Handbook of Smoke Control Engineering

ABOUT THE AUTHORS

John H. Klote

Dr. John Klote is known throughout the world as an expert in smoke control due to his many books on the topic and his 19 years of fire research conducted at the U.S. National Institute of Standards and Technology (NIST) in Gaithersburg, Maryland. For 11 years, he operated his own consulting company specializing in analysis of smoke control systems. Klote developed a series of smoke control seminars that he teaches for the Society of Fire Protection Engineers. The primary author of the 2007 ICC book *A Guide to Smoke Control in the 2006 IBC* and the 2002 ASHRAE book *Principles of Smoke Management*, Dr. Klote is also the primary author of two other ASHRAE books about smoke control, and he has written chapters about smoke control in a number of books, as well as over 80 papers and articles on smoke control, smoke movement, CFD fire simulations, and other aspects of fire protection. He is a licensed professional engineer in Washington, DC. Klote earned his doctorate in mechanical engineering from George Washington University. Klote is a member of NFPA, a fellow of SFPE and a fellow of ASHRAE. He is a member and past chair of ASHRAE Technical Committee 5.6, Fire and Smoke Control, and a member of the NFPA Smoke Management Committee.

James A. Milke

Professor Milke is the chairman of the Department of Fire Protection Engineering at the University of Maryland. He earned his doctorate in aerospace engineering from the University of Maryland. Milke is an author of the ASHRAE book *Principles of Smoke Management*, and of the chapters "Smoke Movement in Buildings" and "Fundamentals of Fire Detection" in the 2008 *NFPA Fire Protection Handbook*. He is also an author of the chapters "Analytical Methods for Determining Fire Resistance of Steel Members," "Smoke Management in Covered Malls and Atria," and "Conduction of Heat in Solids" in the 2008 *SFPE Handbook*. Milke is a licensed professional engineer in Delaware, a member of NFPA and American Society of Civil Engineers (ASCE), a fellow of SFPE, and a past chairman of the NFPA Smoke Management Committee.

Paul G. Turnbull

Paul Turnbull has been actively involved in the development of codes and standards for smoke control systems for over 24 years. He began his career as a hardware developer, designing RFI power line filters, and later moved into development of control products and accessories for building control systems. He then spent 10 years responsible for safety certifications of building controls, HVAC, fire alarm, and smoke control equipment. For the past 15 years, he has specialized in the development and application of gateways that enable fire alarm, security, and lighting control systems to be integrated with building controls in order to provide coordinated operations between these systems. He is an active member in several professional associations focused on control of fire and smoke. Turnbull has a baccalaureate degree in electrical engineering and a master's degree in computer science. He is a member of ASHRAE Technical Committee 5.6, Fire and Smoke Control, and the NFPA Smoke Management Committee. He is an instructor for the SFPE smoke control seminars.

Ahmed Kashef

Dr. Kashef is a group leader of Fire Resistance and Risk Management in the Fire Research Program at the Institute for Research in Construction, National Research Council of Canada. He holds a PhD in civil engineering and has more than 20 years research and practical experience. Dr. Kashef's expertise involves applying numerical and experimental techniques in a wide range of engineering applications including fire risk analysis, fire dynamics, tenability, heat transfer, and smoke transport in the built environment and transportation systems. He has authored and co-authored more than 180 publications. He has managed a broad range of projects involving modeling and full-scale fire experiments to address fire related issues. This includes projects that investigated the ventilation strategies and detection systems in road and subway tunnels. He is the technical secretary of the ASHRAE Technical Committee 5.6, Fire and Smoke Control, and the chair of the research subprogram of ASHRAE Technical Committee 5.9, Enclosed Vehicular Facilities. Dr. Kashef is a registered professional engineer in the province of Ontario, and a member of the NFPA Technical Committee 502 on Road Tunnel and Highway Fire Protection. He is an associate member of the World Road Association (PIARC), Working Group 4, Ventilation and Fire Control and a corresponding member of the Technical Committee 4 Road Tunnel Operations.

Michael J. Ferreira

Michael Ferreira is a senior fire protection engineer and project manager at Hughes Associates, a fire science and engineering consulting company. He has been primarily involved with smoke management system design projects for the past 17 years and has published several articles on the innovative use of computer models for these systems. Ferreira has extensive experience in performing smoke control commissioning testing and calibrating computer models using field data. He was the lead investigator responsible for evaluating smoke control system performance in NIST's investigation of the World Trade Center disaster. He has also conducted a performance-based analysis of the smoke control system at the Statue of Liberty. Ferreira is a professional engineer and holds a BS in Mechanical Engineering and an MS in Fire Protection Engineering from Worcester Polytechnic Institute. He is a member of the NFPA Smoke Management Systems Committee, and is an instructor for the NFPA and SFPE smoke control seminars.

ASHRAE

ASHRAE, founded in 1894, is a building technology society with more than 50,000 members worldwide. The Society and its members focus on building systems, energy efficiency, indoor air quality and sustainability within the industry. Through research, standards writing, publishing, and continuing education, ASHRAE shapes tomorrow's built environment today.

1791 Tullie Circle, NE
Atlanta, GA 30329
1-800-527-4723
www.ashrae.org

International Code Council

International Code Council is a member-focused association dedicated to helping the building safety community and construction industry provide safe, sustainable, and affordable construction through the development of codes and standards used in the design, build, and compliance process. Most U.S. communities and many global markets choose the International Codes. ICC Evaluation Service (ICC-ES), a subsidiary of the International Code Council, has been the industry leader in performing technical evaluations for code compliance fostering safe and sustainable design and construction.

Headquarters: 500 New Jersey Avenue, NW, 6th Floor, Washington, DC 20001-2070
District Offices: Birmingham, AL; Chicago. IL; Los Angeles, CA
1-888-422-7233
www.iccsafe.org

Society of Fire Protection Engineers

Organized in 1950, the Society of Fire Protection Engineers (SFPE) is the professional organization that represents engineers engaged in fire protection worldwide. Through its membership of over 5000 professionals and 65 international chapters, SFPE advances the science and practice of fire protection engineering while maintaining a high ethical standard. SFPE and its members serve to make the world a safer place by reducing the burden of unwanted fire through the application of science and technology. To become a member, go to www.sfpe.org.

7315 Wisconsin Ave., #620E
Bethesda, MD 20814
1-301-718-2910
www.sfpe.org

National Fire Protection Association

The National Fire Protection Association (NFPA) is an international nonprofit organization that was established in 1896. The company's mission is to reduce the worldwide burden of fire and other hazards on the quality of life by providing and advocating consensus codes and standards, research, training, and education.

1 Batterymarch Park
Quincy, MA 02169-7471
1-617-770-3000
www.nfpa.org

Handbook of Smoke Control Engineering

John H. Klote
James A. Milke
Paul G. Turnbull
Ahmed Kashef
Michael J. Ferreira

ISBN 978-1-936504-24-4

Published in cooperation with International Code Council, Inc.,
National Fire Protection Association, and Society of Fire Protection Engineers.

ASHRAE
1791 Tullie Circle, N.E.
Atlanta, GA 30329
www.ashrae.org

Printed in the United States of America

Printed on 30% post-consumer waste using soy-based inks.

Illustrations by John H. Klote, unless otherwise credited.

Library of Congress Cataloging-in-Publication Data

Handbook of smoke control engineering / John H. Klote, editor and chief ; James A. Milke, Paul G. Turnbull.
p. cm.
Includes bibliographical references and index.
ISBN 978-1-936504-24-4 (hardcover : alk. paper) 1. Buildings--Smoke control systems--Handbooks, manuals, etc. 2. Smoke prevention--Handbooks, manuals, etc. 3. Ventilation--Handbooks, manuals, etc. 4. Fire testing--Handbooks, manuals, etc. I. Klote, John H. II. Milke, J. A. (James A.) III. Turnbull, Paul G., 1961- IV. American Society of Heating, Refrigerating and Air-Conditioning Engineers.
TH1088.5.H36 2012
693.8--dc23

2012009054

Any updates/errata to this publication will be posted on the ASHRAE Web site at www.ashrae.org/publicationupdates.

DEDICATION

This book is dedicated to the memory of Harold (Bud) Nelson. Because of his many significant contributions when he worked at the General Services Administration (GSA) and the National Institute of Standards and Technology (NIST), Bud Nelson was recognized as one of the great pioneers of fire protection engineering. Bud Nelson also was the first chairman of the NFPA Smoke Management Committee.

HOW TO USE THIS BOOK

This book is organized in the classic handbook format to help engineers and other professionals who need to get information about a topic quickly. The Table of Contents and the Index can be used so readers can go directly to their topic of interest. The handbook format has no introductory chapter, and the most fundamental material is in the first chapters and applied material is in later chapters. To help readers get information quickly, the chapters do not include derivations of equations. Unlike textbooks, some redundancy is needed in handbooks so that the chapters can be relatively independent. This redundancy is minimized, and in some places readers are referred to another section or chapter for more information. This book includes all the information in my earlier smoke control books plus a number of other topics, and there are many example calculations. This handbook can be used as a textbook with the teacher selecting the chapters and parts of chapters to be taught. The only departure from the handbook format is that derivations of equations are in an appendix included to make the book more useful to scholars, teachers, and students.

John H. Klote

TABLE OF CONTENTS

PREFACE

In 1983, ASHRAE published *Design of Smoke Control Systems for Buildings* by John Fothergill and me. This book was the first attempt to consolidate and present practical information about smoke control design. Judging by the many favorable comments and suggestions about this first book, I feel that it was a success. The first publication was limited to systems that control smoke by means of the physical mechanisms of pressurization and airflow.

In 1992, ASHRAE and SFPE jointly published *Design of Smoke Management Systems* by James Milke and me. The term *smoke management* was used in the title of this publication to indicate that the physical mechanisms were expanded from pressurization and airflow to include compartmentation, dilution, and buoyancy. Based on heightened concerns about supplying combustion air to the fire, a caution was added about the use of airflow for smoke management.

In 2002, ASHRAE and SFPE jointly published *Principles of Smoke Management* by James Milke and me. This publication included the material of the two earlier books plus people movement in fire, hazard analysis, scale modeling, and computational fluid dynamics.

This new publication is in handbook form that is intended to make the book more useful to practicing engineers. The earlier books were aimed at both practicing engineers and students, and derivations of equations were included in many of the chapters. To make the handbook easier to use for engineers who want information on a specific topic quickly, the derivations are not included in the chapters. However, to make the book useful to students and teachers, the derivations are in an appendix.

This new book addresses the material of the earlier books plus (1) controls, (2) fire and smoke control in transport tunnels, and (3) full scale fire testing. For those getting started with the computer models CONTAM and CFAST, there are simplified instructions with examples. As with the other books, this new book is primarily intended for designers, but it is expected that it will be of interest to other professionals (architects, code officials, researchers, etc.).

In this book, the term *smoke control system* is used to mean an engineered system that includes all methods that can be used singly or in combination to modify smoke movement. This usage is consistent with that of the 2009 NFPA 92A, 2012 NFPA 92, and most codes including the International Building Code. This usage is a departure from the earlier ASHRAE smoke control books and earlier versions of NFPA 92A. The meaning of the term *smoke management system* was completely changed in the 2009 NFPA 92A, and this term is almost never used in this handbook. Because these terms have different meanings in many publications, readers are cautioned to be careful about this terminology when reading different books, research papers, and articles.

This book and its predecessors are different from other design books in a number of respects. This book is written in both English units (also called I-P for inch-pound) and SI units so that it can be used by a wide audience. Physical descriptions are worked into the text as simple explanations of how particular mechanisms, processes or events happen. Many example calculations are included. As with the earlier book, I hope that this book is of value to the engineering community. Further, I invite readers to mail their suggestions and comments to me at the address below.

John H. Klote, D.Sc., P.E.
19355 Cypress Ridge Terrace
Unit 502
Leesburg, VA 22101

ACKNOWLEDGMENTS

This project would not have been possible without the support of ASHRAE. In addition to publishing books about smoke control, ASHRAE has funded a considerable body of smoke control research from the 1980s to the present time. A debt is owed to my coauthors: James A. Milke, Paul G. Turnbull, Ahmed Kashef, and Michael J. Ferreira. Each of them has authored a chapter or more, and they have provided valuable advice during development of this handbook.

Acknowledgement is made to the members of the ASHRAE Smoke Control Monitoring Committee for their generous support and constructive criticism. The members of this subcommittee are: William A. Webb (Chair), Jeffrey S. Tubbs, and Douglas Evans. Gary D. Lougheed, Paul G. Turnbull, John A. Clark, John Breen, and W. Stuart Dols also provided constructive criticism.

Special thanks are due to Gary Lougheed for his insightful comments regarding fluid flow, design fires, and full scale fire testing. Paul Turnbull made valuable comments about practically every aspect of the book. John Clark provided helpful comments in a number of areas. John Breen, who is a student at the Department of Fire Protection Engineering at the University of Maryland, made valuable comments regarding the computer program CONTAM. W. Stuart Dols, who is in charge of the development of CONTAM at NIST, made helpful comments about a number of aspects of CONTAM. In addition to chairing the review subcommittee, Bill Webb made practical comments on subjects in every chapter of the book.

Acknowledgement must be made to the many engineers and scientists who have conducted the research that is the foundation of modern smoke control technology. These researchers are too many to mention here, but many of their efforts are referenced in the text. It should be mentioned that I personally owe much to the National Institute of Standards and Technology in Gaithersburg, MD for the opportunity of being able to do fire research there for nineteen years.

The content of this book is heavily dependent on extensive smoke control research conducted at the National Research Council of Canada (NRCC). Much of this research has been conducted at NRCC's Experimental Fire Tower near Ottawa.

John H. Klote

NOTE ON SUSTAINABILITY

Sustainability has attracted considerable attention in recent years, and the design of green buildings requires ingenuity and understanding of the technology. This handbook does not explicitly address sustainability, but it can be thought of as a treatment of sustainability to the extent that designers can develop sustainable smoke control systems based on information provided herein.

In one sense, smoke control systems can be thought of as sustainable systems in that they can minimize the extent of smoke damage to building components during fires. However, the amount of materials used in some smoke control systems can be minimized or even eliminated.

The use of natural smoke venting for smoke control in atria and other large volume spaces eliminates the fans and ductwork used in conventional smoke exhaust systems. The only equipment needed for this kind of venting is a roof vent that opens in the event of a fire. Natural smoke venting has been used for many decades in the United Kingdom, Australia, and Japan. An algebraic equation in Chapter 15 can be used as a starting point for analysis of a natural venting system. Wind effects are a special concern with natural smoke venting, and these systems should be analyzed with computational fluid dynamic (CFD) modeling (Chapter 20).

Smoke filling is the simplest form of smoke control for atria and other large volume spaces, because it eliminates the need for any equipment. This approach consists of allowing smoke to fill the large volume space without any smoke exhaust or other smoke removal. For very large spaces, the smoke filling time can be long enough for evacuation. Smoke filling time can be calculated by algebraic equations or with the use of computer models as discussed in Chapter 15. It is essential that calculations of evacuation time include the times needed for recognition, validation, and premovement as discussed in Chapter 4.

For some applications, passive smoke control using smoke barriers has the potential to be used in place of pressurization smoke control systems. This can reduce or eliminate the fans and ductwork of the pressurization systems. Such systems need to provide equivalent life-safety protection as that of the pressurization systems. The tenability of such passive systems can be analyzed with CFD modeling or with a combination of CONTAM and zone fire modeling as discussed in Chapter 19.

Stairwell ventilation systems have the potential to maintain tenability in stairwells at reduced fan capacity compared to stairwell pressurization. The idea of these ventilation systems is to supply air to and exhaust air from the stairwell so that any smoke leaking into the stairwell is diluted to maintain tenable conditions in the stairwell. The amount of air needed for stairwell pressurization is proportional to the number of floors served by the stairwell, but the amount of air needed for stairwell ventilation, is almost independent of the number of floors. This means that the greatest savings in fan capacity are for stairwells in very tall buildings. For stairwell ventilation the most important location is the landing of the fire floor, and tenability here can be analyzed by CFD modeling as discussed in Chapter 20.

The extent to which smoke control systems can be more sustainable depends on the ingenuity, creativity, and knowledge of the design team. Some old ideas (such as smoke shafts and smoke venting with exterior wall vents) may be reevaluated and revised to become sustainable systems or parts of sustainable systems. It is essential that the alternate smoke control systems provide protection that is equivalent to that of conventional systems.

CHAPTER 1

Units and Properties

John H. Klote

The international system (SI) of units is used for almost all applications outside the U.S. and for many applications inside the U.S. In the U.S., a collection of mostly old English units are used for many applications. These old style units are referred to here as inch-pound (I-P) units. This chapter deals with units of measurement and physical properties.

DUAL UNITS

Most equations in this handbook are presented in dual units, but exceptions are noted at the beginning of some chapters. The equation below for the Reynolds number is an example of these dual units.

$$R_e = \frac{1.39 \times 10^{-3} D_h U}{\nu}$$
$$R_e = \frac{D_h U}{\nu} \text{ for SI} \qquad (1.1)$$

where

R_e = Reynolds number, dimensionless,
D_h = hydraulic diameter of flow path, in. (m),
U = average velocity in flow path, fpm (m/s),
ν = kinematic viscosity, ft^2/s (m^2/s).

This equation consists of an I-P version followed by an SI version. The "where" list below the equation contains the variable names, followed by the I-P units with the SI units in parentheses. For example, the I-P units of average velocity in flow path are fpm, and the SI units for this variable are m/s.

The I-P units are used in the following systems: (1) the pound-mass and pound-force system, (2) the slug and pound system, and (3) the pound-mass and poundal system. Each version has its own rules for dealing with units, but these are not discussed here. The approach taken here is to focus on the SI system, and to provide conversions between the I-P units and SI units.

THE SI SYSTEM

Today's SI system is based on the metric system that was first adopted in France in 1791. This section is a general discussion of the SI system. More detailed information is available from NIST (Thompson and Taylor 2008) and IEEE/ASTM (IEEE/ASTM 2002). The NIST publication can be downloaded over the Internet at no cost.

The SI system consists of base units and derived units which together form what is called a coherent system of SI units. Such a coherent system needs no additional factors in equations to adjust for the units, and the advantage of this is illustrated later. The seven base quantities upon which the SI system is founded are length, mass, time, thermodynamic temperature, electric current, amount of substance, and luminous intensity. Table 1.1 lists the names and symbols of the units for these base quantities.

Derived units are expressed algebraically in terms of base units or other derived units. The symbols for derived units are obtained by means of the mathematical operations of multiplication and division. For example, the derived unit for the derived quantity mass flow (mass divided by time) is the kilogram per second, and the symbol for mass flow is kg/s. Other examples of derived units expressed in terms of SI base units are given in Table 1.2.

There are a number of coherent derived units that have special names and symbols. For example, the pascal is the special unit for pressure, and the symbol Pa is the special symbol for the pascal. Table 1.3 lists some of these units with special names and symbols. When it is stated that an equation is valid for the SI system, it is meant that the equation is valid for variables that are the coherent units of the SI system.

Prefixes are listed in Table 1.4. For example, the prefix kilo (k) means a multiplication factor of one thousand, and a kilometer (km) is a thousand meters (m). Conversions between I-P and SI units are listed in Table 1.5.

Chapters in SI Only

Some of the chapters in this handbook are only in SI units. This was done because the equations in these chapters are intended primarily for explanation. These equations can also be used to write computer programs, and most computer programs are written in SI units because they are based on equations from research done in SI units. All of the variables in an SI equation are in base units or coherent derived units (Tables 1.1 to 1.3). Care needs to be taken because units with a prefix are not coherent except for the kilogram, which is an exception. For example, the following is an SI equation for the pressure difference between two nodes:

$$\Delta p_{ij} = p_i - p_j + p_i g(z_i - z_j) \tag{1.2}$$

where

Δp_{ij} = pressure difference from node *i* to node *j*,

p_i = pressure at node *i*,

p_j = pressure at node *j*,

r_i = density of gas at node *i*,

z_i = elevation of node *i*,

z_j = elevation of node *j*,

g = acceleration of gravity.

It can be seen from Table 1.3 that the pressures and the pressure difference are in the units of pascals (Pa). Elevations are quantities of length, and they are in meters (m) as can be seen from Table 1.1. From Table 1.2, it can be seen that the acceleration term has units of meter per second squared (m/s^2).

Table 1.1: Base Units of the SI System

Base Quantity	Unit	Symbol
Length	meter	m
Mass	kilogram	kg
Time	second	s
Thermodynamic temperature[1]	kelvin	K
Electric current	ampere	A
Amount of substance	mole	mole
Luminous intensity	candela	cd

[1]This is also called absolute temperature. Kelvin is also the unit for temperature difference and temperature rise.

Table 1.2: Some Coherent Derived Units

Quantity	Name	Symbol
Acceleration	meter per second squared	m/s^2
Area	square meter	m^2
Density	kilogram per cubic meter	kg/m^3
Mass flow	mass per second	kg/s
Velocity	meter per second	m/s
Volume	cubic meter	m^3
Volumetric flow	cubic meter per second	m^3/s

Table 1.3: Some Coherent Derived Units with Special Names and Symbols

Quantity	Special Name	Special Symbol	Expression in other SI Units	Expression in SI Base Units
Electrical charge	coulomb	C	–	s A
Electric potential difference	volt	V	W/A	$m^2\ kg\ s^{-3}\ A^{-1}$
Energy, heat, and work	joule	J	N m	$m^2\ kg\ s^{-3}$
Force	newton	N	–	$m\ kg\ s^{-2}$
Frequency	hertz	Hz	–	s^{-1}
Power, heat release rate	watt	W	J/s	$m^2\ kg\ s^{-3}$
Pressure, pressure difference	pascal	Pa	N/m^2	$m^{-1}\ kg\ s^{-2}$

TEMPERATURE CONVERSION

The SI unit of absolute temperature is kelvin, and the I-P unit of absolute temperature is Rankine. In addition, temperature is frequently measured in the Celsius or the Fahrenheit scale. The following equations can be used to convert between temperature scales:

$$\begin{aligned} T_F &= T_R - 459.67 \\ T_R &= T_F + 459.67 \\ T_C &= T_K - 273.15 \\ T_K &= T_C + 273.15 \\ T_F &= 1.8T_C + 32 \\ T_C &= \frac{T_F - 32}{1.8} \end{aligned} \quad (1.3)$$

where

T_F = temperature in degrees Fahrenheit,

T_R = temperature in degrees Rankin,

T_C = temperature in degrees Celsius,

T_K = temperature in kelvin.

Temperature Difference

This section deals with temperature difference, temperature rise, and temperature drop. All of these are handled the same way, and they are referred to here in a generic sense as temperature difference. The following equations can be used for temperature difference conversions:

$$\begin{aligned} \Delta T_F &= 1.8\Delta T_C \\ \Delta T_F &= \Delta T_R \\ \Delta T_C &= \frac{\Delta T_F}{1.8} \\ \Delta T_C &= \Delta T_K \end{aligned} \quad (1.4)$$

where

ΔT_F = temperature difference in degrees Fahrenheit,

ΔT_C = temperature difference in degrees Celsius,

ΔT_K = temperature difference in kelvin,

ΔT_R = temperature difference in degrees Rankine.

SOFT AND HARD CONVERSIONS

Many people are confused by the terms soft conversion and hard conversion, because the terms seem backwards. Regarding conversions, *soft* means exact or nearly so, and *hard* means approximate. An example of a soft conversion is 810 ft equals exactly 246.888 m. What is hard about a hard conversion is deciding how many digits should be kept in the rounded number. Should 810 ft be rounded to 250 m, 247 m, or something else? The answer depends on numerous considerations, some of which are unique to specific areas of engineering.

In this handbook, hard conversions are used. Often, values are rounded to three significant digits because calculations based on such rounding are equivalent for engineering purposes in both systems. Rounding is sometimes based on accuracy considerations of the original value. With most research work and some standards, the original value is in SI units. For consistency in this handbook, I-P units are listed first, followed by SI units in parentheses, regardless of the source of the data.

UNIT CONVERSIONS FOR EQUATIONS

Because almost all research is conducted in SI units, there is a need to convert SI equations to I-P equations. This section discusses a method that can be used for such conversions. For SI equations with temperature in degrees Celsius, the equation needs to be converted to one with temperature in kelvin.

The following is an equation in functional form:

$$y = f(x_1, x_2, \ldots, x_n) \quad (1.5)$$

where y is a dependent variable, and x_i from $i = 1$ to n are independent variables. Equation 1.5 is in SI units, and it is desired to convert it to I-P units. The variables in this equation are related to the ones in the other unit system as follows:

$$\begin{aligned} y &= ay' \\ x_i &= b_i x_i' \end{aligned} \quad (1.6)$$

Table 1.4: SI Prefixes

Prefix	Symbol	Multiplication Factor
giga	G	$10^9 = 1\ 000\ 000\ 000$
mega	M	$10^6 = 1\ 000\ 000$
kilo	k	$10^3 = 1\ 000$
centi[1]	c	$10^{-2} = 0.01$
milli	m	$10^{-3} = 0.001$
micro	μ	$10^{-6} = 0.000\ 001$
nano	n	$10^{-9} = 0.000\ 000\ 001$

[1]The prefix centi is to be avoided where possible.

Table 1.5: Factors for Unit Conversions

TO CONVERT FROM	TO	MULTIPLY BY
Acceleration		
foot per second squared (ft/s^2)	meter per second squared (m/s^2)	0.3048
meter per second squared (m/s^2)	foot per second squared (ft/s^2)	3.2808
standard gravity (g)	meter per second^2 (m/s^2)	9.80665
standard gravity (g)	foot per second (ft/s^2)	32.174
Area		
foot squared (ft^2)	meter^2 (m^2)	0.09290
foot squared (ft^2)	inch squared ($in.^2$)	144
meter squared (m^2)	foot squared (ft^2)	10.76
meter squared (m^2)	inch squared (in^2)	1550
meter squared (m^2)	yard squared (yd^2)	1.196
yard squared (yd^2)	meter^2 (m^2)	0.8361
yard squared (yd^2)	foot squared (ft2)	9
yard squared (yd^2)	inch squared ($in.^2$)	1296
Density		
gram per cubic meter (g/m^3)	kilogram per cubic meter (kg/m^3)	0.001
kilogram per cubic meter (kg/m^3)	gram per cubic meter (g/m^3)	1000
gram per cubic meter (g/m^3)	pound per cubic foot (lb/ft^3)	6.2428E-5
kilogram per cubic meter (kg/m^3)	pound per cubic foot (lb/ft^3)	0.062428
pound per cubic foot (lb/ft^3)	kilogram per cubic meter (kg/m^3)	16.018
pound per cubic foot (lb/ft^3)	gram per cubic meter (g/m^3)	16,018
Energy (also Heat and Work)		
British thermal unit (Btu)	joule (J)	1055
British thermal unit (Btu)	foot pound (ft lb)	778
erg	joule (J)	1.000E-7
foot pound (ft lb)	joule (J)	1.356
joule (J)	British thermal unit (Btu)	9.479E-4
Flow, Mass		
kilogram per second (kg/s)	pound per hour (lb/h)	7937
kilogram per second (kg/s)	pound per minute (lb/min)	132.3
kilogram per second (kg/s)	pound per second (lb/s)	2.205
kilogram per second (kg/s)	standard cubic feet per min (scfm) at 68°F	1760
pound per hour (lb/h)	kilogram per second (kg/s)	0.0001260
pound per minute (lb/min)	kilogram per second (kg/s)	0.007560
pound per second (lb/s)	kilogram per second (kg/s)	0.4536
pound per second (lb/s)	standard cubic feet per min (scfm) at 68°F	798.5
standard cubic feet per min (scfm) at 68°F	kilogram per second (kg/s)	0.005680
standard cubic feet per min (scfm) at 68°F	pound per second (lb/s)	0.0012523

Table 1.5: Factors for Unit Conversions *(Continued)*

TO CONVERT FROM	TO	MULTIPLY BY
Flow, Volumetric		
foot cubed per minute (ft^3/min or cfm)	meter cubed per second (m^3/s)	4.719E-04
foot cubed per second (ft^3/s)	meter cubed per second (m^3/s)	0.02832
gallon per minute (gal/min or gpm)	meter cubed per second (m^3/s)	6.309E-05
meter cubed per second (m^3/s)	foot cubed per minute (ft^3/min or cfm)	2119
meter cubed per second (m^3/s)	foot cubed per second (ft^3/s)	35.31
meter cubed per second (m^3/s)	gallon per minute (gal/min or gpm)	15850
gallon per minute (gal/min or gpm)	foot cubed per minute (ft^3/min or cfm)	0.1337
foot cubed per minute (ft^3/min or cfm)	gallon per minute (gal/min or gpm)	7.481
Force		
kilogram-force (at sea level)	newton (N)	9.80665
pound-force (lb)	newton (N)	4.448
newton (N)	pound-force (lb)	0.2248
Heat (See Energy)		
Heat Release Density		
Btu/s ft^2	kW/m^2	11.36
kW/m^2	Btu/s ft^2	0.08806
Heat Release Rate (see Power)		
Length		
foot (ft)	meter (m)	0.3048
foot (ft)	inch (in.)	12
inch (in.)	meter (m)	0.02540
inch (in.)	centimeter (cm)	2.54
inch (in.)	foot (ft)	0.08333
meter (m)	foot (ft)	3.2808
meter (m)	inch (in)	39.3701
meter (m)	nautical mile (U.S.)	0.0005
meter (m)	mile	6.214E-4
meter (m)	yard	1.0936
mile	meter (m)	1609.3
mile	foot (ft)	5280
nautical mile (U.S.)	meter (m)	1852
yard	meter (m)	0.9144
yard	foot (ft)	3
yard	meter (m)	0.9144
Light		
footcandle	lux (lx)	10.764
lux (lx)	footcandle	0.0929
Mass		
gram (g)	kilogram (kg)	0.001

Table 1.5: Factors for Unit Conversions ***(Continued)***

TO CONVERT FROM	TO	MULTIPLY BY
gram (g)	pound (lb)	0.002205
kilogram (kg)	gram (g)	1000
kilogram (kg)	pound (lb)	2.205
ounce (avoirdupois)	kilogram (kg)	0.03110
pound (lb)	kilogram (kg)	0.4536
pound (lb)	gram (g)	453.6
pound (lb)	slug	0.03108
slug	kilogram (kg)	14.60
slug	pound (lb)	32.174
ton (long, 2240 lb)	kilogram (kg)	1016
ton (metric)	kilogram (kg)	1000
ton (short, 2000 lb)	kilogram (kg)	907.2
Mass Flow (see Flow, Mass)		
Temperature (see equations in the text)		
Power (also Heat Release Rate)		
British thermal unit per hour (Btu/h)	kilowatt (kW)	2.931E-04
British thermal unit per hour (Btu/h)	watt (W)	0.293
British thermal unit per minute (Btu/min)	watt (W)	17.58
British thermal unit per minute (Btu/min)	kilowatt (kW)	0.01758
British thermal unit per second (Btu/s)	watt (W)	1055
British thermal unit per second (Btu/s)	kilowatt (kW)	1.055
horsepower	watt (W)	745.7
horsepower	foot pound per second (ft lb/s)	550.0
horsepower	kilowatt (kW)	0.7457
ton (refrigeration)	watt (W)	3517
ton (refrigeration)	kilowatt (kW)	3.517
Pressure		
atmosphere, standard (atm)	pascal (Pa)	101325
atmosphere, standard (atm)	pound per square inch (lb/in.^{2} or psi)	14.696
atmosphere, standard (atm)	pound per square foot (lb/ft^2)	2116.2
atmosphere, standard (atm)	inch of water (in. H_20) at 60 °F	407.19
atmosphere, standard (atm)	foot of water (ft H_20) at 60 °F	33.932
centimeter of mercury (cm Hg) at 0°C	pascal (Pa)	1333.22
centimeter of water (cm H_2O) 60°C	pascal (Pa)	97.97
foot of water (ft H_20) at 60°F	pascal (Pa)	2986
inch of mercury (in. Hg)	pascal (Pa)	3386
inch of water (in. H_20) at 60°F	pascal (Pa)	248.84
pascal (Pa)	inch of mercury (in. Hg)	2.953E-04
pascal (Pa)	inch of water (in. H_20) at 60°F	0.004019
pascal (Pa)	foot of water (ft H_20) at 60°F	3.349E-04
pascal (Pa)	centimeter of mercury (cm Hg) at 0°C	7.501E-04

Table 1.5: Factors for Unit Conversions *(Continued)*

TO CONVERT FROM	TO	MULTIPLY BY
pascal (Pa)	centimeter of water (cm H_2O) 60° C	0.01021
pascal (Pa)	pound per square foot (lbf/ft^2)	0.02089
pascal (Pa)	pound per square inch (lbf/in^2 or psi)	1.450E-04
pound per square foot (lbf/ft^2)	pascal (Pa)	47.88
pound per square inch (lbf/in.^{2} or psi)	pascal (Pa)	6895
Velocity (also Speed)		
foot per hour (ft/h)	meter per second (m/s)	8.467E-05
foot per minute (ft/min or fpm)	meter per second (m/s)	0.005080
foot per second (ft/s)	meter per second (m/s)	0.3048
kilometer per hour (km/h)	meter per second (m/s)	0.2778
knot	meter per second (m/s)	0.5144
meter per second (m/s)	foot per minute (ft/min or fpm)	196.9
meter per second (m/s)	foot per second (ft/s)	3.281
meter per second (m/s)	foot per hour (ft/h)	11811
meter per second (m/s)	kilometer per hour (km/h)	3.600
meter per second (m/s)	knot	1.944
meter per second (m/s)	mile per hour (mi/h or mph)	2.237
mile per hour (mi/h or mph)	kilometer per hour (km/h)	1.609
Volume		
foot cubed (ft^3)	meter cubed (m^3)	0.02832
foot cubed (ft^3)	inch cubed (in.^{3})	1728
foot cubed (ft^3)	gallon (U.S.)	7.4805428
foot cubed (ft^3)	yard cubed (yd^3)	0.03704
gallon (U.S.)	meter cubed (m^3)	0.003785412
gallon (U.S.)	foot cubed (ft^3)	0.1337
inch cubed (in.^{3})	meter cubed (m^3)	1.639×10^{-5}
inch cubed (in.^{3})	foot cubed (ft^3)	0.0005787
liter	meter cubed (m^3)	0.001
liter	gallon (U.S.)	0.2642
meter cubed (m^3)	foot cubed (ft^3)	35.31
meter cubed (m^3)	inch cubed (in.^{3})	61013
meter cubed (m^3)	gallon (U.S.)	264.2
meter cubed (m^3)	liter	1000
meter cubed (m^3)	yard cubed (yd^3)	1.308
yard cubed (yd^3)	meter cubed (m^3)	0.7646
yard cubed (yd^3)	foot cubed (ft^3)	27
Volumetric Flow (see Flow, Volumetric)		
Work (see Energy)		

where y' and x_i' are corresponding variables in I-P units, and a and b_i are conversion constants. Table 1.5 lists some conversion factors. Substituting Equations 1.6 into Equation 1.5 results in

$$ay' = f(b_1x_1', b_2x_2', ..., b_nx_n') \quad (1.7)$$

This equation is equivalent to Equation 1.6, but it is in I-P units. Equation 1.7 demonstrates that an alternate form of any equation can be developed. In practice, the coefficients of a function in the form of Equation 1.7 would be rearranged and rounded off. The resulting equation can be written as

$$y' = f'(x_1', x_2', ..., x_n') \quad (1.8)$$

where f' is a new function with rounded off coefficients. The level of agreement between Equations 1.7 and 1.8 can be expressed as

$$\varepsilon = \frac{af'(x_1', x_2', ..., x_n') - f(x_1, x_2, ..., x_n)}{f(x_1, x_2, ..., x_n)} \quad (1.9)$$

where ε is the error in the function, f', due to rounding. A positive value of ε means that f' is overpredicting in comparison to the predictions of f.

When rounding off the coefficients, the temptation of using a simple rule based on the accuracy of the original research needs to be avoided. For example, a person might mistakenly think that because the original research has an accuracy of only two significant figures, all the coefficients should be rounded to two places. Some constants in a function can have a much greater impact than others, and using such a simple approach can result in error values ε, that are unacceptably high.

A more appropriate rule is to round coefficients to the smallest values that will result in values of ε that are within a predetermined limit. For many engineering applications, a value ε of 1% would be reasonable, and this value is used in Example 1.1.

Table 1.6: Some Physical Constants

Acceleration of gravity at sea level, g	9.80665 m/s^2
	32.174 ft/s^2
Gas constant of air, R	287.0 J/kg K
	53.34 ft lbf/lbm/°R
	1716. ft lbf/slug/°R
	0.06858 Btu/lbm/°R
Standard atmospheric pressure, P_{atm}	101,325 Pa
	14.696 psi
	2116.2 lb/ft^2
	407.19 in. H_2O (60°F)
	33.932 ft H_2O (60°F)
	1033.3 cm H_2O (4°C)
	30.006 inch mercury (60°F)
	760.00 mm mercury (0°C)

PHYSICAL DATA

The values of some physical constants are listed in Table 1.6. The properties of air are listed in Tables 1.7 and 1.8. The thermal properties of a number of materials are listed in Tables 1.9 and 1.10.

U.S. STANDARD ATMOSPHERE

The barometric pressure and temperature of the air vary with altitude, local geographic conditions, and weather conditions. Altitude is the elevation above sea level. The standard atmosphere is a standard of reference for properties at various altitudes. At sea level, the standard temperature is 59°F (15°C) and the standard barometric pressure is 14.6959 psi (101.325 kPa). The barometric pressure and temperature decrease with increasing altitude. The temperature is considered to decrease linearly throughout the troposphere, which is the lowest portion of the earth's atmosphere. The standard barometric pressure varies with altitude as

$$\begin{aligned} p &= 14.6959(1 - 6.87559 \times 10^{-6}z)^{5.2559} \\ p &= 101.325(1 - 2.25577 \times 10^{-5}z)^{5.2559} \text{ for SI.} \end{aligned} \quad (1.10)$$

The standard temperature varies with altitude as

$$\begin{aligned} T &= 59 - 0.00357z \\ T &= 15 - 0.0065z \text{ for SI} \end{aligned} \quad (1.11)$$

where

p = barometric pressure, psi (kPa),

T = temperature, °F (°C),

z = altitude, ft (m).

Example 1.2 shows how to calculate the standard barometric pressure. The climatic data listed in Chapter 2 lists the standard barometric pressure calculated from Equation 1.10 for locations throughout the world. The above equations for barometric pressure and temperature are accurate from –16,400 to 36,000 ft (–5000 to 11,000 m). For higher altitudes, see NASA (1976).

Example 1.1. Equivalent I-P Equation

For the following SI equation, develop an equivalent I-P version. The SI equation is

$$m = 0.59Q_c^{1/3}W^{1/5}(z_b + 0.17W^{7/15}H + 10.35W^{7/15} - 15)$$

where

m = mass flow rate in plume (kg/sec),
Q_c = convective heat release rate of the fire (kW),
W = length of the spill (m),
z_b = height of the plume above the balcony edge (m),
H = height of balcony above fuel (m).

This equation is applicable for $z_b < 15$ m and $W < 10$ m. It is desired to convert this equation to another one with mass flow in pounds per second, heat release in Btu/s, and length in feet. The variables are related between the two systems as

$$m = 0.4536m' ; Q_c = 1.055Q_c' ; W = 0.3048W' ; z_p = 0.3048z_p' ; H = 0.3048H' .$$

Substituting the relations between the two unit systems into the SI version of the equation, rearranging, and rounding coefficients to four places yields

$$0.4536m' = 0.59(1.055Q_c')^{1/3}(0.3048W')^{1/5}$$

$$[0.3048z_b' + 0.17(0.3048W')^{7/15}0.3048H' + 10.35(0.3048W')^{7/15} - 15] .$$

Next, the coefficients in this equation were rearranged and calculated to four places

$$m' = 0.3182(Q_c')^{1/3}(W')^{1/5}[z_b' + 0.09764(W')^{7/15}H' + 19.50(W')^{7/15} - 49.21] .$$

These coefficients need to be rounded down further. The first attempt will be to round the coefficients to two places and calculate the error. A spread sheet program was used to evaluate a version of the equation with coefficients rounded to two places. Errors were calculated over a range of useful values which is: 350 Btu/s < Q_c < 1400 Btu/s, 3 ft < z_b < 50 ft, 7 ft < W < 32.8 ft, 8 ft < H < 18 ft. It was found that ε is independent of Q_c, but it depends on the other variables. Over this range, the error, ε, varied from 0.8% to 5.9%.

On inspection, the last coefficient in the equation appears to have the most impact on the predicted results. The spread sheet data was modified so that this last coefficient was to three places and the others unchanged. With these coefficients, ε varied from 0.6% to 0.8%. Because these errors are less than the predetermined limit of 1%, the coefficients are acceptable.

Based on this analysis, the equation in I-P units can be written without the prime notation as

$$m = 0.32Q_c^{1/3}W^{1/5}(z_b + 0.098W^{7/15}H + 19.5W^{7/15} - 49.2)$$

where

m = mass flow rate in plume (lb/s),
Q = heat release rate of the fire (Btu/s),
W = length of the spill (ft),
z_b = height of the plume above the balcony edge (ft),
H = height of balcony above fuel (ft).

Table 1.7: Properties of air in I-P units

T, °F	ρ, lbm/ft³	C_p, Btu/lbm·°F	μ, lbm/ft·s	ν, ft²/s)	k, Btu/h·ft·°F
0	0.086	0.239	1.110×10^{-5}	0.130×10^{-3}	0.0133
32	0.081	0.240	1.165×10^{-5}	0.145×10^{-3}	0.0140
100	0.071	0.240	1.285×10^{-5}	0.180×10^{-3}	0.0154
200	0.060	0.241	1.440×10^{-5}	0.239×10^{-3}	0.0174
300	0.052	0.243	1.610×10^{-5}	0.306×10^{-3}	0.0193
400	0.046	0.245	1.750×10^{-5}	0.378×10^{-3}	0.0212
500	0.0412	0.247	1.890×10^{-5}	0.455×10^{-3}	0.0231
600	0.0373	0.250	2.000×10^{-5}	0.540×10^{-3}	0.0250
700	0.0341	0.253	2.14×10^{-5}	0.625×10^{-3}	0.0268
800	0.0314	0.256	2.25×10^{-5}	0.717×10^{-3}	0.0286
900	0.0291	0.259	2.36×10^{-5}	0.815×10^{-3}	0.0303
1000	0.0271	0.262	2.47×10^{-5}	0.917×10^{-3}	0.0319
1500	0.0202	0.276	3.00×10^{-5}	1.47×10^{-3}	0.0400
2000	0.0161	0.286	3.45×10^{-5}	2.14×10^{-3}	0.0471
2500	0.0133	0.292	3.69×10^{-5}	2.80×10^{-3}	0.051
3000	0.0114	0.297	3.86×10^{-5}	3.39×10^{-3}	0.054

Notation: T = temperature, ρ = density, C_p = constant pressure specific heat,
μ = dynamic (absolute) viscosity, ν = kinematic viscosity ($\nu = \mu/\rho$), k = thermal conductivity

Table 1.8: Properties of Air in SI Units

T, K	ρ, kg/m³	C_p, J/kg·K	μ, kg/m·s	ν, m²/s	k, W/m·K
200	1.7684	1.0061×10^{3}	1.3289×10^{-5}	7.514×10^{-6}	0.01809
250	1.4128	1.0053×10^{3}	1.488×10^{-5}	10.5×10^{-6}	0.02227
300	1.1774	1.0057×10^{3}	1.983×10^{-5}	16.8×10^{-6}	0.02624
350	0.9980	1.0090×10^{3}	2.075×10^{-5}	20.8×10^{-6}	0.03003
400	0.8826	1.0140×10^{3}	2.286×10^{-5}	25.9×10^{-6}	0.03365
500	0.7048	1.0295×10^{3}	2.671×10^{-5}	37.9×10^{-6}	0.04038
600	0.5879	1.0551×10^{3}	3.018×10^{-5}	51.3×10^{-6}	0.04659
700	0.5030	1.0752×10^{3}	3.332×10^{-5}	66.3×10^{-6}	0.05230
800	0.4405	1.0978×10^{3}	3.625×10^{-5}	82.3×10^{-6}	0.05779
900	0.3925	1.1212×10^{3}	3.899×10^{-5}	99.3×10^{-6}	0.06279
1000	0.3524	1.1417×10^{3}	4.152×10^{-5}	117.8×10^{-6}	0.06752
1200	0.3204	1.160×10^{3}	4.44×10^{-5}	138.6×10^{-6}	0.0732
1400	0.2515	1.214×10^{3}	5.17×10^{-5}	205.5×10^{-6}	0.0891
1600	0.2211	1.248×10^{3}	5.63×10^{-5}	254.5×10^{-6}	0.100
1800	0.1970	1.287×10^{3}	6.07×10^{-5}	308.1×10^{-6}	0.111
2000	0.1762	1.338×10^{3}	6.50×10^{-5}	369.0×10^{-6}	0.124

Note: Notation listed at bottom of Table 1.7.

Table 1.9: Thermal Properties of Solid Materials in I-P Units

Material	Density ρ, lb/ft^3	Specific Heat C_p, Btu/lb·°F	Thermal Conductivity k, Btu/h·ft·°F	Thermal Inertia $k\rho C$, Btu^2/ft^4·h·°F
Aluminum (pure)	169	0.21	119	4300
Steel (0.5% carbon)	490	0.11	31	1700
Copper (pure)	558	0.091	169	8490
Concrete	150	0.18	0.92	25
Brick	162	0.19	0.46	14
Glass, plate	169	0.19	0.44	14
Brick/concrete block	119	0.20	0.42	10
Gypsum wallboard	59.9	0.26	0.10	1.5
Plywood	33.7	0.60	0.07	1.4
Fiber insulation board	15.0	0.30	0.31	1.4
Chipboard	49.9	0.30	0.087	1.3
Aerated concrete	31.2	0.23	0.15	1.1
Plasterboard	59.3	0.20	0.092	1.1
Calcium silicate board	43.7	0.27	0.064–0.081	0.74–0.95
Alumina silicate block	16.2	0.24	0.081	0.31
Glass fiber insulation	3.75	0.19	0.021	0.015
Expanded polystyrene	1.25	0.36	0.020	0.0088

Table 1.10: Thermal Properties of Solid Materials in SI Units

Material	Density ρ, kg/m^3	Specific Heat C_p, kJ/kg K	Thermal Conductivity $k \times 10,^3$ kW/m K	Thermal Inertia $k\rho C$, kW^2 /m^4·K^2·s
Aluminum (pure)	2710	0.895	206	500
Steel (0.5% carbon)	7850	0.465	54	197
Copper (pure)	8940	0.381	293	998
Concrete	2400	0.75	1.6	2.9
Brick	2600	0.8	0.8	1.7
Glass, plate	2710	0.8	0.76	1.6
Brick/concrete block	1900	0.84	0.73	1.2
Gypsum wallboard	960	1.1	0.17	0.18
Plywood	540	2.5	0.12	0.16
Fiber insulation board	240	1.25	0.53	0.16
Chipboard	800	1.25	0.15	0.15
Aerated concrete	500	0.96	0.26	0.12
Plasterboard	950	0.84	0.16	0.12
Calcium silicate board	700	1.12	0.11–0.14	0.086–0.11
Alumina silicate block	260	1	0.14	0.036
Glass fiber insulation	60	0.8	0.037	0.0018
Expanded polystyrene	20	1.5	0.034	0.0010

Example 1.2. Standard Barometric Pressure

What is the standard barometric pressure at Pikes Peak, Colorado? The elevation there is $z = 14{,}115$ ft (4302 m).

$$p = 14.6959[1 - 6.87559 \times (10^{-6}14,115)]^{5.2559} = 8.59 \text{ psi (59.3 kPa) standard barometric pressure}$$

NOMENCLATURE

D_h = hydraulic diameter of flow path, in (m)

g = acceleration of gravity

p = barometric pressure, psi (kPa)

p_i = pressure at node *i*

p_j = pressure at node *j*

R_e = Reynolds number, dimensionless

T = temperature, °F (°C)

T_C = temperature in degrees Celsius

T_K = temperature in kelvin

T_F = temperature in degrees Fahrenheit

T_R = temperature in degrees Rankine

V = average velocity in flow path, fpm (m/s)

z = altitude, ft (m)

z_i = elevation of node *i*

z_j = elevation of node *j*

ΔT_C = temperature difference in degrees Celsius

ΔT_F = temperature difference in degrees Fahrenheit

ΔT_K = temperature difference in kelvin

ΔT_R = temperature difference in degrees Rankine

ν = kinematic viscosity, ft^2/s (m^2/s)

r_i = density of gas at node *i*

Δp_{ij} = pressure difference from node i to node *j*

REFERENCES

IEEE/ASTM. 2002. *Standard for Use of the International System of Units (SI): The Modern Metric System.* New York: Institute of Electrical and Electronic Engineers.

NASA. 1976. U.S. Standard Atmosphere. National Oceanic and Atmospheric Administration, National Aeronautics and Space Administration, and the United States Air Force. Available from the National Geophysical Data Center, Bolder CO.

Thompson, A., and B.N. Taylor. 2008. *Guide for the Use of the International System of Units (SI), NIST Special Publication 811,* 2nd ed. Gaithersburg, MD: National Institute of Standards and Technology.

CHAPTER 2

Climatic Design Data

John H. Klote

Outdoor temperature and wind data are needed for the design and analysis of smoke control systems, and this chapter provides such data for locations in the U.S., Canada, and many other countries. Standard barometric pressures are also provided.

CLIMATIC DATA

Climatic data in IP and SI units for winter and summer design temperatures plus extreme wind speeds are listed in Tables 2.1 and 2.2. This climatic data was from a study by Thevenard (2009), which was funded by ASHRAE. For information about the source data used for Thevenard's project, see Lott, Baldwin, and Jones (2001) and *Data Documentation for Data Set 3505* (NCDC 2003). The design values of extreme wind speed are based on work of Lamming and Salmon (1998). The data in Tables 2.1 and 2.2 are the same as corresponding temperature and wind data in *ASHRAE Handbook—Fundamentals* (ASHRAE 2009).

Thevenard's study was for the period from 1982 to 2006. This 25-year period of weather data was a compromise between trying to derive design conditions from the longest possible period and using the most recent data to capture the effects of climate change. The actual amount of data used for a station depended on the amount of missing data. While most stations had 25 years of usable data, some stations had as few as eight years.

Earlier climatic design data were compiled by Hubbard et al. (2004) based on weather data from 1972 to 2001. For most weather stations, the more recent data of Thevenard had small increases in temperatures as compared to that of Hubbard et al. For example, the 99.6% winter design temperatures increased 0.20°F (0.11°C) on average, and the summer design temperatures increased 0.25°F (0.14°C) on average.

Tables 2.1 and 2.2 have data for 1663 weather stations around the world. Of these stations, 726 are in the U.S. and 136 are in Canada. These stations include all North American cities and towns in of Thevenard's study that have populations of 10,000 or more plus locations of special interest, such as resorts. For climatic design data for locations not included in Table 2.1, see the CD-ROM that accompanies the 2009 *ASHRAE Handbook—Fundamentals.*

As a convenience, the station names used in Table 2.1 and 2.2 are the commonly used names of the locations and not the World Meteorological Organization (WMO) station identifiers. The WMO station identifiers are: (1) all capital letters, (2) of inconsistent format, and (3) do not always correspond to current station names. For these reasons, the WMO identifiers are not used in Table 2.1.

Most of the stations in Tables 2.1 and 2.2 are airports, and most of these stations have a term such as airport, field, or air field in their name to identify them. Stations at many military sites are also airports, and many of these have abbreviations such as AAF (Army Airfield), AFB (Air Force Base), ANGB (Air National Guard Base), ARB (Air Reserve Base), MCAS (Marine Corps Air Station), NAS (Naval Air Station), and RAAF (Royal Australian Air Force).

The names of many of the civilian airports and military airports indicate their geographical location. For other airports, the name is not indicative of location, and for these the stations name in the table consists of the name of a nearby city followed by the airport name. For example, Andrews AFB is located in Maryland, but it is part of the Washington, DC metropolitan area. The station name used in the table for this airport is *Washington, DC, Andrews AFB*. Some weather stations are not at airports, and the names of such stations are those of the cities they are in or near.

Standard Barometric Pressure

As a convenience, standard barometric pressures are included in Table 2.1. These pressures were calculated from the station elevation using the equation for pressure of the U.S. Standard Atmosphere (Chapter 1) (NASA 1976). For example, the elevation of Denver Stapleton International Airport in Colorado is 5285 ft (1611 m), and the standard barometric pressure at this altitude is 12.1 psi (83.4 kPa). This differs from the standard barometric pressure at sea level, which is 14.696 psi (101.325 kPa).

Winter Design Temperature

Tables 2.1 and 2.2 list winter design temperatures. These are the dry bulb temperatures corresponding to 99.6% annual cumulative frequency of occurrence. For example, the 99.6% winter design temperature for Washington Dulles International Airport in Virginia is 10.7°F (–11.8°C). This means that the temperature at Dulles Airport can be thought of as being above 10.7°F (–11.8°C) for 99.6% of the year.

Summer Design Temperature

Tables 2.1 and 2.2 list summer design temperatures, which are the dry bulb temperatures corresponding to 0.4% annual cumulative frequency of occurrence. For example, the 0.4% summer design value at Dulles Airport is 93.5°F (34.1°C). The temperature at Dulles Airport can be thought of as being above 93.5°F (34.1°C) for only 0.4% of the year.

Design Wind

Tables 2.1 and 2.2 list extreme wind speeds corresponding to 1% annual cumulative frequency of occurrence. This is the same as the 1% extreme wind speed in *ASHRAE Handbook—Fundamentals* (2009). For example, the 1% extreme wind at Dulles Airport is 20.5 mph (9.2 m/s), which means that the wind at Dulles Airport is above 20.5 mph (9.2 m/s) for only 1% of the year. NOAA (1998) provides data regarding prevailing winds for a number of locations in the U.S. Some readers may notice that the design wind speed for smoke control systems is much lower than that for structures. This is because smoke control systems need to withstand the wind for the relatively short duration of system operation, but structures need to withstand the wind over the entire life of the structure.

Table 2.1: Climatic Data in I-P Units

Station	Latitude, degrees	Longitude, degrees	Elevation, ft	St. Br. Pressure, psi	Winter Temp., °F	Summer Temp., °F	Wind, mph
United States of America							
Alabama							
Anniston Metropolitan Airport	33.59N	85.86W	600	14.38	20.1	94.5	16.8
Auburn Opelika Airport	32.62N	85.43W	774	14.29	23.7	93.0	17.5
Birmingham Municipal Airport	33.56N	86.75W	630	14.36	19.6	95.0	18.3
Cairns AAF/Ozark	31.28N	85.72W	299	14.54	26.9	95.4	16.9
Dothan Municipal Airport	31.32N	85.45W	322	14.53	27.3	95.3	19.2
Gadsen Municipal Airport	33.97N	86.08W	568	14.40	18.6	93.4	16.7
Huntsville Intl, Jones Field	34.64N	86.79W	643	14.36	17.0	94.6	21.5
Mobile Regional Airport	30.69N	88.25W	220	14.58	26.9	93.5	20.6
Montgomery, Dannelly Field	32.30N	86.39W	203	14.59	23.7	96.2	18.6
Montgomery, Maxwell AFB	32.38N	86.37W	174	14.60	27.9	97.2	18.0
Muscle Shoals Regional Airport	34.75N	87.61W	561	14.40	17.8	95.5	18.7
Tuscaloosa Municipal Airport	33.21N	87.62W	187	14.60	20.6	95.7	17.5
Alaska							
Anchorage International Airport	61.18N	149.99W	131	14.63	–8.9	71.4	20.7
Anchorage, Elmendorf AFB	61.25N	149.80W	194	14.59	–14.8	73.7	18.8

Table 2.1: Climatic Data in I-P Units ***(Continued)***

Station	Latitude, degrees	Longitude, degrees	Elevation, ft	St. Br. Pressure, psi	Winter Temp., °F	Summer Temp., °F	Wind, mph
Anchorage, Lake Hood Seaplane	61.18N	149.96W	131	14.63	–8.7	74.5	18.7
Bethel Airport	60.79N	161.83W	151	14.62	–27.1	72.4	30.6
Fairbanks International Airport	64.82N	147.86W	453	14.46	–43.3	81.2	17.7
Fort Richardson	61.27N	149.65W	377	14.50	–18.6	73.9	19.0
Juneau International Airport	58.36N	134.58W	23	14.68	3.8	73.5	26.6
Kenai Municipal Airport	60.58N	151.24W	92	14.65	–22.3	69.9	24.1
Ketchikan International Airport	55.36N	131.71W	95	14.65	12.6	71.4	24.6
Kodiak Airport	57.75N	152.49W	112	14.64	8.6	68.9	33.4
Nome Municipal Airport	64.51N	165.44W	23	14.68	–29.0	68.5	28.1
Palmer Municipal Airport	61.60N	149.09W	249	14.56	–16.4	74.7	31.3
Sitka, Japonski Airport	57.05N	135.36W	66	14.66	16.4	66.7	24.5
Arizona							
Casa Granda Municipal Airport	32.95N	111.77W	1463	13.94	32.0	108.5	20.3
Douglas, Bisbee Intl Airport	31.47N	109.60W	4101	12.64	22.6	99.5	23.9
Flagstaff, Pulliam Airport	35.13N	111.67W	7005	11.34	3.9	85.5	20.9
Nogales International Airport	31.42N	110.85W	4055	12.67	26.9	99.9	19.4
Phoenix Sky Harbor Intl Airport	33.44N	111.99W	1106	14.12	38.6	110.2	18.3
Phoenix, Luke AFB	33.53N	112.38W	1086	14.13	35.3	110.8	19.7
Prescott, Love Field	34.65N	112.42W	5052	12.20	17.5	94.3	20.8
Safford	32.82N	109.68W	3117	13.11	25.4	104.3	22.2
Tucson International Airport	32.13N	110.96W	2556	13.39	31.7	105.9	21.4
Tucson, Davis Monthan AFB	32.17N	110.88W	2654	13.34	32.9	105.4	19.6
Winslow Municipal Airport	35.02N	110.72W	4882	12.28	11.1	97.3	27.2
Yuma International Airport	32.65N	114.60W	207	14.59	41.8	110.8	20.7
Yuma MCAS	32.65N	114.62W	213	14.58	41.7	110.9	20.8
Arkansas							
Bentonville Municipal Airport	36.35N	94.22W	1296	14.02	10.1	93.5	19.5
Blytheville, Eaker AFB	35.97N	89.95W	262	14.56	11.7	97.0	22.8
El Dorado, Goodwin Field	33.22N	92.81W	285	14.54	21.7	98.8	17.0
Fayetteville, Drake Field	36.01N	94.17W	1260	14.04	8.0	95.2	20.5
Flippin	36.30N	92.47W	1148	14.10	12.1	97.2	16.8
Fort Smith Regional Airport	35.33N	94.37W	463	14.45	14.7	99.1	20.4
Harrison Airport	36.26N	93.16W	1385	13.98	9.8	94.4	20.5
Jonesboro Municipal Airport	35.83N	90.63W	269	14.55	12.4	96.4	20.9
Little Rock AFB	34.92N	92.15W	338	14.52	15.3	99.3	17.6
Little Rock, Adams Field	34.75N	92.23W	256	14.56	17.2	98.0	18.6
North Little Rock Municipal Airport	34.83N	92.25W	1152	14.09	16.4	95.2	18.4
Pine Bluff Airport	34.18N	91.94W	213	14.58	21.1	97.1	18.5
Rogers Field	36.37N	94.10W	1362	13.99	10.0	93.4	20.8
Siloam Spring	36.18N	94.48W	1194	14.07	10.3	95.5	22.8
Texarkana Municipal Airport	33.45N	94.01W	400	14.48	21.6	98.5	18.8

Table 2.1: Climatic Data in I-P Units *(Continued)*

Station	Latitude, degrees	Longitude, degrees	Elevation, ft	St. Br. Pressure, psi	Winter Temp., °F	Summer Temp., °F	Wind, mph
California							
Alameda NAS	37.73N	122.32W	13	14.69	40.3	83.4	20.6
Bakersfield, Meadows Field	35.43N	119.06W	492	14.44	32.2	103.1	18.2
Marysville, Beale AFB	39.13N	121.43W	125	14.63	32.1	100.7	21.4
Blythe Airport	33.62N	114.72W	394	14.49	36.3	112.7	24.3
Burbank Glendale Pasadena Apt	34.20N	118.36W	732	14.31	39.0	98.3	18.4
Oxnard AFB	34.22N	119.08W	75	14.66	37.3	84.6	24.9
Camp Pendleton MCAS	33.30N	117.35W	79	14.65	32.8	92.0	16.9
Carlsbad, Palomar Airport	33.13N	117.28W	328	14.52	43.0	81.7	14.1
Merced, Castle AFB	37.38N	120.57W	197	14.59	30.6	99.5	18.2
Crescent City, McNamara Field	41.78N	124.24W	56	14.67	35.6	68.1	28.1
El Toro MCAS	33.68N	117.73W	384	14.49	40.3	92.1	14.4
Eureka	40.80N	124.17W	59	14.66	34.2	68.2	21.9
Fresno Air Terminal	36.78N	119.72W	328	14.52	31.5	103.6	18.1
Fullerton Municipal Airport	33.87N	117.98W	95	14.65	39.2	92.7	13.4
Victorville, George AFB	34.58N	117.38W	2874	13.23	27.6	100.7	22.7
Hayward Executive Airport	37.66N	122.12W	46	14.67	37.1	87.6	19.8
Imperial County Airport	32.83N	115.58W	–56	14.73	35.6	111.1	25.9
Hawthorne Municipal Airport	33.92N	118.33W	69	14.66	44.7	85.9	16.3
Lancaster, Gen Wm Fox Field	34.74N	118.22W	2339	13.50	21.5	102.3	29.7
Lemoore NAS	36.33N	119.95W	233	14.57	29.7	103.0	19.0
Livermore Municipal Airport	37.69N	121.82W	397	14.49	30.3	98.8	19.6
Lompoc Airport	34.67N	120.47W	89	14.65	32.4	81.4	20.3
Long Beach Municipal Airport	33.83N	118.16W	39	14.68	41.2	91.2	17.5
Los Angeles Intl Airport	33.94N	118.41W	325	14.52	44.4	83.7	20.0
Riverside, March AFB	33.88N	117.27W	1516	13.91	34.2	100.6	17.9
Sacramento, McClellan AFB	38.67N	121.40W	82	14.65	31.7	101.9	20.5
Modesto Municipal Airport	37.63N	120.95W	98	14.64	31.0	101.1	18.9
Monterey Peninsula	36.58N	121.85W	164	14.61	36.7	77.3	17.0
Mount Shasta	41.33N	122.33W	3537	12.91	17.5	90.8	12.8
Mountain View, Moffett NAS	37.42N	122.05W	33	14.68	36.4	88.2	18.8
Napa County Airport	38.21N	122.28W	56	14.67	30.0	91.2	21.3
San Bernardino, Norton AFB	34.10N	117.23W	1158	14.09	33.9	102.9	16.7
Oakland International Airport	37.76N	122.22W	89	14.65	37.2	81.8	23.1
Ontario	34.07N	117.65W	997	14.17	36.0	100.6	23.1
Palm Springs Intl Airport	33.83N	116.50W	476	14.44	42.7	111.2	22.9
Palm Springs, Thermal Airport	33.63N	116.16W	–118	14.76	31.0	111.3	19.2
Paso Robles Municipal Airport	35.67N	120.63W	817	14.27	26.7	102.1	21.8
Point Arguello	34.57N	120.63W	112	14.64	45.6	71.2	42.4
Point Mugu NAS	34.12N	119.12W	13	14.69	38.9	82.0	22.8
Porterville Airport	36.03N	119.07W	443	14.46	30.4	100.4	12.8

Table 2.1: Climatic Data in I-P Units *(Continued)*

Station	Latitude, degrees	Longitude, degrees	Elevation, ft	St. Br. Pressure, psi	Winter Temp., °F	Summer Temp., °F	Wind, mph
Redding Municipal Airport	40.52N	122.31W	502	14.43	28.4	105.9	24.9
Riverside Municipal Airport	33.95N	117.43W	840	14.26	36.0	99.8	20.7
Sacramento Executive Airport	38.51N	121.49W	26	14.68	31.4	100.1	20.4
Sacramento, Mather Airport	38.55N	121.29W	95	14.65	29.8	101.3	20.4
Sacramento Metropolitan Airport	38.70N	121.59W	33	14.68	31.5	100.4	23.3
Salinas Municipal Airport	36.66N	121.61W	79	14.65	33.8	82.9	20.9
San Diego International Airport	32.74N	117.17W	30	14.68	44.8	84.1	17.6
San Diego, Miramar MCAS	32.87N	117.15W	479	14.44	38.9	90.3	14.6
San Diego, North Island NAS	32.70N	117.20W	26	14.68	44.6	82.3	18.8
San Diego, Brown Field	32.57N	116.98W	522	14.42	38.9	88.3	16.4
San Diego, Montgomery Field	32.82N	117.13W	423	14.47	40.9	88.3	15.9
San Francisco Intl Airport	37.62N	122.40W	20	14.69	38.8	83.0	28.6
San Jose International Airport	37.36N	121.93W	49	14.67	35.7	92.3	19.8
San Luis Obispo County Airport	35.23N	120.63W	217	14.58	34.0	88.3	25.7
Santa Barbara Municipal Airport	34.43N	119.84W	20	14.69	34.7	82.9	19.4
Santa Maria Public Airport	34.92N	120.47W	240	14.57	32.2	84.2	23.8
Sonoma County Airport	38.51N	122.81W	148	14.62	29.7	95.3	17.0
Stockton Metropolitan Airport	37.89N	121.24W	26	14.68	30.5	100.8	22.3
Fairfield, Travis AFB	38.27N	121.93W	59	14.66	31.8	98.8	28.3
Truckee Tahoe Airport	39.32N	120.13W	5899	11.82	0.9	88.5	22.0
Tustin MCAF	33.70N	117.83W	56	14.67	38.9	93.5	17.0
Ukiah Municipal Airport	39.13N	123.20W	627	14.37	29.6	99.9	16.0
Visalia Municipal Airport	36.32N	119.40W	292	14.54	29.9	99.8	14.7
Colorado							
Alamosa Municipal Airport	37.44N	105.87W	7543	11.11	–15.4	85.2	27.0
Aspen-Pitkin County Airport	39.22N	106.87W	8018	10.91	–2.4	84.3	18.3
Colorado Springs Airport	38.81N	104.71W	6171	11.70	–0.7	90.3	27.7
Cortez Montezuma Co Airport	37.30N	108.63W	5915	11.81	3.3	93.2	23.0
Craig Moffat Airport	40.50N	107.53W	6283	11.65	–15.4	90.1	24.8
Denver International Airport	39.83N	104.66W	5430	12.03	0.7	94.3	26.7
Denver Stapleton Intl Airport	39.77N	104.87W	5285	12.10	–4.0	93.5	24.3
Denver, Buckley AFB	39.72N	104.75W	5663	11.93	–0.2	93.0	23.4
Denver, Centennial Airport	39.57N	104.85W	5883	11.83	–1.8	91.4	24.8
Fort Collins	40.58N	105.08W	5003	12.23	–4.8	89.8	20.8
Fort Collins Loveland Airport	40.45N	105.02W	5016	12.22	0.1	93.5	25.6
Grand Junction, Walker Field	39.13N	108.54W	4839	12.30	6.0	97.4	23.1
Greeley, Weld County Airport	40.43N	104.63W	4659	12.38	–5.5	95.4	27.7
Lamar Municipal Airport	38.07N	102.68W	3704	12.83	2.6	99.8	28.4
Montrose County Airport	38.50N	107.90W	5758	11.88	6.7	93.4	22.6
Pueblo Memorial Airport	38.29N	104.50W	4721	12.36	–2.0	98.4	28.8
Rifle, Garfield County Airport	39.53N	107.72W	5548	11.98	1.4	96.6	22.5

Table 2.1: Climatic Data in I-P Units *(Continued)*

Station	Latitude, degrees	Longitude, degrees	Elevation, ft	St. Br. Pressure, psi	Winter Temp., °F	Summer Temp., °F	Wind, mph
Trinidad, Las Animas Co Airport	37.26N	104.34W	5745	11.89	1.3	92.9	27.3
Connecticut							
Bridgeport, Sikorsky Airport	41.18N	73.15W	16	14.69	10.7	87.4	24.5
Hartford, Bradley Intl Airport	41.94N	72.68W	180	14.60	3.1	91.5	22.3
Hartford Brainard Field	41.74N	72.65W	20	14.69	6.4	90.5	20.0
Waterbury, Oxford Airport	41.48N	73.13W	728	14.31	3.1	87.7	19.7
Willimantic	41.73N	72.18W	249	14.56	3.1	89.9	19.6
Delaware							
Dover AFB	39.13N	75.47W	23	14.68	13.9	92.1	24.2
Greater Wilmington Airport	39.67N	75.60W	79	14.65	11.7	91.9	24.4
Florida							
Crestview, Bob Sikes Airport	30.78N	86.52W	184	14.60	24.1	95.3	17.5
Daytona Beach Intl Airport	29.18N	81.06W	43	14.67	34.7	92.7	20.3
Destin-Fort Walton Beach Apt	30.40N	86.47W	23	14.68	34.1	91.0	18.8
Fort Lauderdale-Hollywood Apt	26.07N	80.15W	10	14.69	46.7	91.8	22.4
Fort Myers Metropolitan Airport	26.59N	81.86W	20	14.69	42.4	93.7	18.9
Fort Myers, SW Florida Airport	26.53N	81.75W	30	14.68	41.2	93.8	20.7
Gainesville Regional Airport	29.69N	82.27W	164	14.61	29.7	93.5	18.5
Homestead AFB	25.48N	80.38W	16	14.69	46.4	91.2	18.9
Hurlburt Field	30.43N	86.68W	39	14.68	29.2	92.1	18.9
Jacksonville International Airport	30.49N	81.69W	33	14.68	29.2	94.5	20.0
Jacksonville NAS	30.23N	81.68W	23	14.68	32.2	95.5	20.4
Jacksonville, Cecil Field	30.22N	81.87W	89	14.65	29.8	96.3	18.7
Jacksonville, Craig Airport	30.34N	81.52W	43	14.67	31.6	93.4	18.9
Jacksonville, Mayport NS	30.40N	81.42W	13	14.69	34.4	94.5	20.1
Kennedy Space Center	28.62N	80.72W	10	14.69	38.4	91.9	19.0
Key West International Airport	24.55N	81.75W	20	14.69	54.9	90.6	23.0
Marathon Airport	24.73N	81.05W	7	14.69	54.0	91.4	19.6
Melbourne International Airport	28.10N	80.65W	26	14.68	39.0	92.6	20.9
Miami International Airport	25.82N	80.30W	30	14.68	47.7	91.8	20.6
Miami, Kendall Tamiami Airport	25.65N	80.43W	10	14.69	45.4	92.4	20.6
Naples Municipal Airport	26.15N	81.78W	23	14.68	43.8	90.9	18.8
Ocala Municipal Airport	29.17N	82.22W	89	14.65	29.7	93.2	17.8
Orlando Executive Airport	28.55N	81.33W	112	14.64	40.0	93.6	19.0
Orlando International Airport	28.43N	81.33W	105	14.64	37.7	93.7	20.1
Orlando, Central Florida Airport	28.78N	81.24W	56	14.67	38.6	94.9	20.4
Panama City Bay County Airport	30.20N	85.68W	20	14.69	32.1	92.7	18.8
Pensacola NAS	30.35N	87.32W	30	14.68	28.3	93.2	23.5
Pensacola Regional Airport	30.47N	87.19W	118	14.63	29.5	93.7	20.0
Sarasota Bradenton Airport	27.38N	82.55W	33	14.68	39.3	92.2	21.7
St Petersburg Clearwater Airport	27.90N	82.68W	10	14.69	42.6	93.2	20.9

Table 2.1: Climatic Data in I-P Units *(Continued)*

Station	Latitude, degrees	Longitude, degrees	Elevation, ft	St. Br. Pressure, psi	Winter Temp., °F	Summer Temp., °F	Wind, mph
Tallahassee Municipal Airport	30.39N	84.35W	69	14.66	25.4	95.5	17.9
Tampa International Airport	27.96N	82.54W	10	14.69	38.4	92.4	18.1
Tampa, MacDill AFB	27.85N	82.52W	26	14.68	40.3	93.1	18.7
Tyndall AFB	30.07N	85.58W	23	14.68	31.9	91.2	19.8
Valparaiso, Eglin AFB	30.48N	86.53W	66	14.66	28.4	92.6	20.0
Venice Municipal Airport	27.07N	82.45W	16	14.69	42.3	87.6	28.2
Vero Beach Municipal Airport	27.66N	80.42W	30	14.68	38.9	91.7	20.2
West Palm Beach Intl Airport	26.69N	80.10W	20	14.69	44.2	91.4	23.1
Georgia							
Albany, Dougherty Co Airport	31.54N	84.19W	194	14.59	26.4	96.7	18.6
Athens, Ben Epps Airport	33.95N	83.33W	801	14.28	21.8	95.2	18.4
Atlanta	33.37N	84.55W	971	14.19	19.1	92.8	17.6
Atlanta, Dekalb Peachtree Apt	33.87N	84.30W	1027	14.16	22.6	93.3	18.3
Atlanta. Hartsfield Intl Airport	33.64N	84.43W	1027	14.16	20.7	93.8	22.0
Augusta, Bush Field	33.37N	81.97W	148	14.62	22.2	97.1	18.8
Augusta, Daniel Field	33.47N	82.03W	420	14.47	27.5	96.8	16.8
Brunswick, M McKinnon Airport	31.25N	81.39W	23	14.68	30.1	92.7	18.6
Columbus Metropolitan Airport	32.52N	84.94W	394	14.49	25.1	96.3	18.2
Fort Benning, Lawson AAF	32.33N	85.00W	289	14.54	23.0	96.7	16.6
Fulton County Airport Brown	33.77N	84.52W	863	14.24	19.8	93.6	18.0
Gainesville, Lee Gilmer Airport	34.27N	83.83W	1276	14.03	23.0	91.3	19.1
Macon, Middle Georgia Reg Apt	32.69N	83.65W	361	14.51	23.5	96.7	18.2
Mariett, Dobbins AFB	33.92N	84.52W	1083	14.13	19.5	93.5	18.7
Rome, R B Russell Airport	34.35N	85.16W	643	14.36	17.2	96.4	15.1
Savannah International Airport	32.12N	81.20W	52	14.67	27.1	95.5	18.8
Savannah, Hunter AAF	32.00N	81.13W	43	14.67	28.0	95.4	18.8
Valdosta Regional Airport	30.78N	83.28W	197	14.59	27.5	95.4	17.1
Valdosta, Moody AFB	30.97N	83.20W	233	14.57	29.9	95.7	16.2
Warner Robins AFB	32.63N	83.60W	302	14.54	25.0	97.3	19.0
Waycross Ware County Airport	31.25N	82.40W	141	14.62	27.9	96.2	16.7
Hawaii							
Hilo International Airport	19.72N	155.05W	36	14.68	61.5	85.6	17.6
Honolulu International Airport	21.33N	157.94W	16	14.69	61.2	89.9	21.8
Kahului Airport	20.90N	156.43W	49	14.67	58.8	89.7	26.7
Kalaeloa Airport (Barbers Point)	21.30N	158.07W	33	14.68	59.4	90.9	19.6
Kaneohe Bay MCAS	21.45N	157.77W	20	14.69	64.3	85.3	19.3
Keahol, Kona Intl Airport	19.73N	156.03W	49	14.67	65.7	88.2	19.8
Lihue Airport	21.98N	159.34W	148	14.62	60.3	85.3	26.1
Idaho							
Boise Air Terminal	43.57N	116.22W	2867	13.24	2.7	98.1	21.8
Caldwell	43.63N	116.63W	2428	13.45	11.6	97.0	21.4

Table 2.1: Climatic Data in I-P Units *(Continued)*

Station	Latitude, degrees	Longitude, degrees	Elevation, ft	St. Br. Pressure, psi	Winter Temp., °F	Summer Temp., °F	Wind, mph
Coeur d'Alene Airport	47.77N	116.82W	2320	13.50	6.7	91.4	22.4
Idaho Falls, Fanning Field	43.52N	112.07W	4744	12.35	–8.7	91.7	27.2
Lewiston Nez Perce Co Airport	46.38N	117.01W	1437	13.95	10.2	97.8	20.6
Mountain Home AFB	43.05N	115.87W	2992	13.17	2.0	99.5	25.4
Pocatello Regional Airport	42.92N	112.57W	4478	12.47	–4.9	94.4	28.1
Salmon Airport	45.12N	113.88W	4045	12.67	–5.5	90.4	20.1
Twin Falls, Magic Valley Airport	42.48N	114.48W	4255	12.57	9.4	94.5	27.5
Illinois							
Aurora Municipal Airport	41.77N	88.47W	705	14.33	–2.0	90.8	25.9
Belleville, Scott AFB	38.55N	89.85W	443	14.46	4.5	95.1	21.5
Cahokia/St. Louis Airport	38.57N	90.15W	413	14.48	9.5	93.5	20.6
Charleston/Mattoon, Coles Apt	39.48N	88.28W	722	14.32	3.2	91.1	24.2
Chicago Midway Airport	41.79N	87.75W	617	14.37	–1.6	92.1	24.4
Chicago O'Hare Intl Airport	41.99N	87.91W	673	14.34	–4.0	91.9	24.8
Decatur Airport	39.83N	88.87W	699	14.33	–0.4	93.0	24.8
Glenview NAS	42.08N	87.82W	653	14.35	–4.6	93.2	21.0
Lawrenceville Municipal Airport	38.77N	87.60W	430	14.47	6.8	93.0	24.3
Moline, Quad City Airport	41.47N	90.52W	594	14.38	–6.2	93.5	24.6
Mount Vernon Airport	38.32N	88.87W	479	14.44	4.6	93.3	20.5
Peoria, Greater Peoria Airport	40.67N	89.68W	663	14.35	–3.6	92.7	23.8
Quincy Municipal Airport	39.94N	91.19W	768	14.29	–2.0	93.3	24.7
Rockford, Greater Rockford Apt	42.20N	89.09W	745	14.30	–8.4	91.5	24.6
Springfield, Capital Airport	39.85N	89.68W	614	14.37	–2.1	92.9	24.9
Urbana/Champaign, Willard Apt	40.03N	88.27W	774	14.29	–1.0	92.5	27.4
W Chicago, Du Page Airport	41.92N	88.25W	758	14.30	–5.4	90.5	24.5
Indiana							
Bloomington, Monroe Co Apt	39.13N	86.62W	866	14.24	4.8	90.6	19.5
Evansville Regional Airport	38.04N	87.54W	387	14.49	5.6	93.9	20.7
Fort Wayne Municipal Airport	41.01N	85.21W	827	14.26	–2.6	91.1	24.6
Grissom ARB	40.65N	86.15W	830	14.26	–3.9	92.4	24.9
Indianapolis International Airport	39.71N	86.27W	807	14.27	–0.5	91.1	24.5
Lafayette, Purdue Univ Airport	40.41N	86.94W	636	14.36	–2.3	92.4	23.1
South Bend, Michiana Airport	41.71N	86.33W	774	14.29	–1.5	90.9	24.1
Terre Haute, Hulman Airport	39.45N	87.32W	574	14.39	–0.3	92.1	22.8
Iowa							
Ames Municipal Airport	42.00N	93.62W	955	14.20	–5.8	90.5	26.5
Ankeny Regional Airport	41.68N	93.55W	902	14.22	–5.4	94.9	20.7
Boone Municipal Airport	42.05N	93.85W	1161	14.09	–5.8	91.3	26.4
Burlington Municipal Airport	40.78N	91.13W	702	14.33	–3.5	93.2	21.4
Carroll, Neu Airport	42.05N	94.78W	1230	14.05	–6.1	93.0	28.9
Cedar Rapids Municipal Airport	41.88N	91.71W	873	14.24	–9.6	91.6	26.3

Table 2.1: Climatic Data in I-P Units *(Continued)*

Station	Latitude, degrees	Longitude, degrees	Elevation, ft	St. Br. Pressure, psi	Winter Temp., °F	Summer Temp., °F	Wind, mph
Clinton Municipal Airport	41.83N	90.33W	709	14.32	–6.3	90.5	28.0
Davenport Municipal Airport	41.62N	90.58W	850	14.25	–3.4	90.3	26.4
Des Moines International Airport	41.54N	93.67W	965	14.19	–6.9	93.4	25.6
Dubuque Regional Airport	42.40N	90.70W	1079	14.13	–8.2	89.3	25.6
Fort Dodge Regional Airport	42.55N	94.18W	1165	14.09	–11.0	91.0	25.1
Keokuk Municipal Airport	40.47N	91.43W	673	14.34	1.7	94.7	22.0
Marshalltown Municipal Airport	42.10N	92.92W	971	14.19	–7.7	90.2	27.6
Mason City Municipal Airport	43.16N	93.33W	1224	14.06	–13.8	91.0	28.3
Ottumwa Industrial Airport	41.11N	92.45W	846	14.25	–4.7	93.3	28.0
Sioux City, Sioux Gateway Apt	42.39N	96.38W	1102	14.12	–8.8	93.4	28.7
Spencer	43.17N	95.15W	1339	14.00	–13.6	91.2	28.2
Storm Lake Municipal Airport	42.60N	95.23W	1490	13.92	–8.4	91.4	29.6
Waterloo Municipal Airport	42.55N	92.40W	879	14.24	–12.7	91.8	26.2
Kansas							
Coffeyville Municipal Airport	37.10N	95.57W	755	14.30	10.3	99.6	24.5
Dodge City Regional Airport	37.77N	99.97W	2592	13.37	2.0	100.1	30.9
Fort Riley, Marshall Airport	39.05N	96.77W	1063	14.14	–1.4	99.6	20.7
Garden City Municipal Airport	37.93N	100.73W	2881	13.23	–0.1	99.4	29.9
Great Bend Municipal Airport	38.35N	98.87W	1886	13.72	3.1	99.8	30.1
Hays Municipal Airport	38.85N	99.27W	1998	13.67	2.6	100.4	29.7
Lawrence Municipal Airport	39.00N	95.22W	833	14.26	5.1	99.1	25.0
Liberal Municipal Airport	37.03N	100.95W	2956	13.19	7.2	99.7	31.8
Manhattan Municipal Airport	39.13N	96.67W	1083	14.13	2.5	99.9	24.2
Newton City County Airport	38.05N	97.28W	1532	13.90	3.4	99.0	28.7
Olathe, Johnson County Airport	38.85N	94.73W	1096	14.12	5.2	96.7	23.4
Salina Municipal Airport	38.81N	97.66W	1283	14.03	0.9	101.3	27.6
Topeka, Billard Muni Airport	39.07N	95.63W	886	14.23	0.6	97.1	23.8
Topeka, Forbes Field	38.95N	95.66W	1079	14.13	1.0	96.7	25.4
Wichita Mid Continent Airport	37.65N	97.43W	1339	14.00	4.0	100.4	28.0
Wichita, Col. Jabara Airport	37.75N	97.22W	1421	13.96	6.9	99.7	27.3
Wichita, McConnel AFB	37.62N	97.27W	1358	13.99	5.2	99.9	26.4
Winfield	37.30N	97.00W	1158	14.09	9.8	100.3	26.2
Kentucky							
Bowling Green Warren Co Apt	36.98N	86.44W	538	14.41	8.5	93.2	20.1
Cincinnati International Airport	39.04N	84.67W	883	14.23	3.1	91.4	22.3
Fort Campbell AAF	36.67N	87.50W	568	14.40	9.9	94.4	19.0
Fort Knox, Godman AAF	37.90N	85.97W	784	14.28	7.3	93.0	17.8
Frankfort, Capital City Airport	38.18N	84.90W	804	14.27	9.6	91.3	18.8
Henderson City Airport	37.82N	87.68W	384	14.49	7.3	92.9	21.0
Lexington, Blue Grass Airport	38.04N	84.61W	988	14.18	6.0	91.7	20.1
Louisville, Bowman Field	38.23N	85.66W	558	14.40	7.1	93.3	18.9

Table 2.1: Climatic Data in I-P Units *(Continued)*

Station	Latitude, degrees	Longitude, degrees	Elevation, ft	St. Br. Pressure, psi	Winter Temp., °F	Summer Temp., °F	Wind, mph
Louisville, Standiford Field	38.18N	85.73W	489	14.44	8.0	93.4	21.0
Paducah, Barkley Reg Airport	37.06N	88.77W	413	14.48	8.6	94.6	20.3
Louisiana							
Alexandria International Airport	31.33N	92.55W	89	14.65	26.7	96.6	18.5
Alexandria, Esler Reg Airport	31.40N	92.30W	118	14.63	26.5	97.3	16.5
Baton Rouge Metropolitan Apt	30.54N	91.15W	75	14.66	27.6	94.2	18.7
Fort Polk, Polk AAF	31.05N	93.20W	335	14.52	26.7	95.8	17.3
Lafayette Regional Airport	30.21N	91.99W	43	14.67	28.6	94.2	20.5
Lake Charles Regional Airport	30.13N	93.23W	10	14.69	29.6	94.1	20.7
Monroe Regional Airport	32.51N	92.04W	82	14.65	23.6	96.9	18.9
New Orleans Lakefront Airport	30.04N	90.03W	10	14.69	35.6	93.4	24.0
New Orleans NAS, Callender Fld	29.83N	90.03W	0	14.70	30.1	92.5	17.7
New Orleans, Armstrong Intl Apt	29.99N	90.25W	20	14.69	31.5	93.6	20.6
Patterson, H P Williams Airport	29.72N	91.33W	10	14.69	31.0	93.2	19.0
Shreveport Downtown Airport	32.54N	93.74W	180	14.60	27.0	99.3	18.7
Shreveport Regional Airport	32.45N	93.82W	259	14.56	23.8	97.8	19.6
Shreveport, Barksdale AFB	32.50N	93.67W	177	14.60	22.5	97.3	19.0
Maine							
Auburn/Lewiston Muni Airport	44.05N	70.28W	289	14.54	–6.8	87.9	20.9
Augusta State Airport	44.32N	69.80W	361	14.51	–3.9	87.0	23.5
Bangor International Airport	44.81N	68.82W	194	14.59	–7.4	87.8	23.5
Brunswick NAS	43.90N	69.93W	75	14.66	–2.5	86.4	22.2
Portland International Jetport	43.64N	70.30W	62	14.66	–0.3	86.8	23.2
Presque Isle Municipal Airport	46.69N	68.03W	535	14.41	–18.8	84.7	23.4
Sanford Municipal Airport	43.40N	70.72W	243	14.57	–6.8	88.4	20.9
Waterville Airport	44.53N	69.68W	331	14.52	–6.6	86.1	20.5
Maryland							
Baltimore Washington Intl Apt	39.17N	76.68W	154	14.61	12.9	93.9	22.4
Salisbury-Ocean City Airport	38.34N	75.51W	59	14.66	13.7	92.8	20.8
Washington DC, Andrews AFB	38.82N	76.87W	282	14.55	13.9	94.0	23.5
Massachusetts							
Barnstable Airport	41.67N	70.27W	56	14.67	9.6	84.0	24.8
Boston, Logan International Apt	42.36N	71.01W	30	14.68	7.4	90.8	26.8
Chatham Municipal Airport	41.68N	70.00W	62	14.66	10.4	81.8	23.3
Lawrence Municipal Airport	42.72N	71.12W	151	14.62	2.8	90.2	20.2
Marthas Vineyard Airport	41.40N	70.62W	69	14.66	8.9	82.4	26.1
New Bedford Municipal Airport	41.67N	70.95W	82	14.65	7.4	88.1	23.2
Norwood Memorial Airport	42.18N	71.18W	49	14.67	2.6	90.4	20.5
Plymouth Municipal Airport	41.92N	70.73W	148	14.62	4.6	88.4	23.3
Provincetown Municipal Airport	42.07N	70.22W	7	14.69	12.4	82.3	30.2
South Weymouth NAS	42.15N	70.93W	161	14.61	5.9	91.2	18.5

Table 2.1: Climatic Data in I-P Units *(Continued)*

Station	Latitude, degrees	Longitude, degrees	Elevation, ft	St. Br. Pressure, psi	Winter Temp., °F	Summer Temp., °F	Wind, mph
Worcester Municipal Airport	42.27N	71.88W	1017	14.16	1.6	85.7	25.8
Michigan							
Alpena County Airport	45.07N	83.58W	692	14.33	–5.6	88.1	20.0
Battle Creek Intl Airport	42.23N	85.55W	896	14.23	3.0	90.1	21.7
Benton Harbor, Ross Field	42.13N	86.43W	643	14.36	7.1	89.9	23.2
Cadillac, Wexford Co Airport	44.28N	85.42W	1306	14.02	–9.0	87.9	20.6
Detroit City Airport	42.41N	83.01W	627	14.37	4.1	90.8	20.4
Detroit Metropolitan Airport	42.22N	83.35W	663	14.35	1.4	90.3	25.6
Detroit Willow Run Airport	42.23N	83.53W	715	14.32	0.8	90.4	24.6
Flint, Bishop Intl Airport	42.97N	83.75W	768	14.29	–1.0	89.4	24.2
Grand Rapids, Kent Co Intl Apt	42.88N	85.52W	804	14.27	1.4	89.3	24.8
Grosse Ile, Municipal Airport	42.10N	83.15W	577	14.39	7.3	89.8	20.8
Holland, Tulip City Airport	42.75N	86.10W	689	14.33	7.4	88.3	25.9
Howell, Livingston Co Airport	42.63N	83.98W	961	14.19	1.2	90.4	20.3
Jackson Reynolds Muni Airport	42.26N	84.46W	1020	14.16	–1.0	88.5	20.3
Lansing, Capital City Airport	42.78N	84.58W	873	14.24	–2.5	89.4	24.7
Mackinac Island Airport	45.87N	84.63W	984	14.18	–10.8	84.6	23.4
Marquette, Sawyer Intl Airport	46.53N	87.55W	1414	13.96	–14.4	84.9	21.3
Menominee Airport	45.13N	87.63W	627	14.37	–7.6	86.1	21.3
Muskegon County Airport	43.17N	86.24W	633	14.36	4.0	86.2	25.6
Oakland Pontiac Airport	42.67N	83.42W	1004	14.17	1.1	89.6	24.5
Saginaw, Tri City Intl Airport	43.53N	84.08W	669	14.34	0.0	89.9	24.2
Saint Clair County Intl Airport	42.92N	82.53W	650	14.35	0.6	90.3	18.7
Sault Ste Marie Muni Airport	46.47N	84.37W	725	14.31	–11.3	83.7	21.6
Selfridge ANGB	42.62N	82.83W	581	14.39	2.7	90.3	21.0
Sturgis, Kirsch Municipal	41.82N	85.43W	925	14.21	1.4	89.9	21.1
Traverse City Airport	44.74N	85.58W	630	14.36	–0.4	89.3	20.7
Minnesota							
Albert Lea Municipal Airport	43.68N	93.37W	1257	14.04	–12.3	88.8	24.6
Alexandria, Chandler Field	45.88N	95.39W	1430	13.95	–19.8	89.1	25.8
Austin Municipal Airport	43.67N	92.93W	1230	14.05	–12.4	89.4	24.5
Bemidji Beltrami County Airport	47.50N	94.93W	1378	13.98	–22.7	87.7	24.2
Brainerd, Crow Wing Co Airport	46.40N	94.13W	1227	14.06	–18.4	88.3	19.8
Cloquet, Carlton County Airport	46.70N	92.50W	1280	14.03	–18.1	84.4	20.9
Duluth Harbor	46.77N	92.08W	610	14.37	–12.1	86.1	27.7
Duluth International Airport	46.84N	92.19W	1417	13.96	–19.5	84.5	24.8
Fairmont Municipal Airport	43.65N	94.42W	1161	14.09	–11.1	89.8	30.9
Faribault Municipal Airport	44.33N	93.32W	1056	14.14	–14.6	90.0	23.5
Fergus Falls Municipal Airport	46.28N	96.15W	1184	14.08	–18.7	90.0	27.9
Hibbing, Chisholm-Hibbing Apt	47.39N	92.84W	1358	13.99	–24.7	85.5	20.5
Hutchinson Municipal Airport	44.87N	94.38W	1060	14.14	–16.5	90.2	24.9

Table 2.1: Climatic Data in I-P Units *(Continued)*

Station	Latitude, degrees	Longitude, degrees	Elevation, ft	St. Br. Pressure, psi	Winter Temp., °F	Summer Temp., °F	Wind, mph
Mankato Municipal Airport	44.22N	93.92W	1020	14.16	–13.2	89.8	26.7
Marshall, Ryan Field	44.45N	95.82W	1178	14.08	–12.8	90.2	31.5
Minneapolis St Paul Intl Airport	44.88N	93.23W	837	14.26	–13.4	91.0	24.8
Minneapolis, Anoka Co Airport	45.15N	93.22W	912	14.22	–8.7	90.3	23.0
Minneapolis, Crystal Airport	45.07N	93.35W	869	14.24	–8.6	90.6	21.5
Minneapolis, Flying Cloud Apt	44.82N	93.45W	928	14.21	–8.6	90.8	22.2
New Ulm Municipal Airport	44.32N	94.50W	1010	14.17	–13.3	90.3	27.4
Owatonna Municipal Airport	44.12N	93.25W	1148	14.10	–14.5	89.7	26.5
Red Wing Municipal Airport	44.58N	92.48W	784	14.28	–12.8	90.5	24.9
Rochester Municipal Airport	43.90N	92.49W	1319	14.01	–15.2	88.2	28.9
South St Paul Municipal Airport	44.85N	93.15W	820	14.27	–9.3	90.7	18.3
St Cloud Municipal Airport	45.55N	94.05W	1024	14.16	–19.3	90.0	22.7
St Paul Downtown Airport	44.93N	93.05W	712	14.32	–12.0	90.4	22.9
Willmar Municipal Airport	45.12N	95.08W	1129	14.11	–19.1	88.5	25.7
Winona Municipal Airport	44.08N	91.70W	656	14.35	–9.5	91.0	18.6
Worthington Municipal Airport	43.65N	95.58W	1575	13.88	–11.4	88.4	31.4
Mississippi							
Biloxi, Keesler AFB	30.42N	88.92W	26	14.68	30.3	93.5	17.6
Columbus AFB	33.65N	88.45W	223	14.58	20.3	96.8	18.6
Golden Triangle Reg Airport	33.45N	88.58W	262	14.56	20.5	97.0	17.8
Greenwood Leflore Airport	33.50N	90.08W	135	14.62	20.5	96.1	18.9
Hattiesburg/Laurel, Pine Belt Apt	31.47N	89.33W	305	14.53	25.3	95.8	15.4
Jackson International Airport	32.32N	90.08W	331	14.52	22.0	95.6	18.9
Meridian NAS	32.55N	88.57W	318	14.53	20.8	97.2	15.4
Meridian Regional Airport	32.33N	88.75W	312	14.53	21.9	95.7	18.4
Natchez, Hardy Anders Field	31.62N	91.30W	272	14.55	25.4	93.2	18.5
Tupelo Municipal Airport	34.26N	88.77W	361	14.51	18.3	95.9	18.9
Missouri							
Cape Girardeau Muni Airport	37.23N	89.57W	351	14.51	7.4	94.8	21.2
Columbia Regional Airport	38.82N	92.22W	899	14.22	1.3	94.7	24.2
Jefferson City Memorial Airport	38.58N	92.15W	548	14.41	7.4	95.2	20.9
Joplin Regional Airport	37.15N	94.50W	984	14.18	5.8	96.6	24.6
Kansas City Downtown Airport	39.12N	94.59W	751	14.30	2.6	97.2	22.6
Kansas City International Airport	39.30N	94.72W	1024	14.16	–0.1	96.2	25.8
Kirksville Municipal Airport	40.10N	92.54W	965	14.19	–3.5	94.1	24.4
Poplar Bluff	36.77N	90.47W	479	14.44	9.5	93.8	19.0
Sedalia Memorial Airport	38.70N	93.18W	909	14.22	6.9	96.9	23.3
Springfield Regional Airport	37.24N	93.39W	1270	14.03	4.3	95.0	22.8
St Louis, Lambert Intl Airport	38.75N	90.37W	709	14.32	4.1	95.6	24.1
St Louis, Spirit of St Louis Apt	38.66N	90.66W	463	14.45	5.1	95.2	20.7
Montana							

Table 2.1: Climatic Data in I-P Units ***(Continued)***

Station	Latitude, degrees	Longitude, degrees	Elevation, ft	St. Br. Pressure, psi	Winter Temp., °F	Summer Temp., °F	Wind, mph
Billings Logan Intl Airport	45.81N	108.54W	3570	12.90	–11.6	94.9	26.8
Bozeman, Gallatin Field	45.79N	111.15W	4449	12.48	–18.1	91.3	20.7
Butte, Mooney Airport	45.95N	112.51W	5535	11.98	–20.4	87.8	23.0
Great Falls	47.45N	111.38W	3707	12.83	–12.5	90.5	N/A
Great Falls International Airport	47.47N	111.38W	3658	12.85	–17.7	92.1	31.3
Great Falls, Malmstrom AFB	47.52N	111.18W	3465	12.95	–16.9	92.9	29.0
Helena Regional Airport	46.61N	111.96W	3868	12.75	–15.4	92.7	24.3
Kalispell, Glacier Park Intl Apt	48.30N	114.26W	2972	13.18	–8.6	89.8	20.5
Missoula International Airport	46.92N	114.09W	3189	13.08	–6.0	92.5	21.6
Nebraska							
Beatrice Municipal Airport	40.30N	96.75W	1322	14.01	–1.0	97.3	27.9
Bellevue, Offutt AFB	41.12N	95.92W	1047	14.15	–3.2	95.0	24.4
Columbus Municipal Airport	41.45N	97.33W	1480	13.93	–5.6	93.0	27.0
Fremont Municipal Airport	41.45N	96.52W	1243	14.05	–2.1	93.2	26.8
Grand Island, Central NE Airport	40.96N	98.31W	1857	13.74	–6.1	96.2	28.3
Hastings Municipal Airport	40.60N	98.43W	1955	13.69	–2.0	96.8	28.0
Kearney Municipal Airport	40.73N	99.00W	2129	13.60	–2.0	93.5	31.1
Lincoln Municipal Airport	40.83N	96.76W	1188	14.08	–5.4	97.2	26.9
Norfolk, Karl Stefan Airport	41.98N	97.44W	1552	13.89	–8.5	94.6	28.1
North Platte, Lee Bird Field	41.12N	100.67W	2789	13.27	–7.7	96.3	27.8
Omaha, Eppley Airfield	41.31N	95.90W	981	14.18	–6.3	95.0	25.9
Omaha, North Omaha Airport	41.37N	96.02W	1309	14.01	–7.8	93.4	23.5
Scotts Bluff County Airport	41.87N	103.60W	3957	12.71	–9.3	96.5	28.9
Nevada							
Elko Municipal Airport	40.83N	115.79W	5075	12.19	–3.9	94.6	20.6
Las Vegas, McCarran Intl Apt	36.08N	115.16W	2182	13.57	30.5	108.3	26.9
Las Vegas, Nellis AFB	36.23N	115.03W	1880	13.72	28.2	109.2	26.2
Reno / Tahoe Intl Airport	39.48N	119.77W	4400	12.50	11.9	95.7	25.9
New Hampshire							
Concord Municipal Airport	43.20N	71.50W	348	14.51	–4.8	89.8	20.9
Keene, Dillant Hopkins Airport	42.90N	72.27W	502	14.43	–6.4	88.3	18.4
Laconia Municipal Airport	43.57N	71.42W	545	14.41	–4.2	88.1	19.8
Lebanon Regional Airport	43.63N	72.30W	597	14.38	–8.6	88.6	17.9
Manchester Municipal Airport	42.93N	71.44W	233	14.57	1.0	91.2	19.0
Portsmout, Pease AFB	43.08N	70.82W	102	14.64	2.6	89.3	22.6
New Jersey							
Atlantic City Intl Airport	39.46N	74.46W	66	14.66	9.9	92.3	24.8
Belmar/Farmingdale Airport	40.18N	74.13W	85	14.65	10.7	90.7	25.4
Cape May County Airport	39.00N	74.90W	23	14.68	15.7	89.6	23.1
Lakehurst, McGuire AFB	40.02N	74.60W	148	14.62	10.3	92.9	23.3
Millville Municipal Airport	39.37N	75.08W	75	14.66	10.3	91.9	19.8

Table 2.1: Climatic Data in I-P Units ***(Continued)***

Station	Latitude, degrees	Longitude, degrees	Elevation, ft	St. Br. Pressure, psi	Winter Temp., °F	Summer Temp., °F	Wind, mph
Newark International Airport	40.72N	74.17W	30	14.68	11.0	94.0	25.0
Teterboro Airport	40.85N	74.06W	7	14.69	9.9	92.4	20.6
Trenton, Mercer County Airport	40.28N	74.81W	213	14.58	9.8	92.6	23.5
New Mexico							
Alamogordo White Sands Apt	32.83N	105.98W	4308	12.55	21.1	100.0	21.5
Alamogordo, Holloman AFB	32.85N	106.10W	4094	12.65	19.0	99.1	22.9
Albuquerque Intl Airport	35.04N	106.62W	5315	12.08	17.7	95.2	28.2
Carlsbad Airport	32.34N	104.26W	3261	13.04	20.3	102.3	26.8
Clovis Municipal Airport	34.43N	103.08W	4213	12.59	13.8	97.1	31.7
Clovis, Cannon AFB	34.38N	103.32W	4295	12.55	11.8	97.4	27.9
Farmington Municipal Airport	36.74N	108.23W	5502	12.00	7.5	94.9	24.7
Gallup Municipal Airport	35.51N	108.79W	6467	11.57	0.6	91.3	24.7
Roswell International Air Center	33.31N	104.54W	3668	12.85	16.3	99.8	25.0
New York							
Albany International Airport	42.75N	73.80W	292	14.54	–1.9	89.0	23.9
Binghamton, Greater Airport	42.21N	75.98W	1637	13.85	–1.0	85.5	21.5
Buffalo, Greater Buffalo Airport	42.94N	78.74W	705	14.33	2.7	86.5	28.1
Dunkirk	42.48N	79.35W	646	14.36	7.4	81.7	44.1
Elmira/Corning Regional Airport	42.16N	76.89W	955	14.20	–1.8	89.9	20.5
Farmingdale, Republic Airport	40.72N	73.40W	85	14.65	11.9	90.2	24.9
Glens Falls, Bennett Airport	43.34N	73.61W	335	14.52	–10.2	88.0	18.3
Islip, Long Island MacArthur Apt	40.79N	73.10W	108	14.64	10.6	88.4	23.7
Jamestown Airport	42.15N	79.27W	1722	13.80	1.0	82.4	21.5
New York, Kennedy Intl Airport	40.66N	73.80W	23	14.68	12.8	89.7	27.3
New York, Laguardia Airport	40.78N	73.88W	30	14.68	12.6	92.2	27.3
Newburgh, Stewart Intl Airport	41.50N	74.10W	581	14.39	3.5	89.8	24.4
Niagara Falls Intl Airport	43.11N	78.95W	587	14.39	2.5	87.9	26.2
Plattsburgh AFB	44.65N	73.47W	236	14.57	–9.6	86.5	20.6
Poughkeepsie, Dutchess Co Apt	41.63N	73.88W	161	14.61	0.5	91.4	18.4
Rochester, Monroe Co Airport	43.12N	77.68W	554	14.40	2.1	88.4	25.2
Rome, Griffiss AFB	43.23N	75.40W	518	14.42	–6.0	88.7	22.1
Syracuse Hancock Intl Airport	43.11N	76.10W	417	14.48	–2.7	88.9	24.2
Utica, Oneida County Airport	43.15N	75.38W	745	14.30	–5.0	87.5	20.7
Watertown International Airport	43.99N	76.02W	335	14.52	–13.0	85.2	22.4
White Plains, Westchester Apt	41.07N	73.71W	397	14.49	7.7	90.0	20.4
North Carolina							
Asheville Regional Airport	35.43N	82.54W	2169	13.58	13.6	88.1	23.2
Charlotte Douglas Intl Airport	35.21N	80.94W	768	14.29	20.5	94.1	18.7
Fayetteville Municipal Airport	34.98N	78.88W	194	14.59	21.4	95.5	19.8
Fayetteville, Simmons AAF	35.13N	78.93W	305	14.53	21.7	96.6	17.8
Fayetteville, Pope AFB	35.17N	79.02W	217	14.58	21.2	97.0	18.4

Table 2.1: Climatic Data in I-P Units *(Continued)*

Station	Latitude, degrees	Longitude, degrees	Elevation, ft	St. Br. Pressure, psi	Winter Temp., °F	Summer Temp., °F	Wind, mph
Goldsboro, S Johnson AFB	35.33N	77.97W	108	14.64	21.7	96.5	18.3
Greensboro, Piedmont Triad Apt	36.10N	79.94W	886	14.23	17.1	92.2	19.3
Greenville (Pitt County) Airport	35.63N	77.40W	26	14.68	20.9	95.2	18.8
Havelock, Cherry Point MCAS	34.90N	76.88W	30	14.68	23.6	93.2	20.9
Hickory Regional Airport	35.74N	81.39W	1188	14.08	18.9	92.5	17.6
Jacksonville, Albert J Ellis Apt	34.83N	77.62W	95	14.65	20.0	93.5	20.0
Jacksonville, New River MCAF	34.70N	77.43W	26	14.68	22.9	93.2	19.8
New Bern, Craven Co Reg Apt	35.07N	77.05W	23	14.68	22.4	93.7	18.6
Raleigh Durham Intl Airport	35.87N	78.79W	436	14.47	18.8	94.1	18.9
Southern Pines, Moore Co Apt	35.23N	79.40W	463	14.45	18.9	94.6	17.9
Wilmington International Airport	34.27N	77.91W	39	14.68	24.0	93.3	20.7
Winston-Salem, Reynolds Apt	36.13N	80.22W	971	14.19	18.2	92.4	18.7
North Dakota							
Bismarck Municipal Airport	46.77N	100.75W	1660	13.84	–20.0	93.9	27.3
Dickinson Municipal Airport	46.80N	102.80W	2582	13.38	–18.3	93.2	29.1
Fargo, Hector Intl Airport	46.93N	96.81W	899	14.22	–20.4	91.0	28.3
Grand Forks AFB	47.97N	97.40W	906	14.22	–20.4	90.8	28.3
Grand Forks Intl Airport	47.95N	97.18W	833	14.26	–22.2	90.0	26.9
Jamestown Municipal Airport	46.93N	98.68W	1493	13.92	–19.9	91.6	28.5
Minot AFB	48.42N	101.35W	1631	13.85	–22.2	93.2	28.9
Minot International Airport	48.26N	101.28W	1713	13.81	–19.9	91.4	27.5
Williston, Sloulin Intl Airport	48.20N	103.64W	1906	13.71	–23.5	94.9	27.1
Ohio							
Akron Canton Regional Airport	40.92N	81.44W	1237	14.05	1.8	88.7	23.4
Cincinnati Muni Lunken Airport	39.10N	84.42W	499	14.43	6.3	92.8	20.2
Cleveland Hopkins Intl Airport	41.41N	81.85W	804	14.27	2.5	89.4	24.7
Columbus, Ohio State Univ Apt	40.07N	83.07W	928	14.21	6.9	90.4	21.6
Columbus, Port Columbus Apt	39.99N	82.88W	817	14.27	3.2	91.1	21.9
Columbus, Rickenbacker Airport	39.82N	82.93W	755	14.30	4.3	92.5	22.1
Dayton International Airport	39.91N	84.22W	1004	14.17	0.6	90.3	24.4
Dayton, Wright Paterson AFB	39.83N	84.05W	820	14.27	1.4	91.3	21.5
Findlay Airport	41.01N	83.67W	814	14.27	–0.4	90.4	24.4
Lancaster, Fairfield Co Airport	39.75N	82.65W	866	14.24	3.1	90.5	20.2
Mansfield Lahm Muni Airport	40.82N	82.52W	1312	14.01	0.1	88.0	24.6
Toledo Express Airport	41.59N	83.80W	692	14.33	–0.3	91.2	24.3
Youngstown Municipal Airport	41.25N	80.67W	1188	14.08	1.8	88.5	21.9
Zanesville Municipal Airport	39.94N	81.89W	902	14.22	2.8	90.2	19.4
Oklahoma							
Enid, Vance AFB	36.33N	97.92W	1339	14.00	6.5	100.6	26.7
Lawton Municipal Airport	34.57N	98.42W	1109	14.12	17.9	102.4	26.0
Lawton, Fort Sill	34.65N	98.40W	1211	14.06	12.6	100.5	24.8

Table 2.1: Climatic Data in I-P Units *(Continued)*

Station	Latitude, degrees	Longitude, degrees	Elevation, ft	St. Br. Pressure, psi	Winter Temp., °F	Summer Temp., °F	Wind, mph
McAlester Regional Airport	34.90N	95.78W	771	14.29	13.0	99.3	20.4
Midwest City, Tinker AFB	35.42N	97.38W	1260	14.04	12.1	99.3	25.5
Oklahoma City, Wiley Post Apt	35.53N	97.65W	1299	14.02	12.1	99.5	26.4
Oklahoma City, Will Rogers Apt	35.39N	97.60W	1306	14.02	11.4	99.5	27.2
Ponca City Municipal Airport	36.73N	97.10W	997	14.17	9.5	100.3	22.3
Stillwater Municipal Airport	36.15N	97.08W	1010	14.17	13.6	101.8	24.6
Tulsa International Airport	36.20N	95.89W	676	14.34	10.9	99.4	24.5
Tulsa, R L Jones Airport	36.03N	95.98W	633	14.36	15.8	100.1	19.8
Oregon							
Astoria, Port of Astoria Airport	46.16N	123.88W	23	14.68	26.8	76.7	24.6
Canby, Aurora State Airport	45.25N	122.77W	197	14.59	27.5	91.2	18.2
Corvallis Municipal Airport	44.48N	123.28W	253	14.56	25.0	92.9	19.7
Eugene, Mahlon Sweet Field	44.13N	123.21W	374	14.50	22.4	91.4	19.6
Klamath Falls, Kingsley Field	42.15N	121.72W	4091	12.65	6.2	91.1	24.7
La Grande Union County Airport	45.29N	118.01W	2717	13.31	9.4	93.0	27.0
McMinnville Municipal Airport	45.18N	123.13W	167	14.61	27.6	91.4	20.9
Medford Jackson County Airport	42.39N	122.87W	1329	14.00	22.9	98.9	18.4
Newport State Beach	44.62N	124.07W	36	14.68	30.9	65.7	41.0
North Bend Municipal Airport	43.42N	124.25W	16	14.69	30.6	71.5	26.6
Pendleton Municipal Airport	45.70N	118.83W	1516	13.91	7.0	96.4	26.0
Portland Hillsboro Airport	45.53N	122.95W	203	14.59	21.8	91.8	18.9
Portland International Airport	45.59N	122.60W	108	14.64	23.9	91.2	23.8
Redmond, Roberts Field	44.25N	121.15W	3084	13.13	5.4	92.8	20.6
Salem, McNary Field	44.91N	123.00W	200	14.59	21.9	92.0	20.8
Pennsylvania							
Allentown, Lehigh Valley Apt	40.65N	75.45W	384	14.49	7.0	91.0	24.2
Altoona Blair County Airport	40.30N	78.32W	1470	13.93	4.7	88.5	21.9
Butler, Butler County Airport	40.78N	79.95W	1247	14.05	3.1	88.0	17.8
Erie International Airport	42.08N	80.18W	738	14.31	5.2	86.4	24.7
Harrisburg International Airport	40.19N	76.76W	312	14.53	10.7	92.6	25.3
Harrisburg, Capital City Airport	40.22N	76.85W	348	14.51	8.7	92.4	20.5
Johnstown Cambria Co Airport	40.30N	78.83W	2283	13.52	2.5	85.1	23.8
Philadelphia Intl Airport	39.87N	75.23W	30	14.68	12.6	93.2	24.4
Philadelphia, NE Phil Apt	40.08N	75.01W	118	14.63	11.0	93.1	21.1
Pittsburgh International Airport	40.50N	80.23W	1204	14.07	3.7	89.5	23.4
Pittsburgh, Allegheny Co Airport	40.36N	79.92W	1273	14.03	4.3	89.9	20.3
Reading Spaatz Field	40.37N	75.96W	354	14.51	9.4	92.4	22.6
Washington Airport	40.13N	80.28W	1184	14.08	2.7	88.4	19.5
Wilkes-Barre Scranton Intl Apt	41.34N	75.73W	961	14.19	3.5	88.9	20.3
Williamsport Lycoming Co Apt	41.24N	76.92W	525	14.42	4.5	90.8	21.6
Willow Grove NAS	40.20N	75.15W	361	14.51	10.2	92.7	18.3

Table 2.1: Climatic Data in I-P Units *(Continued)*

Station	Latitude, degrees	Longitude, degrees	Elevation, ft	St. Br. Pressure, psi	Winter Temp., °F	Summer Temp., °F	Wind, mph
Rhode Island							
Pawtucket, North Central Airport	41.92N	71.50W	440	14.46	3.1	88.1	19.5
Providence, T F Green State Apt	41.72N	71.43W	62	14.66	7.2	90.1	24.4
South Carolina							
Anderson County Airport	34.50N	82.71W	774	14.29	22.0	94.9	18.9
Beaufort MCAS	32.48N	80.72W	39	14.68	27.6	94.9	18.2
Charleston International Airport	32.90N	80.04W	49	14.67	26.9	94.3	20.3
Columbia Metropolitan Airport	33.94N	81.12W	226	14.58	22.0	97.0	19.1
Florence Regional Airport	34.19N	79.73W	151	14.62	23.4	95.9	19.1
Greenville Spartanburg Airport	34.90N	82.22W	971	14.19	20.5	94.0	19.5
Myrtle Beach AFB	33.68N	78.92W	26	14.68	25.3	92.4	19.8
Sumter, Shaw AFB	33.97N	80.47W	243	14.57	24.2	95.4	18.9
South Dakota							
Aberdeen Regional Airport	45.45N	98.42W	1299	14.02	–17.9	92.2	28.2
Brookings Municipal Airport	44.30N	96.82W	1647	13.84	–15.2	89.5	26.5
Huron Regional Airport	44.39N	98.23W	1289	14.02	–15.0	94.5	27.4
Mitchell Municipal Airport	43.77N	98.03W	1302	14.02	–11.0	93.4	28.2
Pierre Municipal Airport	44.38N	100.29W	1726	13.80	–11.8	99.3	28.7
Rapid City Regional Airport	44.05N	103.05W	3169	13.09	–10.5	96.9	34.7
Rapid City, Ellsworth AFB	44.15N	103.10W	3215	13.07	–9.2	96.5	34.3
Sioux Falls, Joe Foss Field	43.58N	96.75W	1427	13.95	–13.7	92.8	27.6
Watertown Municipal Airport	44.93N	97.15W	1739	13.80	–16.9	90.2	28.2
Yankton, Chan Gurney Airport	42.92N	97.38W	1339	14.00	–8.2	92.7	29.8
Tennessee							
Bristol, Tri City Airport	36.48N	82.40W	1526	13.90	11.4	89.8	18.9
Chattanooga, Lovell Field	35.03N	85.20W	689	14.33	17.7	94.5	17.9
Dyersburg Municipal Airport	36.02N	89.32W	338	14.52	14.6	93.4	18.9
Jackson, McKellar Sipes Airport	35.59N	88.92W	423	14.47	13.5	94.7	19.6
Knoxville, McGhee Tyson Apt	35.82N	83.99W	981	14.18	15.0	92.6	20.2
Memphis International Airport	35.06N	89.99W	331	14.52	17.0	96.0	20.3
Millington Municipal Airport	35.35N	89.87W	322	14.53	14.0	97.3	18.6
Nashville International Airport	36.12N	86.69W	604	14.38	12.9	94.4	19.9
Texas							
Abilene Regional Airport	32.41N	99.68W	1791	13.77	17.9	99.3	25.9
Abilene, Dyess AFB	32.43N	99.85W	1788	13.77	16.2	101.4	24.0
Alice International Airport	27.74N	98.03W	180	14.60	34.3	99.7	21.9
Amarillo International Airport	35.22N	101.71W	3606	12.88	7.2	97.3	28.8
Austin, Camp Mabry	30.32N	97.77W	659	14.35	26.7	99.4	20.2
Austin-Bergstrom Intl Airport	30.18N	97.68W	495	14.43	25.2	99.7	20.9
Beeville, Chase Field NAS	28.37N	97.67W	190	14.60	27.9	100.2	23.3
Brownsville, South Padre Is. Apt	25.91N	97.43W	23	14.68	36.4	95.5	26.2

Table 2.1: Climatic Data in I-P Units *(Continued)*

Station	Latitude, degrees	Longitude, degrees	Elevation, ft	St. Br. Pressure, psi	Winter Temp., °F	Summer Temp., °F	Wind, mph
College Station, Easterwood Fld	30.59N	96.36W	328	14.52	26.0	99.2	20.0
Corpus Christi Intl Airport	27.77N	97.51W	43	14.67	32.6	95.8	26.9
Corpus Christi NAS	27.70N	97.28W	20	14.69	35.6	93.2	24.8
Dallas Love Field	32.85N	96.85W	489	14.44	21.6	100.3	21.9
Dallas NAS	32.73N	96.97W	495	14.43	18.5	99.7	20.8
Dallas, Redbird Airport	32.68N	96.87W	659	14.35	26.5	100.2	22.6
Dallas/Fort Worth Intl Airport	32.90N	97.04W	597	14.38	20.3	100.4	26.1
Del Rio International Airport	29.37N	100.92W	1027	14.16	30.7	101.5	20.6
Del Rio, Laughlin AFB	29.37N	100.78W	1073	14.13	29.4	104.2	22.1
El Paso International Airport	31.81N	106.38W	3917	12.73	22.6	100.6	25.4
Fort Hood, Robert Gray AAF	31.07N	97.83W	1024	14.16	23.7	99.9	22.0
Fort Worth Alliance Airport	32.98N	97.32W	741	14.31	21.5	101.6	24.0
Fort Worth Meacham Airport	32.82N	97.36W	705	14.33	19.9	100.9	21.8
Fort Worth NAS	32.77N	97.44W	650	14.35	18.8	100.2	23.5
Galveston, Scholes Field	29.27N	94.86W	10	14.69	33.0	91.4	25.6
Georgetown Municipal Airport	30.68N	97.68W	787	14.28	26.5	99.1	21.0
Harlingen, Rio Grande Apt	26.23N	97.65W	36	14.68	36.6	98.8	27.8
Houston, Bush Intl Airport	29.99N	95.36W	105	14.64	29.1	96.8	19.2
Houston, D W Hooks Airport	30.07N	95.55W	151	14.62	30.2	98.7	17.6
Houston, Ellington Field	29.60N	95.17W	39	14.68	32.0	96.1	19.8
Houston, W P Hobby Airport	29.65N	95.28W	46	14.67	31.3	95.1	20.9
Killeen Municipal Airport	31.08N	97.68W	846	14.25	25.9	99.5	22.0
Kingsville NAS	27.50N	97.82W	59	14.66	32.0	97.7	23.8
Laredo International Airport	27.55N	99.47W	509	14.43	33.9	102.5	24.7
Longview, Gregg County Airport	32.39N	94.71W	374	14.50	21.6	99.1	19.4
Lubbock International Airport	33.67N	101.82W	3241	13.05	13.3	98.8	28.9
Lubbock, Reese AFB	33.60N	102.05W	3327	13.01	12.1	100.0	27.1
Lufkin, Angelina County Airport	31.23N	94.75W	315	14.53	25.4	97.5	17.8
McAllen, Miller Intl Airport	26.18N	98.24W	112	14.64	36.5	100.1	24.7
McGregor Municipal Airport	31.48N	97.32W	591	14.38	25.2	99.9	22.7
McKinney Municipal Airport	33.18N	96.58W	577	14.39	21.4	100.3	22.5
Midland International Airport	31.93N	102.21W	2861	13.24	18.3	100.3	26.7
Nacogdoches, Mangham Airport	31.58N	94.72W	354	14.51	25.3	97.4	18.2
Paris, Cox Field	33.63N	95.45W	561	14.40	20.7	99.2	24.3
Port Aransas	27.82N	97.05W	20	14.69	36.6	85.9	39.4
Port Arthur, Jefferson Co Airport	29.95N	94.02W	16	14.69	30.5	94.2	21.3
Rockport, Aransas Co Airport	28.08N	97.05W	26	14.68	36.4	92.8	22.4
Sabine	29.67N	94.05W	20	14.69	32.2	88.4	35.8
San Angelo, Mathis Field	31.35N	100.49W	1893	13.72	20.1	100.2	24.8
San Antonio Intl Airport	29.53N	98.46W	810	14.27	27.4	98.5	20.2
San Antonio, Kelly AFB	29.38N	98.58W	682	14.34	28.4	100.2	19.9

Table 2.1: Climatic Data in I-P Units *(Continued)*

Station	Latitude, degrees	Longitude, degrees	Elevation, ft	St. Br. Pressure, psi	Winter Temp., °F	Summer Temp., °F	Wind, mph
San Antonio, Stinson Muni Apt	29.34N	98.47W	577	14.39	30.5	100.0	18.8
San Marcos Municipal Airport	29.88N	97.87W	597	14.38	27.9	99.1	24.3
Stephenville, Clark Muni Airport	32.22N	98.18W	1322	14.01	18.8	97.6	19.0
Temple, D Miller Muni Airport	31.15N	97.40W	699	14.33	24.9	99.5	24.8
Universal City, Randolph AFB	29.53N	98.28W	761	14.30	27.5	99.4	20.6
Victoria Regional Airport	28.86N	96.93W	118	14.63	29.8	96.6	24.3
Waco Municipal Airport	31.61N	97.23W	509	14.43	22.9	100.8	24.7
Wichita Falls Municipal Airport	33.98N	98.49W	1030	14.16	15.3	102.5	26.9
Utah							
Cedar City Municipal Airport	37.70N	113.10W	5617	11.95	3.4	93.4	26.0
Logan-Cache Airport	41.79N	111.85W	4455	12.48	–6.3	95.0	19.6
Ogden, Hill AFB	41.12N	111.97W	4787	12.33	8.6	93.3	22.8
Price, Carbon County Airport	39.62N	110.75W	5902	11.82	4.1	92.6	22.5
Provo Municipal Airport	40.22N	111.72W	4491	12.46	9.0	94.6	24.1
Salt Lake City Intl Airport	40.79N	111.97W	4226	12.59	9.3	97.4	25.0
St George Municipal Airport	37.08N	113.60W	2940	13.20	26.8	106.2	26.7
Vernal Airport	40.43N	109.52W	5276	12.10	1.0	93.1	20.3
Vermont							
Burlington International Airport	44.47N	73.15W	341	14.52	–8.3	88.3	23.8
Montpelier, Knapp State Airport	44.20N	72.58W	1122	14.11	–10.7	85.1	20.4
Virginia							
Abington, VA Highlands Airport	36.68N	82.03W	2070	13.63	11.8	89.6	17.4
Blacksburg, Virginia Tech Apt	37.22N	80.42W	2133	13.60	10.3	88.4	20.7
Chesapeake	36.90N	75.72W	75	14.66	22.9	84.1	50.3
Danville Regional Airport	36.57N	79.34W	591	14.38	18.4	93.2	18.5
Fort Belvoir, Davison AAF	38.72N	77.18W	89	14.65	12.6	96.2	20.5
Fredericksburg, Shannon Airport	38.27N	77.45W	85	14.65	15.5	95.2	18.9
Hampton, Langley AFB	37.08N	76.37W	10	14.69	20.6	92.9	23.4
Leesburg Executive Airport	39.08N	77.57W	390	14.49	14.2	94.8	22.3
Lynchburg Municipal Airport	37.34N	79.21W	938	14.20	14.4	92.1	18.3
Manassas Municipal Airport	38.72N	77.52W	194	14.59	11.9	92.8	21.5
Martinsville, Blue Ridge Airport	36.63N	80.02W	942	14.20	14.6	93.4	16.9
Newport News, P Henry Intl Apt	37.13N	76.49W	52	14.67	18.7	94.5	19.9
Norfolk International Airport	36.90N	76.19W	30	14.68	21.7	93.7	25.0
Norfolk NAS	36.93N	76.28W	16	14.69	22.7	93.8	25.3
Petersburg Municipal Airport	37.18N	77.50W	197	14.59	15.9	97.4	18.1
Quantico MCAS	38.50N	77.30W	13	14.69	16.2	92.6	19.1
Richmond International Airport	37.51N	77.32W	164	14.61	16.8	94.8	20.7
Roanoke Regional Airport	37.32N	79.97W	1175	14.08	14.2	92.1	22.5
Staunton, Shenandoah Reg Apt	38.27N	78.90W	1201	14.07	12.1	93.3	17.6
Virginia Beach, Oceana NAS	36.82N	76.03W	23	14.68	21.1	92.9	24.0

Table 2.1: Climatic Data in I-P Units *(Continued)*

Station	Latitude, degrees	Longitude, degrees	Elevation, ft	St. Br. Pressure, psi	Winter Temp., °F	Summer Temp., °F	Wind, mph
Washington DC, Reagan Airport	38.87N	77.03W	66	14.66	16.3	94.3	23.2
Washington Dulles Intl Airport	38.94N	77.45W	325	14.52	10.7	93.5	20.5
Winchester Regional Airport	39.15N	78.15W	728	14.31	14.3	93.2	22.4
Washington							
Arlington Municipal Airport	48.17N	122.17W	138	14.62	20.8	82.1	20.9
Bellingham International Airport	48.79N	122.54W	151	14.62	18.0	79.4	25.0
Bremerton National Airport	47.48N	122.75W	449	14.46	23.7	85.6	18.7
Fort Lewis, Gray AAF	47.08N	122.58W	302	14.54	19.3	86.4	17.5
Hoquiam, Bowerman Airport	46.97N	123.94W	16	14.69	28.1	77.6	25.9
Kelso Longview Airport	46.12N	122.89W	20	14.69	21.6	88.0	17.7
Olympia Airport	46.97N	122.90W	200	14.59	20.1	87.3	18.8
Paine Field	47.90N	122.28W	620	14.37	27.7	79.4	24.2
Pasco, Tri Cities Airport	46.27N	119.12W	404	14.48	6.8	99.5	24.6
Port Angeles, Fairchild Airport	48.12N	123.50W	299	14.54	24.2	79.4	17.9
Pullman-Moscow Reg Airport	46.75N	117.12W	2552	13.39	9.1	92.5	23.5
Seattle Tacoma Intl Airport	47.46N	122.31W	433	14.47	24.5	84.9	20.2
Seattle, Boeing Field	47.53N	122.30W	30	14.68	24.0	86.1	18.7
Spokane International Airport	47.62N	117.53W	2365	13.48	2.9	92.8	25.6
Spokane, Fairchild AFB	47.62N	117.65W	2438	13.45	4.6	91.4	23.6
Spokane, Felts Field	47.68N	117.32W	2001	13.66	6.9	94.7	19.7
Tacoma Narrows Airport	47.27N	122.57W	299	14.54	29.5	83.7	19.6
Tacoma, McChord AFB	47.13N	122.48W	289	14.54	20.6	86.2	19.7
Vancouver, Pearson Airpark	45.62N	122.65W	26	14.68	25.3	90.7	16.9
Walla Walla City County Airport	46.10N	118.29W	1204	14.07	8.1	98.9	23.7
Wenatchee, Pangborn Airport	47.40N	120.21W	1230	14.05	6.1	95.3	23.2
Whidbey Island NAS	48.35N	122.65W	46	14.67	22.1	72.3	28.6
Yakima Air Terminal	46.56N	120.53W	1066	14.14	6.3	95.7	23.5
West Virginia							
Beckley Airport	37.80N	81.13W	2513	13.41	5.5	84.6	19.7
Bluefield, Mercer County Airport	37.30N	81.21W	2867	13.24	6.0	84.4	17.4
Bridgeport, Benedum Airport	39.28N	80.22W	1247	14.05	9.4	89.8	19.2
Charleston, Yeager Airport	38.38N	81.59W	981	14.18	8.5	91.3	17.3
Huntington, Tri-State Airport	38.38N	82.56W	837	14.26	8.4	91.9	17.1
Martinsburg, Shepherd Field	39.40N	77.98W	535	14.41	9.7	93.3	21.0
Morgantown Municipal Airport	39.64N	79.92W	1253	14.04	5.9	89.5	17.0
Parkersburg, Wood Co Airport	39.35N	81.44W	863	14.24	5.4	90.8	18.2
Wisconsin							
Appleton, Outagamie Co Airport	44.25N	88.52W	919	14.21	–5.8	88.5	24.7
Eau Claire County Airport	44.87N	91.49W	896	14.23	–15.6	90.8	19.8
Fond Du Lac County Airport	43.77N	88.49W	807	14.27	–3.9	88.5	23.7
Green Bay, Austin Straubel Field	44.51N	88.12W	702	14.33	–9.9	88.4	23.9

Table 2.1: Climatic Data in I-P Units *(Continued)*

Station	Latitude, degrees	Longitude, degrees	Elevation, ft	St. Br. Pressure, psi	Winter Temp., °F	Summer Temp., °F	Wind, mph
Juneau, Dodge County Airport	43.43N	88.70W	935	14.21	–1.9	88.1	23.7
Kenosha Municipal Airport	42.58N	87.92W	761	14.30	0.5	90.3	24.9
La Crosse Municipal Airport	43.75N	91.26W	656	14.35	–12.3	92.1	23.0
Madison, Dane County Airport	43.14N	89.35W	866	14.24	–9.1	89.8	23.6
Manitowoc Municipal Airport	44.13N	87.68W	650	14.35	–4.2	84.5	24.2
Marshfield Municipal Airport	44.63N	90.18W	1276	14.03	–12.9	88.1	22.7
Milwaukee, Mitchell Intl Airport	42.95N	87.90W	692	14.33	–4.0	90.3	25.7
Rice Lake Municipal Airport	45.48N	91.72W	1138	14.10	–11.3	89.6	19.9
Sheboygan	43.75N	87.68W	620	14.37	–2.2	83.2	41.8
Sheboygan County Airport	43.78N	87.85W	748	14.30	–1.9	88.2	24.6
Watertown Municipal Airport	43.17N	88.72W	833	14.26	–4.2	89.7	20.8
Wausau Municipal Airport	44.93N	89.63W	1198	14.07	–13.6	88.1	19.3
Wisconsin Rapids, Alexander Fld	44.35N	89.83W	1010	14.17	–8.2	89.6	20.4
Wittman Regional Airport	43.98N	88.55W	830	14.26	–5.6	88.4	23.3
Wyoming							
Casper Airport	42.90N	106.47W	5289	12.10	–10.3	93.8	32.2
Cheyenne Airport	41.16N	104.81W	6142	11.71	–6.4	89.2	33.7
Gillette Campbell County Airport	44.35N	105.53W	4035	12.68	–11.8	95.1	28.7
Jackson Hole Airport	43.60N	110.73W	6614	11.51	–15.2	84.1	24.3
Laramie, General Brees Field	41.31N	105.67W	7274	11.22	–9.2	84.8	31.9
Rock Springs Sweetwater Airport	41.59N	109.07W	6745	11.45	–6.1	88.0	31.4
Sheridan County Airport	44.77N	106.98W	3967	12.71	–12.7	95.5	27.0
Canada							
Alberta							
Banff	51.19N	115.55W	4583	12.42	–21.9	81.8	14.4
Bowness	51.08N	114.22W	4052	12.67	–18.3	82.7	23.4
Brooks	50.55N	111.85W	2451	13.44	–23.6	89.2	24.3
Calgary International Airport	51.11N	114.02W	3556	12.90	–20.4	83.5	27.3
Calgary, Springbank Airport	51.10N	114.37W	3940	12.72	–25.0	80.3	24.8
Camrose Airport	53.03N	112.81W	2425	13.45	–27.8	82.9	23.1
Cold Lake Airport	54.42N	110.28W	1775	13.78	–30.9	82.2	20.9
Edmonton City Centre Airport	53.57N	113.52W	2201	13.56	–21.5	82.5	21.7
Edmonton International Airport	53.32N	113.58W	2372	13.48	–26.5	81.5	23.0
Edmonton Namao Airport	53.67N	113.47W	2257	13.54	–23.1	80.9	23.2
Fort McMurray Airport	56.65N	111.22W	1211	14.06	–34.2	84.1	18.5
Grande Prairie Airport	55.18N	118.88W	2195	13.57	–33.5	81.4	25.0
Lacombe	52.45N	113.76W	2822	13.26	–27.2	82.9	21.1
Lethbridge	49.70N	112.78W	3022	13.16	–19.0	89.5	29.7
Lethbridge Airport	49.63N	112.80W	3048	13.15	–21.6	88.5	36.2
Lloydminster Airport	53.31N	110.07W	2192	13.57	–28.1	83.1	23.0
Medicine Hat Airport	50.02N	110.72W	2352	13.49	–24.1	90.4	24.9

Table 2.1: Climatic Data in I-P Units *(Continued)*

Station	Latitude, degrees	Longitude, degrees	Elevation, ft	St. Br. Pressure, psi	Winter Temp., °F	Summer Temp., °F	Wind, mph
Red Deer Industrial Airport	52.18N	113.89W	2969	13.19	–26.8	82.3	20.6
British Columbia							
Abbotsford Airport	49.03N	122.36W	194	14.59	16.5	85.4	19.0
Agassiz	49.25N	121.77W	49	14.67	18.6	86.2	23.2
Ballenas Island	49.35N	124.16W	43	14.67	30.7	75.1	35.5
Comox Airport	49.72N	124.90W	85	14.65	22.4	80.1	30.1
Cranbrook International Airport	49.61N	115.78W	3084	13.13	–12.0	88.7	18.7
Entrance Island	49.22N	123.80W	16	14.69	29.1	75.1	32.0
Esquimalt Harbour	48.43N	123.44W	10	14.69	28.2	72.6	21.6
Fort St John Airport	56.24N	120.74W	2280	13.52	–27.5	79.1	23.2
Kamloops Airport	50.70N	120.44W	1132	14.10	–5.5	92.7	22.2
Kelowna International Airport	49.96N	119.38W	1411	13.96	–1.0	90.9	16.8
Nelson	49.49N	117.31W	1755	13.79	10.5	89.0	15.5
Penticton Airport	49.46N	119.60W	1129	14.11	6.4	90.6	22.8
Pitt Meadows	49.21N	122.69W	16	14.69	18.9	87.0	12.4
Port Alberni Airport	49.32N	124.93W	249	14.56	21.2	88.6	13.7
Prince George Airport	53.89N	122.68W	2267	13.53	–23.9	81.8	21.0
Prince Rupert Airport	54.29N	130.44W	115	14.64	11.3	66.5	25.7
Salmon Arm	50.70N	119.29W	1152	14.09	0.7	90.1	14.1
Squamish Airport	49.78N	123.16W	171	14.61	18.7	84.5	14.4
Summerland	49.56N	119.64W	1490	13.92	5.9	91.3	18.7
Terrace Airport	54.47N	128.58W	712	14.32	–1.3	81.9	25.8
Vancouver Harbour	49.30N	123.12W	10	14.69	26.8	78.3	N/A
Vancouver International Airport	49.20N	123.18W	13	14.69	19.4	77.0	23.1
Vancouver, Point Atkinson	49.33N	123.26W	115	14.63	31.2	77.0	30.7
Vancouver, West	49.35N	123.19W	551	14.41	21.6	80.6	11.4
Vernon	50.22N	119.19W	1581	13.88	5.4	91.6	15.4
Victoria International Airpor	48.65N	123.43W	62	14.66	23.6	79.9	19.0
Victoria University	48.46N	123.30W	197	14.59	28.2	80.5	13.2
Victoria, Gonzales	48.41N	123.33W	230	14.57	27.5	76.4	27.4
Victoria, Hartland	48.53N	123.46W	505	14.43	27.0	83.4	21.8
White Rock Campbell Scientific	49.02N	122.78W	43	14.67	22.4	76.7	14.3
Williams Lake Airport	52.18N	122.05W	3084	13.13	–19.8	82.5	21.6
Manitoba							
Brandon Airport	49.91N	99.95W	1342	14.00	–28.3	87.2	25.3
Winnipeg International Airport	49.92N	97.23W	784	14.28	–25.9	87.4	27.9
New Brunswick							
Fredericton Airport	45.87N	66.53W	69	14.66	–10.5	85.5	22.0
Miramichi Airport	47.01N	65.47W	108	14.64	–10.2	86.0	24.5
Miramichi RCS	47.01N	65.46W	108	14.64	–11.3	87.1	25.2
Moncton International Airport	46.10N	64.69W	233	14.57	–8.7	83.2	27.8

Table 2.1: Climatic Data in I-P Units *(Continued)*

Station	Latitude, degrees	Longitude, degrees	Elevation, ft	St. Br. Pressure, psi	Winter Temp., °F	Summer Temp., °F	Wind, mph
Point Lepreau	45.07N	66.45W	20	14.69	–1.5	74.2	23.2
Saint John Airport	45.32N	65.89W	358	14.51	–8.3	78.9	26.6
Newfoundland and Labrador							
Corner Brook	48.93N	57.92W	499	14.43	–1.6	79.1	20.9
Gander International Airport	48.95N	54.58W	495	14.43	–3.5	78.8	31.3
Goose Bay Airport	53.32N	60.42W	161	14.61	–22.4	81.9	25.6
St. John's International Airport	47.62N	52.74W	463	14.45	3.7	76.3	36.1
Wabush Airport	52.93N	66.87W	1808	13.76	–32.4	76.6	21.1
Northwest Territories							
Yellowknife Airport	62.46N	114.44W	676	14.34	–41.7	77.4	21.3
Nova Scotia							
Amherst	45.85N	64.27W	75	14.66	–5.6	77.9	29.3
Halifax International Airport	44.88N	63.52W	476	14.44	–1.4	81.7	26.9
Halifax Shearwater Airport	44.63N	63.50W	167	14.61	1.4	78.7	26.7
Shearwater Jetty	44.63N	63.52W	20	14.69	7.0	78.5	27.2
Sydney Airport	46.17N	60.05W	203	14.59	–0.4	81.1	28.1
Nunavut							
Iqaluit Airport	63.75N	68.55W	112	14.64	–40.2	61.7	34.0
Ontario							
Beausoleil	44.85N	79.87W	600	14.38	–11.0	85.9	14.0
Belle River	42.30N	82.70W	604	14.38	5.7	88.9	29.0
Burlington Piers	43.30N	79.80W	253	14.56	4.5	86.5	23.3
Cobourg	43.95N	78.17W	256	14.56	–4.0	78.3	25.5
Collingwood	44.50N	80.22W	591	14.39	–1.6	86.4	24.8
Erieau	42.25N	81.90W	584	14.39	5.7	80.4	29.2
Kenora Airport	49.79N	94.37W	1345	14.00	–25.4	84.7	19.4
Lagoon City	44.55N	79.22W	725	14.31	–11.1	81.6	28.5
London International Airport	43.03N	81.15W	912	14.22	–0.9	86.2	24.1
North Bay Airport	46.36N	79.42W	1214	14.06	–18.1	82.0	21.3
Ottawa International Airport	45.32N	75.67W	374	14.50	–12.0	87.1	22.4
Petawawa Airport	45.95N	77.32W	427	14.47	–20.5	86.9	20.2
Peterborough Airport	44.23N	78.37W	627	14.37	–11.0	85.2	19.8
Port Colborne	42.87N	79.25W	604	14.38	4.1	78.9	40.9
Port Weller	43.25N	79.22W	259	14.56	7.7	84.7	32.3
Sault Ste Marie Airport	46.48N	84.51W	630	14.36	–13.2	83.1	24.6
Sudbury Airport	46.62N	80.80W	1142	14.10	–18.7	84.5	23.2
Thunder Bay Intl Airport	48.37N	89.33W	653	14.35	–21.9	84.3	22.2
Timmins Victor M Power Apt	48.57N	81.38W	968	14.19	–28.6	84.9	18.8
Toronto Buttonville Airport	43.86N	79.37W	650	14.35	–4.2	89.1	21.6
Toronto Island Airport	43.63N	79.40W	253	14.56	3.0	83.3	29.7
Toronto Pearson Intl Airport	43.68N	79.63W	568	14.40	–1.8	88.1	26.5

Table 2.1: Climatic Data in I-P Units *(Continued)*

Station	Latitude, degrees	Longitude, degrees	Elevation, ft	St. Br. Pressure, psi	Winter Temp., °F	Summer Temp., °F	Wind, mph
Trenton Airport	44.12N	77.53W	282	14.55	–7.3	84.5	24.1
Windsor Airport	42.28N	82.96W	623	14.37	2.4	89.4	25.4
Prince Edward Island							
Charlottetown Airport	46.29N	63.13W	161	14.61	–4.9	80.1	25.7
Quebec							
Bagotville Airport	48.33N	71.00W	522	14.42	–22.2	84.6	26.5
Charlevoix	47.28N	70.64W	2372	13.48	–19.4	76.1	16.7
Gaspe Airport	48.78N	64.48W	112	14.64	–12.8	82.0	21.4
Iles-de-la-Madeleine Airport	47.43N	61.77W	26	14.68	2.1	74.8	36.1
La Baie	48.30N	70.92W	499	14.43	–22.8	84.4	23.1
La Tuque Airport	47.41N	72.79W	554	14.40	–22.6	86.6	16.9
Lac Saint-Pierre	46.18N	72.92W	52	14.67	–13.9	81.7	29.6
L'Acadie	45.29N	73.35W	144	14.62	–11.3	86.2	23.2
L'Assomption	45.81N	73.43W	69	14.66	–14.7	86.7	19.1
Lennoxville	45.37N	71.82W	594	14.38	–14.8	85.0	20.2
Mont-Joli Airport	48.60N	68.22W	171	14.61	–11.0	80.1	28.2
Mont-Orford	45.31N	72.24W	2776	13.28	–19.2	77.2	35.1
Montreal Mirabel Intl Airport	45.67N	74.03W	269	14.55	–15.6	85.0	19.8
Montreal St Hubert Airport	45.52N	73.42W	89	14.65	–11.7	86.2	25.4
Montreal Trudeau Intl Airport	45.47N	73.75W	118	14.63	–10.6	86.0	24.8
Montreal-Est	45.63N	73.55W	164	14.61	–10.0	86.9	19.3
Nicolet	46.23N	72.66W	26	14.68	–14.1	83.8	21.5
Pointe-Au-Pere	48.51N	68.47W	16	14.69	–7.8	73.6	29.4
Quebec Airport	46.80N	71.38W	243	14.57	–15.2	84.0	24.9
Riviere-Du-Loup	47.81N	69.55W	482	14.44	–12.1	79.2	18.4
Roberval Airport	48.52N	72.27W	587	14.39	–22.5	83.7	23.0
Rouyn	48.25N	79.03W	1043	14.15	–23.7	84.2	20.4
Sainte-Agathe-des-Monts	46.05N	74.28W	1296	14.02	–18.5	81.5	20.2
Sainte-Anne-de-Bellevue	45.43N	73.93W	128	14.63	–11.5	86.1	20.1
Sainte-Foy	46.78N	71.29W	299	14.54	–12.9	84.5	21.1
Sept-Îles Airport	50.22N	66.27W	180	14.60	–18.3	73.0	26.5
Sherbrooke Airport	45.43N	71.68W	791	14.28	–19.0	83.5	20.3
Trois-Rivieres	46.35N	72.52W	20	14.69	–11.4	81.4	23.9
Val D Or Airport	48.06N	77.79W	1106	14.12	–27.7	83.5	21.1
Varennes	45.72N	73.38W	59	14.66	–10.9	86.6	24.5
Victoria Hartland	48.53N	123.46W	505	14.43	–20.6	84.2	23.8
Saskatchewan							
Estevan Airport	49.22N	102.97W	1906	13.71	–26.8	89.9	29.2
Moose Jaw	50.33N	105.56W	1893	13.72	–21.5	90.0	27.0
Moose Jaw Airport	50.33N	105.55W	1893	13.72	–27.9	90.3	29.5
North Battleford Airport	52.77N	108.26W	1798	13.77	–31.5	85.6	25.0

Table 2.1: Climatic Data in I-P Units *(Continued)*

Station	Latitude, degrees	Longitude, degrees	Elevation, ft	St. Br. Pressure, psi	Winter Temp., °F	Summer Temp., °F	Wind, mph
Prince Albert Airport	53.22N	105.67W	1404	13.97	–33.8	84.7	21.2
Regina Airport	50.43N	104.67W	1893	13.72	–29.2	88.5	29.9
Saskatoon Diefenbaker Intl Apt	52.17N	106.72W	1654	13.84	–31.0	87.5	25.1
Saskatoon, Kernen Farm	52.15N	106.55W	1673	13.83	–28.3	87.2	24.0
Swift Current Airport	50.30N	107.68W	2684	13.33	–24.7	88.2	28.9
Yorkton Airport	51.27N	102.47W	1634	13.85	–30.2	85.5	23.7
Yukon Territory							
Whitehorse Airport	60.71N	135.07W	2316	13.51	–40.0	78.1	23.2
Albania							
Tirana	41.33N	19.78E	295	14.54	28.1	93.3	15.1
Algeria							
Constantine Airport	36.28N	6.62E	2277	13.53	31.2	100.8	22.9
Dar-El-Beida Airport	36.68N	3.22E	95	14.65	35.4	95.4	23.9
Es Senia Airport	35.63N	0.60W	299	14.54	35.8	93.0	27.3
Argentina							
Buenos Aires, J Newbery Airport	34.57S	58.42W	20	14.69	39.5	87.8	24.8
Cordoba Airport	31.32S	64.22W	1555	13.89	31.8	94.2	26.4
Corrientes International Airport	27.45S	58.77W	203	14.59	40.6	97.2	23.1
Ezeiza International Airport	34.82S	58.53W	66	14.66	31.7	92.6	22.5
Mar del Plata Airport	37.93S	57.58W	69	14.66	30.0	87.8	25.7
Mendoza, El Plumerillo Airport	32.83S	68.78W	2310	13.51	31.1	95.7	19.7
Parana, General Urquiza Airport	31.78S	60.48W	256	14.56	36.4	93.3	24.6
Posadas Airport	27.37S	55.97W	410	14.48	40.6	96.8	19.7
Resistencia International Airport	27.45S	59.05W	171	14.61	35.5	98.4	20.2
Rosario International Airport	32.92S	60.78W	82	14.65	30.8	93.5	26.6
Salta Airport	24.85S	65.48W	4006	12.69	30.4	91.0	17.8
San Juan Airport	31.40S	68.42W	1962	13.68	28.6	100.2	31.2
Santiago del Estero Airport	27.77S	64.30W	653	14.35	31.0	102.0	23.4
Sauce Viejo Airport	31.70S	60.82W	59	14.66	33.0	94.7	33.9
Tucuman, Matienzo Airport	26.85S	65.10W	1476	13.93	37.8	96.8	20.4
Armenia							
Yerevan Airport	40.13N	44.47E	3740	12.82	7.0	96.9	22.3
Australia							
Adelaide Airport	34.95S	138.53E	26	14.68	39.1	96.5	26.3
Adelaide International Airport	34.92S	138.62E	167	14.61	40.6	97.4	18.9
Brisbane International Airport	27.38S	153.13E	33	14.68	42.6	88.0	21.8
Brisbane, Archerfield Airport	27.57S	153.00E	43	14.67	41.4	91.6	20.3
Canberra International Airport	35.30S	149.20E	1903	13.71	26.0	91.7	23.8
Canberra, Isabella Plains	35.42S	149.10E	1929	13.70	24.9	92.5	19.0
Coolangatta Airport	28.17S	153.50E	20	14.69	43.7	84.6	21.8
Gold Coast Seaway	27.93S	153.43E	10	14.69	49.4	87.3	28.2

Table 2.1: Climatic Data in I-P Units ***(Continued)***

Station	Latitude, degrees	Longitude, degrees	Elevation, ft	St. Br. Pressure, psi	Winter Temp., °F	Summer Temp., °F	Wind, mph
Laverton Airport	37.87S	144.75E	66	14.66	35.2	93.4	27.2
Melbourne	37.82S	144.97E	105	14.64	40.4	94.3	17.1
Melbourne International Airport	37.67S	144.85E	390	14.49	37.1	94.3	30.9
Moorabbin Airport	37.98S	145.10E	43	14.67	36.5	92.9	26.3
Newcastle Nobbys Si	32.92S	151.78E	108	14.64	45.9	86.6	40.7
Perth International Airport	31.93S	115.97E	66	14.66	39.5	98.8	24.5
Perth, Jandakot Airport	32.10S	115.88E	102	14.64	35.2	96.7	23.5
Perth, Mount Lawley	31.92S	115.87E	82	14.65	39.1	97.1	18.9
Perth, Swanbourne	31.95S	115.77E	66	14.66	43.5	94.4	30.3
Scoresby Research	37.87S	145.25E	295	14.54	36.1	92.4	18.7
Sydney	33.85S	151.20E	131	14.63	45.0	87.9	N/A
Sydney International Airport	33.93S	151.18E	16	14.69	42.8	91.0	28.3
Sydney, Bankstown Airport	33.92S	150.98E	26	14.68	37.7	92.9	22.0
Sydney, Homebush	33.85S	151.07E	92	14.65	42.7	92.8	21.7
Williamtown Airport (RAAF)	32.80S	151.83E	26	14.68	39.5	93.1	27.6
Austria							
Gumpoldskirchen	48.03N	16.28E	764	14.29	14.2	87.7	17.8
Tulln	48.32N	16.12E	577	14.39	9.0	87.6	26.4
Vienna Downtown	48.20N	16.37E	561	14.40	17.3	88.8	20.2
Vienna International Airport	48.12N	16.57E	623	14.37	11.2	87.5	27.2
Vienna, Hohe Warte	48.25N	16.37E	656	14.35	13.4	87.1	22.2
Belarus							
Brest	52.12N	23.68E	479	14.44	–1.4	85.2	17.0
Gomel	52.40N	30.95E	413	14.48	–6.0	84.6	18.5
Grodno Southeast Airport	53.60N	24.05E	440	14.46	–4.4	82.9	24.2
Minsk	53.93N	27.63E	758	14.30	–5.1	82.8	18.0
Mogilev	53.95N	30.07E	630	14.36	–8.9	81.8	22.5
Vitebsk	55.17N	30.22E	577	14.39	–8.5	81.3	18.5
Belgium							
Antwerp Deurne Airport	51.20N	4.47E	46	14.67	18.5	84.5	22.8
Brussels National Airport	50.90N	4.53E	190	14.60	18.2	84.1	25.8
Brussels, Uccle	50.80N	4.35E	341	14.52	18.8	83.7	21.6
Benin							
Cotonou Cadjehoun Airport	6.35N	2.38E	30	14.68	71.4	90.7	18.3
Bolivia							
Cochabamba Intl Airport	17.42S	66.18W	8360	10.77	35.2	85.8	21.8
La Paz, El Alto Intel Airport	16.52S	68.18W	13248	8.90	23.8	63.4	19.5
Santa Cruz, Viru Viru Intl Apt	17.63S	63.13W	1224	14.06	48.5	93.6	29.3
Bosnia and Herzegovina							
Bjelasnica	43.72N	18.27E	6791	11.43	–3.7	65.5	76.1
Sarajevo Butmir Intl Airport	43.82N	18.33E	1677	13.83	7.5	89.7	18.6

Table 2.1: Climatic Data in I-P Units *(Continued)*

Station	Latitude, degrees	Longitude, degrees	Elevation, ft	St. Br. Pressure, psi	Winter Temp., °F	Summer Temp., °F	Wind, mph
Sarajevo, Bjelave	43.87N	18.43E	2093	13.62	9.7	88.3	11.9
Brazil							
Anapolis AFB	16.23S	48.97W	3730	12.82	55.0	89.1	14.8
Aracaju, Santa Maria Airport	10.98S	37.07W	30	14.68	69.9	89.9	16.3
Belem, International Airport	1.38S	48.48W	52	14.67	73.0	91.6	18.8
Belo Horizonte	19.93S	43.93W	2789	13.27	51.8	89.8	17.6
Belo Horizonte Airport	19.85S	43.95W	2575	13.38	52.6	91.4	13.9
Brasilia International Airport	15.87S	47.93W	3481	12.94	49.7	89.8	16.6
Campo Grande Intl Airport	20.47S	54.67W	1860	13.73	47.3	97.2	23.3
Cuiaba, Marechal Rondon Apt	15.65S	56.10W	614	14.37	55.6	100.4	17.2
Curitiba International Airport	25.52S	49.17W	2979	13.18	36.9	87.7	19.0
Florianopolis, Hercílio Luz Apt	27.67S	48.55W	16	14.69	46.0	89.9	18.6
Fortaleza, Pinto Martins Airport	3.78S	38.53W	82	14.65	73.1	90.0	20.8
Goiania, Santa Genoveva Airport	16.63S	49.22W	2451	13.44	53.4	95.0	18.2
Guarulhos International Airport	23.43S	46.47W	2520	13.41	45.0	91.0	17.0
Londrina Airport	23.33S	51.13W	1867	13.73	46.1	93.0	14.6
Macapa	0.03N	51.05W	49	14.67	73.0	93.2	18.8
Maceio Airport	9.52S	35.78W	384	14.49	66.4	91.4	17.0
Manaus, Eduardo Gomes Airport	3.03S	60.05W	7	14.69	71.2	96.7	13.1
Manaus, Ponta Pelada Airport	3.15S	59.98W	276	14.55	71.7	94.4	13.8
Natal International Airport	5.92S	35.25W	171	14.61	69.7	91.2	22.3
Porto Alegre, Salgado Filho Apt	30.00S	51.18W	10	14.69	39.4	94.8	20.7
Porto Velho Airport	8.77S	63.92W	335	14.52	64.4	95.8	13.7
Recife, Guararapes Intl Airport	8.07S	34.85W	62	14.66	71.2	93.2	18.2
Rio de Janeiro Intl Airport	22.82S	43.25W	20	14.69	58.8	100.1	18.5
Rio de Janeiro, Dumont Airport	22.90S	43.17W	10	14.69	61.1	93.3	18.8
Salvador International Airport	12.90S	38.33W	20	14.69	68.3	90.3	20.4
Sao Luis International Airport	2.60S	44.23W	174	14.60	73.1	93.3	20.2
Sao Paulo Campinas Intl Airport	23.00S	47.13W	2169	13.58	47.9	91.8	25.2
Sao Paulo, Congonhas Intl Apt	23.62S	46.65W	2635	13.35	48.0	89.8	16.8
Teresina Airport	5.05S	42.82W	226	14.58	71.3	100.5	11.6
Vitoria, Goiabeiras Airport	20.27S	40.28W	13	14.69	61.9	93.2	22.5
Bulgaria							
Plovdiv	42.13N	24.75E	607	14.38	13.6	93.5	27.0
Sofia	42.65N	23.38E	1939	13.69	9.6	89.6	21.1
Varna	43.20N	27.92E	141	14.62	15.9	87.7	30.3
Burkina Faso							
Bobo Dioulasso Airport	11.17N	4.32W	1509	13.91	64.0	100.4	16.5
Ouagadougou Airport	12.35N	1.52W	1004	14.17	61.0	105.0	17.1
Chad							
Ndjamena Airport	12.13N	15.03E	968	14.19	55.3	109.3	21.2

Table 2.1: Climatic Data in I-P Units *(Continued)*

Station	Latitude, degrees	Longitude, degrees	Elevation, ft	St. Br. Pressure, psi	Winter Temp., °F	Summer Temp., °F	Wind, mph
Chile							
Antofagasta, Cerro Moreno Airport	23.43S	70.45W	459	14.45	50.1	76.7	20.9
Santiago, Pudahuel Intl Airport	33.38S	70.78W	1555	13.89	30.0	89.2	18.7
China							
Anqing	30.53N	117.05E	66	14.66	28.6	96.1	18.1
Anyang	36.05N	114.40E	210	14.58	17.5	95.0	16.9
Baoding	38.85N	115.57E	62	14.66	14.0	95.3	14.4
Baoji	34.35N	107.13E	2001	13.66	21.3	94.5	13.8
Beijing	39.93N	116.28E	180	14.60	12.5	94.7	21.4
Bengbu	32.95N	117.37E	72	14.66	22.8	96.1	16.2
Benxi	41.32N	123.78E	607	14.38	–9.1	88.6	15.3
Cangzhou	38.33N	116.83E	36	14.68	13.7	93.7	20.4
Changchun, Dafang Shen Airport	43.90N	125.22E	781	14.29	–12.6	87.2	26.8
Changde	29.05N	111.68E	115	14.63	30.3	97.0	13.5
Changsha	28.23N	112.87E	223	14.58	30.2	96.7	16.1
Chaoyang	41.55N	120.45E	577	14.39	–3.5	92.7	20.9
Chengde	40.98N	117.95E	1266	14.04	–0.2	91.4	14.3
Chengdu	30.67N	104.02E	1667	13.83	32.6	91.6	12.1
Chifeng	42.27N	118.97E	1877	13.73	–4.3	90.9	18.5
Chongqing	29.58N	106.47E	853	14.25	37.7	98.4	11.5
Dalian	38.90N	121.63E	318	14.53	10.1	88.0	25.1
Dandong	40.05N	124.33E	46	14.67	2.5	85.7	20.1
Datong	40.10N	113.33E	3507	12.93	–5.6	88.8	22.1
Dezhou	37.43N	116.32E	72	14.66	15.5	93.5	16.6
Fuzhou	26.08N	119.28E	279	14.55	40.0	96.0	18.6
Ganyu	34.83N	119.13E	33	14.68	19.4	91.5	16.7
Gaoyao	23.05N	112.47E	39	14.68	43.6	95.0	15.5
Guangzhou	23.17N	113.33E	138	14.62	42.4	95.1	15.2
Guilin	25.33N	110.30E	545	14.41	34.4	94.4	18.2
Guiyang	26.58N	106.73E	4012	12.69	27.9	86.6	14.5
Haikou Airport	20.03N	110.35E	79	14.65	51.3	95.0	14.5
Hangzhou	30.23N	120.17E	141	14.62	28.1	97.0	16.1
Harbin	45.75N	126.77E	469	14.45	–19.1	87.9	20.3
Hefei	31.87N	117.23E	118	14.63	24.5	95.3	17.1
Hohhot	40.82N	111.68E	3494	12.93	–7.4	88.9	19.1
Hong Kong	22.30N	114.17E	203	14.59	49.3	90.0	19.3
Hong Kong International Airport	22.32N	113.92E	26	14.68	48.2	92.8	22.7
Hong Kong, Taipa Macau	22.15N	113.60E	20	14.69	45.0	91.3	24.5
Jiangling	30.33N	112.18E	108	14.64	29.3	94.6	15.8
Jinan	36.60N	117.05E	554	14.40	17.3	95.0	20.5
Jingdezhen	29.30N	117.20E	197	14.59	28.9	97.0	13.2

Table 2.1: Climatic Data in I-P Units *(Continued)*

Station	Latitude, degrees	Longitude, degrees	Elevation, ft	St. Br. Pressure, psi	Winter Temp., °F	Summer Temp., °F	Wind, mph
Jinzhou	41.13N	121.12E	230	14.57	2.5	88.9	22.7
Jixi	45.28N	130.95E	768	14.29	–13.3	87.0	23.5
Kunming	25.02N	102.68E	6207	11.69	32.6	80.8	18.0
Lanzhou	36.05N	103.88E	4980	12.24	11.0	90.0	9.9
Lingxian	37.33N	116.57E	62	14.66	12.2	95.2	19.3
Liuzhou	24.35N	109.40E	318	14.53	38.1	95.3	12.5
Mengjin	34.82N	112.43E	1093	14.12	19.8	94.5	20.7
Mudanjiang	44.57N	129.60E	794	14.28	–15.5	88.0	20.9
Nanchang	28.60N	115.92E	164	14.61	30.6	96.2	14.4
Nanjing	32.00N	118.80E	23	14.68	23.4	95.2	17.1
Nanning	22.82N	108.35E	413	14.48	41.3	94.9	13.9
Neijiang	29.58N	105.05E	1171	14.08	36.1	95.3	11.9
Qingdao	36.07N	120.33E	253	14.56	18.2	89.7	26.3
Qingjiang	33.60N	119.03E	62	14.66	20.9	92.5	15.7
Qiqihar	47.38N	123.92E	486	14.44	–17.8	89.1	22.1
Shanghai	31.40N	121.47E	13	14.69	28.5	94.8	17.4
Shanghai, Hongqiao	31.17N	121.43E	23	14.68	26.5	95.1	20.2
Shantou	23.40N	116.68E	10	14.69	44.7	92.2	18.4
Shaoguan	24.80N	113.58E	223	14.58	36.7	95.8	14.9
Shenyang	41.73N	123.52E	141	14.62	–7.7	88.8	21.8
Shenyang Taoxian Intl Airport	41.63N	123.48E	203	14.59	–13.1	89.7	24.1
Shenzhen	22.55N	114.10E	59	14.66	44.4	92.8	18.1
Shijiazhuang	38.03N	114.42E	266	14.56	15.7	96.3	14.7
Siping	43.18N	124.33E	548	14.41	–10.0	87.8	20.8
Taishan	36.25N	117.10E	5039	12.21	1.7	72.4	41.2
Taiyuan	37.78N	112.55E	2556	13.39	4.8	91.3	20.7
Tangshan	39.67N	118.15E	95	14.65	8.8	91.7	18.4
Tianjin	39.10N	117.17E	16	14.69	13.6	93.6	18.8
Tianjin, Binhai Intl Airport	39.12N	117.33E	7	14.69	12.3	93.5	22.8
Urumqi Diwopu Intl Airport	43.90N	87.47E	2178	13.58	–9.8	95.7	16.3
Weifang	36.77N	119.18E	72	14.66	12.0	93.7	22.5
Wenzhou	28.02N	120.67E	23	14.68	34.0	92.9	14.5
Wuhan	30.62N	114.13E	75	14.66	28.1	96.3	15.6
Wuhu	31.33N	118.35E	52	14.67	26.1	96.4	17.6
Wulumuqi	43.80N	87.65E	3107	13.12	–8.8	92.1	17.5
Xiamen	24.48N	118.08E	456	14.46	43.3	93.3	20.0
Xian	34.30N	108.93E	1306	14.02	20.7	96.7	17.5
Xihua	33.78N	114.52E	174	14.60	21.4	95.0	15.0
Xingtai	37.07N	114.50E	256	14.56	17.9	96.0	13.2
Xining	36.62N	101.77E	7533	11.11	2.8	81.3	13.8
Xinyang	32.13N	114.05E	377	14.50	23.8	94.1	18.8

Table 2.1: Climatic Data in I-P Units *(Continued)*

Station	Latitude, degrees	Longitude, degrees	Elevation, ft	St. Br. Pressure, psi	Winter Temp., °F	Summer Temp., °F	Wind, mph
Xuzhou	34.28N	117.15E	138	14.62	20.2	94.3	15.4
Yangjiang	21.87N	111.97E	72	14.66	44.9	91.5	18.1
Yanji	42.87N	129.50E	584	14.39	–9.2	87.9	22.6
Yichang	30.70N	111.30E	440	14.46	30.6	96.0	10.5
Yinchuan	38.47N	106.20E	3648	12.86	1.4	89.8	19.8
Yingkou	40.67N	122.20E	13	14.69	0.2	87.0	23.9
Yueyang	29.38N	113.08E	171	14.61	30.4	93.9	16.2
Yuncheng	35.05N	111.05E	1198	14.07	17.4	97.4	21.0
Zhangjiakou	40.78N	114.88E	2382	13.47	2.0	90.7	16.2
Zhanjiang	21.22N	110.40E	92	14.65	46.1	93.0	17.8
Zhengzhou	34.72N	113.65E	364	14.50	19.9	95.2	18.8
Zunyi	27.70N	106.88E	2772	13.28	29.9	90.6	10.8
Colombia							
Barranquilla Intl Airport	10.88N	74.78W	98	14.64	73.0	93.5	29.6
Bogota, Eldorado Intl Airport	4.70N	74.13W	8353	10.77	37.1	70.2	18.8
Cali, Aragon Intl Airport	3.55N	76.38W	3179	13.08	63.9	89.7	18.9
Cartagena, Rafael Nunez Airport	10.45N	75.52W	39	14.68	73.5	90.2	20.5
Medellín, J M Cordova Airport	6.13N	75.43W	7028	11.33	49.9	75.1	20.4
Congo							
Brazzaville, Maya-Maya Airport	4.25S	15.25E	1037	14.15	64.4	93.2	13.1
Costa Rica							
Juan Santamaria Intl Airport	9.98N	84.22W	3064	13.14	61.8	87.6	28.3
Cote d'Ivoire							
Abidjan Port Bouet Airpot	5.25N	3.93W	26	14.68	69.8	91.0	16.1
Croatia							
Zagreb, Maksimir	45.82N	16.03E	420	14.47	12.0	89.0	13.4
Zagreb, Pleso Airport	45.73N	16.07E	351	14.51	10.0	89.4	19.0
Cuba							
Havana, Jose Marti Intl Airport	22.98N	82.40W	246	14.57	51.9	91.5	23.3
Camaguey Intl Airport	21.42N	77.85W	387	14.49	59.4	92.0	23.3
Santiago de Cuba Airport	19.97N	75.85W	180	14.60	65.8	89.4	23.3
Czech Republic							
Brno, Turany Airport	49.15N	16.70E	807	14.27	7.8	86.2	22.9
Ostrava, Mosnov Airport	49.68N	18.12E	853	14.25	3.0	86.0	22.7
Prague, Kbely Airport	50.12N	14.53E	942	14.20	8.6	85.2	20.4
Prague, Libus	50.02N	14.45E	994	14.18	7.6	86.2	19.4
Prague, Ruzyne Airport	50.10N	14.25E	1198	14.07	5.7	84.6	27.3
Denmark							
Copenhagen Kastrup Airport	55.62N	12.65E	16	14.69	15.4	77.9	28.4
Copenhagen Roskilde Airport	55.58N	12.13E	141	14.62	13.8	78.6	28.0
Copenhagen, Drogden	55.53N	12.72E	0	14.70	17.5	71.7	40.7

Table 2.1: Climatic Data in I-P Units *(Continued)*

Station	Latitude, degrees	Longitude, degrees	Elevation, ft	St. Br. Pressure, psi	Winter Temp., °F	Summer Temp., °F	Wind, mph
Vaerloese Airport	55.77N	12.33E	102	14.64	10.2	79.2	27.5
Dominican Republic							
Santo Domingo	18.43N	69.88W	46	14.67	67.2	90.6	16.6
Santo Domingo, Americas Apt	18.43N	69.67W	59	14.66	65.3	90.7	16.4
Ecuador							
Guayaquil International Airport	2.15S	79.88W	30	14.68	65.9	91.6	16.2
Quito International Airport	0.13S	78.48W	9226	10.41	44.3	71.4	17.3
Egypt							
Alexandria, Nouzha	31.20N	29.95E	23	14.68	44.4	91.4	22.9
Assiut Airport	27.05N	31.02E	230	14.57	39.7	105.3	23.6
Cairo International Airport	30.13N	31.40E	243	14.57	45.9	100.6	21.0
Luxor International Airport	25.67N	32.70E	325	14.52	41.0	109.7	16.1
Port Said	31.27N	32.30E	20	14.69	48.7	89.8	24.2
Port Said Airport	31.28N	32.23E	20	14.69	49.4	89.0	26.6
Estonia							
Tallinn	59.47N	24.82E	112	14.64	–2.3	78.7	20.6
Finland							
Helsinki Vantaa Airport	60.32N	24.97E	184	14.60	–9.1	80.1	22.3
Isosaari	60.10N	25.07E	16	14.69	–4.3	73.0	35.2
France							
Cap Couronne	43.33N	5.05E	89	14.65	26.8	87.3	38.3
Cap Ferrat	43.68N	7.33E	472	14.45	37.8	84.3	30.0
Cap Pomegues	43.27N	5.30E	230	14.57	29.4	83.5	52.5
Le Bourget Airport	48.97N	2.43E	171	14.61	24.5	88.0	22.8
Lyon, Bron Airport	45.72N	4.93E	663	14.35	22.3	92.5	25.8
Lyon, Satolas Airport	45.73N	5.08E	787	14.28	19.7	90.4	24.1
Marignane	43.45N	5.23E	105	14.64	26.5	90.9	36.8
Nice	43.65N	7.20E	89	14.65	35.3	85.2	26.2
Paris, Charles de Gaulle Intl Apt	49.02N	2.53E	367	14.50	20.9	87.1	26.0
Paris, Montsouris	48.82N	2.33E	253	14.56	27.4	88.8	16.3
Paris, Orly International Airport	48.72N	2.38E	295	14.54	21.3	87.7	24.7
Toulouse, Blagnac Airport	43.63N	1.37E	505	14.43	24.2	91.6	23.4
Trappes	48.77N	2.00E	551	14.41	24.7	86.6	15.6
Villacoublay, Velizy Airport	48.77N	2.20E	587	14.39	21.4	85.7	21.8
Gabon							
Libreville International Airport	0.45N	9.42E	49	14.67	71.5	88.5	15.9
Gambia							
Banjul/Yundum	13.20N	16.63W	108	14.64	61.2	100.1	20.0
Georgia							
Tbilisi International Airport	41.68N	44.95E	1470	13.93	21.8	93.9	46.6

Table 2.1: Climatic Data in I-P Units *(Continued)*

Station	Latitude, degrees	Longitude, degrees	Elevation, ft	St. Br. Pressure, psi	Winter Temp., °F	Summer Temp., °F	Wind, mph
Germany							
Berlin, Dahlem	52.47N	13.30E	167	14.61	10.4	84.7	16.5
Berlin, Schonefeld	52.38N	13.52E	154	14.61	7.1	85.3	24.9
Berlin, Tegel Airport	52.57N	13.32E	121	14.63	9.7	86.1	23.3
Berlin, Tempelhof Airport	52.47N	13.40E	164	14.61	10.8	86.0	23.2
Bremen Airport	53.05N	8.80E	10	14.69	12.3	83.2	25.5
Celle Airport	52.60N	10.02E	171	14.61	10.8	86.2	20.8
Dresden, Klotzsche Airport	51.13N	13.77E	755	14.30	7.5	85.0	21.5
Dusseldorf Airport	51.28N	6.78E	148	14.62	14.1	85.3	23.3
Essen/Mulheim	51.40N	6.97E	505	14.43	14.3	82.8	21.6
Frankfurt International Airport	50.05N	8.60E	367	14.50	13.0	87.4	22.6
Furstenfeldbruck Airport	48.20N	11.27E	1755	13.79	4.8	84.3	24.8
Guetersloh Airport	51.93N	8.32E	236	14.57	14.3	85.8	22.5
Hamburg Fuhlsbuettel Airport	53.63N	10.00E	52	14.67	11.1	82.1	22.7
Hannover Airport	52.47N	9.70E	180	14.60	9.2	84.0	22.7
Heidelberg AAF	49.40N	8.65E	358	14.51	14.4	89.6	17.6
Koln Bonn Airport	50.87N	7.17E	299	14.54	12.9	85.9	20.1
Leipzig	51.32N	12.42E	495	14.43	14.9	86.7	15.3
Leipzig Airport	51.42N	12.23E	436	14.47	8.0	85.7	27.9
Munich	48.13N	11.55E	1706	13.81	10.5	85.2	17.5
Munich, Riem	48.13N	11.70E	1736	13.80	6.5	85.0	25.8
Norvenich Airport	50.83N	6.67E	443	14.46	15.4	86.4	22.9
Nuremberg Airport	49.50N	11.08E	1047	14.15	6.1	86.3	20.7
Potsdam	52.38N	13.07E	266	14.56	8.7	84.8	24.1
Quickborn	53.73N	9.88E	56	14.67	14.6	83.1	20.0
Roth Airport	49.22N	11.10E	1296	14.02	6.5	87.5	18.8
Stuttgart Echterdingen Airport	48.68N	9.22E	1299	14.02	9.1	84.8	20.9
Stuttgart/Schnarren	48.83N	9.20E	1033	14.16	11.3	85.2	20.3
Wunstorf	52.47N	9.43E	167	14.61	11.8	86.3	24.1
Greece							
Athens, Ellinikon Airport	37.90N	23.73E	49	14.67	34.8	95.1	22.4
Elefsis Airport	38.07N	23.55E	102	14.64	33.4	97.1	22.7
Thessaloniki, Makedonia Airport	40.52N	22.97E	13	14.69	26.2	93.3	27.8
Guatemala							
Guatemala International Airport	14.58N	90.52W	4885	12.28	51.4	82.4	27.4
Honduras							
San Pedro Sula, La Mesa Airport	15.45N	87.93W	102	14.64	63.0	98.6	19.9
Tegucigalpa, Toncontín Airport	14.05N	87.22W	3304	13.02	52.6	89.7	20.9
Hungary							
Budaors Airport	47.45N	18.97E	433	14.47	11.8	87.8	31.2
Budapest, Ferihegy Airport	47.43N	19.27E	607	14.38	9.2	89.9	31.1

Table 2.1: Climatic Data in I-P Units ***(Continued)***

Station	Latitude, degrees	Longitude, degrees	Elevation, ft	St. Br. Pressure, psi	Winter Temp., °F	Summer Temp., °F	Wind, mph
Budapest, Pestszentl	47.43N	19.18E	456	14.46	12.9	90.1	17.2
India							
Ahmadabad International Airport	23.07N	72.63E	180	14.60	51.6	107.7	14.4
Akola Airport	20.70N	77.07E	1014	14.17	55.3	109.8	12.8
Aurangabad Chikalthan	19.85N	75.40E	1900	13.71	50.9	104.0	20.7
Bangalore	12.97N	77.58E	3022	13.16	59.2	93.6	12.8
Belgaum, Sambra	15.85N	74.62E	2451	13.44	56.0	97.3	18.7
Bhopal Airport	23.28N	77.35E	1716	13.81	50.3	107.1	20.5
Bhubaneswar Airport	20.25N	85.83E	151	14.62	57.2	101.4	23.1
Bikaner	28.00N	73.30E	735	14.31	42.1	111.5	15.7
Bombay, Santacruz	19.12N	72.85E	46	14.67	61.7	96.4	15.5
Calcutta, Dum Dum	22.65N	88.45E	20	14.69	52.8	98.9	12.7
Coimbatore, Peelamedu Airport	11.03N	77.05E	1309	14.01	64.4	98.0	22.8
CWC Vishakhapatnam	17.70N	83.30E	217	14.58	68.1	92.7	18.8
Guwahati Airport	26.10N	91.58E	177	14.60	51.5	94.0	11.3
Gwalior	26.23N	78.25E	679	14.34	42.7	110.6	10.8
Hyderabad Airport	17.45N	78.47E	1788	13.77	57.1	104.4	18.4
Indore Airport	22.72N	75.80E	1860	13.73	48.3	105.4	25.0
Jabalpur	23.20N	79.95E	1289	14.02	47.1	108.2	9.4
Jaipur Sanganer Airport	26.82N	75.80E	1280	14.03	44.8	108.3	15.8
Jamshedpur	22.82N	86.18E	466	14.45	50.0	108.2	8.1
Jodhpur	26.30N	73.02E	735	14.31	48.0	108.7	13.1
Kozhikode	11.25N	75.78E	16	14.69	71.9	92.7	15.0
Lucknow Amausi	26.75N	80.88E	420	14.47	44.3	107.8	16.6
Madras Chennai Airport	13.00N	80.18E	52	14.67	67.9	101.3	18.6
Mangalore Bajpe	12.92N	74.88E	335	14.52	69.1	93.8	17.9
Nagpur, Sonegaon Airport	21.10N	79.05E	1017	14.16	53.1	110.7	17.8
Nellore	14.45N	79.98E	66	14.66	68.6	105.1	12.1
New Delhi/Palam Intl Airport	28.57N	77.12E	764	14.29	43.2	110.8	18.1
New Delhi/Safdarjung	28.58N	77.20E	709	14.32	43.4	107.6	15.8
Pataliputra	25.60N	85.10E	197	14.59	46.7	105.8	14.2
Patiala	30.33N	76.47E	823	14.26	40.9	106.5	9.8
Pune	18.53N	73.85E	1834	13.75	49.5	100.6	11.6
Rajkot Airport	22.30N	70.78E	453	14.46	53.3	105.8	24.5
Sholapur	17.67N	75.90E	1572	13.88	60.7	105.7	7.6
Surat	21.20N	72.83E	39	14.68	57.9	100.0	12.6
Thiruvananthapuram	8.48N	76.95E	210	14.58	71.8	92.6	13.6
Tiruchirapalli Airport	10.77N	78.72E	289	14.54	68.0	102.1	26.0
Indonesia							
Denpasar, Ngurah Rai Intl Apt	8.75S	115.17E	3	14.69	71.2	90.7	18.0
Jakarta, Soekarno–Hatta Airport	6.12S	106.65E	26	14.68	71.7	92.9	21.7

Table 2.1: Climatic Data in I-P Units *(Continued)*

Station	Latitude, degrees	Longitude, degrees	Elevation, ft	St. Br. Pressure, psi	Winter Temp., °F	Summer Temp., °F	Wind, mph
Makassar, Hasanuddin Intl Apt	5.07S	119.55E	46	14.67	68.8	93.3	16.3
Medan, Polonia Intl Airport	3.57N	98.68E	82	14.65	72.1	93.5	13.6
Menado, Sam Ratulangi Intl Apt	1.53N	124.92E	262	14.56	68.5	91.3	17.0
Padang, Tabing Airport	0.88S	100.35E	10	14.69	70.5	90.0	12.0
Pekanbaru, SSQ II Intl Airport	0.47N	101.45E	102	14.64	71.0	94.0	13.4
Surabaya, Juanda Intl Airport	7.37S	112.77E	10	14.69	69.6	93.5	18.5
Iran, Islamic Republic of							
Abadan Airport	30.37N	48.25E	20	14.69	40.8	117.8	23.5
Ahwaz	31.33N	48.67E	72	14.66	41.0	117.7	20.8
Anzali	37.47N	49.47E	–85	14.74	35.5	86.5	24.8
Arak	34.10N	49.77E	5604	11.95	3.8	97.2	18.8
Bandar Abbas Intel Airport	27.22N	56.37E	33	14.68	48.8	107.3	19.1
Esfahan	32.47N	51.67E	5085	12.19	19.5	102.3	22.8
Hamedan	34.85N	48.53E	5738	11.89	1.7	95.9	23.0
Kashan	33.98N	51.45E	3222	13.06	27.0	107.1	13.9
Kerman Airport	30.25N	56.97E	5755	11.89	19.7	100.4	25.8
Kermanshah	34.27N	47.12E	4337	12.53	18.8	103.2	21.7
Mashhad	36.27N	59.63E	3278	13.04	18.4	98.8	20.4
Orumieh	37.53N	45.08E	4318	12.54	12.7	91.1	20.1
Shiraz	29.53N	52.53E	4859	12.29	28.7	102.4	22.0
Tabriz	38.08N	46.28E	4465	12.47	12.3	95.4	23.2
Tehran, Mehrabad Intl Airport	35.68N	51.32E	3907	12.74	26.9	101.3	25.5
Zahedan	29.47N	60.88E	4495	12.46	23.8	102.4	26.8
Zanjan	36.68N	48.48E	5456	12.02	8.0	93.1	22.0
Ireland							
Casement Airport	53.30N	6.43W	305	14.53	27.4	73.5	34.0
Dublin Airport	53.43N	6.25W	279	14.55	28.5	71.8	30.0
Israel							
Tel Aviv, Ben Gurion Intl Apt	32.00N	34.90E	161	14.61	41.1	94.9	22.5
Tel Aviv, Sde Dov Airport	32.10N	34.78E	13	14.69	44.8	88.1	26.8
Italy							
Bari, Palese Macchie Airport	41.13N	16.78E	161	14.61	33.5	92.8	21.3
Bologna, G Marconi Airport	44.53N	11.30E	161	14.61	23.3	93.3	16.2
Catania, Fontanarossa Airport	37.47N	15.05E	56	14.67	35.0	95.0	22.1
Catania, Sigonella Airport	37.40N	14.92E	72	14.66	35.6	96.9	27.1
Florence Peretola Airport	43.80N	11.20E	125	14.63	26.4	95.1	19.4
Genoa Cristoforo Colombo Apt	44.42N	8.85E	10	14.69	33.7	85.9	26.1
Grazzanise Airport	41.05N	14.07E	33	14.68	30.3	89.9	22.4
Milan, Linate Airport	45.43N	9.28E	338	14.52	22.9	91.4	16.4
Naples	40.90N	14.30E	305	14.53	35.6	92.3	18.3
Naples Capodichino Airport	40.85N	14.30E	236	14.57	33.1	91.6	21.1

Table 2.1: Climatic Data in I-P Units *(Continued)*

Station	Latitude, degrees	Longitude, degrees	Elevation, ft	St. Br. Pressure, psi	Winter Temp., °F	Summer Temp., °F	Wind, mph
Palermo Punta Raisi Airport	38.18N	13.10E	69	14.66	44.2	92.4	29.3
Rome, Ciampino	41.78N	12.58E	344	14.51	29.9	91.8	25.2
Rome, Fiumicino Airport	41.80N	12.23E	10	14.69	30.6	87.8	25.7
Rome, Pratica Di Mare AFB	41.65N	12.45E	69	14.66	33.5	87.5	22.5
Turin, Bric Della Croce	45.03N	7.73E	2329	13.50	24.2	82.8	19.5
Turin-Caselle Airport	45.22N	7.65E	942	14.20	21.2	87.7	14.0
Jamaica							
Kingston, Norman Manley Apt	17.93N	76.78W	46	14.67	71.7	92.0	32.2
Japan							
Akita	39.72N	140.10E	69	14.66	22.2	88.8	27.4
Asahikawa	43.77N	142.37E	381	14.49	–1.1	85.8	16.0
Ashiya Air Field	33.88N	130.65E	108	14.64	30.1	90.0	23.4
Atsugi NAS	35.45N	139.45E	213	14.58	29.8	91.3	23.3
Chiba	35.60N	140.10E	62	14.66	32.6	90.2	27.6
Fukuoka	33.58N	130.38E	49	14.67	33.2	92.1	18.7
Fukuoka Airport	33.58N	130.45E	39	14.68	30.7	93.0	20.8
Fukuyama	34.45N	133.25E	10	14.69	27.2	92.9	13.6
Fushiki	36.80N	137.05E	43	14.67	27.7	92.1	16.7
Futenma MCAS	26.27N	127.75E	276	14.55	51.5	89.9	25.1
Gifu	35.40N	136.77E	56	14.67	29.4	95.2	17.9
Gifu Air Field	35.38N	136.87E	138	14.62	26.2	93.5	18.0
Hamamatsu	34.72N	137.72E	108	14.64	32.2	90.9	19.9
Hamamatsu AB	34.75N	137.70E	157	14.61	30.0	91.1	21.7
Himeji	34.83N	134.67E	131	14.63	27.9	92.1	19.2
Hiroshima	34.40N	132.47E	174	14.60	30.0	92.1	20.9
Iizuka	33.65N	130.70E	125	14.63	28.8	92.4	16.4
Iruma AB	35.83N	139.42E	305	14.53	24.7	93.0	21.8
Kadena Air Base	26.35N	127.77E	148	14.62	49.9	91.7	25.9
Kagoshima	31.55N	130.55E	105	14.64	33.4	91.7	19.9
Kanazawa	36.58N	136.63E	108	14.64	28.8	91.6	26.1
Kansai International Airport	34.43N	135.25E	26	14.68	35.2	91.4	28.7
Kobe	34.70N	135.22E	98	14.64	31.1	91.2	21.1
Kochi	33.57N	133.55E	16	14.69	30.2	90.7	13.0
Komatsu Airport	36.40N	136.40E	30	14.68	28.0	91.5	24.7
Kumagaya	36.15N	139.38E	102	14.64	28.1	95.4	17.8
Kumamoto	32.82N	130.70E	128	14.63	28.5	93.6	16.0
Kure	34.23N	132.55E	16	14.69	31.9	90.2	15.7
Kyoto	35.02N	135.73E	151	14.62	30.1	94.2	11.7
Matsuyama	33.85N	132.78E	112	14.64	31.5	91.5	13.2
Miyazaki	31.93N	131.42E	49	14.67	31.3	92.6	20.4
Nagano	36.67N	138.20E	1375	13.98	19.6	90.7	17.8

Table 2.1: Climatic Data in I-P Units *(Continued)*

Station	Latitude, degrees	Longitude, degrees	Elevation, ft	St. Br. Pressure, psi	Winter Temp., °F	Summer Temp., °F	Wind, mph
Nagasaki	32.73N	129.87E	115	14.63	33.2	90.6	17.5
Nagoya	35.17N	136.97E	184	14.60	29.3	93.7	19.3
Nagoya Airport	35.25N	136.92E	56	14.67	28.0	94.1	21.9
Naha	26.20N	127.68E	174	14.60	53.0	89.8	29.3
Naha Airport	26.20N	127.65E	20	14.69	53.6	89.9	29.8
NARA	34.70N	135.83E	348	14.51	28.1	93.4	10.2
Niigata	37.92N	139.05E	20	14.69	28.1	91.1	23.3
Nyutabaru AB	32.08N	131.45E	269	14.55	28.8	90.0	22.0
Oita	33.23N	131.62E	43	14.67	30.8	91.7	16.1
Okayama	34.67N	133.92E	59	14.66	29.9	93.9	22.1
Onahama	36.95N	140.90E	16	14.69	27.1	84.0	18.5
Osaka	34.68N	135.52E	272	14.55	32.6	93.5	20.6
Osaka International Airport	34.78N	135.43E	49	14.67	28.7	93.6	18.8
Otaru	43.18N	141.02E	85	14.65	14.5	82.4	18.0
Ozuki Air Field	34.05N	131.05E	23	14.68	30.4	89.9	24.6
Sapporo	43.07N	141.33E	85	14.65	12.8	84.6	20.3
Sendai	38.27N	140.90E	141	14.62	24.5	87.5	23.1
Shimofusa AB	35.80N	140.02E	108	14.64	26.8	91.7	23.6
Shimonoseki	33.95N	130.93E	62	14.66	33.9	89.3	23.0
Shizuhama AB	34.82N	138.30E	33	14.68	30.4	91.1	24.1
Shizuoka	34.98N	138.40E	49	14.67	31.5	91.4	14.1
Sumoto	34.33N	134.90E	367	14.50	31.9	89.3	16.0
Tadotsu	34.28N	133.75E	16	14.69	31.9	92.5	16.7
Takamatsu	34.32N	134.05E	33	14.68	30.4	93.3	18.1
Tokyo	35.68N	139.77E	118	14.63	33.0	91.7	19.3
Tokyo International Airport	35.55N	139.78E	30	14.68	32.1	91.1	27.9
Toyama	36.72N	137.20E	56	14.67	26.9	91.8	20.5
Tsuiki Airport	33.68N	131.05E	66	14.66	28.0	89.8	22.4
Utsunomiya	36.55N	139.87E	459	14.45	23.9	91.4	20.5
Wakayama	34.23N	135.17E	59	14.66	32.7	91.7	24.2
Yokohama	35.43N	139.65E	138	14.62	32.6	89.7	21.2
Yokosuka	35.28N	139.67E	174	14.60	33.6	91.8	30.1
Yokota AB	35.75N	139.35E	456	14.46	24.8	92.9	20.3
Jordan							
Amman Airport	31.98N	35.98E	2556	13.39	33.8	95.6	22.9
Irbid	32.55N	35.85E	2021	13.65	35.6	93.8	20.1
Queen Alia International Airport	31.72N	35.98E	2369	13.48	30.6	98.3	27.2
Kazakhstan							
Almaty	43.23N	76.93E	2792	13.27	–3.8	93.0	13.3
Astana	51.13N	71.37E	1148	14.10	–23.5	89.4	25.6
Karaganda Southeast Airport	49.80N	73.15E	1814	13.76	–21.0	89.6	24.7

Table 2.1: Climatic Data in I-P Units *(Continued)*

Station	Latitude, degrees	Longitude, degrees	Elevation, ft	St. Br. Pressure, psi	Winter Temp., °F	Summer Temp., °F	Wind, mph
Pavlodar	52.30N	76.93E	400	14.48	–26.2	90.7	21.0
Shymkent	42.32N	69.70E	1982	13.67	7.1	98.8	17.8
Taraz (was Zhambyl)	42.85N	71.38E	2149	13.59	–3.2	96.1	25.6
Kenya							
Mombasa	4.03S	39.62E	180	14.60	67.8	91.3	19.4
Nairobi, Kenyatta International Apt	1.32S	36.92E	5328	12.08	49.7	84.1	21.8
Korea, North							
Chongjin	41.78N	129.82E	141	14.62	7.9	81.5	16.4
Hamheung	39.93N	127.55E	72	14.66	7.9	88.5	18.4
Kaesong	37.97N	126.57E	230	14.57	8.2	87.7	18.6
Namp'o	38.72N	125.38E	154	14.61	8.8	86.1	22.4
Pyongyang	39.03N	125.78E	118	14.63	4.1	88.1	15.0
Sinuiju	40.10N	124.38E	23	14.68	3.7	87.5	17.3
Wonsan	39.18N	127.43E	118	14.63	13.0	88.8	17.3
Korea, South							
Busan	35.10N	129.03E	230	14.57	21.9	88.1	22.8
Cheongju	36.63N	127.45E	194	14.59	10.8	91.1	14.9
Cheongju International Airport	36.72N	127.50E	197	14.59	6.8	91.7	16.1
Daegu	35.88N	128.62E	194	14.59	18.3	93.3	18.8
Daegu International Airport	35.90N	128.67E	115	14.63	15.7	95.0	19.1
Daejeon	36.37N	127.37E	236	14.57	12.3	90.8	15.9
Gimhae International Airport	35.18N	128.93E	16	14.69	19.7	91.1	20.7
Gwangju	35.17N	126.90E	243	14.57	19.7	90.4	16.9
Gwangju, Kwangju Intl Airport	35.12N	126.82E	43	14.67	17.9	93.4	16.5
Incheon	37.47N	126.63E	230	14.57	12.7	88.0	21.1
Jeju	33.52N	126.53E	75	14.66	32.0	89.1	24.8
Jeju International Airport	33.52N	126.50E	79	14.65	30.5	89.5	27.6
Jeonju	35.82N	127.15E	180	14.60	15.9	91.6	12.6
Jinju	35.20N	128.12E	75	14.66	16.6	91.1	16.1
Masan	35.18N	128.57E	13	14.69	23.5	90.4	15.2
Osan AB	37.10N	127.03E	39	14.68	8.3	91.4	18.3
Pohang	36.03N	129.38E	13	14.69	20.0	92.5	18.9
Pohang Airport	35.98N	129.42E	66	14.66	19.2	93.2	21.5
Pyongtaek Airport A-511	36.97N	127.03E	46	14.67	8.4	91.3	17.6
Sachon Airport	35.08N	128.08E	26	14.68	13.9	91.6	15.6
Seogwipo	33.25N	126.57E	167	14.61	31.7	88.4	18.8
Seoul	37.57N	126.97E	282	14.55	11.2	89.7	16.1
Seoul, Korea AF HQ	37.50N	126.93E	161	14.61	10.8	91.7	14.1
Seoul AB	37.43N	127.12E	66	14.66	5.4	91.7	13.9
Seoul, Gimpo Intl Airport	37.57N	126.78E	56	14.67	7.2	89.6	18.7
Suwon	37.27N	126.98E	115	14.63	10.1	89.7	14.3

Table 2.1: Climatic Data in I-P Units *(Continued)*

Station	Latitude, degrees	Longitude, degrees	Elevation, ft	St. Br. Pressure, psi	Winter Temp., °F	Summer Temp., °F	Wind, mph
Ulsan	35.55N	129.32E	118	14.63	20.8	91.6	15.7
Yeosu	34.73N	127.75E	220	14.58	22.6	86.9	27.1
Kyrgyzstan							
Bishkek	42.85N	74.53E	2493	13.42	–4.7	95.4	19.0
Latvia							
Riga	56.97N	24.05E	85	14.65	–1.4	81.3	22.3
Riga International Airport	56.92N	23.97E	36	14.68	–0.8	84.1	20.5
Lebanon							
Beirut International Airport	33.82N	35.48E	62	14.66	45.7	89.2	25.0
Libyan Arab Jamahiriya							
Benina International Airport	32.10N	20.27E	433	14.47	44.3	98.9	32.5
Misurata	32.42N	15.05E	105	14.64	46.4	97.7	29.4
Tripoli International Airport	32.70N	13.08E	207	14.59	39.6	107.5	23.3
Lithuania							
Kaunas	54.88N	23.83E	2526	13.40	–3.4	82.0	22.0
Vilnius Airport	54.63N	25.28E	512	14.43	–4.3	82.4	23.0
Macedonia (Yugoslav)							
Skopje Airport	41.97N	21.65E	784	14.28	9.2	96.0	19.9
Madagascar							
Antananarivo Ivato Airport	18.80S	47.48E	4186	12.60	46.2	84.5	18.2
Malaysia							
Kota Kinabalu Intl Airport	5.93N	116.05E	10	14.69	72.9	91.9	15.6
Kuala Lumpur, Subang Airport	3.12N	101.55E	72	14.66	71.6	93.6	14.4
Kuantan Airport	3.78N	103.22E	52	14.67	70.3	92.9	14.7
Kuching Airport	1.48N	110.33E	89	14.65	71.4	93.0	11.9
Sandakan Airport	5.90N	118.07E	43	14.67	73.2	92.3	15.8
Tawau Airport	4.27N	117.88E	66	14.66	71.6	90.2	13.4
Mali							
Bamako Senou Airport	12.53N	7.95W	1250	14.04	59.4	104.2	19.0
Mauritania							
Nouakchott Airport	18.10N	15.95W	10	14.69	55.2	106.2	22.7
Mexico							
Acapulco, Gen J N Alvarez Apt	16.75N	99.75W	16	14.69	67.6	92.7	19.5
Apodaca, Gen M Escobedo Apt	25.77N	100.10W	1309	14.01	37.7	102.0	31.5
Cancun International Airport	21.03N	86.87W	20	14.69	56.8	93.4	23.4
De Guanajuato, Del Bajío Apt	20.98N	101.48W	6106	11.73	39.2	93.2	28.2
Guadalajara International Airport	20.52N	103.30W	5138	12.16	35.5	92.5	23.1
Mazatlan, Gen R Buelna Intl Apt	23.15N	106.25W	16	14.69	51.6	93.9	23.2
Merida	20.98N	89.65W	30	14.68	56.7	100.7	23.5
Mexico City	19.43N	99.13W	7333	11.20	39.3	84.2	47.3
Mexico City Intl Airport	19.43N	99.07W	7500	11.13	37.4	84.5	25.9

Table 2.1: Climatic Data in I-P Units *(Continued)*

Station	Latitude, degrees	Longitude, degrees	Elevation, ft	St. Br. Pressure, psi	Winter Temp., °F	Summer Temp., °F	Wind, mph
Monterrey	25.73N	100.30W	1690	13.82	39.8	100.7	12.4
San Luis Potosi	22.18N	100.98W	6178	11.70	31.6	90.1	22.3
Tampico, Gen F J Mina Airport	22.28N	97.87W	82	14.65	50.2	93.4	33.3
Tijuana, Gen A L Rodriguez Apt	32.53N	116.97W	512	14.43	42.6	89.7	18.9
Toluca, Uruapan Intl Airport	19.33N	99.57W	8671	10.64	28.3	79.2	20.3
Veracruz, Gen H Jara Airport	19.13N	96.18W	108	14.64	58.8	95.3	44.9
Moldova, Republic of							
Kishinev	47.02N	28.98E	568	14.40	6.3	88.0	14.4
Mongolia							
Ulaanbaatar	47.92N	106.87E	4285	12.56	–29.5	85.8	23.1
Morocco							
Agadir	30.38N	9.57W	75	14.66	41.0	95.3	23.6
Agadir Al Massira Airport	30.32N	9.40W	75	14.66	41.2	99.6	21.2
Casablanca	33.57N	7.67W	187	14.60	42.8	85.1	17.4
Casablanca Airport	33.37N	7.58W	676	14.34	37.7	95.4	22.2
Fes, Saiss Airport	33.93N	4.98W	1900	13.71	33.4	102.3	22.2
Marrakech	31.62N	8.03W	1529	13.90	39.4	106.1	18.8
Meknes, Bassatine Airport	33.88N	5.53W	1837	13.75	35.9	101.4	18.8
Oujda	34.78N	1.93W	1542	13.90	32.4	98.9	27.0
Rabat, Sale	34.05N	6.77W	259	14.56	40.9	90.0	18.4
Tanger, Boukhalf Airport	35.73N	5.90W	69	14.66	39.4	91.6	39.9
Tetouan, Sania Ramel Airport	35.58N	5.33W	33	14.68	43.3	91.0	28.1
Mozambique							
Maputo International Airport	25.92S	32.57E	144	14.62	53.5	95.3	35.8
Netherlands							
Amsterdam Schiphol Airport	52.30N	4.77E	–13	14.70	18.9	82.0	30.4
Hoek Van Holland	51.98N	4.10E	46	14.67	20.7	80.9	36.5
IJmuiden	52.47N	4.57E	43	14.67	19.8	77.9	41.5
Rotterdam Hague Airport	51.95N	4.45E	–13	14.70	18.8	82.3	28.6
Valkenburg	52.18N	4.42E	7	14.69	19.0	80.8	30.9
Woensdrecht AB	51.45N	4.33E	56	14.67	19.0	85.4	21.8
New Zealand							
Auckland Aero AWS	37.00S	174.80E	23	14.68	40.1	77.5	29.1
Auckland Airport	37.02S	174.80E	20	14.69	35.3	77.3	29.2
Christchurch Aero A	43.48S	172.52E	121	14.63	27.4	81.8	25.6
Christchurch Airport	43.48S	172.55E	98	14.64	27.2	82.4	25.6
Nicaragua							
Managua, A C Sandino Intl Apt	12.15N	86.17W	184	14.60	67.6	96.8	17.7
Niger							
Niamey, Diori Hamani Airport	13.48N	2.17E	745	14.30	60.3	107.9	21.8
Norway							
Oslo Fornebu Airport	59.90N	10.62E	56	14.67	1.0	80.3	18.9

Table 2.1: Climatic Data in I-P Units *(Continued)*

Station	Latitude, degrees	Longitude, degrees	Elevation, ft	St. Br. Pressure, psi	Winter Temp., °F	Summer Temp., °F	Wind, mph
Oslo-Blindern	59.95N	10.72E	315	14.53	7.0	79.4	18.0
Oman							
Buraimi Airport	24.23N	55.78E	981	14.18	48.9	113.7	18.7
Pakistan							
Islamabad, Benazir Bhutto Apt	33.62N	73.10E	1667	13.83	35.9	105.9	26.9
Karachi, Rudra Mata Airport	24.90N	67.13E	72	14.66	50.0	101.9	20.3
Lahore, Allama Iqbal Airport	31.52N	74.40E	712	14.32	39.0	109.7	18.0
Palestinian Ter, Occupied							
Atarot Airport	31.87N	35.22E	2490	13.42	33.5	91.1	22.2
Panama							
Panama City, Albrook Intl Airport	8.97N	79.55W	33	14.68	73.0	94.7	17.4
Tocumen, Panama City Metro Apt	9.05N	79.37W	108	14.64	58.8	95.3	44.9
Paraguay							
Asuncion, Silvio Pettirossi Apt	25.25S	57.52W	331	14.52	41.1	98.4	23.3
Peru							
Arequipa, Rodriguez Ballon Apt	16.33S	71.57W	8268	10.80	42.5	75.4	22.7
Chiclayo Airport	6.78S	79.82W	98	14.64	59.1	90.0	23.4
Cuzco	13.53S	71.93W	10659	9.85	31.9	73.3	20.5
Iquitos Airport	3.78S	73.30W	413	14.48	66.2	93.4	13.9
Lima-Callao Airport	12.00S	77.12W	43	14.67	57.1	84.8	21.0
Piura	5.20S	80.60W	180	14.60	60.6	93.4	19.8
Pucallpa Airport	8.37S	74.57W	489	14.44	63.7	94.6	14.8
Trujillo Airport	8.08S	79.10W	98	14.64	58.4	84.1	19.3
Philippines							
Cagayan de Oro	8.48N	124.63E	20	14.69	71.6	94.2	11.0
Davao, F Bangoy Intl Airport	7.12N	125.65E	59	14.66	72.3	92.7	17.9
General Santo City	6.12N	125.18E	49	14.67	72.6	95.2	13.3
Iloilo City	10.70N	122.57E	26	14.68	73.0	94.5	17.0
Mactan-Cebu Intl Airport	10.30N	123.97E	79	14.65	73.4	91.9	18.4
Manila	14.58N	120.98E	43	14.67	73.6	94.1	22.1
Ninoy Aquino Intl Airport	14.52N	121.00E	49	14.67	69.6	94.9	36.4
Sangley Point AB	14.50N	120.92E	7	14.69	73.6	94.5	20.8
Science Garden	14.63N	121.02E	151	14.62	68.4	95.4	13.3
Zamboanga City	6.90N	122.07E	20	14.69	72.3	93.3	13.0
Poland							
Gdansk, Rebiechowo Airport	54.38N	18.47E	453	14.46	3.0	80.7	28.4
Gdansk, Swibno	54.33N	18.93E	23	14.68	1.4	78.5	22.9
Hel	54.60N	18.82E	10	14.69	14.7	77.4	23.0
Katowice Airport	50.23N	19.03E	932	14.21	4.3	84.4	18.6
Krakow	50.08N	19.80E	778	14.29	1.6	85.5	20.8
Lodz Lublinek Airport	51.73N	19.40E	623	14.37	3.5	85.0	20.5

Table 2.1: Climatic Data in I-P Units *(Continued)*

Station	Latitude, degrees	Longitude, degrees	Elevation, ft	St. Br. Pressure, psi	Winter Temp., °F	Summer Temp., °F	Wind, mph
Lublin Radawiec Airport	51.22N	22.40E	787	14.28	0.2	83.1	19.2
Poznan, Lawica Airport	52.42N	16.85E	276	14.55	6.2	86.0	21.9
Raciborz	50.05N	18.20E	676	14.34	3.2	85.2	22.5
Szczecin	53.40N	14.62E	23	14.68	8.8	84.1	21.3
Terespol	52.07N	23.62E	449	14.46	–3.0	84.5	16.6
Warsaw Chopin Airport	52.17N	20.97E	348	14.51	2.1	85.2	23.1
Wrocław Strachowice Airport	51.10N	16.88E	407	14.48	4.6	85.7	20.0
Portugal							
Lisbon Portela Airport	38.77N	9.13W	374	14.50	39.6	93.5	23.2
Puerto Rico							
San Juan International Airport	18.42N	66.00W	62	14.66	69.1	91.4	19.0
San Juan, Luis Muñoz Marín Apt	18.43N	66.00W	10	14.69	69.5	91.1	20.6
Qatar							
Doha International Airport	25.25N	51.57E	33	14.68	52.0	110.6	24.2
Romania							
Bucharest, Afumati	44.48N	26.18E	295	14.54	6.8	91.7	23.1
Bucharest inmh-bane	44.48N	26.12E	299	14.54	8.7	92.8	19.8
Cluj-Napoca	46.78N	23.57E	1355	13.99	4.3	86.1	18.6
Constanta	44.22N	28.65E	46	14.67	15.2	85.4	30.0
Craiova Airport	44.32N	23.87E	640	14.36	9.8	92.2	30.5
Iasi North Airport	47.17N	27.63E	341	14.52	3.3	89.6	22.0
Kogalniceanu Airport	44.33N	28.43E	335	14.52	11.7	89.4	25.3
Timisoara	45.77N	21.25E	289	14.54	9.9	92.2	19.0
Russian Federation							
Adler	43.43N	39.90E	43	14.67	27.9	85.4	17.7
Arhangelsk	64.55N	40.58E	13	14.69	–28.9	80.6	18.0
Astrahan	46.28N	48.05E	–75	14.74	–1.0	94.6	23.1
Barnaul	53.43N	83.52E	604	14.38	–25.9	85.8	25.2
Bryansk	53.25N	34.32E	709	14.32	–8.6	82.6	21.1
Chelyabinsk Balandino Airport	55.30N	61.53E	745	14.30	–20.3	86.2	25.1
Cherepovec	59.25N	37.97E	374	14.50	–23.7	81.2	20.9
Chita	52.08N	113.48E	2201	13.56	–33.9	86.7	22.8
Elabuga	55.77N	52.07E	630	14.36	–20.3	85.3	30.0
Irkutsk	52.27N	104.32E	1539	13.90	–31.1	82.6	22.7
Izhevsk Southeast Airport	56.83N	53.45E	522	14.42	–22.1	84.7	22.7
Kaliningrad	54.72N	20.55E	69	14.66	–0.8	81.8	19.0
Kaluga	54.57N	36.40E	659	14.35	–13.4	81.7	20.7
Kazan International Airport	55.60N	49.28E	381	14.49	–19.2	85.4	25.6
Kemerovo	55.23N	86.12E	853	14.25	–27.2	83.7	24.4
Khabarovsk Novy Airport	48.52N	135.17E	249	14.56	–22.1	86.4	23.3
Kirov	58.65N	49.62E	538	14.41	–27.0	82.0	22.1

Table 2.1: Climatic Data in I-P Units *(Continued)*

Station	Latitude, degrees	Longitude, degrees	Elevation, ft	St. Br. Pressure, psi	Winter Temp., °F	Summer Temp., °F	Wind, mph
Kirov	58.60N	49.63E	518	14.42	–21.4	84.4	14.1
Krasnodar	45.03N	39.15E	112	14.64	5.2	92.4	23.2
Krasnoyarsk	56.00N	92.88E	909	14.22	–28.6	83.1	22.7
Krasnoyarsk Opytnoe	56.03N	92.75E	906	14.22	–35.5	84.6	16.4
Kurgan Airport	55.47N	65.40E	259	14.56	–26.5	87.7	25.7
Kursk	51.77N	36.17E	810	14.27	–9.2	84.2	22.7
Magnitogorsk	53.35N	59.08E	1253	14.04	–20.9	86.1	22.7
Makhachkala	43.02N	47.48E	105	14.64	12.1	88.1	25.0
Moscow	55.83N	37.62E	512	14.43	–9.6	83.1	15.1
Moscow Sheremtyevo Airport	55.97N	37.42E	646	14.36	–13.3	82.7	20.8
Murmansk	68.97N	33.05E	167	14.61	–25.4	75.4	25.0
Nizhniy Novgorod, Gorkij Apt	56.22N	43.82E	269	14.55	–17.2	83.2	20.9
Nizhny Novgorod	56.27N	44.00E	515	14.42	–16.8	84.7	16.1
Nizhnyj Tagil	57.88N	60.07E	846	14.25	–25.9	83.1	16.5
Novokuznetsk Airport	53.82N	86.88E	1010	14.17	–25.5	84.2	27.1
Novosibirsk	55.08N	82.90E	577	14.39	–29.9	84.5	23.9
Omsk	55.02N	73.38E	400	14.48	–26.8	87.8	23.7
Orel Southwest Airport	52.93N	36.00E	666	14.35	–11.5	83.5	23.6
Orenburg	51.68N	55.10E	384	14.49	–20.5	92.9	24.2
Penza South Airport	53.12N	45.02E	571	14.40	–16.7	86.7	23.4
Perm, Bakharerka Airport	57.95N	56.20E	558	14.40	–23.5	85.0	22.3
Rostov-on-Don, Rostov E Apt	47.25N	39.82E	253	14.56	1.3	91.6	28.5
Ryazan	54.62N	39.72E	525	14.42	–10.3	83.1	21.7
Ryazan	54.63N	39.70E	518	14.42	–14.1	84.3	14.9
Samara	53.25N	50.45E	131	14.63	–17.0	88.7	23.6
Saratov Airport	51.57N	46.03E	512	14.43	–9.5	89.0	23.8
Smolensk	54.75N	32.07E	784	14.28	–9.5	80.5	16.6
St Petersburg	59.97N	30.30E	20	14.69	–9.8	81.1	19.7
Stavropol	45.12N	42.08E	1483	13.93	1.2	91.8	28.9
Surgut	61.25N	73.50E	184	14.60	–41.4	83.0	22.8
Tomsk	56.50N	84.92E	456	14.46	–33.2	83.0	22.3
Tula	54.23N	37.62E	669	14.34	–13.1	84.2	16.4
Tver	56.90N	35.88E	479	14.44	–15.2	83.0	21.0
Tyumen	57.12N	65.43E	341	14.52	–26.3	84.9	14.6
Ufa	54.72N	55.83E	341	14.52	–24.7	87.4	23.1
Ulan-Ude	51.83N	107.60E	1690	13.82	–32.9	87.4	26.3
Ulyanovsk	54.32N	48.33E	417	14.48	–19.1	86.5	25.2
Vladikavkaz	43.05N	44.65E	2306	13.51	6.4	85.9	11.5
Vladimir	56.12N	40.35E	558	14.40	–16.0	82.8	20.8
Vladivostok	43.12N	131.93E	600	14.38	–12.1	82.5	30.6
Vnukovo	55.58N	37.25E	702	14.33	–11.4	82.5	22.1

Table 2.1: Climatic Data in I-P Units *(Continued)*

Station	Latitude, degrees	Longitude, degrees	Elevation, ft	St. Br. Pressure, psi	Winter Temp., °F	Summer Temp., °F	Wind, mph
Volgograd, Gumrak Airport	48.78N	44.37E	440	14.46	–7.3	92.8	28.7
Voronezh	51.70N	39.22E	489	14.44	–10.6	87.7	18.2
Voronezh East Airport	51.65N	39.25E	341	14.52	–11.0	84.8	24.5
Yekaterinburg	56.83N	60.63E	928	14.21	–22.4	84.3	20.0
Saudi Arabia							
Abha Airport	18.23N	42.65E	6867	11.40	42.6	87.7	21.8
Buraidah, Gassim Airport	26.30N	43.77E	2126	13.60	37.7	111.6	20.5
Dhahran International Airport	26.27N	50.17E	56	14.67	46.1	111.6	24.9
Jeddah, King Abdulaziz Airport	21.70N	39.18E	56	14.67	59.2	105.6	22.3
Khamis Mushait Airport	18.30N	42.80E	6745	11.45	42.4	89.3	21.0
Mecca	21.43N	39.77E	787	14.28	60.9	113.2	14.1
Medina Airport	24.55N	39.70E	2087	13.62	48.2	113.0	20.8
Riyadh	24.70N	46.73E	2034	13.65	42.7	111.6	21.3
Tabuk Airport	28.38N	36.60E	2520	13.41	35.2	105.5	23.4
Senegal							
Dakar Yoff International Airport	14.73N	17.50W	79	14.65	61.7	89.9	22.1
Serbia							
Belgrade	44.80N	20.47E	433	14.47	15.9	92.7	16.5
Belgrade Nikola Tesla Airport	44.82N	20.28E	325	14.52	12.2	92.8	23.2
Singapore							
Singapore, Changi Airport	1.37N	103.98E	52	14.67	73.4	91.7	16.7
Slovakia							
Bratislava Letisko	48.20N	17.20E	440	14.46	10.6	89.6	22.3
South Africa							
Bloemfontein International Airport	29.10S	26.30E	4442	12.49	23.7	93.1	20.9
Cape Town International Airport	33.97S	18.60E	138	14.62	38.8	87.8	31.2
Durban International Airport	29.97S	30.95E	46	14.67	49.0	86.4	25.2
East London Airport	33.03S	27.83E	410	14.48	46.5	86.4	28.7
Johannesburg, Tambo Intl Apt	26.15S	28.23E	5643	11.94	32.4	84.1	20.7
Port Elizabeth Airport	33.98S	25.62E	207	14.59	41.7	84.6	32.4
Pretoria, Eendracht	25.73S	28.18E	4350	12.53	37.4	89.8	12.7
Pretoria, Irene	25.92S	28.22E	4997	12.23	36.8	87.1	19.9
Spain							
Alicante Airport	38.28N	0.55W	102	14.64	37.9	91.0	22.8
Barcelona Airport	41.28N	2.07E	20	14.69	33.6	86.3	21.2
Madrid, Barajas Airport	40.45N	3.55W	1909	13.71	24.7	97.1	21.6
Madrid, Torrejon	40.48N	3.45W	2005	13.66	24.4	98.2	20.6
Malaga Airport	36.67N	4.48W	23	14.68	39.0	94.7	25.2
Murcia	38.00N	1.17W	203	14.59	36.2	96.4	17.9
Palma Mallorca Airport	39.55N	2.73E	23	14.68	31.9	91.7	22.8
Palmas de Gran Canaria	27.93N	15.38W	154	14.61	55.8	86.3	32.2

Table 2.1: Climatic Data in I-P Units ***(Continued)***

Station	Latitude, degrees	Longitude, degrees	Elevation, ft	St. Br. Pressure, psi	Winter Temp., °F	Summer Temp., °F	Wind, mph
Seville, San Pablo Airport	37.42N	5.90W	102	14.64	34.3	103.8	19.8
Sondika, Bilbao Airport	43.30N	2.90W	128	14.63	31.7	90.8	22.0
Valencia	39.50N	0.47W	203	14.59	33.7	91.6	25.5
Valladolid	41.65N	4.77W	2411	13.46	24.5	93.8	18.6
Zaragoza AB	41.67N	1.05W	863	14.24	28.1	96.9	27.9
Zaragoza Airport	41.67N	1.00W	846	14.25	26.8	97.0	29.9
Sri Lanka							
Katunayake, Bandaranaike Apt	7.17N	79.88E	26	14.68	69.7	91.7	20.1
Sweden							
Gothenburg	57.72N	12.00E	7	14.69	10.9	80.0	20.2
Gothenburg City Airport	57.78N	11.88E	52	14.67	5.3	78.4	25.3
Gothenburg, Landvetter Airport	57.67N	12.30E	554	14.40	6.6	78.8	25.1
Stockholm, Bromma Airport	59.37N	17.90E	46	14.67	1.1	80.6	20.2
Switzerland							
Laegern	47.48N	8.40E	2766	13.29	11.0	79.0	28.2
Zurich, Kloten	47.48N	8.53E	1417	13.96	13.9	86.1	18.9
Zurich, MeteoSwiss	47.38N	8.57E	1867	13.73	15.3	83.8	20.0
Syrian Arab Republic							
Aleppo International Airport	36.18N	37.20E	1260	14.04	28.4	101.9	23.1
Damascus International Airport	33.42N	36.52E	1998	13.67	25.8	102.3	27.1
Daraa	32.60N	36.10E	1781	13.77	33.9	96.4	18.5
Hama	35.12N	36.75E	994	14.18	29.5	102.0	15.9
Latakia	35.53N	35.77E	23	14.68	39.4	90.0	22.1
Taiwan, Province of China							
Chi-lung	25.15N	121.80E	10	14.69	50.3	92.8	20.8
Chinmem / Shatou AFB	24.43N	118.37E	30	14.68	44.4	91.5	21.8
Hsinchu AFB	24.82N	120.93E	26	14.68	48.2	91.6	29.8
Hsinchu City	24.83N	120.93E	89	14.65	47.8	93.0	22.0
Kangshan AFB	22.78N	120.27E	33	14.68	50.0	91.7	18.8
Kaohsiung	22.63N	120.28E	95	14.65	54.2	91.1	16.8
Kaohsiung International Airport	22.58N	120.35E	30	14.68	53.3	91.8	19.8
Pingtung North Airport	22.70N	120.48E	95	14.65	51.9	93.6	16.2
Pingtung South Airport	22.68N	120.47E	79	14.65	53.2	94.8	16.3
Taichung AFB	24.18N	120.65E	367	14.50	46.2	93.6	20.5
Tainan	23.00N	120.22E	46	14.67	51.2	92.2	19.0
Tainan AFB	22.95N	120.20E	62	14.66	50.4	91.9	20.5
Taipei	25.03N	121.52E	30	14.68	49.3	94.8	17.7
Taipei, Sungshan Airport	25.07N	121.55E	20	14.69	48.3	94.9	19.8
Taiwan Taoyuan Intl Airport	25.08N	121.22E	108	14.64	48.0	93.5	28.7
Taizhong	24.15N	120.68E	256	14.56	49.1	92.1	11.2
Taoyuan AB	25.07N	121.23E	148	14.62	47.4	93.0	26.6

Table 2.1: Climatic Data in I-P Units *(Continued)*

Station	Latitude, degrees	Longitude, degrees	Elevation, ft	St. Br. Pressure, psi	Winter Temp., °F	Summer Temp., °F	Wind, mph
Wu-Chi Observatory	24.25N	120.52E	16	14.69	49.8	91.1	35.7
Wuchia Observatory	24.27N	120.62E	16	14.69	46.2	90.0	26.7
Tajikistan							
Dushanbe	38.55N	68.78E	2625	13.35	19.3	99.4	14.2
Tanzania							
Dar Es Salaam Airport	6.87S	39.20E	174	14.60	63.8	91.7	19.3
Thailand							
Bangkok	13.73N	100.57E	13	14.69	66.3	96.5	14.1
Bangkok International Airport	13.92N	100.60E	39	14.68	66.2	98.9	18.0
Togo							
Lome Tokoin Airport	6.17N	1.25E	82	14.65	69.8	91.7	18.5
Tunisia							
Tunis-Carthage Airport	36.83N	10.23E	13	14.69	40.9	99.2	26.4
Turkey							
Adana	36.98N	35.30E	66	14.66	34.1	98.0	17.8
Adana, Incirlik AB	37.00N	35.43E	240	14.57	31.9	98.3	18.7
Adana, Incirlik AFB	37.00N	35.42E	249	14.56	32.3	97.1	18.9
Antalya	36.87N	30.73E	177	14.60	34.5	100.3	23.5
Bursa	40.18N	29.07E	328	14.52	25.1	93.4	16.5
Diyarbakır	37.88N	40.18E	2221	13.55	15.8	104.2	20.0
Erzurum Airport	39.95N	41.17E	5768	11.88	–21.2	86.0	23.2
Esenboga Airport	40.12N	33.00E	3114	13.12	3.8	91.5	20.0
Eskisehir Airport	39.78N	30.57E	2579	13.38	12.3	91.4	19.6
Etimesgut Airport	39.95N	32.68E	2644	13.35	11.8	93.3	20.6
Gaziantep	37.08N	37.37E	2300	13.51	23.4	101.7	18.2
Istanbul, Ataturk Airport	40.97N	28.82E	121	14.63	27.3	88.1	24.7
Izmir, Cigli	38.52N	27.02E	16	14.69	28.8	97.1	24.4
Kayseri, Erkilet	38.82N	35.43E	3458	12.95	3.2	92.8	20.9
Konya Airport	37.97N	32.55E	3383	12.99	8.7	92.8	25.6
Malatya Erhac Airport	38.43N	38.08E	2785	13.28	11.3	99.0	22.7
Menderes, Izmir	38.27N	27.15E	394	14.49	27.0	98.5	26.9
Samsun	41.28N	36.30E	13	14.69	29.7	82.5	18.5
Van Airport	38.45N	43.32E	5453	12.02	6.9	84.2	18.4
Turkmenistan							
Ashgabat Keshi	37.92N	58.33E	1024	14.16	20.0	104.2	21.0
Ukraine							
Chernihiv	51.47N	31.25E	463	14.45	–5.3	85.3	20.5
Dnipropetrovsk Oblast	48.37N	35.08E	469	14.45	–0.1	89.4	26.0
Donetsk Airport	48.07N	37.77E	738	14.31	–2.0	88.3	27.8
Kharkiv	49.97N	36.13E	509	14.43	–3.2	87.5	21.5
Kherson	46.63N	32.57E	177	14.60	3.6	90.9	22.0

Table 2.1: Climatic Data in I-P Units *(Continued)*

Station	Latitude, degrees	Longitude, degrees	Elevation, ft	St. Br. Pressure, psi	Winter Temp., °F	Summer Temp., °F	Wind, mph
Kiev	50.40N	30.57E	548	14.41	–0.6	85.0	19.8
Kryvyi Rih Airport	48.03N	33.22E	407	14.48	–0.4	89.0	26.0
Luhansk	48.57N	39.25E	203	14.59	–5.0	91.3	26.4
Lviv, Sknilov Airport	49.82N	23.95E	1060	14.14	0.8	82.7	21.7
Mariupol	47.03N	37.50E	230	14.57	4.2	86.7	32.3
Odessa	46.43N	30.77E	138	14.62	6.9	88.1	25.2
Poltava	49.60N	34.55E	525	14.42	–2.8	86.7	22.6
Simferopol	45.02N	33.98E	594	14.38	9.1	90.0	27.5
Vinnytsia	49.23N	28.60E	978	14.18	–2.3	83.2	25.6
Zaporizhia	47.80N	35.02E	367	14.50	0.1	90.1	22.6
United Arab Emirates							
Abu Dhabi Bateen Airport	24.43N	54.47E	10	14.69	55.5	109.4	21.2
Abu Dhabi International Airport	24.43N	54.65E	89	14.65	52.8	112.8	21.0
Al Ain International Airport	24.27N	55.60E	860	14.24	51.9	114.7	23.3
Dubai International Airport	25.25N	55.33E	16	14.69	54.9	109.0	20.6
Sharjah International Airport	25.33N	55.52E	108	14.64	49.9	111.1	18.6
United Kingdom							
Aughton	53.55N	2.92W	184	14.60	26.0	76.0	25.8
Bingley	53.82N	1.87W	876	14.24	24.6	74.8	28.3
Birmingham Intl Airport	52.45N	1.73W	325	14.52	22.7	80.0	22.4
Bristol Airport	51.38N	2.72W	636	14.36	26.3	78.5	27.0
Bristol Weather Center	51.47N	2.60W	36	14.68	27.0	79.9	23.2
Cardiff Airport	51.40N	3.35W	220	14.58	26.6	76.6	29.7
Cardiff Weather Center	51.48N	3.18W	171	14.61	30.2	79.2	26.1
Church Lawford	52.37N	1.33W	348	14.51	23.6	79.8	22.5
Cilfynydd	51.63N	3.30W	636	14.36	23.7	78.1	25.8
Crosby	53.50N	3.07W	30	14.68	25.9	76.1	39.4
East Midlands Airport	52.83N	1.32W	315	14.53	24.8	80.2	28.7
Edinburgh Turnhouse Airport	55.95N	3.35W	135	14.62	22.5	72.4	27.8
Emley Moor	53.62N	1.67W	850	14.25	26.3	75.2	33.1
Glasgow International Airport	55.87N	4.43W	26	14.68	21.4	74.7	29.3
Gravesend Broadness	51.47N	0.30E	10	14.69	28.9	83.1	24.8
Hawarden Airport	53.17N	2.98W	30	14.68	24.3	77.6	23.3
Kenley Airfield	51.30N	0.08W	558	14.40	27.2	80.0	24.5
Leconfield	53.87N	0.43W	23	14.68	25.3	76.9	28.1
Leeds Bradford Airport	53.87N	1.65W	699	14.33	26.6	75.6	28.8
Leeds Weather Center	53.80N	1.55W	154	14.61	27.9	79.1	29.0
Liverpool Lennon Intl Airport	53.33N	2.85W	82	14.65	26.3	77.4	29.7
London Heathrow Airport	51.48N	0.45W	82	14.65	26.5	83.0	22.5
London Weather Center	51.52N	0.12W	75	14.66	27.3	80.9	26.2
London Weather Center	51.52N	0.10W	141	14.62	30.9	83.2	20.7

Table 2.1: Climatic Data in I-P Units *(Continued)*

Station	Latitude, degrees	Longitude, degrees	Elevation, ft	St. Br. Pressure, psi	Winter Temp., °F	Summer Temp., °F	Wind, mph
Manchester Airport	53.35N	2.28W	226	14.58	25.3	78.5	24.8
Northolt Airport	51.55N	0.42W	128	14.63	24.6	82.9	23.3
Uruguay							
Carrasco	34.83S	56.00W	105	14.64	34.5	88.3	29.5
Prado	34.85S	56.20W	52	14.67	37.4	88.8	22.8
Uzbekistan							
Namangan Airport	40.98N	71.58E	1555	13.89	18.7	97.5	16.5
Samarkand	39.57N	66.95E	2375	13.48	14.4	96.8	23.2
Tashkent	41.27N	69.27E	1529	13.90	15.1	100.8	13.8
Venezuela							
Caracas, Simon Bolivar Intl Apt	10.60N	66.98W	157	14.61	69.3	92.2	11.6
San Antonio del Tachira	7.85N	72.45W	1240	14.05	67.9	94.9	27.4
Viet Nam							
Da Nang	16.07N	108.35E	23	14.68	61.7	96.8	16.6
Hanoi	21.03N	105.80E	20	14.69	50.1	96.3	15.5
Ho Chi Minh City, Tan Son Hoa	10.82N	106.67E	16	14.69	68.1	95.4	25.7
Phu Lien	20.80N	106.63E	381	14.49	49.6	93.2	15.6
Zimbabwe							
Harare International Airport	17.92S	31.13E	4856	12.29	44.1	87.2	20.2

Table 2.2: Climatic Data in SI Units

Station	Latitude, degrees	Longitude, degrees	Elevation, m	St. Br. Pressure, kPa	Winter Temp., °C	Summer Temp., °C	Wind, m/s
United States of America							
Alabama							
Anniston Metropolitan Airport	33.59N	85.86W	183	99.2	–6.6	34.7	7.5
Auburn Opelika Airport	32.62N	85.43W	236	98.5	–4.6	33.9	7.8
Birmingham Municipal Airport	33.56N	86.75W	192	99.0	–6.9	35.0	8.2
Cairns AAF/Ozark	31.28N	85.72W	91	100.2	–2.8	35.2	7.6
Dothan Municipal Airport	31.32N	85.45W	98	100.2	–2.6	35.2	8.6
Gadsen Municipal Airport	33.97N	86.08W	173	99.3	–7.5	34.1	7.5
Huntsville Intl, Jones Field	34.64N	86.79W	196	99.0	–8.3	34.8	9.6
Mobile Regional Airport	30.69N	88.25W	67	100.5	–2.8	34.2	9.2
Montgomery, Dannelly Field	32.30N	86.39W	62	100.6	–4.6	35.7	8.3
Montgomery, Maxwell AFB	32.38N	86.37W	53	100.7	–2.3	36.2	8.0
Muscle Shoals Regional Airport	34.75N	87.61W	171	99.3	–7.9	35.3	8.3
Tuscaloosa Municipal Airport	33.21N	87.62W	57	100.6	–6.4	35.4	7.8
Alaska							
Anchorage International Airport	61.18N	149.99W	40	100.9	–22.7	21.9	9.2
Anchorage, Elmendorf AFB	61.25N	149.80W	59	100.6	–26.0	23.2	8.4
Anchorage, Lake Hood Seaplane	61.18N	149.96W	40	100.9	–22.6	23.6	8.4
Bethel Airport	60.79N	161.83W	46	100.8	–32.8	22.4	13.7
Fairbanks International Airport	64.82N	147.86W	138	99.7	–41.9	27.4	7.9
Fort Richardson	61.27N	149.65W	115	100.0	–28.1	23.3	8.5
Juneau International Airport	58.36N	134.58W	7	101.2	–15.7	23.1	11.9
Kenai Municipal Airport	60.58N	151.24W	28	101.0	–30.2	21.0	10.8
Ketchikan International Airport	55.36N	131.71W	29	101.0	–10.8	21.9	11.0
Kodiak Airport	57.75N	152.49W	34	100.9	–13.0	20.5	14.9
Nome Municipal Airport	64.51N	165.44W	7	101.2	–33.9	20.3	12.6
Palmer Municipal Airport	61.60N	149.09W	76	100.4	–26.9	23.7	14.0
Sitka, Japonski Airport	57.05N	135.36W	20	101.1	–8.7	19.3	11.0
Arizona							
Casa Granda Municipal Airport	32.95N	111.77W	446	96.1	0.0	42.5	9.1
Douglas, Bisbee Intl Airport	31.47N	109.60W	1250	87.2	–5.2	37.5	10.7
Flagstaff, Pulliam Airport	35.13N	111.67W	2135	78.2	–15.6	29.7	9.3
Nogales International Airport	31.42N	110.85W	1236	87.3	–2.8	37.7	8.7
Phoenix Sky Harbor Intl Airport	33.44N	111.99W	337	97.3	3.7	43.4	8.2
Phoenix, Luke AFB	33.53N	112.38W	331	97.4	1.8	43.8	8.8
Prescott, Love Field	34.65N	112.42W	1540	84.1	–8.1	34.6	9.3
Safford	32.82N	109.68W	950	90.4	–3.6	40.2	9.9
Tucson International Airport	32.13N	110.96W	779	92.3	–0.2	41.1	9.6
Tucson, Davis Monthan AFB	32.17N	110.88W	809	92.0	0.5	40.8	8.7
Winslow Municipal Airport	35.02N	110.72W	1488	84.7	–11.6	36.3	12.2
Yuma International Airport	32.65N	114.60W	63	100.6	5.5	43.8	9.3

Table 2.2: Climatic Data in SI Units *(Continued)*

Station	Latitude, degrees	Longitude, degrees	Elevation, m	St. Br. Pressure, kPa	Winter Temp., °C	Summer Temp., °C	Wind, m/s
Yuma MCAS	32.65N	114.62W	65	100.6	5.4	43.8	9.3
Arkansas							
Bentonville Municipal Airport	36.35N	94.22W	395	96.7	–12.1	34.2	8.7
Blytheville, Eaker AFB	35.97N	89.95W	80	100.4	–11.3	36.1	10.2
El Dorado, Goodwin Field	33.22N	92.81W	87	100.3	–5.7	37.1	7.6
Fayetteville, Drake Field	36.01N	94.17W	384	96.8	–13.3	35.1	9.2
Flippin	36.30N	92.47W	350	97.2	–11.1	36.2	7.5
Fort Smith Regional Airport	35.33N	94.37W	141	99.6	–9.6	37.3	9.1
Harrison Airport	36.26N	93.16W	422	96.4	–12.3	34.7	9.2
Jonesboro Municipal Airport	35.83N	90.63W	82	100.3	–10.9	35.8	9.3
Little Rock AFB	34.92N	92.15W	103	100.1	–9.3	37.4	7.9
Little Rock, Adams Field	34.75N	92.23W	78	100.4	–8.2	36.7	8.3
North Little Rock Municipal Airport	34.83N	92.25W	351	97.2	–8.7	35.1	8.2
Pine Bluff Airport	34.18N	91.94W	65	100.6	–6.1	36.2	8.3
Rogers Field	36.37N	94.10W	415	96.4	–12.2	34.1	9.3
Siloam Spring	36.18N	94.48W	364	97.0	–12.1	35.3	10.2
Texarkana Municipal Airport	33.45N	94.01W	122	99.9	–5.8	36.9	8.4
California							
Alameda NAS	37.73N	122.32W	4	101.3	4.6	28.5	9.2
Bakersfield, Meadows Field	35.43N	119.06W	150	99.5	0.1	39.5	8.2
Marysville, Beale AFB	39.13N	121.43W	38	100.9	0.1	38.1	9.6
Blythe Airport	33.62N	114.72W	120	99.9	2.4	44.8	10.9
Burbank Glendale Pasadena Apt	34.20N	118.36W	223	98.7	3.9	36.8	8.2
Oxnard AFB	34.22N	119.08W	23	101.1	3.0	29.2	11.1
Camp Pendleton MCAS	33.30N	117.35W	24	101.0	0.4	33.3	7.6
Carlsbad, Palomar Airport	33.13N	117.28W	100	100.1	6.1	27.6	6.3
Merced, Castle AFB	37.38N	120.57W	60	100.6	–0.8	37.5	8.1
Crescent City, McNamara Field	41.78N	124.24W	17	101.1	2.0	20.0	12.6
El Toro MCAS	33.68N	117.73W	117	99.9	4.6	33.4	6.4
Eureka	40.80N	124.17W	18	101.1	1.2	20.1	9.8
Fresno Air Terminal	36.78N	119.72W	100	100.1	–0.3	39.8	8.1
Fullerton Municipal Airport	33.87N	117.98W	29	101.0	4.0	33.7	6.0
Victorville, George AFB	34.58N	117.38W	876	91.2	–2.5	38.2	10.2
Hayward Executive Airport	37.66N	122.12W	14	101.2	2.8	30.9	8.9
Imperial County Airport	32.83N	115.58W	–17	101.5	2.0	43.9	11.6
Hawthorne Municipal Airport	33.92N	118.33W	21	101.1	7.0	29.9	7.3
Lancaster, Gen Wm Fox Field	34.74N	118.22W	713	93.1	–5.9	39.0	13.3
Lemoore NAS	36.33N	119.95W	71	100.5	–1.3	39.5	8.5
Livermore Municipal Airport	37.69N	121.82W	121	99.9	–1.0	37.1	8.8
Lompoc Airport	34.67N	120.47W	27	101.0	0.2	27.4	9.1
Long Beach Municipal Airport	33.83N	118.16W	12	101.2	5.1	32.9	7.8

Table 2.2: Climatic Data in SI Units *(Continued)*

Station	Latitude, degrees	Longitude, degrees	Elevation, m	St. Br. Pressure, kPa	Winter Temp., °C	Summer Temp., °C	Wind, m/s
Los Angeles Intl Airport	33.94N	118.41W	99	100.1	6.9	28.7	8.9
Riverside, March AFB	33.88N	117.27W	462	95.9	1.2	38.1	8.0
Sacramento, McClellan AFB	38.67N	121.40W	25	101.0	–0.2	38.8	9.2
Modesto Municipal Airport	37.63N	120.95W	30	101.0	–0.5	38.4	8.4
Monterey Peninsula	36.58N	121.85W	50	100.7	2.6	25.2	7.6
Mount Shasta	41.33N	122.33W	1078	89.0	–8.1	32.6	5.7
Mountain View, Moffett NAS	37.42N	122.05W	10	101.2	2.5	31.2	8.4
Napa County Airport	38.21N	122.28W	17	101.1	–1.1	32.9	9.5
San Bernardino, Norton AFB	34.10N	117.23W	353	97.2	1.1	39.4	7.5
Oakland International Airport	37.76N	122.22W	27	101.0	2.9	27.7	10.3
Ontario	34.07N	117.65W	304	97.7	2.2	38.1	10.3
Palm Springs Intl Airport	33.83N	116.50W	145	99.6	5.9	44.0	10.2
Palm Springs, Thermal Airport	33.63N	116.16W	–36	101.8	–0.6	44.1	8.6
Paso Robles Municipal Airport	35.67N	120.63W	249	98.4	–2.9	38.9	9.8
Point Arguello	34.57N	120.63W	34	100.9	7.6	21.8	19.0
Point Mugu NAS	34.12N	119.12W	4	101.3	3.8	27.8	10.2
Porterville Airport	36.03N	119.07W	135	99.7	–0.9	38.0	5.7
Redding Municipal Airport	40.52N	122.31W	153	99.5	–2.0	41.0	11.1
Riverside Municipal Airport	33.95N	117.43W	256	98.3	2.2	37.6	9.3
Sacramento Executive Airport	38.51N	121.49W	8	101.2	–0.4	37.9	9.1
Sacramento, Mather Airport	38.55N	121.29W	29	101.0	–1.2	38.5	9.1
Sacramento Metropolitan Airport	38.70N	121.59W	10	101.2	–0.3	38.0	10.4
Salinas Municipal Airport	36.66N	121.61W	24	101.0	1.0	28.3	9.4
San Diego International Airport	32.74N	117.17W	9	101.2	7.1	28.9	7.9
San Diego, Miramar MCAS	32.87N	117.15W	146	99.6	3.8	32.4	6.5
San Diego, North Island NAS	32.70N	117.20W	8	101.2	7.0	28.0	8.4
San Diego, Brown Field	32.57N	116.98W	159	99.4	3.8	31.3	7.3
San Diego, Montgomery Field	32.82N	117.13W	129	99.8	4.9	31.3	7.1
San Francisco Intl Airport	37.62N	122.40W	6	101.3	3.8	28.3	12.8
San Jose International Airport	37.36N	121.93W	15	101.2	2.1	33.5	8.9
San Luis Obispo County Airport	35.23N	120.63W	66	100.5	1.1	31.3	11.5
Santa Barbara Municipal Airport	34.43N	119.84W	6	101.3	1.5	28.3	8.7
Santa Maria Public Airport	34.92N	120.47W	73	100.5	0.1	29.0	10.6
Sonoma County Airport	38.51N	122.81W	45	100.8	–1.3	35.2	7.6
Stockton Metropolitan Airport	37.89N	121.24W	8	101.2	–0.8	38.2	10.0
Fairfield, Travis AFB	38.27N	121.93W	18	101.1	–0.1	37.1	12.6
Truckee Tahoe Airport	39.32N	120.13W	1798	81.5	–17.3	31.4	9.8
Tustin MCAF	33.70N	117.83W	17	101.1	3.8	34.2	7.6
Ukiah Municipal Airport	39.13N	123.20W	191	99.1	–1.3	37.7	7.2
Visalia Municipal Airport	36.32N	119.40W	89	100.3	–1.1	37.7	6.6
Colorado							
Alamosa Municipal Airport	37.44N	105.87W	2299	76.6	–26.3	29.6	12.1

Table 2.2: Climatic Data in SI Units *(Continued)*

Station	Latitude, degrees	Longitude, degrees	Elevation, m	St. Br. Pressure, kPa	Winter Temp., °C	Summer Temp., °C	Wind, m/s
Aspen-Pitkin County Airport	39.22N	106.87W	2444	75.2	–19.1	29.0	8.2
Colorado Springs Airport	38.81N	104.71W	1881	80.7	–18.2	32.4	12.4
Cortez Montezuma Co Airport	37.30N	108.63W	1803	81.5	–15.9	34.0	10.3
Craig Moffat Airport	40.50N	107.53W	1915	80.3	–26.4	32.3	11.1
Denver International Airport	39.83N	104.66W	1655	83.0	–17.4	34.6	11.9
Denver Stapleton Intl Airport	39.77N	104.87W	1611	83.4	–20.0	34.1	10.9
Denver, Buckley AFB	39.72N	104.75W	1726	82.2	–17.9	33.9	10.5
Denver, Centennial Airport	39.57N	104.85W	1793	81.6	–18.8	33.0	11.1
Fort Collins	40.58N	105.08W	1525	84.3	–20.4	32.1	9.3
Fort Collins Loveland Airport	40.45N	105.02W	1529	84.3	–17.7	34.2	11.5
Grand Junction, Walker Field	39.13N	108.54W	1475	84.8	–14.5	36.3	10.3
Greeley, Weld County Airport	40.43N	104.63W	1420	85.4	–20.8	35.2	12.4
Lamar Municipal Airport	38.07N	102.68W	1129	88.5	–16.3	37.7	12.7
Montrose County Airport	38.50N	107.90W	1755	81.9	–14.1	34.1	10.1
Pueblo Memorial Airport	38.29N	104.50W	1439	85.2	–18.9	36.9	12.9
Rifle, Garfield County Airport	39.53N	107.72W	1691	82.6	–17.0	35.9	10.1
Trinidad, Las Animas Co Airport	37.26N	104.34W	1751	82.0	–17.1	33.8	12.2
Connecticut							
Bridgeport, Sikorsky Airport	41.18N	73.15W	5	101.3	–11.8	30.8	10.9
Hartford, Bradley Intl Airport	41.94N	72.68W	55	100.7	–16.0	33.1	10.0
Hartford Brainard Field	41.74N	72.65W	6	101.3	–14.2	32.5	9.0
Waterbury, Oxford Airport	41.48N	73.13W	222	98.7	–16.1	30.9	8.8
Willimantic	41.73N	72.18W	76	100.4	–16.1	32.2	8.8
Delaware							
Dover AFB	39.13N	75.47W	7	101.2	–10.1	33.4	10.8
Greater Wilmington Airport	39.67N	75.60W	24	101.0	–11.3	33.3	10.9
Florida							
Crestview, Bob Sikes Airport	30.78N	86.52W	56	100.7	–4.4	35.1	7.8
Daytona Beach Intl Airport	29.18N	81.06W	13	101.2	1.5	33.7	9.1
Destin-Fort Walton Beach Apt	30.40N	86.47W	7	101.2	1.2	32.8	8.4
Fort Lauderdale-Hollywood Apt	26.07N	80.15W	3	101.3	8.2	33.2	10.0
Fort Myers Metropolitan Airport	26.59N	81.86W	6	101.3	5.8	34.3	8.5
Fort Myers, SW Florida Airport	26.53N	81.75W	9	101.2	5.1	34.3	9.3
Gainesville Regional Airport	29.69N	82.27W	50	100.7	–1.3	34.1	8.3
Homestead AFB	25.48N	80.38W	5	101.3	8.0	32.9	8.4
Hurlburt Field	30.43N	86.68W	12	101.2	–1.5	33.4	8.5
Jacksonville International Airport	30.49N	81.69W	10	101.2	–1.5	34.7	8.9
Jacksonville NAS	30.23N	81.68W	7	101.2	0.1	35.3	9.1
Jacksonville, Cecil Field	30.22N	81.87W	27	101.0	–1.2	35.7	8.3
Jacksonville, Craig Airport	30.34N	81.52W	13	101.2	–0.2	34.1	8.4
Jacksonville, Mayport NS	30.40N	81.42W	4	101.3	1.3	34.7	9.0

Table 2.2: Climatic Data in SI Units *(Continued)*

Station	Latitude, degrees	Longitude, degrees	Elevation, m	St. Br. Pressure, kPa	Winter Temp., °C	Summer Temp., °C	Wind, m/s
Kennedy Space Center	28.62N	80.72W	3	101.3	3.5	33.3	8.5
Key West International Airport	24.55N	81.75W	6	101.3	12.7	32.5	10.3
Marathon Airport	24.73N	81.05W	2	101.3	12.2	33.0	8.8
Melbourne International Airport	28.10N	80.65W	8	101.2	3.9	33.7	9.3
Miami International Airport	25.82N	80.30W	9	101.2	8.7	33.2	9.2
Miami, Kendall Tamiami Airport	25.65N	80.43W	3	101.3	7.4	33.6	9.2
Naples Municipal Airport	26.15N	81.78W	7	101.2	6.6	32.7	8.4
Ocala Municipal Airport	29.17N	82.22W	27	101.0	–1.3	34.0	8.0
Orlando Executive Airport	28.55N	81.33W	34	100.9	4.4	34.2	8.5
Orlando International Airport	28.43N	81.33W	32	100.9	3.2	34.3	9.0
Orlando, Central Florida Airport	28.78N	81.24W	17	101.1	3.7	34.9	9.1
Panama City Bay County Airport	30.20N	85.68W	6	101.3	0.0	33.7	8.4
Pensacola NAS	30.35N	87.32W	9	101.2	–2.1	34.0	10.5
Pensacola Regional Airport	30.47N	87.19W	36	100.9	–1.4	34.3	9.0
Sarasota Bradenton Airport	27.38N	82.55W	10	101.2	4.0	33.4	9.7
St Petersburg Clearwater Airport	27.90N	82.68W	3	101.3	5.9	34.0	9.3
Tallahassee Municipal Airport	30.39N	84.35W	21	101.1	–3.7	35.3	8.0
Tampa International Airport	27.96N	82.54W	3	101.3	3.6	33.6	8.1
Tampa, MacDill AFB	27.85N	82.52W	8	101.2	4.6	33.9	8.4
Tyndall AFB	30.07N	85.58W	7	101.2	0.0	32.9	8.8
Valparaiso, Eglin AFB	30.48N	86.53W	20	101.1	–2.0	33.7	8.9
Venice Municipal Airport	27.07N	82.45W	5	101.3	5.7	30.9	12.6
Vero Beach Municipal Airport	27.66N	80.42W	9	101.2	3.8	33.2	9.0
West Palm Beach Intl Airport	26.69N	80.10W	6	101.3	6.8	33.0	10.3
Georgia							
Albany, Dougherty Co Airport	31.54N	84.19W	59	100.6	–3.1	36.0	8.3
Athens, Ben Epps Airport	33.95N	83.33W	244	98.4	–5.7	35.1	8.2
Atlanta	33.37N	84.55W	296	97.8	–7.2	33.8	7.9
Atlanta, Dekalb Peachtree Apt	33.87N	84.30W	313	97.6	–5.2	34.0	8.2
Atlanta. Hartsfield Intl Airport	33.64N	84.43W	313	97.6	–6.3	34.4	9.8
Augusta, Bush Field	33.37N	81.97W	45	100.8	–5.4	36.2	8.4
Augusta, Daniel Field	33.47N	82.03W	128	99.8	–2.5	36.0	7.5
Brunswick, M McKinnon Airport	31.25N	81.39W	7	101.2	–1.0	33.7	8.3
Columbus Metropolitan Airport	32.52N	84.94W	120	99.9	–3.8	35.7	8.2
Fort Benning, Lawson AAF	32.33N	85.00W	88	100.3	–5.0	36.0	7.4
Fulton County Airport Brown	33.77N	84.52W	263	98.2	–6.8	34.2	8.1
Gainesville, Lee Gilmer Airport	34.27N	83.83W	389	96.7	–5.0	32.9	8.5
Macon, Middle Georgia Reg Apt	32.69N	83.65W	110	100.0	–4.7	35.9	8.1
Mariett, Dobbins AFB	33.92N	84.52W	330	97.4	–6.9	34.2	8.3
Rome, R B Russell Airport	34.35N	85.16W	196	99.0	–8.2	35.8	6.8
Savannah International Airport	32.12N	81.20W	16	101.1	–2.7	35.3	8.4

Table 2.2: Climatic Data in SI Units ***(Continued)***

Station	Latitude, degrees	Longitude, degrees	Elevation, m	St. Br. Pressure, kPa	Winter Temp., °C	Summer Temp., °C	Wind, m/s
Savannah, Hunter AAF	32.00N	81.13W	13	101.2	–2.2	35.2	8.4
Valdosta Regional Airport	30.78N	83.28W	60	100.6	–2.5	35.2	7.6
Valdosta, Moody AFB	30.97N	83.20W	71	100.5	–1.2	35.4	7.3
Warner Robins AFB	32.63N	83.60W	92	100.2	–3.9	36.3	8.5
Waycross Ware County Airport	31.25N	82.40W	43	100.8	–2.3	35.7	7.5
Hawaii							
Hilo International Airport	19.72N	155.05W	11	101.2	16.4	29.8	7.9
Honolulu International Airport	21.33N	157.94W	5	101.3	16.2	32.2	9.8
Kahului Airport	20.90N	156.43W	15	101.1	14.9	32.1	11.9
Kalaeloa Airport (Barbers Point)	21.30N	158.07W	10	101.2	15.2	32.7	8.8
Kaneohe Bay MCAS	21.45N	157.77W	6	101.3	17.9	29.6	8.6
Keahol, Kona Intl Airport	19.73N	156.03W	15	101.1	18.7	31.2	8.9
Lihue Airport	21.98N	159.34W	45	100.8	15.7	29.6	11.7
Idaho							
Boise Air Terminal	43.57N	116.22W	874	91.3	–16.3	36.7	9.7
Caldwell	43.63N	116.63W	740	92.7	–11.3	36.1	9.6
Coeur d'Alene Airport	47.77N	116.82W	707	93.1	–14.0	33.0	10.0
Idaho Falls, Fanning Field	43.52N	112.07W	1446	85.1	–22.6	33.2	12.2
Lewiston Nez Perce Co Airport	46.38N	117.01W	438	96.2	–12.1	36.6	9.2
Mountain Home AFB	43.05N	115.87W	912	90.8	–16.7	37.5	11.4
Pocatello Regional Airport	42.92N	112.57W	1365	86.0	–20.5	34.7	12.5
Salmon Airport	45.12N	113.88W	1233	87.4	–20.8	32.5	9.0
Twin Falls, Magic Valley Airport	42.48N	114.48W	1297	86.7	–12.5	34.7	12.3
Illinois							
Aurora Municipal Airport	41.77N	88.47W	215	98.8	–18.9	32.7	11.6
Belleville, Scott AFB	38.55N	89.85W	135	99.7	–15.3	35.0	9.6
Cahokia/St. Louis Airport	38.57N	90.15W	126	99.8	–12.5	34.2	9.2
Charleston/Mattoon, Coles Apt	39.48N	88.28W	220	98.7	–16.0	32.8	10.8
Chicago Midway Airport	41.79N	87.75W	188	99.1	–18.7	33.4	10.9
Chicago O'Hare Intl Airport	41.99N	87.91W	205	98.9	–20.0	33.3	11.1
Decatur Airport	39.83N	88.87W	213	98.8	–18.0	33.9	11.1
Glenview NAS	42.08N	87.82W	199	99.0	–20.3	34.0	9.4
Lawrenceville Municipal Airport	38.77N	87.60W	131	99.8	–14.0	33.9	10.9
Moline, Quad City Airport	41.47N	90.52W	181	99.2	–21.2	34.1	11.0
Mount Vernon Airport	38.32N	88.87W	146	99.6	–15.2	34.0	9.2
Peoria, Greater Peoria Airport	40.67N	89.68W	202	98.9	–19.8	33.7	10.6
Quincy Municipal Airport	39.94N	91.19W	234	98.5	–18.9	34.0	11.0
Rockford, Greater Rockford Apt	42.20N	89.09W	227	98.6	–22.5	33.1	11.0
Springfield, Capital Airport	39.85N	89.68W	187	99.1	–19.0	33.8	11.1
Urbana/Champaign, Willard Apt	40.03N	88.27W	236	98.5	–18.3	33.6	12.3
W Chicago, Du Page Airport	41.92N	88.25W	231	98.6	–20.8	32.5	11.0

Table 2.2: Climatic Data in SI Units *(Continued)*

Station	Latitude, degrees	Longitude, degrees	Elevation, m	St. Br. Pressure, kPa	Winter Temp., °C	Summer Temp., °C	Wind, m/s
Indiana							
Bloomington, Monroe Co Apt	39.13N	86.62W	264	98.2	–15.1	32.5	8.7
Evansville Regional Airport	38.04N	87.54W	118	99.9	–14.7	34.4	9.2
Fort Wayne Municipal Airport	41.01N	85.21W	252	98.3	–19.2	32.8	11.0
Grissom ARB	40.65N	86.15W	253	98.3	–20.0	33.5	11.2
Indianapolis International Airport	39.71N	86.27W	246	98.4	–18.0	32.8	11.0
Lafayette, Purdue Univ Airport	40.41N	86.94W	194	99.0	–19.1	33.6	10.3
South Bend, Michiana Airport	41.71N	86.33W	236	98.5	–18.6	32.7	10.8
Terre Haute, Hulman Airport	39.45N	87.32W	175	99.2	–17.9	33.4	10.2
Iowa							
Ames Municipal Airport	42.00N	93.62W	291	97.9	–21.0	32.5	11.8
Ankeny Regional Airport	41.68N	93.55W	275	98.1	–20.8	34.9	9.3
Boone Municipal Airport	42.05N	93.85W	354	97.2	–21.0	32.9	11.8
Burlington Municipal Airport	40.78N	91.13W	214	98.8	–19.7	34.0	9.6
Carroll, Neu Airport	42.05N	94.78W	375	96.9	–21.1	33.9	12.9
Cedar Rapids Municipal Airport	41.88N	91.71W	266	98.2	–23.1	33.1	11.8
Clinton Municipal Airport	41.83N	90.33W	216	98.8	–21.3	32.5	12.5
Davenport Municipal Airport	41.62N	90.58W	259	98.3	–19.6	32.4	11.8
Des Moines International Airport	41.54N	93.67W	294	97.8	–21.6	34.1	11.5
Dubuque Regional Airport	42.40N	90.70W	329	97.4	–22.3	31.8	11.5
Fort Dodge Regional Airport	42.55N	94.18W	355	97.1	–23.9	32.8	11.2
Keokuk Municipal Airport	40.47N	91.43W	205	98.9	–16.9	34.8	9.8
Marshalltown Municipal Airport	42.10N	92.92W	296	97.8	–22.1	32.3	12.3
Mason City Municipal Airport	43.16N	93.33W	373	96.9	–25.4	32.8	12.7
Ottumwa Industrial Airport	41.11N	92.45W	258	98.3	–20.4	34.1	12.5
Sioux City, Sioux Gateway Apt	42.39N	96.38W	336	97.4	–22.7	34.1	12.8
Spencer	43.17N	95.15W	408	96.5	–25.3	32.9	12.6
Storm Lake Municipal Airport	42.60N	95.23W	454	96.0	–22.4	33.0	13.2
Waterloo Municipal Airport	42.55N	92.40W	268	98.2	–24.8	33.2	11.7
Kansas							
Coffeyville Municipal Airport	37.10N	95.57W	230	98.6	–12.1	37.5	11.0
Dodge City Regional Airport	37.77N	99.97W	790	92.2	–16.7	37.8	13.8
Fort Riley, Marshall Airport	39.05N	96.77W	324	97.5	–18.5	37.6	9.3
Garden City Municipal Airport	37.93N	100.73W	878	91.2	–17.8	37.4	13.4
Great Bend Municipal Airport	38.35N	98.87W	575	94.6	–16.1	37.6	13.5
Hays Municipal Airport	38.85N	99.27W	609	94.2	–16.3	38.0	13.3
Lawrence Municipal Airport	39.00N	95.22W	254	98.3	–14.9	37.3	11.2
Liberal Municipal Airport	37.03N	100.95W	901	91.0	–13.8	37.6	14.2
Manhattan Municipal Airport	39.13N	96.67W	330	97.4	–16.4	37.7	10.8
Newton City County Airport	38.05N	97.28W	467	95.8	–15.9	37.2	12.8
Olathe, Johnson County Airport	38.85N	94.73W	334	97.4	–14.9	35.9	10.5

Table 2.2: Climatic Data in SI Units *(Continued)*

Station	Latitude, degrees	Longitude, degrees	Elevation, m	St. Br. Pressure, kPa	Winter Temp., °C	Summer Temp., °C	Wind, m/s
Salina Municipal Airport	38.81N	97.66W	391	96.7	–17.3	38.5	12.3
Topeka, Billard Muni Airport	39.07N	95.63W	270	98.1	–17.4	36.2	10.6
Topeka, Forbes Field	38.95N	95.66W	329	97.4	–17.2	35.9	11.3
Wichita Mid Continent Airport	37.65N	97.43W	408	96.5	–15.6	38.0	12.5
Wichita, Col. Jabara Airport	37.75N	97.22W	433	96.2	–14.0	37.6	12.2
Wichita, McConnel AFB	37.62N	97.27W	414	96.5	–14.9	37.7	11.8
Winfield	37.30N	97.00W	353	97.2	–12.3	38.0	11.7
Kentucky							
Bowling Green Warren Co Apt	36.98N	86.44W	164	99.4	–13.1	34.0	9.0
Cincinnati International Airport	39.04N	84.67W	269	98.1	–16.1	33.0	10.0
Fort Campbell AAF	36.67N	87.50W	173	99.3	–12.3	34.7	8.5
Fort Knox, Godman AAF	37.90N	85.97W	239	98.5	–13.7	33.9	8.0
Frankfort, Capital City Airport	38.18N	84.90W	245	98.4	–12.4	32.9	8.4
Henderson City Airport	37.82N	87.68W	117	99.9	–13.7	33.8	9.4
Lexington, Blue Grass Airport	38.04N	84.61W	301	97.8	–14.4	33.2	9.0
Louisville, Bowman Field	38.23N	85.66W	170	99.3	–13.9	34.1	8.4
Louisville, Standiford Field	38.18N	85.73W	149	99.6	–13.3	34.1	9.4
Paducah, Barkley Reg Airport	37.06N	88.77W	126	99.8	–13.0	34.8	9.1
Louisiana							
Alexandria International Airport	31.33N	92.55W	27	101.0	–3.0	35.9	8.3
Alexandria, Esler Reg Airport	31.40N	92.30W	36	100.9	–3.0	36.3	7.4
Baton Rouge Metropolitan Apt	30.54N	91.15W	23	101.1	–2.5	34.6	8.4
Fort Polk, Polk AAF	31.05N	93.20W	102	100.1	–2.9	35.5	7.7
Lafayette Regional Airport	30.21N	91.99W	13	101.2	–1.9	34.6	9.2
Lake Charles Regional Airport	30.13N	93.23W	3	101.3	–1.3	34.5	9.3
Monroe Regional Airport	32.51N	92.04W	25	101.0	–4.7	36.1	8.5
New Orleans Lakefront Airport	30.04N	90.03W	3	101.3	2.0	34.1	10.7
New Orleans NAS, Callender Fld	29.83N	90.03W	0	101.3	–1.1	33.6	7.9
New Orleans, Armstrong Intl Apt	29.99N	90.25W	6	101.3	–0.3	34.2	9.2
Patterson, H P Williams Airport	29.72N	91.33W	3	101.3	–0.5	34.0	8.5
Shreveport Downtown Airport	32.54N	93.74W	55	100.7	–2.8	37.4	8.4
Shreveport Regional Airport	32.45N	93.82W	79	100.4	–4.6	36.6	8.8
Shreveport, Barksdale AFB	32.50N	93.67W	54	100.7	–5.3	36.3	8.5
Maine							
Auburn/Lewiston Muni Airport	44.05N	70.28W	88	100.3	–21.5	31.0	9.3
Augusta State Airport	44.32N	69.80W	110	100.0	–19.9	30.6	10.5
Bangor International Airport	44.81N	68.82W	59	100.6	–21.9	31.0	10.5
Brunswick NAS	43.90N	69.93W	23	101.1	–19.1	30.2	9.9
Portland International Jetport	43.64N	70.30W	19	101.1	–18.0	30.4	10.4
Presque Isle Municipal Airport	46.69N	68.03W	163	99.4	–28.2	29.3	10.5
Sanford Municipal Airport	43.40N	70.72W	74	100.4	–21.6	31.4	9.3

Table 2.2: Climatic Data in SI Units *(Continued)*

Station	Latitude, degrees	Longitude, degrees	Elevation, m	St. Br. Pressure, kPa	Winter Temp., °C	Summer Temp., °C	Wind, m/s
Waterville Airport	44.53N	69.68W	101	100.1	–21.4	30.1	9.1
Maryland							
Baltimore Washington Intl Apt	39.17N	76.68W	47	100.8	–10.6	34.4	10.0
Salisbury-Ocean City Airport	38.34N	75.51W	18	101.1	–10.2	33.8	9.3
Washington DC, Andrews AFB	38.82N	76.87W	86	100.3	–10.1	34.4	10.5
Massachusetts							
Barnstable Airport	41.67N	70.27W	17	101.1	–12.5	28.9	11.1
Boston, Logan International Apt	42.36N	71.01W	9	101.2	–13.6	32.7	12.0
Chatham Municipal Airport	41.68N	70.00W	19	101.1	–12.0	27.6	10.4
Lawrence Municipal Airport	42.72N	71.12W	46	100.8	–16.2	32.3	9.0
Marthas Vineyard Airport	41.40N	70.62W	21	101.1	–12.8	28.0	11.7
New Bedford Municipal Airport	41.67N	70.95W	25	101.0	–13.7	31.2	10.4
Norwood Memorial Airport	42.18N	71.18W	15	101.2	–16.3	32.4	9.2
Plymouth Municipal Airport	41.92N	70.73W	45	100.8	–15.2	31.3	10.4
Provincetown Municipal Airport	42.07N	70.22W	2	101.3	–10.9	27.9	13.5
South Weymouth NAS	42.15N	70.93W	49	100.7	–14.5	32.9	8.3
Worcester Municipal Airport	42.27N	71.88W	310	97.7	–16.9	29.8	11.5
Michigan							
Alpena County Airport	45.07N	83.58W	211	98.8	–20.9	31.2	9.0
Battle Creek Intl Airport	42.23N	85.55W	273	98.1	–16.1	32.3	9.7
Benton Harbor, Ross Field	42.13N	86.43W	196	99.0	–13.8	32.1	10.4
Cadillac, Wexford Co Airport	44.28N	85.42W	398	96.6	–22.8	31.0	9.2
Detroit City Airport	42.41N	83.01W	191	99.1	–15.5	32.7	9.1
Detroit Metropolitan Airport	42.22N	83.35W	202	98.9	–17.0	32.4	11.5
Detroit Willow Run Airport	42.23N	83.53W	218	98.7	–17.3	32.4	11.0
Flint, Bishop Intl Airport	42.97N	83.75W	234	98.5	–18.3	31.9	10.8
Grand Rapids, Kent Co Intl Apt	42.88N	85.52W	245	98.4	–17.0	31.8	11.1
Grosse Ile, Municipal Airport	42.10N	83.15W	176	99.2	–13.7	32.1	9.3
Holland, Tulip City Airport	42.75N	86.10W	210	98.8	–13.7	31.3	11.6
Howell, Livingston Co Airport	42.63N	83.98W	293	97.9	–17.1	32.4	9.1
Jackson Reynolds Muni Airport	42.26N	84.46W	311	97.7	–18.3	31.4	9.1
Lansing, Capital City Airport	42.78N	84.58W	266	98.2	–19.1	31.9	11.0
Mackinac Island Airport	45.87N	84.63W	300	97.8	–23.5	29.2	10.5
Marquette, Sawyer Intl Airport	46.53N	87.55W	431	96.3	–25.8	29.4	9.5
Menominee Airport	45.13N	87.63W	191	99.1	–22.0	30.0	9.5
Muskegon County Airport	43.17N	86.24W	193	99.0	–15.5	30.1	11.4
Oakland Pontiac Airport	42.67N	83.42W	306	97.7	–17.2	32.0	10.9
Saginaw, Tri City Intl Airport	43.53N	84.08W	204	98.9	–17.8	32.2	10.8
Saint Clair County Intl Airport	42.92N	82.53W	198	99.0	–17.4	32.4	8.3
Sault Ste Marie Muni Airport	46.47N	84.37W	221	98.7	–24.0	28.7	9.6
Selfridge ANGB	42.62N	82.83W	177	99.2	–16.3	32.4	9.4

Table 2.2: Climatic Data in SI Units *(Continued)*

Station	Latitude, degrees	Longitude, degrees	Elevation, m	St. Br. Pressure, kPa	Winter Temp., °C	Summer Temp., °C	Wind, m/s
Sturgis, Kirsch Municipal	41.82N	85.43W	282	98.0	–17.0	32.2	9.4
Traverse City Airport	44.74N	85.58W	192	99.0	–18.0	31.8	9.3
Minnesota							
Albert Lea Municipal Airport	43.68N	93.37W	383	96.8	–24.6	31.6	11.0
Alexandria, Chandler Field	45.88N	95.39W	436	96.2	–28.8	31.7	11.5
Austin Municipal Airport	43.67N	92.93W	375	96.9	–24.7	31.9	11.0
Bemidji Beltrami County Airport	47.50N	94.93W	420	96.4	–30.4	31.0	10.8
Brainerd, Crow Wing Co Airport	46.40N	94.13W	374	96.9	–28.0	31.3	8.8
Cloquet, Carlton County Airport	46.70N	92.50W	390	96.7	–27.8	29.1	9.4
Duluth Harbor	46.77N	92.08W	186	99.1	–24.5	30.1	12.4
Duluth International Airport	46.84N	92.19W	432	96.2	–28.6	29.2	11.1
Fairmont Municipal Airport	43.65N	94.42W	354	97.1	–24.0	32.1	13.8
Faribault Municipal Airport	44.33N	93.32W	322	97.5	–25.9	32.2	10.5
Fergus Falls Municipal Airport	46.28N	96.15W	361	97.1	–28.2	32.2	12.5
Hibbing, Chisholm-Hibbing Apt	47.39N	92.84W	414	96.5	–31.5	29.7	9.2
Hutchinson Municipal Airport	44.87N	94.38W	323	97.5	–26.9	32.4	11.1
Mankato Municipal Airport	44.22N	93.92W	311	97.7	–25.1	32.1	11.9
Marshall, Ryan Field	44.45N	95.82W	359	97.1	–24.9	32.3	14.1
Minneapolis St Paul Intl Airport	44.88N	93.23W	255	98.3	–25.2	32.8	11.1
Minneapolis, Anoka Co Airport	45.15N	93.22W	278	98.0	–22.6	32.4	10.3
Minneapolis, Crystal Airport	45.07N	93.35W	265	98.2	–22.6	32.6	9.6
Minneapolis, Flying Cloud Apt	44.82N	93.45W	283	98.0	–22.6	32.7	9.9
New Ulm Municipal Airport	44.32N	94.50W	308	97.7	–25.2	32.4	12.2
Owatonna Municipal Airport	44.12N	93.25W	350	97.2	–25.8	32.1	11.8
Red Wing Municipal Airport	44.58N	92.48W	239	98.5	–24.9	32.5	11.1
Rochester Municipal Airport	43.90N	92.49W	402	96.6	–26.2	31.2	12.9
South St Paul Municipal Airport	44.85N	93.15W	250	98.4	–22.9	32.6	8.2
St Cloud Municipal Airport	45.55N	94.05W	312	97.6	–28.5	32.2	10.2
St Paul Downtown Airport	44.93N	93.05W	217	98.7	–24.4	32.4	10.2
Willmar Municipal Airport	45.12N	95.08W	344	97.3	–28.4	31.4	11.5
Winona Municipal Airport	44.08N	91.70W	200	99.0	–23.0	32.8	8.3
Worthington Municipal Airport	43.65N	95.58W	480	95.7	–24.1	31.3	14.0
Mississippi							
Biloxi, Keesler AFB	30.42N	88.92W	8	101.2	–0.9	34.1	7.9
Columbus AFB	33.65N	88.45W	68	100.5	–6.5	36.0	8.3
Golden Triangle Reg Airport	33.45N	88.58W	80	100.4	–6.4	36.1	8.0
Greenwood Leflore Airport	33.50N	90.08W	41	100.8	–6.4	35.6	8.5
Hattiesburg/Laurel, Pine Belt Apt	31.47N	89.33W	93	100.2	–3.7	35.4	6.9
Jackson International Airport	32.32N	90.08W	101	100.1	–5.6	35.3	8.4
Meridian NAS	32.55N	88.57W	97	100.2	–6.2	36.2	6.9
Meridian Regional Airport	32.33N	88.75W	95	100.2	–5.6	35.4	8.2

Table 2.2: Climatic Data in SI Units *(Continued)*

Station	Latitude, degrees	Longitude, degrees	Elevation, m	St. Br. Pressure, kPa	Winter Temp., °C	Summer Temp., °C	Wind, m/s
Natchez, Hardy Anders Field	31.62N	91.30W	83	100.3	–3.7	34.0	8.3
Tupelo Municipal Airport	34.26N	88.77W	110	100.0	–7.6	35.5	8.4
Missouri							
Cape Girardeau Muni Airport	37.23N	89.57W	107	100.1	–13.7	34.9	9.5
Columbia Regional Airport	38.82N	92.22W	274	98.1	–17.0	34.8	10.8
Jefferson City Memorial Airport	38.58N	92.15W	167	99.3	–13.7	35.1	9.3
Joplin Regional Airport	37.15N	94.50W	300	97.8	–14.5	35.9	11.0
Kansas City Downtown Airport	39.12N	94.59W	229	98.6	–16.3	36.2	10.1
Kansas City International Airport	39.30N	94.72W	312	97.6	–17.8	35.7	11.5
Kirksville Municipal Airport	40.10N	92.54W	294	97.8	–19.7	34.5	10.9
Poplar Bluff	36.77N	90.47W	146	99.6	–12.5	34.3	8.5
Sedalia Memorial Airport	38.70N	93.18W	277	98.0	–13.9	36.1	10.4
Springfield Regional Airport	37.24N	93.39W	387	96.8	–15.4	35.0	10.2
St Louis, Lambert Intl Airport	38.75N	90.37W	216	98.8	–15.5	35.4	10.8
St Louis, Spirit of St Louis Apt	38.66N	90.66W	141	99.6	–15.0	35.1	9.3
Montana							
Billings Logan Intl Airport	45.81N	108.54W	1088	88.9	–24.2	34.9	12.0
Bozeman, Gallatin Field	45.79N	111.15W	1356	86.1	–27.8	32.9	9.2
Butte, Mooney Airport	45.95N	112.51W	1687	82.6	–29.1	31.0	10.3
Great Falls	47.45N	111.38W	1130	88.5	–24.7	32.5	N/A
Great Falls International Airport	47.47N	111.38W	1115	88.6	–27.6	33.4	14.0
Great Falls, Malmstrom AFB	47.52N	111.18W	1056	89.3	–27.2	33.8	13.0
Helena Regional Airport	46.61N	111.96W	1179	87.9	–26.3	33.7	10.9
Kalispell, Glacier Park Intl Apt	48.30N	114.26W	906	90.9	–22.6	32.1	9.2
Missoula International Airport	46.92N	114.09W	972	90.2	–21.1	33.6	9.7
Nebraska							
Beatrice Municipal Airport	40.30N	96.75W	403	96.6	–18.3	36.3	12.5
Bellevue, Offutt AFB	41.12N	95.92W	319	97.6	–19.6	35.0	10.9
Columbus Municipal Airport	41.45N	97.33W	451	96.0	–20.9	33.9	12.0
Fremont Municipal Airport	41.45N	96.52W	379	96.9	–18.9	34.0	12.0
Grand Island, Central NE Airport	40.96N	98.31W	566	94.7	–21.2	35.7	12.7
Hastings Municipal Airport	40.60N	98.43W	596	94.4	–18.9	36.0	12.5
Kearney Municipal Airport	40.73N	99.00W	649	93.8	–18.9	34.2	13.9
Lincoln Municipal Airport	40.83N	96.76W	362	97.1	–20.8	36.2	12.0
Norfolk, Karl Stefan Airport	41.98N	97.44W	473	95.8	–22.5	34.8	12.5
North Platte, Lee Bird Field	41.12N	100.67W	850	91.5	–22.1	35.7	12.4
Omaha, Eppley Airfield	41.31N	95.90W	299	97.8	–21.3	35.0	11.6
Omaha, North Omaha Airport	41.37N	96.02W	399	96.6	–22.1	34.1	10.5
Scotts Bluff County Airport	41.87N	103.60W	1206	87.7	–22.9	35.8	12.9
Nevada							
Elko Municipal Airport	40.83N	115.79W	1547	84.1	–20.0	34.8	9.2

Table 2.2: Climatic Data in SI Units *(Continued)*

Station	Latitude, degrees	Longitude, degrees	Elevation, m	St. Br. Pressure, kPa	Winter Temp., °C	Summer Temp., °C	Wind, m/s
Las Vegas, McCarran Intl Apt	36.08N	115.16W	665	93.6	–0.8	42.4	12.0
Las Vegas, Nellis AFB	36.23N	115.03W	573	94.6	–2.1	42.9	11.7
Reno / Tahoe Intl Airport	39.48N	119.77W	1341	86.2	–11.2	35.4	11.6
New Hampshire							
Concord Municipal Airport	43.20N	71.50W	106	100.1	–20.4	32.1	9.3
Keene, Dillant Hopkins Airport	42.90N	72.27W	153	99.5	–21.3	31.3	8.2
Laconia Municipal Airport	43.57N	71.42W	166	99.4	–20.1	31.2	8.8
Lebanon Regional Airport	43.63N	72.30W	182	99.2	–22.5	31.5	8.0
Manchester Municipal Airport	42.93N	71.44W	71	100.5	–17.2	32.9	8.5
Portsmout, Pease AFB	43.08N	70.82W	31	101.0	–16.3	31.8	10.1
New Jersey							
Atlantic City Intl Airport	39.46N	74.46W	20	101.1	–12.3	33.5	11.1
Belmar/Farmingdale Airport	40.18N	74.13W	26	101.0	–11.8	32.6	11.3
Cape May County Airport	39.00N	74.90W	7	101.2	–9.1	32.0	10.3
Lakehurst, McGuire AFB	40.02N	74.60W	45	100.8	–12.1	33.8	10.4
Millville Municipal Airport	39.37N	75.08W	23	101.1	–12.1	33.3	8.9
Newark International Airport	40.72N	74.17W	9	101.2	–11.6	34.5	11.2
Teterboro Airport	40.85N	74.06W	2	101.3	–12.3	33.5	9.2
Trenton, Mercer County Airport	40.28N	74.81W	65	100.6	–12.3	33.6	10.5
New Mexico							
Alamogordo White Sands Apt	32.83N	105.98W	1313	86.5	–6.1	37.8	9.6
Alamogordo, Holloman AFB	32.85N	106.10W	1248	87.2	–7.2	37.3	10.2
Albuquerque Intl Airport	35.04N	106.62W	1620	83.3	–7.9	35.1	12.6
Carlsbad Airport	32.34N	104.26W	994	89.9	–6.5	39.0	12.0
Clovis Municipal Airport	34.43N	103.08W	1284	86.8	–10.1	36.2	14.2
Clovis, Cannon AFB	34.38N	103.32W	1309	86.6	–11.2	36.4	12.5
Farmington Municipal Airport	36.74N	108.23W	1677	82.7	–13.6	34.9	11.0
Gallup Municipal Airport	35.51N	108.79W	1971	79.8	–17.4	33.0	11.0
Roswell International Air Center	33.31N	104.54W	1118	88.6	–8.7	37.7	11.2
New York							
Albany International Airport	42.75N	73.80W	89	100.3	–18.8	31.7	10.7
Binghamton, Greater Airport	42.21N	75.98W	499	95.5	–18.3	29.7	9.6
Buffalo, Greater Buffalo Airport	42.94N	78.74W	215	98.8	–16.3	30.3	12.6
Dunkirk	42.48N	79.35W	197	99.0	–13.6	27.6	19.7
Elmira/Corning Regional Airport	42.16N	76.89W	291	97.9	–18.8	32.2	9.2
Farmingdale, Republic Airport	40.72N	73.40W	26	101.0	–11.2	32.3	11.1
Glens Falls, Bennett Airport	43.34N	73.61W	102	100.1	–23.4	31.1	8.2
Islip, Long Island MacArthur Apt	40.79N	73.10W	33	100.9	–11.9	31.4	10.6
Jamestown Airport	42.15N	79.27W	525	95.2	–17.2	28.0	9.6
New York, Kennedy Intl Airport	40.66N	73.80W	7	101.2	–10.7	32.1	12.2
New York, Laguardia Airport	40.78N	73.88W	9	101.2	–10.8	33.5	12.2

Table 2.2: Climatic Data in SI Units *(Continued)*

Station	Latitude, degrees	Longitude, degrees	Elevation, m	St. Br. Pressure, kPa	Winter Temp., °C	Summer Temp., °C	Wind, m/s
Newburgh, Stewart Intl Airport	41.50N	74.10W	177	99.2	–15.8	32.1	10.9
Niagara Falls Intl Airport	43.11N	78.95W	179	99.2	–16.4	31.0	11.7
Plattsburgh AFB	44.65N	73.47W	72	100.5	–23.1	30.3	9.2
Poughkeepsie, Dutchess Co Apt	41.63N	73.88W	49	100.7	–17.5	33.0	8.2
Rochester, Monroe Co Airport	43.12N	77.68W	169	99.3	–16.6	31.3	11.3
Rome, Griffiss AFB	43.23N	75.40W	158	99.4	–21.1	31.5	9.9
Syracuse Hancock Intl Airport	43.11N	76.10W	127	99.8	–19.3	31.6	10.8
Utica, Oneida County Airport	43.15N	75.38W	227	98.6	–20.6	30.9	9.3
Watertown International Airport	43.99N	76.02W	102	100.1	–25.0	29.6	10.0
White Plains, Westchester Apt	41.07N	73.71W	121	99.9	–13.5	32.2	9.1
North Carolina							
Asheville Regional Airport	35.43N	82.54W	661	93.6	–10.2	31.2	10.4
Charlotte Douglas Intl Airport	35.21N	80.94W	234	98.5	–6.4	34.5	8.3
Fayetteville Municipal Airport	34.98N	78.88W	59	100.6	–5.9	35.3	8.9
Fayetteville, Simmons AAF	35.13N	78.93W	93	100.2	–5.7	35.9	8.0
Fayetteville, Pope AFB	35.17N	79.02W	66	100.5	–6.0	36.1	8.2
Goldsboro, S Johnson AFB	35.33N	77.97W	33	100.9	–5.7	35.8	8.2
Greensboro, Piedmont Triad Apt	36.10N	79.94W	270	98.1	–8.3	33.4	8.6
Greenville (Pitt County) Airport	35.63N	77.40W	8	101.2	–6.2	35.1	8.4
Havelock, Cherry Point MCAS	34.90N	76.88W	9	101.2	–4.7	34.0	9.3
Hickory Regional Airport	35.74N	81.39W	362	97.1	–7.3	33.6	7.8
Jacksonville, Albert J Ellis Apt	34.83N	77.62W	29	101.0	–6.7	34.1	8.9
Jacksonville, New River MCAF	34.70N	77.43W	8	101.2	–5.0	34.0	8.9
New Bern, Craven Co Reg Apt	35.07N	77.05W	7	101.2	–5.3	34.3	8.3
Raleigh Durham Intl Airport	35.87N	78.79W	133	99.7	–7.4	34.5	8.5
Southern Pines, Moore Co Apt	35.23N	79.40W	141	99.6	–7.3	34.8	8.0
Wilmington International Airport	34.27N	77.91W	12	101.2	–4.5	34.0	9.2
Winston-Salem, Reynolds Apt	36.13N	80.22W	296	97.8	–7.7	33.6	8.3
North Dakota							
Bismarck Municipal Airport	46.77N	100.75W	506	95.4	–28.9	34.4	12.2
Dickinson Municipal Airport	46.80N	102.80W	787	92.2	–28.0	34.0	13.0
Fargo, Hector Intl Airport	46.93N	96.81W	274	98.1	–29.1	32.8	12.7
Grand Forks AFB	47.97N	97.40W	276	98.1	–29.1	32.6	12.7
Grand Forks Intl Airport	47.95N	97.18W	254	98.3	–30.1	32.2	12.0
Jamestown Municipal Airport	46.93N	98.68W	455	96.0	–28.8	33.1	12.8
Minot AFB	48.42N	101.35W	497	95.5	–30.1	34.0	12.9
Minot International Airport	48.26N	101.28W	522	95.2	–28.8	33.0	12.3
Williston, Sloulin Intl Airport	48.20N	103.64W	581	94.5	–30.9	34.9	12.1
Ohio							
Akron Canton Regional Airport	40.92N	81.44W	377	96.9	–16.8	31.5	10.4
Cincinnati Muni Lunken Airport	39.10N	84.42W	152	99.5	–14.3	33.8	9.1

Table 2.2: Climatic Data in SI Units *(Continued)*

Station	Latitude, degrees	Longitude, degrees	Elevation, m	St. Br. Pressure, kPa	Winter Temp., °C	Summer Temp., °C	Wind, m/s
Cleveland Hopkins Intl Airport	41.41N	81.85W	245	98.4	–16.4	31.9	11.0
Columbus, Ohio State Univ Apt	40.07N	83.07W	283	98.0	–14.0	32.4	9.7
Columbus, Port Columbus Apt	39.99N	82.88W	249	98.4	–16.0	32.8	9.8
Columbus, Rickenbacker Airport	39.82N	82.93W	230	98.6	–15.4	33.6	9.9
Dayton International Airport	39.91N	84.22W	306	97.7	–17.4	32.4	10.9
Dayton, Wright Paterson AFB	39.83N	84.05W	250	98.4	–17.0	33.0	9.6
Findlay Airport	41.01N	83.67W	248	98.4	–18.0	32.4	10.9
Lancaster, Fairfield Co Airport	39.75N	82.65W	264	98.2	–16.0	32.5	9.0
Mansfield Lahm Muni Airport	40.82N	82.52W	400	96.6	–17.7	31.1	11.0
Toledo Express Airport	41.59N	83.80W	211	98.8	–18.0	32.9	10.9
Youngstown Municipal Airport	41.25N	80.67W	362	97.1	–16.8	31.4	9.8
Zanesville Municipal Airport	39.94N	81.89W	275	98.1	–16.2	32.4	8.7
Oklahoma							
Enid, Vance AFB	36.33N	97.92W	408	96.5	–14.1	38.1	11.9
Lawton Municipal Airport	34.57N	98.42W	338	97.3	–7.8	39.1	11.6
Lawton, Fort Sill	34.65N	98.40W	369	97.0	–10.8	38.0	11.1
McAlester Regional Airport	34.90N	95.78W	235	98.5	–10.5	37.4	9.1
Midwest City, Tinker AFB	35.42N	97.38W	384	96.8	–11.1	37.4	11.4
Oklahoma City, Wiley Post Apt	35.53N	97.65W	396	96.7	–11.0	37.5	11.8
Oklahoma City, Will Rogers Apt	35.39N	97.60W	398	96.6	–11.4	37.5	12.1
Ponca City Municipal Airport	36.73N	97.10W	304	97.7	–12.5	37.9	10.0
Stillwater Municipal Airport	36.15N	97.08W	308	97.7	–10.2	38.8	11.0
Tulsa International Airport	36.20N	95.89W	206	98.9	–11.7	37.5	10.9
Tulsa, R L Jones Airport	36.03N	95.98W	193	99.0	–9.0	37.8	8.8
Oregon							
Astoria, Port of Astoria Airport	46.16N	123.88W	7	101.2	–2.9	24.8	11.0
Canby, Aurora State Airport	45.25N	122.77W	60	100.6	–2.5	32.9	8.1
Corvallis Municipal Airport	44.48N	123.28W	77	100.4	–3.9	33.8	8.8
Eugene, Mahlon Sweet Field	44.13N	123.21W	114	100.0	–5.3	33.0	8.8
Klamath Falls, Kingsley Field	42.15N	121.72W	1247	87.2	–14.3	32.8	11.1
La Grande Union County Airport	45.29N	118.01W	828	91.8	–12.5	33.9	12.0
McMinnville Municipal Airport	45.18N	123.13W	51	100.7	–2.4	33.0	9.3
Medford Jackson County Airport	42.39N	122.87W	405	96.6	–5.1	37.2	8.2
Newport State Beach	44.62N	124.07W	11	101.2	–0.6	18.7	18.3
North Bend Municipal Airport	43.42N	124.25W	5	101.3	–0.8	22.0	11.9
Pendleton Municipal Airport	45.70N	118.83W	462	95.9	–13.9	35.8	11.6
Portland Hillsboro Airport	45.53N	122.95W	62	100.6	–5.7	33.2	8.4
Portland International Airport	45.59N	122.60W	33	100.9	–4.5	32.9	10.6
Redmond, Roberts Field	44.25N	121.15W	940	90.5	–14.8	33.8	9.2
Salem, McNary Field	44.91N	123.00W	61	100.6	–5.6	33.3	9.3

Table 2.2: Climatic Data in SI Units *(Continued)*

Station	Latitude, degrees	Longitude, degrees	Elevation, m	St. Br. Pressure, kPa	Winter Temp., °C	Summer Temp., °C	Wind, m/s
Pennsylvania							
Allentown, Lehigh Valley Apt	40.65N	75.45W	117	99.9	–13.9	32.8	10.8
Altoona Blair County Airport	40.30N	78.32W	448	96.1	–15.2	31.4	9.8
Butler, Butler County Airport	40.78N	79.95W	380	96.8	–16.1	31.1	8.0
Erie International Airport	42.08N	80.18W	225	98.7	–14.9	30.2	11.0
Harrisburg International Airport	40.19N	76.76W	95	100.2	–11.8	33.7	11.3
Harrisburg, Capital City Airport	40.22N	76.85W	106	100.1	–13.0	33.5	9.2
Johnstown Cambria Co Airport	40.30N	78.83W	696	93.2	–16.4	29.5	10.6
Philadelphia Intl Airport	39.87N	75.23W	9	101.2	–10.8	34.0	10.9
Philadelphia, NE Phil Apt	40.08N	75.01W	36	100.9	–11.7	33.9	9.4
Pittsburgh International Airport	40.50N	80.23W	367	97.0	–15.7	31.9	10.5
Pittsburgh, Allegheny Co Airport	40.36N	79.92W	388	96.8	–15.4	32.2	9.1
Reading Spaatz Field	40.37N	75.96W	108	100.0	–12.6	33.5	10.1
Washington Airport	40.13N	80.28W	361	97.1	–16.3	31.3	8.7
Wilkes-Barre Scranton Intl Apt	41.34N	75.73W	293	97.9	–15.9	31.6	9.1
Williamsport Lycoming Co Apt	41.24N	76.92W	160	99.4	–15.3	32.7	9.6
Willow Grove NAS	40.20N	75.15W	110	100.0	–12.1	33.7	8.2
Rhode Island							
Pawtucket, North Central Airport	41.92N	71.50W	134	99.7	–16.0	31.2	8.7
Providence, T F Green State Apt	41.72N	71.43W	19	101.1	–13.8	32.3	10.9
South Carolina							
Anderson County Airport	34.50N	82.71W	236	98.5	–5.5	35.0	8.5
Beaufort MCAS	32.48N	80.72W	12	101.2	–2.5	34.9	8.2
Charleston International Airport	32.90N	80.04W	15	101.2	–2.8	34.6	9.1
Columbia Metropolitan Airport	33.94N	81.12W	69	100.5	–5.5	36.1	8.5
Florence Regional Airport	34.19N	79.73W	46	100.8	–4.8	35.5	8.5
Greenville Spartanburg Airport	34.90N	82.22W	296	97.8	–6.4	34.4	8.7
Myrtle Beach AFB	33.68N	78.92W	8	101.2	–3.7	33.5	8.8
Sumter, Shaw AFB	33.97N	80.47W	74	100.4	–4.4	35.2	8.4
South Dakota							
Aberdeen Regional Airport	45.45N	98.42W	396	96.7	–27.7	33.5	12.6
Brookings Municipal Airport	44.30N	96.82W	502	95.4	–26.2	31.9	11.8
Huron Regional Airport	44.39N	98.23W	393	96.7	–26.1	34.7	12.2
Mitchell Municipal Airport	43.77N	98.03W	397	96.7	–23.9	34.1	12.6
Pierre Municipal Airport	44.38N	100.29W	526	95.2	–24.3	37.4	12.9
Rapid City Regional Airport	44.05N	103.05W	966	90.3	–23.6	36.1	15.5
Rapid City, Ellsworth AFB	44.15N	103.10W	980	90.1	–22.9	35.8	15.3
Sioux Falls, Joe Foss Field	43.58N	96.75W	435	96.2	–25.4	33.8	12.3
Watertown Municipal Airport	44.93N	97.15W	530	95.1	–27.2	32.3	12.6
Yankton, Chan Gurney Airport	42.92N	97.38W	408	96.5	–22.3	33.7	13.3
Tennessee							
Bristol, Tri City Airport	36.48N	82.40W	465	95.9	–11.5	32.1	8.4

Table 2.2: Climatic Data in SI Units ***(Continued)***

Station	Latitude, degrees	Longitude, degrees	Elevation, m	St. Br. Pressure, kPa	Winter Temp., °C	Summer Temp., °C	Wind, m/s
Chattanooga, Lovell Field	35.03N	85.20W	210	98.8	–7.9	34.7	8.0
Dyersburg Municipal Airport	36.02N	89.32W	103	100.1	–9.7	34.1	8.5
Jackson, McKellar Sipes Airport	35.59N	88.92W	129	99.8	–10.3	34.8	8.7
Knoxville, McGhee Tyson Apt	35.82N	83.99W	299	97.8	–9.4	33.7	9.0
Memphis International Airport	35.06N	89.99W	101	100.1	–8.3	35.6	9.1
Millington Municipal Airport	35.35N	89.87W	98	100.2	–10.0	36.3	8.3
Nashville International Airport	36.12N	86.69W	184	99.1	–10.6	34.7	8.9
Texas							
Abilene Regional Airport	32.41N	99.68W	546	94.9	–7.8	37.4	11.6
Abilene, Dyess AFB	32.43N	99.85W	545	95.0	–8.8	38.6	10.7
Alice International Airport	27.74N	98.03W	55	100.7	1.3	37.6	9.8
Amarillo International Airport	35.22N	101.71W	1099	88.8	–13.8	36.3	12.9
Austin, Camp Mabry	30.32N	97.77W	201	98.9	–3.0	37.4	9.0
Austin-Bergstrom Intl Airport	30.18N	97.68W	151	99.5	–3.8	37.6	9.4
Beeville, Chase Field NAS	28.37N	97.67W	58	100.6	–2.3	37.9	10.4
Brownsville, South Padre Is. Apt	25.91N	97.43W	7	101.2	2.4	35.3	11.7
College Station, Easterwood Fld	30.59N	96.36W	100	100.1	–3.3	37.3	8.9
Corpus Christi Intl Airport	27.77N	97.51W	13	101.2	0.4	35.5	12.0
Corpus Christi NAS	27.70N	97.28W	6	101.3	2.0	34.0	11.1
Dallas Love Field	32.85N	96.85W	149	99.6	–5.8	38.0	9.8
Dallas NAS	32.73N	96.97W	151	99.5	–7.5	37.6	9.3
Dallas, Redbird Airport	32.68N	96.87W	201	98.9	–3.0	37.9	10.1
Dallas/Fort Worth Intl Airport	32.90N	97.04W	182	99.2	–6.5	38.0	11.6
Del Rio International Airport	29.37N	100.92W	313	97.6	–0.7	38.6	9.2
Del Rio, Laughlin AFB	29.37N	100.78W	327	97.5	–1.5	40.1	9.9
El Paso International Airport	31.81N	106.38W	1194	87.8	–5.2	38.1	11.4
Fort Hood, Robert Gray AAF	31.07N	97.83W	312	97.6	–4.6	37.7	9.8
Fort Worth Alliance Airport	32.98N	97.32W	226	98.6	–5.8	38.7	10.7
Fort Worth Meacham Airport	32.82N	97.36W	215	98.8	–6.7	38.3	9.7
Fort Worth NAS	32.77N	97.44W	198	99.0	–7.4	37.9	10.5
Galveston, Scholes Field	29.27N	94.86W	3	101.3	0.6	33.0	11.4
Georgetown Municipal Airport	30.68N	97.68W	240	98.5	–3.0	37.3	9.4
Harlingen, Rio Grande Apt	26.23N	97.65W	11	101.2	2.5	37.1	12.4
Houston, Bush Intl Airport	29.99N	95.36W	32	100.9	–1.6	36.0	8.6
Houston, D W Hooks Airport	30.07N	95.55W	46	100.8	–1.0	37.0	7.9
Houston, Ellington Field	29.60N	95.17W	12	101.2	0.0	35.6	8.9
Houston, W P Hobby Airport	29.65N	95.28W	14	101.2	–0.4	35.0	9.4
Killeen Municipal Airport	31.08N	97.68W	258	98.3	–3.4	37.5	9.8
Kingsville NAS	27.50N	97.82W	18	101.1	0.0	36.5	10.6
Laredo International Airport	27.55N	99.47W	155	99.5	1.0	39.2	11.0
Longview, Gregg County Airport	32.39N	94.71W	114	100.0	–5.8	37.3	8.7

Table 2.2: Climatic Data in SI Units *(Continued)*

Station	Latitude, degrees	Longitude, degrees	Elevation, m	St. Br. Pressure, kPa	Winter Temp., °C	Summer Temp., °C	Wind, m/s
Lubbock International Airport	33.67N	101.82W	988	90.0	–10.4	37.1	12.9
Lubbock, Reese AFB	33.60N	102.05W	1014	89.7	–11.0	37.8	12.1
Lufkin, Angelina County Airport	31.23N	94.75W	96	100.2	–3.7	36.4	8.0
McAllen, Miller Intl Airport	26.18N	98.24W	34	100.9	2.5	37.8	11.1
McGregor Municipal Airport	31.48N	97.32W	180	99.2	–3.8	37.7	10.1
McKinney Municipal Airport	33.18N	96.58W	176	99.2	–5.9	38.0	10.1
Midland International Airport	31.93N	102.21W	872	91.3	–7.6	37.9	11.9
Nacogdoches, Mangham Airport	31.58N	94.72W	108	100.0	–3.7	36.4	8.1
Paris, Cox Field	33.63N	95.45W	171	99.3	–6.3	37.3	10.9
Port Aransas	27.82N	97.05W	6	101.3	2.5	30.0	17.6
Port Arthur, Jefferson Co Airport	29.95N	94.02W	5	101.3	–0.8	34.6	9.5
Rockport, Aransas Co Airport	28.08N	97.05W	8	101.2	2.4	33.8	10.0
Sabine	29.67N	94.05W	6	101.3	0.1	31.3	16.0
San Angelo, Mathis Field	31.35N	100.49W	577	94.6	–6.6	37.9	11.1
San Antonio Intl Airport	29.53N	98.46W	247	98.4	–2.6	37.0	9.0
San Antonio, Kelly AFB	29.38N	98.58W	208	98.9	–2.0	37.9	8.9
San Antonio, Stinson Muni Apt	29.34N	98.47W	176	99.2	–0.8	37.8	8.4
San Marcos Municipal Airport	29.88N	97.87W	182	99.2	–2.3	37.3	10.8
Stephenville, Clark Muni Airport	32.22N	98.18W	403	96.6	–7.3	36.5	8.5
Temple, D Miller Muni Airport	31.15N	97.40W	213	98.8	–4.0	37.5	11.1
Universal City, Randolph AFB	29.53N	98.28W	232	98.6	–2.5	37.5	9.2
Victoria Regional Airport	28.86N	96.93W	36	100.9	–1.2	35.9	10.9
Waco Municipal Airport	31.61N	97.23W	155	99.5	–5.1	38.2	11.0
Wichita Falls Municipal Airport	33.98N	98.49W	314	97.6	–9.3	39.2	12.0
Utah							
Cedar City Municipal Airport	37.70N	113.10W	1712	82.4	–15.9	34.1	11.6
Logan-Cache Airport	41.79N	111.85W	1358	86.0	–21.3	35.0	8.8
Ogden, Hill AFB	41.12N	111.97W	1459	85.0	–13.0	34.1	10.2
Price, Carbon County Airport	39.62N	110.75W	1799	81.5	–15.5	33.7	10.0
Provo Municipal Airport	40.22N	111.72W	1369	85.9	–12.8	34.8	10.8
Salt Lake City Intl Airport	40.79N	111.97W	1288	86.8	–12.6	36.3	11.2
St George Municipal Airport	37.08N	113.60W	896	91.0	–2.9	41.2	11.9
Vernal Airport	40.43N	109.52W	1608	83.4	–17.2	33.9	9.1
Vermont							
Burlington International Airport	44.47N	73.15W	104	100.1	–22.4	31.3	10.6
Montpelier, Knapp State Airport	44.20N	72.58W	342	97.3	–23.7	29.5	9.1
Virginia							
Abington, VA Highlands Airport	36.68N	82.03W	631	94.0	–11.2	32.0	7.8
Blacksburg, Virginia Tech Apt	37.22N	80.42W	650	93.8	–12.1	31.3	9.2
Chesapeake	36.90N	75.72W	23	101.1	–5.1	29.0	22.5
Danville Regional Airport	36.57N	79.34W	180	99.2	–7.5	34.0	8.3

Table 2.2: Climatic Data in SI Units ***(Continued)***

Station	Latitude, degrees	Longitude, degrees	Elevation, m	St. Br. Pressure, kPa	Winter Temp., °C	Summer Temp., °C	Wind, m/s
Fort Belvoir, Davison AAF	38.72N	77.18W	27	101.0	–10.8	35.6	9.2
Fredericksburg, Shannon Airport	38.27N	77.45W	26	101.0	–9.1	35.1	8.4
Hampton, Langley AFB	37.08N	76.37W	3	101.3	–6.4	33.9	10.5
Leesburg Executive Airport	39.08N	77.57W	119	99.9	–9.9	34.9	10.0
Lynchburg Municipal Airport	37.34N	79.21W	286	97.9	–9.8	33.4	8.2
Manassas Municipal Airport	38.72N	77.52W	59	100.6	–11.2	33.8	9.6
Martinsville, Blue Ridge Airport	36.63N	80.02W	287	97.9	–9.7	34.1	7.6
Newport News, P Henry Intl Apt	37.13N	76.49W	16	101.1	–7.4	34.7	8.9
Norfolk International Airport	36.90N	76.19W	9	101.2	–5.7	34.3	11.2
Norfolk NAS	36.93N	76.28W	5	101.3	–5.1	34.3	11.3
Petersburg Municipal Airport	37.18N	77.50W	60	100.6	–8.9	36.3	8.1
Quantico MCAS	38.50N	77.30W	4	101.3	–8.8	33.7	8.6
Richmond International Airport	37.51N	77.32W	50	100.7	–8.4	34.9	9.3
Roanoke Regional Airport	37.32N	79.97W	358	97.1	–9.9	33.4	10.1
Staunton, Shenandoah Reg Apt	38.27N	78.90W	366	97.0	–11.0	34.0	7.9
Virginia Beach, Oceana NAS	36.82N	76.03W	7	101.2	–6.0	33.8	10.7
Washington DC, Reagan Airport	38.87N	77.03W	20	101.1	–8.7	34.6	10.4
Washington Dulles Intl Airport	38.94N	77.45W	99	100.1	–11.8	34.1	9.2
Winchester Regional Airport	39.15N	78.15W	222	98.7	–9.8	34.0	10.0
Washington							
Arlington Municipal Airport	48.17N	122.17W	42	100.8	–6.2	27.9	9.3
Bellingham International Airport	48.79N	122.54W	46	100.8	–7.8	26.3	11.2
Bremerton National Airport	47.48N	122.75W	137	99.7	–4.6	29.8	8.3
Fort Lewis, Gray AAF	47.08N	122.58W	92	100.2	–7.1	30.2	7.8
Hoquiam, Bowerman Airport	46.97N	123.94W	5	101.3	–2.2	25.3	11.6
Kelso Longview Airport	46.12N	122.89W	6	101.3	–5.8	31.1	7.9
Olympia Airport	46.97N	122.90W	61	100.6	–6.6	30.7	8.4
Paine Field	47.90N	122.28W	189	99.1	–2.4	26.3	10.8
Pasco, Tri Cities Airport	46.27N	119.12W	123	99.9	–14.0	37.5	11.0
Port Angeles, Fairchild Airport	48.12N	123.50W	91	100.2	–4.3	26.4	8.0
Pullman-Moscow Reg Airport	46.75N	117.12W	778	92.3	–12.7	33.6	10.5
Seattle Tacoma Intl Airport	47.46N	122.31W	132	99.8	–4.2	29.4	9.0
Seattle, Boeing Field	47.53N	122.30W	9	101.2	–4.5	30.0	8.4
Spokane International Airport	47.62N	117.53W	721	93.0	–16.2	33.8	11.4
Spokane, Fairchild AFB	47.62N	117.65W	743	92.7	–15.2	33.0	10.6
Spokane, Felts Field	47.68N	117.32W	610	94.2	–13.9	34.8	8.8
Tacoma Narrows Airport	47.27N	122.57W	91	100.2	–1.4	28.7	8.8
Tacoma, McChord AFB	47.13N	122.48W	88	100.3	–6.3	30.1	8.8
Vancouver, Pearson Airpark	45.62N	122.65W	8	101.2	–3.7	32.6	7.5
Walla Walla City County Airport	46.10N	118.29W	367	97.0	–13.3	37.2	10.6
Wenatchee, Pangborn Airport	47.40N	120.21W	375	96.9	–14.4	35.2	10.4

Table 2.2: Climatic Data in SI Units *(Continued)*

Station	Latitude, degrees	Longitude, degrees	Elevation, m	St. Br. Pressure, kPa	Winter Temp., °C	Summer Temp., °C	Wind, m/s
Whidbey Island NAS	48.35N	122.65W	14	101.2	–5.5	22.4	12.8
Yakima Air Terminal	46.56N	120.53W	325	97.5	–14.3	35.4	10.5
West Virginia							
Beckley Airport	37.80N	81.13W	766	92.5	–14.7	29.2	8.8
Bluefield, Mercer County Airport	37.30N	81.21W	874	91.3	–14.5	29.1	7.8
Bridgeport, Benedum Airport	39.28N	80.22W	380	96.8	–12.5	32.1	8.6
Charleston, Yeager Airport	38.38N	81.59W	299	97.8	–13.0	32.9	7.7
Huntington, Tri-State Airport	38.38N	82.56W	255	98.3	–13.1	33.3	7.6
Martinsburg, Shepherd Field	39.40N	77.98W	163	99.4	–12.4	34.1	9.4
Morgantown Municipal Airport	39.64N	79.92W	382	96.8	–14.5	31.9	7.6
Parkersburg, Wood Co Airport	39.35N	81.44W	263	98.2	–14.8	32.7	8.1
Wisconsin							
Appleton, Outagamie Co Airport	44.25N	88.52W	280	98.0	–21.0	31.4	11.0
Eau Claire County Airport	44.87N	91.49W	273	98.1	–26.4	32.7	8.8
Fond Du Lac County Airport	43.77N	88.49W	246	98.4	–20.0	31.4	10.6
Green Bay, Austin Straubel Field	44.51N	88.12W	214	98.8	–23.3	31.3	10.7
Juneau, Dodge County Airport	43.43N	88.70W	285	98.0	–18.8	31.2	10.6
Kenosha Municipal Airport	42.58N	87.92W	232	98.6	–17.5	32.4	11.1
La Crosse Municipal Airport	43.75N	91.26W	200	99.0	–24.6	33.4	10.3
Madison, Dane County Airport	43.14N	89.35W	264	98.2	–22.8	32.1	10.5
Manitowoc Municipal Airport	44.13N	87.68W	198	99.0	–20.1	29.2	10.8
Marshfield Municipal Airport	44.63N	90.18W	389	96.7	–24.9	31.2	10.1
Milwaukee, Mitchell Intl Airport	42.95N	87.90W	211	98.8	–20.0	32.4	11.5
Rice Lake Municipal Airport	45.48N	91.72W	347	97.2	–24.1	32.0	8.9
Sheboygan	43.75N	87.68W	189	99.1	–19.0	28.4	18.7
Sheboygan County Airport	43.78N	87.85W	228	98.6	–18.8	31.2	11.0
Watertown Municipal Airport	43.17N	88.72W	254	98.3	–20.1	32.0	9.3
Wausau Municipal Airport	44.93N	89.63W	365	97.0	–25.3	31.2	8.6
Wisconsin Rapids, Alexander Fld	44.35N	89.83W	308	97.7	–22.3	32.0	9.1
Wittman Regional Airport	43.98N	88.55W	253	98.3	–20.9	31.3	10.4
Wyoming							
Casper Airport	42.90N	106.47W	1612	83.4	–23.5	34.3	14.4
Cheyenne Airport	41.16N	104.81W	1872	80.8	–21.3	31.8	15.1
Gillette Campbell County Airport	44.35N	105.53W	1230	87.4	–24.3	35.1	12.8
Jackson Hole Airport	43.60N	110.73W	2016	79.3	–26.2	28.9	10.9
Laramie, General Brees Field	41.31N	105.67W	2217	77.4	–22.9	29.3	14.2
Rock Springs Sweetwater Airport	41.59N	109.07W	2056	78.9	–21.2	31.1	14.0
Sheridan County Airport	44.77N	106.98W	1209	87.6	–24.8	35.3	12.1
Canada							
Alberta							
Banff	51.19N	115.55W	1397	85.6	–29.9	27.6	6.4
Bowness	51.08N	114.22W	1235	87.3	–28.0	28.2	10.5

Table 2.2: Climatic Data in SI Units *(Continued)*

Station	Latitude, degrees	Longitude, degrees	Elevation, m	St. Br. Pressure, kPa	Winter Temp., °C	Summer Temp., °C	Wind, m/s
Brooks	50.55N	111.85W	747	92.7	–30.9	31.8	10.9
Calgary International Airport	51.11N	114.02W	1084	89.0	–29.1	28.6	12.2
Calgary, Springbank Airport	51.10N	114.37W	1201	87.7	–31.7	26.8	11.1
Camrose Airport	53.03N	112.81W	739	92.8	–33.2	28.3	10.3
Cold Lake Airport	54.42N	110.28W	541	95.0	–34.9	27.9	9.3
Edmonton City Centre Airport	53.57N	113.52W	671	93.5	–29.7	28.1	9.7
Edmonton International Airport	53.32N	113.58W	723	92.9	–32.5	27.5	10.3
Edmonton Namao Airport	53.67N	113.47W	688	93.3	–30.6	27.2	10.4
Fort McMurray Airport	56.65N	111.22W	369	97.0	–36.8	28.9	8.3
Grande Prairie Airport	55.18N	118.88W	669	93.5	–36.4	27.5	11.2
Lacombe	52.45N	113.76W	860	91.4	–32.9	28.3	9.4
Lethbridge	49.70N	112.78W	921	90.7	–28.3	32.0	13.3
Lethbridge Airport	49.63N	112.80W	929	90.7	–29.8	31.4	16.2
Lloydminster Airport	53.31N	110.07W	668	93.6	–33.4	28.4	10.3
Medicine Hat Airport	50.02N	110.72W	717	93.0	–31.2	32.5	11.1
Red Deer Industrial Airport	52.18N	113.89W	905	90.9	–32.7	27.9	9.2
British Columbia							
Abbotsford Airport	49.03N	122.36W	59	100.6	–8.6	29.6	8.5
Agassiz	49.25N	121.77W	15	101.2	–7.5	30.1	10.4
Ballenas Island	49.35N	124.16W	13	101.2	–0.7	23.9	15.9
Comox Airport	49.72N	124.90W	26	101.0	–5.4	26.7	13.4
Cranbrook International Airport	49.61N	115.78W	940	90.5	–24.4	31.5	8.4
Entrance Island	49.22N	123.80W	5	101.3	–1.6	23.9	14.3
Esquimalt Harbour	48.43N	123.44W	3	101.3	–2.1	22.6	9.6
Fort St John Airport	56.24N	120.74W	695	93.3	–33.0	26.1	10.4
Kamloops Airport	50.70N	120.44W	345	97.3	–20.8	33.7	9.9
Kelowna International Airport	49.96N	119.38W	430	96.3	–18.3	32.7	7.5
Nelson	49.49N	117.31W	535	95.1	–11.9	31.7	6.9
Penticton Airport	49.46N	119.60W	344	97.3	–14.2	32.5	10.2
Pitt Meadows	49.21N	122.69W	5	101.3	–7.3	30.5	5.5
Port Alberni Airport	49.32N	124.93W	76	100.4	–6.0	31.5	6.1
Prince George Airport	53.89N	122.68W	691	93.3	–31.0	27.7	9.4
Prince Rupert Airport	54.29N	130.44W	35	100.9	–11.5	19.1	11.5
Salmon Arm	50.70N	119.29W	351	97.2	–17.4	32.3	6.3
Squamish Airport	49.78N	123.16W	52	100.7	–7.4	29.2	6.4
Summerland	49.56N	119.64W	454	96.0	–14.5	33.0	8.3
Terrace Airport	54.47N	128.58W	217	98.8	–18.5	27.7	11.5
Vancouver Harbour	49.30N	123.12W	3	101.3	–2.9	25.7	N/A
Vancouver International Airport	49.20N	123.18W	4	101.3	–7.0	25.0	10.3
Vancouver, Point Atkinson	49.33N	123.26W	35	100.9	–0.4	25.0	13.7
Vancouver, West	49.35N	123.19W	168	99.3	–5.8	27.0	5.1

Table 2.2: Climatic Data in SI Units *(Continued)*

Station	Latitude, degrees	Longitude, degrees	Elevation, m	St. Br. Pressure, kPa	Winter Temp., °C	Summer Temp., °C	Wind, m/s
Vernon	50.22N	119.19W	482	95.7	–14.8	33.1	6.9
Victoria International Airpor	48.65N	123.43W	19	101.1	–4.7	26.6	8.5
Victoria University	48.46N	123.30W	60	100.6	–2.1	27.0	5.9
Victoria, Gonzales	48.41N	123.33W	70	100.5	–2.5	24.7	12.3
Victoria, Hartland	48.53N	123.46W	154	99.5	–2.8	28.5	9.8
White Rock Campbell Scientific	49.02N	122.78W	13	101.2	–5.3	24.9	6.4
Williams Lake Airport	52.18N	122.05W	940	90.5	–28.8	28.1	9.6
Manitoba							
Brandon Airport	49.91N	99.95W	409	96.5	–33.5	30.6	11.3
Winnipeg International Airport	49.92N	97.23W	239	98.5	–32.2	30.8	12.5
New Brunswick							
Fredericton Airport	45.87N	66.53W	21	101.1	–23.6	29.7	9.8
Miramichi Airport	47.01N	65.47W	33	100.9	–23.4	30.0	11.0
Miramichi RCS	47.01N	65.46W	33	100.9	–24.1	30.6	11.3
Moncton International Airport	46.10N	64.69W	71	100.5	–22.6	28.4	12.4
Point Lepreau	45.07N	66.45W	6	101.3	–18.6	23.5	10.4
Saint John Airport	45.32N	65.89W	109	100.0	–22.4	26.0	11.9
Newfoundland and Labrador							
Corner Brook	48.93N	57.92W	152	99.5	–18.7	26.2	9.4
Gander International Airport	48.95N	54.58W	151	99.5	–19.7	26.0	14.0
Goose Bay Airport	53.32N	60.42W	49	100.7	–30.2	27.7	11.4
St. John's International Airport	47.62N	52.74W	141	99.6	–15.7	24.6	16.1
Wabush Airport	52.93N	66.87W	551	94.9	–35.8	24.8	9.4
Northwest Territories							
Yellowknife Airport	62.46N	114.44W	206	98.9	–40.9	25.2	9.5
Nova Scotia							
Amherst	45.85N	64.27W	23	101.1	–20.9	25.5	13.1
Halifax International Airport	44.88N	63.52W	145	99.6	–18.6	27.6	12.0
Halifax Shearwater Airport	44.63N	63.50W	51	100.7	–17.0	26.0	11.9
Shearwater Jetty	44.63N	63.52W	6	101.3	–13.9	25.8	12.2
Sydney Airport	46.17N	60.05W	62	100.6	–18.0	27.3	12.6
Nunavut							
Iqaluit Airport	63.75N	68.55W	34	100.9	–40.1	16.5	15.2
Ontario							
Beausoleil	44.85N	79.87W	183	99.2	–23.9	29.9	6.2
Belle River	42.30N	82.70W	184	99.1	–14.6	31.6	12.9
Burlington Piers	43.30N	79.80W	77	100.4	–15.3	30.3	10.4
Cobourg	43.95N	78.17W	78	100.4	–20.0	25.7	11.4
Collingwood	44.50N	80.22W	180	99.2	–18.7	30.2	11.1
Erieau	42.25N	81.90W	178	99.2	–14.6	26.9	13.0
Kenora Airport	49.79N	94.37W	410	96.5	–31.9	29.3	8.7

Table 2.2: Climatic Data in SI Units *(Continued)*

Station	Latitude, degrees	Longitude, degrees	Elevation, m	St. Br. Pressure, kPa	Winter Temp., °C	Summer Temp., °C	Wind, m/s
Lagoon City	44.55N	79.22W	221	98.7	–23.9	27.5	12.7
London International Airport	43.03N	81.15W	278	98.0	–18.3	30.1	10.8
North Bay Airport	46.36N	79.42W	370	97.0	–27.8	27.8	9.5
Ottawa International Airport	45.32N	75.67W	114	100.0	–24.5	30.6	10.0
Petawawa Airport	45.95N	77.32W	130	99.8	–29.2	30.5	9.0
Peterborough Airport	44.23N	78.37W	191	99.1	–23.9	29.5	8.9
Port Colborne	42.87N	79.25W	184	99.1	–15.5	26.1	18.3
Port Weller	43.25N	79.22W	79	100.4	–13.5	29.3	14.5
Sault Ste Marie Airport	46.48N	84.51W	192	99.0	–25.1	28.4	11.0
Sudbury Airport	46.62N	80.80W	348	97.2	–28.2	29.2	10.4
Thunder Bay Intl Airport	48.37N	89.33W	199	99.0	–30.0	29.0	9.9
Timmins Victor M Power Apt	48.57N	81.38W	295	97.8	–33.7	29.4	8.4
Toronto Buttonville Airport	43.86N	79.37W	198	99.0	–20.1	31.7	9.7
Toronto Island Airport	43.63N	79.40W	77	100.4	–16.1	28.5	13.3
Toronto Pearson Intl Airport	43.68N	79.63W	173	99.3	–18.8	31.2	11.9
Trenton Airport	44.12N	77.53W	86	100.3	–21.8	29.2	10.8
Windsor Airport	42.28N	82.96W	190	99.1	–16.4	31.9	11.4
Prince Edward Island							
Charlottetown Airport	46.29N	63.13W	49	100.7	–20.5	26.7	11.5
Quebec							
Bagotville Airport	48.33N	71.00W	159	99.4	–30.1	29.2	11.9
Charlevoix	47.28N	70.64W	723	92.9	–28.5	24.5	7.5
Gaspe Airport	48.78N	64.48W	34	100.9	–24.9	27.8	9.5
Iles-de-la-Madeleine Airport	47.43N	61.77W	8	101.2	–16.6	23.8	16.2
La Baie	48.30N	70.92W	152	99.5	–30.4	29.1	10.3
La Tuque Airport	47.41N	72.79W	169	99.3	–30.3	30.4	7.6
Lac Saint-Pierre	46.18N	72.92W	16	101.1	–25.5	27.6	13.2
L'Acadie	45.29N	73.35W	44	100.8	–24.1	30.1	10.4
L'Assomption	45.81N	73.43W	21	101.1	–25.9	30.4	8.5
Lennoxville	45.37N	71.82W	181	99.2	–26.0	29.5	9.0
Mont-Joli Airport	48.60N	68.22W	52	100.7	–23.9	26.7	12.6
Mont-Orford	45.31N	72.24W	846	91.6	–28.4	25.1	15.7
Montreal Mirabel Intl Airport	45.67N	74.03W	82	100.3	–26.4	29.4	8.9
Montreal St Hubert Airport	45.52N	73.42W	27	101.0	–24.3	30.1	11.4
Montreal Trudeau Intl Airport	45.47N	73.75W	36	100.9	–23.7	30.0	11.1
Montreal-Est	45.63N	73.55W	50	100.7	–23.4	30.5	8.6
Nicolet	46.23N	72.66W	8	101.2	–25.6	28.8	9.6
Pointe-Au-Pere	48.51N	68.47W	5	101.3	–22.1	23.1	13.2
Quebec Airport	46.80N	71.38W	74	100.4	–26.2	28.9	11.1
Riviere-Du-Loup	47.81N	69.55W	147	99.6	–24.5	26.2	8.2
Roberval Airport	48.52N	72.27W	179	99.2	–30.3	28.7	10.3

Table 2.2: Climatic Data in SI Units *(Continued)*

Station	Latitude, degrees	Longitude, degrees	Elevation, m	St. Br. Pressure, kPa	Winter Temp., °C	Summer Temp., °C	Wind, m/s
Rouyn	48.25N	79.03W	318	97.6	–31.0	29.0	9.1
Sainte-Agathe-des-Monts	46.05N	74.28W	395	96.7	–28.1	27.5	9.0
Sainte-Anne-de-Bellevue	45.43N	73.93W	39	100.9	–24.2	30.0	9.0
Sainte-Foy	46.78N	71.29W	91	100.2	–25.0	29.2	9.4
Sept-Îles Airport	50.22N	66.27W	55	100.7	–28.0	22.8	11.9
Sherbrooke Airport	45.43N	71.68W	241	98.5	–28.3	28.6	9.1
Trois-Rivieres	46.35N	72.52W	6	101.3	–24.1	27.4	10.7
Val D Or Airport	48.06N	77.79W	337	97.3	–33.2	28.6	9.4
Varennes	45.72N	73.38W	18	101.1	–23.8	30.3	11.0
Victoria Hartland	48.53N	123.46W	154	99.5	–29.2	29.0	10.6
Saskatchewan							
Estevan Airport	49.22N	102.97W	581	94.5	–32.7	32.2	13.1
Moose Jaw	50.33N	105.56W	577	94.6	–29.7	32.2	12.1
Moose Jaw Airport	50.33N	105.55W	577	94.6	–33.3	32.4	13.2
North Battleford Airport	52.77N	108.26W	548	94.9	–35.3	29.8	11.2
Prince Albert Airport	53.22N	105.67W	428	96.3	–36.6	29.3	9.5
Regina Airport	50.43N	104.67W	577	94.6	–34.0	31.4	13.4
Saskatoon Diefenbaker Intl Apt	52.17N	106.72W	504	95.4	–35.0	30.8	11.2
Saskatoon, Kernen Farm	52.15N	106.55W	510	95.4	–33.5	30.6	10.7
Swift Current Airport	50.30N	107.68W	818	91.9	–31.5	31.2	12.9
Yorkton Airport	51.27N	102.47W	498	95.5	–34.5	29.7	10.6
Yukon Territory							
Whitehorse Airport	60.71N	135.07W	706	93.1	–40.0	25.6	10.4
Albania							
Tirana	41.33N	19.78E	90	100.3	–2.2	34.1	6.8
Algeria							
Constantine Airport	36.28N	6.62E	694	93.3	–0.5	38.2	10.2
Dar-El-Beida Airport	36.68N	3.22E	29	101.0	1.9	35.2	10.7
Es Senia Airport	35.63N	0.60W	91	100.2	2.1	33.9	12.2
Argentina							
Buenos Aires, J Newbery Airport	34.57S	58.42W	6	101.3	4.2	31.0	11.1
Cordoba Airport	31.32S	64.22W	474	95.8	–0.1	34.5	11.8
Corrientes International Airport	27.45S	58.77W	62	100.6	4.8	36.2	10.3
Ezeiza International Airport	34.82S	58.53W	20	101.1	–0.1	33.7	10.0
Mar del Plata Airport	37.93S	57.58W	21	101.1	–1.1	31.0	11.5
Mendoza, El Plumerillo Airport	32.83S	68.78W	704	93.2	–0.5	35.4	8.8
Parana, General Urquiza Airport	31.78S	60.48W	78	100.4	2.5	34.0	11.0
Posadas Airport	27.37S	55.97W	125	99.8	4.8	36.0	8.8
Resistencia International Airport	27.45S	59.05W	52	100.7	2.0	36.9	9.0
Rosario International Airport	32.92S	60.78W	25	101.0	–0.6	34.1	11.9
Salta Airport	24.85S	65.48W	1221	87.5	–0.9	32.8	8.0

Table 2.2: Climatic Data in SI Units *(Continued)*

Station	Latitude, degrees	Longitude, degrees	Elevation, m	St. Br. Pressure, kPa	Winter Temp., °C	Summer Temp., °C	Wind, m/s
San Juan Airport	31.40S	68.42W	598	94.3	–1.9	37.9	13.9
Santiago del Estero Airport	27.77S	64.30W	199	99.0	–0.6	38.9	10.5
Sauce Viejo Airport	31.70S	60.82W	18	101.1	0.5	34.8	15.2
Tucuman, Matienzo Airport	26.85S	65.10W	450	96.0	3.2	36.0	9.1
Armenia							
Yerevan Airport	40.13N	44.47E	1140	88.4	–13.9	36.0	10.0
Australia							
Adelaide Airport	34.95S	138.53E	8	101.2	4.0	35.8	11.7
Adelaide International Airport	34.92S	138.62E	51	100.7	4.8	36.3	8.4
Brisbane International Airport	27.38S	153.13E	10	101.2	5.9	31.1	9.8
Brisbane, Archerfield Airport	27.57S	153.00E	13	101.2	5.2	33.1	9.1
Canberra International Airport	35.30S	149.20E	580	94.6	–3.3	33.2	10.6
Canberra, Isabella Plains	35.42S	149.10E	588	94.5	–4.0	33.6	8.5
Coolangatta Airport	28.17S	153.50E	6	101.3	6.5	29.2	9.7
Gold Coast Seaway	27.93S	153.43E	3	101.3	9.7	30.7	12.6
Laverton Airport	37.87S	144.75E	20	101.1	1.8	34.1	12.2
Melbourne	37.82S	144.97E	32	100.9	4.7	34.6	7.6
Melbourne International Airport	37.67S	144.85E	119	99.9	2.8	34.6	13.8
Moorabbin Airport	37.98S	145.10E	13	101.2	2.5	33.8	11.8
Newcastle Nobbys Si	32.92S	151.78E	33	100.9	7.7	30.3	18.2
Perth International Airport	31.93S	115.97E	20	101.1	4.2	37.1	11.0
Perth, Jandakot Airport	32.10S	115.88E	31	101.0	1.8	36.0	10.5
Perth, Mount Lawley	31.92S	115.87E	25	101.0	3.9	36.2	8.4
Perth, Swanbourne	31.95S	115.77E	20	101.1	6.4	34.7	13.6
Scoresby Research	37.87S	145.25E	90	100.3	2.3	33.6	8.4
Sydney	33.85S	151.20E	40	100.9	7.2	31.1	N/A
Sydney International Airport	33.93S	151.18E	5	101.3	6.0	32.8	12.6
Sydney, Bankstown Airport	33.92S	150.98E	8	101.2	3.2	33.8	9.8
Sydney, Homebush	33.85S	151.07E	28	101.0	5.9	33.8	9.7
Williamtown Airport (RAAF)	32.80S	151.83E	8	101.2	4.1	33.9	12.3
Austria							
Gumpoldskirchen	48.03N	16.28E	233	98.6	–9.9	30.9	7.9
Tulln	48.32N	16.12E	176	99.2	–12.8	30.9	11.8
Vienna Downtown	48.20N	16.37E	171	99.3	–8.2	31.5	9.0
Vienna International Airport	48.12N	16.57E	190	99.1	–11.5	30.9	12.2
Vienna, Hohe Warte	48.25N	16.37E	200	99.0	–10.3	30.6	9.9
Belarus							
Brest	52.12N	23.68E	146	99.6	–18.6	29.6	7.6
Gomel	52.40N	30.95E	126	99.8	–21.1	29.2	8.3
Grodno Southeast Airport	53.60N	24.05E	134	99.7	–20.2	28.3	10.8
Minsk	53.93N	27.63E	231	98.6	–20.6	28.2	8.0

Table 2.2: Climatic Data in SI Units *(Continued)*

Station	Latitude, degrees	Longitude, degrees	Elevation, m	St. Br. Pressure, kPa	Winter Temp., °C	Summer Temp., °C	Wind, m/s
Mogilev	53.95N	30.07E	192	99.0	–22.7	27.7	10.0
Vitebsk	55.17N	30.22E	176	99.2	–22.5	27.4	8.3
Belgium							
Antwerp Deurne Airport	51.20N	4.47E	14	101.2	–7.5	29.2	10.2
Brussels National Airport	50.90N	4.53E	58	100.6	–7.7	29.0	11.5
Brussels, Uccle	50.80N	4.35E	104	100.1	–7.3	28.7	9.7
Benin							
Cotonou Cadjehoun Airport	6.35N	2.38E	9	101.2	21.9	32.6	8.2
Bolivia							
Cochabamba Intl Airport	17.42S	66.18W	2548	74.2	1.8	29.9	9.7
La Paz, El Alto Intel Airport	16.52S	68.18W	4038	61.3	–4.5	17.4	8.7
Santa Cruz, Viru Viru Intl Apt	17.63S	63.13W	373	96.9	9.2	34.2	13.1
Bosnia and Herzegovina							
Bjelasnica	43.72N	18.27E	2070	78.8	–19.8	18.6	34.0
Sarajevo Butmir Intl Airport	43.82N	18.33E	511	95.3	–13.6	32.1	8.3
Sarajevo, Bjelave	43.87N	18.43E	638	93.9	–12.4	31.3	5.3
Brazil							
Anapolis AFB	16.23S	48.97W	1137	88.4	12.8	31.7	6.6
Aracaju, Santa Maria Airport	10.98S	37.07W	9	101.2	21.1	32.1	7.3
Belem, International Airport	1.38S	48.48W	16	101.1	22.8	33.1	8.4
Belo Horizonte	19.93S	43.93W	850	91.5	11.0	32.1	7.9
Belo Horizonte Airport	19.85S	43.95W	785	92.2	11.5	33.0	6.2
Brasilia International Airport	15.87S	47.93W	1061	89.2	9.8	32.1	7.4
Campo Grande Intl Airport	20.47S	54.67W	567	94.7	8.5	36.2	10.4
Cuiaba, Marechal Rondon Apt	15.65S	56.10W	187	99.1	13.1	38.0	7.7
Curitiba International Airport	25.52S	49.17W	908	90.9	2.7	30.9	8.5
Florianopolis, Hercílio Luz Apt	27.67S	48.55W	5	101.3	7.8	32.2	8.3
Fortaleza, Pinto Martins Airport	3.78S	38.53W	25	101.0	22.8	32.2	9.3
Goiania, Santa Genoveva Airport	16.63S	49.22W	747	92.7	11.9	35.0	8.1
Guarulhos International Airport	23.43S	46.47W	768	92.4	7.2	32.8	7.6
Londrina Airport	23.33S	51.13W	569	94.7	7.8	33.9	6.5
Macapa	0.03N	51.05W	15	101.2	22.8	34.0	8.4
Maceio Airport	9.52S	35.78W	117	99.9	19.1	33.0	7.6
Manaus, Eduardo Gomes Airport	3.03S	60.05W	2	101.3	21.8	35.9	5.8
Manaus, Ponta Pelada Airport	3.15S	59.98W	84	100.3	22.1	34.7	6.2
Natal International Airport	5.92S	35.25W	52	100.7	21.0	32.9	10.0
Porto Alegre, Salgado Filho Apt	30.00S	51.18W	3	101.3	4.1	34.9	9.2
Porto Velho Airport	8.77S	63.92W	102	100.1	18.0	35.4	6.1
Recife, Guararapes Intl Airport	8.07S	34.85W	19	101.1	21.8	34.0	8.1
Rio de Janeiro Intl Airport	22.82S	43.25W	6	101.3	14.9	37.9	8.3
Rio de Janeiro, Dumont Airport	22.90S	43.17W	3	101.3	16.2	34.1	8.4

Table 2.2: Climatic Data in SI Units *(Continued)*

Station	Latitude, degrees	Longitude, degrees	Elevation, m	St. Br. Pressure, kPa	Winter Temp., °C	Summer Temp., °C	Wind, m/s
Salvador International Airport	12.90S	38.33W	6	101.3	20.2	32.4	9.1
Sao Luis International Airport	2.60S	44.23W	53	100.7	22.8	34.1	9.0
Sao Paulo Campinas Intl Airport	23.00S	47.13W	661	93.6	8.8	33.2	11.3
Sao Paulo, Congonhas Intl Apt	23.62S	46.65W	803	92.0	8.9	32.1	7.5
Teresina Airport	5.05S	42.82W	69	100.5	21.9	38.1	5.2
Vitoria, Goiabeiras Airport	20.27S	40.28W	4	101.3	16.6	34.0	10.1
Bulgaria							
Plovdiv	42.13N	24.75E	185	99.1	–10.2	34.2	12.1
Sofia	42.65N	23.38E	591	94.4	–12.5	32.0	9.4
Varna	43.20N	27.92E	43	100.8	–8.9	31.0	13.5
Burkina Faso							
Bobo Dioulasso Airport	11.17N	4.32W	460	95.9	17.8	38.0	7.4
Ouagadougou Airport	12.35N	1.52W	306	97.7	16.1	40.5	7.6
Chad							
Ndjamena Airport	12.13N	15.03E	295	97.8	13.0	42.9	9.5
Chile							
Antofagasta, Cerro Moreno Airport	23.43S	70.45W	140	99.7	10.0	24.8	9.4
Santiago, Pudahuel Intl Airport	33.38S	70.78W	474	95.8	–1.1	31.8	8.4
China							
Anqing	30.53N	117.05E	20	101.1	–1.9	35.6	8.1
Anyang	36.05N	114.40E	64	100.6	–8.0	35.0	7.5
Baoding	38.85N	115.57E	19	101.1	–10.0	35.2	6.4
Baoji	34.35N	107.13E	610	94.2	–6.0	34.7	6.2
Beijing	39.93N	116.28E	55	100.7	–10.8	34.9	9.6
Bengbu	32.95N	117.37E	22	101.1	–5.1	35.6	7.2
Benxi	41.32N	123.78E	185	99.1	–22.8	31.5	6.8
Cangzhou	38.33N	116.83E	11	101.2	–10.1	34.3	9.1
Changchun, Dafang Shen Airport	43.90N	125.22E	238	98.5	–24.8	30.7	12.0
Changde	29.05N	111.68E	35	100.9	–0.9	36.1	6.0
Changsha	28.23N	112.87E	68	100.5	–1.0	36.0	7.2
Chaoyang	41.55N	120.45E	176	99.2	–19.7	33.7	9.3
Chengde	40.98N	117.95E	386	96.8	–17.9	33.0	6.4
Chengdu	30.67N	104.02E	508	95.4	0.4	33.1	5.4
Chifeng	42.27N	118.97E	572	94.6	–20.2	32.7	8.3
Chongqing	29.58N	106.47E	260	98.2	3.2	36.9	5.1
Dalian	38.90N	121.63E	97	100.2	–12.2	31.1	11.2
Dandong	40.05N	124.33E	14	101.2	–16.4	29.8	9.0
Datong	40.10N	113.33E	1069	89.1	–20.9	31.5	9.9
Dezhou	37.43N	116.32E	22	101.1	–9.2	34.2	7.4
Fuzhou	26.08N	119.28E	85	100.3	4.4	35.5	8.3
Ganyu	34.83N	119.13E	10	101.2	–7.0	33.1	7.5

Table 2.2: Climatic Data in SI Units *(Continued)*

Station	Latitude, degrees	Longitude, degrees	Elevation, m	St. Br. Pressure, kPa	Winter Temp., °C	Summer Temp., °C	Wind, m/s
Gaoyao	23.05N	112.47E	12	101.2	6.5	35.0	6.9
Guangzhou	23.17N	113.33E	42	100.8	5.8	35.0	6.8
Guilin	25.33N	110.30E	166	99.4	1.3	34.7	8.1
Guiyang	26.58N	106.73E	1223	87.5	–2.3	30.3	6.5
Haikou Airport	20.03N	110.35E	24	101.0	10.7	35.0	6.5
Hangzhou	30.23N	120.17E	43	100.8	–2.2	36.1	7.2
Harbin	45.75N	126.77E	143	99.6	–28.4	31.1	9.1
Hefei	31.87N	117.23E	36	100.9	–4.1	35.2	7.6
Hohhot	40.82N	111.68E	1065	89.2	–21.9	31.6	8.6
Hong Kong	22.30N	114.17E	62	100.6	9.6	32.2	8.6
Hong Kong International Airport	22.32N	113.92E	8	101.2	9.0	33.8	10.2
Hong Kong, Taipa Macau	22.15N	113.60E	6	101.3	7.2	32.9	11.0
Jiangling	30.33N	112.18E	33	100.9	–1.5	34.8	7.1
Jinan	36.60N	117.05E	169	99.3	–8.2	35.0	9.2
Jingdezhen	29.30N	117.20E	60	100.6	–1.7	36.1	5.9
Jinzhou	41.13N	121.12E	70	100.5	–16.4	31.6	10.1
Jixi	45.28N	130.95E	234	98.5	–25.2	30.5	10.5
Kunming	25.02N	102.68E	1892	80.6	0.3	27.1	8.0
Lanzhou	36.05N	103.88E	1518	84.4	–11.7	32.2	4.4
Lingxian	37.33N	116.57E	19	101.1	–11.0	35.1	8.6
Liuzhou	24.35N	109.40E	97	100.2	3.4	35.2	5.6
Mengjin	34.82N	112.43E	333	97.4	–6.8	34.7	9.3
Mudanjiang	44.57N	129.60E	242	98.5	–26.4	31.1	9.3
Nanchang	28.60N	115.92E	50	100.7	–0.8	35.7	6.4
Nanjing	32.00N	118.80E	7	101.2	–4.8	35.1	7.6
Nanning	22.82N	108.35E	126	99.8	5.2	35.0	6.2
Neijiang	29.58N	105.05E	357	97.1	2.3	35.2	5.3
Qingdao	36.07N	120.33E	77	100.4	–7.7	32.0	11.8
Qingjiang	33.60N	119.03E	19	101.1	–6.2	33.6	7.0
Qiqihar	47.38N	123.92E	148	99.6	–27.7	31.7	9.9
Shanghai	31.40N	121.47E	4	101.3	–1.9	34.9	7.8
Shanghai, Hongqiao	31.17N	121.43E	7	101.2	–3.0	35.1	9.0
Shantou	23.40N	116.68E	3	101.3	7.1	33.5	8.2
Shaoguan	24.80N	113.58E	68	100.5	2.6	35.5	6.7
Shenyang	41.73N	123.52E	43	100.8	–22.1	31.6	9.7
Shenyang Taoxian Intl Airport	41.63N	123.48E	62	100.6	–25.0	32.1	10.8
Shenzhen	22.55N	114.10E	18	101.1	6.9	33.8	8.1
Shijiazhuang	38.03N	114.42E	81	100.4	–9.1	35.7	6.6
Siping	43.18N	124.33E	167	99.3	–23.3	31.0	9.3
Taishan	36.25N	117.10E	1536	84.2	–16.8	22.5	18.4
Taiyuan	37.78N	112.55E	779	92.3	–15.1	32.9	9.3

Table 2.2: Climatic Data in SI Units *(Continued)*

Station	Latitude, degrees	Longitude, degrees	Elevation, m	St. Br. Pressure, kPa	Winter Temp., °C	Summer Temp., °C	Wind, m/s
Tangshan	39.67N	118.15E	29	101.0	–12.9	33.2	8.2
Tianjin	39.10N	117.17E	5	101.3	–10.2	34.2	8.4
Tianjin, Binhai Intl Airport	39.12N	117.33E	2	101.3	–10.9	34.2	10.2
Urumqi Diwopu Intl Airport	43.90N	87.47E	664	93.6	–23.2	35.4	7.3
Weifang	36.77N	119.18E	22	101.1	–11.1	34.3	10.1
Wenzhou	28.02N	120.67E	7	101.2	1.1	33.8	6.5
Wuhan	30.62N	114.13E	23	101.1	–2.2	35.7	7.0
Wuhu	31.33N	118.35E	16	101.1	–3.3	35.8	7.9
Wulumuqi	43.80N	87.65E	947	90.5	–22.7	33.4	7.8
Xiamen	24.48N	118.08E	139	99.7	6.3	34.0	8.9
Xian	34.30N	108.93E	398	96.6	–6.3	35.9	7.8
Xihua	33.78N	114.52E	53	100.7	–5.9	35.0	6.7
Xingtai	37.07N	114.50E	78	100.4	–7.9	35.5	5.9
Xining	36.62N	101.77E	2296	76.6	–16.2	27.4	6.2
Xinyang	32.13N	114.05E	115	100.0	–4.6	34.5	8.4
Xuzhou	34.28N	117.15E	42	100.8	–6.6	34.6	6.9
Yangjiang	21.87N	111.97E	22	101.1	7.2	33.0	8.1
Yanji	42.87N	129.50E	178	99.2	–22.9	31.0	10.1
Yichang	30.70N	111.30E	134	99.7	–0.8	35.6	4.7
Yinchuan	38.47N	106.20E	1112	88.7	–17.0	32.1	8.9
Yingkou	40.67N	122.20E	4	101.3	–17.7	30.6	10.7
Yueyang	29.38N	113.08E	52	100.7	–0.9	34.4	7.2
Yuncheng	35.05N	111.05E	365	97.0	–8.1	36.3	9.4
Zhangjiakou	40.78N	114.88E	726	92.9	–16.7	32.6	7.3
Zhanjiang	21.22N	110.40E	28	101.0	7.8	33.9	8.0
Zhengzhou	34.72N	113.65E	111	100.0	–6.7	35.1	8.4
Zunyi	27.70N	106.88E	845	91.6	–1.2	32.5	4.8
Colombia							
Barranquilla Intl Airport	10.88N	74.78W	30	101.0	22.8	34.1	13.2
Bogota, Eldorado Intl Airport	4.70N	74.13W	2546	74.3	2.8	21.2	8.4
Cali, Aragon Intl Airport	3.55N	76.38W	969	90.2	17.7	32.1	8.4
Cartagena, Rafael Nunez Airport	10.45N	75.52W	12	101.2	23.0	32.3	9.2
Medellín, J M Cordova Airport	6.13N	75.43W	2142	78.1	10.0	23.9	9.1
Congo							
Brazzaville, Maya-Maya Airport	4.25S	15.25E	316	97.6	18.0	34.0	5.9
Costa Rica							
Juan Santamaria Intl Airport	9.98N	84.22W	934	90.6	16.6	30.9	12.7
Cote d'Ivoire							
Abidjan Port Bouet Airpot	5.25N	3.93W	8	101.2	21.0	32.8	7.2
Croatia							
Zagreb, Maksimir	45.82N	16.03E	128	99.8	–11.1	31.6	6.0

Table 2.2: Climatic Data in SI Units *(Continued)*

Station	Latitude, degrees	Longitude, degrees	Elevation, m	St. Br. Pressure, kPa	Winter Temp., °C	Summer Temp., °C	Wind, m/s
Zagreb, Pleso Airport	45.73N	16.07E	107	100.1	–12.2	31.9	8.5
Cuba							
Havana, Jose Marti Intl Airport	22.98N	82.40W	75	100.4	11.0	33.1	10.4
Camaguey Intl Airport	21.42N	77.85W	118	99.9	15.2	33.3	10.4
Santiago de Cuba Airport	19.97N	75.85W	55	100.7	18.8	31.9	10.4
Czech Republic							
Brno, Turany Airport	49.15N	16.70E	246	98.4	–13.5	30.1	10.2
Ostrava, Mosnov Airport	49.68N	18.12E	260	98.2	–16.1	30.0	10.2
Prague, Kbely Airport	50.12N	14.53E	287	97.9	–13.0	29.6	9.1
Prague, Libus	50.02N	14.45E	303	97.7	–13.5	30.1	8.7
Prague, Ruzyne Airport	50.10N	14.25E	365	97.0	–14.6	29.2	12.2
Denmark							
Copenhagen Kastrup Airport	55.62N	12.65E	5	101.3	–9.2	25.5	12.7
Copenhagen Roskilde Airport	55.58N	12.13E	43	100.8	–10.1	25.9	12.5
Copenhagen, Drogden	55.53N	12.72E	0	101.3	–8.0	22.1	18.2
Vaerloese Airport	55.77N	12.33E	31	101.0	–12.1	26.2	12.3
Dominican Republic							
Santo Domingo	18.43N	69.88W	14	101.2	19.5	32.5	7.4
Santo Domingo, Americas Apt	18.43N	69.67W	18	101.1	18.5	32.6	7.3
Ecuador							
Guayaquil International Airport	2.15S	79.88W	9	101.2	18.8	33.1	7.3
Quito International Airport	0.13S	78.48W	2812	71.8	6.8	21.9	7.7
Egypt							
Alexandria, Nouzha	31.20N	29.95E	7	101.2	6.9	33.0	10.2
Assiut Airport	27.05N	31.02E	70	100.5	4.3	40.7	10.6
Cairo International Airport	30.13N	31.40E	74	100.4	7.7	38.1	9.4
Luxor International Airport	25.67N	32.70E	99	100.1	5.0	43.1	7.2
Port Said	31.27N	32.30E	6	101.3	9.3	32.1	10.8
Port Said Airport	31.28N	32.23E	6	101.3	9.7	31.7	11.9
Estonia							
Tallinn	59.47N	24.82E	34	100.9	–19.1	25.9	9.2
Finland							
Helsinki Vantaa Airport	60.32N	24.97E	56	100.7	–22.8	26.7	10.0
Isosaari	60.10N	25.07E	5	101.3	–20.2	22.8	15.7
France							
Cap Couronne	43.33N	5.05E	27	101.0	–2.9	30.7	17.1
Cap Ferrat	43.68N	7.33E	144	99.6	3.2	29.0	13.4
Cap Pomegues	43.27N	5.30E	70	100.5	–1.4	28.6	23.5
Le Bourget Airport	48.97N	2.43E	52	100.7	–4.2	31.1	10.2
Lyon, Bron Airport	45.72N	4.93E	202	98.9	–5.4	33.6	11.5
Lyon, Satolas Airport	45.73N	5.08E	240	98.5	–6.9	32.4	10.8

Table 2.2: Climatic Data in SI Units *(Continued)*

Station	Latitude, degrees	Longitude, degrees	Elevation, m	St. Br. Pressure, kPa	Winter Temp., °C	Summer Temp., °C	Wind, m/s
Marignane	43.45N	5.23E	32	100.9	–3.1	32.7	16.5
Nice	43.65N	7.20E	27	101.0	1.8	29.5	11.7
Paris, Charles de Gaulle Intl Apt	49.02N	2.53E	112	100.0	–6.2	30.6	11.6
Paris, Montsouris	48.82N	2.33E	77	100.4	–2.5	31.5	7.3
Paris, Orly International Airport	48.72N	2.38E	90	100.3	–5.9	30.9	11.0
Toulouse, Blagnac Airport	43.63N	1.37E	154	99.5	–4.3	33.1	10.4
Trappes	48.77N	2.00E	168	99.3	–4.1	30.3	7.0
Villacoublay, Velizy Airport	48.77N	2.20E	179	99.2	–5.9	29.8	9.8
Gabon							
Libreville International Airport	0.45N	9.42E	15	101.2	21.9	31.4	7.1
Gambia							
Banjul/Yundum	13.20N	16.63W	33	100.9	16.2	37.8	8.9
Georgia							
Tbilisi International Airport	41.68N	44.95E	448	96.1	–5.7	34.4	20.8
Germany							
Berlin, Dahlem	52.47N	13.30E	51	100.7	–12.0	29.3	7.4
Berlin, Schonefeld	52.38N	13.52E	47	100.8	–13.9	29.6	11.1
Berlin, Tegel Airport	52.57N	13.32E	37	100.9	–12.4	30.0	10.4
Berlin, Tempelhof Airport	52.47N	13.40E	50	100.7	–11.8	30.0	10.4
Bremen Airport	53.05N	8.80E	3	101.3	–10.9	28.4	11.4
Celle Airport	52.60N	10.02E	52	100.7	–11.8	30.1	9.3
Dresden, Klotzsche Airport	51.13N	13.77E	230	98.6	–13.6	29.4	9.6
Dusseldorf Airport	51.28N	6.78E	45	100.8	–9.9	29.6	10.4
Essen/Mulheim	51.40N	6.97E	154	99.5	–9.9	28.2	9.7
Frankfurt International Airport	50.05N	8.60E	112	100.0	–10.5	30.8	10.1
Furstenfeldbruck Airport	48.20N	11.27E	535	95.1	–15.1	29.1	11.1
Guetersloh Airport	51.93N	8.32E	72	100.5	–9.9	29.9	10.1
Hamburg Fuhlsbuettel Airport	53.63N	10.00E	16	101.1	–11.6	27.8	10.2
Hannover Airport	52.47N	9.70E	55	100.7	–12.7	28.9	10.2
Heidelberg AAF	49.40N	8.65E	109	100.0	–9.8	32.0	7.8
Koln Bonn Airport	50.87N	7.17E	91	100.2	–10.6	29.9	9.0
Leipzig	51.32N	12.42E	151	99.5	–9.5	30.4	6.8
Leipzig Airport	51.42N	12.23E	133	99.7	–13.3	29.8	12.5
Munich	48.13N	11.55E	520	95.2	–11.9	29.5	7.8
Munich, Riem	48.13N	11.70E	529	95.1	–14.2	29.4	11.5
Norvenich Airport	50.83N	6.67E	135	99.7	–9.2	30.2	10.2
Nuremberg Airport	49.50N	11.08E	319	97.6	–14.4	30.2	9.2
Potsdam	52.38N	13.07E	81	100.4	–12.9	29.3	10.8
Quickborn	53.73N	9.88E	17	101.1	–9.6	28.4	8.9
Roth Airport	49.22N	11.10E	395	96.7	–14.2	30.8	8.4
Stuttgart Echterdingen Airport	48.68N	9.22E	396	96.7	–12.7	29.3	9.4

Table 2.2: Climatic Data in SI Units *(Continued)*

Station	Latitude, degrees	Longitude, degrees	Elevation, m	St. Br. Pressure, kPa	Winter Temp., °C	Summer Temp., °C	Wind, m/s
Stuttgart/Schnarren	48.83N	9.20E	315	97.6	–11.5	29.6	9.1
Wunstorf	52.47N	9.43E	51	100.7	–11.2	30.2	10.8
Greece							
Athens, Ellinikon Airport	37.90N	23.73E	15	101.2	1.6	35.1	10.0
Elefsis Airport	38.07N	23.55E	31	101.0	0.8	36.2	10.2
Thessaloniki, Makedonia Airport	40.52N	22.97E	4	101.3	–3.2	34.1	12.4
Guatemala							
Guatemala International Airport	14.58N	90.52W	1489	84.7	10.8	28.0	12.2
Honduras							
San Pedro Sula, La Mesa Airport	15.45N	87.93W	31	101.0	17.2	37.0	8.9
Tegucigalpa, Toncontín Airport	14.05N	87.22W	1007	89.8	11.5	32.0	9.3
Hungary							
Budaors Airport	47.45N	18.97E	132	99.8	–11.2	31.0	13.9
Budapest, Ferihegy Airport	47.43N	19.27E	185	99.1	–12.7	32.2	13.9
Budapest, Pestszentl	47.43N	19.18E	139	99.7	–10.6	32.3	7.7
India							
Ahmadabad International Airport	23.07N	72.63E	55	100.7	10.9	42.0	6.4
Akola Airport	20.70N	77.07E	309	97.7	12.9	43.2	5.7
Aurangabad Chikalthan	19.85N	75.40E	579	94.6	10.5	40.0	9.3
Bangalore	12.97N	77.58E	921	90.7	15.1	34.2	5.7
Belgaum, Sambra	15.85N	74.62E	747	92.7	13.3	36.3	8.4
Bhopal Airport	23.28N	77.35E	523	95.2	10.2	41.7	9.2
Bhubaneswar Airport	20.25N	85.83E	46	100.8	14.0	38.5	10.3
Bikaner	28.00N	73.30E	224	98.7	5.6	44.2	7.0
Bombay, Santacruz	19.12N	72.85E	14	101.2	16.5	35.8	6.9
Calcutta, Dum Dum	22.65N	88.45E	6	101.3	11.6	37.2	5.7
Coimbatore, Peelamedu Airport	11.03N	77.05E	399	96.6	18.0	36.7	10.2
CWC Vishakhapatnam	17.70N	83.30E	66	100.5	20.1	33.7	8.4
Guwahati Airport	26.10N	91.58E	54	100.7	10.8	34.5	5.1
Gwalior	26.23N	78.25E	207	98.9	6.0	43.7	4.8
Hyderabad Airport	17.45N	78.47E	545	95.0	13.9	40.2	8.2
Indore Airport	22.72N	75.80E	567	94.7	9.1	40.8	11.2
Jabalpur	23.20N	79.95E	393	96.7	8.4	42.4	4.2
Jaipur Sanganer Airport	26.82N	75.80E	390	96.7	7.1	42.4	7.1
Jamshedpur	22.82N	86.18E	142	99.6	10.0	42.3	3.6
Jodhpur	26.30N	73.02E	224	98.7	8.9	42.6	5.8
Kozhikode	11.25N	75.78E	5	101.3	22.1	33.7	6.7
Lucknow Amausi	26.75N	80.88E	128	99.8	6.8	42.1	7.4
Madras Chennai Airport	13.00N	80.18E	16	101.1	20.0	38.5	8.3
Mangalore Bajpe	12.92N	74.88E	102	100.1	20.6	34.3	8.0
Nagpur, Sonegaon Airport	21.10N	79.05E	310	97.7	11.7	43.7	8.0

Table 2.2: Climatic Data in SI Units *(Continued)*

Station	Latitude, degrees	Longitude, degrees	Elevation, m	St. Br. Pressure, kPa	Winter Temp., °C	Summer Temp., °C	Wind, m/s
Nellore	14.45N	79.98E	20	101.1	20.3	40.6	5.4
New Delhi/Palam Intl Airport	28.57N	77.12E	233	98.6	6.2	43.8	8.1
New Delhi/Safdarjung	28.58N	77.20E	216	98.8	6.3	42.0	7.1
Pataliputra	25.60N	85.10E	60	100.6	8.2	41.0	6.4
Patiala	30.33N	76.47E	251	98.4	5.0	41.4	4.4
Pune	18.53N	73.85E	559	94.8	9.7	38.1	5.2
Rajkot Airport	22.30N	70.78E	138	99.7	11.8	41.0	11.0
Sholapur	17.67N	75.90E	479	95.7	16.0	40.9	3.4
Surat	21.20N	72.83E	12	101.2	14.4	37.8	5.6
Thiruvananthapuram	8.48N	76.95E	64	100.6	22.1	33.7	6.1
Tiruchirapalli Airport	10.77N	78.72E	88	100.3	20.0	38.9	11.6
Indonesia							
Denpasar, Ngurah Rai Intl Apt	8.75S	115.17E	1	101.3	21.8	32.6	8.0
Jakarta, Soekarno–Hatta Airport	6.12S	106.65E	8	101.2	22.0	33.8	9.7
Makassar, Hasanuddin Intl Apt	5.07S	119.55E	14	101.2	20.4	34.1	7.3
Medan, Polonia Intl Airport	3.57N	98.68E	25	101.0	22.3	34.1	6.1
Menado, Sam Ratulangi Intl Apt	1.53N	124.92E	80	100.4	20.3	32.9	7.6
Padang, Tabing Airport	0.88S	100.35E	3	101.3	21.4	32.2	5.4
Pekanbaru, SSQ II Intl Airport	0.47N	101.45E	31	101.0	21.7	34.4	6.0
Surabaya, Juanda Intl Airport	7.37S	112.77E	3	101.3	20.9	34.2	8.3
Iran, Islamic Republic of							
Abadan Airport	30.37N	48.25E	6	101.3	4.9	47.7	10.5
Ahwaz	31.33N	48.67E	22	101.1	5.0	47.6	9.3
Anzali	37.47N	49.47E	–26	101.6	1.9	30.3	11.1
Arak	34.10N	49.77E	1708	82.4	–15.7	36.2	8.4
Bandar Abbas Intel Airport	27.22N	56.37E	10	101.2	9.3	41.8	8.6
Esfahan	32.47N	51.67E	1550	84.0	–7.0	39.0	10.2
Hamedan	34.85N	48.53E	1749	82.0	–16.8	35.5	10.3
Kashan	33.98N	51.45E	982	90.1	–2.8	41.7	6.2
Kerman Airport	30.25N	56.97E	1754	82.0	–6.8	38.0	11.5
Kermanshah	34.27N	47.12E	1322	86.4	–7.3	39.6	9.7
Mashhad	36.27N	59.63E	999	89.9	–7.5	37.1	9.1
Orumieh	37.53N	45.08E	1316	86.5	–10.7	32.8	9.0
Shiraz	29.53N	52.53E	1481	84.8	–1.8	39.1	9.8
Tabriz	38.08N	46.28E	1361	86.0	–10.9	35.2	10.4
Tehran, Mehrabad Intl Airport	35.68N	51.32E	1191	87.8	–2.8	38.5	11.4
Zahedan	29.47N	60.88E	1370	85.9	–4.6	39.1	12.0
Zanjan	36.68N	48.48E	1663	82.9	–13.4	34.0	9.8
Ireland							
Casement Airport	53.30N	6.43W	93	100.2	–2.6	23.1	15.2
Dublin Airport	53.43N	6.25W	85	100.3	–1.9	22.1	13.4

Table 2.2: Climatic Data in SI Units ***(Continued)***

Station	Latitude, degrees	Longitude, degrees	Elevation, m	St. Br. Pressure, kPa	Winter Temp., °C	Summer Temp., °C	Wind, m/s
Israel							
Tel Aviv, Ben Gurion Intl Apt	32.00N	34.90E	49	100.7	5.0	34.9	10.1
Tel Aviv, Sde Dov Airport	32.10N	34.78E	4	101.3	7.1	31.2	12.0
Italy							
Bari, Palese Macchie Airport	41.13N	16.78E	49	100.7	0.9	33.8	9.5
Bologna, G Marconi Airport	44.53N	11.30E	49	100.7	–4.8	34.1	7.2
Catania, Fontanarossa Airport	37.47N	15.05E	17	101.1	1.6	35.0	9.9
Catania, Sigonella Airport	37.40N	14.92E	22	101.1	2.0	36.1	12.1
Florence Peretola Airport	43.80N	11.20E	38	100.9	–3.1	35.1	8.7
Genoa Cristoforo Colombo Apt	44.42N	8.85E	3	101.3	1.0	29.9	11.7
Grazzanise Airport	41.05N	14.07E	10	101.2	–0.9	32.1	10.0
Milan, Linate Airport	45.43N	9.28E	103	100.1	–5.1	33.0	7.3
Naples	40.90N	14.30E	93	100.2	2.0	33.5	8.2
Naples Capodichino Airport	40.85N	14.30E	72	100.5	0.6	33.1	9.4
Palermo Punta Raisi Airport	38.18N	13.10E	21	101.1	6.8	33.5	13.1
Rome, Ciampino	41.78N	12.58E	105	100.1	–1.2	33.2	11.3
Rome, Fiumicino Airport	41.80N	12.23E	3	101.3	–0.8	31.0	11.5
Rome, Pratica Di Mare AFB	41.65N	12.45E	21	101.1	0.8	30.9	10.1
Turin, Bric Della Croce	45.03N	7.73E	710	93.1	–4.4	28.2	8.7
Turin-Caselle Airport	45.22N	7.65E	287	97.9	–6.0	31.0	6.3
Jamaica							
Kingston, Norman Manley Apt	17.93N	76.78W	14	101.2	22.1	33.3	14.4
Japan							
Akita	39.72N	140.10E	21	101.1	–5.5	31.5	12.3
Asahikawa	43.77N	142.37E	116	99.9	–18.4	29.9	7.1
Ashiya Air Field	33.88N	130.65E	33	100.9	–1.1	32.2	10.4
Atsugi NAS	35.45N	139.45E	65	100.6	–1.2	32.9	10.4
Chiba	35.60N	140.10E	19	101.1	0.4	32.3	12.3
Fukuoka	33.58N	130.38E	15	101.2	0.6	33.4	8.4
Fukuoka Airport	33.58N	130.45E	12	101.2	–0.7	33.9	9.3
Fukuyama	34.45N	133.25E	3	101.3	–2.7	33.8	6.1
Fushiki	36.80N	137.05E	13	101.2	–2.4	33.4	7.5
Futenma MCAS	26.27N	127.75E	84	100.3	10.8	32.2	11.2
Gifu	35.40N	136.77E	17	101.1	–1.4	35.1	8.0
Gifu Air Field	35.38N	136.87E	42	100.8	–3.2	34.1	8.0
Hamamatsu	34.72N	137.72E	33	100.9	0.1	32.7	8.9
Hamamatsu AB	34.75N	137.70E	48	100.8	–1.1	32.8	9.7
Himeji	34.83N	134.67E	40	100.9	–2.3	33.4	8.6
Hiroshima	34.40N	132.47E	53	100.7	–1.1	33.4	9.3
Iizuka	33.65N	130.70E	38	100.9	–1.8	33.6	7.3
Iruma AB	35.83N	139.42E	93	100.2	–4.1	33.9	9.7

Table 2.2: Climatic Data in SI Units *(Continued)*

Station	Latitude, degrees	Longitude, degrees	Elevation, m	St. Br. Pressure, kPa	Winter Temp., °C	Summer Temp., °C	Wind, m/s
Kadena Air Base	26.35N	127.77E	45	100.8	10.0	33.2	11.6
Kagoshima	31.55N	130.55E	32	100.9	0.8	33.2	8.9
Kanazawa	36.58N	136.63E	33	100.9	–1.8	33.1	11.7
Kansai International Airport	34.43N	135.25E	8	101.2	1.8	33.0	12.8
Kobe	34.70N	135.22E	30	101.0	–0.5	32.9	9.4
Kochi	33.57N	133.55E	5	101.3	–1.0	32.6	5.8
Komatsu Airport	36.40N	136.40E	9	101.2	–2.2	33.1	11.1
Kumagaya	36.15N	139.38E	31	101.0	–2.2	35.2	7.9
Kumamoto	32.82N	130.70E	39	100.9	–1.9	34.2	7.2
Kure	34.23N	132.55E	5	101.3	–0.1	32.4	7.0
Kyoto	35.02N	135.73E	46	100.8	–1.1	34.6	5.2
Matsuyama	33.85N	132.78E	34	100.9	–0.3	33.1	5.9
Miyazaki	31.93N	131.42E	15	101.2	–0.4	33.7	9.1
Nagano	36.67N	138.20E	419	96.4	–6.9	32.6	7.9
Nagasaki	32.73N	129.87E	35	100.9	0.6	32.5	7.8
Nagoya	35.17N	136.97E	56	100.7	–1.5	34.3	8.6
Nagoya Airport	35.25N	136.92E	17	101.1	–2.2	34.5	9.8
Naha	26.20N	127.68E	53	100.7	11.7	32.1	13.1
Naha Airport	26.20N	127.65E	6	101.3	12.0	32.2	13.3
NARA	34.70N	135.83E	106	100.1	–2.2	34.1	4.6
Niigata	37.92N	139.05E	6	101.3	–2.2	32.8	10.4
Nyutabaru AB	32.08N	131.45E	82	100.3	–1.8	32.2	9.9
Oita	33.23N	131.62E	13	101.2	–0.7	33.2	7.2
Okayama	34.67N	133.92E	18	101.1	–1.2	34.4	9.9
Onahama	36.95N	140.90E	5	101.3	–2.7	28.9	8.3
Osaka	34.68N	135.52E	83	100.3	0.3	34.2	9.2
Osaka International Airport	34.78N	135.43E	15	101.2	–1.8	34.2	8.4
Otaru	43.18N	141.02E	26	101.0	–9.7	28.0	8.0
Ozuki Air Field	34.05N	131.05E	7	101.2	–0.9	32.2	11.0
Sapporo	43.07N	141.33E	26	101.0	–10.6	29.2	9.1
Sendai	38.27N	140.90E	43	100.8	–4.2	30.8	10.3
Shimofusa AB	35.80N	140.02E	33	100.9	–2.9	33.2	10.6
Shimonoseki	33.95N	130.93E	19	101.1	1.1	31.8	10.3
Shizuhama AB	34.82N	138.30E	10	101.2	–0.9	32.8	10.8
Shizuoka	34.98N	138.40E	15	101.2	–0.3	33.0	6.3
Sumoto	34.33N	134.90E	112	100.0	–0.1	31.8	7.2
Tadotsu	34.28N	133.75E	5	101.3	–0.1	33.6	7.5
Takamatsu	34.32N	134.05E	10	101.2	–0.9	34.0	8.1
Tokyo	35.68N	139.77E	36	100.9	0.5	33.2	8.6
Tokyo International Airport	35.55N	139.78E	9	101.2	0.1	32.9	12.5
Toyama	36.72N	137.20E	17	101.1	–2.9	33.2	9.2

Table 2.2: Climatic Data in SI Units *(Continued)*

Station	Latitude, degrees	Longitude, degrees	Elevation, m	St. Br. Pressure, kPa	Winter Temp., °C	Summer Temp., °C	Wind, m/s
Tsuiki Airport	33.68N	131.05E	20	101.1	–2.2	32.1	10.0
Utsunomiya	36.55N	139.87E	140	99.7	–4.5	33.0	9.2
Wakayama	34.23N	135.17E	18	101.1	0.4	33.1	10.8
Yokohama	35.43N	139.65E	42	100.8	0.3	32.1	9.5
Yokosuka	35.28N	139.67E	53	100.7	0.9	33.2	13.5
Yokota AB	35.75N	139.35E	139	99.7	–4.0	33.9	9.1
Jordan							
Amman Airport	31.98N	35.98E	779	92.3	1.0	35.3	10.2
Irbid	32.55N	35.85E	616	94.1	2.0	34.3	9.0
Queen Alia International Airport	31.72N	35.98E	722	93.0	–0.8	36.8	12.2
Kazakhstan							
Almaty	43.23N	76.93E	851	91.5	–19.9	33.9	5.9
Astana	51.13N	71.37E	350	97.2	–30.8	31.9	11.4
Karaganda Southeast Airport	49.80N	73.15E	553	94.9	–29.4	32.0	11.1
Pavlodar	52.30N	76.93E	122	99.9	–32.4	32.6	9.4
Shymkent	42.32N	69.70E	604	94.3	–13.9	37.1	7.9
Taraz (was Zhambyl)	42.85N	71.38E	655	93.7	–19.6	35.6	11.5
Kenya							
Mombasa	4.03S	39.62E	55	100.7	19.9	33.0	8.7
Nairobi, Kenyatta International Apt	1.32S	36.92E	1624	83.3	9.8	29.0	9.8
Korea, North							
Chongjin	41.78N	129.82E	43	100.8	–13.4	27.5	7.3
Hamheung	39.93N	127.55E	22	101.1	–13.4	31.4	8.2
Kaesong	37.97N	126.57E	70	100.5	–13.2	30.9	8.3
Namp'o	38.72N	125.38E	47	100.8	–12.9	30.1	10.0
Pyongyang	39.03N	125.78E	36	100.9	–15.5	31.2	6.7
Sinuiju	40.10N	124.38E	7	101.2	–15.7	30.8	7.8
Wonsan	39.18N	127.43E	36	100.9	–10.5	31.5	7.7
Korea, South							
Busan	35.10N	129.03E	70	100.5	–5.6	31.2	10.2
Cheongju	36.63N	127.45E	59	100.6	–11.8	32.8	6.7
Cheongju International Airport	36.72N	127.50E	60	100.6	–14.0	33.2	7.2
Daegu	35.88N	128.62E	59	100.6	–7.6	34.1	8.4
Daegu International Airport	35.90N	128.67E	35	100.9	–9.1	35.0	8.5
Daejeon	36.37N	127.37E	72	100.5	–11.0	32.7	7.1
Gimhae International Airport	35.18N	128.93E	5	101.3	–6.8	32.8	9.2
Gwangju	35.17N	126.90E	74	100.4	–6.8	32.4	7.6
Gwangju, Kwangju Intl Airport	35.12N	126.82E	13	101.2	–7.8	34.1	7.4
Incheon	37.47N	126.63E	70	100.5	–10.7	31.1	9.5
Jeju	33.52N	126.53E	23	101.1	0.0	31.7	11.1
Jeju International Airport	33.52N	126.50E	24	101.0	–0.8	31.9	12.3

Table 2.2: Climatic Data in SI Units *(Continued)*

Station	Latitude, degrees	Longitude, degrees	Elevation, m	St. Br. Pressure, kPa	Winter Temp., °C	Summer Temp., °C	Wind, m/s
Jeonju	35.82N	127.15E	55	100.7	–8.9	33.1	5.6
Jinju	35.20N	128.12E	23	101.1	–8.6	32.8	7.2
Masan	35.18N	128.57E	4	101.3	–4.7	32.4	6.8
Osan AB	37.10N	127.03E	12	101.2	–13.1	33.0	8.2
Pohang	36.03N	129.38E	4	101.3	–6.7	33.6	8.5
Pohang Airport	35.98N	129.42E	20	101.1	–7.1	34.0	9.6
Pyongtaek Airport A-511	36.97N	127.03E	14	101.2	–13.1	32.9	7.9
Sachon Airport	35.08N	128.08E	8	101.2	–10.0	33.1	7.0
Seogwipo	33.25N	126.57E	51	100.7	–0.1	31.3	8.4
Seoul	37.57N	126.97E	86	100.3	–11.6	32.1	7.2
Seoul, Korea AF HQ	37.50N	126.93E	49	100.7	–11.8	33.2	6.3
Seoul AB	37.43N	127.12E	20	101.1	–14.8	33.2	6.2
Seoul, Gimpo Intl Airport	37.57N	126.78E	17	101.1	–13.8	32.0	8.4
Suwon	37.27N	126.98E	35	100.9	–12.2	32.1	6.4
Ulsan	35.55N	129.32E	36	100.9	–6.2	33.1	7.0
Yeosu	34.73N	127.75E	67	100.5	–5.2	30.5	12.1
Kyrgyzstan							
Bishkek	42.85N	74.53E	760	92.5	–20.4	35.2	8.5
Latvia							
Riga	56.97N	24.05E	26	101.0	–18.5	27.4	10.0
Riga International Airport	56.92N	23.97E	11	101.2	–18.2	28.9	9.1
Lebanon							
Beirut International Airport	33.82N	35.48E	19	101.1	7.6	31.8	11.2
Libyan Arab Jamahiriya							
Benina International Airport	32.10N	20.27E	132	99.8	6.8	37.2	14.5
Misurata	32.42N	15.05E	32	100.9	8.0	36.5	13.1
Tripoli International Airport	32.70N	13.08E	63	100.6	4.2	41.9	10.4
Lithuania							
Kaunas	54.88N	23.83E	770	92.4	–19.7	27.8	9.8
Vilnius Airport	54.63N	25.28E	156	99.5	–20.2	28.0	10.3
Macedonia (Yugoslav)							
Skopje Airport	41.97N	21.65E	239	98.5	–12.7	35.6	8.9
Madagascar							
Antananarivo Ivato Airport	18.80S	47.48E	1276	86.9	7.9	29.2	8.1
Malaysia							
Kota Kinabalu Intl Airport	5.93N	116.05E	3	101.3	22.7	33.3	7.0
Kuala Lumpur, Subang Airport	3.12N	101.55E	22	101.1	22.0	34.2	6.4
Kuantan Airport	3.78N	103.22E	16	101.1	21.3	33.9	6.6
Kuching Airport	1.48N	110.33E	27	101.0	21.9	33.9	5.3
Sandakan Airport	5.90N	118.07E	13	101.2	22.9	33.5	7.1
Tawau Airport	4.27N	117.88E	20	101.1	22.0	32.3	6.0

Table 2.2: Climatic Data in SI Units *(Continued)*

Station	Latitude, degrees	Longitude, degrees	Elevation, m	St. Br. Pressure, kPa	Winter Temp., °C	Summer Temp., °C	Wind, m/s
Mali							
Bamako Senou Airport	12.53N	7.95W	381	96.8	15.2	40.1	8.5
Mauritania							
Nouakchott Airport	18.10N	15.95W	3	101.3	12.9	41.2	10.2
Mexico							
Acapulco, Gen J N Alvarez Apt	16.75N	99.75W	5	101.3	19.8	33.7	8.7
Apodaca, Gen M Escobedo Apt	25.77N	100.10W	399	96.6	3.2	38.9	14.1
Cancun International Airport	21.03N	86.87W	6	101.3	13.8	34.1	10.5
De Guanajuato, Del Bajío Apt	20.98N	101.48W	1861	80.9	4.0	34.0	12.6
Guadalajara International Airport	20.52N	103.30W	1566	83.9	1.9	33.6	10.3
Mazatlan, Gen R Buelna Intl Apt	23.15N	106.25W	5	101.3	10.9	34.4	10.4
Merida	20.98N	89.65W	9	101.2	13.7	38.2	10.5
Mexico City	19.43N	99.13W	2235	77.2	4.1	29.0	21.1
Mexico City Intl Airport	19.43N	99.07W	2286	76.7	3.0	29.2	11.6
Monterrey	25.73N	100.30W	515	95.3	4.3	38.2	5.5
San Luis Potosi	22.18N	100.98W	1883	80.7	–0.2	32.3	10.0
Tampico, Gen F J Mina Airport	22.28N	97.87W	25	101.0	10.1	34.1	14.9
Tijuana, Gen A L Rodriguez Apt	32.53N	116.97W	156	99.5	5.9	32.0	8.4
Toluca, Uruapan Intl Airport	19.33N	99.57W	2643	73.4	–2.0	26.2	9.1
Veracruz, Gen H Jara Airport	19.13N	96.18W	33	100.9	14.9	35.2	20.1
Moldova, Republic of							
Kishinev	47.02N	28.98E	173	99.3	–14.3	31.1	6.4
Mongolia							
Ulaanbaatar	47.92N	106.87E	1306	86.6	–34.1	29.9	10.3
Morocco							
Agadir	30.38N	9.57W	23	101.1	5.0	35.2	10.6
Agadir Al Massira Airport	30.32N	9.40W	23	101.1	5.1	37.6	9.5
Casablanca	33.57N	7.67W	57	100.6	6.0	29.5	7.8
Casablanca Airport	33.37N	7.58W	206	98.9	3.2	35.2	9.9
Fes, Saiss Airport	33.93N	4.98W	579	94.6	0.8	39.1	9.9
Marrakech	31.62N	8.03W	466	95.9	4.1	41.2	8.4
Meknes, Bassatine Airport	33.88N	5.53W	560	94.8	2.2	38.6	8.4
Oujda	34.78N	1.93W	470	95.8	0.2	37.2	12.1
Rabat, Sale	34.05N	6.77W	79	100.4	5.0	32.2	8.2
Tanger, Boukhalf Airport	35.73N	5.90W	21	101.1	4.1	33.1	17.8
Tetouan, Sania Ramel Airport	35.58N	5.33W	10	101.2	6.3	32.8	12.6
Mozambique							
Maputo International Airport	25.92S	32.57E	44	100.8	12.0	35.2	16.0
Netherlands							
Amsterdam Schiphol Airport	52.30N	4.77E	–4	101.4	–7.3	27.8	13.6
Hoek Van Holland	51.98N	4.10E	14	101.2	–6.3	27.2	16.3

Table 2.2: Climatic Data in SI Units *(Continued)*

Station	Latitude, degrees	Longitude, degrees	Elevation, m	St. Br. Pressure, kPa	Winter Temp., °C	Summer Temp., °C	Wind, m/s
IJmuiden	52.47N	4.57E	13	101.2	–6.8	25.5	18.6
Rotterdam Hague Airport	51.95N	4.45E	–4	101.4	–7.3	27.9	12.8
Valkenburg	52.18N	4.42E	2	101.3	–7.2	27.1	13.8
Woensdrecht AB	51.45N	4.33E	17	101.1	–7.2	29.7	9.8
New Zealand							
Auckland Aero AWS	37.00S	174.80E	7	101.2	4.5	25.3	13.0
Auckland Airport	37.02S	174.80E	6	101.3	1.8	25.2	13.0
Christchurch Aero A	43.48S	172.52E	37	100.9	–2.5	27.6	11.5
Christchurch Airport	43.48S	172.55E	30	101.0	–2.6	28.0	11.4
Nicaragua							
Managua, A C Sandino Intl Apt	12.15N	86.17W	56	100.7	19.8	36.0	7.9
Niger							
Niamey, Diori Hamani Airport	13.48N	2.17E	227	98.6	15.7	42.2	9.7
Norway							
Oslo Fornebu Airport	59.90N	10.62E	17	101.1	–17.2	26.8	8.5
Oslo-Blindern	59.95N	10.72E	96	100.2	–13.9	26.4	8.1
Oman							
Buraimi Airport	24.23N	55.78E	299	97.8	9.4	45.4	8.4
Pakistan							
Islamabad, Benazir Bhutto Apt	33.62N	73.10E	508	95.4	2.2	41.1	12.0
Karachi, Rudra Mata Airport	24.90N	67.13E	22	101.1	10.0	38.9	9.1
Lahore, Allama Iqbal Airport	31.52N	74.40E	217	98.7	3.9	43.2	8.1
Palestinian Ter, Occupied							
Atarot Airport	31.87N	35.22E	759	92.5	0.8	32.9	9.9
Panama							
Panama City, Albrook Intl Airport	8.97N	79.55W	10	101.2	22.8	34.8	7.8
Tocumen, Panama City Metro Apt	9.05N	79.37W	45	100.8	20.0	34.0	7.6
Paraguay							
Asuncion, Silvio Pettirossi Apt	25.25S	57.52W	101	100.1	5.1	36.9	10.4
Peru							
Arequipa, Rodriguez Ballon Apt	16.33S	71.57W	2520	74.5	5.8	24.1	10.2
Chiclayo Airport	6.78S	79.82W	30	101.0	15.0	32.2	10.5
Cuzco	13.53S	71.93W	3249	67.9	0.0	22.9	9.2
Iquitos Airport	3.78S	73.30W	126	99.8	19.0	34.1	6.2
Lima-Callao Airport	12.00S	77.12W	13	101.2	14.0	29.3	9.4
Piura	5.20S	80.60W	55	100.7	15.9	34.1	8.9
Pucallpa Airport	8.37S	74.57W	149	99.6	17.6	34.8	6.6
Trujillo Airport	8.08S	79.10W	30	101.0	14.7	29.0	8.6
Philippines							
Cagayan de Oro	8.48N	124.63E	6	101.3	22.0	34.6	4.9
Davao, F Bangoy Intl Airport	7.12N	125.65E	18	101.1	22.4	33.7	8.0

Table 2.2: Climatic Data in SI Units *(Continued)*

Station	Latitude, degrees	Longitude, degrees	Elevation, m	St. Br. Pressure, kPa	Winter Temp., °C	Summer Temp., °C	Wind, m/s
General Santo City	6.12N	125.18E	15	101.2	22.6	35.1	6.0
Iloilo City	10.70N	122.57E	8	101.2	22.8	34.7	7.6
Mactan-Cebu Intl Airport	10.30N	123.97E	24	101.0	23.0	33.3	8.2
Manila	14.58N	120.98E	13	101.2	23.1	34.5	9.9
Ninoy Aquino Intl Airport	14.52N	121.00E	15	101.2	20.9	34.9	16.3
Sangley Point AB	14.50N	120.92E	2	101.3	23.1	34.7	9.3
Science Garden	14.63N	121.02E	46	100.8	20.2	35.2	5.9
Zamboanga City	6.90N	122.07E	6	101.3	22.4	34.1	5.8
Poland							
Gdansk, Rebiechowo Airport	54.38N	18.47E	138	99.7	–16.1	27.1	12.7
Gdansk, Swibno	54.33N	18.93E	7	101.2	–17.0	25.8	10.2
Hel	54.60N	18.82E	3	101.3	–9.6	25.2	10.3
Katowice Airport	50.23N	19.03E	284	98.0	–15.4	29.1	8.3
Krakow	50.08N	19.80E	237	98.5	–16.9	29.7	9.3
Lodz Lublinek Airport	51.73N	19.40E	190	99.1	–15.8	29.5	9.2
Lublin Radawiec Airport	51.22N	22.40E	240	98.5	–17.6	28.4	8.6
Poznan, Lawica Airport	52.42N	16.85E	84	100.3	–14.3	30.0	9.8
Raciborz	50.05N	18.20E	206	98.9	–16.0	29.6	10.1
Szczecin	53.40N	14.62E	7	101.2	–12.9	28.9	9.5
Terespol	52.07N	23.62E	137	99.7	–19.5	29.2	7.4
Warsaw Chopin Airport	52.17N	20.97E	106	100.1	–16.6	29.6	10.3
Wrocław Strachowice Airport	51.10N	16.88E	124	99.8	–15.2	29.8	8.9
Portugal							
Lisbon Portela Airport	38.77N	9.13W	114	100.0	4.2	34.2	10.4
Puerto Rico							
San Juan International Airport	18.42N	66.00W	19	101.1	20.6	33.0	8.5
San Juan, Luis Muñoz Marín Apt	18.43N	66.00W	3	101.3	20.9	32.8	9.2
Qatar							
Doha International Airport	25.25N	51.57E	10	101.2	11.1	43.7	10.8
Romania							
Bucharest, Afumati	44.48N	26.18E	90	100.3	–14.0	33.2	10.3
Bucharest inmh-bane	44.48N	26.12E	91	100.2	–12.9	33.8	8.9
Cluj-Napoca	46.78N	23.57E	413	96.5	–15.4	30.0	8.3
Constanta	44.22N	28.65E	14	101.2	–9.3	29.7	13.4
Craiova Airport	44.32N	23.87E	195	99.0	–12.3	33.4	13.6
Iasi North Airport	47.17N	27.63E	104	100.1	–15.9	32.0	9.8
Kogalniceanu Airport	44.33N	28.43E	102	100.1	–11.3	31.9	11.3
Timisoara	45.77N	21.25E	88	100.3	–12.3	33.5	8.5
Russian Federation							
Adler	43.43N	39.90E	13	101.2	–2.3	29.7	7.9
Arhangelsk	64.55N	40.58E	4	101.3	–33.8	27.0	8.1

Table 2.2: Climatic Data in SI Units *(Continued)*

Station	Latitude, degrees	Longitude, degrees	Elevation, m	St. Br. Pressure, kPa	Winter Temp., °C	Summer Temp., °C	Wind, m/s
Astrahan	46.28N	48.05E	–23	101.6	–18.3	34.8	10.3
Barnaul	53.43N	83.52E	184	99.1	–32.2	29.9	11.3
Bryansk	53.25N	34.32E	216	98.8	–22.6	28.1	9.4
Chelyabinsk Balandino Airport	55.30N	61.53E	227	98.6	–29.0	30.1	11.2
Cherepovec	59.25N	37.97E	114	100.0	–30.9	27.3	9.3
Chita	52.08N	113.48E	671	93.5	–36.6	30.4	10.2
Elabuga	55.77N	52.07E	192	99.0	–29.0	29.6	13.4
Irkutsk	52.27N	104.32E	469	95.8	–35.0	28.1	10.1
Izhevsk Southeast Airport	56.83N	53.45E	159	99.4	–30.1	29.3	10.2
Kaliningrad	54.72N	20.55E	21	101.1	–18.2	27.7	8.5
Kaluga	54.57N	36.40E	201	98.9	–25.2	27.6	9.2
Kazan International Airport	55.60N	49.28E	116	99.9	–28.4	29.6	11.5
Kemerovo	55.23N	86.12E	260	98.2	–32.9	28.7	10.9
Khabarovsk Novy Airport	48.52N	135.17E	76	100.4	–30.0	30.2	10.4
Kirov	58.65N	49.62E	164	99.4	–32.8	27.8	9.9
Kirov	58.60N	49.63E	158	99.4	–29.7	29.1	6.3
Krasnodar	45.03N	39.15E	34	100.9	–14.9	33.6	10.4
Krasnoyarsk	56.00N	92.88E	277	98.0	–33.7	28.4	10.1
Krasnoyarsk Opytnoe	56.03N	92.75E	276	98.1	–37.5	29.2	7.3
Kurgan Airport	55.47N	65.40E	79	100.4	–32.5	30.9	11.5
Kursk	51.77N	36.17E	247	98.4	–22.9	29.0	10.2
Magnitogorsk	53.35N	59.08E	382	96.8	–29.4	30.1	10.2
Makhachkala	43.02N	47.48E	32	100.9	–11.1	31.2	11.2
Moscow	55.83N	37.62E	156	99.5	–23.1	28.4	6.8
Moscow Sheremtyevo Airport	55.97N	37.42E	197	99.0	–25.2	28.2	9.3
Murmansk	68.97N	33.05E	51	100.7	–31.9	24.1	11.2
Nizhniy Novgorod, Gorkij Apt	56.22N	43.82E	82	100.3	–27.4	28.4	9.3
Nizhny Novgorod	56.27N	44.00E	157	99.5	–27.1	29.3	7.2
Nizhnyj Tagil	57.88N	60.07E	258	98.3	–32.1	28.4	7.4
Novokuznetsk Airport	53.82N	86.88E	308	97.7	–31.9	29.0	12.1
Novosibirsk	55.08N	82.90E	176	99.2	–34.4	29.2	10.7
Omsk	55.02N	73.38E	122	99.9	–32.7	31.0	10.6
Orel Southwest Airport	52.93N	36.00E	203	98.9	–24.2	28.6	10.6
Orenburg	51.68N	55.10E	117	99.9	–29.2	33.8	10.8
Penza South Airport	53.12N	45.02E	174	99.3	–27.0	30.4	10.4
Perm, Bakharerka Airport	57.95N	56.20E	170	99.3	–30.8	29.5	10.0
Rostov-on-Don, Rostov E Apt	47.25N	39.82E	77	100.4	–17.0	33.1	12.7
Ryazan	54.62N	39.72E	160	99.4	–23.5	28.4	9.7
Ryazan	54.63N	39.70E	158	99.4	–25.6	29.1	6.7
Samara	53.25N	50.45E	40	100.9	–27.2	31.5	10.6
Saratov Airport	51.57N	46.03E	156	99.5	–23.1	31.6	10.6

Table 2.2: Climatic Data in SI Units *(Continued)*

Station	Latitude, degrees	Longitude, degrees	Elevation, m	St. Br. Pressure, kPa	Winter Temp., °C	Summer Temp., °C	Wind, m/s
Smolensk	54.75N	32.07E	239	98.5	–23.1	26.9	7.4
St Petersburg	59.97N	30.30E	6	101.3	–23.2	27.3	8.8
Stavropol	45.12N	42.08E	452	96.0	–17.1	33.2	12.9
Surgut	61.25N	73.50E	56	100.7	–40.8	28.3	10.2
Tomsk	56.50N	84.92E	139	99.7	–36.2	28.4	10.0
Tula	54.23N	37.62E	204	98.9	–25.1	29.0	7.3
Tver	56.90N	35.88E	146	99.6	–26.2	28.4	9.4
Tyumen	57.12N	65.43E	104	100.1	–32.4	29.4	6.5
Ufa	54.72N	55.83E	104	100.1	–31.5	30.8	10.3
Ulan-Ude	51.83N	107.60E	515	95.3	–36.1	30.8	11.8
Ulyanovsk	54.32N	48.33E	127	99.8	–28.4	30.3	11.3
Vladikavkaz	43.05N	44.65E	703	93.2	–14.2	29.9	5.1
Vladimir	56.12N	40.35E	170	99.3	–26.7	28.2	9.3
Vladivostok	43.12N	131.93E	183	99.2	–24.5	28.1	13.7
Vnukovo	55.58N	37.25E	214	98.8	–24.1	28.1	9.9
Volgograd, Gumrak Airport	48.78N	44.37E	134	99.7	–21.8	33.8	12.8
Voronezh	51.70N	39.22E	149	99.6	–23.7	30.9	8.1
Voronezh East Airport	51.65N	39.25E	104	100.1	–23.9	29.4	10.9
Yekaterinburg	56.83N	60.63E	283	98.0	–30.2	29.1	9.0
Saudi Arabia							
Abha Airport	18.23N	42.65E	2093	78.6	5.9	30.9	9.7
Buraidah, Gassim Airport	26.30N	43.77E	648	93.8	3.2	44.2	9.2
Dhahran International Airport	26.27N	50.17E	17	101.1	7.8	44.2	11.1
Jeddah, King Abdulaziz Airport	21.70N	39.18E	17	101.1	15.1	40.9	10.0
Khamis Mushait Airport	18.30N	42.80E	2056	79.0	5.8	31.8	9.4
Mecca	21.43N	39.77E	240	98.5	16.0	45.1	6.3
Medina Airport	24.55N	39.70E	636	93.9	9.0	45.0	9.3
Riyadh	24.70N	46.73E	620	94.1	5.9	44.2	9.5
Tabuk Airport	28.38N	36.60E	768	92.4	1.8	40.8	10.4
Senegal							
Dakar Yoff International Airport	14.73N	17.50W	24	101.0	16.5	32.1	9.9
Serbia							
Belgrade	44.80N	20.47E	132	99.8	–8.9	33.7	7.4
Belgrade Nikola Tesla Airport	44.82N	20.28E	99	100.1	–11.0	33.8	10.4
Singapore							
Singapore, Changi Airport	1.37N	103.98E	16	101.1	23.0	33.2	7.5
Slovakia							
Bratislava Letisko	48.20N	17.20E	134	99.7	–11.9	32.0	10.0
South Africa							
Bloemfontein International Airport	29.10S	26.30E	1354	86.1	–4.6	33.9	9.4
Cape Town International Airport	33.97S	18.60E	42	100.8	3.8	31.0	14.0

Table 2.2: Climatic Data in SI Units ***(Continued)***

Station	Latitude, degrees	Longitude, degrees	Elevation, m	St. Br. Pressure, kPa	Winter Temp., °C	Summer Temp., °C	Wind, m/s
Durban International Airport	29.97S	30.95E	14	101.2	9.4	30.2	11.3
East London Airport	33.03S	27.83E	125	99.8	8.1	30.2	12.8
Johannesburg, Tambo Intl Apt	26.15S	28.23E	1720	82.3	0.2	29.0	9.3
Port Elizabeth Airport	33.98S	25.62E	63	100.6	5.4	29.2	14.5
Pretoria, Eendracht	25.73S	28.18E	1326	86.4	3.0	32.1	5.7
Pretoria, Irene	25.92S	28.22E	1523	84.3	2.7	30.6	8.9
Spain							
Alicante Airport	38.28N	0.55W	31	101.0	3.3	32.8	10.2
Barcelona Airport	41.28N	2.07E	6	101.3	0.9	30.2	9.5
Madrid, Barajas Airport	40.45N	3.55W	582	94.5	–4.1	36.2	9.7
Madrid, Torrejon	40.48N	3.45W	611	94.2	–4.2	36.8	9.2
Malaga Airport	36.67N	4.48W	7	101.2	3.9	34.8	11.3
Murcia	38.00N	1.17W	62	100.6	2.4	35.8	8.0
Palma Mallorca Airport	39.55N	2.73E	7	101.2	–0.1	33.2	10.2
Palmas de Gran Canaria	27.93N	15.38W	47	100.8	13.2	30.1	14.4
Seville, San Pablo Airport	37.42N	5.90W	31	101.0	1.3	39.9	8.8
Sondika, Bilbao Airport	43.30N	2.90W	39	100.9	–0.2	32.7	9.8
Valencia	39.50N	0.47W	62	100.6	1.0	33.1	11.4
Valladolid	41.65N	4.77W	735	92.8	–4.2	34.3	8.3
Zaragoza AB	41.67N	1.05W	263	98.2	–2.2	36.1	12.5
Zaragoza Airport	41.67N	1.00W	258	98.3	–2.9	36.1	13.4
Sri Lanka							
Katunayake, Bandaranaike Apt	7.17N	79.88E	8	101.2	20.9	33.1	9.0
Sweden							
Gothenburg	57.72N	12.00E	2	101.3	–11.7	26.7	9.0
Gothenburg City Airport	57.78N	11.88E	16	101.1	–14.8	25.8	11.3
Gothenburg, Landvetter Airport	57.67N	12.30E	169	99.3	–14.1	26.0	11.2
Stockholm, Bromma Airport	59.37N	17.90E	14	101.2	–17.1	27.0	9.0
Switzerland							
Laegern	47.48N	8.40E	843	91.6	–11.7	26.1	12.6
Zurich, Kloten	47.48N	8.53E	432	96.2	–10.1	30.1	8.5
Zurich, MeteoSwiss	47.38N	8.57E	569	94.7	–9.3	28.8	8.9
Syrian Arab Republic							
Aleppo International Airport	36.18N	37.20E	384	96.8	–2.0	38.8	10.3
Damascus International Airport	33.42N	36.52E	609	94.2	–3.5	39.1	12.1
Daraa	32.60N	36.10E	543	95.0	1.0	35.8	8.3
Hama	35.12N	36.75E	303	97.7	–1.4	38.9	7.1
Latakia	35.53N	35.77E	7	101.2	4.1	32.2	9.9
Taiwan, Province of China							
Chi-lung	25.15N	121.80E	3	101.3	10.2	33.8	9.3
Chinmem / Shatou AFB	24.43N	118.37E	9	101.2	6.9	33.1	9.7

Table 2.2: Climatic Data in SI Units ***(Continued)***

Station	Latitude, degrees	Longitude, degrees	Elevation, m	St. Br. Pressure, kPa	Winter Temp., °C	Summer Temp., °C	Wind, m/s
Hsinchu AFB	24.82N	120.93E	8	101.2	9.0	33.1	13.3
Hsinchu City	24.83N	120.93E	27	101.0	8.8	33.9	9.8
Kangshan AFB	22.78N	120.27E	10	101.2	10.0	33.2	8.4
Kaohsiung	22.63N	120.28E	29	101.0	12.4	32.8	7.5
Kaohsiung International Airport	22.58N	120.35E	9	101.2	11.8	33.2	8.9
Pingtung North Airport	22.70N	120.48E	29	101.0	11.1	34.2	7.2
Pingtung South Airport	22.68N	120.47E	24	101.0	11.8	34.9	7.3
Taichung AFB	24.18N	120.65E	112	100.0	7.9	34.2	9.2
Tainan	23.00N	120.22E	14	101.2	10.7	33.5	8.5
Tainan AFB	22.95N	120.20E	19	101.1	10.2	33.3	9.2
Taipei	25.03N	121.52E	9	101.2	9.6	34.9	7.9
Taipei, Sungshan Airport	25.07N	121.55E	6	101.3	9.0	34.9	8.8
Taiwan Taoyuan Intl Airport	25.08N	121.22E	33	100.9	8.9	34.2	12.8
Taizhong	24.15N	120.68E	78	100.4	9.5	33.4	5.0
Taoyuan AB	25.07N	121.23E	45	100.8	8.5	33.9	11.9
Wu-Chi Observatory	24.25N	120.52E	5	101.3	9.9	32.8	16.0
Wuchia Observatory	24.27N	120.62E	5	101.3	7.9	32.2	11.9
Tajikistan							
Dushanbe	38.55N	68.78E	800	92.1	–7.1	37.4	6.3
Tanzania							
Dar Es Salaam Airport	6.87S	39.20E	53	100.7	17.7	33.1	8.6
Thailand							
Bangkok	13.73N	100.57E	4	101.3	19.0	35.8	6.3
Bangkok International Airport	13.92N	100.60E	12	101.2	19.0	37.2	8.1
Togo							
Lome Tokoin Airport	6.17N	1.25E	25	101.0	21.0	33.1	8.3
Tunisia							
Tunis-Carthage Airport	36.83N	10.23E	4	101.3	5.0	37.3	11.8
Turkey							
Adana	36.98N	35.30E	20	101.1	1.2	36.6	8.0
Adana, Incirlik AB	37.00N	35.43E	73	100.5	–0.1	36.8	8.4
Adana, Incirlik AFB	37.00N	35.42E	76	100.4	0.2	36.1	8.5
Antalya	36.87N	30.73E	54	100.7	1.4	38.0	10.5
Bursa	40.18N	29.07E	100	100.1	–3.8	34.1	7.4
Diyarbakır	37.88N	40.18E	677	93.5	–9.0	40.1	8.9
Erzurum Airport	39.95N	41.17E	1758	81.9	–29.6	30.0	10.4
Esenboga Airport	40.12N	33.00E	949	90.4	–15.7	33.0	8.9
Eskisehir Airport	39.78N	30.57E	786	92.2	–11.0	33.0	8.8
Etimesgut Airport	39.95N	32.68E	806	92.0	–11.2	34.1	9.2
Gaziantep	37.08N	37.37E	701	93.2	–4.8	38.7	8.1
Istanbul, Ataturk Airport	40.97N	28.82E	37	100.9	–2.6	31.1	11.1

Table 2.2: Climatic Data in SI Units *(Continued)*

Station	Latitude, degrees	Longitude, degrees	Elevation, m	St. Br. Pressure, kPa	Winter Temp., °C	Summer Temp., °C	Wind, m/s
Izmir, Cigli	38.52N	27.02E	5	101.3	–1.8	36.2	10.9
Kayseri, Erkilet	38.82N	35.43E	1054	89.3	–16.0	33.8	9.3
Konya Airport	37.97N	32.55E	1031	89.5	–13.0	33.8	11.5
Malatya Erhac Airport	38.43N	38.08E	849	91.5	–11.5	37.2	10.1
Menderes, Izmir	38.27N	27.15E	120	99.9	–2.8	37.0	12.0
Samsun	41.28N	36.30E	4	101.3	–1.3	28.0	8.3
Van Airport	38.45N	43.32E	1662	82.9	–14.0	29.0	8.2
Turkmenistan							
Ashgabat Keshi	37.92N	58.33E	312	97.6	–6.7	40.1	9.4
Ukraine							
Chernihiv	51.47N	31.25E	141	99.6	–20.7	29.6	9.2
Dnipropetrovsk Oblast	48.37N	35.08E	143	99.6	–17.8	31.9	11.6
Donetsk Airport	48.07N	37.77E	225	98.7	–18.9	31.3	12.4
Kharkiv	49.97N	36.13E	155	99.5	–19.6	30.8	9.6
Kherson	46.63N	32.57E	54	100.7	–15.8	32.7	9.8
Kiev	50.40N	30.57E	167	99.3	–18.1	29.4	8.9
Kryvyi Rih Airport	48.03N	33.22E	124	99.8	–18.0	31.7	11.6
Luhansk	48.57N	39.25E	62	100.6	–20.6	33.0	11.8
Lviv, Sknilov Airport	49.82N	23.95E	323	97.5	–17.3	28.1	9.7
Mariupol	47.03N	37.50E	70	100.5	–15.4	30.4	14.4
Odessa	46.43N	30.77E	42	100.8	–13.9	31.2	11.2
Poltava	49.60N	34.55E	160	99.4	–19.4	30.4	10.1
Simferopol	45.02N	33.98E	181	99.2	–12.7	32.2	12.3
Vinnytsia	49.23N	28.60E	298	97.8	–19.1	28.4	11.4
Zaporizhia	47.80N	35.02E	112	100.0	–17.7	32.3	10.1
United Arab Emirates							
Abu Dhabi Bateen Airport	24.43N	54.47E	3	101.3	13.0	43.0	9.5
Abu Dhabi International Airport	24.43N	54.65E	27	101.0	11.5	44.9	9.4
Al Ain International Airport	24.27N	55.60E	262	98.2	11.1	45.9	10.4
Dubai International Airport	25.25N	55.33E	5	101.3	12.7	42.8	9.2
Sharjah International Airport	25.33N	55.52E	33	100.9	9.9	44.0	8.3
United Kingdom							
Aughton	53.55N	2.92W	56	100.7	–3.4	24.4	11.5
Bingley	53.82N	1.87W	267	98.2	–4.1	23.8	12.7
Birmingham Intl Airport	52.45N	1.73W	99	100.1	–5.2	26.7	10.0
Bristol Airport	51.38N	2.72W	194	99.0	–3.2	25.8	12.1
Bristol Weather Center	51.47N	2.60W	11	101.2	–2.8	26.6	10.4
Cardiff Airport	51.40N	3.35W	67	100.5	–3.0	24.8	13.3
Cardiff Weather Center	51.48N	3.18W	52	100.7	–1.0	26.2	11.7
Church Lawford	52.37N	1.33W	106	100.1	–4.7	26.6	10.0
Cilfynydd	51.63N	3.30W	194	99.0	–4.6	25.6	11.6

Table 2.2: Climatic Data in SI Units *(Continued)*

Station	Latitude, degrees	Longitude, degrees	Elevation, m	St. Br. Pressure, kPa	Winter Temp., °C	Summer Temp., °C	Wind, m/s
Crosby	53.50N	3.07W	9	101.2	–3.4	24.5	17.6
East Midlands Airport	52.83N	1.32W	96	100.2	–4.0	26.8	12.8
Edinburgh Turnhouse Airport	55.95N	3.35W	41	100.8	–5.3	22.4	12.4
Emley Moor	53.62N	1.67W	259	98.3	–3.2	24.0	14.8
Glasgow International Airport	55.87N	4.43W	8	101.2	–5.9	23.7	13.1
Gravesend Broadness	51.47N	0.30E	3	101.3	–1.7	28.4	11.1
Hawarden Airport	53.17N	2.98W	9	101.2	–4.3	25.3	10.4
Kenley Airfield	51.30N	0.08W	170	99.3	–2.7	26.7	11.0
Leconfield	53.87N	0.43W	7	101.2	–3.7	25.0	12.6
Leeds Bradford Airport	53.87N	1.65W	213	98.8	–3.0	24.2	12.9
Leeds Weather Center	53.80N	1.55W	47	100.8	–2.3	26.1	13.0
Liverpool Lennon Intl Airport	53.33N	2.85W	25	101.0	–3.1	25.2	13.3
London Heathrow Airport	51.48N	0.45W	25	101.0	–3.1	28.3	10.1
London Weather Center	51.52N	0.12W	23	101.1	–2.6	27.2	11.7
London Weather Center	51.52N	0.10W	43	100.8	–0.6	28.5	9.3
Manchester Airport	53.35N	2.28W	69	100.5	–3.7	25.8	11.1
Northolt Airport	51.55N	0.42W	39	100.9	–4.1	28.3	10.4
Uruguay							
Carrasco	34.83S	56.00W	32	100.9	1.4	31.3	13.2
Prado	34.85S	56.20W	16	101.1	3.0	31.6	10.2
Uzbekistan							
Namangan Airport	40.98N	71.58E	474	95.8	–7.4	36.4	7.4
Samarkand	39.57N	66.95E	724	92.9	–9.8	36.0	10.4
Tashkent	41.27N	69.27E	466	95.9	–9.4	38.2	6.2
Venezuela							
Caracas, Simon Bolivar Intl Apt	10.60N	66.98W	48	100.8	20.7	33.4	5.2
San Antonio del Tachira	7.85N	72.45W	378	96.9	19.9	35.0	12.3
Viet Nam							
Da Nang	16.07N	108.35E	7	101.2	16.5	36.0	7.4
Hanoi	21.03N	105.80E	6	101.3	10.0	35.7	6.9
Ho Chi Minh City, Tan Son Hoa	10.82N	106.67E	5	101.3	20.1	35.2	11.5
Phu Lien	20.80N	106.63E	116	99.9	9.8	34.0	7.0
Zimbabwe							
Harare International Airport	17.92S	31.13E	1480	84.8	6.7	30.7	9.0

REFERENCES

ASHRAE. 2009. *ASHRAE Handbook—Fundamentals*. Atlanta: ASHRAE.

Hubbard, K., K. Kunkel, A. DeGaetano, and K. Redmond. 2004. Sources of uncertainty in the calculation of the design weather conditions. RP-1171, ASHRAE, Atlanta.

Lamming, S.D., and J.R. Salmon. 1998. Wind data for design of smoke control systems. *ASHRAE Transactions*, 104(1):742–751.

Lott, J.N., R. Baldwin, and P. Jones. 2001. The FCC Integrated Surface Hourly Database, a new resource of global climate data. NCDC Technical Report 2001–01, National Climatic Data Center, Asheville, NC.

NASA. 1976. U.S. Standard Atmosphere. National Oceanic and Atmospheric Administration, National Aeronautics and Space Administration, and the United States Air Force. Availably from the National Geophysical Data Center, Bolder CO.

NCDC. 2003. Data documentation for data set 3505 (DSI-3505) integrated surface hourly (ISH) data. National Climatic Data Center, Asheville, NC.

NOAA. 1998. Climatic Wind Data for the United States. National Oceanic and Atmospheric Administration, National Climatic Data Center, Asheville, NC.

Thevenard, D. 2009. Updating the ASHRAE climatic data for design and standards. RP-1453, ASHRAE, Atlanta.

CHAPTER 3

Flow of Air and Smoke

John H. Klote

In building fires, smoke often travels through shafts to locations remote from the fire to threaten life and damage property. This chapter discusses equations for air and smoke flow, effective areas, symmetry, and the driving forces of smoke movement. These driving forces are buoyancy of combustion gases, expansion of combustion gases, fan-powered ventilation systems, elevator piston effect, stack effect, and wind. With the exception of the first two, these driving forces also apply to airflow in the absence of a fire.

FLOW EQUATIONS

Many equations have been used to express the relation between fluid flow and pressure difference. The characteristics of this flow depend on the geometry of the flow path and the Reynolds number, which is the ratio of kinetic forces to viscous forces.

The Reynolds number is

$$R_e = \frac{1.39 \times 10^{-3} D_h U}{\nu}$$
$$R_e = \frac{D_h U}{\nu} \text{ for SI} \tag{3.1}$$

where

R_e = Reynolds number, dimensionless,

D_h = hydraulic diameter of flow path, in. (m),

U = average velocity in flow path, fpm (m/s),

ν = kinematic viscosity, ft²/s (m²/s).

See Chapter 1 for values of kinematic viscosity. The hydraulic diameter is

$$D_h = \frac{4A}{P} \tag{3.2}$$

where

A = cross-sectional area of the path, ft² (m²),

P = perimeter of the path, ft (m).

For flow paths with rectangular cross sections, the hydraulic diameter is $D_h = 2ab/(a + b)$ where a and b are the sides of the rectangle in ft (m). For very long and thin rectangular gaps, the length is much greater than the width ($b >> a$) and the hydraulic diameter is twice the width ($D_h = 2a$). The hydraulic diameter of a circle is the diameter of the circle, and the hydraulic diameter of a square is the side of the square.

At Reynolds numbers greater than about 2000 or 4000, the flow is dominated by kinetic forces. At these Reynolds numbers, the flow becomes turbulent, and the flow fluctuates rapidly at a given point in what seems like a random manner. For such turbulent flow, the orifice equation is discussed next.

Orifice Flow Equation

The primary equation used for analysis of pressurization smoke control systems is the orifice equation.

$$m = 12.9CA\sqrt{2\rho\Delta p}$$
$$m = CA\sqrt{2\rho\Delta p} \text{ for SI} \tag{3.3}$$

For a standard temperature of 70°F (21°C) and standard atmospheric pressure, Equation 3.3 becomes

$$m = 4.99CA\sqrt{\rho}$$
$$m = 1.41CA\sqrt{\rho} \text{ for SI} \tag{3.3a}$$

For mass flow in alternative units at 70°F (21°C) and standard atmospheric pressure, Equation 3.3 becomes

$$m_{sv} = 2610CA\sqrt{\Delta\rho}$$
$$m_{sv} = 0.839CA\sqrt{\Delta\rho} \text{ for SI} \tag{3.3b}$$

where

m = mass flow through the path, lb/s (kg/s),

m_{sv} = mass flow through the path, scfm (standard m^3/s),

C = flow coefficient, dimensionless,

A = flow area (or leakage area), ft^2 (m^2),

Δp = pressure difference across path, in. H_2O (Pa),

ρ = density gas in path, lb/ft^3 (kg/m^3).

One standard cubic foot per minute, scfm, equals 0.00125 pounds per second, and one standard cubic meter per second (standard m^3/s) equals 1.2 kilograms per second (kg/s). Alternatively, the orifice equation can be expressed in terms of volumetric flow.

$$V = 776CA\sqrt{\frac{2\Delta p}{\rho}}$$
$$V = CA\sqrt{\frac{2\Delta p}{\rho}} \text{ for SI} \tag{3.4}$$

where V is volumetric flow through the path in cfm (m^3/s).

Equations 3.3 and 3.4 are equivalent forms of the same equation, and name orifice equation applies to both. The orifice equation gets its name because it is used to calculate the flow through orifice flow meters. For these flow meters, the area term above is called the cross-sectional area, and the flow coefficient is called the discharge coefficient. As discussed in Chapter 14, the network flow program CONTAM uses the flow meter terminology. Flow areas and flow coefficients for building components are discussed later, and Idelchik (1986) is also a source of flow data for many items. Example 3.1 illustrates use of the orifice equation.

As mentioned, the orifice equation is for turbulent flow, but it can be extended to flow with somewhat lower Reynolds numbers by appropriate selection of the flow coefficient or the flow area.

Depending on the kind of pressurization smoke control system, the boundaries of systems and smoke control zones are at stairwell doors, elevator doors, or doors at other locations. The flow in the gaps around the doors of these boundaries is expected to be turbulent, but the flow through the construction cracks may not be turbulent. Considering that there is so much more flow through the gaps around the doors than through the construction cracks, it is appropriate to use the orifice equation for analysis of pressurization smoke control systems.

For buildings with extremely tight leakage, including gasketed doors, the orifice equation may not be appropriate. Systems that rely on compartmentation without pressurization may have low Reynolds numbers such that the orifice equation may not be appropriate.

Density of Gases

The density of air and smoke are expressed by the ideal gas law,

$$\rho = \frac{144p}{RT}$$
$$\rho = \frac{p}{RT} \text{ for SI} \tag{3.5}$$

where

ρ = density, lb/ft^3 (kg/m^3),

p = pressure, lb/in^2 (Pa),

R = gas constant, 53.34 ft·lbf·lbm/°R (287 J/kg·K),

T = absolute temperature, °R (K).

Exponential Flow

The exponential flow equation is

$$V = C_e(\Delta p)^n \tag{3.6}$$

where

V = volumetric flow, cfm (m^3/s),

C_e = flow coefficient for exponential flow equation, ft^3/min·(in. H_2O) [m^3/s·Pa^n],

Example 3.1. Flow Using the Orifice Equation

Calculate the volumetric flow through an orifice with the following values:

$C = 0.65$, $A = 0.5$ ft^2 (0.046 m^2), $\Delta p = 0.10$ in. H_2O (25 Pa) and $\rho = 0.075$ lb/ft^3 (1.20 kg/m^3).

The volumetric flow is $V = 776CA\sqrt{\frac{2\Delta p}{\rho}} = 776(0.65)(0.5)\sqrt{\frac{2(0.10)}{0.075}} = 412 \text{ cfm } (0.194 \text{ m}^3/\text{s})$.

Δp = pressure difference across the path, in. H_2O (Pa),

n = flow exponent, dimensionless.

This equation can be used to simulate flows in the entire range from viscous-dominated to kinetic-dominated. For viscous-dominated forces, the flow exponent is 1. For kinetic-dominated flows, the flow exponent is 1/2, which is the same as the orifice equation.

Equation 3.6 has proven useful for evaluation of flows through small cracks in buildings at low levels of pressure difference. However, this equation is not directly related to the geometry of the flow path. The exponents for leakage areas (cracks) in exterior walls are sometimes considered to be about 0.6 or 0.65.

Gap Method

Gross and Haberman (1988) developed the gap method of determining the airflow through slots of different geometry such as those of door assemblies. They correlated considerable experimental data over a wide range of Reynolds numbers. The gap method is cumbersome for design calculations, but it is useful for calculating flow values that can be used in other flow equations, and this was done for the door leakage tables discussed later in this chapter. For projects where the published flow areas and flow coefficients are inappropriate, the gap method may be helpful in calculating values that can be used in either the orifice equation or the exponential equation.

There is a relationship between the dimensionless variables NQ and NP.

$$NQ = R_e\left(\frac{a}{x}\right) \tag{3.7}$$

and

$$NP = \frac{1.16\Delta p D_h^2}{\rho \nu^2}\left(\frac{D_h}{x}\right)^2$$

$$NP = \frac{\Delta p D_h^2}{\rho \nu^2}\left(\frac{D_h}{x}\right)^2 \text{ for SI} \tag{3.8}$$

where

NQ = dimensionless flow,

NP = dimensionless pressure difference,

R_e = Reynolds number, dimensionless (Equation 3.1),

a = thickness of gap in direction perpendicular to flow, in. (m),

x = depth of gap in flow direction, in. (m),

Δp = pressure difference across gap, in. H_2O (Pa),

D_h = hydraulic diameter, in. (m), $D_h = 2a$,

ρ = density of gas in gap, lb/ft^3 (kg/m^3),

ν = kinematic viscosity, ft^2/s (m^2/s).

After some entrance length in a slot, the flow becomes fully developed. The gap method accounts for the developing flow in this entrance length. For a straight-through gap, the relationship for flow versus pressure difference is shown in Figure 3.1. The regions of flow for a straight-through slot are:

Region A (Viscous dominated region: for $NP \leq 250$):

$$NQ = 0.0104NP \tag{3.9}$$

Region B (Transition region: for $250 < NP < 10^6$):

$$NQ = 0.016984NP^{\alpha} \tag{3.10}$$

where $\alpha = 1.01746 - 0.044181 \text{ Log}_{10}(NP)$

Region C (Kinetic dominated region: for $NP \geq 10^6$):

$$NQ = 0.555NP^{1/2} \tag{3.11}$$

The equation for region B was developed by Forney (1989) as a modification of the original gap method. Forney's equation is particularly attractive for computer applications, because it is continuous with the expressions for the other two regions.

The volumetric flow V through a straight-through slot is

$$V = \frac{30\nu\, xL\, NQ}{a}$$

$$V = \frac{\nu\, xL\, NQ}{2a} \text{ for SI} \tag{3.12}$$

where L is the length of the gap in ft (m). For single- and double-bend slots, the dimensionless flow NQ can be

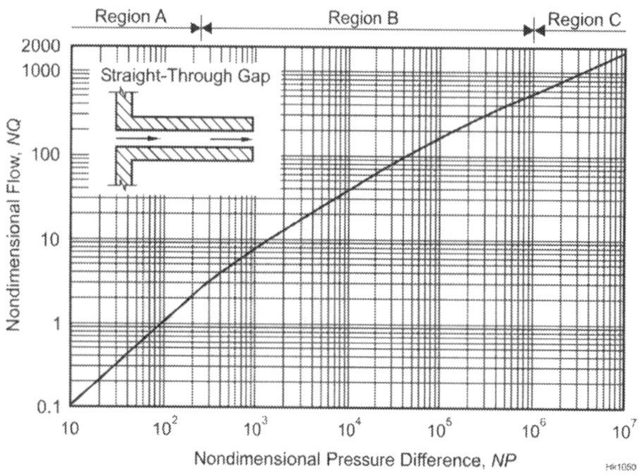

Figure 3.1 *Flow relationship for straight-through gaps.*

Table 3.1: Flow Factors for Single- and Double-Bend Gaps

Dimensionless Pressure Difference, *NP*	Flow Factor for Single-Bend Slot, F_1	Flow Factor for Double-Bend Slot, F_2
≤ 4000	1.000	1.000
7000	0.981	0.939
10,000	0.972	0.908
15,000	0.960	0.880
20,000	0.952	0.862
40,000	0.935	0.826
100,000	0.910	0.793
200,000	0.890	0.772
400,000	0.872	0.742
1,000,000	0.848	0.720
2,000,000	0.827	0.700

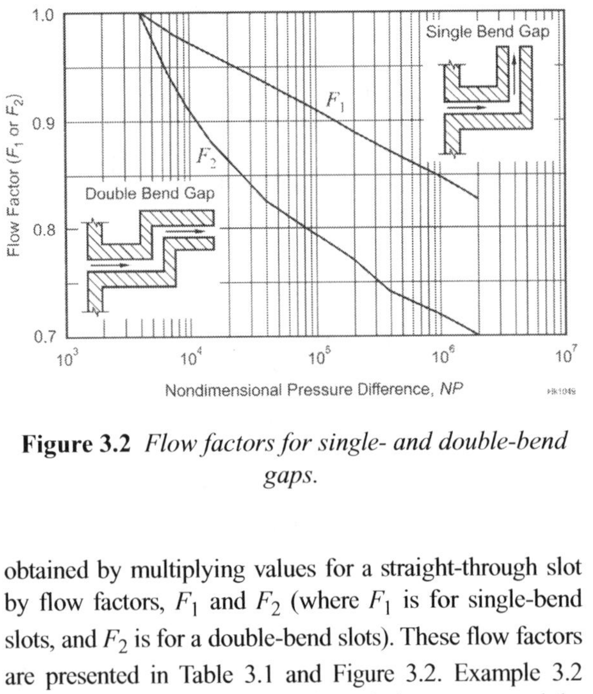

Figure 3.2 *Flow factors for single- and double-bend gaps.*

obtained by multiplying values for a straight-through slot by flow factors, F_1 and F_2 (where F_1 is for single-bend slots, and F_2 is for a double-bend slots). These flow factors are presented in Table 3.1 and Figure 3.2. Example 3.2 illustrates calculating the flow through the gaps around the door in Figure 3.3.

Use in Orifice Equation

Orifice equation calculations for a specific slot can be based on the gap method. The first approach bases the orifice equation on one pressure difference. This approach consists of the following steps:

1. Selecting a pressure difference,
2. Calculating the flow through the slot at the selected pressure difference using equations of the gap method,
3. Calculating either the flow area or the flow coefficient from the orifice equation.

The flow area or the flow coefficient evaluated previously can be used in the orifice equation for flow calculations, which produce nearly the same flows as the gap method in the vicinity of the selected pressure difference. Pressure difference selection is important. For analysis of a system where the relevant pressure differences range from 0.10 to 0.35 in. H_2O, (25 to 87 Pa) selecting a pressure difference in step 1 of 0.15 in. H_2O (37 Pa) would be reasonable. Example 3.3 illustrates calculation of the flow area for use in the orifice equation from the gap method calculations of Example 3.2.

An alternate approach uses regression analysis to obtain a least squares fit value of either the flow area or the flow coefficient. This approach requires that a number of pairs of flow and pressure difference be calculated. Because of the nature of the orifice equation, it is not possible for the regression analysis approach to be significantly better than using the first approach.

Use in Exponential Equation

Exponential flow calculations for a slot also can be based on the gap method, using the following steps:

1. Selecting two pressure differences,
2. Calculating the two flows through the slot at the selected pressure differences using Equations 3.8 to 3.12,
3. Calculating the flow coefficient and flow exponent for exponential flow equation from the following equations.

$$n = \frac{\log_e(V_1/V_2)}{\log_e(\Delta p_1/\Delta p_2)} \tag{3.13}$$

and

Example 3.2. Flow Using the Gap Method

The single door shown in Figure 3.3 has the following dimensions:

Dimension	Bottom Slot	Top and Side Slots
Thickness of gap in direction perpendicular to flow, a	0.25 in. (0.00635 m)	0.1 in. (0.00254 m)
Depth of gap in flow direction, x	1.75 in. (0.0445 m)	2.37 in. (0.0602 m)
Length of gap, L	3 ft (0.914 m)	17 ft (5.18 m)
Hydraulic diameter, $D_h = 2a$	0.5 in. (0.0127 m)	0.2 in. (0.00508)

The air density is 0.075 lb/ft^3 (1.20 kg/m^3), and the kinematic viscosity is 1.64×10^{-4} ft^2/s (1.52×10^{-5} m^2/s).

What is the volumetric flow with a pressure difference, $\Delta p = 0.15$ in H_2O (37 Pa)?

Bottom Slot

$$NP = \frac{1.16\Delta p D_h^2}{\rho v^2}\left(\frac{D_h}{x}\right)^2 = \frac{1.16(0.15)(0.5)^2}{0.075(1.64\times10^{-4})^2}\left(\frac{0.5}{1.75}\right)^2 = 1.76\times10^2$$

Because this is in Region C, the following equation is used

$$NQ = 0.555(NP)^{1/2} = 0.555(1.76\times10^6)^{1/2} = 736$$

The volumetric flow is

$$V = \frac{30vxLNQ}{a} = \frac{30(1.64\times10^{-4})(1.75)(3)(736)}{0.25} = 76 \text{ cfm } (0.036 \text{ m}^3/\text{s})$$

Top and Side Slots

$$NP = \frac{1.16\Delta p D_h^2}{\rho v^2}\left(\frac{D_h}{x}\right)^2 = \frac{1.16(0.15)(0.2)^2}{0.075(1.64\times10^{-4})^2}\left(\frac{0.2}{2.37}\right)^2 = 24,600$$

Because this is in Region B, the following equation is used

$$\alpha = 1.01746 - 0.044181\ \text{Log}_{10}(NP) = 1.01746 - 0.044181\ \text{Log}_{10}(24{,}600) = 0.8235$$

$$NQ = 0.016984NP^{\alpha} = 0.016984(24,000)^{0.8235} = 70$$

$$V = \frac{30vxLNQ}{a} = \frac{30(1.64\times10^{-4})(2.37)(17)(70)}{0.10} = 139 \text{ cfm } (0.066 \text{ m}^3/\text{s})$$

For a single bend slot, from Table 3.1 or Figure 3.2: $F_1 = 0.95$

$$V = 0.95(139) = 132 \text{ cfm } (0.062 \text{ m}^3/\text{s})$$

Total flow is 76 + 132 = 208 cfm (0.098 m^3/s)

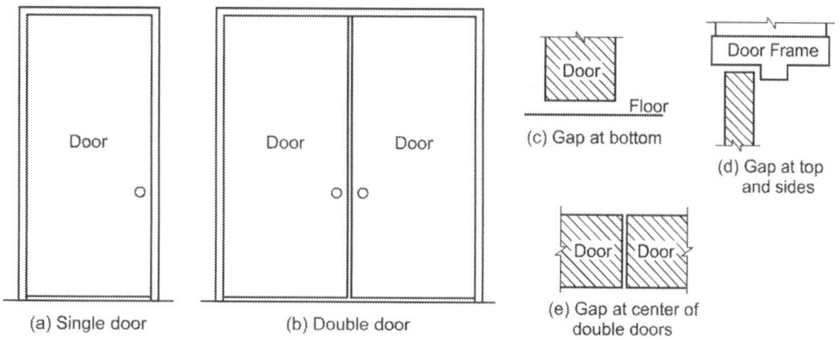

Figure 3.3 *Gaps around single and double doors.*

Example 3.3. Orifice Equation Based on the Gap Method

Calculate the flow area of the orifice equation that would have the same flow as the door of Example 3.2.

Use a flow coefficient of $C = 0.65$.

From Example 3.2, $V = 132$ cfm (0.062 m^3/s), $\rho = 0.075$ lb/ft^3 (1.20 kg/m^3) and $\Delta\rho = 0.15$ in. H_2O (37 Pa).

The orifice equation is

$$V = 776CA\sqrt{2\Delta\frac{p}{\rho}}$$

Rearrange this equation to solve for the flow area

$$A = \frac{V}{776CA}\sqrt{\frac{\rho}{2\Delta p}} = \frac{132}{776(0.65)}\sqrt{\frac{0.075}{2(0.15)}} = 0.13\ \text{ft}^2(0.012\ \text{m}^2)$$

This flow area can be used in the orifice equation to make approximate flow calculations for the door.

$$C_e = \frac{V_1}{(\Delta p_1)^n} \tag{3.14}$$

where

C_e = flow coefficient for exponential flow equation, ft^3/min·(in. H_2O) [m^3/s·Pa^n],

n = flow exponent, dimensionless,

V_1 = volumetric flow at Δp_1, cfm (m^3/s),

V_2 = volumetric flow at Δp_1, cfm (m^3/s),

Δp_1 = pressure difference across the path at V_1, in. H_2O (Pa),

Δp_2 = pressure difference across the path at V_2, in. H_2O (Pa).

BIDIRECTIONAL FLOW

Bidirectional flow happens through an opening or openings between two spaces that are at different temperatures. These spaces can be (1) two rooms, (2) a fire compartment and the surroundings, and (3) a stairwell or other shaft and the outdoors. When bidirectional flow is between a stairwell or other shaft and the outdoors, it is referred to as stack effect. The analysis that follows considers the temperatures in each space to be uniform, and this is often a reasonable approximation for two rooms connected to each other and for stack effect. When the shaft temperature varies, stack effect can be analyzed by a network model (Chapter 14). Further aspects of stack effect are discussed later in this chapter.

Because the temperatures in fire compartments are often far from uniform, the equations of this section need to be used with care for fire compartments. For fully developed fires (Chapter 5), the temperatures in fire compartment can be roughly approximated as uniform, and the analysis of bidirectional flow that follows has some utility. For other compartment fires, the flow between rooms can be complicated by the presence of a high-temperature smoke layer and a lower nonsmoke layer. For an analysis of flows involving smoke and nonsmoke layers, readers are referred to Jones and Bodart (1986). For further discussion of compartment fires, see the section on Buoyancy of Combustion Gases in this chapter. In the following discussion, the term *space* is used in a generic sense to mean any space inside a building or even the outdoors.

Pressure Difference

As shown on Figure 3.4, a neutral plane forms between the two spaces. The reason this plane is called the neutral plane is because the pressure in both spaces are the same at this plane. It follows that there is no horizontal flow along this plane.

The pressure difference related to bidirectional flow is

$$\Delta p_{12} = \frac{0.00598 g p_{atm}}{R}\left(\frac{1}{T_2} - \frac{1}{T_1}\right)z$$
$$\Delta p_{12} = \frac{g p_{atm}}{R}\left(\frac{1}{T_2} - \frac{1}{T_1}\right)z \text{ for SI} \tag{3.15}$$

and at standard atmospheric pressure this is

$$\Delta p_{12} = 7.63\left(\frac{1}{T_2} - \frac{1}{T_1}\right)z$$
$$\Delta p_{12} = 3460\left(\frac{1}{T_2} - \frac{1}{T_1}\right)z \text{ for SI} \tag{3.16}$$

where

Δp_{12} = pressure difference from space 1 to space 2, in. H_2O (Pa),

g = acceleration due to gravity, ft/s^2 (m/s^2),

p_{atm} = absolute atmospheric pressure, lb/ft^2 (Pa),
R = gas constant of air, 53.34 ft·lbf/lbm·°R (287.0 J/kg·K),
T_1 = absolute temperature of space 1, °R (K),
T_2 = absolute temperature of space 2, °R (K),
z = distance above the neutral plane, ft (m).

Continuous Opening

The opening between the spaces has constant width, and it is shown in a general way in Figure 3.4a. The opening could be small, it could be an open doorway, or it could extend from the floor to the ceiling. Above the neutral plane, the flow is from the warmer space to the cooler space. Below the neutral plane, the flow is in the opposite direction.

The flows between the spaces are

$$m_{12} = \frac{2}{3}12.9Cw(H-H_n)\sqrt{2\rho_1 g(\rho_2-\rho_1)}$$
$$m_{12} = \frac{2}{3}Cw(H-H_n)\sqrt{2\rho_1 g(\rho_2-\rho_1)} \text{ for SI} \tag{3.17}$$

and

$$m_{21} = \frac{2}{3}12.9CwH_n\sqrt{2\rho_2 g(\rho_2-\rho_1)}$$
$$m_{21} = \frac{2}{3}CwH_n\sqrt{2\rho_2 g(\rho_2-\rho_1)} \text{ for SI} \tag{3.18}$$

where

m_{12} = mass flow from Space 1 to Space 2, lb/s (kg/s),
m_{21} = mass flow from Space 2 to Space 1, lb/s (kg/s),
C = flow coefficient, dimensionless,
w = width of the opening, ft (m),
H = height of the opening, ft (m),
H_n = height of neutral plane, ft (m),
ρ_1 = density in Space 1, lb/ft^3 (kg/m^3),
ρ_2 = density in Space 2, lb/ft^3 (kg/m^3).

Equations 3.17 and 3.18 apply when space 1 is warmer than space 2 ($T_1 > T_2$). The height of the neutral plane, H_n, depends on a number of factors including flow paths to other spaces and the presence of forced air or exhaust. If m_{12} and m_{21} are the only flow into or out of the two spaces, the height of the neutral plane is

$$H_n = \frac{H}{1+(T_1/T_2)^{1/3}} \tag{3.19}$$

where

H = height of the opening, ft (m),
H_n = height of neutral plane, ft (m),
T_1 = absolute temperature of space 1, °R (K),

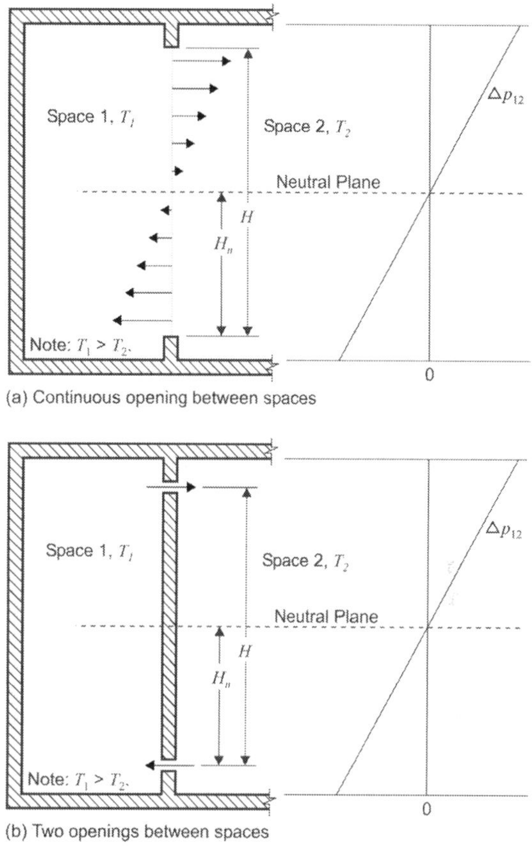

Figure 3.4 *Bidirectional flow between two spaces.*

T_2 = absolute temperature of space 2, °R (K).

Two Openings

Bidirectional flow with two openings is shown in Figure 3.4b. These openings are considered small enough such that the pressure difference across each opening is constant with respect to height. The flow through these openings can be calculated from the orifice equation. If the flows through these two openings are the only flow into or out of the two spaces, the location of the neutral plane is

$$H_n = \frac{H}{1+(T_1/T_2)(A_b/A_a)^2} \tag{3.20}$$

where

H_n = distance from the bottom opening to the neutral plane, ft (m),
H = distance between the two openings, ft (m),
A_a = area above neutral plane, ft^2 (m^2),
A_b = area below neutral plane, ft^2 (m^2).

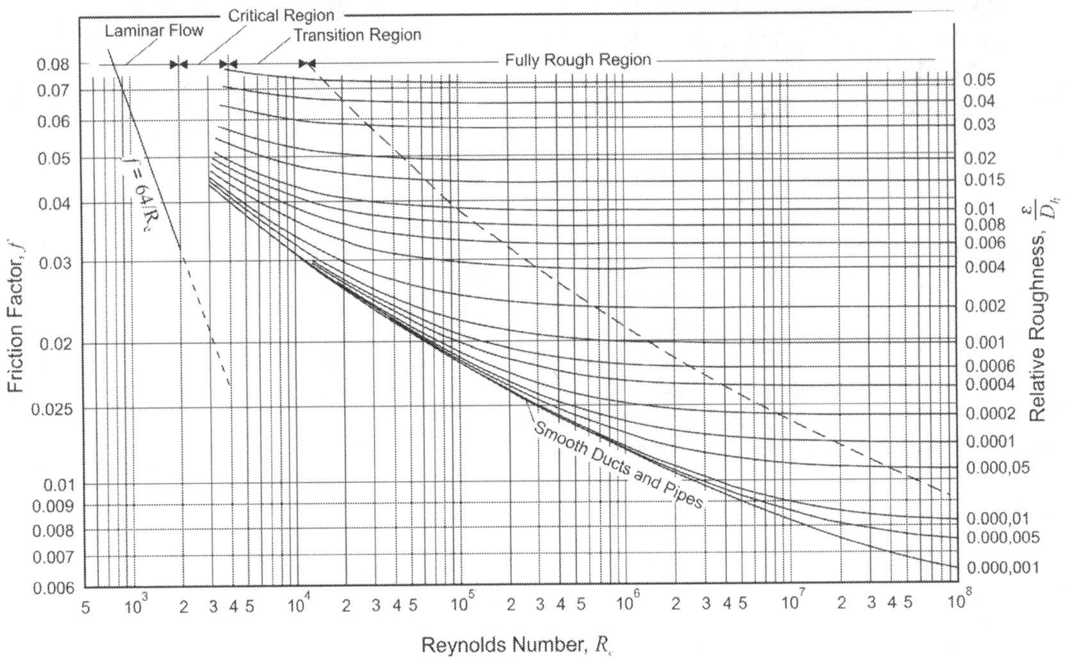

Figure 3.5 *Moody diagram for the friction factor for flow in ducts and pipes.*

Table 3.2: Duct and Shaft Roughness

Duct and Shaft	Roughness, e	
	in.	mm
Average galvanized steel duct	0.006	0.15
Concrete duct or shaft	0.12	3.0
Elevator shaft	4.0	100

PRESSURE LOSSES IN SHAFTS

Ducts and Shafts

The pressure losses due to friction in ducts and shafts is represented by the Darcy-Weisbach equation:

$$\Delta p_f = 1.66 \times 10^{-6} f \frac{L}{D_h} \frac{\rho U^2}{2}$$

$$\Delta p_f = f \frac{L}{D_h} \frac{\rho U^2}{2} \text{ for SI} \tag{3.21}$$

where

Δp_f = pressure loss in shaft or duct due to friction, in. H_2O (Pa),

f = dimensionless friction factor of shaft or duct,

L = shaft or duct length, ft (m),

D_h = hydraulic diameter of shaft or duct, ft (m),

ρ = density of gas inside shaft or duct, lb/ft^3 (kg/m^3),

U = average velocity inside shaft or duct, fpm (m/s).

The hydraulic diameter of a shaft or duct can be calculated from Equation 3.2. For ducts and pipes, the friction factor can be obtained from the Moody diagram of Figure 3.5. This figure provides friction factors for the laminar region, critical region, transitional region, and fully rough region. The transitional region and fully rough region are for turbulent flow.

For the transitional region and fully rough region the friction factor can be calculated from the Colebrook equation:

$$\frac{1}{\sqrt{f}} = -2 \log_{10}\left(\frac{\varepsilon}{3.7 D_h} + \frac{2.51}{R_e \sqrt{f}}\right) \tag{3.22}$$

where ε is the roughness of the inside surface of the duct in ft (m). Because of the form of Equation 3.22, it is suggested that it be solved with the aid of a computer program using the Newton Raphson method[1]. The roughness of some ducts and shafts are listed in Table 3.2.

1. The efficiency of this method is improved by substituting $x = f^{1/2}$, and solving for x.

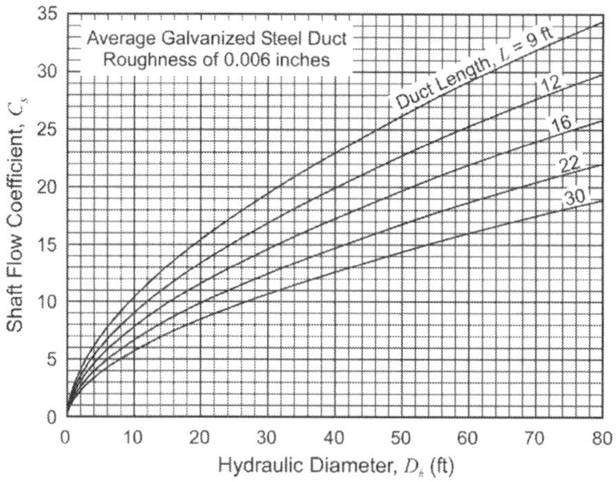

Figure 3.6 *Flow coefficient for average galvanized steel duct (I-P units).*

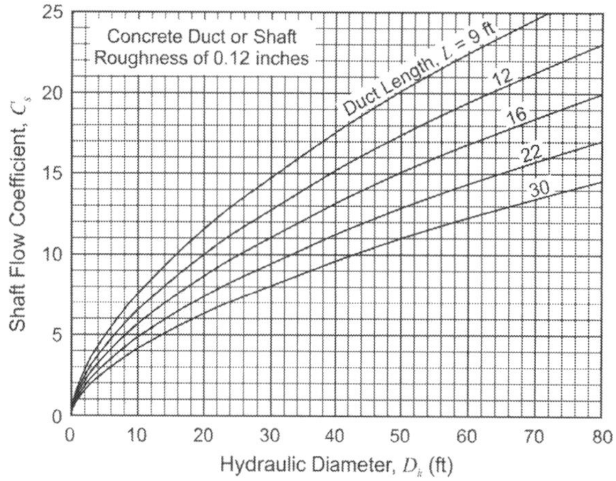

Figure 3.7 *Flow coefficient for concrete duct or shaft (I-P units).*

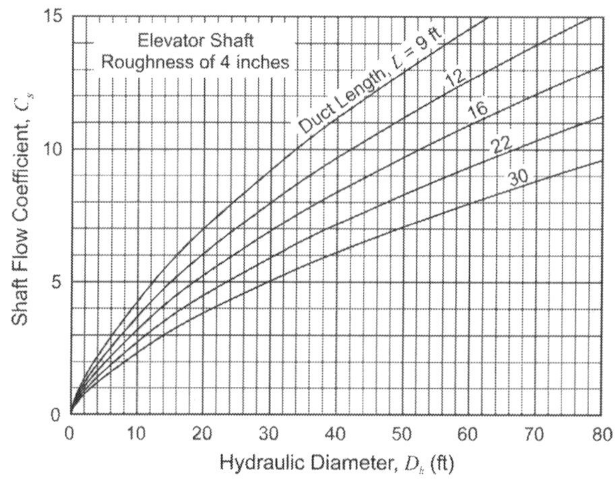

Figure 3.8 *Flow coefficient for elevator shaft (I-P units).*

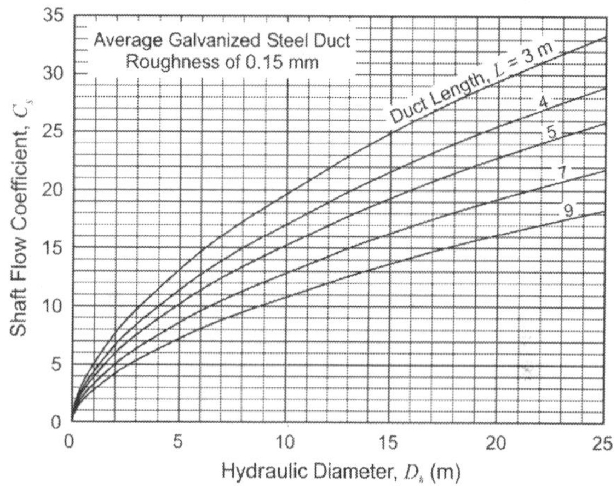

Figure 3.9 *Flow coefficient for average galvanized steel duct (SI units).*

For the fully rough flow region the friction factor can be calculated from

$$f = \frac{1}{\left[1.14 + 2\log_{10}\left(\frac{D_h}{\varepsilon}\right)\right]^2} \tag{3.23}$$

The orifice equation can be used to account for the friction losses due to flow in shafts and vertical ducts. This can be done by using the cross-sectional area and the following shaft or duct flow coefficient.

$$C_s = \sqrt{\frac{D_h}{fL}} \tag{3.24}$$

where C_s is the dimensionless shaft flow coefficient.

In network flow models (Chapter 14), vertical ducts and shafts are represented as a series of vertical spaces (rooms) with openings between them. The pressure losses of these ducts and shafts are calculated in the computer model using the orifice equation where the area of the opening is the cross-sectional area of the duct or shaft, and the flow coefficient, C_s. The flow coefficient, C_s, can be calculated from Equation 3.24 or it can be obtained from Figures 3.6 to 3.11. These values of C_s are for the fully rough flow region with the friction factor calculated from Equation 3.23. Example 3.4 illustrates calculation of the shaft flow coefficient for an elevator shaft.

Stairwell Flow

The pressure losses due to friction in stairwells are similar to those in shafts. The orifice equation can be used to account for the friction losses due to flow in stairwells. This can be done by using a flow coefficient of 0.60 and the equivalent orifice area for the stairwell below.

$$\frac{A_o}{A_s} = \frac{1}{C}\sqrt{\frac{D_h}{hK_{sw}}} \tag{3.25}$$

where

C = flow coefficient of 0.60, dimensionless,

A_o = equivalent orifice area for the stairwell, ft^2 (m^2),

A_s = cross-sectional area of the stairwell (or leakage area), ft^2 (m^2),

D_h = hydraulic diameter of shaft, ft (m),

h = height of floor, ft (m),

K_{sw} = dimensionless friction factor of stairwell.

Tamura and Shaw (1976b) and Achakji and Tamura (1988) conducted tests of pressure loss in stairwells and determined values for the friction factor, K_{sw}. Tamura and Shaw's tests were field tests conducted without people in the stairs. The tests of Achakji and Tamura were conducted in laboratory conditions at the National Research Council's fire tower near Ottawa, and the tests included the effects of people and stair treads. Table 3.3 lists friction factors of conventional stairwells based on the work of Achakji and Tamura. Figure 3.12 shows a conventional stairwell with one of four possible door locations. For scissor stairs with closed treads and without people, a friction factor, K_{sw}, of 15 can be used based on research of Tamura and Shaw. Example 3.5 illustrates calculation of the equivalent area of a stairwell.

FLOW AREAS & COEFFICIENTS

The flow areas and flow coefficients for the gaps around doors can be calculated by the gap method.

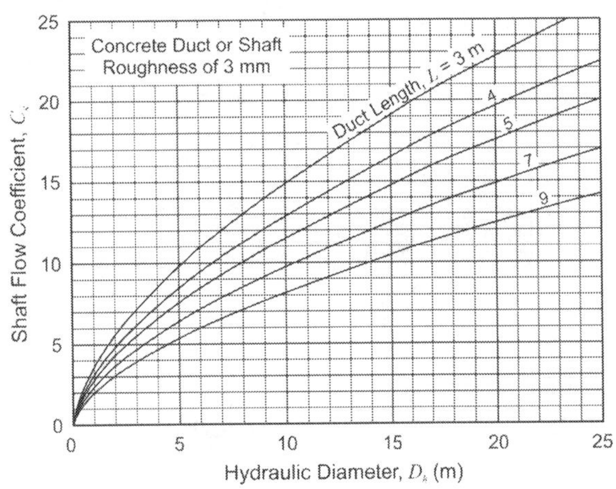

Figure 3.10 *Flow coefficient for concrete duct or shaft (SI units).*

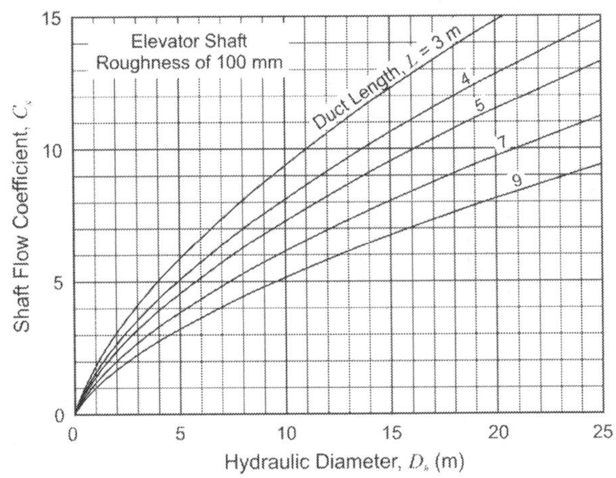

Figure 3.11 *Flow coefficient for elevator shaft (SI units).*

Example 3.4. Shaft Flow Coefficient

Calculate the shaft flow coefficient of an elevator shaft that is 10.5 by 12 ft (3.20 by 3.66 m) with a floor-to-floor height of 12 ft (3.66 m). In a network model, the shafts are simulated on a floor-to-floor basis. This makes the length of duct in this analysis the same as the floor-to-floor height (L = 12 ft).

The cross-sectional area of the shaft is 10.5(12) = 126 ft^2, and the perimeter is 2(10.5 + 12) = 45 ft.

The hydraulic diameter of shaft is $D_h = \frac{4A}{P} = \frac{4(126)}{45} = 11.2$ ft (3.4 m).

From Figure 3.8, the shaft flow coefficient is C_s = 4.0. The friction pressure loss in the elevator shaft can be simulated in a network model by the orifice equation using an area of 126 ft^2 (11.7 m^2) and a flow coefficient of 4.0.

Table 3.4 lists cross-sectional areas and flow coefficients for single doors that can be used in the orifice equation. Table 3.5 lists flow areas for single doors that can be used in the orifice equation with a flow coefficient of 0.65. Table 3.6 lists cross-sectional areas and flow coefficients for double doors that can be used in the orifice equation. Table 3.7 lists flow areas for double doors that can be used in the orifice equation with a flow coefficient of 0.65.

The open doors of pressurized stairwells commonly have stationary vortices which reduce the flow significantly (Cresci 1973, Klote and Bardot 1985). These vortices are thought to be caused by the asymmetric flow from the stairs, and stationary vortices can be expected at many open doors in other locations of smoke control systems. For open doors in stairwells, the geometric area of the opening should be used for the flow area along with a flow coefficient of 0.35.

Because the geometry of the leakage paths through elevator doors is so complex, the gap method is inappropriate for analysis of open or closed elevator doors. Figure 3.13 shows the air leakage of closed elevator doors measured by Tamura and Shaw (1976b). Tamura and Shaw indicate that the average value of the flow exponent is about 0.55 for the exponential flow equation. For this exponent, the flow is similar to that of the orifice equation. Table 3.8 lists flow areas for closed elevator doors with a flow coefficient of 0.65. This table lists values for tight, average, and loose leakage which are based on the results of the field tests. The flow areas of this table were evaluated at 0.2 in. H_2O (50 Pa) which is appropriate for applications with and without pressurized elevators. The values of this table are somewhat larger than those of an earlier table by Klote and Milke (2002) which was based on 0.1 in. H_2O (25 Pa).

Table 3.3: Typical Friction Factors for Conventional Stairwells[1]

Stair Treads	People Density	Friction Factors, K_{sw}
Closed	None	70
Closed	High[2]	170
Open	None	60
Open	High[2]	100

[1]Based on data from Achakji and Tamura (1988).
[2]High is high density of 0.18 person/ft^2 (2.0 persons/m^2).

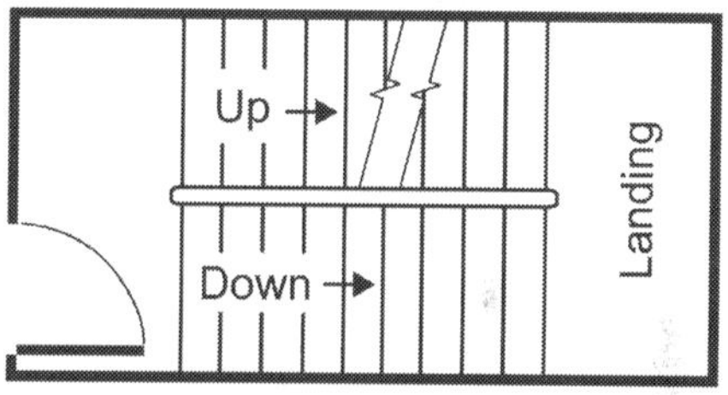

Figure 3.12 *Conventional stairwell.*

Example 3.5. Equivalent Area of a Stairwell

What is the equivalent area of a conventional stairwell 8.5 by 17.5 ft (2.59 by 5.33 m) that has a floor height of 10 ft? The stair treads are closed, and there are no people in the stairwell.

From Table 3.3, the friction coefficient, K_{sw}, is 71. The cross-sectional area of the stairwell is (8.5)(17.5) = 148.8 ft^2, and the perimeter is 2(8.5 + 17.5) = 52 ft.

The hydraulic diameter of shaft is $D_h = \frac{4A}{P} = \frac{4(148.8)}{52} = 11.4 \text{ ft } (3.47 \text{ m})$.

The ratio of the equivalent area to the cross-sectional area is $\frac{A_o}{A_s} = \frac{1}{C}\sqrt{\frac{D_h}{hK_{sw}}} = \frac{1}{0.6}\sqrt{\frac{11.4}{10(71)}} = 0.21$.

The equivalent area is A_o = 0.21(148.8) = 31 ft^2 (9.4 m^2). The friction pressure loss in this stairwell can be simulated in a network model by the orifice equation using an area of 31 ft^2 (9.4 m^2) and a flow coefficient of 0.6.

Table 3.4: Flow Coefficients for the Gaps around Single Doors

Door Width		Gap Thickness at Bottom (Figure 3.3c)		Gap Thickness at Top and Sides (Figure 3.3d)		Cross-Sectional Area		Flow Coefficient
in.	m	in.	mm	in.	mm	ft^2	m^2	
36	0.914	0.25	6.36	0.02	0.508	0.090	0.0084	0.57
36	0.914	0.25	6.36	0.08	2.032	0.175	0.0163	0.63
36	0.914	0.25	6.36	0.12	3.048	0.233	0.0216	0.68
36	0.914	0.25	6.36	0.16	4.064	0.290	0.0269	0.70
36	0.914	0.50	12.70	0.02	0.508	0.153	0.0142	0.66
36	0.914	0.50	12.70	0.08	2.032	0.238	0.0221	0.67
36	0.914	0.50	12.70	0.12	3.048	0.295	0.0274	0.70
36	0.914	0.50	12.70	0.16	4.064	0.352	0.0327	0.71
36	0.914	0.75	19.05	0.02	0.508	0.216	0.0201	0.69
36	0.914	0.75	19.05	0.08	2.032	0.300	0.0279	0.69
36	0.914	0.75	19.05	0.12	3.048	0.357	0.0332	0.72
36	0.914	0.75	19.05	0.16	4.064	0.414	0.0385	0.73
44	1.118	0.25	6.36	0.02	0.508	0.105	0.0098	0.59
44	1.118	0.25	6.36	0.08	2.032	0.194	0.0180	0.64
44	1.118	0.25	6.36	0.12	3.048	0.253	0.0235	0.68
44	1.118	0.25	6.36	0.16	4.064	0.312	0.0290	0.70
44	1.118	0.50	12.70	0.02	0.508	0.182	0.0169	0.67
44	1.118	0.50	12.70	0.08	2.032	0.270	0.0251	0.68
44	1.118	0.50	12.70	0.12	3.048	0.329	0.0306	0.71
44	1.118	0.50	12.70	0.16	4.064	0.389	0.0361	0.72
44	1.118	0.75	19.05	0.02	0.508	0.258	0.0240	0.71
44	1.118	0.75	19.05	0.08	2.032	0.347	0.0322	0.70
44	1.118	0.75	19.05	0.12	3.048	0.406	0.0377	0.72
44	1.118	0.75	19.05	0.16	4.064	0.465	0.0432	0.73

Note: The data in this table are for use with the orifice equation. Flow coefficients and cross-sectional areas are for doors 7 ft (2.13 m) high, 1.75 in. (44.5 mm) thick, and with a doorstop protruding 0.62 in. (15.7 mm) from the frame. The flow coefficients were evaluated by the gap method at 0.15 in. H_2O (37.3 Pa).

Table 3.5: Flow Areas of Gaps Around Single Doors with $C = 0.65$

Door Width		Gap Thickness at Bottom (Figure 3.3c)		Gap Thickness at Top and Sides (Figure 3.3d)		Flow Area[2]	
in.	m	in.	mm	in.	mm	ft^2	m^2
36	0.914	0.25	6.35	0.02	0.508	0.079	0.0073
36	0.914	0.25	6.35	0.08	2.032	0.169	0.0157
36	0.914	0.25	6.35	0.12	3.048	0.242	0.0225
36	0.914	0.25	6.35	0.16	4.064	0.310	0.0288
36	0.914	0.50	12.70	0.02	0.508	0.155	0.0144
36	0.914	0.50	12.70	0.08	2.032	0.244	0.0227
36	0.914	0.50	12.70	0.12	3.048	0.317	0.0295
36	0.914	0.50	12.70	0.16	4.064	0.385	0.0358
36	0.914	0.75	19.05	0.02	0.508	0.230	0.0214
36	0.914	0.75	19.05	0.08	2.032	0.320	0.0297
36	0.914	0.75	19.05	0.12	3.048	0.392	0.0364
36	0.914	0.75	19.05	0.16	4.064	0.461	0.0428
44	1.118	0.25	6.35	0.02	0.508	0.096	0.0089
44	1.118	0.25	6.35	0.08	2.032	0.186	0.0173
44	1.118	0.25	6.35	0.12	3.048	0.259	0.0241
44	1.118	0.25	6.35	0.16	4.064	0.327	0.0304
44	1.118	0.50	12.70	0.02	0.508	0.188	0.0175
44	1.118	0.50	12.70	0.08	2.032	0.278	0.0258
44	1.118	0.50	12.70	0.12	3.048	0.351	0.0326
44	1.118	0.50	12.70	0.16	4.064	0.419	0.0389
44	1.118	0.75	19.05	0.02	0.508	0.280	0.026
44	1.118	0.75	19.05	0.08	2.032	0.370	0.0344
44	1.118	0.75	19.05	0.12	3.048	0.443	0.0412
44	1.118	0.75	19.05	0.16	4.064	0.511	0.0475

Note: The data in this table are for use with the orifice equation. Flow areas are for doors 7 ft (2.13 m) high, 1.75 in. (44.5 mm) thick, and with a doorstop protruding 0.62 in. (15.7 mm) from the frame. The flow areas were evaluated by the gap method at 0.15 in. H_2O (37.3 Pa).

Table 3.6: Flow Coefficients for the Gaps around Double Doors 6 ft (1.83 m) Wide

Gap Thickness at Bottom (Figure 3.3c)		Gap Thickness at Top and Sides (Figure 3.3d)		Gap Thickness at Center (Figure 3.3e)		Cross-Sectional Area		Flow Coefficient
in.	mm	in.	mm	in.	mm	ft^2	m^2	
0.25	6.36	0.02	0.508	0.08	2.032	0.205	0.0190	0.63
0.25	6.36	0.02	0.508	0.12	3.048	0.228	0.0212	0.67
0.25	6.36	0.02	0.508	0.16	4.064	0.252	0.0234	0.69
0.25	6.36	0.08	2.032	0.08	2.032	0.305	0.0283	0.65
0.25	6.36	0.08	2.032	0.12	3.048	0.328	0.0305	0.67
0.25	6.36	0.08	2.032	0.16	4.064	0.352	0.0327	0.69
0.25	6.36	0.12	3.048	0.08	2.032	0.372	0.0345	0.68
0.25	6.36	0.12	3.048	0.12	3.048	0.395	0.0367	0.70
0.25	6.36	0.12	3.048	0.16	4.064	0.418	0.0389	0.71
0.25	6.36	0.16	4.064	0.08	2.032	0.438	0.0407	0.70
0.25	6.36	0.16	4.064	0.12	3.048	0.462	0.0429	0.71
0.25	6.36	0.16	4.064	0.16	4.064	0.485	0.0451	0.72
0.50	12.70	0.02	0.508	0.08	2.032	0.330	0.0307	0.69
0.50	12.70	0.02	0.508	0.12	3.048	0.353	0.0328	0.71
0.50	12.70	0.02	0.508	0.16	4.064	0.377	0.0350	0.72
0.50	12.70	0.08	2.032	0.08	2.032	0.430	0.0399	0.69
0.50	12.70	0.08	2.032	0.12	3.048	0.453	0.0421	0.70
0.50	12.70	0.08	2.032	0.16	4.064	0.477	0.0443	0.71
0.50	12.70	0.12	3.048	0.08	2.032	0.497	0.0461	0.71
0.50	12.70	0.12	3.048	0.12	3.048	0.520	0.0483	0.72
0.50	12.70	0.12	3.048	0.16	4.064	0.543	0.0505	0.73
0.50	12.70	0.16	4.064	0.08	2.032	0.563	0.0523	0.72
0.50	12.70	0.16	4.064	0.12	3.048	0.587	0.0545	0.73
0.50	12.70	0.16	4.064	0.16	4.064	0.610	0.0567	0.74
0.75	19.05	0.02	0.508	0.08	2.032	0.455	0.0423	0.72
0.75	19.05	0.02	0.508	0.12	3.048	0.478	0.0444	0.73
0.75	19.05	0.02	0.508	0.16	4.064	0.502	0.0466	0.74
0.75	19.05	0.08	2.032	0.08	2.032	0.555	0.0516	0.71
0.75	19.05	0.08	2.032	0.12	3.048	0.578	0.0537	0.72
0.75	19.05	0.08	2.032	0.16	4.064	0.602	0.0559	0.73
0.75	19.05	0.12	3.048	0.08	2.032	0.622	0.0578	0.72
0.75	19.05	0.12	3.048	0.12	3.048	0.645	0.0599	0.73
0.75	19.05	0.12	3.048	0.16	4.064	0.668	0.0621	0.74
0.75	19.05	0.16	4.064	0.08	2.032	0.688	0.0639	0.73
0.75	19.05	0.16	4.064	0.12	3.048	0.712	0.0661	0.74
0.75	19.05	0.16	4.064	0.16	4.064	0.735	0.0683	0.74

Note: The data in this table are for use with the orifice equation. The cross-sectional areas and flow coefficients are for double doors 7 ft (2.13 m) high, 6 ft (1.83 m) wide, 1.75 in. (44.5 mm) thick, and with a doorstop protruding 0.62 in. (15.7 mm) from the frame. The flow coefficients were evaluated by the gap method at 0.15 in. H_2O (37.3 Pa).

Table 3.7: Flow Areas for the Gaps around Double Doors 6 ft (1.83 m) Wide with $C = 0.65$

Gap Thickness at Bottom (Figure 3.3c)		Gap Thickness at Top and Sides (Figure 3.3d)		Gap Thickness at Center (Figure 3.3e)		Flow Area	
in.	mm	in.	mm	in.	mm	ft^2	m^2
0.25	6.36	0.02	0.508	0.08	2.032	0.200	0.0185
0.25	6.36	0.02	0.508	0.12	3.048	0.234	0.0217
0.25	6.36	0.02	0.508	0.16	4.064	0.267	0.0248
0.25	6.36	0.08	2.032	0.08	2.032	0.305	0.0284
0.25	6.36	0.08	2.032	0.12	3.048	0.340	0.0316
0.25	6.36	0.08	2.032	0.16	4.064	0.373	0.0346
0.25	6.36	0.12	3.048	0.08	2.032	0.391	0.0363
0.25	6.36	0.12	3.048	0.12	3.048	0.426	0.0396
0.25	6.36	0.12	3.048	0.16	4.064	0.459	0.0426
0.25	6.36	0.16	4.064	0.08	2.032	0.472	0.0438
0.25	6.36	0.16	4.064	0.12	3.048	0.506	0.0471
0.25	6.36	0.16	4.064	0.16	4.064	0.540	0.0501
0.50	12.70	0.02	0.508	0.08	2.032	0.350	0.0325
0.50	12.70	0.02	0.508	0.12	3.048	0.385	0.0358
0.50	12.70	0.02	0.508	0.16	4.064	0.418	0.0388
0.50	12.70	0.08	2.032	0.08	2.032	0.456	0.0424
0.50	12.70	0.08	2.032	0.12	3.048	0.491	0.0456
0.50	12.70	0.08	2.032	0.16	4.064	0.524	0.0487
0.50	12.70	0.12	3.048	0.08	2.032	0.542	0.0504
0.50	12.70	0.12	3.048	0.12	3.048	0.577	0.0536
0.50	12.70	0.12	3.048	0.16	4.064	0.610	0.0566
0.50	12.70	0.16	4.064	0.08	2.032	0.623	0.0579
0.50	12.70	0.16	4.064	0.12	3.048	0.657	0.0611
0.50	12.70	0.16	4.064	0.16	4.064	0.690	0.0641
0.75	19.05	0.02	0.508	0.08	2.032	0.501	0.0466
0.75	19.05	0.02	0.508	0.12	3.048	0.536	0.0498
0.75	19.05	0.02	0.508	0.16	4.064	0.569	0.0528
0.75	19.05	0.08	2.032	0.08	2.032	0.607	0.0564
0.75	19.05	0.08	2.032	0.12	3.048	0.641	0.0596
0.75	19.05	0.08	2.032	0.16	4.064	0.675	0.0627
0.75	19.05	0.12	3.048	0.08	2.032	0.693	0.0644
0.75	19.05	0.12	3.048	0.12	3.048	0.727	0.0676
0.75	19.05	0.12	3.048	0.16	4.064	0.760	0.0706
0.75	19.05	0.16	4.064	0.08	2.032	0.774	0.0719
0.75	19.05	0.16	4.064	0.12	3.048	0.808	0.0751
0.75	19.05	0.16	4.064	0.16	4.064	0.841	0.0781

Note: The data in this table are for use with the orifice equation. Flow areas are for double doors 7 ft (2.13 m) high, 6 ft (1.83 m) wide, 1.75 in. (44.5 mm) thick, and with a doorstop protruding 0.62 in. (15.7 mm) from the frame. The flow areas were evaluated by the gap method at 0.15 in. H_2O (37.3 Pa).

Table 3.9 lists the area ratios of walls and floors of commercial buildings. This table lists values for tight, average, and loose leakage which are based on the results of the field tests by Tamura and Wilson (1966), Tamura and Shaw (1976a; 1976b; 1978), and Shaw, Reardon, and Cheung (1993). Floor leakage does not account for gaps that sometimes exist between the floor and curtain walls.

For a large opening such as an open doorway or window to a fire compartment, bidirectional flow is common. For such an opening, the flow area is the geometric area of the opening, and the flow coefficient of 0.7 is commonly used.

EFFECTIVE AREAS

Effective flow areas were essential in the early days of smoke control to simplify flow networks. With today's software (Chapter 14), there is little need for network simplification. The effective flow area concept is still used (1) with the equation approach for analysis of pressurized stairwells, (2) with analysis of elevator piston effect, (3) to reduce data input to network models, and (4) to solve some problems without calculations.

The effective area of a system of flow areas is the area that results in the same flow as the system when it is subjected to the same pressure difference over the total system of flow paths. The effective area concept is an inexact analogy to the analysis of systems of electric resistances. The paths in a system can be in parallel with one another, in series, or a combination of parallel and series paths. Three parallel flow paths from a pressurized space are shown in Figure 3.14, and three paths in series are shown in Figure 3.15.

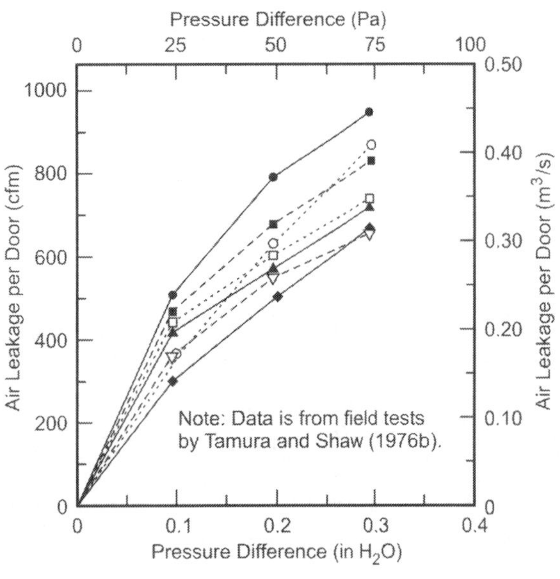

Figure 3.13 *Air leakage of closed elevator doors.*

The effective area of any number of flow paths in parallel is

$$A_e = \sum_{i=1}^{n} A_i \tag{3.26}$$

and the effective area of any number of paths in series is

$$A_e = \left(\sum_{i=1}^{n} \frac{1}{A_i^2}\right)^{-1/2} \tag{3.27}$$

where

A_e = effective flow area, ft^2 (m^2),

A_i = flow area of path i, ft^2 (m^2).

Equations 3.26 and 3.27 are based on having the same flow coefficients and temperatures for all the paths

Table 3.8: Flow Areas for Closed Elevator Doors with $C = 0.65$

Width of Door Opening		Leakage	Flow Area	
ft	m		ft^2	m^2
3.0	0.914	Tight	0.41	0.038
		Average	0.50	0.047
		Loose	0.65	0.060
3.5	1.07	Tight	0.43	0.040
		Average	0.53	0.049
		Loose	0.68	0.063
4.0	1.22	Tight	0.45	0.042
		Average	0.55	0.052
		Loose	0.72	0.066
4.5	1.37	Tight	0.47	0.044
		Average	0.58	0.054
		Loose	0.75	0.069
5.0	1.52	Tight	0.49	0.046
		Average	0.60	0.056
		Loose	0.78	0.072

Note: The data in this table are for use with the orifice equation. The elevator doors are 7 ft (2.13 m) high. The flow areas were evaluated at 0.2 in H_2O (50 Pa) from data from Tamura and Shaw (1976b). For an open elevator doors, Tamura and Shaw evaluated a flow area of 6 ft^2 (0.56 m^2).

Table 3.9: Flow Areas of Walls and Floors of Commercial Buildings[1]

Construction Element	Leakage	Area Ratio[2] Leakage Area per Unit Wall Area in²/ft²	ft²/ft²	m²/m²
Exterior building walls (includes construction crack and cracks around windows and doors)	Tight	7.2×10^{-3}	5.0×10^{-5}	5.0×10^{-5}
	Average	2.5×10^{-2}	1.7×10^{-4}	1.7×10^{-4}
	Loose	5.0×10^{-2}	3.5×10^{-4}	3.5×10^{-4}
	Very Loose	1.7×10^{-1}	1.2×10^{-3}	1.2×10^{-3}
Stairwell walls (includes construction cracks but not cracks around windows or doors)	Tight	2.0×10^{-3}	1.4×10^{-5}	1.4×10^{-5}
	Average	1.6×10^{-2}	1.1×10^{-4}	1.1×10^{-4}
	Loose	5.0×10^{-2}	3.5×10^{-4}	3.5×10^{-4}
Elevator shaft walls (includes construction cracks but not cracks around doors)	Tight	2.6×10^{-2}	1.8×10^{-4}	1.8×10^{-4}
	Average	1.2×10^{-1}	8.4×10^{-4}	8.4×10^{-4}
	Loose	2.6×10^{-1}	1.8×10^{-3}	1.8×10^{-3}
Construction Element	**Leakage**	**Area Ratio[2] Leakage Area per Unit Floor Area in²/ft²**	**ft²/ft²**	**m²/m²**
Floors[3] (includes construction cracks and gaps around penetrations)	Tight[4]	9.5×10^{-4}	6.6×10^{-6}	6.6×10^{-6}
	Average	7.5×10^{-3}	5.2×10^{-5}	5.2×10^{-5}
	Loose[4]	2.4×10^{-2}	1.7×10^{-4}	1.7×10^{-4}

[1]The data in this table are for use with the orifice equation. Flow area ratios for flow coefficient of $C = 0.65$ at 0.3 in. H_2O (75 Pa).

[2]Values of area ratios based on pressurization measurements in buildings by Tamura and Wilson (1966), Tamura and Shaw (1976a; 1976b; 1978) and Shaw, Reardon, and Cheung (1993).

[3]Floor leakage does not account for gaps that can be between a floor and a curtain wall.

[4]Values extrapolated from the average floor tightness based on range of tightness of other construction components.

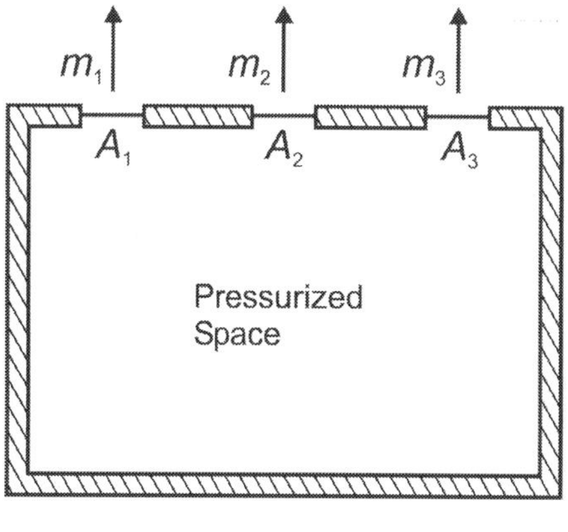

Figure 3.14 *Flow paths in parallel.*

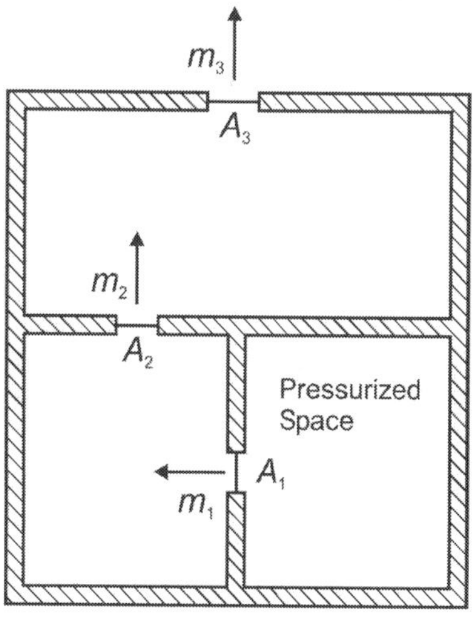

Figure 3.15 *Flow paths in series.*

Example 3.6. Effective Flow Area for Parallel Paths

In Figure 3.14, for A_1 of 1.0 ft² (0.093 m²) and A_2 and A_3 both of 0.5 ft² (0.046 m²), what is the effective flow area of the system?

$$A_e = A_1 + A_2 + A_3 = 1.0 + 0.5 + 0.5 = 2.0 \text{ ft}^2 \ (0.19 \text{ m}^2)$$

Example 3.7. Effective Flow Area for Series Paths

Part 1: Calculate the effective leakage area of two equal paths of 1.0 ft² (0.093 m²) in series.

$$A_e = \left(\frac{1}{A_1^2} + \frac{1}{A_2^2}\right)^{-1/2} = \left(\frac{1}{1} + \frac{1}{1}\right)^{-1/2} = 0.707 \text{ ft}^2 (0.215 \text{ m}^2)$$

Part 2: Calculate the effective leakage area for A_1 of 1.0 ft² (0.093 m²) and A_2 of 10 ft² (0.93 m²).

$$A_e = \left(\frac{1}{A_1^2} + \frac{1}{A_2^2}\right)^{-1/2} = \left(\frac{1}{1} + \frac{1}{10^2}\right)^{-1/2} = 0.995 \text{ ft}^2 (0.924 \text{ m}^2)$$

This shows that when one path is much smaller than the other, the effective flow path is a little less than the smaller one.

Example 3.8. Effective Flow Area for Combination of Paths

In Figure 3.16, what is the effective flow area of the system with the flow areas below?

$A_1 = A_2 = 1.0$ ft² (0.093 m²) and $A_3 = A_4 = A_5 = 2.0$ ft² (0.19 m²).

All the effective area of the parallel path is calculated first.

The effective area of A_2 and A_3 is $A_{e23} = A_2 + A_3 = 1 + 2 = 3$ ft² (0.279 m²).

The effective area of A_4 and A_5 is $A_{e45} = A_4 + A_5 = 2 + 2 = 4$ ft² (0.372 m²).

The effective area of the system is $A_e = \left(\frac{1}{A_1^2} + \frac{1}{A_{e23}^2} + \frac{1}{A_{e45}^2}\right)^{-1/2} = \left(\frac{1}{1} + \frac{1}{3^2} + \frac{1}{4^2}\right)^{-1/2} = 0.92 \text{ ft}^2 (0.085 \text{ m}^2)$.

Note: This example only applies if the only source of pressurization air is the space shown in Figure 3.14. If air is supplied to or exhausted from other spaces, the calculations of this example are not appropriate.

in the total system of flow paths. Example 3.6 illustrates parallel paths, and Example 3.7 illustrates series paths. Figure 3.16 shows a system of both parallel and series paths, and the method of developing an effective area for such a system is to combine parallel paths first and then series paths. This is illustrated in Example 3.8.

SYMMETRY

As with effective areas, symmetry was essential in the early days of smoke control to simplify flow networks. The symmetry concept is used with the equation approach for smoke control applications such as stairwell pressurization. Figure 3.17 is a floor plan of a multistory building that can be divided in half by a plane of symmetry. The flow areas on one side of this plane are equal to corresponding areas on the other side. It is not necessary that the building be geometrically symmetric as shown in this figure, but it must be symmetric with respect to flow.

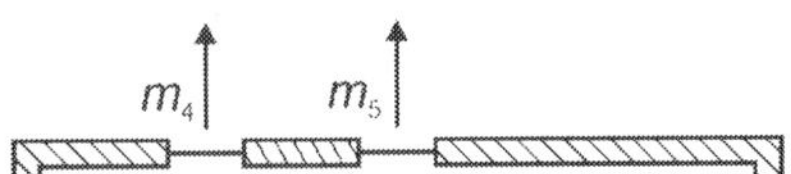

Figure 3.16 *Flow paths in parallel and series.*

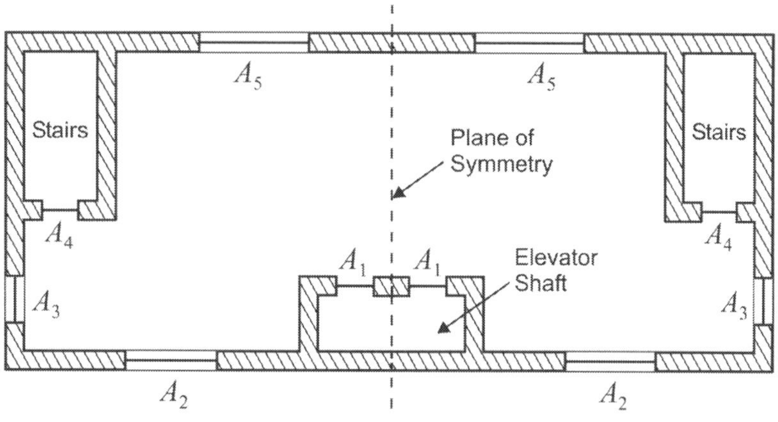

Figure 3.17 *Floor plan of building floor illustration flow symmetry.*

To apply symmetry to a building, every floor must be such that it can be divided in the same manner by the plane of symmetry. Consider that the stairwells in Figure 3.17 are pressurized. Only one side of the plane of symmetry needs to be analyzed, because the flows and pressure differences are the same on the other side. This means that when calculations are made for an open stairwell door, the corresponding door on the other side of the plane of symmetry also is open. If wind effects are included in the analysis, the wind direction can only be parallel to the plane of symmetry.

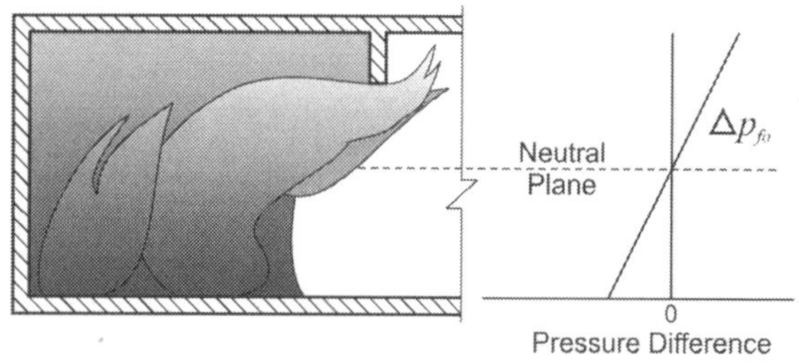

Figure 3.18 *Pressure difference due to a fully developed room fire.*

DRIVING FORCES

The driving forces of smoke movement are

- buoyancy of combustion gases,
- expansion of combustion gases,
- fan-powered ventilation systems,
- elevator piston effect,
- stack effect, and
- wind.

All but the first two are also driving forces of air movement under nonfire conditions. These forces are discussed in the following text.

Buoyancy of Combustion Gases

Hot smoke from a fire has a buoyancy force due to reduced density. Figure 3.18 shows the pressure difference due to buoyancy resulting from a fully developed room fire. The temperature of the fire room is greater near the ceiling than lower in the room. By using an average fire room temperature for T_I in the equations for bidirectional flow, rough estimates for pressure differences and flows due to buoyancy can be obtained. Fang (1980) studied pressure differences caused by the buoyancy of combustion gases in a room fire during a series of full scale fire tests. During these tests, the maximum pressure difference reached was 0.064 in. H_2O (16 Pa) across the burn room wall near the ceiling. Example 3.9 shows how to calculate the pressure difference due to buoyancy of a fully developed fire.

Expansion of Combustion Gases

The release of energy in a fire causes gases to expand. In a fire room with only one opening to the building, air will flow into the room and hot smoke (or fire gases) will flow out of the room. The thermal properties of smoke can be considered to be the same as those of air. The mass of the fuel is small compared to the airflow, and this small mass can be neglected here. For a fully developed fire, the flows are quasi-steady, and the ratio of the volumetric flows can be expressed as a ratio of absolute temperatures:

$$\frac{V_{out}}{V_{in}} = \frac{T_{out}}{T_{in}} \tag{3.28}$$

where

V_{out} = volumetric flow of smoke out of the fire compartment, cfm (m^3/s),

V_{in} = volumetric flow of air into the fire compartment, cfm (m^3/s),

Example 3.9. Buoyancy Pressure Difference

In Figure 3.18, the average temperature of a room with a fully developed fire is 2700°F (1480°C), and the temperature away from the fire is 70°F (21°C). What is the pressure difference 4.5 ft (1.37 m) above the neutral plane? Consider the room to be at standard atmospheric pressure. The fire room temperature is $T_1 = 2700 + 460 = 3160$°R, and the temperature of the surrounding space is $T_2 = 70 + 460 = 530$°R.

$$\Delta p_{12} = 7.63\left(\frac{1}{T_2} - \frac{1}{T_1}\right)z = 7.63\left(\frac{1}{530} - \frac{1}{3160}\right)(4.5) = 0.054 \text{ in. } H_2O \text{ (13 Pa)}$$

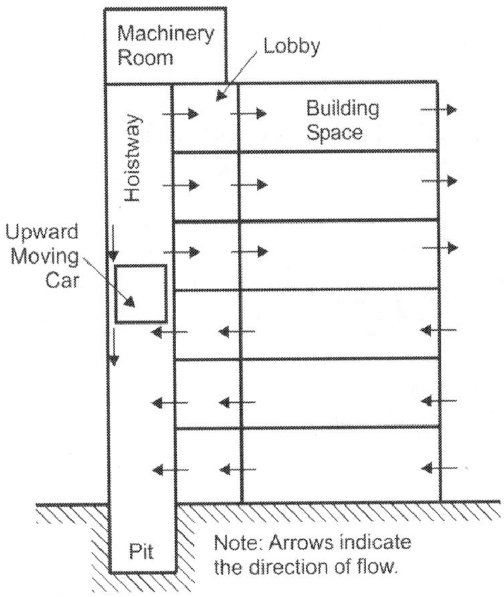

Figure 3.19 *Airflow due to an ascending elevator car.*

T_{out} = absolute temperature of smoke leaving the fire compartment, °R (K),

T_{in} = absolute temperature of air entering the fire compartment, °R (K).

For fire gas temperature of 2200°F (1260°C), the gas will expand to about five times its original volume. For a fire room with open doors or windows, the pressure difference across these openings due to expansion is negligible because of the large flow areas involved. However, for a fire space without open doors or windows, the pressure differences due to expansion may be important, provided there is sufficient oxygen to support combustion for a significant time.

Fan-Powered Ventilation Systems

The current code requirements for heating, ventilating, and air-conditioning (HVAC) systems started with a 1939 report by the National Board of Fire Underwriters (NBFU 1939). The NBFU examined NFPA fire data from January 1936 to April 1938. Of 25 fires recorded, 19 involved combustion of parts of the air-moving system. In five cases of no fire in the HVAC system, smoke was distributed by the system. Modern HVAC systems are built of materials intended to withstand fires. Also, modern HVAC systems either shut down in the event of a fire or go into a smoke control mode of operation. This mode of operation is called zoned smoke control (Chapter 13).

Elevator Piston Effect

The motion of an elevator car causes transient pressures that are called elevator piston effect. Figure 3.19 shows the airflows resulting from an upward-moving elevator car. A joint US/Canadian project developed analytical equations for elevator piston effect in shafts with and without pressurization (Klote and Tamura 1986,1987; Klote 1988).

The joint US/Canadian project also included experiments of piston effects of an elevator in a hotel in Mississauga, Ontario, Canada. This elevator served each floor of the 15-story building, and the hoistway was pressurized by a vaneaxial fan. Figure 3.20 is a comparison of measured and calculated pressure differences due to an elevator car ascending from the ground floor to the top floor. The general trends of the calculations are in agreement with the measurements. Without elevator car motion, the background pressurization of the elevator shaft is slightly less than 0.15 in. H_2O (37 Pa). It can be seen from this figure that piston effect caused a rapid drop in pressure on the ground floor followed by a gradual pressure increase as the car moved upward. On the top floor (Floor 15), the piston effect caused a gradual pressure increase. On a middle floor, the pressure increases as the car approaches, and it drops suddenly as the car passes and increases after it travels away.

The upper limit of piston effect is the largest maximum pressure difference produced by piston effect. For the calculated values on Floor 15 shown in Figure 3.20, the upper limit is about 0.12 in. H_2O (30 Pa) above the background pressurization at Floor 15.

The upper limit of piston effect for an elevator without enclosed lobbies is

$$\Delta p_{u,si} = \frac{1.66 \times 10^{-6}\rho}{2}\left(\frac{A_s A_e U}{A_a A_{si} C_c}\right)^2$$

$$\Delta p_{u,si} = \frac{\rho}{2}\left(\frac{A_s A_e U}{A_a A_{si} C_c}\right)^2 \text{ for SI} \quad (3.29)$$

and the upper limit of piston effect for an elevator with enclosed lobbies is

$$\Delta p_{u,ir} = \frac{1.66 \times 10^{-6}\rho}{2}\left(\frac{A_s A_e U}{A_a A_{ir} C_c}\right)^2$$

$$\Delta p_{u,ir} = \frac{\rho}{2}\left(\frac{A_s A_e U}{A_a A_{ir} C_c}\right)^2 \text{ for SI} \quad (3.30)$$

where

$\Delta p_{u,si}$ = upper limit pressure difference from the shaft to the building, in. H_2O (Pa),

$\Delta p_{u,ir}$ = upper limit pressure difference from the lobby to the building, in. H_2O (Pa),

ρ = air density in hoistway, lb/ft^3 (kg/m^3),

A_s = cross-sectional area of shaft, ft^2 (m^2),

A_{si} = leakage area between shaft and building, ft^2 (m^2),

A_{ir} = leakage area between building and lobby, ft^2 (m^2),

A_a = free area around the elevator car, ft^2 (m^2),

A_e = effective area, ft^2 (m^2),

U = elevator car velocity, fpm (m/s),

C_c = flow coefficient for flow around car, dimensionless.

The flow coefficient C_c was determined experimentally (Klote and Tamura 1986) at about 0.94 for a multiple car hoistway and 0.83 for a single car hoistway. The free area around the elevator car is the cross-sectional area of the shaft less the cross-sectional area of the car. For an elevator without enclosed lobbies, the effective area is

$$A_e = \left(\frac{1}{A_{si}^2} + \frac{1}{A_{io}^2}\right)^{-1/2} \quad (3.31)$$

and the effective area for an elevator with enclosed lobbies is

$$A_e = \left(\frac{1}{A_{sr}^2} + \frac{1}{A_{ir}^2} + \frac{1}{A_{io}^2}\right)^{-1/2} \quad (3.32)$$

where

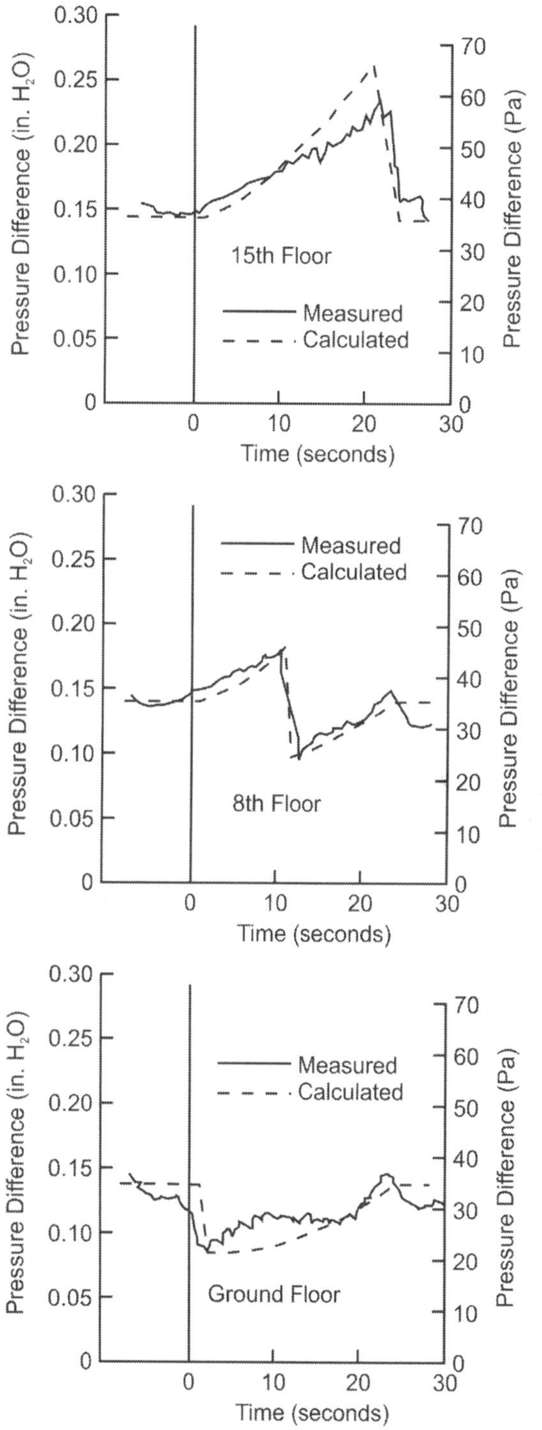

Figure 3.20 *Pressure differences due to piston effect of an ascending elevator car.*

Example 3.10. Upper Limit of Piston Effect with Enclosed Lobbies

Determine the upper limit of piston effect for a two car elevator shaft with the following parameters:

A_{sr} = 1.60 ft^2 (0.149 m^2), A_{ir} = 0.42 ft^2 (0.039 m^2), A_{io} = 0.54 ft^2 (0.0502 m^2), A_s = 121 ft^2 (11.2 m^2), A_a = 79.8 ft^2 (7.43 m^2), ρ = 0.075 lb/ft^3 (1.20 kg/m^3), U = 500 ft/min (2.54 m/s), C_c = 0.94.

The effective area for an elevator with enclosed lobbies is

$$A_e = \left(\frac{1}{A_{sr}^2} + \frac{1}{A_{ir}^2} + \frac{1}{A_{io}^2}\right)^{-1/2} = \left(\frac{1}{1.6^2} + \frac{1}{0.42^2} + \frac{1}{0.54^2}\right)^{-1/2} = 0.325 \text{ ft}^2 \ (0.302 \text{ m}^2).$$

The upper limit of piston effect for an elevator with enclosed lobbies is

$$\Delta p_{u,\,ir} = \frac{1.66 \times 10^{-6}\rho}{2}\left(\frac{A_s A_e U}{A_a A_{ir} C_c}\right)^2 = \frac{1.66 \times 10^{-6}(0.075)}{2}\left(\frac{121(0.325)(500)}{79.8(0.42)(0.94)}\right)^2 = 0.024 \text{ in. H}_2\text{O (6 Pa)}.$$

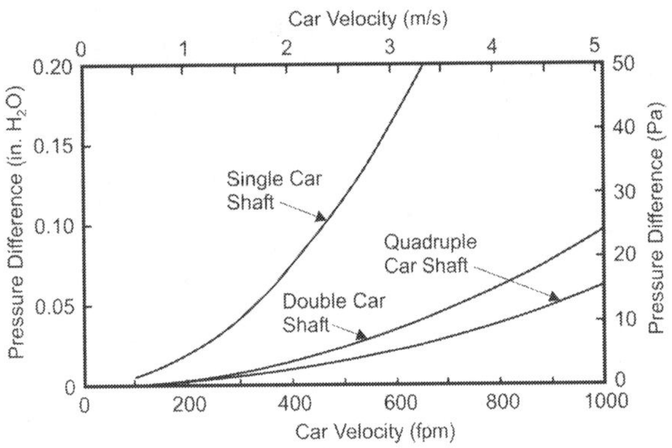

Figure 3.21 *Calculated upper limit of piston effect across elevator lobby doors.*

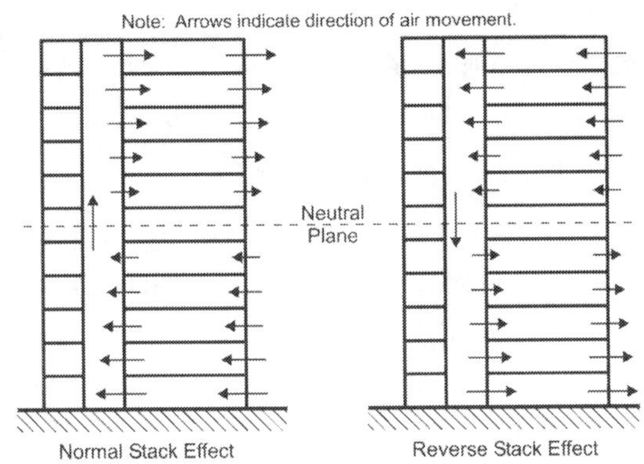

Figure 3.22 *Airflow due to normal and reverse stack effect.*

A_{sr} = leakage area between shaft and lobby, ft^2 (m^2),

A_{io} = leakage area between the building and the outdoors, ft^2 (m^2).

Example 3.10 illustrates calculation of the upper limit of piston effect for an elevator with lobbies. Figure 3.21 shows the upper limit of piston effect from the lobby to the building for car velocities from 100 to 1000 fpm (1 to 5 m/s). All elevator velocities are in this range with the exception of those of extremely tall buildings. These curves were calculated from Equation 3.30. The parameters used for the curve of double car shaft are the same as those of Example 3.10. The other curves were calculated with compatible parameters. At 1000 fpm (5 m/s), the upper limit of the double car shaft is about 0.1 in. H_2O (25 Pa). If the upper limit of piston effect is less than or equal to the minimum design pressure difference of a pressurized elevator, piston effect will not be a problem. Because the upper limit is greater for the single car shaft than multiple car shafts, piston effect is of more concern with single car shafts.

Stack Effect

It is common to have an upward flow of air in building shafts during winter. These shafts include stairwells, elevator shafts, dumbwaiters, and mechanical shafts. The upward flow is caused by the buoyancy of warm air relative to the cold outdoor air. This upward flow is similar to the upward flow in smoke stacks, and it is from this analogy that the upward flow in shafts got the name stack effect. In summer, there is a downward flow in shafts. The upward flow in shafts is called normal stack effect, and the downward flow in shafts is called reverse stack effect.

Figure 3.22 shows both kinds of stack effect. In normal stack effect, air flows into the building below the neutral plane, flows up building shafts, and out of the building above the neutral plane. As already stated, the neutral plane is a horizontal plane where the pressure

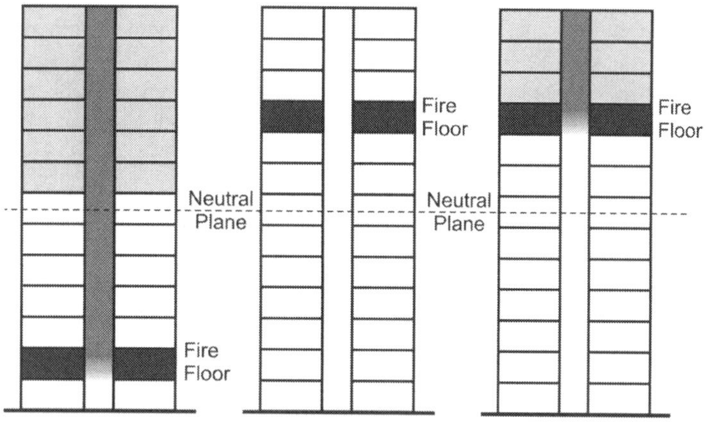

Figure 3.23 *Smoke movement in a high-rise building due to stack effect.*

inside the shaft equals the outdoor pressure. The flows due the reverse stack effect are also as shown in Figure 3.22.

Stack effect can have a significant impact on smoke flow during building fires. When it is cold outdoors, the upward airflow in shafts can be enhanced by the buoyancy of the smoke. Figure 3.23 shows smoke flows in a building subjected to normal stack effect. For a fire below the neutral plane, smoke tends to enter and flow up shafts, and above the neutral plane this smoke flows from the shaft to building spaces (Figure 3.23a). Smoke from a fire above the neutral plane can flow through cracks and gaps in the floor to the floor above the fire, but the forces of stack effect work to prevent smoke from entering shafts as shown in Figure 3.23b. If the smoke from a fire above the neutral plane has sufficient buoyancy to overcome stack effect and flow into a shaft, it will flow up the shaft and infiltrate floors above the fire floor as can be seen in Figure 3.23c.

The smoke flow shown in Figure 2.23 is for a building without pressurized stairwells or any other pressurization system. Smoke flow in buildings with pressurization systems can be analyzed by the methods described in Chapter 19.

Pressure Difference

The equations for bidirectional flow discussed apply to stack effect. Equations 3.15 and 3.16 can be rewritten for stack effect as

$$\Delta p_{SO} = \frac{0.00598 g p_{atm}}{R}\left(\frac{1}{T_O} - \frac{1}{T_S}\right)z$$

$$\Delta p_{SO} = \frac{g p_{atm}}{R}\left(\frac{1}{T_O} - \frac{1}{T_S}\right)z \text{ for SI} \tag{3.33}$$

and at standard atmospheric pressure this is

$$\Delta p_{SO} = 7.63\left(\frac{1}{T_O} - \frac{1}{T_S}\right)z$$

$$\Delta p_{SO} = 3460\left(\frac{1}{T_O} - \frac{1}{T_S}\right)z \text{ for SI} \tag{3.34}$$

where

Δp_{SO} = pressure difference from shaft to the outdoors, in. H_2O (Pa),

g = acceleration due to gravity, ft/s^2 (m/s^2),

p_{atm} = absolute atmospheric pressure, lb/ft^2 (Pa),

R = gas constant of air, 53.34 ft·lbf/lbm·°R (287.0 J/kg·K),

T_S = absolute temperature of the shaft, °R (K),

T_O = absolute temperature of the outdoors, °R (K),

z = distance above the neutral plane, ft (m).

Equations 3.33 and 3.34 have been verified by field tests on several buildings (Tamura and Wilson 1966, 1967a, 1967b), and full scale tests of stack effect were conducted at the Canadian Research Tower near Ottawa (Tamura and Klote 1988). Because of unique features of this experimental facility, it was possible to do these tests for building leakage area ratios, A_{si} / A_{io}, of 1.7, 3.4, and 7 (where A_{si} is the flow area between the shaft and the building and A_{io} is the flow area between the building and the outdoors). Figure 3.24 shows pressure differences due to stack effect for A_{si} / A_{io} of seven and from these tests at various outdoor temperatures, and the agreement between the measured and calculated values can be seen.

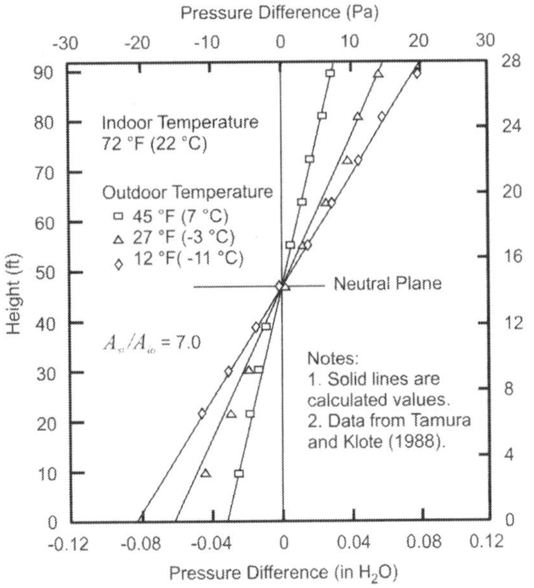

Figure 3.24 *Pressure differences due to stack effect at the Canadian Research Tower.*

For conditions of steady flow, the location of the neutral plane can be determined by an analysis that includes conservation of mass such as is done with a network computer model such as CONTAM (Chapter14). For the simple situation of a continuous opening and of two flow paths, the mass flows and the location of the neutral plane can be evaluated as discussed in the section on bidirectional flow. However, the neutral plane is often located near the mid height of a building.

Example 3.11 illustrates calculation of the pressure difference due to stack effect, and Example 3.12 illustrates calculation of the neutral plane. The pressure difference due to stack effect is nearly proportional to (1) the building height, and (2) the temperature difference from the shaft to the outdoors.

Myth: It is a myth that the pressure difference due to stack effect is nearly proportional to the temperature difference between the building and the outdoors.

Fact: This pressure difference is nearly proportional to the temperature difference between a shaft and the outdoors.

When looking at Figure 3.22, it is easy to see that people might consider the shaft temperature to be the same as the building temperature, and often it is. However, shafts that have one or more walls on the outside of the building tend to be relatively cold in winter and warm in summer, and this can have a major impact on stack effect.

Example 3.11. Pressure due to Stack Effect

For a building with a stairwell 1000 ft (305 m) tall, what is the pressure difference due to stack effect when the temperature of the stairwell is 73°F (23°C), and the outdoor temperature is 20°F (–7°C)? The shaft temperature is $T_S = 73 + 460 = 533$°R (296 K), and the outdoor temperature is $T_O = 20 + 460 = 480$°R (267 K). Consider the neutral plane to be at the mid height of the building which is $z = 500$ ft.

The pressure difference due to stack effect is $\Delta p_{SO} = 7.63\left(\frac{1}{T_O} - \frac{1}{T_S}\right)z = 7.63\left(\frac{1}{480} - \frac{1}{533}\right)(500) = 0.79$ in. H_2O (197 Pa).

This illustrates how large stack effect can be in extremely tall buildings.

Example 3.12. Location of Neutral Plane

For the building of Example 3.11, what is the location of the neutral plane if the building leakage is continuous? $H = 1000$ ft (305 m)?

For stack effect, Equation 3.19 is written as $H_n = \frac{H}{1 + (T_S/T_O)^{1/3}} = \frac{1000}{1 + (533/480)^{1/3}} = 491$ ft (150 m)

It can be seen that this is nearly at the midheight of the shaft.

Equations 3.33 and 3.34 give good predictions for buildings with normal floor leakage provided that all of the building shafts are similar. By similar it is meant that (1) the shafts are at the same temperature, (2) the shafts serve the same floors, and (3) the leakage of each shaft is about the same from floor to floor.

For example, consider a 20-story building with two stairwells, an elevator shaft, and a mechanical shaft. Each shaft is at the same temperature. Each shaft serves Floors 1 to 20. Each shaft has flow areas between it and the rest of the building. All walls have some flow areas in the form of construction cracks and gaps or cracks around penetrations. It is well known that the leakage of walls depends more on workmanship than materials. Elevator doors have relatively large gaps at the top, sides, and bottom. Stairwell doors usually have smaller gaps around them. The leakage areas from the shafts to the building are: largest for the elevator shaft, next largest for the stairwells, and smallest for the mechanical shaft. For each shaft, the leakage from the shaft to the building is about the same from floor to floor. For this building, Equations 3.33 and 3.34 would be applicable. Also, for this building, the basic ideas of stack effect are applicable.

In the 20-story example, all the shafts could be located inside the building, and they could all be at the temperature of the building. The shafts could also all be located on exterior walls, but if the temperatures in the shafts were the same or nearly so, Equations 3.33 and 3.34 would still be applicable.

For a building with shafts of various heights and different shaft temperatures, the flows become very complicated. These flows would not look like those of Figure 3.22. Each shaft could have its own neutral plane with respect to the outdoors, and sometimes a shaft may have more than one neutral plane. Equations 3.33 and 3.34 are not applicable for such complicated buildings, but the flows and pressures in such buildings can be analyzed by a network flow model like CONTAM.

Another Meaning of Stack Effect

The term stack effect is often used in a different way from that just discussed. Sometimes engineers will say that a pressurized stairwell (or elevator) needs to be designed to account for the impact of stack effect. If the stairwell (or elevator) is properly pressurized, there is no neutral plane, and the flows do not look like those in Figure 3.22. Strictly speaking, there is no stack effect in the pressurized stairwell (or elevator). What is meant when an engineer says "a pressurized stairwell (or elevator) needs to be designed to account for the impact of stack effect" is that it needs to be designed to account for the temperature differences that cause stack effect.

Myth: It is a myth that stack effect is the major factor impacting stairwell and elevator pressurization.

Fact: Today the impact of stack effect is a minor factor for most pressurized stairwells and elevators. The pressurization air for many stairwells and elevators is untreated outdoor air that is not heated or cooled. The temperature of these shafts is often nearly the same as the outdoor temperature, and the impact of stack effect is significantly reduced as compared to shafts pressurized with air treated to the building temperature.

Wind

Above the surface of the earth, the wind velocity increases until it reaches the gradient winds. This layer of increasing wind speed is referred to as the wind boundary layer, and the roughness of the terrain impacts the boundary layer as shown in Figure 3.25. In the absence of local obstructions to the wind, the relationship between velocity and height in the boundary layer is often approximated by the power law equation:

$$U = U_o\left(\frac{z}{z_o}\right)^a \qquad (3.35)$$

where

U = wind velocity, mph (m/s),

U_o = velocity at reference elevation, mph (m/s),

z = elevation of velocity, U, ft (m),

z_o = reference elevation, ft (m),

a = wind exponent, dimensionless.

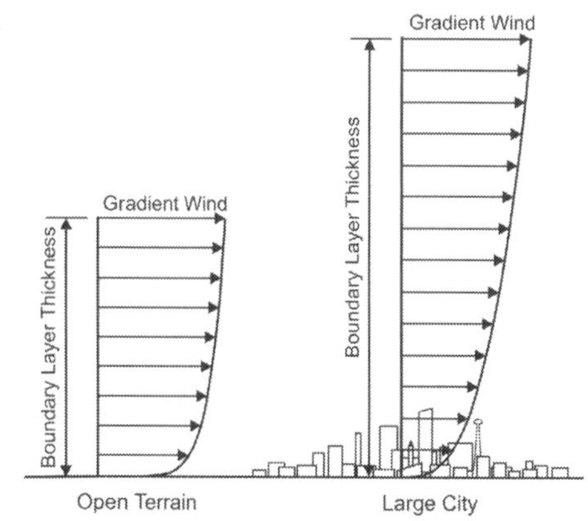

Figure 3.25 *Effect of terrain on wind velocity profile.*

Table 3.10: Boundary Layer Thickness and Exponents

Terrain Category	Description	Boundary Layer Thickness δ (ft)	Boundary Layer Thickness δ (m)	Wind Exponent *a*
1	Large city center with 50% of buildings higher than 70 ft (2 m), and over an upwind distance of at least 2600 ft (800 m) or 10 times the height of the structure, whichever is greater	1500	460	0.33
2	Urban and suburban areas, wooded areas, or other terrain with many closely spaced obstructions of the size of single-family residences, over an upwind distance of at least 1500 ft (460 m) or 10 times the height of the structure, whichever is greater	1200	370	0.22
3	Open terrain such as airports with scattered obstructions having heights generally less than 30 ft (9.1 m), including flat, open country	890	270	0.14
4	Flat, unobstructed areas exposed to wind flowing over water for at least 1 mi (1.6 km)	690	210	0.10

Note: Table adapted from ASHRAE (2009).

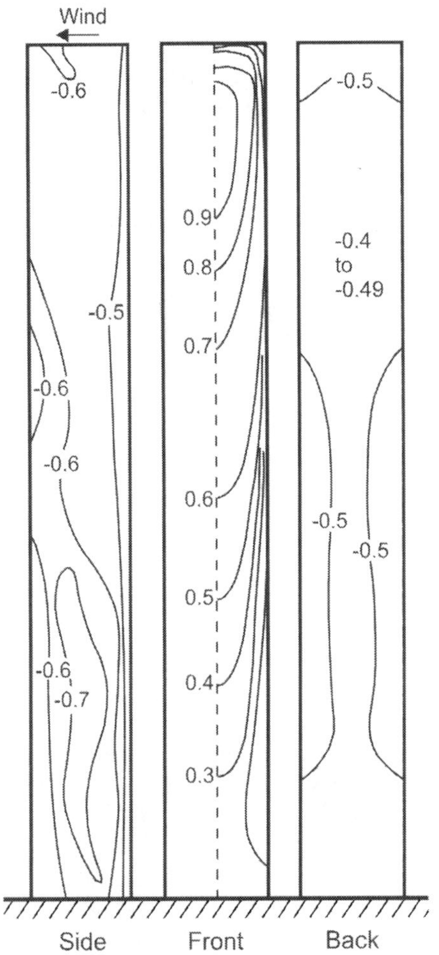

Figure 3.26 *Typical distribution of pressure coefficient over a tall building free of obstructions.*

A number of approaches have been developed for categorizing terrain and the wind exponent, and two of these approaches are discussed here. Table 3.10 lists the wind exponent a and the boundary layer thickness for four terrain categories. The data in this table are based on those from ASHRAE (2009). The wind exponent a can be different for different wind directions. For example, a building on the edge of a large city may be Category 1 for the wind blowing across the city, but it could be Category 2 for wind blowing from the opposite direction.

CONTAM uses different values of wind exponent and the boundary layer thickness than those in Table 3.10. CONTAM is a network model that is discussed in Chapter 14, and the wind exponent values and the boundary layer thickness in CONTAM also are acceptable for smoke control calculations.

The pressure that wind exerts on a wall is

$$p_w = 0.00645 C_w \rho_o U_H^2$$
$$p_w = \frac{1}{2} C_w \rho_o U_H^2 \text{ for SI} \qquad (3.36)$$

where

p_w = wind pressure, in. H_2O (Pa),

C_w = pressure coefficient, dimensionless,

ρ_o = outdoor air density, lb/ft^3 (kg/m^3),

U_H = velocity at the upwind wall of height H, mph (m/s).

Wind is constantly fluctuating, but it is the mean average wind velocity that is relevant for smoke control applications. For this handbook, the term wind velocity is used to indicate the mean average velocity.

The pressure coefficient C_w depends on wind direction, building geometry, and local obstructions to the wind. For a tall building without local wind obstructions, typical distribution of the pressure coefficient is shown in Figure 3.26. The wind is blowing directly at the front wall, and the pressure coefficients on the wall are symmetrical. The wind coefficients on this wall range from 0.3 to 0.9, and the lower values tend to be lower on the wall. The values on the side wall range from –0.7 to –0.5, and the values on the back range from 0.5 to –0.4. For rectangular buildings which are free of local obstructions, average pressure coefficients $\bar{C}_w$ are listed in Figure 3.27.

The local wind speed, U_H, at the top of a wall is needed for wind calculations.

$$U_H = U_{met}\left(\frac{\delta_{met}}{H_{met}}\right)^{a_{met}}\left(\frac{H}{\delta}\right)^{a} \tag{3.37}$$

where

U_H = velocity at the upwind wall of height H, mph (m/s),

U_{met} = measured velocity, mph (m/s),

H_{met} = height of wind measurement, ft (m),

δ_{met} = boundary layer height in the vicinity of the wind anemometer, ft (m),

a_{met} = wind exponent in the vicinity of the wind anemometer, dimensionless,

H = height of wall, ft (m),

δ = boundary layer height at wall, ft (m),

a = wind exponent at wall.

Wind is usually measured at airports, and airports generally have terrain like that of Category 3 (Table 3.10). The standard height for measuring velocity and direction is 33 ft (10 m). Example 3.13 shows the calculation of local wind speed in a large city from data measured at a nearby airport.

For additional information about wind, see Shaw and Tamura (1977), Kandola (1986a, 1986b, 1986c), Aynsley (1989), and Klote (1995). Some civil engineering texts have useful information about wind (MacDonald 1975; Liu 1991; Simiu and Scanlan 1996; Dyrbye and Hansen 1997).

Building Height Ratio	Building Plan Ratio	Elevation	Plan	Wind Angle α	$\bar{C}_w$ for Surface A	B	C	D
$\frac{h}{w} \le \frac{1}{2}$	$1 < \frac{l}{w} \le \frac{3}{2}$			0°	+0.7	−0.2	−0.5	−0.5
				90°	−0.5	−0.5	+0.7	−0.2
	$\frac{3}{2} < \frac{l}{w} \le 4$			0°	+0.7	−0.25	−0.6	−0.6
				90°	−0.5	−0.5	+0.7	−0.1
$\frac{1}{2} < \frac{h}{w} \le \frac{3}{2}$	$1 < \frac{l}{w} \le \frac{3}{2}$			0°	+0.7	−0.25	−0.6	−0.6
				90°	−0.6	−0.6	+0.7	−0.25
	$\frac{3}{2} < \frac{l}{w} \le 4$			0°	+0.7	−0.3	−0.7	−0.7
				90°	−0.5	−0.5	+0.7	−0.1
$\frac{3}{2} < \frac{h}{w} \le 6$	$1 < \frac{l}{w} \le \frac{3}{2}$			0°	+0.8	−0.25	−0.8	−0.8
				90°	−0.8	−0.8	+0.8	−0.25
	$\frac{3}{2} < \frac{l}{w} \le 4$			0°	+0.7	−0.4	−0.7	−0.7
				90°	−0.5	−0.5	+0.8	−0.1

Note: h = height to eaves or parapet; l = length (greater horizontal dimension of a building); w = width (lesser horizontal dimension of a building).

Figure 3.27 *Average pressure coefficients for walls of rectangular buildings free of local obstructions (adapted from MacDonald [1975]).*

Example 3.13. Wind Velocity in a Large City

For a building in a large city that is 470 ft (143 m) tall, what is the design wind velocity at the building corresponding to a design wind velocity based of 24 mph (10.7 m/s)? $H = 470$ ft, $H_{met} = 33$ ft, and $U_{met} = 24$ mph.

The design wind velocity is based on weather data from an airport that is terrain Category 3, and the large city is Category 1. From Table 3.10, $a = 0.33$, $\delta = 1500$ ft, $a_{met} = 0.14$, and $\delta_{met} = 890$ ft.

$$U_H = U_{met}\left(\frac{\delta_{met}}{H_{met}}\right)^{a_{met}}\left(\frac{H}{\delta}\right)^{a} = 24\left(\frac{890}{33}\right)^{0.14}\left(\frac{470}{1500}\right)^{0.33} = 26 \text{ mph } (11.6 \text{ m/s})$$

NOMENCLATURE

A = flow area (or leakage area), ft^2 (m^2); or cross-sectional area of the path, ft^2 (m^2)
a = thickness of gap in direction perpendicular to flow, in. (m); or wind exponent, dimensionless
A_a = area above neutral plane, ft^2 (m^2); or free area around the elevator car, ft^2 (m^2)
A_b = area below neutral plane, ft^2 (m^2)
A_e = effective flow area, ft^2 (m^2)
A_i = flow area of path i, ft^2 (m^2)
A_{io} = leakage area between the building and the outdoors, ft^2 (m^2)
A_{ir} = leakage area between building and lobby, ft^2 (m^2)
a_{met} = wind exponent in the vicinity of the wind anemometer, dimensionless
A_s = cross-sectional area of shaft, ft^2 (m^2)
A_{si} = leakage area between shaft and building, ft^2 (m^2)
A_{sr} = leakage area between shaft and lobby, ft^2 (m^2)
C = flow coefficient, dimensionless
C_c = flow coefficient for flow around car, dimensionless
C_e = flow coefficient for exponential flow equation, ft^3/min/(in. H_2O) [m^3/s /Pa^n]
C_w = pressure coefficient, dimensionless
D_h = hydraulic diameter, in. (m), $D_h = 2a$
f = friction factor of shaft or duct, dimensionless
g = acceleration due to gravity, ft/s^2 (m/s^2)
H = distance between the two openings, ft (m); height of the opening, ft (m); or height of wall, ft (m)
H_{met} = height of wind measurement, ft (m)
H_n = height of neutral plane, ft (m); or distance from the bottom opening to the neutral plane, ft (m)
L = length of gap, ft (m); or shaft or duct length, ft (m)
m = mass flow through the path, lb/s (kg/s)
m_{12} = mass flow from space 1 to space 2, lb/s (kg/s)
m_{21} = mass flow from space 2 to space 1, lb/s (kg/s)
m_{sv} = mass flow through the path, scfm (standard m^3/s)
n = flow exponent, dimensionless
NP = dimensionless pressure difference
NQ = dimensionless flow
P = perimeter of the path, ft (m)
p = pressure, lb/in^2 (Pa)
p_{atm} = absolute atmospheric pressure, lb/ft^2 (Pa)
p_w = wind pressure, in. H_2O (Pa)
R = gas constant, 53.34 ft lbf/lbm/°R (287 J/kg K)
R_e = Reynolds number, dimensionless
T = absolute temperature, °R (K)
T_1 = absolute temperature of space 1, °R (K)
T_2 = absolute temperature of space 2, °R (K)
T_{in} = absolute temperature of air entering the fire compartment, °R (K)
T_O = absolute temperature of the outdoors, °R (K)
T_{out} = absolute temperature of smoke leaving the fire compartment, °R (K)
T_S = absolute temperature of the shaft, °R (K)
U = average velocity, fpm (m/s); wind velocity, mph (m/s); or elevator car velocity, fpm (m/s)
U_H = velocity at the upwind wall of height H, mph (m/s)
U_{met} = measured velocity, mph (m/s)
U_o = velocity at reference elevation, mph (m/s)
V = volumetric flow, cfm (m^3/s)

V_1 = volumetric flow at Δp_1, cfm (m^3/s)

V_2 = volumetric flow at Δp_1, cfm (m^3/s)

V_{in} = volumetric flow of air into the fire compartment, cfm (m^3/s)

V_{out} = volumetric flow of smoke out of the fire compartment, cfm (m^3/s)

w = width of the opening, ft (m)

x = depth of gap in flow direction, in. (m)

z = distance above the neutral plane, ft (m); or elevation of velocity, U, ft (m)

z_o = reference elevation, ft (m)

ρ = density, lb/ft^3 (kg/m^3)

ν = kinematic viscosity, ft^2/s (m^2/s)

δ = boundary layer height at wall, ft (m)

ρ_1 = density in space 1, lb/ft^3 (kg/m^3)

ρ_2 = density in space 2, lb/ft^3 (kg/m^3)

ρ_o = outdoor air density, lb/ft^3 (kg/m^3)

δ_{met} = boundary layer height in the vicinity of the wind anemometer, ft (m)

Δp = pressure difference, in. H_2O (Pa)

Δp_1 = pressure difference across the path at V_1, in. H_2O (Pa)

Δp_{12} = pressure difference from space 1 to space 2, in. H_2O (Pa)

Δp_2 = pressure difference across the path at V_2, in. H_2O (Pa)

Δp_f = pressure loss in shaft or duct due to friction, in. H_2O (Pa)

Δp_{SO} = pressure difference from shaft to the outdoors, in. H_2O (Pa)

$\Delta p_{u,ir}$ = upper limit pressure difference from the lobby to the building, in. H_2O (Pa)

$\Delta p_{u,si}$ = upper limit pressure difference from the shaft to the building, in. H_2O (Pa)

REFERENCES

Achakji, G.Y., and G.T. Tamura. 1988. Pressure drop characteristics of typical stairshafts in high-rise buildings. *ASHRAE Transactions*, 94(1):1223–1237.

ASHRAE. 2009. *ASHRAE Handbook—Fundamentals*. Atlanta: ASHRAE.

Aynsley, R.M. 1989. The estimation of wind pressures at ventilation inlets and outlets on buildings. *ASHRAE Transactions*, 95(2).

Cresci, R.J. 1973. Smoke and fire control in high-rise office buildings—Part II: Analysis of stair pressurization systems. *Symposium on Experience and Applications on Smoke and Fire Control at the ASHRAE Annual Meeting, Louisville, KY, Atlanta, GA*, pp. 16–23.

Dyrbye, C., and S.O. Hansen. 1997. *Wind Loads on Structures*. New York, NY: Wiley.

Fang, J.B. 1980. Static pressures produced by room fires. NBSIR 80-1984, National Bureau of Standards, Gaithersburg, MD.

Forney, G.P. 1989. Personal Communications at the Center for Fire Research, National Institute of Standards and Technology, Gaithersburg, MD.

Gross, D. and W.L. Haberman. 1988. Analysis and prediction of air leakage through door assemblies. *Fire Safety Science, Proceedings of the 2nd International Symposium*, Tokyo, Japan, pp. 129–131.

Idelchik, I.E. 1986. *Handbook of Hydraulic Resistance*, 2nd ed. New York: Hemisphere Publishing.

Jones, W.W., and X.E. Bodart. 1986. Buoyancy driven flow as the forcing function of smoke transport models. NBSIR 86-3329, National Bureau of Standards, Gaithersburg, MD.

Kandola, B.S. 1986a. A wind tunnel building model for the investigation of smoke movement problems. *Fire Safety Journal*, 10(3).

Kandola, B.S. 1986b. Comparison of wind tunnel pressure measurements and smoke movement computer predictions inside a five-story model building. *Fire Safety Journal*, 10(3).

Kandola, B. S. 1986c. The effects of simulated pressure and outside wind on the internal pressure distribution in a five-story building. *Fire Safety Journal*, 10(3).

Klote, J.H. 1988. An analysis of the influence of piston effect on elevator smoke control. NBSIR-88-3751, National Bureau of Standards, Gaithersburg, MD.

Klote, J.H. 1995. Design of smoke control systems for elevator fire evacuation including wind effects. 2nd Symposium on Elevators, Fire, and Accessibility, April 19–21, Baltimore, MD.

Klote, J.H., and X. Bodart. 1985. Validation of network models for smoke control analysis. *ASHRAE Transactions* 91(2b).

Klote, J.H., and J.A. Milke. 2002. *Principles of Smoke Management*. Atlanta: ASHRAE.

Klote, J.H., and G.T. Tamura. 1986. Elevator piston effect and the smoke problem. *Fire Safety Journal*, (11)3.

Klote, J.H., and G.T. Tamura. 1987. Experiments of piston effect on elevator smoke control. *ASHRAE Transactions*, 93(2a).

Liu, H. 1991. *Wind Engineering—A Handbook for Structural Engineers*. Englewood, NJ: Prentice Hall.

MacDonald, A.J. 1975. *Wind Loading on Buildings*. New York: Wiley.

NBFU. 1939. Smoke hazards of air-conditioning systems. *NFPA Quarterly*, 33(2).

Shaw, C.T. and G.T. Tamura. 1977. The calculation of air infiltration rates caused by wind and stack action for tall buildings. *ASHRAE Transactions*, 83(2).

Shaw, C.Y., J.T. Reardon, and M.S. Cheung. 1993. Changes in air leakage levels of six canadian office buildings. *ASHRAE Journal*, 35(2).

Simiu, E., and R.H. Scanlan. 1996. *Wind Effects on Structures: Fundamentals and Application to Design*, 3rd ed. New York: Wiley.

Tamura, G.T. and J.H. Klote. 1988. Experimental fire tower studies on adverse pressures caused by stack and wind action: studies on smoke movement and control. ASTM International Symposium on Characterization and Toxicity of Smoke, December 5, Phoenix, AZ.

Tamura, G.T., and C.Y. Shaw. 1976a. Studies on exterior wall air tightness and air infiltration of tall buildings. *ASHRAE Transactions*, 82(1).

Tamura, G.T., and C.Y. Shaw. 1976b. Air leakage data for the design of elevator and stair shaft pressurization systems. *ASHRAE Transactions*, 82(2).

Tamura, G.T., and C.Y. Shaw. 1978. Experimental studies of mechanical venting for smoke control in tall office buildings. *ASHRAE Transactions*, 86(1).

Tamura, G.T., and A.G. Wilson. 1966. Pressure differences for a nine-story building as a result of chimney effect and ventilation system operation. *ASHRAE Transactions*, 72(1).

Tamura, G.T., and A.G. Wilson. 1967a. Building pressures caused by chimney action and mechanical ventilation. *ASHRAE Transactions*, 73(2).

Tamura, G.T., and A.G. Wilson. 1967b. Pressure differences caused by chimney effect in three high buildings, ASHRAE Transactions, 73(2).

CHAPTER 4

Timed Egress Analysis

James E. Milke

Some smoke management systems are explicitly designed to maintain tenability for the period required for evacuation by building occupants, while others may be designed for other time periods and other purposes. For the former, timed egress analyses need to be conducted to determine the time required for evacuation. This chapter provides an overview of the considerations to address when conducting a timed egress analysis, along with the scope and limitations of computational methods which may be employed. The results of a timed egress analysis may be used to determine if occupants can evacuate prior to the smoke layer descending to their position or may be part of an overall tenability analysis to assess the amount of time individuals might be exposed to smoke.

Elevators are not normally used for evacuation, but the idea of using them to speed up evacuation has gained considerable attention. Chapter 12 has information about elevator evacuation including calculation of egress time by the use of elevators.

TIMELINE

A timeline depicting the steps that a building occupant follows in order to evacuate is presented in Figure 4.1. Following ignition of a fire, occupants will require some time to perceive that a fire has been initiated in the building or be alerted to it by hearing the audible fire alarm in the building, seeing the flashing strobe lights, or being told of the fire by another building occupant. This delay is noted in the timeline as *recognition*, where the person needs to recognize the need to evacuate.

An occupant may not immediately initiate their evacuation even after they become aware of a fire, or hear the fire alarm in the building. The reasons for

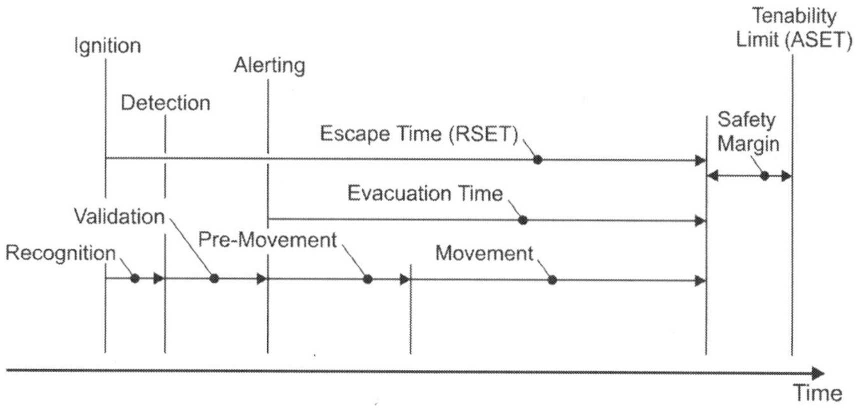

Figure 4.1 *Timeline for occupant response.*

delays are described in the section on Human Behavior near the end of this chapter. Even where the cues are obvious, such as in the DuPont Plaza fire where occupants in the casino could see smoke billowing into the casino area, individuals may continue with their activities until the urgency of the situation becomes apparent to them. Even so, the timelines of their reactions as well as the particular actions taken will be based on their perception of the severity of the incident and time available, which may be different from reality (Bryan 2008).

Once recognizing the need to evacuate, occupants may take an additional amount of time to prepare for evacuation. This may include getting dressed, finding a coat, gathering family members or other activities (Bryan 1977). This is labeled *premovement* in the timeline. In other presentations of the evacuation process, some authors may combine the *recognition*, *validation,* and *premovement* periods presented in the timeline in Figure 4.1 and incorporate it into a single premovement period.

Following the premovement period, individuals are now potentially ready to move and can begin their movement toward the exit. This movement time is usually the focus of any calculations done in timed egress studies.

Timed egress studies generally consider that once occupants become aware of the fire, the only actions they take are those associated with evacuation. Notably absent from these analyses are times associated with other actions that the individual may undertake, such as attempting extinguishment, assisting others, or calling the fire department. Further, the possibility of occupants making "wrong turns" while evacuating is typically ignored. While it's difficult to estimate the amount of time that an individual might be engaged in the various activities or spend by following a mistaken path, these times are sometimes accounted for indirectly by expanding the premovement time or applying a factor of safety.

ANALYSIS APPROACHES

The approaches followed in conducting timed egress studies for engineering purposes can be divided into two groups: (1) algebraic equation-based methods and (2) computer-based models.

The basis for any of these methods relies on data from observations of people movement during normal, everyday activities and fire drills (Proulx 2008; Bryan, 2008). Data sources of people movement from normal use activities included movement in situations such as subway stations during rush hour (London Transport Board 1958), an indoor arena (Fruin 1971), and an outdoor arena (Pauls 1980). Fire drills monitored to collect data were either announced or unannounced.

The relevance of using data from such activities for emergency movement analyses has been highly debated. Some argue that movement with fire effects provides motivation to move faster than in fire drills, so that drill data should provide a lower bound for evacuation estimates (Proulx 2008). However, in fires incidents in large buildings, people may evacuate without any secondary cues of the fire (e.g., visible smoke or odors, such that movement in a fire drill would be directly relevant). Without data from actual fire incidents, the correct answer to the debate is unknown.

ALGEBRAIC EQUATION-BASED METHODS

There are two versions of the methods involving the application of algebraic equations: *simplified method* and *component by component analysis*. The simplified version requires that a controlling element in the egress system be identified. A controlling element is one where the greatest normalized flow is expected (the normalized flow is defined as the flow rate along a path divided by a characteristic width for the path as described later in this section). The simplified version consists of three calculations: (1) time to reach controlling element, (2) time to travel through controlling element, and (3) time to travel from controlling element to outdoors (or place of safety).

These three time periods listed are determined by adopting a hydraulic analogy to assess the flows associated with evacuating building occupants. In this respect, the movement of occupants is described in terms of velocities and flow rates. The velocity is defined as expected (i.e., the distance traveled by the occupant per unit time[1]). The flow rate is defined as the number of persons per unit time who pass a particular point in the egress component (e.g., the number of persons per minute who pass through a doorway). One other useful parameter is termed the *specific flow.* The specific flow is the flow rate normalized by the effective width of the egress component[2].

In either version, the evacuation time is estimated using a global perspective; i.e., the egress time for the entire group is determined without distinguishing between occupants within the group.

1. The velocity on stairs refers to the rate of travel along a diagonal path obtained by connecting the tips of the stairs.
2. The effective width will be defined later in this section, though refers to the portion of the width of the egress components in which occupants actually travel.

Table 4.1: Proportionality Constant *k* for Egress Components

Egress Component				*k*, ft/min	*k*, m/s
Corridor, aisle, ramp, doorway				275	1.40
Stair Riser		**Stair Tread**			
in.	**mm**	**in.**	**mm**		
7.5	191	10	254	196	1.00
7.0	178	11	279	212	1.08
6.5	165	12	305	229	1.16
6.5	165	13	330	242	1.23

Velocity

The velocity that an occupant travels is dependent on several factors, including the physical abilities of the individual, the proximity of other people, and the type of egress component on which the individual is traveling (i.e., components which involve a change in elevation, such as stairs, and ramps and level walkways such as corridors and aisles) (Gwynne et al. 1999; Nelson and MacLennan 1988; Predtechenskii and Milinskii 1978). The abilities of the individual are dependent on the gender, age, and overall physical capabilities of the individual (Fruin 1987; Proulx 2008).

A correlation for the velocity as a function of density of mobile occupants is presented in the *SFPE Handbook of Fire Protection Engineering* (Gwynne and Rosenbaum 2008). This correlation is based on research by Pauls who observed evacuation drills in high-rise office buildings in the 1970s.

The occupant density parameter included in the correlation is defined as the number of individuals within a particular area of the egress component. The correlation presented in the *Handbook* applies to the situation where the density of people surrounding the individual of interest is sufficiently high so as to impede the ability of the person to move at their maximum velocity. This occurs at a density of at least 0.051 p/ft^2 (0.55 p/m^2).[3] The correlation is

$$v = k - 2.86kD$$
$$v = k - 0.266kD \text{ for SI} \tag{4.1}$$

where

v = velocity, ft/min (m/s),

k = velocity factor, ft/min (m/s),

D = density of occupant flow, p/ft^2 (p/m^2)

Velocity factors k associated with travel on various surfaces are listed in Table 4.1. The velocity determined from Equation 4.1 applies for the average, mobile individual. Children and the elderly have been observed to walk at velocities which are approximately half that of young and middle-aged adults (Proulx 2008). Very little information has been reported by Pauls or others to indicate the range of velocities that have been observed for building occupants without disabilities. The dependence of the velocity on density is presented in Figure 4.2.

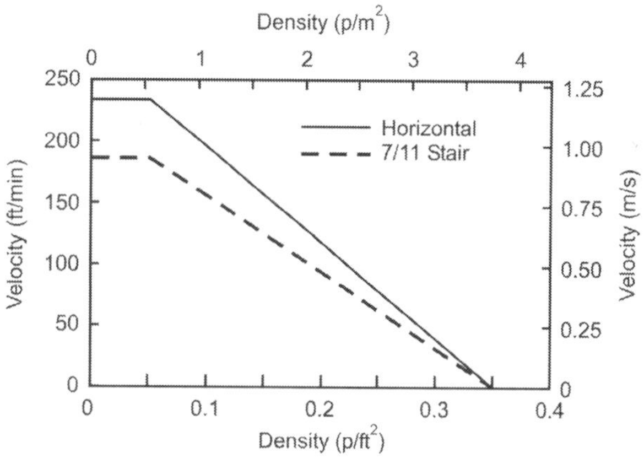

Figure 4.2 *Influence of density on mean walking velocity.*

A limited amount of data is available to estimate the velocity of mobility-impaired individuals if unimpeded by others (i.e., for low-density situations [Proulx 2008]). The mean velocity for such individuals depends on the type and extent of impairment.

Density

As noted in the previous section, density has a significant influence on walking speeds. As such, it is an important parameter in evacuation studies. The density of moving groups of individuals is typically between 0.093 p/ft^2 and 0.196 p/ft^2 (1.0 p/m^2 and 2.0 p/m^2) (Predtechenskii and Milinskii 1978; Frantzich 1996; Pauls 1980; Fruin 1987). However, a wide range of densities are feasible. A minimum density would be

3. The symbol p is for persons.

Table 4.2: Recommended Occupant Load Factors*

Space Use	Occupant Load Factor**	
	p/ft^2	p/m^2
Assembly		
Less concentrated use without fixed seating	15 net	1.4 net
Concentrated use without fixed seating	7 net	0.65 net
Waiting space	3 net	0.28 net
Library stack areas	100 gross	9.3 gross
Library reading areas	50 net	4.6 net
Mercantile		
Street floor and sales basement	30 gross	2.8 gross
Multiple street floors	40 gross	3.7 gross
Other floors	60 gross	5.6 gross
Storage, shipping	300 gross	27.9 gross
Educational		
Classroom area	20 net	1.9 net
Shops	50 net	4.6 net
Daycare centers	35 net	3.3 net
Business (offices), industrial	100 gross	9.3 gross
Hotel and apartment	200 gross	18.6 gross
Healthcare		
Sleeping departments	120 gross	11.1 gross
Impatient treatment departments	240 gross	22.3 gross
Detention and correctional	120 gross	11.1 gross

*The load factors of this table are based on experience in the United States.

**The population of a space is the product of the load factor and the net area or gross area of that space as indicated above. See NFPA 101 for a definition of the space uses and the terms of net and gross areas.

associated with situations where only one individual is located in a large egress component. Conversely, a maximum density is associated with crowd flows where individuals are virtually in contact with one another.

Rosenbaum and Gwynne express the density of a flow as the ratio of the number of people in a group in an egress component divided by the total floor area occupied by the group (including the area between individuals). Other references may express the density in terms of the portion of floor area occupied by individuals (Predtechenskii and Milinskii 1978).

In the algebraic equation-based methods and even some of the computer-based methods, the density is an input to the analysis. In some methods, the value of the density is selected which maximizes the flow rate through the component. In other methods, the initial density is based on the expected number of occupants per unit floor area for the space. The IBC provides occupant load factors.Where the building codes do not specify occupant load factors for calculating the number of people expected to occupy spaces, the factors in Table 4.2 are recommended.

Specific Flow

The flow rate of occupants along a particular egress path has been found to be linearly proportional to the portion of the width of the path that people use. The portion of the path that individuals actually use is referred to as the *effective width*. This parameter was initially identified by Pauls (1980).

Figure 4.3 depicts the effective width as compared to the clear width, which is the term typically used in building code analyses of the adequacy of the means of

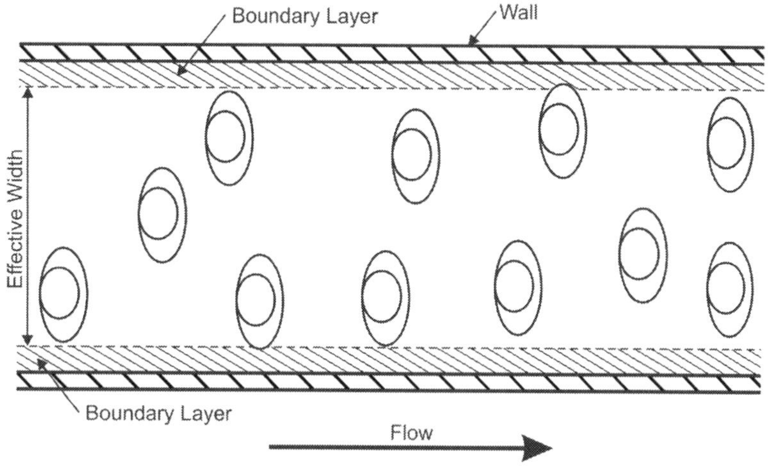

Figure 4.3 *Illustration of effective width when people move in a corridor.*

Table 4.3: Boundary Layer Width

Component	Boundary Layer in.	Boundary Layer m
Theater chairs, stadium benches	0	0
Railings, handrails[1]	3.5	0.089
Obstacles	4	0.10
Stairways, doors, archways	6	0.15
Corridor and ramp walls	8	0.20

[1]Where there are handrails, Nelson and MacLennan suggest that the boundary layer width for handrails should be used if the boundary layer width for handrails is less than that for the egress component where the handrail is present.

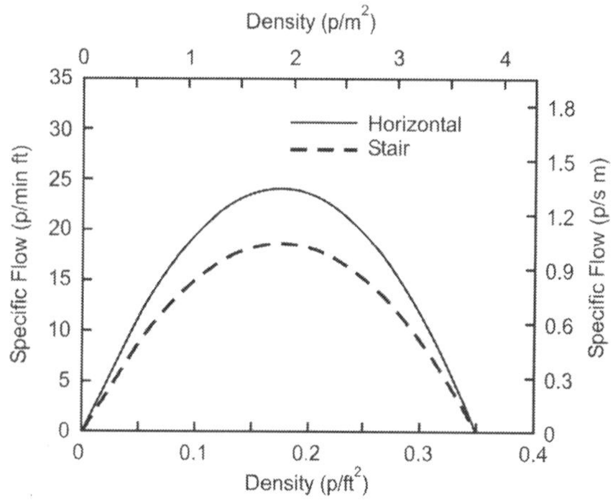

Figure 4.4 *Specific flow versus density.*

egress component. The effective width is smaller than the clear width because a small portion of the egress component is not often occupied by individuals as they tend to maintain a small distance between themselves and the wall or handrail situated at the edge of the egress component. The small distance on either side of the effective width was termed the *boundary layer* by Nelson and MacLennan, following the hydraulics analogy for evacuation analyses (1988). The width of the boundary layer for various types of egress components is presented in Table 4.3.

Following the hydraulics analogy for evacuation analyses, the specific flow is comparable to the mass flux in hydraulic systems and is defined as the product of the velocity and density of the flow:

$$F_s = Dv \tag{4.2}$$

where F_s is the specific flow in p/min·ft (p/s·m). For a density greater than 0.051 p/ft^2 (0.55 p/m^2), the velocity of Equation 4.1 can be substituted into Equation 4.2 to yield

$$F_s = (1 - aD)kD \tag{4.3}$$

The specific flow predicted by Equation 4.3 is presented in Figure 4.4. As indicated in this figure, a maximum value of the specific flow, F_{sm}, occurs at a density of 0.175 p/ft^2 (1.88 p/m^2). This density is independent of the type of egress component.

Flow

The flow rate of occupants is the product of the specific flow and effective width of the egress component.

The flow rate is an important parameter in several considerations.

The flow rate parameter may be used in a simplified method for determining the egress time in buildings. This method is described in the following section. Flow rates are also used to determine if queues form and the amount of time for dissipating queues. Queues form whenever the flow rate approaching a particular point in the egress system exceeds the maximum flow rate possible from that point. This is relevant where two egress paths merge (e.g., two corridors, or in stairwells where people entering a stairwell merge with those travelling in the stairwell from other floor levels). Queues dissipate whenever the flow rate leaving the front of the queue exceeds the flow rate into the back of the queue.

SIMPLIFIED METHOD

The simplified method, developed by Nelson and MacLennan (1988), is based on determining the controlling element along a path of travel that occupants might travel along in order to evacuate. The controlling element in the egress system is the component that has the smallest value of the maximum flow rate for each of the components in the egress system that a particular group of individuals might travel along in order to evacuate. The maximum flow rate occurs when the specific flow is maximized. After identifying the controlling component, the method would then estimate the evacuation time for the building as the sum of the following times:

- Time to reach the controlling component, t_1
- Time to use the controlling component, t_2
- Time to travel from controlling component to the point of safety, t_3

This method assumes that all occupants start their evacuation simultaneously. In cases with high-rise buildings, the exterior stairwell door is often the controlling element, in which case the estimated evacuation time, t_e, is determined as

$$t_e = t_1 + t_2 + t_3 \tag{4.4}$$

where

t_e = estimated evacuation time, s,

t_1 = time to reach the controlling component, s,

t_2 = time to use controlling component, s,

t_3 = time to travel from controlling component to the point of safety, s.

The simplified method is illustrated in Example 4.1.

INDIVIDUAL COMPONENT ANALYSIS

Another method of analysis of the evacuation time with the use of algebraic equations considers the time to use each component in the means of egress along a particular path of travel. As part of this approach, determining the velocity of travel along a component or the flow rate through a doorway will require that the occupant density needs to be determined at each component. The density of the people may be expected to change as a result of three types of transitions:

- merging flows (e.g., at corridor intersections or where people entering a stairwell merge with people traveling in the stairwell from other floors,
- changes in the width of the egress component, and
- changes in the type of egress component to another, (e.g., a corridor to a stair).

In the analysis of any of these transitions, the analysis needs to consider whether the flow capacity of the downstream component can accommodate the flow(s) entering the transition. A queue is expected if the flow rate downstream from the point of the change exceeds the maximum capacity for that component, $F_{sm}w_e$. Consequently, when addressing transitions, two possibilities exist:

- If the incoming flow(s) is less than the flow capacity for the downstream component, then a conservation principle applies where the flow rate leaving the transition is equal to the flow rate entering the transition.
- If the incoming flow(s) is greater than the flow capacity for the downstream component, then a queue forms and the outgoing flow is set equal to the flow capacity for the downstream component.

The overall evacuation time estimated by this approach is not dependent upon the details of the merger if a queue forms. The details of the merger may be important in some cases, as in the situation where a group of people entering a stairwell merges with occupants already in the stairwell. If the people entering the stairwell are attempting to leave the floor of fire origin and thus are potentially threatened by the fire, an important detail could be whether the group in the stairwell yields to the group entering the stairwell. However, there is a lack of data indicating whether yielding would be more or less likely in that situation.

The density of a group of occupants downstream from a transition is determined by initially applying the conservation principle noted. The conservation principle is used to determine a possible flow rate of people leaving the transition. If this flow rate yields a specific flow that is

Example 4.1. Simplified Method

Determine the evacuation time for a seven-story building with the following characteristics. There are 300 people on each floor. Each floor is served by two 48 in. (1.22 m) wide stairways. The doors leading into and from the stairway are 36 in. wide (0.91 m). The stair design includes 7/11 risers and treads. The floor-to-floor distance is 14 ft (4.27 m) and the landing between floors is 4 × 8 ft (1.22 × 2.44 m). Handrails are provided on both sides of the stairways.

Solution:

Component	Effective Width, ft (m)	Specific Flow, p/ft·min (p/m·s)	Flow Rate, p/min (p/s)
Door into stairway	2.00 (0.61)	24.0 (1.32)	48 (0.81)
Stairway	3.00 (0.91)	18.5 (1.01)	55.5 (0.92)
Door from stairway	2.00 (0.61)	24.0 (1.32)	48 (0.81)

The controlling component is selected as the door leading from the stairway. The time required for the half of the building occupants on the upper floors (900 persons) to pass through this doorway is estimated to be 18.8 min (900/48). The time required for the first person traveling at a velocity associated with the maximum density is given by the time to travel down one flight of stairways and two landings.

Time to travel down one flight of stairways:

The hypotenuse of 7/11 stair is 13 in. Thus, to travel a vertical distance of 14 ft (4.27 m) requires traveling a diagonal distance of 26 ft (8.54 m). The occupant density in the stairs is considered to be the maximum of D_{max} which is 0.175 p/ft^2 (1.88 p/m^2). From Table 4.1, the velocity factor on the 7/11 stairs is $k = 212$ ft/min (1.08 m/s). The velocity on the stairs is

$$v = k - 2.86kD = 212 - 2.86(212)(0.175) = 106 \text{ ft/min (54 m/s).}$$

The length of travel along each landing is 8 ft (2.4 m) (assuming an average length of travel on the middle of the landing). Because the velocity on a stairway is less than that for a horizontal component such as a landing, the velocity on the landing is limited to that achieved on the stairway. As such, the length of travel on the landing can be added to that for the stairway, giving a total length of travel of 42 ft (13.3 m). The time required to traverse this distance at the velocity achieved on the stairways is 0.40 min (24 s).

Thus the total time is 18.8 + 0.4 min. or 19.2 min.

Such an analysis is most relevant in situations where a queue is expected to form at the controlling egress component. Generally, these situations consist of cases where an appreciable number of people occupy the area of the building being modeled. Conversely, in buildings with low occupant loads, a queue is unlikely. In cases with low occupant loads, a more complex analysis is needed to examine the occupant flow on a component by component basis. These analyses also may be applied to provide a more accurate assessment in cases where queuing is likely.

less than the maximum specific flow permitted for the egress component, then the density is determined by solving Equation 4.3 for the density D. Alternatively, if the flow rate yields a specific flow greater than the maximum specific flow permitted for the egress component, then a queue is assumed and the density is set equal to 0.175 p/ft^2 (1.88 p/m^2).

For cases where Equation 4.3 applies, two possible solutions are obtained from solving the quadratic equation. The lesser value for density should be selected as the correct value.

If people traveling down a corridor reach an intersection with two possible choices, the analyst will need to estimate what proportion of the crowd should be allocated to each choice. There is little research that can be used as a basis for determining the respective proportions of occupants choosing the two paths. Predtechenskii and Milinskii (1978) suggest that the number of occupants choosing each path should be proportional to the respective capacities of the available paths. This individual component approach is illustrated in Example 4.2.

COMPUTER-BASED EVACUATION MODELS

Reviews of the numerous evacuation models are provided by Peacock and Kuligowski (2005) and Gwynne and Galea (1999). Evident in this review is the wide range of capabilities that are included in these models. Based on the characteristics of the models, they can be grouped as follows: (1) egress system, (2) human behavior, (3) individual tracking, and (4) uncertainty.

Example 4.2. Component-by-Component Analysis

Determine the evacuation time for the same seven-story building as in Example 4.1. The corridor on the upper floors is 4 ft wide (1.22 m). Assume the beginning density of the people in the corridor is 0.125 p/ft^2 (1.35 p/m^2).

Solution:

Assume that all occupants initiate movement simultaneously and half of the building occupants are located in the corridor at a distance of at least 100 ft (15.2 m) from the stair door. Other occupants are in the spaces adjacent to the corridor and are assumed to join the people in the corridor promptly upon notification. Assume the occupants distribute themselves evenly to the two stairs; i.e., half of the occupants use one stair, the other half use the other stair.

The solution is begun by considering the initial movement of people in the corridor. The thickness of the boundary layer in the corridor is 0.67 ft (20 mm). Thus, the effective width of the corridor is 2.67 ft (0.82 m). Considering the initial density of the occupants, the initial specific flow of the people in the corridor is: 22.1 p/ft·min (1.21 p/m-s) $< F_{sm}$.

The velocity in the corridor is 177 ft/min (0.90 m/s).

The flow rate in the corridor is 59 p/min (0.99 p/s).

Time to reach stairway: 100/177 = 0.56 min (34 s).

The maximum flow of the door leading into the stairway is 48 p/min (see Example 1)(0.81 p/s). Because the flow rate leading up to the doorway is greater than the maximum flow rate that can be accommodated by the doorway, a queue forms at the doorway.

The queue builds at a rate of 11 p/min (0.18 p/s), the difference between the incoming flow rate and the maximum flow rate for the doorway.

With the flow rate in the stairway limited by the doorway to 48 p/min (0.81 p/s), the specific flow in the stairway is determined to be 48/3 = 16 p/ft-min (0.81/0.91 = 0.89 p/m-s). Using Equation 5, the density associated with that specific flow is 0.074 p/ft^2 (0.79 p/m^2). Hence, the velocity moving down the stairs from the seventh floor approaching the sixth floor is determined from Equation 3 as 167 ft/min (0.85 m/s).

The time to travel 42 ft (13.3 m) to reach the sixth floor: 0.25 min (15 s).
At this point, flows from the sixth and seventh floors merge at the landing of the fourth floor, as well as every other floor level.

The total time required for the last person from the seventh floor to enter the stair at that floor level is: 3.69 min (221 s) (this time is determined as the number of occupants per stair [150] divided by the flow rate into the stair, plus the total time to reach the stair, 0.56 min.). The time required for the last person from the seventh floor to reach the sixth floor: 3.94 min (236 s).

With a flow proceeding down the stairs from the seventh floor of 48 p/min (0.81 p/s) and the same flow rate entering the stairway from the sixth floor, the outflow from the point of merger would be 96 p/min (1.62 p/s) if no queue occurs. However, because the flow capacity in the stairway is 55.5 p/min (0.925 p/s), the flow in the stairway will be limited to that maximum value. Priority of flow in the stairway is assumed to be given to occupants from the top floor level (though no empirical evidence is available to justify that assumption).

The time for a queue to form in a stairway is the time to reach the stairway plus the time to travel one story in the stairway, which is 0.56 min + 0.25 min = 0.81 min (49 s). Prior to the queue forming in the stairway, 39 people exit from each of the lower floors. Because the flow capacity in the stairway is limited to 48 p/min (0.81 p/s), the flow from all lower floors is stopped. Once the last person from the seventh floor reaches the sixth floor, the flow of the remaining people from the sixth floor commences.

The time required for the last person from the sixth floor to enter the stair at that floor level is: (150-39)/48 + 3.94= 6.25 min (375 s).

The time required for the last person from the sixth floor to reach the fifth floor: 6.50 min (390 s).

Similarly:

The time required for the last person from the fifth floor to enter the stair at that floor level is: 8.81 min (529 s).

The time required for the last person from the fifth floor to reach the fourth floor: 9.06 min (544 s).

The time required for the last person from the fourth floor to enter the stair at that floor level is: 11.4 min (684 s).

The time required for the last person from the fourth floor to reach the third floor: 11.62 min (697 s).

The time required for the last person from the third floor to enter the stair at that floor level is: 13.93 min (836 s).

The time required for the last person from the third floor to reach the second floor: 14.2 min (851 s).

The time required for the last person from the second floor to enter the stair at that floor level is: 16.5 min (989 s).

The time required for the last person from the second floor to reach the first floor: 16.7 min (1004 s).

Egress System

The egress system can be approximated either as discrete parts, either using a coarse or fine grid, or as a continuous path. A coarse grid may consist of each room being represented by a single node with links (or arcs) used to connect the nodes. In contrast, a fine grid divides each room into several small squares which could be small enough to allow only one person to occupy the particular square. Individuals move from grid square to an adjacent square in order to evacuate. In the continuous model, occupants move along egress path without being limited to discrete steps.

Human Behavior Modeling

Some of the models, especially the earliest models developed, follow the hydraulic analogy, essentially automating the approaches using the algebraic equations such as those described in the previous section. Recent research has indicated that the homogeneous flow assumed in the hydraulic analogy does not occur in building evacuation (Leahy 2011). Instead, people travel in groups or in platoons, with the speed of the platoon being that associated with the slowest member of the group (Proulx 2008).

Most of the more recently developed models account for various aspects of human behavior. Some of these recently developed models allow occupants to travel in groups, start at different times, have a preference to use exits that they are familiar with, account for patience (for how long to wait in a queue) and drive (affecting which of two occupants vying for a particular space move to that space), among other capabilities.

For the models that include various behavioral considerations, users need to be cautious about the level of confidence that should be placed in the results. The state-of-the-art in human behavior in fire is limited. Behavioral patterns in fire have only been systematically collected in a limited number of studies. The conclusions from those studies may indicate some trends in behavior, but these trends are highly subject to change, based on the particular conditions associated with the fire scenario, the social structure of the occupants, experience and training of individuals in fire emergencies, and capabilities of the individuals (both mental and physical).

Individual Tracking

The earlier models used a global perspective to model the evacuation of the entire population in the building. Consequently, the results of evacuation time from a particular area or space were for the population at large, without knowing where occupants evacuating in a particular time period originated. Most of the more recently developed models have an individual perspective allowing the analyst to track a particular individual as that individual makes their way along the egress path.

Uncertainty Reference

As with other areas of engineering, some evacuation models are deterministic. A deterministic model will provide the same results for a particular set of inputs. In contrast, stochastic evacuation models allow for variability that may occur in an evacuation so that the same output is not achieved every time for a particular input data set. The evacuation models that provide a stochastic analysis do not necessarily consider the uncertainty in the same set of variables. Some may consider the variability only in movement speeds, human behavior, premovement times, or perhaps all of the above.

Summary

Considering the wide range in capabilities of the evacuation models, appreciable differences in results obtained from different models should be expected. For example, one model, EVACNET4, is an optimization model. As such, it determines the distribution of occupants to the various exits necessary in order to minimize the evacuation time.

As with any model used in engineering, the results are strongly dependent on the input provided. For evacuation models, the technical validity of data to support the various input parameters is relatively thin. As noted in a previous section, mean velocities are reported in the literature, but little information is available on the range of velocities at which people move. Similarly, for models that include behavioral factors, the supporting data are relatively limited (e.g., to justify a particular patience level or drive).

Kuligowski (2003) conducted a study of the evacuation times acquired from two evacuation models, EVACNET4 and Simulex, for a relatively simple high-rise hotel. The evacuation times ranged from 730 to 960 s. Where possible, default values were applied. Otherwise, a consistent set of assumptions were applied when identifying the input for the two models where differences existed in the input to be provided. It should be noted that this significant range of evacuation times were achieved while attempting to provide as similar a set of inputs as possible and did not attempt to provide an analysis of the greatest variation of times that could be achieved for the same building design and number of occupants.

Considering the state of knowledge of people movement and human behavior factors, results from evacuation models should be treated as rough approximations. It is especially incumbent upon users to appreciate the assumptions imbedded in the model and relevance of input data used for a particular application (Proulx 2002; Fahy 2002). The uncertainty associated with the assumptions and input data should affect the selection of safety factors. For example, those models without human behavior aspects are likely to underestimate the evacuation times, without making other adjustments. Given the inexact nature of the simulations, rather than apply the models to predict a particular evacuation time, their best application is to compare evacuation times of multiple evacuation strategies.

HUMAN BEHAVIOR

Bryan (2008) divides the decision process for individuals responding to a fire into the following six steps:

1. Recognition: observation of cues that indicate something different than usual is occurring. In many cases, the cues are ambiguous (e.g., strange odors, slight haze, or abnormal sounds). The time required for individuals to note the cues will depend on their alertness, proximity to the fire, and whether automatic detectors are present. Solely hearing the building fire alarm or seeing flashing strobes may not necessarily be interpreted as a fire, depending on the experience of the individual with nuisance alarms.
2. Validation: realization that the cues are associated with a fire. Where an individual senses a strange odor or sees a haze, they may choose to investigate to determine the source of the odor or haze. This realization may come following a search for the source of the cues, other individuals communicating their observations, receiving additional cues, etc.
3. Definition: determination of the severity of the incident and time available for safe egress.
4. Evaluation: identification of possible initial actions and assessment of which action should be carried out. This is done considering the likelihood of success and the challenge of completing the action.
5. Commitment: implementation of the first action.
6. Reassessment: continuous analysis of feasible actions, depending on the success of the initial action and the observation of changing conditions.

All of these steps are performed based on the individuals' perceptions and understanding of the relative risk posed by the incident. Their perceptions and understanding will be affected by their training and experience relative to fire and their mental capabilities. Thus, a trained emergency responder would be expected to respond differently than someone who has received no training and has no experience with fires. Finally, these steps assume that an individual has sufficient cognitive abilities to understand the nature of the threat and the need to evacuate or otherwise seek to mitigate the threat.

While building occupants reacting to a significant fire may feel fear or anxiety, they rarely panic or react irrationally. The term *panic* is used by researchers of human behavior in fire to denote irrational behavior. Considering this definition, in a significant majority of cases, people react rationally in fire events by working through the event via a logical decision process (Quarantelli 1979; Proulx and Sime 1991). Thus, the common perception that people regularly panic in fire incidents is incorrect. The common use of the term *panic* is generally intended to refer to people being afraid or anxious about the situation. Such anxiety or fear is actually quite rational, especially for serious fires.

PREMOVEMENT

A significant amount of time may be required prior to occupants initiating their evacuation attempt from a building. This may be due to their activities to validate and define the incident, attempts to suppress the fire, prepare to evacuate (i.e., get dressed or put on a coat), or to gather family members, friends, pets, or belongings (Bryan 2008). Delay times in fire incidents and drills may be a few minutes to an hour (or more if the individual sleeps through most of the event) (Proulx 2008; Proulx and Fahy 1997).

Proulx's observations suggest that premovement time is dependent on the occupancy of the building. For example, premovement times tend to be less for office buildings than for private residences.

NOMENCLATURE

a = constant for units conversion ft^2/p, (m^2/p)
D = density of occupant flow, p/ft^2 (p/m^2)
F_s = specific flow, p/min-ft (p/s-m)
F_{sm} = maximum value of specific flow, p/min-ft (p/s-m)
k = velocity factor, ft/min (m/s)
P = population using the stair, p
T = evacuation time, min
t_1 = time to reach controlling component, s
t_2 = time to use controlling component, s
t_3 = time to travel from controlling component to the point of safety, s
t_e = estimated evacuation time, s
v = velocity, ft/min (m/s)
w_e = effective width of stair, in (m)

Note: The symbol p is for persons.

REFERENCES

Bryan, J.L. 2008. *SFPE Handbook of Fire Protection Engineering,* Behavioral Response to Fire and Smoke. Quincy, MA: National Fire Protection Association.

Bryan, J.L. 1977. Smoke as a determinant of human behavior in fire situations (Project People). NBS-GCR-77-94, NBS, Center for Fire Research, Gaithersburg, MD.

Fahy, R.F. 2002. Tools for the simulation of human behavior. *Fire Protection Engineering* 16(Fall).

Frantzich, H. 1996. *Study of Movement on Stairs During Evacuation Using Video Analysing Techniques.* Sweden: Lund Institute of Technology.

Fruin, J.J. 1971. *Pedestrian Planning and Design,* revised ed. Mobile, AL: Elevator World Educational Services Division.

Fruin, J.J. 1987. *Pedestrian Planning and Design,* revised ed. Mobile, AL: Elevator World Educational Services Division.

Gwynne, S., and E. Rosenbaum. 2008. Employing the hydraulic model in assessing emergency movement. *SFPE Handbook on Fire Protection Engineering,* P. DiNenno, 4th ed. Quincy, MA: National Fire Protection Association.

Gwynne, S., and E.R. Galea. 1999. *A Review of the Methodologies and Critical Appraisal of Computer Models Used in the Simulation of Evacuation from the Built Environment.* Bethesda, MD: Society of. Fire Protection Engineers.

Gwynne, S., E.R. Galea, M. Owen, and P.J. Lawrence. 1999. Escape as a social response. Research Report, Society of Fire Protection Engineers, Bethesda, MD.

Kuligowski, E.D., 2003. The evaluation of a performance-based design process for a hotel building: the comparison of two egress models. Master's Thesis, University of Maryland, College Park, MD.

Kuligowski, E.D., and R.D. Peacock. 2005. review of building evacuation models. NIST TN 1471, National Institute of Standards and Technology, Gaithersburg, MD.

Leahy, A., 2011, Observed trends in human behavior phenomena within high-rise stairwells. Master's Thesis, Department of Fire Protection Engineering, University of Maryland, College Park, MD.

London Transport Board. 1958. Second report of the operational research team on capacity of footways. Research Report, Issue 95, London Transport Board, London.

Nelson, and MacLennan. 1988. *SFPE Handbook of Fire Protection Engineering,* Emergency Movement. Quincy, MA: National Fire Protection Association.

Pauls, J., 1980. *Building Evacuation: Research Findings and Recommendations, Fires and Human Behaviour,* D. Canter, ed. New York: John Wiley.

Predtechenskii, and Milinskii. 1978. *Planning for Foot Traffic Flow in Buildings.* New Delhi: Amerind Publishing Co.

Proulx, G. and R. Fahy. 1997. The time delay to start evacuation: review of five case studies. *5th International Symposium on Fire Safety Science,* pp. 783–794.

Proulx, G. and Sime, J., 1991. To prevent panic in an underground emergency, why not tell people the truth. 3rd International Symposium on Fire Safety Science, July 8–12, University of Edinburgh, Scotland.

Proulx, G. 2002. Cool under fire. *Fire Protection Engineering* 16(Fall).

Proulx, G. 2008. SFPE *Handbook of Fire Protection Engineering,* Evacuation Time. Quincy, MA: National Fire Protection Association.

Quarantelli, E.L. 1979. *Panic Behavior in Fire Situations: Findings and a Model from the English Language Research Literature.* Columbus, OH: Ohio State University.

CHAPTER 5

Fire Science and Design Fires

John H. Klote

In an analysis of a smoke control system, the design fire is an important part of each design scenario. The heat release rate (HRR) is probably the most important aspect of a design fire. Analysis of design fires requires an understanding of the stages of fire development, the impact of sprinklers on HRR, the HRR of various objects, and radiant ignition.

DESIGN FIRES

Often, steady fires are used as design fires, because they simplify design calculations. By nature, fire is an unsteady process, and much of the focus of this chapter is on unsteady fires. When steady design fires are based on test data, it is generally accepted that HRR of the steady fire is taken as the maximum HRR of the test data. For example, test data of a sofa burn starts out small and grows to a maximum of about 3000 Btu/s (3200 kW) followed by a decrease in HRR as the fuel burns out. A sofa design fire could be unsteady based on the fire test data or it could be a steady 3000 Btu/s (3200 kW).

Avoid Wishful Thinking

Professionals involved with the analysis of design fires must avoid wishful thinking, because such thinking can lead to the blunder of significantly underestimating design fire size.

An example of wishful thinking is the designer who foolishly proposed a wastebasket-size design fire for an atrium smoke control system. The erroneous reasoning went something like "the atrium is designed to have almost no materials that can burn, so the fire size should be very small." This reasoning does not take into account either (1) changes in space usage or (2) transient fuels. It may not be practical to design systems for uses that greatly exceed the planned use, but design fires must take into account transient fuels which are discussed in the next section.

Transient Fuels

Transient fuels are materials that are in a space temporarily. A few examples of transient fuels are Christmas decorations, paint and solvents in stairwells during redecorating, unpacked foam cups in cardboard boxes after delivery, cut-up cardboard boxes awaiting removal, upholstered furniture after delivery, and stacked folding chairs. Sometimes transient fuels remain in place for long periods. Some examples are (1) a number of polyurethane mattresses delivered to a dormitory and waiting for distribution in the next school year, (2) automobiles on display in a shopping mall, (3) boats and campers on display in an arena, and (4) a two-story wood frame house built for display inside a shopping mall.

Transient fuel is likely to accumulate at most locations in a building except where it would block the usual paths of heavy traffic. It is unlikely that a commonly used building entrance would be blocked by transient fuel, but there could be transient fuel next to a wall near such an entrance. It is also unlikely that a frequently used corridor would be blocked with transient fuel, but there could be some transient fuel in the corridor.

Location can play a key role in transient fuels. Consider the sofa with polyurethane foam padding that is delivered for the office of the corporate president. Because the sofa is new and clean, it is decided to temporarily leave it in the nearby atrium until it can be moved to the president's office. In a corridor of an office building, the fuel could be trash consisting of any number of things such as an old upholstered chair or cardboard boxes with packing materials. A minimum value for the

HRR of transient fuels of 2000 Btu/s (2100 kW) is suggested, but the HRR of transient fuels can be much higher.

Decision Tree

Figure 5.1 is a decision tree that can be used to help develop design fires. Fires are either sprinklered or unsprinklered. Most new buildings in the U.S. are fully sprinklered, and the following discussion focuses on fully sprinklered buildings.

Sprinklered or Not

Design fires are considered to stop growing on sprinkler activation with two exceptions: (1) fires for which the owner wants extra smoke protection and (2) fires in spaces with high ceilings such as atria. An owner may require a high level of protection, because of needs such as security or continuity of operation. Such an owner may want a smoke control system that can withstand large fires in the event of sprinkler failure. Sprinkler reliability and failure have been estimated for different occupancies based on data from the National Fire Incident Reporting System (Koffel 2005; Hall 2006, 2011).

When smoke rises above a fire, it entrains air from the surrounding, such that the temperature of the smoke drops as it rises. In spaces with high ceilings, the temperature of the smoke can drop so much that sprinklers may not activate or activation may be so delayed that the spray may evaporate before it reaches the fire. For more information about smoke temperature, see Chapter 16.

There is a critical ceiling height in large spaces above which the design fire should not take into account any benefits of the sprinkler system, but this is not to say that sprinklers will not work above this critical height. Many professionals believe that this critical height is in the range of 35 to 45 ft (10.7 to 13.7 m). While little research has been done on this subject, this critical height depends on the HRR curve, the type of fuel and the sprinkler system.

Even though a building is fully sprinklered, it is possible that some spaces in the buildings may not be sprinklered, and the design fires for such spaces need to take this into account. An example would be a museum display area where an exception for sprinklers was obtained based on concerns about water damage that may not realistically recognize resulting risks of fire damage.

Shielded Fires

When an object keeps sprinkler spray from directly reaching a fire, the fire is said to be shielded. A common shielded fire scenario consists of burning boxes of paper under a table or desk. Suggested design fires for shielded fires are discussed later in this chapter.

HRR per Unit Area

This approach to determining a design fire is rough, and it should be reserved mostly as a reality check for design fires based on more realistic approaches. In the HRR per unit area approach, the area of a design fire is estimated, and that area is multiplied by a heat release density to get the HRR of the design fire.

For retail spaces in sprinklered buildings, Morgan et al. (1999) suggest 25 to 57 Btu/s·ft^2 (280 to 650 kW/m^2). They suggest 24 Btu/s·ft^2 (270 kW/m^2) for office fires in sprinklered buildings. For library spaces in sprinklered buildings, Morgan et al. suggest 13 to 57 Btu/s·ft^2 (150 to 650 kW/m^2).

A fire occurring over 100 ft^2 (9.3 m^2) of floor space has often been used for estimating design fires. For smoke control applications, a heat release density of 20 Btu/s·ft^2 (230 kW/m^2) might be considered for a space where the major fuels would be transient fuels. Using this fire area and heat release density, a design fire of

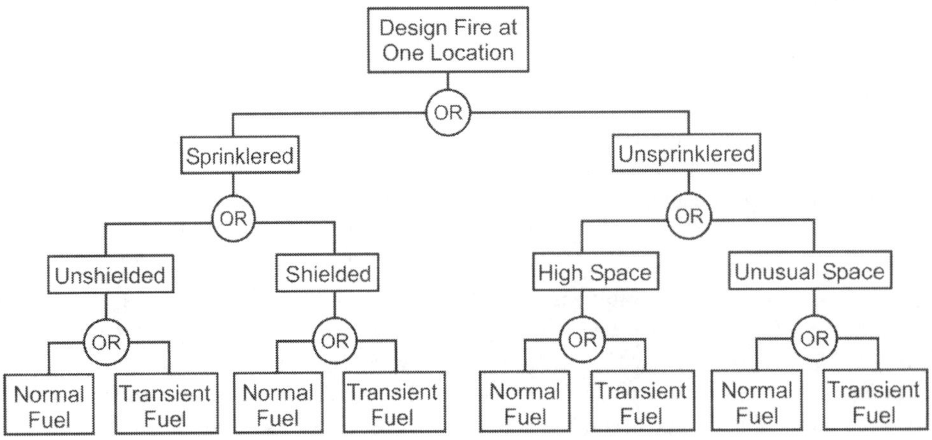

Figure 5.1 *Design fire decision tree.*

2000 Btu/s (2100 kW) would result for spaces where transient fuels dominate. A value of 57 Btu/s ft^2 (650 kW/m^2) might be considered for spaces with furniture or other combustible materials, and using this with the above fire area results in a design fire of 5700 Btu/s (6050 kW).

It should be remembered that this approach is rough, and more detailed analysis can result in design fires that are smaller or larger. When design fires are smaller than the values mentioned here, the possibility of analysis error or wishful thinking should be considered. The heat release densities of some materials are listed in Table 5.1.

STAGES OF FIRE DEVELOPMENT

The stages of fire development are useful when discussing fires. These stages are (1) growth, (2) flashover, (3) fully developed fire and (4) decay. Figure 5.2 shows these stages, but not all fires go through all the stages.

Fire Growth

The growth stage follows ignition, and the early part of the growth stage is characterized by an abundance of air for the fire. During the growth stage, the fire often spreads from one object to another. For example, a fire in a living room that started in a sofa could spread to a lamp shade as shown in Figure 5.3. In this figure, smoke plumes are rising from the burning sofa and lamp shade to form a smoke layer under the ceiling. Also smoke is flowing out of the doorway. During the growth stage, the flames are pulsating, and the smoke plume and smoke layer are in a state of turbulence. Even with this motion, the growth of this kind of fire often seems gradual.

Fire spreads by convection, radiation, or by both. Fire spread by convection happens when flame comes into contact with another object and ignites that object. For example, in Figure 5.3, fire could spread to the paint on the wall behind the sofa by convection. Fire spread by radiation happens when the thermal radiation from flames and hot smoke are sufficient to ignite an item. In Figure 5.3, fire was spread to the lamp shade by radiation.

The HRR curve in the growth stage can take many shapes depending on the object burning. Figure 5.4a shows a fluctuating growth curve that has an increasing trend, and this kind of growth stage can be represented by the exponential curve shown in Figure 5.4b.

$$Q = \alpha(t - t_o)^n \tag{5.1}$$

where

Q = heat release rate (HRR), Btu/s (kW),

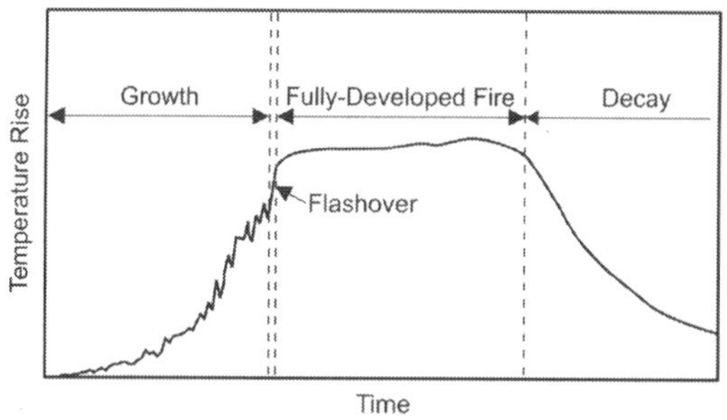

Figure 5.2 *Stages of fire development.*

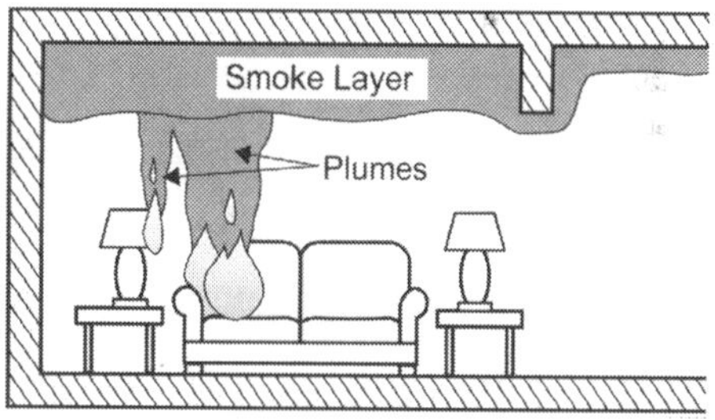

Figure 5.3 *Growth stage fire with a smoke layer.*

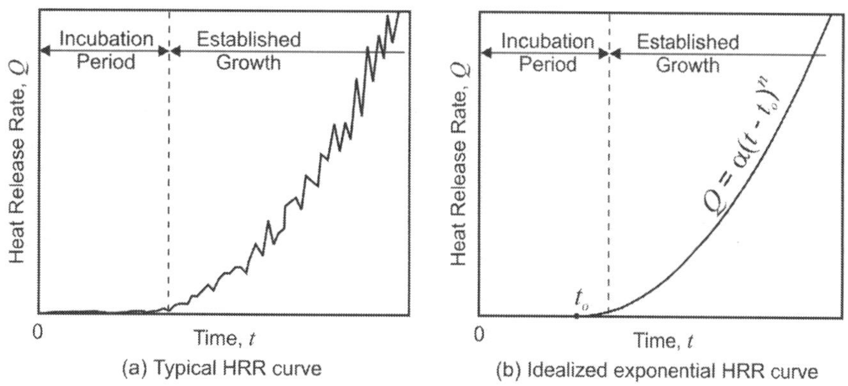

Figure 5.4 *Fire growth curves.*

Table 5.1: Heat Release Density of Some Materials

	Material Burned	Heat Release Density	
		Btu/s·ft^2	kW/m^2
1.	Wood pallets, stacked 1.5 ft (0.46 m) high (6%–12% moisture)	125	1400
2.	Wood pallets, stacked 5 ft (1.52 m) high (6%–12% moisture)	350	4000
3.	Wood pallets, stacked 10 ft (3.05 m) high (6%–12% moisture)	600	6800
4.	Wood pallets, stacked 16 ft (4.88 m) high (6%–12% moisture)	900	10,000
5.	Mail bags, filled, stored 5 ft (1.52 m) high	35	400
6.	Cartons, compartmented, stacked 15 ft (4.57 m) high	150	1700
7.	PE letter trays, filled, stacked 5 ft (1.52 m) high on cart	750	8500
8.	PE trash barrels in cartons, stacked 15 ft (4.57 m) high	175	2000
9.	PE fiberglass shower stalls in cartons, stacked 15 ft (4.57 m) high	125	1400
10.	PE bottles packed in item 6	550	6200
11.	PE bottles in cartons, stacked 15 ft (4.57 m) high	175	2000
12.	PU insulation board, rigid foam, stacked 15 ft (4.57 m) high	170	1900
13.	PS jars packed in item 6	1250	14,000
14.	PS tubes nested in cartons, stacked 14 ft (4.27 m) high	475	5400
15.	PS toy parts in cartons, stacked 15 ft (4.57 m) high	180	2000
16.	PS insulation board, rigid foam, stacked 14 ft (4.27 m) high	290	3300
17.	PVC bottles packed in item 6	300	3400
18.	PP tubes packed in item 6	390	4400
19.	PP & PE film in rolls, stacked 14 ft (4.27 m) high	550	6200
20.	Methanol pool, 0.52 ft (0.16 m) diameter	180	2000
21.	Methanol pool, 4.0 ft (1.22 m) diameter	35	400
22.	Methanol pool, 5.7 ft (1.74 m) diameter	35	400
23.	Methanol pool, 8 ft (2.44 m) diameter	37	420
24.	Methanol pool, 3.2 ft (0.97 m) square	66	745
25.	Silicone transformer fluid pool, 5.7 ft (1.74 m) diameter	8	90
26.	Silicone transformer fluid pool, 8 ft (2.44 m) diameter	8	90
27.	Hydrocarbon transformer fluid pool, 4.0 ft (1.22 m) diameter	83	940
28.	Hydrocarbon transformer fluid pool, 5.7 ft (1.74 m) diameter	80	900
29.	Heptane pool, 4 ft (1.22 m) diameter	270	3000
30.	Heptane pool, 5.7 ft (1.74 m) diameter	280	3200
31.	Work station (four sided)	160	1800

Notes:

1. Abbreviations are: PE = polyethylene, PS = polystyrene, PVC = polyvinyl chloride, PP = polypropylene, PU = polyurethane.
2. Items 1 through 19 from NFPA 92 (2012).
3. Items 20 through 30 from Heskestad (1984).
4. Item 31 based on Madrzykowski (1996).
4. Items 25 through 28 are proprietary products.

α = fire growth coefficient, Btu/s^3 (kW/s^2),

t = time from ignition, s,

t_o = effective ignition time, s,

n = growth exponent, dimensionless.

During experimental work on fire plumes, Heskestad (1984) showed that for some fires, a growth exponent of two was applicable. Using an exponent of $n = 2$, and neglecting much of the fire incubation time, Equation 5.1 becomes

$$Q = \alpha t^2 \quad (5.2)$$

For I-P units, the following form of the growth equation is often used

$$Q = 1000\left(\frac{1}{t_g}\right)^2 \quad (5.3)$$

where

Q = heat release rate (HRR), Btu/s,

t = time from ignition, s,

t_g = growth time, s.

In the equations above, the time, t, is referred to as the time from ignition, which is an appropriate simplification for smoke control applications. Strictly speaking, this time is the time from effective ignition. The fires described by Equations 5.2 and 5.3 are called the t-squared fires. Table 5.2 lists values of growth coefficients and growth times from NFPA 92 (NFPA 2012), and these values are used in this handbook. Figure 5.5 shows the relation of the t-squared fire to some test data (Nelson 1987). The NFPA 72 (NFPA 2007) has other growth times that are appropriate for detection applications.

When $t = t_g$, Equation 5.3 gives a HRR of 1000 Btu/s. While Equation 5.3 is for I-P units, it is sometimes used for SI units along with the growth times of Table 5.2. This alternate SI use of Equation 5.3 results in a HRR of 1000 kW at $t = t_g$. Because 1000 kW equals 948 Btu/s, this alternate use of equation results in HRR that are lower by about 5%.

Table 5.2: Fire Growth Constants for T-Squared Fires

	α, Btu/s^3	α, kW/s^2	t_g, s
Slow	0.002778	0.002931	600
Medium	0.01111	0.01127	300
Fast	0.04444	0.04689	150
Ultra Fast	0.1778	0.1878	75

Flashover

Flashover is a rapid change from a growth stage fire to a fully developed fire. Flashover is an exception to most processes dealt with in engineering that are gradual. A person observing a growing fire like that of Figure 5.3 might mistakenly think that they have plenty of time to retrieve a valued item from the fire room. Because of the rapid nature of flashover, going into such a room could be a fatal mistake. In a fully developed fire, everything in the

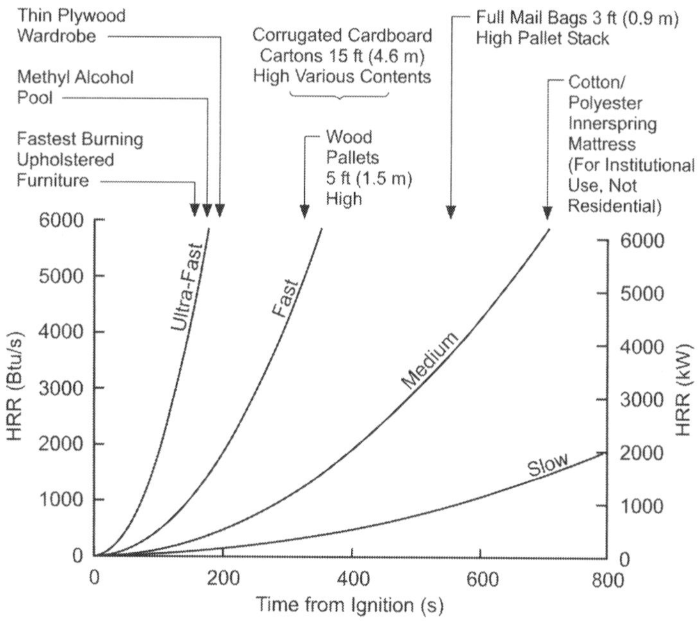

Figure 5.5 *Relation of t-squared fires to some test data.*

room that can support fire is burning, and a person in such a fire would almost certainly die.

Flashover is due primarily to fire spread by thermal radiation. This radiation is from the flames, the smoke plume and the hot smoke layer below the ceiling. Thin easy-to-ignite materials (newspapers, draperies etc.) near the fire are the first to burst into flame, and this is followed by ignition of the rest of the materials in the room that are capable of burning.

At the end of flashover, flames generally extend from the doorways or open windows of the fire room. Flashover generally happens when the smoke layer is in the range of 930°F to 1300°F (500°C to 700°C). Peacock et al. (1999) suggest criteria for flashover of a smoke layer temperature of 1100°F (600°C) or a radiant heat flux of 1.8 Btu/ft^2·s (20 kW/m^2) at the floor of the fire room. In an extremely large room, like an open office floor plan, only part of the floor may flashover.

Fully Developed Fire

In a room with a fully developed fire, every thing that can burn is burning. A fully developed fire also is called a *ventilation controlled* fire, because the HRR depends on the amount of air that reaches the fire. During a fully developed fire, flames generally extend from the doorways or open windows of the fire room. A fully developed fire is characterized by inefficient combustion resulting in high CO production. Based on the research of Pitts (1994) and Mulholland (1995), approximate CO yields of fires are listed in Table 5.3.

For a fully developed fire in room with one opening, the HRR within the fire room can be expressed as

Table 5.3: Approximate CO Yield for Room Fires

	CO Yield*
Flaming fires in "free air"	0.04
Fully involved fire (in a room without cellulosic materials on ceiling or upper portion of walls)	0.2

*Yield is in lb (g) CO produced per lb (g) of fuel burne

$$Q = 61.2 A_w H_w^{1/2}$$
$$Q = 1260 A_w H_w^{1/2} \text{ for SI} \qquad (5.4)$$

where

Q = heat release rate of a fully developed fire, Btu/s (kW),

A_w = area of ventilation opening, ft^2 (m^2),

H_w = height of ventilation opening, ft (m).

The opening to the fire room can be a doorway or window to the outdoors or another space in a building. For a room with more than one opening all with the same top and bottom elevations, Equation 5.4 can be used by setting A_w to the sum of the areas of all the openings.

Example 5.1 illustrates calculation of the HRR of a fully developed fire. Figure 5.6 shows the HRR for various sizes of openings. The previous equation is for rooms of normal construction (drywall, brick, concrete, etc.), but it is not appropriate for metal rooms such as those on many naval ships.

The temperature in the fire room can be calculated by computer models (Chapters 18 and 20). For information about the temperatures of room fires, see Thomas (2008).

Fire Decay

The decay stage is a decrease in the HRR, which is the result of either fuel consumption or fire suppression. As the fuel is consumed, the fire may change from ventilation controlled to fuel controlled.

SPRINKLERS

The extensive use of sprinklers is due to the success with which they suppress fires. Figure 5.7 shows the possible responses to sprinkler spray: (1) HRR decay, (2) constant HRR, and (3) an increase in HRR. The first two responses might be considered successful suppression, but the third consists of the sprinkler spray being overpowered by the fire, which can occasionally happen. A sprinkler can be overpowered when the fire grows to such an extent before sprinkler activation that the sprinkler spray is inadequate for suppression. This

Example 5.1. A Fully Developed Fire

What is the HRR of a fully developed fire in a room with one open doorway 3.5 ft (1.07 m) wide by 7 ft (2.13 m) high?

$$Q = 61.2 A_w H_w^{1/2} = 61.2(3.5 \times 7)(7)^{1/2} = 3970 \text{ Btu/s } (4190 \text{ kW})$$

can happen with an extremely fast growing fire that exceeds the sprinkler design. Also, it can happen in spaces with high ceilings such that the smoke plumes cools to an extent that significantly delays sprinkler activation. Equations for calculating the temperature of smoke plumes are in Chapter 16.

HRR Decay

The HRR after sprinkler actuation can be expressed as

$$Q = Q_{act} e^{-(t - t_{act})/\tau} \quad (5.5)$$

where

Q = post sprinkler actuation HRR, Btu/s (kW),
Q_{act} = HRR at sprinkler actuation, Btu/s (kW),
t = time from ignition, s,
t_{act} = time of sprinkler actuation, s
τ = time constant for fire decay, s.

Madrzykowski and Vettori (1992) conducted a study at National Institute of Standards and Technology (NIST) of the decay after sprinkler activation. For these tests, the spray density was 0.10 gpm/ft^2 (0.07 mm/s) of water. The tests were for arrangements of office furnishings that were not heavily shielded. The ratio of Q/Q_{act} for these tests is shown in Figure 5.8, and a time constant of 435 s in Equation 5.5 results in a larger HRR than most of fires of these tests. For fuels and spray densities similar to those of the tests by Madrzykowski and Vettori, a time constant of 435 s is recommended.

Evans (1993) used data from wood crib fires with spray densities of 0.06 gpm/ft^2 (0.041 mm/s) and 0.097 gpm/ft^2 (0.066 mm/s) to develop the correlation $\tau = C_\tau w^{-1.85}$ where C_τ is 5.15 (3.0) and w is the spray density in gpm/ft^2 (mm/s). While this correlation for the time constant has not been experimentally verified for other materials, it allows the use of Equation 5.5 for other spray densities. Because Equation 5.5 was developed for office fires, it has limited applicability to fires in other occupancies.

Sprinkler Actuation

This section applies to unshielded fires. Sprinkler actuation depends on the temperature and velocity of the gases flowing by the sprinkler and on the responsiveness of the sprinkler. The responsiveness of a sprinkler is characterized by the response time index (RTI). In a fire, a ceiling jet of hot gases flows in a radial direction from where the smoke plume contacts the ceiling.

To test a sprinkler's responsiveness, the plunge test is used where the sprinkler is suddenly put into an oven in which heated air is circulated (Heskestad and Smith 1976). From the plunge test, the RTI of the sprinkler is determined. The larger the RTI, the longer it takes for a sprinkler to respond to a particular fire exposure. Sprinkler manufacturers do not provide RTI values for their sprinklers.

NFPA 13, *Installation for the Sprinkler Systems*, (NFPA 2010) defines standard response sprinklers as those that have an RTI equal to or less than 145 ft$^{1/2}$ s$^{1/2}$

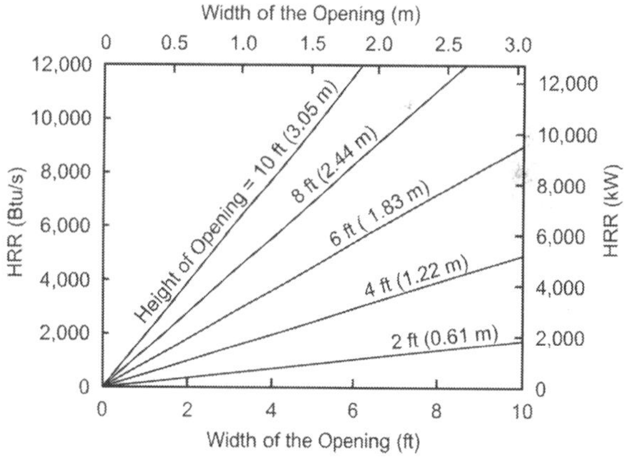

Figure 5.6 *HRR of a fully developed fire in a room with an opening.*

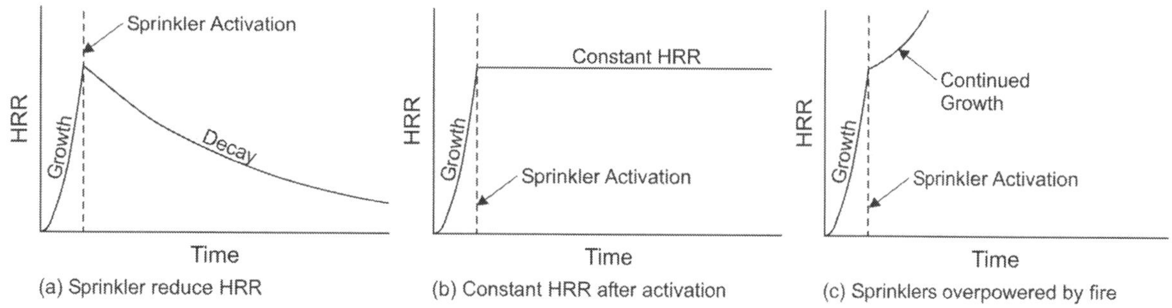

Figure 5.7 *Possible results to sprinkler action.*

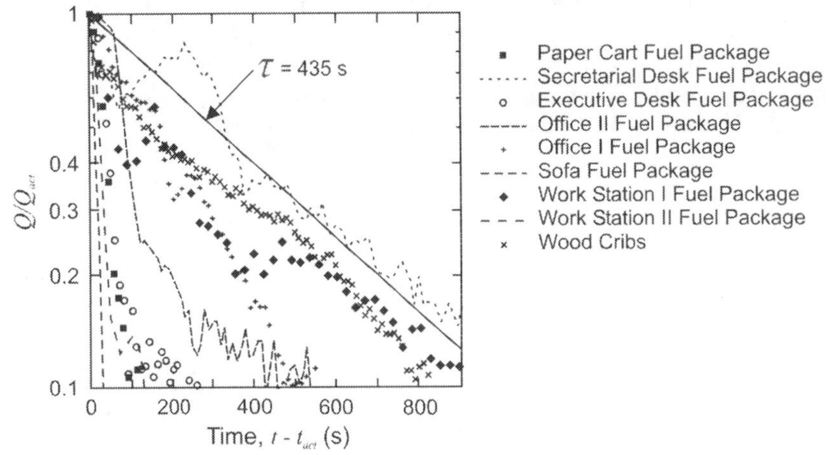

Figure 5.8 *Fire decay due to sprinklers (Adapted from Madrzykowski and Vettori [1992)]).*

(80 $m^{1/2}$ $s^{1/2}$), and fast response sprinklers as those that have an RTI equal to or less than 90 $ft^{1/2}$ $s^{1/2}$ (50 $m^{1/2}$ $s^{1/2}$). For more information about RTI, see Budnick, Hunt, and Wright (2008) and Custer, Meacham, and Schifiliti (2008).

Computer programs have been developed that use the RTI and correlations for the ceiling jet to predict sprinkler actuation time. The ceiling jet correlations are only appropriate for a plume that forms above a single burning object and flows upward without interruption to the ceiling. This kind of plume is called an axisymmetric plume. These programs also can be used to predict actuation times of detectors and fusible links of heat and smoke vents. Chapter 18 discusses computer models that can use the RTI, axisymmetric plume correlations, and ceiling jet correlations to calculate activation times. That chapter also has an example calculation using the model, CFAST, to calculate sprinkler actuation time.

Shielded Fires

A fire can be shielded from the sprinkler spray by any surface that is between it and the sprinkler. For a fire to be shielded, the orientation of the obstruction needs to be such that it is between the sprinkler and the fire. Not only does the obstruction shield the fire from the water spray, but it also prevents the usual formation of a smoke plume. Because the smoke plume of a shielded fire can be very different from that of an unshielded fire, the sprinkler actuation time of shielded fires must not be calculated by the computer programs mentioned in the previous section.

Generally, it is expected that the temperature of the smoke layer would be cooler for a shielded fire than an unshielded fire. This means that the computer models discussed previously would generally result in underpredicting actuation time. Thus, these computer models discussed should not be used for predicting actuation time with shielded fire.

The National Research Council of Canada (NRCC) conducted a study of shielded fires. Part 1 of this project included surveys to quantify typical shielded fire scenarios in high-rise office buildings in the U.S. and Canada (Lougheed and Mawhinney 1996). The most frequently observed scenario was a table or desk shielding fuel with a load ranging from 66 to 331 lb (30 to 150 kg). This fuel consisted primarily of cellulosic material such as cardboard boxes, files, journals, and reports. Varying amounts of plastic materials (such as ring binders, bubble wrap, and foam plastic) and textiles (such as clothing and duffel bags) were noted in many scenarios. Also, a chair was often partly positioned under the shielded area.

Another scenario observed in Part 1 of the NRCC study was rooms or other areas dedicated to miscellaneous storage such as equipment, supplies, documents, and records. The storage areas in single-tenant buildings were frequently isolated from the main portion of the building (such as in basements) with limited access by personnel. There were situations of large records storage systems in open-plan office areas. In multitenant buildings, storage areas were found throughout buildings. Part 1 of this study also noted shielded fire scenarios in mercantile portions of mixed-occupancy buildings and building complexes.

The surveys of Part 1 of the NRCC study identified code nonconformance as a critical issue in many of the shielded fires. In a number of cases, the use of the space by occupants did not conform to codes and

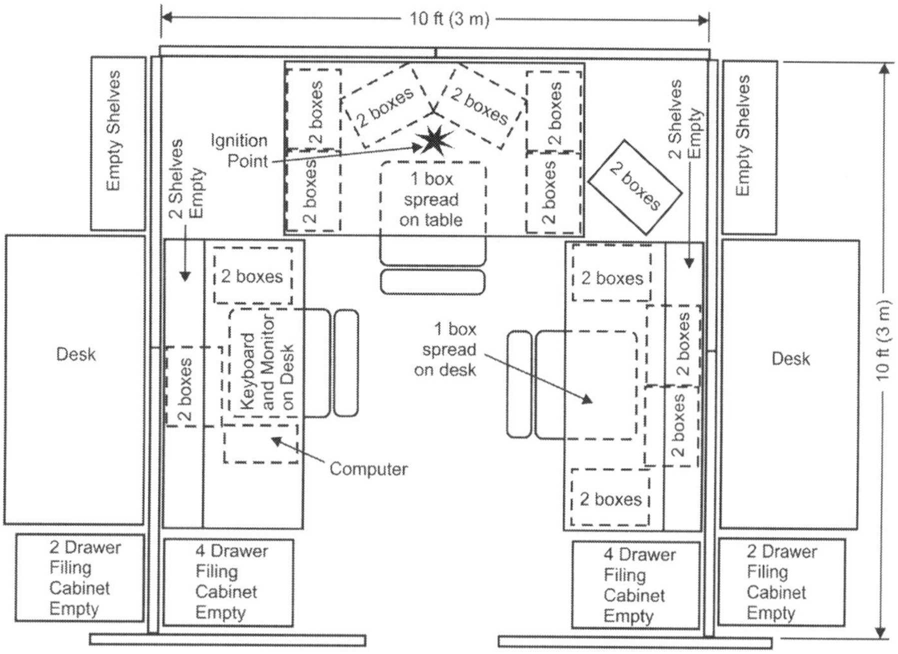

Figure 5.9 *Workspace layout for Test 28 of NRCC shielded fire study.*

standards and thus created or exacerbated the potential fire challenge to the sprinkler system. While there were many examples in which the assumed good performance of safety systems was compromised, the NRCC study stated that this does not mean that fire protection systems should be adjusted to take into account all the extremes that can occur. The full scale fire tests of Part 2 of the NRCC study were in conformance with relevant standards in force at the time of the fire tests.

The full scale fire experiments of Part 2 of the NRCC study focused on open-plan office fires where the fuel load was shielded by tables and desks (Lougheed 1997). In addition, a shielded fire test of document storage and two single office shielded fire tests were conducted. The HRRs of these tests posed less of a challenge than those of the open-plan scenarios.

The facility for the open-plan tests had a ceiling height of 9 ft (2.74 m) with four standard response 165°F (74°C) pendent sprinklers. Oxygen consumption calorimetry was used to measure HRR as described later in this chapter. One open-plan office arrangement of furniture was used with variations of fuel loadings, locations of sprinklers, and locations of ignition source. Figure 5.9 shows the workspace layout for Test 28, which had a heavy fuel load. The HRR of these fires are shown in Figure 5.10.

Based on these tests, NRCC recommended the design HRR curve for shielded fires shown in Figure 5.10. This design can be considered for normal applications where there are no special circumstances that would limit the accumulation of shielded materials. This normal shielded design fire consists of a growth stage, a constant HRR stage and a decay stage. The growth stage of this design fire is for time less than 150 s:

$$Q = 948\left(\frac{t}{150}\right)^2$$
$$Q = 1000\left(\frac{t}{150}\right)^2 \text{ for SI} \qquad (5.6)$$

The constant HRR stage consists 948 Btu/s (1000 kW) from 150 to 240 s, and the decay stage is for time greater than 240 s:

$$Q = 948e^{-0.00155(t-240)}$$
$$Q = 1000e^{-0.00155(t-240)} \text{ for SI} \qquad (5.7)$$

The NIST fires of Madrzykowski and Vettori (1992) had lower shielded fuel loads than those of the NRCC study. The maximum HRR from the NIST sprinklered fires was 474 Btu/s (500 kW). Based on this data, a light-load shielded fire is suggested for locations where the fuel load is limited. This light shielded fire consists of a growth stage and a decay stage. The growth stage is according to Equation 5.6 for time less than 106.1 s. Based on this, the decay stage is for time greater than 106.1 s:

$$Q = 474e^{-0.0023(t - 106.1)}$$
$$Q = 500e^{-0.0023(t - 106.1)} \text{ for SI} \quad (5.8)$$

where

Q = postsprinkler actuation HRR, Btu/s (kW),

t = time from ignition, s.

This light shielded fire is appropriate for locations where the fuel is limited such as in a showplace like the office of the president of a large corporation. Figure 5.11 shows both the normal and light recommended shielded design fires.

MEASUREMENT OF HRR

Until the development of oxygen consumption calorimetry in the 1980s, the mass loss method was used to determine the HRR of burning objects. Materials were often burned on a load cell to measure the weight during burning. The HRR was estimated from the mass loss and the heat of combustion of the material. Often the load cell was water cooled to ensure that elevated temperatures did not invalidate data.

Ideally, the mass loss method is for only one heat of combustion. Further, the heat of combustion of a material is not always constant during a fire. Many objects are made of different materials each with its own heat of combustion. In spite of these difficulties, some useful data has been obtained from this mass loss method.

Oxygen Consumption Calorimetry

This technology is based on the fact that the heat released per unit oxygen consumed is almost a constant for most materials. This constant is 5630 Btu per lb of oxygen consumed, or 13.1 MJ per kg of oxygen consumed (Huggett 1980), and it has an uncertainty of about 6%.

Figure 5.12 shows an open air calorimeter where furniture or other objects are burned under a hood that is

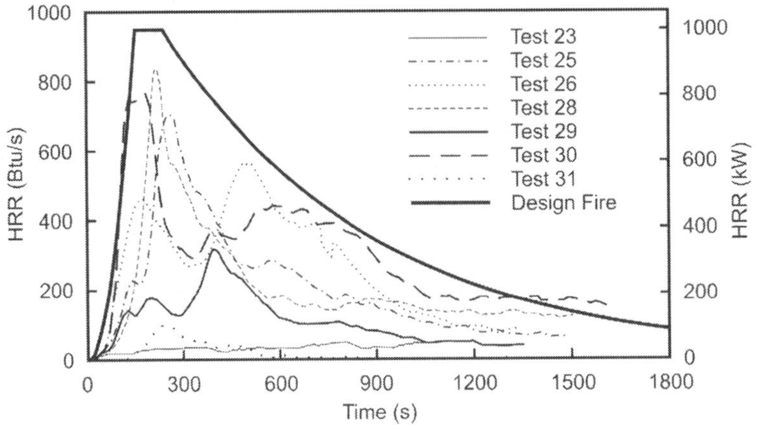

Figure 5.10 *HRR of shielded fires from the NRCC study.*

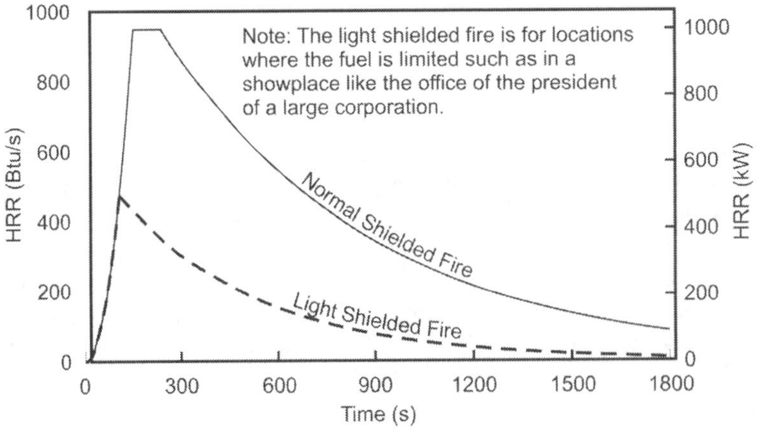

Figure 5.11 *Recommended shielded design fires.*

connected to a high temperature exhaust fan. This kind of burning also is called burning in *free air*, and sometimes an open air calorimeter is called a *furniture calorimeter*. All the smoke is pulled into a hood. From measurements of the temperature, flow rate and O_2 concentration, the HRR can be calculated. Measurements of CO and CO_2 are also made to account for incomplete combustion. Equations for the HRR using O_2, CO, and CO_2 concentrations are provided by Parker (1982). A burner using a gas such as propane or methane at a measured flow rate is used to calibrate an oxygen consumption calorimeter. A well-calibrated calorimeter that collects all the smoke from a burning object may have an uncertainty of about 10%. For more information about the uncertainty of these devices, see Stroup et al. (2000). Open air calorimeters often have a load cell, but this is not used to determine HRR.

Burning in an open air calorimeter is like that in a very large space away from the walls and ceiling. For example, a sofa fire in an atrium away from walls and ceilings would burn much like it would in open air. If the same sofa were burning in a room, it would initially burn as if it were in open air until room effects start to take over. Room effects are due to radiation from fire gases in the plume, under the ceiling, as well as from other burning objects in the room. These room effects are strongly dependent on the room size and the other objects in the room. These room effects can lead to flashover.

Figure 5.13 shows a room calorimeter that also is an oxygen consumption calorimeter. The opening shown in this figure is a doorway, but the opening can also be a window. A room calorimeter can measure the HRR of a fire going through all the stages of fire development. Burning in rooms depends on the location and orientation of the fuel in the room, the size of the room, and the size of the opening. Small scale laboratory calorimeters have been developed. For more information about calorimeters, see Babrauskas (2008a, 2008b) and Janssens (2008).

Experiments using a room calorimeter with a room 12.5 by 13.8 by 7.9 ft high (3.8 by 4.2 by 2.4 m high) that had a window opening of 4.9 by 4.9 ft (1.5 by 1.5 m) were conducted at the National Research Council of Canada (Bwalya et al. 2010). In these experiments, the HRR of many household items were measured, including upholstered furniture, mattress with beddings, clothing, toys, shoes, books, and a computer workstation.

HRR of Objects

This section discusses the HRR of objects burning in free air. The most apparent use of these data is for the analysis of design fires for smoke control applications, but the data have also been used for smoke control research.

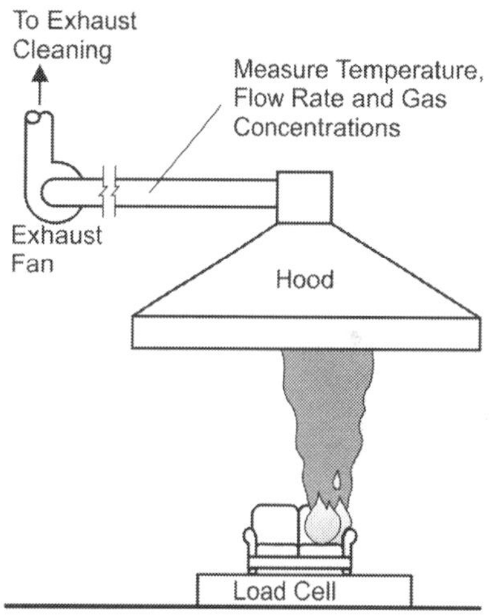

Figure 5.12 *Open burning calorimeter.*

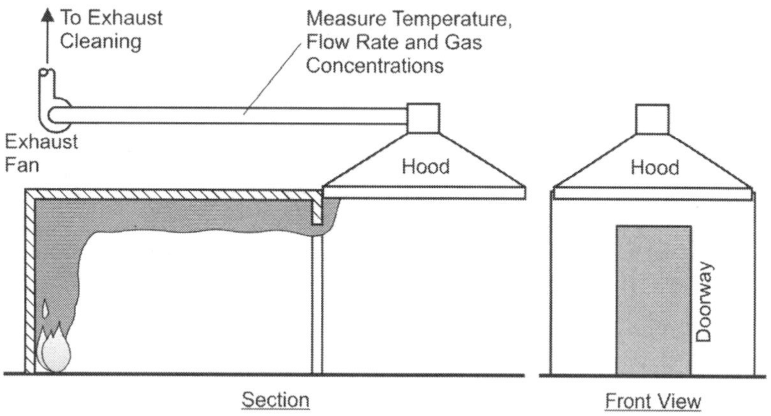

Figure 5.13 *Room calorimeter.*

Unless otherwise noted, the following data and equations were obtained using oxygen consumption calorimetry.

Wood Cribs

Wood cribs consisting of geometrically arranged sticks have been used for fire research for many decades. The burning of wood cribs is reproducible and fairly well understood (Gross 1962; Block 1971; Babrauskas 2008a). The crib of Figure 5.14 was one of the fuel packages for the NIST study of HRR decay after sprinkler activation (Madrzykowski and Vettori 1992). This crib was made of fir sticks 1.5 in. (38 mm) square and 2 ft (0.61 m) long, and it had a peak HRR of about 470 Btu/s (500 kW). A crib like this one but twice as tall was used for the full scale smoke control tests conducted at the Plaza Hotel (Klote 1990).

Wood Pallets

Wood pallets are used for shipping a wide range of materials, but the concern with pallets is when they are stacked in piles as shown in Figure 5.15. The peak HRR of a stack of wood pallets can be estimated from

$$Q = 81LW(1 + 0.65H)(1 - 0.03M)$$
$$Q = 919LW(1 + 2.14H)(1 - 0.03M) \text{ for SI} \quad (5.9)$$

where

Q = peak HRR, Btu/s (kW),
L = pallet length, ft (m),
W = pallet width, ft (m),
H = stack height, ft (m),
M = moisture, %.

This equation is based on an equation of Babrauskas (2008a), and it is for net heat of combustion of 5200 Btu/lb (12,000 kJ/kg). Example 5.2 illustrates calculation of the HRR of a stack of wood pallets. Table 5.4 lists peak HRRs calculated from Equation 5.9 for a number of pallet sizes.

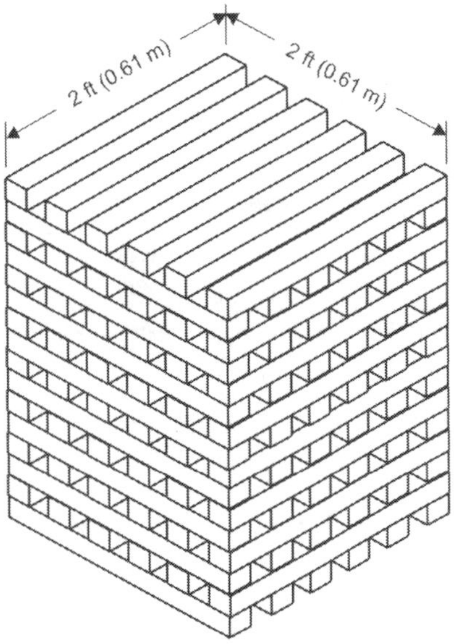

Figure 5.14 *Wood crib made of geometrically arranged sticks.*

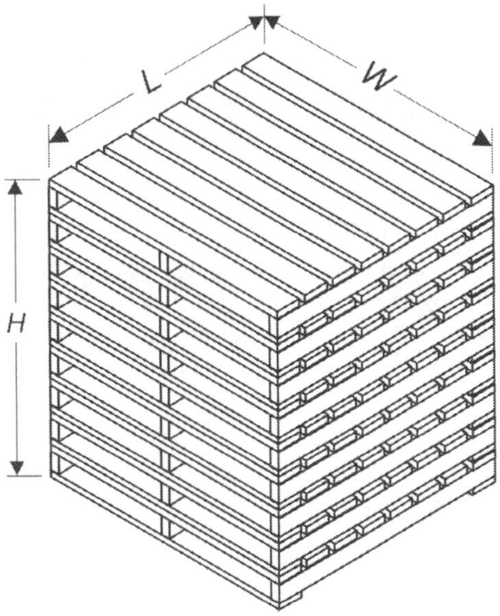

Figure 5.15 *Stack of wood pallets.*

Example 5.2. Wood Pallet Fire

What is the peak HRR of a stack of 12 wood pallets that are 48 by 40 by 5 in. (1.22 by 1.02 by 127 mm)? The wood has a moisture content of 15%.

The height of the stack is $H = 12$ (5 in.)(1 ft/12 in.) = 5 ft; $L = 4$ ft; $W = 3.33$ ft

$$Q = 81LW(1 + 0.65H)(1 - 0.03M)$$
$$Q = 81(4)(3.33)(1 + 0.65 \times 5)(1 - 0.03 \times 15) = 2520 \text{ Btu/s } (2660 \text{ kW})$$

Upholstered Furniture

The use of polyurethane cushions in most modern upholstered furniture is a significant improvement in comfort and durability over the natural materials (such as cotton and horse hair) of the past. Figure 5.16 shows HRRs for furniture with polyurethane padding and wood frames from tests in an open air calorimeter by Lawson et al. (1984). The peak HRR of the chair was about 2000 Btu/s (2100 kW), and that of the sofa was about 3000 Btu/s (3200 kW).

Mattresses

Figure 5.17 shows the HRR of a box spring mattress filled with polyurethane foam burning in open air (Lawson et al. 1984). This mattress had a peak HRR of about 1600 Btu/s (1700 kW). Bedding items such as pillows, pillow cases, and sheets add relatively little to the HRR, but can have an impact on ignitability. However, comforters and duvets for queen and king size beds can produce peak HRR in the neighborhood of 1000 Btu/s (1000 kW) (Bwalya et al. 2010).

After the initial part of the growth stage, room effects typically have a major impact on bed fires making them differ significantly from HRR data for open air burning. This also applies to many other objects burning in rooms. The introduction of new designs including pillow top

Table 5.4: Peak HRR of Stacks of Wood Pallets

Length		Width		Height		HRR at 16% Moisture		HRR at 5% Moisture	
in.	m	in.	m	ft	m	Btu/s	kW	Btu/s	kW
36	0.91	36	0.91	2.5	0.76	1000	1050	1630	1720
				5.0	1.52	1610	1700	2640	2780
				7.5	2.29	2230	2350	3650	3850
				10.0	3.05	2850	3010	4660	4910
				15.0	4.57	4080	4310	6680	7040
				17.5	5.33	4700	4960	7690	8110
42	1.07	42	1.07	2.5	0.76	1360	1430	2220	2340
				5.0	1.52	2200	2320	3590	3790
				7.5	2.29	3040	3200	4960	5240
				10.0	3.05	3880	4090	6340	6690
				15.0	4.57	5560	5860	9090	9590
				17.5	5.33	6400	6750	10,460	11,040
44	1.12	44	1.12	2.5	0.76	1490	1570	2430	2570
				5.0	1.52	2410	2540	3940	4160
				7.5	2.29	3330	3520	5450	5750
				10.0	3.05	4260	4490	6960	7340
				15.0	4.57	6100	6440	9970	10,520
				17.5	5.33	7020	7410	11,480	12,110
48	1.22	40	1.02	2.5	0.76	1480	1560	2410	2550
				5.0	1.52	2390	2520	3910	4120
				7.5	2.29	3310	3490	5400	5700
				10.0	3.05	4220	4450	6900	7280
				15.0	4.57	6050	6380	9890	10,430
				17.5	5.33	6970	7350	11,390	12,010
48	1.22	48	1.22	2.5	0.76	1770	1870	2900	3050
				5.0	1.52	2870	3030	4690	4950
				7.5	2.29	3970	4190	6480	6840
				10.0	3.05	5070	5340	8280	8730
				15.0	4.57	7260	7660	11,870	12,520
				17.5	5.33	8360	8820	13,660	14,420

Note: Pallets are 5 in. (127 mm) thick.

mattresses substantially increased the heat output for modern mattress sets. Bwalya et al. using a room calorimeter measured peak heat release rates of more than 3000 Btu/s (3000 kW) for queen-sized beds.

The recent introduction of CFR 1633 requirements for mattresses by the U.S. Consumer Products Safety Commission requires the use of fire barrier technology (thin fabric barrier systems) that limits the initial heat release from the mattress to less than 200 kW. This will delay peak heat release, but the HRR can still be significant. It is expected to take about 10 to 15 years before the majority of the existing mattresses have been replaced.

Wardrobes

Wardrobes are common in patient rooms in health care facilities, and the HRRs of a number of wardrobes with cotton and polyester garments were measured by NIST (Lawson et al. 1984). Various cotton and polyester fabrics were used to simulate the kind of clothes that would be found in a wardrobe. The fabrics were hung in the wardrobe.

Paper in a cardboard box was the ignition source for the fabrics and the wardrobe. The box was placed in the left rear corner of the wardrobe under the hanging fabrics. The door next to the box was closed, and the other door at the other end of the wardrobe was opened 7 in. (180 mm) to allow for ventilation. Figure 5.18 shows the HRR for a factory-built wardrobe and one made by technicians out of 0.5 in. (12.7 mm) plywood. The factory-built wardrobe was made of 0.125 in. (3.2 mm) plywood except for the bottom and back, which were of hardboard. The factory-built wardrobe had a peak HRR of about 6000 Btu/s (6300 kW), and the one made of 0.5 in. (12.7 mm) plywood had a peak HRR of about 3300 Btu/s (3500 kW). A metal wardrobe (not shown) had a peak of only about 730 Btu/s (770 kW).

Christmas Trees

Moisture content is probably the most important factor regarding the HRR of Christmas trees, and some dry Christmas trees were burned at NIST (Stroup et al. 1999; Madrzykowski 2008). Dry trees burn very rapidly

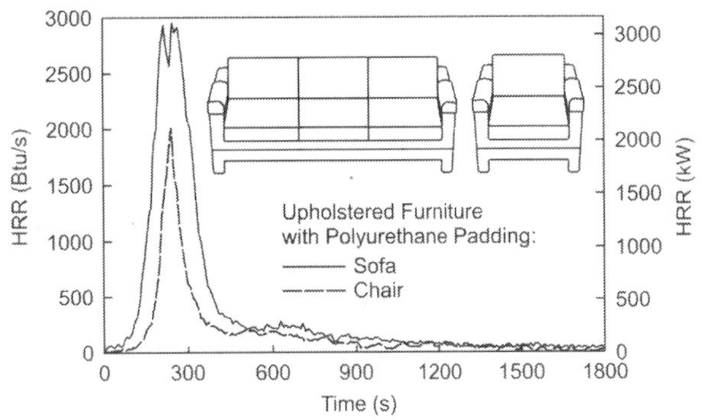

Figure 5.16 *HRR of upholstered furniture with polyurethane foam padding.*

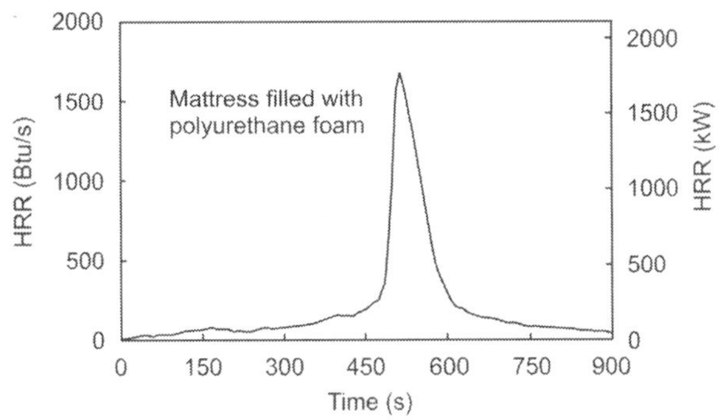

Figure 5.17 *HRR of mattress filled with polyurethane foam.*

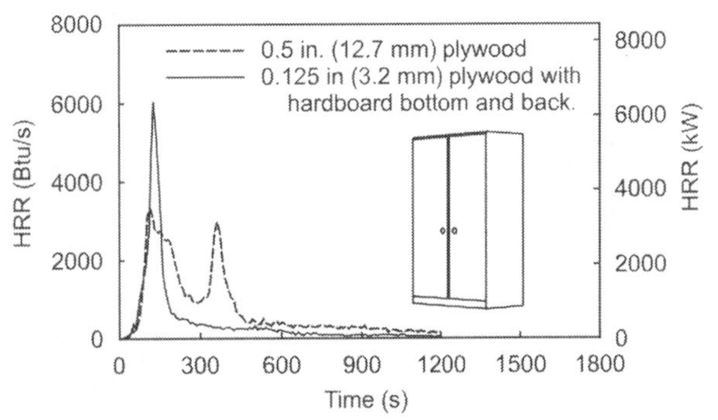

Figure 5.18 *HRR of plywood wardrobes with cotton and polyester garments.*

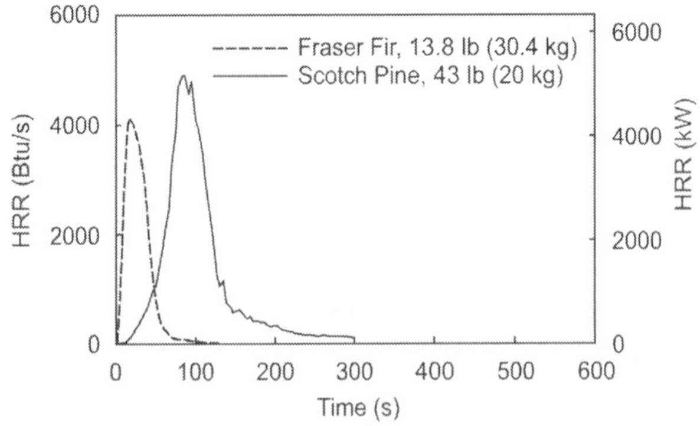

Figure 5.19 *HRR of dry Christmas trees.*

and reach high HRR peaks as can be seen from Figure 5.19.

Kiosks

Mitler (1996) conducted burns of small kiosks for the sale of t-shirts. Figure 5.20 shows the kiosk, with the shelves down and t-shirts placed on them. The bottom of the kiosk is basically a box, with two hinged doors in the front to provide access to the interior. The inside of the box is for storage, and it also contains electric cables for lights and electric outlets in the kiosk. Three cardboard boxes of cotton t-shirts were placed in the box. The fire started inside the box similar to an electrical fire, and the doors were not completely closed providing air for the fire. The HRR of these kiosks were obtained by the mass loss method.

Figure 5.21 shows the HRR of three burns that were as identical as possible for such tests. The first peak HRRs for Tests 2, 3, and 4 were 1364, 1440, and 1668 Btu/s (1440, 1519, and 1760 kW), respectively. Also, these first peaks happened at 1086, 1218, and 1620 s, respectively. Minor differences of arrangement and composition of materials can be causes for such deviations. In hydrodynamics, a small cause can have a large effect, and slight differences in the openings of the doors and the cardboard boxes might have been major contributors to the deviations. However, the shapes and peaks of the HRR curves have similar trends.

Work Stations

NIST conducted burns of office cubicles (work stations) divided by fabric-covered partitions (Madrzykowski and Vettori 1992; Madrzykowski 1996). The layouts of the work stations burned are shown in Figure 5.22; the furniture in these work stations is typical of that used in the offices of the U.S. General Services Administration (GSA) and the fuel loading was consistent with that observed in GSA offices. The HRRs of the burns are shown in Figure 5.23. The work stations with more sides tend to have higher peak HRRs. The lowest peak was at about 1600 Btu/s (1700 kW) for the two-sided (A) layout. The three-sided (A) layout and the four-sided one had peaks of about 6400 Btu/s (6800 kW). Fire tests of work stations conducted at NIST by Ohlemiller et al. (2005) had peak HRRs as high as 7100 Btu/s (7500 kW).

Automobiles

Over the last four decades, there has been a very significant increase in the HRR of cars. This increase is probably due to increased use of polymers and

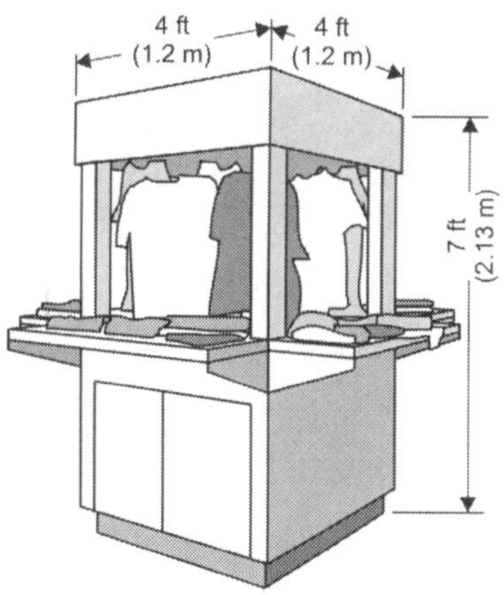

Figure 5.20 *Small kiosk.*

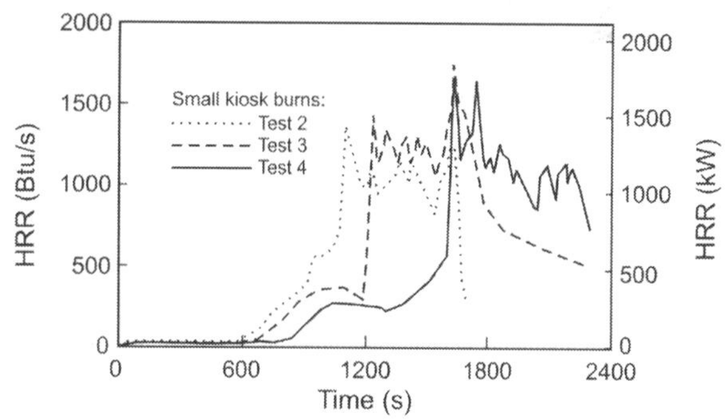

Figure 5.21 *HRR of small kiosk.*

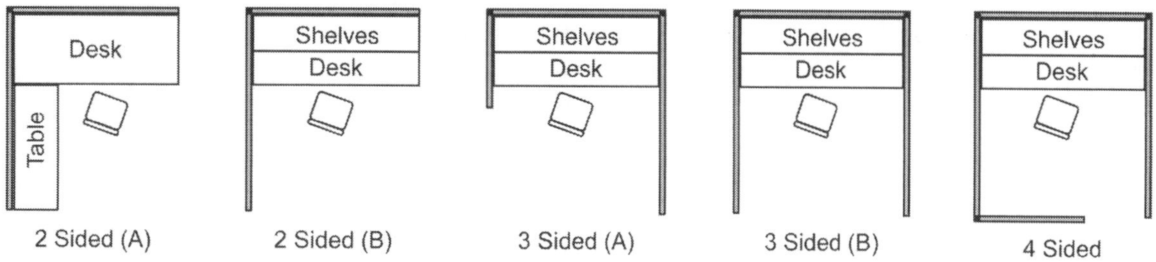

Figure 5.22 *Layouts of NIST work stations.*

other nonmetallic materials in cars. In a 1972 test sponsored by the American Iron and Steel Institute, a 1964 model car burning in an open parking garage did not spread to the cars in adjacent parking places (Cohn 1973). In tests at the Centre Technique Industriel de la Construction Metallique (CTICM) in France (Joyeux 1997), the HRR of a burning car in a parking garage ignited a car in an adjacent parking place. In tests conducted at the British Research Establishment (BRE) in the UK (Shipp et al. 2006; BRE 2010), the fires were generally larger than those of the tests at CTICM.

The BRE tests were done in 2006, and all cars were either less than five years old or a model still in production. The fuel tank in each car contained 5.3 gal (20 liters) of gasoline. HRR of these tests is shown in Figure 5.24.

Three Car Burns: In the BRE study, Tests 1, 2, and 3 were with three cars arranged as shown in Figure 5.25. Cars 1 and 2 were next to each other in parking places, and Car 3 was separated by a parking space. For all of these tests, the fire initiation simulated arson. Car 1 driver's window and passenger window were open. All other car windows were closed. The fire was started using a No. 7 crib on the driver's seat of Car 1. This crib is commonly used in the U.K. for fire tests of upholstered seating, and it is roughly 2.5 in. (60 mm) on a side.

Tests 1 and 3 were unsprinklered. In both tests, the initial fire in Car 1 ignited Car 2, and when the fire spread to Car 3, the fire was terminated to avoid overloading the calorimeter. The peak HRR for Test 1 was 15,200 Btu/s (16,000 kW) at 21 minutes after ignition, and the peak HRR for Test 2 was 10,400 Btu/s (11,000 kW) at 10 minutes after ignition (Figure 5.24). The higher peak HRR of Test 1 is probably due to the larger cars burned in that test.

Test 2 was a sprinklered fire, and the HRR was below 280 Btu/s (300 kW) for the first 40 minutes of the test. After 40 minutes, the HRR grows to a peak of 6400 Btu/s (7000 kW) 55 minutes into the test.

Engine Fire: Test 7 was an engine fire, and for the first 40 minutes of this test the HRR was less than 1900 Btu/s (2000 kW). This test peaked 45 minutes into the test with a HRR of 4500 Btu/s (4800 kW).

Stacker Fire: A car stacker is a mechanical device parking for two or more cars stacked one on top of another, and stackers in parking garages can be computer controlled. Using a steel and block rig, Test 11 had two cars supported one on top of the other as they would be in a stacker. The fire started in the lower car with the driver's window open. The peak HRR was 8100 Btu/s (8500 kW) at 12 minutes into the test.

Trash Bags and Containers

A common scenario is that a trash fire ignites a larger object such as a bed or a piece of upholstered

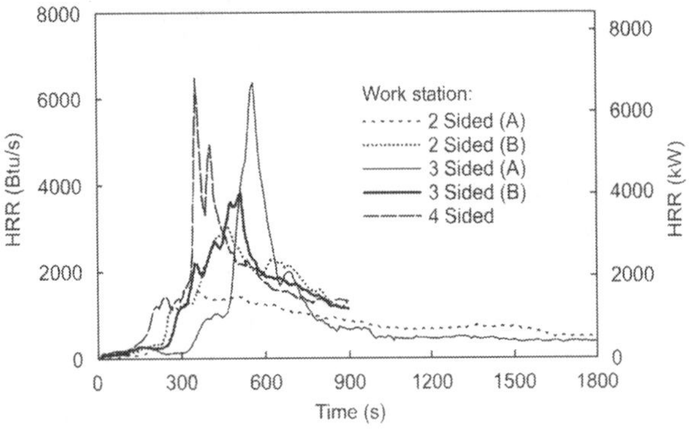

Figure 5.23 *HRR from NIST work station burns.*

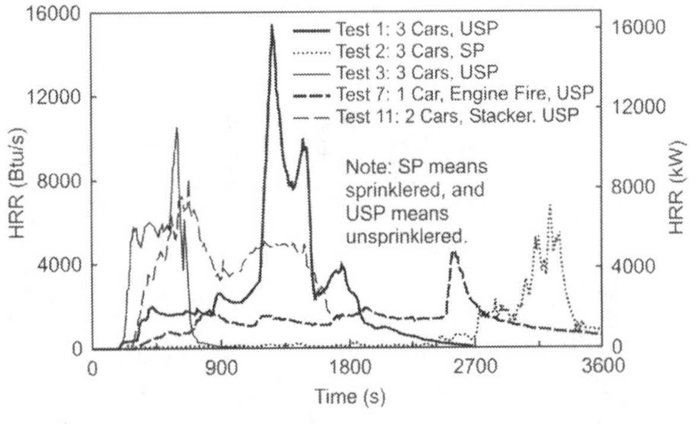

Figure 5.24 *HRR of automobile burns at BRE.*

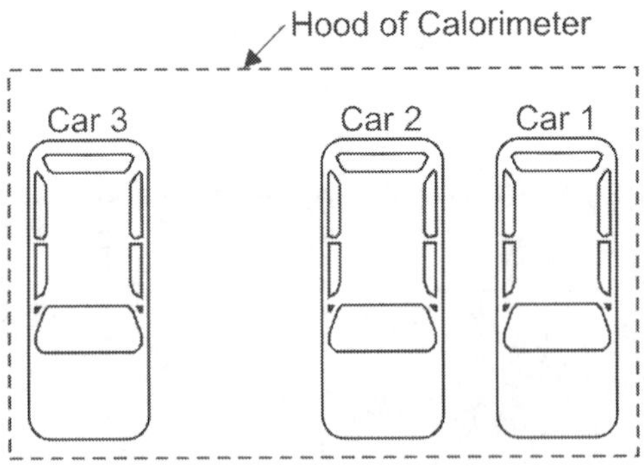

Figure 5.25 *Arrangement of automobile in BRE burn Hall.*

furniture. Many research projects have used this scenario. A seven-liter polyethylene wastebasket filled with 12 paper/polyethylene milk cartons (six upright and six shredded) resulted in a HRR of about 50 kW for about 200 s (Klitgaad and Williamson 1975). This is probably near the upper limit of a fire that can be supported by the contents of a small wastebasket. For repeatability, a gas burner with a HRR of 50 kW has been used to simulate ignition by such a wastebasket (Babrauskas and Krasny 1985).

Tests were done at NIST on the kind of paper trash bags used on Amtrak trains (Peacock et al. 2002). The HHRs for these tests are shown in Figures 5.26 and 5.27. The HRR depended on the weight of the filled bag and how the crumpled paper was packed into the bag. The bag of Test 1 weighed only 4.0 lb (1.8 kg), and it had a peak HRR of about 240 Btu/s (250 kW). Test 2 weighed 20 lb (9.5 kg) and had a peak of only about 40 Btu/s (42 kW). This shows that a less densely packed trash bag can have a higher peak HRR. The trash bags for tests 8, 9, and 10 were packed with the same amount of newspaper in the same manner as much as was possible, but the peaks for these tests ranged from about 130 to 220 Btu/s (140 to 230 kW). This variation in peak values is possibly due to the difficulty in hand packing crumpled newspaper in a repeatable fashion.

Tests were conducted at NIST on high-density polyethylene (HDPE) trash containers filled with cellulosic debris (Stroup and Madrzykowski 2003). Figure 5.28 shows the HRR from these tests. In the first test, the container opened up from the top down and had a peak heat release rate of approximately 300 kW. In the second case, the trash container melted and opened from the midpoint in the container down. This resulted in a container that tended to close up into itself instead of opening up, and thus yielded a peak heat release rate of approximately 150 kW.

RADIANT IGNITION

The point source radiant model can be useful in evaluating radiant ignition. This model considers the flame as a small thermal source such that the intensity of thermal radiation is proportional to the inverse of the square of distance from the source. Thermal radiation also is called radiative heat flux.

The intensity of thermal radiation is

$$q_r'' = \frac{Q_r}{4\pi R^2} \tag{5.10}$$

where

$$Q_r = \chi_r Q \tag{5.11}$$

and

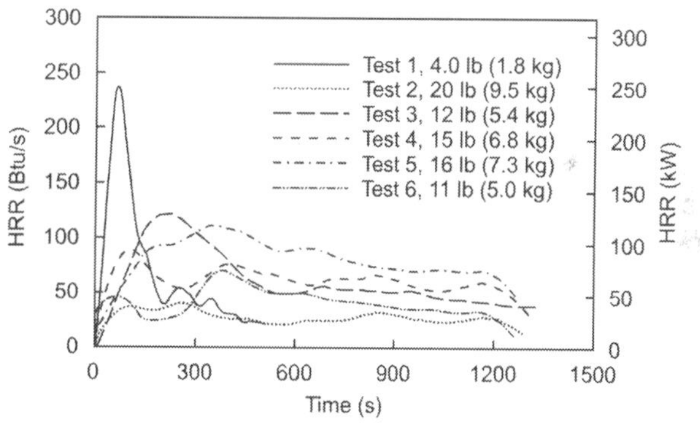

Figure 5.26 *HRR of Amtrak trash bag fire tests 1 to 6.*

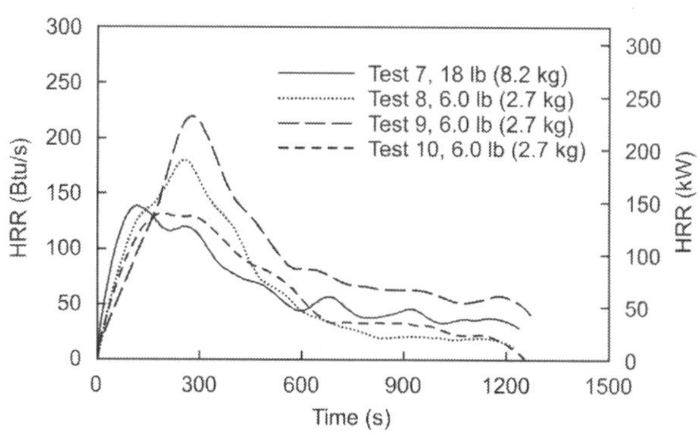

Figure 5.27 *HRR of Amtrak trash bag fire tests 7 to 10.*

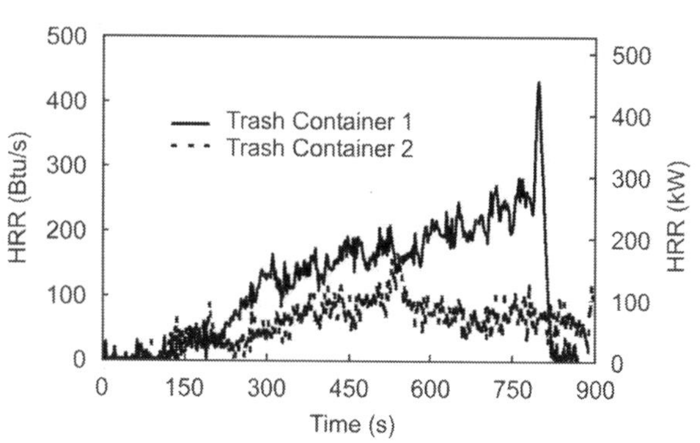

Figure 5.28 *HRR of plastic trash containers measured at NIST.*

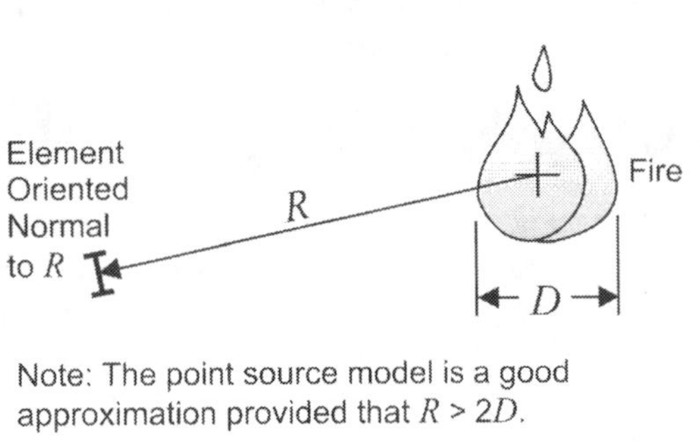

Figure 5.29 *Point source radiation model.*

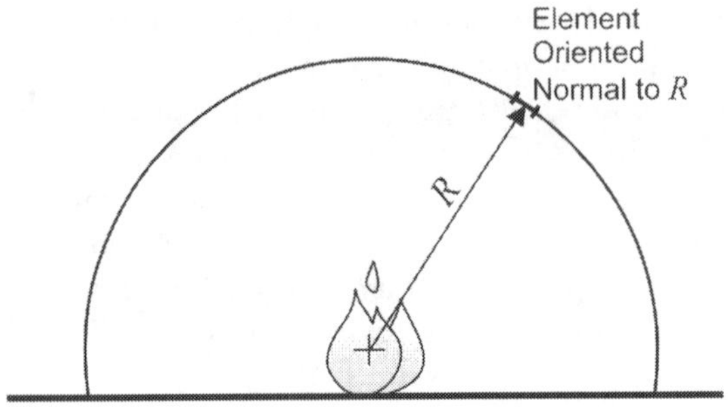

Figure 5.30 *Separation distance.*

q_r'' = intensity of thermal radiation, Btu/ft²·s (kW/m²),

Q = HRR of the fire, Btu/s (kW),

Q_r = radiant HRR of the fire, Btu/s (kW),

R = distance from the center of the fire, ft (m).

χ_r = radiative fraction, dimensionless.

Figure 5.29 illustrates the point source model, which is appropriate provided that the distance from the center of the flame is greater than twice the diameter of the fire.

Heat transfer is by conduction, convection, and radiation. For most fires, conduction heat transfer from the flame is negligible. The radiative fraction depends on the material burned and the diameter of the fire. The radiative fraction varies from about 0.1 to 0.6. Low sooting fuels such as methanol have low radiative fractions, and high sooting materials such as gasoline and polystyrene have high radiative fractions. The radiant fraction can be expressed as

$$\chi_r = (1 - \chi_c) \tag{5.12}$$

where χ_c is the convective fraction. While convective and radiative fractions vary for many materials, the values of $\chi_r = 0.3$ and $\chi_c = 0.7$ are commonly used for many design applications.

Fuel Packages

A fuel package can be made of one or a number of items. For a particular situation, the concept of the separation distance can be used to determine what objects constitute the fuel package of a fire.

For the point source radiant model, the separation distance is

$$R_{SD} = \sqrt{\frac{Q_r}{4\pi q_{r,i}''}} \tag{5.13}$$

where

R_{SD} = separation distance from the center of the fire to a target, ft (m),

Q_r = radiant HRR of the fire, Btu/s (kW),

$q_{r,i}''$ = intensity of thermal radiation needed for nonpiloted ignition, Btu/ft²·s (kW/m²).

Fuel items that are less than R_{SD} away from the fire are expected to ignite, and fuel items further than R_{SD} away are not expected to ignite. Figure 5.30 shows a fire that is the center of a partial sphere of radius, R_{SD}. Items inside the partial sphere are expected to ignite, and items outside it are not be expected to ignite. The radiant flux needed for nonpiloted ignition varies from about 0.9 Btu/ft²·s (10 kW/m²) for thin, easy-to-ignite materials to 1.8 Btu/ft²·s (20 kW/m²) for thick materials.

The HRR needed to ignite a second object can be expressed as

$$Q_{r,i} = 4\pi R^2 q_{r,i}'' \tag{5.14}$$

where

R = distance from the center of the fire to a the second object, ft (m),

$Q_{r,i}$ = radiant HRR of the fire to cause ignition of second item, Btu/s (kW),

$q_{r,i}''$ = intensity of thermal radiation needed for nonpiloted ignition, Btu/ft²·s (kW/m²).

Example 5.3 illustrates the determination of a fuel package for the four chairs shown in Figure 5.31. When the fuel package is determined, the HRR curve for it is determined graphically as shown in Figure 5.32. For the fuel package of Figure 5.33, Example 5.4 calculates the HRR curve graphically (Figure 5.34).

Examples 5.3 and 5.4 use HRR curves for the chairs and sofa from test data for burning in an open air calorimeter. The calculations consider that the objects burn as they would in open air, or burn as they would in a

Example 5.3. Fire Spread by Radiant Ignition

For the arrangement of chairs in Figure 5.31, a fire starts in chair 1. Determine what chairs will be involved in the fire, and what will be the HRR curve. Use the following simplified version of the HRR for the upholstered chair with polyurethane foam padding of Figure 5.16.

HRR of Chair

Time (s)	Btu/s
0	0
120	28
175	250
248	2010
354	270
600	150
930	75
2100	0

Use an intensity of thermal radiation needed for nonpiloted ignition of 0.9 Btu/ft^2 s (10 kW/m^2). The maximum HRR of the chair is 2010 Btu/s. The radiant HRR is

$$Q_r = \chi_r Q = 0.3(2010) = 603 \text{ Btu/s}$$

Calculate the separation of distance from the chair fire.

$$R_{SD} = \sqrt{\frac{Q_r}{4\pi q''_{r,i}}} = \sqrt{\frac{603}{4\pi(0.9)}} = 7.3 \text{ ft (2.2 m)}$$

This means that Chairs 1 and 2 will burn, but Chairs 3 and 4 will not. The distance, R, from the center of the fire in Chair 1 to the edge of Chair 2 is 3 ft (0.91 m). The radiant HRR of the fire to cause ignition of Chair 2 is

$$Q_{r,i} = 4\pi R^2 q''_{r,i} = 4\pi(3)^2(0.9) = 102 \text{ Btu/s (108 kW)}$$

The total HRR at this point is $Q = Q_{r,i}/0.3 = 103/0.3 = 339$ Btu/s (358 kW).

Graph the HRR curve for Chair 1, and locate when this curve reaches 339 Btu/s. This is the point of ignition of Chair 2, which happens at about 179 s. See Figure 5.32a.

Draw the HRR curve of Chair 2 starting at 179 s. Add the HRR curves of the 2 chairs to get the total HRR of the fire as shown in Figure 5.32b to get the total HRR curve. The maximum of this curve is about 2200 Btu/s (2300 kW).

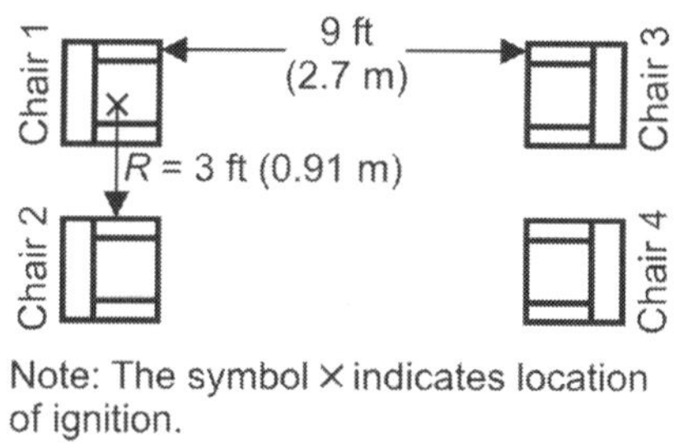

Figure 5.31 *Arrangement of chairs for Example 5.3.*

large volume space away from the ceiling and walls. The calculations do not account for room effects that can become significant in confined spaces.

An additional consideration with the calculation of fuel packages is that furniture is frequently rearranged. A furniture arrangement on an architect's drawing may be followed for the initial stages of occupancy, but occupants tend to move furniture. It is not practical to evaluate fuel packages for all possible furniture arrangements in a space. Further, the selection of the location of ignition can have a major impact on calculated HRR. It is suggested that a location for ignition be selected that results in a worst case scenario or a near worst case scenario.

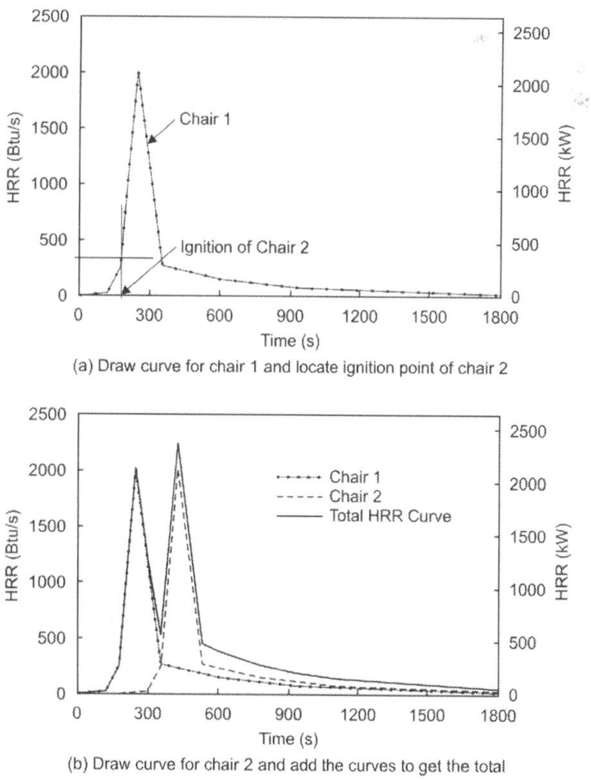

Figure 5.33 *Graphic determination of HRR curves for Example 5.3.*

Example 5.4. Another Fire Spread Example

For the arrangement of furniture in Figure 5.33, a fire starts in the sofa. What will be the HRR curve of the entire fuel package? Use the HRR for the upholstered chair from Example 5.3, and the following simplified curve for the sofa based on that of Figure 5.16.

HRR of Sofa

Time, s	Btu/s
0	0
83	140
130	560
218	2940
265	2960
370	600
410	350
840	133
2250	0

As with Example 5.3, an intensity of thermal radiation needed for nonpiloted ignition of 0.9 $Btu/ft^2 \cdot s$ (10 kW/m^2) will be used. Because of the location of the fire, both chairs will ignite at the same time. The radius for the chairs is $R = 7$ ft (2.1 m).

The radiant HRR of the fire to cause ignition of chairs is

$$Q_{r,i} = 4\pi R^2 q''_{r,i} = 4\pi(7)^2(0.9) = 554 \text{ Btu/s (584 kW)}$$

The total HRR at this point is $Q = Q_{r,i}/0.3 = 554/0.3 = 1850$ Btu/s (1950 kW).

At about 187 s, the sofa gets to this HRR, and the chairs ignite. See Figure 5.34a.

Draw the HRR curves of two chairs starting at 187 s.

Add the HRR curves of the sofa and chairs to get the total HRR of the fire as shown in Figure 5.34b. The maximum of this curve is about 4400 Btu/s (4600 kW).

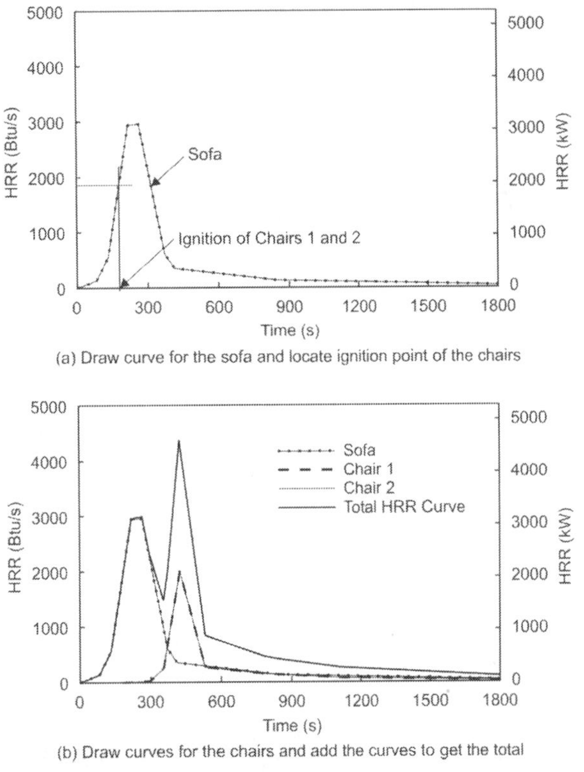

(a) Draw curve for the sofa and locate ignition point of the chairs

(b) Draw curves for the chairs and add the curves to get the total

Figure 5.32 *Graphic determination of HRR curves for Example 5.4.*

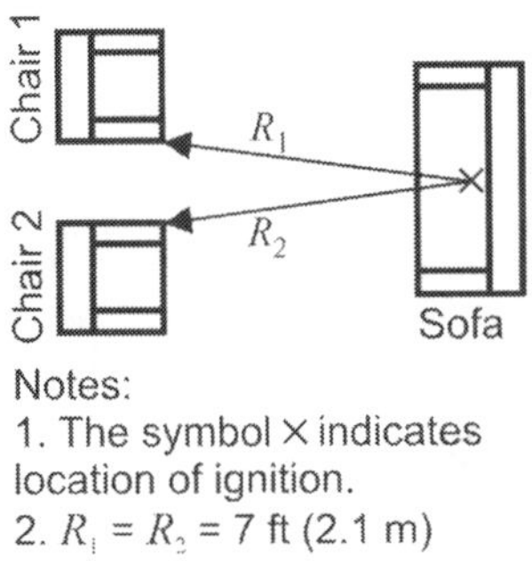

Figure 5.34 *Furniture arrangement for Example 5.4.*

NOMENCLATURE

A_w	=	area of ventilation opening, ft^2 (m^2)
H	=	stack height, ft (m)
H_w	=	height of ventilation opening, ft (m)
L	=	pallet length, ft (m)
M	=	moisture, %
n	=	growth exponent, dimensionless
Q	=	heat release rate, Btu/s (kW); or post sprinkler actuation HRR, Btu/s (kW)
Q_{act}	=	HRR at sprinkler actuation, Btu/s (kW)
Q_r	=	radiant HRR of the fire, Btu/s (kW)

$Q_{r,i}$ = radiant HRR of the fire to cause ignition of second item, Btu/s (kW)

$q''_{r,i}$ = intensity of thermal radiation needed for nonpiloted ignition, Btu/ft^2 s (kW/m^2)

q''_r = intensity of thermal radiation, Btu/ft^2·s (kW/m^2)

R = distance from the center of the fire, ft (m)

R_{SD} = separation distance from the center of the fire to a target, ft (m)

t = time from ignition, s

t_{act} = time of sprinkler actuation, s

t_g = growth time, s

t_o = effective ignition time, s

W = pallet width, ft (m)

α = fire growth coefficient, Btu/s^3 (kW/s^2)

τ = time constant for fire decay, s

χ_c = convective fraction, dimensionless

χ_r = radiative fraction, dimensionless

REFERENCES

Babrauskas, V. 2008a. *SFPE Handbook of Fire Protection Engineering*, Chapter 3-1, Heat Release Rates. Society of Fire Protection Engineers, Bethesda, MD.

Babrauskas, V. 2008b. *SFPE Handbook of Fire Protection Engineering*, Chapter 3-3, The Cone Calorimeter. Society of Fire Protection Engineers, Bethesda, MD.

Babrauskas, V., and J. Krasny. 1985. Fire behavior of upholstered furniture. NBS Monograph 173, National Bureau of Standards, Gaithersburg, MD.

Block, J.A. 1971. A theoretical and experimental study of nonpropagating free-burning fires. Thirteenth Symposium (International) on Combustion, Combustion Institute, August 23–29, Salt Lake City, UT.

BRE. 2010. Fire spread in car parks. Report BD2552, BRE Global Limited, Watford, UK.

Budnick, E.K., S.P. Hunt, and M.T. Wright. 2008. *Fire Protection Handbook*, Vol. I, 20th ed., Chapter 3-9, Closed Form Enclosure Fire Calculations. National Fire Protection Association, Quincy, MA.

Bwalya, A.C., et al. 2010. Characterization of fires in multisuite residential dwellings: phase 1—room fire experiments with individual furnishings. Research Report No. 302, National Research Council, Ottawa.

Cohn, B.M. 1973. Automobile burn-out test in an open-air parking structure, Scranton, PA, October 15, 1972. Report No. 7328, Gage-Babcock, Westchester, IL.

Custer, R.L.P., B.J. Meacham, and R.P. Schifiliti. 2008. *SFPE Handbook of Fire Protection Engineering*, Chapter 4-1, Design of Detection Systems. Society of Fire Protection Engineers, Bethesda, MD.

Evans, D.D. 1993. Sprinkler fire suppression algorithm for HAZARD. NISTIR 5254, National Institute of Standards and Technology, Gaithersburg, MD.

Gross, D. 1962. Experiments on the burning of cross piles of wood. *Journal of Research of NBS* 66C(2).

Hall, J.R. 2006. *An Analysis of Automatic Sprinkler System Reliability Using Current Data.* Quincy, MA: National Fire Protection Association.

Hall, J.R. 2011. *U.S. Experience with Sprinklers.* Quincy, MA: National Fire Protection Association.

Heskestad, G. 1984. Engineering relations for fire plumes. *Fire Safety Journal* 7(1).

Heskestad, G., and H. Smith. 1976. Investigation of a new sprinkler sensitivity approval test: the plunge test. FMRC Serial No. 22485, Factory Mutual Corporation, Norwood, MA.

Huggett, C. 1980. Estimation of rate of heat release by means of oxygen consumption measurements. *Fire and Materials* 4(2).

Joyeux, d. 1997. natural fires in closed car parks: car fire tests. CTICM—Testing and Fire Division, Saint-Rémy-lès-Chevreuse, France.

Klitgaad, P.S. and R.B. Williamson. 1975. The impact of contents on building fires. *Journal of Flammability/Consumer Product Flammability Supplement* (2).

Klote, J.H. 1990. Fire experiments of zoned smoke control at the Plaza Hotel in Washington, DC. *ASHRAE Transactions* 96(2).

Koffel, W.E. 2005. *Reliability of Automatic Sprinkler Systems*. Columbia, MD: Koffel Associates.

Janssens, M. 2008. *SFPE Handbook of Fire Protection Engineering*, Chapter 3-2, Calorimetry, Society of. Fire Protection Engineers, Bethesda, MD.

Lawson, J.R., et al. 1984. Fire Performance of Furnishings as Measured in the NBS Furniture Calorimeter, Part I. National Bureau of Standards, Gaithersburg, MD.

Lougheed, G.D. 1997. Expected size of shielded fires in sprinklered office buildings. *ASHRAE Transactions*, 103(1).

Lougheed, G.D. and J.R. Mawhinney. 1996. Probability of occurrence and expected size of shielded fires in sprinklered buildings, ASHRAE RP-838—Phase 1. Report A4201.5, National Research Council, Ottawa, Canada.

Madrzykowski, D. 1996. Office work station heat release rate study: full scale vs. bench scale, Interfalm 1996, C.A. Franks and S. Grayson, editors. Proceedings of the 7th International Interflam Conference, March 26–28, Cambridge, England.

Madrzykowski, D. 2008. Impact of a residential sprinkler on the heat release rate of a christmas tree fire. NISTIR 7506, National Institute of Standards and Technology, Gaithersburg, MD.

Madrzykowski, D., and R.L. Vettori. 1992. A sprinkler fire suppression algorithm for the GSA engineering fire assessment system. NISTIR 4833, National Institute of Standards and Technology, Gaithersburg, MD.

Mitler, H.E. 1996. Input Data for Fire Modeling. 13th Meeting of the UJNR Panel on Fire Research and Safety, March 13–20. NISTIR 6030, National Institute of Standards and Technology, Gaithersburg, MD.

Morgan, H.P., et al. 1999. *Design Methodologies for Smoke and Heat Exhaust Ventilation*. London: CRC Ltd.

Mulholland, G. 1995. *SFPE Handbook of Fire Protection Engineering*, 2nd ed., Smoke production and properties. Quincy, MA: National Fire Protection Association.

Nelson, H.E. 1987. An engineering analysis of the early stages of fire development—the fire at the DuPont Plaza Hotel and Casino—December 31, 1986. NISTIR 87-3560, National Institute of Standards and Technology, Gaithersburg, MD.

NFPA. 2007. NFPA 72, *National Fire Alarm Code*. Quincy, MA: National Fire Protection Association.

NFPA. 2012. NFPA 92, *Standard for Smoke Control Systems*. Quincy, MA: National Fire Protection Association.

NFPA. 2010. NFPA 13, *Installation for the Sprinkler Systems*. Quincy, MA: National Fire Protection Association.

Ohlemiller, T.J., et al. 2005. Fire tests of single office workstations. NIST NCSTAR 1-5C, National Institute of Standards and Technology, Gaithersburg, MD.

Parker, W.J. 1982. Calculation of heat release rate by oxygen consumption for various applications. NBSIR 82-2427-1, National Bureau of Standards, Gaithersburg, MD.

Peacock, R.D., et al. 2002. Fire safety of passenger trains; Phase II: Application of fire hazard analysis techniques. NISTIR 6525, National Institute of Standards and Technology, Gaithersburg, MD.

Peacock, R.D., et al. 1999. Defining flashover for fire hazard calculations. *Fire Safety Journal* 32.

Pitts, W.M. 1994. The global equivalence ratio concept and production of carbon monoxide in enclosure fires. NIST Monograph 197, National Institute of Standards and Technology, Gaithersburg, MD.

Shipp, M., et al. 2006. Fire spread in car parks—Summary of the CLG/BRE research programme. BRE Global Limited, Watford, UK.

Stroup, W.D., and D. Madrzykowski. 2003. Heat release rate tests of plastic trash containers. Report of Test FR 4018, National Institute of Standards and Technology, Gaithersburg, MD.

Stroup, W.D., et al. 1999. Scotch pine Christmas tree fire tests. Report of Test FR 4010, National Institute of Standards and Technology, Gaithersburg, MD.

Stroup, W.D., et al. 2000. Large fire research facility (building 205) exhaust hood heat release rate measurement system. NISTIR 6509, National Institute of Standards and Technology, Gaithersburg, MD.

Thomas, I. 2008. *SFPE Handbook of Fire Protection Engineering*, Chapter 4-8, Enclosure Fire Temperature-Time Estimates. Society of Fire Protection Engineers, Bethesda, MD.

CHAPTER 6

Human Exposure to Smoke

John H. Klote

Harland and Woolley (1979) and Berl and Halpin (1980) showed that smoke is the major killer in building fires. Smoke is defined as the airborne solid and liquid particulates and gases evolved when a material undergoes pyrolysis or combustion, together with the quantity of air that is entrained or otherwise mixed into the mass. Toxic gases, heat, and thermal radiation are the direct threats to human life from flames and smoke. In thick smoke, people see poorly, walk slowly, and/or become disoriented, which prolongs exposure to smoke. Falls from balconies are an additional threat associated with reduced visibility. In many applications, the primary threat results from reduced visibility, but the other threats still need to be considered. This chapter addresses these threats with respect to smoke control.

TIME EXPOSURE

Haber (1924) postulated that the effect of an exposure to a gas is directly related to the product of the gas concentration and time duration of the exposure. This relationship has become known as Haber's rule. This rule considers a constant ingestion rate of the toxin, but concentrations of toxic gases due to building fires change with time. While not all gases follow this rule, gas concentration and exposure time are important factors for all exposures to toxic gases. The approaches discussed later for evaluation of toxic gas exposure account for the concentrations of gases changing during exposure.

The effect of exposures to heat and thermal radiation also depend on the time duration of the exposure, and there are approaches to evaluate the effects of both of these exposures. Reduced visibility is the exception in that it does not depend on the duration of the exposure.

EXPOSURE TO TOXIC GASES

Carbon monoxide (CO) exposure accounts for the majority of total fire fatalities (Berl and Halpin 1980; Harland and Woolley 1979). However, smoke often includes many other toxic gases. Hyperventilation due to carbon dioxide (CO_2) exposure will increase the rate of intake of CO. Oxygen (O_2) deprivation is a special case, and the reduction in the amount of O_2 available for tissue respiration is referred to as hypoxia. Because of the interaction of these gases, exposure effects discussed below consider the combined effects of these gases. The effect of exposure to toxic gases on a specific individual depends on the physiological characteristics of the individual.

CO and CO_2

Exposure to CO results in carboxyhemoglobin (COHb) uptake in the blood, which results in decreased oxygen-carrying capacity of the blood. Stewart et al. (1973) conducted a series of experiments on humans, and based on this research, COHb uptake can be expressed as

$$C_{COHb} = C_{COHb,0} + 3.317 \times 10^{-5} \sum_{i=1}^{n} C_{CO,i}^{1.036} V \Delta t_i \tag{6.1}$$

where

C_{COHb} = concentration of COHb in the blood, %,

$C_{COHb,0}$ = concentration of COHb in the blood at time zero, %,

$C_{CO,i}$ = concentration of CO, ppm,

V = volume of breathed air per minute, L/min,

Δt_i = time interval, min.

This equation does not include the effects of oxygen depletion, increased breathing rate due to carbon dioxide exposure, or exposure to other toxic gases. The volume of breathed air V is called the respiratory minute volume (RMV). The typical RMV a 150 lb (70 kg) person at rest is about 8.5 L/min. A higher RMV of 18 L/min has been used to account for activity and CO_2 exposure. For calculations, a value of $C_{COHb,0} = 0.75\%$ can be used, and incapacitation and lethality are approximately 25% COHb and 50% COHb, respectively. However, calculation of the COHb level from Equation 6.1 is not a reliable indication of incapacitation or fatality, because it does not include the effects of other gases commonly present in smoke.

Gas Exposure Models

Because of concern for animal rights, the use of animals in toxicity research essentially ended near the end of the 20th century. The fractional effective dose (FED) model and the N-gas model can be used to evaluate fatality for a given exposure. Purser (2008) developed a model to evaluate incapacitation, which is based on experiments with primates.

Animal Tests & the FED Model

The FED animal tests determine the concentration of airborne combustion products that is lethal to 50% of the subjects exposed for a specified time, and this lethal concentration is referred to as the LC_{50}. The specified time for the tests is usually 30 min.

For an exposure at a constant concentration, the FED is

$$\mathrm{FED} = \frac{m_f t}{\mathrm{LC}t_{50}} \tag{6.2}$$

where

FED = fractional effective dose, dimensionless,

m_f = mass concentration of fuel burned, lb/ft^3 (g/m^3),

t = exposure time, min,

$\mathrm{LC}t_{50}$ = lethal exposure dose from test data, lb ft^{-3} min (g m^{-3} min).

An FED greater than or equal to 1 indicates fatality, and the concentration is in mass of the material burned per unit volume. The lethal exposure dose is the product of the LC_{50} and the exposure time, and Table 6.1 lists the lethal exposure doses of some materials. Bukowski et al. (1989) state that a FED of 0.5 can be considered an approximation to the incapacitating dose. It is possible that this approximation is a conservative criterion for smoke management design analysis. Example 6.1 illustrates calculation of the FED.

When the concentration is not constant, the FED is

Table 6.1: Approximate lethal exposure dose, $\mathrm{LC}t_{50}$, for common materials

	Nonflaming Fire		Fuel-Controlled Fire		Fully Developed Fire	
Material	**lb/ft^3·min**	**g/m^3·min**	**lb/ft^3·min**	**g/m^3·min**	**lb/ft^3·min**	**g/m^3·min**
Cellulosics	0.046	730	0.19	3120	0.047	750
C, H, O plastics	0.031	500	0.075	1200	0.033	530
PVC	0.031	500	0.019	300	0.012	200
Wool/Nylon (low N_2)	0.031	500	0.057	920	0.0044	70
Flexible Polyurethane	0.042	680	0.087	1390	0.012	200
Rigid Polyurethane	0.0039	63	0.0062	100	0.0034	54
Modacrylic/PAN[1]	0.010	160	0.0087	140	0.0028	45

[1]PAN is polyacrylonitrile.

$$\mathrm{FED} = \frac{\sum_{i=1}^{n} m_{f,i}\Delta t_i}{\mathrm{LC}t_{50}} \tag{6.3}$$

where

$m_{f,i}$ = mass concentration for time interval i, lb/ft^3 (g/m^3),

Δt_i = time interval i, min,

n = number of discrete concentration time pairs.

N-Gas Model

The N-gas model was developed at the National Institute of Standards and Technology (NIST) and relates fatality with animal test data of exposures to pure gases and mixtures of gases (Levin 1996; Levin et al. 1995; Babrauskas et al. 1991). For mixtures of gases including NO_2, the N-gas model can be stated as

$$\begin{aligned} N_{Gas} = {} & \frac{m[CO]}{[CO_2]-b} + \frac{20.9-[O_2]}{20.9-LC_{50}(O_2)} \\ & + \left(\frac{[HCN]}{LC_{50}(HCN)} \times \frac{9.4[NO_2]}{LC_{50}(NO_2)}\right) \\ & + 0.4\left(\frac{[NO_2]}{LC_{50}(NO_2)}\right) \\ & + \frac{[HCl]}{LC_{50}(HCl)} + \frac{[HBr]}{LC_{50}(HBr)} \end{aligned} \tag{6.4}$$

and for mixtures not including NO_2, the N-gas model can be started as

$$\begin{aligned} N_{Gas} = {} & \frac{m[CO]}{[CO_2]-b} + \frac{20.9-[O_2]}{20.9-LC_{50}(O_2)} \\ & + \frac{[HCN]}{LC_{50}(HCN)} + \frac{[HCl]}{LC_{50}(HCl)} + \frac{[HBr]}{LC_{50}(HBr)}, \end{aligned} \tag{6.5}$$

Example 6.1. FED

Smoke from burning flexible polyurethane foam in a fully developed fire has a mass concentration of 0.001 lb/ft^3 of fuel burned. Calculate the FED for a 20 min exposure to this smoke.

From Table 6.1, the $\mathrm{LC}t_{50}$ is 0.012 lb/ft^3·min (200 g/m^3·min). For a constant concentration,

FED = $m_f t / \mathrm{LC}t_{50}$ where m_f is 0.001 lb/ft^3 of fuel burned, and t is the exposure time of 20 min.

FED = (0.001)(20)/0.012 = 1.7. Because this concentration is greater than 1, fatality is expected.

and

$$\begin{aligned} [CO] &= \frac{1}{t_e}\sum_{i=1}^{n} C_{CO,i}\Delta t_i \\ [CO_2] &= \frac{1}{t_e}\sum_{i=1}^{n} C_{CO2,i}\Delta t_i \\ [O_2] &= \frac{1}{t_e}\sum_{i=1}^{n} C_{O2,i}\Delta t_i \\ [HCN] &= \frac{1}{t_e}\sum_{i=1}^{n} C_{HCN,i}\Delta t_i \\ [NO_2] &= \frac{1}{t_e}\sum_{i=1}^{n} C_{NO2,i}\Delta t_i \\ [HCl] &= \frac{1}{t_e}\sum_{i=1}^{n} C_{HCl,i}\Delta t_i \\ [HBr] &= \frac{1}{t_e}\sum_{i=1}^{n} C_{HBr,i}\Delta t_i \end{aligned} \tag{6.6}$$

where

N_{Gas} = N-Gas model indicator, dimensionless,

m = –18 for $CO_2 \le 5\%$ and 23 for $CO_2 > 5\%$,

b = 122,000 for $CO_2 < 5\%$ and –38,600 for $CO_2 > 5\%$,

$LC_{50}(O_2)$ = lethal concentration of O_2, %,

$LC_{50}(HCN)$ = lethal concentration of HCN, ppm,

$LC_{50}(NO_2)$ = lethal concentration of NO_2, ppm,

$LC_{50}(HCl)$ = lethal concentration of HCl, ppm,

$LC_{50}(HBr)$ = lethal concentration of HBr, ppm,

$C_{CO,i}$ = concentration of CO, ppm,

$C_{CO2,i}$ = concentration of CO_2, ppm,

$C_{O2,i}$ = concentration of O_2, %,

$C_{HCN,i}$ = concentration of HCN, ppm,

$C_{NO2,i}$ = concentration of NO_2, ppm,

$C_{HCl,i}$ = concentration of HCl, ppm,

$C_{HBr,i}$ = concentration of HBr, ppm,

t_e = exposure time, min,

Δt_i = time interval i, min,

n = number of concentration values for each gas and time intervals.

The model incorporates increased breathing rate due to CO_2 exposure. It is apparent that there is a unique interaction between HCN and NO_2. For many of the gases, the contribution to lethality is expressed as the ratio of the gas exposure to the LC_{50}. This is how O_2 is treated, except that it is in terms of oxygen depletion.

The toxicity of CO_2 is not included in the N-gas model, because fire-generated atmospheres do not contain toxic concentrations of CO_2. The LC_{50} of CO_2 is 47%, and the maximum concentration of CO_2 in a fire atmosphere is 20.9% if all the oxygen in the air is converted to CO_2.

For animal tests, it was found that when the N_{Gas} value was approximately 1, some of the animals died. For values below 0.8, there would be no fatalities, and for values above 1.3, all the animals would be expected to die.

Equations 6.4 and 6.5 apply when the exposure time is the same as the duration of the LC_{50} data. Table 6.2 has LC_{50} values for all the gases in this model for many exposure times (Levin et al. 1988, 1989; Levin 1996, 2000; Hartzell, Grand, and Switzer 1990). For exposure times between those listed in this table, LC_{50} values can be interpolated. Example 6.2 demonstrates the use of the N-Gas model for four gases.

EXPOSURE TO HEAT

Heat exposure happens when a person comes into contact with hot gases. Figure 6.1 is a graph of the heat tolerance of naked humans at rest with low air movement (Blockley 1973). This figure shows 250°F (121°C) as a rule of thumb demarcation between skin burns and heat stroke (hyperthermia). Exposures to temperatures above 250°F (121°C) can result in skin pain and burns, and exposures to temperatures below this temperature can result in heat stroke.

Table 6.2: Lethal Concentration LC_{50} of Various Gases

Exposure Time, min	O_2, %	HCN, ppm	HCl, ppm	HBr, ppm	NO_2, ppm
1	–	3000	–	–	–
2	–	1600	–	–	1450
5	4.0	570	15900	12600	830
10	4.8	290	8400	6600	510
15	5.0	230	6900	5400	380
20	5.2	170	6400	5100	320
25	5.3	160	5900	4700	290
30	5.4	150	3800	3000	200
45	5.6	120	3300	2600	150
60	5.8	90	2800	2200	100

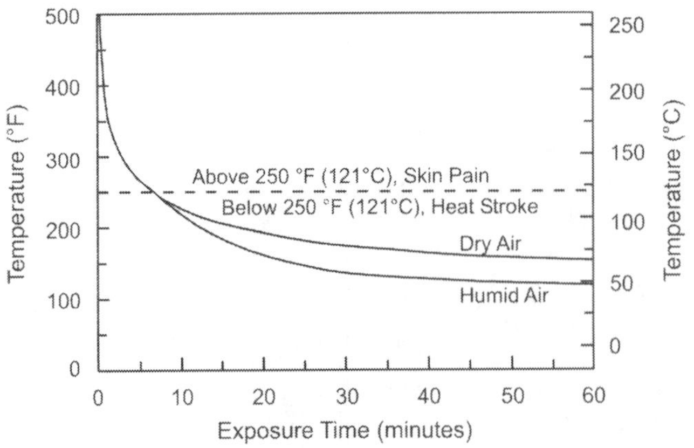

Figure 6.1 *Heat tolerance for humans at rest, naked, with low air movement.*

Example 6.2. N-Gas Model

Evaluate the exposure to smoke with the composition listed below using the N-Gas model.

Time (min)	$C_{O2,i}$ %	$C_{CO2,i}$ ppm	$C_{CO,i}$ ppm	$C_{HCN,i}$ ppm
0	20.9	0	0	0
1	20.8	300	20	1
2	20.7	600	40	2
3	20.5	1200	50	3
4	20.3	2000	60	5
5	20.0	2500	90	8
6	19.8	3200	110	9
7	19.7	3500	120	11
8	19.6	3600	130	12
9	19.5	3700	140	15
10	19.5	3800	170	18
11	19.5	3850	380	25
12	19.5	3850	500	35
13	19.5	3850	600	45
14	19.5	3850	700	45
15	19.5	3850	800	45
16	19.5	3850	900	45
17	19.5	3850	900	45
18	19.5	3850	900	45
19	19.5	3850	900	45
20	19.5	3850	900	45

$[O_2]$ = 19.77; $[CO_2]$ =3145; [CO] = 421; [HCN] = 25.2

Because CO_2 is less than 5% (50,000 ppm), m = –18 and b = 122,000.

From Table 6.2 for a 20 minute exposure, $LC_{50}(O_2)$ = 5.2% and LC_{50}(HCN) = 170 ppm.

Without HCl and HBr:

$$N_{Gas} = \frac{m[CO]}{[CO_2] - b} + \frac{20.9 - [O_2]}{20.9 - LC_{50}(O_2)}$$

$$N_{Gas} = \frac{-18(421)}{3145 - (122,000)} + \frac{20.9 - 19.77}{20.9 - 5.2} + \frac{25.2}{170} = 0.28$$

An exposure of N_{GAS} = 0.28 is not expected to cause fatality.

Figure 6.1 is for naked people, but clothing may inhibit the efficiency of the natural cooling process by perspiration. Conversely, clothing provides insulation from high temperature environments to protect the skin from becoming burned. At temperatures in excess of 250°F (121°C), where pain from skin burns is the dominant effect, the presence of clothing can be considered to be beneficial. However, at the lower temperatures where hyperthermia is the dominant effect, clothing is detrimental.

Humidity of the air has an impact on thermal exposure. Because water vapor is one of the products of combustion, the curve for exposure to humid air in Figure 6.1 is suggested for smoke control applications. Example 6.3 demonstrates evaluation of heat exposure.

EXPOSURE TO THERMAL RADIATION

This exposure happens when a person is subjected to thermal radiation from nearby flames or hot gases. Exposure to thermal radiation is not relevant for the design of most smoke control systems. This can be illustrated by considering both heat and thermal radiation exposures to a gas for the same period of time. If the temperature of the gas is such that heat exposure to that gas can be tolerated, then the exposure to the thermal radiation of the gas can also be tolerated.

Figure 6.2 shows the tolerance of human skin to thermal radiation (Stoll and Chianta 1969). The time to pain and blistering can be expressed as

$$t_{r,p} = \frac{3.20}{(q_r'')^{1.35}}$$

$$t_{r,p} = \frac{85}{(q_r'')^{1.35}} \text{ for IP} \tag{6.7}$$

and

$$t_{r,b} = \frac{8.39}{(q_r'')^{1.35}}$$

$$t_{r,b} = \frac{223}{(q_r'')^{1.35}} \text{ for SI} \tag{6.8}$$

where

$t_{r,p}$ = exposure time to pain, s,

$t_{r,b}$ = exposure time to blister, s,

q_r'' = intensity of thermal radiation, Btu/ft^2 s (kW/m^2).

Based on the point source radiant model discussed in Chapter 5, the minimum distance that an observer can be from a fire for a few minutes without unbearable pain is

$$R = \sqrt{\frac{Q_r}{4\pi q_{r,t}''}} \tag{6.9}$$

where

R = minimum distance from the center of the fire to a person, ft (m),

Q_r = radiant HRR of the fire, Btu/s (kW),

$q_{r,t}''$ = intensity of thermal radiation that can be temporarily tolerated, Btu/ft^2·s (kW/m^2).

A value of $q_{r,t}''$ = 0.22 Btu/ft^2·s (2.5 kW/m^2) is recommended. As discussed in Chapter 5, $Q_r = \chi_r Q$, where Q is the HRR of the fire and χ_r is radiative fraction. The radiative fraction varies from about 0.1 to 0.6, but a value of 0.3

Example 6.3. Heat Exposure

A smoke control system has a duration of operation of 20 min. At a location of concern, the smoke control analysis indicates that the maximum temperature is 140°F (60°C) for part of the 20 min duration. Is there a concern about heat exposure?

From Figure 6.1, it can be seen that for an exposure time of 20 min, a person can tolerate elevated temperatures up to about 160°F (70°C). This means that heat exposure at this location should not be a concern.

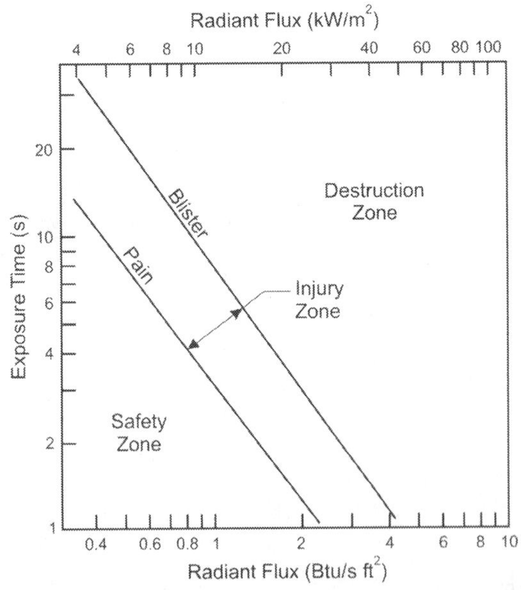

Figure 6.2 *Tolerance of human skin to thermal radiation.*

is commonly used for design applications. Example 6.4 calculates how close a person can approach a fire.

SMOKE OBSCURATION

Figure 6.3 is a diagram of a smoke meter for measuring smoke obscuration. A beam of light passes through a distance x called the path length. The intensity of light at the source is I_o, and the intensity after passing through the path length is I_x. Smoke meters are used in full scale tests and as part of many calorimeters. When used in large scale tests, the path length is often 3.28 ft (1.0 m).

The fraction of light transmitted through the path length is called the *transmittance*, and it is expressed as

$$T = \frac{I_x}{I_o} \tag{6.10}$$

where

T = transmittance, dimensionless,

I_o = intensity of light at the beginning of the path length,

I_x = intensity of light remaining after it has passed through the path length.

The units of light intensity are not given for the above equation because any of a number of units can be used, and the specific units are not needed for the discussion that follows. When the air in the path length is free of smoke, the intensity of light remaining after it has passed through the path length is almost the same as that at the beginning of the path length. When there is smoke in the path length, I_x is less than I_o. It follows that the transmittance through smoke is less than one.

Strictly speaking, this discussion applies to light of one wavelength, such as a laser beam, but incandescent bulbs have been used extensively in smoke meters. Smoke meters are calibrated by placing a translucent filter of known density in the path length, and then the voltage (or current) output of the photo cell can be calibrated to give transmittance directly.

There are a number of terms commonly used to express smoke obscuration. Because some publications use different definitions for obscuration terms, the more common ones are discussed here.

Percent obscuration is defined as

$$\lambda = 100(1 - T) \tag{6.11}$$

where λ is the dimensionless percent obscuration.

The *optical density* δ is expressed as

$$\delta = -\frac{\log_{10} T}{x} \tag{6.12}$$

where

δ = optical density per unit distance, ft^{-1} (m^{-1}),

x = distance of light travel or the path length, ft (m).

The *extinction coefficient* per unit distance is

$$\alpha = -\frac{\log_e T}{x} \tag{6.13}$$

where α is the extinction coefficient per unit distance in units of ft^{-1} (m^{-1}). The relationship between the extinction coefficient and the optical density is $\alpha = 2.303\ \delta$.

The *mass optical density* is

$$\delta_m = \frac{\delta}{m_f} \tag{6.14}$$

where

δ_m = mass optical density, ft^2/lb (m^2/g),

δ = optical density per unit distance, ft^{-1} (m^{-1}),

Example 6.4. Thermal Radiation

How close will a person be able to approach a 2000 Btu/s (2100 kW) fire?

$$Q_r = \chi_r Q = 0.3\,(2000) = 600 \text{ Btu/s (630 kW)}$$

$$q''_{r,t} = 0.22 \text{ Btu/ft}^2\text{·s (2.5 kW/m}^2\text{)}$$

$$R = \sqrt{\frac{Q_r}{4\pi q''_{r,t}}} = \sqrt{\frac{600}{4\pi(0.22)}} = 15 \text{ ft (4.6 m)}$$

This means that a person can be 15 ft (4.6 m) from the center of the 2000 Btu/s fire for a few minutes without unbearable pain.

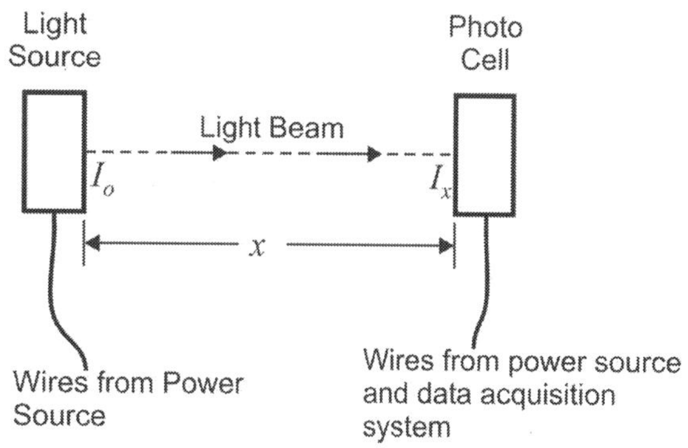

Figure 6.3 *Smoke meter for measuring smoke obscuration.*

m_f = mass concentration of fuel burned in units lb/ft^3 (g/m^3).

In a test chamber, the mass concentration of fuel burned, m_f, is the mass loss of a test sample that is burned divided by the volume of the test chamber.

Reduced Visibility

Tests of the visibility of illuminated and reflecting signs through smoke were conducted at the Fire Research Institute (FRI) in Japan (Jin 1974, 1975, 1985). A chamber was filled with smoke, and participants viewed signs in the chamber through a window. Figure 6.4 shows the results for the illuminated signs. Black smoke was produced by flaming fires, and white smoke was produced by smoldering fires. Visibility through the white smoke was less, probably due to higher light scattering.

Based on the FRI research, the relation between visibility and smoke obscuration is

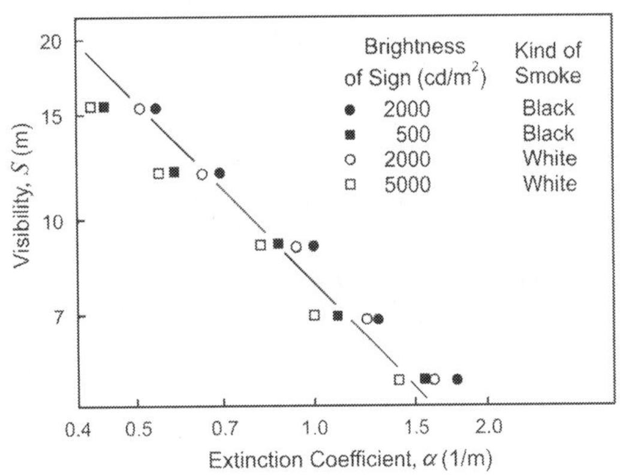

Figure 6.4 *FRI tests of visibility of illuminated signs in smoke.*

$$S = \frac{K}{\alpha}, \tag{6.15}$$

which can also be written as

$$S = \frac{K}{2.303\delta} \tag{6.16}$$

where

S = visibility, ft (m),

α = extinction coefficient ft^{-1} (m^{-1}),

δ = optical density per unit distance, ft^{-1} (m^{-1}),

K = proportionality constant (Table 6.3).

The proportionality constant K is 8 for illuminated signs and 3 for reflecting signs. A value of $K = 3$ is often used for building components that are seen with reflected light. Figures 6.5 and 6.6 are graphs of Equations 6.15 and 6.16 that show how visibility decreases when extinction coefficient and optical density increase. Example 6.5 illustrates calculation of visibility by use of Equation 6.16.

The effect of eye irritation on visibility was studied at FRI. The irritating smoke was white smoke produced by burning wood cribs which are geometrically arranged piles of sticks as described in Chapter 5. Less irritating smoke was produced by burning kerosene.

Table 6.3: Proportionality Constants for Visibility Based on Research at FRI

Situation	*K*
Illuminated signs	8
Reflecting signs	3
Building components in reflected light	3

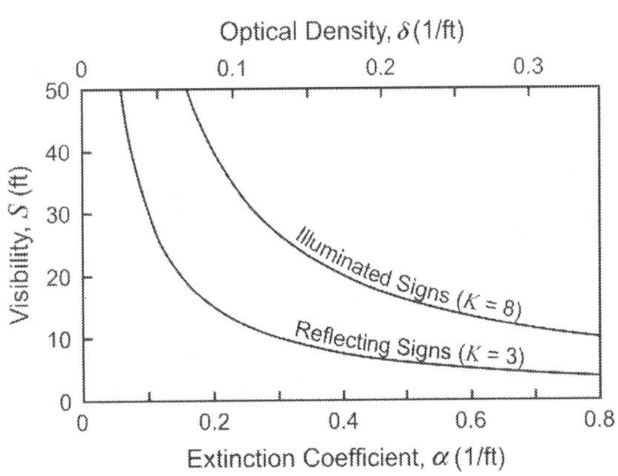

Figure 6.5 *Visibility through smoke in I-P units.*

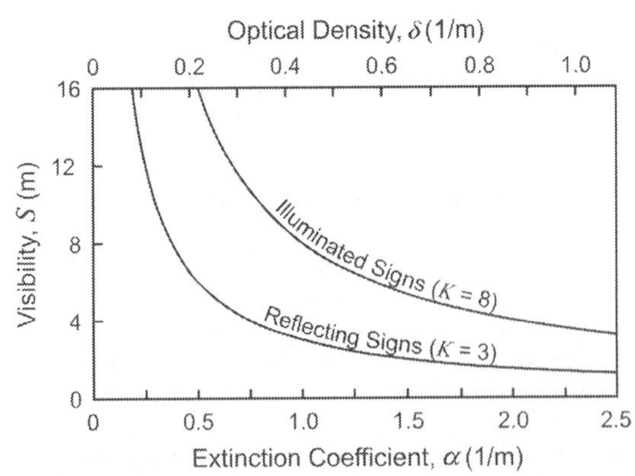

Figure 6.6 *Visibility through smoke in SI units.*

Figure 6.7 shows the relation between smoke obscuration and walking speed of people walking down a corridor in irritating and nonirritating smoke. Both eye irritation and smoke density affect walking speed. Walking speed decreases with extinction coefficient for both smokes, but it is much worse for irritating smoke. For extinction coefficients greater than 0.15 ft^{-1} (0.5 m^{-1}), the walking speed decreased to about 60 fpm (0.3 m/s), the speed of a blindfolded person. The drop in walking speed was because subjects could not keep their eyes open, and they walked in a zigzag or went step by step as they held the side wall.

Calculating Reduced Visibility

Visibility can be calculated from the properties of smoke by the mass optical density method and the specific extinction coefficient method. These methods can be used with computer models capable of simulating mass transport (Chapters 19 and 20).

Mass Optical Density

Equation 6.16 can be written in terms of mass optical density as

$$S = \frac{K}{2.303\delta_m m_f} \tag{6.17}$$

where

S = visibility, ft (m),
K = proportionality constant (Table 6.3),
δ_m = mass optical density, ft^2/lb (m^2/g),
m_f = mass concentration of fuel burned lb/ft^3 (g/m^3).

Computer models can be used to calculate the mass concentration of fuel burned at locations in a smoke control system. Mass optical densities of some natural and synthetic materials are listed in Table 6.4 (Mulholland 2008).

Specific Extinction Coefficient

Equation 6.15 can be expressed in terms of the mass concentration of airborne particulates as

$$S = \frac{K}{\alpha_m m_p} \tag{6.18}$$

where

S = visibility, ft (m),
K = proportionality constant (Table 6.3),
α_m = specific extinction coefficient, ft^2/lb (m^2/g),
m_p = mass concentration of particulate lb/ft^3 (g/m^3).

The specific extinction coefficient is the extinction coefficient divided by the mass concentration of airborne particulates ($\alpha_m = \alpha / m_p$). Commonly used values of the specific extinction coefficient for wood and plastics are

Example 6.5. Visibility Through Smoke

When the optical density of smoke is 0.10 ft^{-1}, what is the visibility for an illuminated sign?

$$K = 8;\ \delta = 0.10\ ft^{-1}$$

$$S = \frac{K}{2.303\delta} = \frac{8}{2.303(0.1)} = 35\ ft\ (11\ m)$$

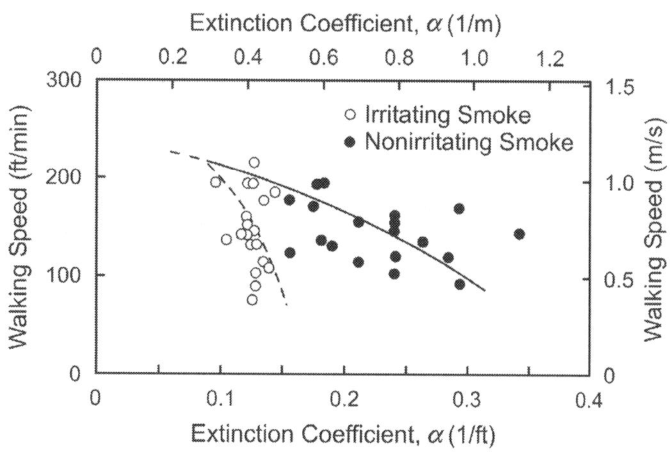

Figure 6.7 *FRI tests of walking speed in irritating and nonirritating smoke.*

Table 6.4: Mass Optical Densities

Material	Mass Optical Density δ_m		Combustion Conditions	Sample Thickness	
	ft^2/lb	m^2/g		in.	cm
Natural Materials:					
Plywood	1400	0.29	Pyrolysis	0.24	0.6
Wood (Douglas fir)	1400	0.28	Pyrolysis	0.24	0.6
Cotton	830	0.17	Flaming[1]		
Cotton	590	0.12	Flaming[2]		
Synthetic Materials:					
Polymethylmethacrylate	730	0.15	Pyrolysis	0.24	0.6
Polyvinylchloride	590	0.12	Pyrolysis	0.24	0.6
Polyvinylchloride (with plasticizer)	3100	0.64	Pyrolysis	0.24	0.6
Neoprene	2700	0.55	Pyrolysis	0.24	0.6
Polypropylene	2600	0.53	Flaming[1]	0.16	0.4
Polyethylene	1400	0.29	Flaming[1]	0.16	0.4
Paraffin wax	1100	0.23	Flaming[1]	0.16	0.4
Polystyrene	6800	1.4	Flaming[1]	0.16	0.4
Styrene	4700	0.96	Flaming[1]	0.16	0.4
Polyvinylchloride	1700	0.34	Flaming[1]	0.16	0.4
Polyurethane	1600	0.33	Flaming[1]		
Polyurethane	1100	0.22	Flaming[2]		
Latex	3200	0.65	Flaming[1]		
Latex	2100	0.44	Flaming[2]		
Neoprene	2000	0.40	Flaming[1]		
Neoprene	1000	0.20	Flaming[2]		
Polystyrene	3900	0.79	Flaming[1]		
Polystyrene	4900	1.0	Flaming[3]		
Polystyrene foam	3900	0.79	Flaming[1]		
Polystyrene foam	4000	0.82	Flaming[3]		
Acrylonitrile-butadiene-styrene (ABS)	2500	0.52	Flaming[1]		
Acrylonitrile-butadiene-styrene (ABS)	2600	0.54	Flaming[3]		

[1]Samples in horizontal configuration (0.005 m^2).
[2]The sample is a mattress.
[3]The sample is a plastic utility table.

3.7×10^4 ft²/lb (7.6 m²/g) for smoke from flaming combustion and 2.1×10^4 ft²/lb (4.4 m²/g) for smoke from pyrolysis (Seader and Einhorn 1976).

Computer models can be used to calculate the mass concentration of particulate, and the soot yield is a key factor in determining this concentration. Values of the soot yield are listed in Table 6.5 from small scale tests of turbulent flaming combustion for a number of materials (Tewarson 1995; Mulholland 2008). As with the mass optical density, it is expected the soot yield will vary with the size of the fire and the orientation of the fuel. The data of Table 6.5 is recommended in the absence of other data.

Visibility in a Well Mixed Space

The two methods of calculating visibility discussed above can be applied to smoke in a well mixed space such as a room.

For a fire with a constant heat release rate, the mass of fuel consumed by a fire can be expressed as

$$M_f = \frac{Q_t}{\Delta H_{ch}}$$
$$M_f = \frac{1000Qt}{\Delta H_{ch}} \text{ for SI} \tag{6.19}$$

where

M_f = mass of fuel burned, lb (g),

Q = total heat release rate Btu/s (kW),

ΔH_{ch} = chemical heat of combustion Btu/lb (kJ/kg),

t = time from ignition, s.

Values of ΔH_{ch} for some materials are listed in Table 6.5. In fires, combustion is never complete. Combustion efficiency is the ratio of the chemical heat of combustion to the net heat of combustion. Using ΔH_{ch} eliminates the need to consider combustion efficiency.

The mass concentration of fuel burned in a well mixed space is

$$n_f = \frac{M_f}{V_c} \tag{6.20}$$

where

m_f = mass concentration of fuel burned lb/ft³ (g/m³),

M_f = mass of fuel burned, lb (g),

V_c = volume of the space, ft³ (m³).

The mass concentration of fuel burned m_f from the above equation can be used in Equation 6.17 to calculate visibility.

The mass concentration of particulate in a well-mixed space is

$$n_p = \frac{y_p M_f}{V_c} \tag{6.21}$$

where

m_p = mass concentration of particulate lb/ft³ (g/m³),

y_p = soot yield, dimensionless,

M_f = mass of fuel burned, lb (g),

V_c = volume of the space, ft³ (m³).

The mass concentration of particulate, m_p, from the above equation is used in Equation 6.18 to calculate visibility. Example 6.6 illustrates calculation of visibility in a room fire.

Nonuniform Smoke

The smoke meter shown in Figure 6.3 measures the average visibility along the path of the light beam. The previous equations for reduced visibility apply to visibility where the smoke properties are uniform from a person to an object being viewed. These equations also apply to the visibility at a point that is an abstract concept, meaning the distance a person could see through smoke that had the same properties as those at the point.

There are many applications where nonuniform smoke can happen, such as smoke on a balcony in an atrium, smoke in a tunnel, and smoke in a hotel corridor. For example, Figure 6.8 shows a small pocket of relatively dense smoke not far from an exit sign. The average visibility for a path with nonuniform smoke is defined as

$$S = \frac{1}{L}\int_0^L S(x)dx \tag{6.22}$$

where

S = visibility over the path, ft (m),

L = length of path, ft (m),

$S(x)$ = visibility as a function of x, ft (m),

x = distance along path, ft (m).

If S is greater than or equal to the length of the path, L, an object can be seen over the path. Because of the lack of detailed information about the function $S(x)$, it is not practical to make calculations based on Equation 6.22. Two approaches for evaluating visibility over a path are discussed here.

Numerical Averaging

This method consists of averaging the visibility at a number of points along the path. The visibility at these points can be calculated by a computational fluid dynamic (CFD) model. The average visibility for a path with nonuniform smoke can be calculated as

Table 6.5: Soot Yield and Heat of Combustion for Well-Ventilated Fires of Solid Fuels

Material	Soot Yield	Chemical Heat of Combustion, ΔH_{ch}	
	y_p	Btu/lb	kJ/kg
Natural Materials:			
Wood (red oak)	0.015	5,340	12,400
Wood (Douglas fir)	0.018	5,590	13,000
Wood (hemlock)	0.015	5,720	13,300
Fiberboard	0.008	6,020	14,000
Wool 100%	0.008	8,390	19,500
Synthetic Materials:			
Acrylonitrile-butadiene-styrene (ABS)	0.105	12,900	30,000
Polymethylmethacrylate	0.022	10,400	24,200
Polypropylene	0.059	16,600	38,600
Polystyrene	0.164	11,600	27,000
Silicone	0.065	4,560	10,600
Polyester[1]	0.090	8,650	20,100
Nylon	0.075	11,700	27,100
Silicone rubber	0.078	4,690	10,900
Polyurethane Foam (flexible)[1]	0.188	7,570	17,600
Polyurethane Foam (rigid)[1]	0.118	7,270	16,900
Polystyrene Foam[3]	0.194	11,000	25,500
Polyethylene Foam[3]	0.076	14,700	34,200
Phenolic Foam	0.002	4,300	10,000
Polyethylene (PE)	0.060	16,500	38,400
PE with 25% chlorine	0.115	9,720	22,600
PE with 36% chlorine	0.139	4,560	10,600
PE with 48% chlorine	0.134	3,100	7,200
Polyvinylchloride (PVC)	0.172	2,450	5,700
PVC 1 (LOI = 0.50)	0.098	3,310	7,700
PVC 2 (LOI = 0.50)	0.076	3,570	8,300
PVC (LOI = 0.20)	0.099	4,860	11,300
PVC (LOI = 0.25)	0.078	4,220	9,800
PVC (LOI = 0.30)	0.098	4,430	10,300
PVC (LOI = 0.35)	0.088	4,650	10,800
Ethylenetetrafluoroethylene	0.042	2,320	5,400
Perfluoroalkoxy	0.002	2,020	4,700
Fluorinated polyethylene-polypropylene	0.003	1,760	4,100
Tetrafluoroethylene	0.003	1,810	4,200

[1]Values listed are an average of a number of different materials under this general name.

$$S_{av} = \frac{1}{L}\sum_{i=1}^{n} S_i \Delta x_i \qquad (6.23)$$

and for evenly spaced intervals ($\Delta x_1 = \Delta x_2 = \Delta x_3$ etc.) this becomes

$$S_{av} = \frac{1}{n}\sum_{i=1}^{n} S_i \qquad (6.24)$$

where

S_{av} = average visibility, ft (m),
L = length of path, ft (m),
S_i = visibility at the center of interval i, ft (m),
Δx_i = length of interval i, ft (m),
n = number of intervals, dimensionless.

If S_{av} is greater than or equal to the length of the path L, an object can be seen over the path. Example 6.7 illustrates calculating visibility of an illuminated sign and a light reflecting sign through nonuniform smoke. Example 6.7 shows how to calculate visibility in non-uniform smoke using numerical averaging.

From Percent Obscuration

Percent obscuration along a path can be simulated by a CFD model. Visibility can also be calculated from percent obscuration as

$$S = -\frac{KL}{\log_e(1 - \lambda/100)} \qquad (6.25)$$

where

S = visibility, ft (m),
K = proportionality constant (Table 6.3),

Example 6.6. Visibility in a Room Fire

In a room 25 by 40 by 12 ft (7.6 by 12 by 3.7 m) high, a 0.4 lb (181 g) polyurethane pillow is burned. If the smoke is well mixed in the room, what is the visibility of for an illuminated sign?

$V_c = (25)(40)(12) = 12{,}000\ \text{ft}^3\ (340\ \text{m}^3)$; $M_f = 0.4$ lb (181 g); From Table 6.3, $K = 8$

Part 1: Mass Optical Density Approach

The mass concentration of fuel burned is $m_f = M_f/V_c = 0.4/12{,}000 = 3.33\times10^{-5}\ \text{lb/ft}^3$

From Table 6.4, the mass optical density, δ_m, of a horizontal sample of polyurethane foam in a flaming fire is 1600 ft²/lb (0.33 m²/g).

$$S = \frac{K}{2.303\delta_m m_f} = \frac{8}{2.303(1600)(3.33\times10^{-5})} = 65 \text{ ft (20 m) visibility of an illuminated sign.}$$

Part 2: Specific Extinction Coefficient Approach

For flaming combustion, $\alpha_m = 3.7\text{x}10^4\ \text{ft}^2/\text{lb}\ (7.6\ \text{m}^2/\text{g})$. From Table 6.5, the soot yield of flexible polyurethane foam is $y_p = 0.188$.

$m_p = y_p M_f/V_c = 0.188\ (0.4)/12{,}000 = 6.27\text{x}10^{-6}\ \text{lb/ft}^3$.

$$S = \frac{K}{\alpha_m m_p} = \frac{8}{(3.7\times10^4)(6.27\times10^{-6})} = 34 \text{ ft (10 m) visibility of an illuminated sign.}$$

This difference from the mass optical density approach illustrates limitations of the current state of this technology.

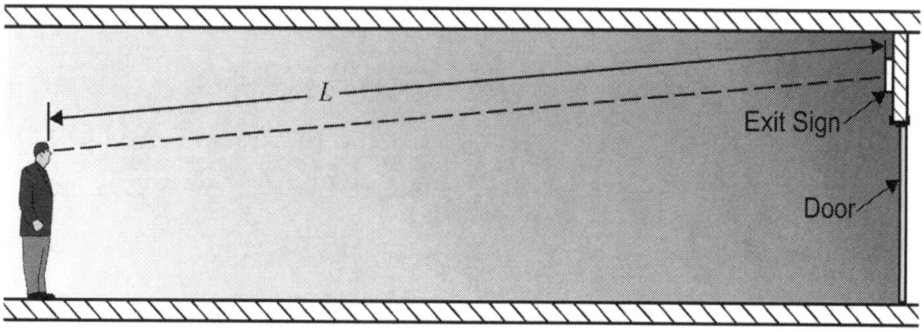

Figure 6.8 *Visibility through nonuniform smoke.*

L = length of path,

λ = percent obscuration, dimensionless.

As previously stated, if S is greater than or equal to the length of the path L, an object can be seen over the path. Example 6.8 shows how to calculate visibility in nonuniform smoke from percent obscuration.

When the path length is the same as the visibility ($L = S$), an object at the end of the path can barely be seen by a person with average eyesight, and if the object were any farther away, such a person could not see it. This is the limit of visibility. At this limit, the obscuration is

$$\lambda = 100(1 - e^{-K}) \tag{6.26}$$

where

λ = percent obscuration at the limit of visibility, dimensionless,

K = proportionality constant (Table 6.3),

It can be seen from Equation 6.26 that the percent obscuration at the limit of visibility does not depend on x or S. For an illuminated sign ($K = 8$), the percent obscuration at the limit of visibility is 99.966%. This means that an illuminated sign can be seen provided that the smoke obscuration is not more than 99.966%. For a nonilluminated sign ($K = 3$), the percent obscuration at this limit is 95.02%. This means that a nonilluminated sign can be seen provided that the smoke obscuration is not more than 95.02%.

It can be stated that the limits of visibility are: (1) 99.966% obscuration for an illuminated sign, and (2) 95.02% obscuration for a nonilluminated sign. This means that an illuminated sign is visible if the percent obscuration is not more than 99.966%, and a nonilluminated sign is visible if the percent obscuration is not more than 95.02%.

TENABILITY

With regard to the tenability of occupants, the objective of a smoke control system is that the atmosphere to which occupants are exposed does not cause fatality for conservatively chosen realistic design fires. A second objective regarding tenability is similar but it regards protection for members of the fire service. Codes such as the International Building Code (ICC 2012) have requirements pertaining to the first objective but not the second. Systems designed to meet the first objective also tend to provide a level of protection for the fire service. For this discussion, the objective will be to maintain a tenable environment for the occupants during evacuation or relocation during a fire.

For many smoke control systems, the intent of the system is to keep smoke away from the occupants. These systems include pressurized stairwells and zoned smoke control systems. Also, atrium smoke exhaust systems that maintain the smoke layer away from the occupants are included. These systems are designed to meet the above objective without need for an analysis of tenability.

Tenability systems are ones where occupants are exposed to some combustion products that are so diluted that the previously stated objective can be maintained. This includes some exposure to toxic gases, heat, thermal radiation, and reduced visibility. In many applications,

Example 6.7. Visibility through Non-Uniform Smoke

i	S_i, ft
1	36
2	36
3	36
4	36
5	36
6	36
7	36
8	30
9	26
10	22
11	19
12	16
13	14
14	11

An exit sign is 16 ft (4.9 m) from an observer, and the smoke has the visibility listed here for 14 evenly spaced intervals. This visibility was calculated for an illuminated sign. $L = 16$ ft (4.9 m).

Part 1: Can the observer see the sign through this smoke?

Use $S_{av} = \frac{1}{n}\sum_{i=1}^{n} S_i$ where $n = 14$. Then $S_{av} = 28$ ft (8.5 m).

Because S_{av} is greater than L, an observer would be expected to see this sign.

Part 2: If the above sign were not illuminated, would the observer be able to see it?

The visibilities S_i listed here were calculated with $K = 8$. For a light reflecting sign, $K = 3$. So for a light reflecting sign, $S_{av} = 28\ (3/8) = 10.5$ ft (3.2 m).

Because S_{av} is greater than L, an observer would not be expected to see the light reflecting sign.

visibility dominates the other exposures such that systems that meet visibility criteria will often not have problems with the other exposures. This is because the products of combustion need to be diluted to a considerable extent so that people can see through them. Design fires that involve very-low-sooting materials can be exceptions.

Exposure Approaches

As previously stated, the effect of exposures to toxic gases, heat, and thermal radiation depend on the time duration of the exposure. Exposures can be evaluated by (1) the egress flow approach, (2) the remain-in-place approach, and (3) logic indicating that detailed calculation of an exposure is unnecessary. This kind of logic can be used for many exposures to toxic gases, heat, and thermal radiation as discussed next.

Egress Flow Approach

This approach is called the egress flow approach because it requires an analysis of people movement during egress. For some applications, people would relocate to building locations remote from the fire, but this approach can still be used. The approach consists of the following steps: (1) simulate the movement of people during the fire, (2) simulate the movement of smoke during the fire, and (3) calculate the exposures to people on their simulated paths during egress or relocation. The egress flow approach requires complex calculations, and the calculations for the remain-in-place approach are simple by comparison.

Remain-in-Place Approach

In the most general form, this approach consists of calculating exposures for the duration of operation of the smoke control system at the locations that are intended to be protected by the smoke control system. This is as if people with disabilities remained at these locations for the duration of operation. This approach is more conservative than the egress flow approach.

The remain-in-place approach might be thought of as having people with mobility limitations at the locations that are intended to be protected. A modified version of this approach can be used where, in unusual design conditions, a space might become untenable but the system objective could still be met.

For example, consider a smoke control system where all spaces intended to be protected are tenable except one refuge area under one extreme wind condition. It can be expected that a person in a wheel chair would not go into the smoke logged refuge area, but would move down the tenable corridor to another refuge area. Thus, a modified version of the approach would allow this smoke logged refuge area provided that the system objective is met.

Protected Locations

Protected locations include the spaces that are intended to be kept tenable, with the exceptions of smoke locations where smoke protection is beyond capability of smoke control.

For example, the ground floor of an atrium would normally be a protected location except in the vicinity of

Example 6.8. Nonuniform Smoke and Percent Obscuration

A person is looking at an illuminated exit sign 30 ft (9.1 m) away. The obscuration along this path is 98%.

Part 1: Can the person see the exit sign?

For an illuminated sign, $K = 8$.

$$S = -\frac{Kx}{\log_e(1 - \lambda/100)} = -\frac{8(30)}{\log_e(1 - 98/100)} = 61 \text{ ft } (19 \text{ m})$$

The visibility is greater than the path length, so the person can see the illuminated sign.

Part 2: If the sign were not illuminated, could the person see it?

For a nonilluminated sign, $K = 3$.

$$S = -\frac{Kx}{\log_e(1 - \lambda/100)} = -\frac{3(30)}{\log_e(1 - 98/100)} = 23 \text{ ft } (7 \text{ m})$$

The visibility is less than the path length, so the person cannot see the nonilluminated sign.

Part 3: Determine the answers to Parts 1 and 2 using the limits of visibility. These limits are: (1) 99.966% obscuration for an illuminated sign, and (2) 95.02% obscuration for a nonilluminated sign.

The obscuration along this path is 98% which is less than 99.966% so an illuminated sign can be seen. The obscuration of 98% is more than 95.02%, so a nonilluminated sign cannot be seen.

a fire. As already discussed, people can approach only so close to a fire, because of thermal radiation. The fire may block an exit, and this cannot be changed by a smoke control system. This is the reason for multiple exits.

Another example is a balcony that is part of the means of egress. This location is to be protected except when it is blocked by a smoke plume from a fire below. It is beyond the capability of smoke control to prevent such smoke blocking in this kind of scenario.

Heat Exposure

Heat exposure happens when a person is in contact with hot air or other gases. For many smoke control applications, the effect of heat exposure can be evaluated by examination of Figure 6.1. This is illustrated by Example 6.3.

Thermal Radiation Exposure

Exposure to thermal radiation happens when a person is near flames or hot gases. Exposure to thermal radiation can often be ruled out on the basis of heat exposure. The reasoning is that if contact with a particular body of gas is an acceptable heat exposure, then the thermal radiation some distance away from the same gas would also be acceptable.

Thermal radiation limits how close a person will approach a flame, and this is illustrated in Example 6.4. It is not possible for smoke control technology to change this, and such areas near flames cannot be protected by a smoke control system. The temperature of the smoke layer in a room fire can be so hot that people cannot withstand the thermal radiation below the layer.

Reduced Visibility

In dense smoke with very low visibility, people can become completely disoriented, which leads to increased smoke exposures and sometimes the possibility of fatal falls. In fire situations, people need to be able to see to the extent necessary for evacuation or relocation. The following discussion of visibility addresses criteria that can be applied to various situations.

Familiarity with the surroundings has a major impact on how far a person needs to see during evacuation. The familiar person needs to see enough to keep their orientation so that he or she can move out of the building or to another safe location. The unfamiliar person needs to be able to see the exit doors or exit signs. If no location in a room is more than 30 ft (9 m) from a door leading out of the room, the unfamiliar person in that room needs to be able to see for 30 ft (9 m). If each of these doors has an illuminated exit sign, the visibility distance can be calculated for an illuminated sign. Otherwise, visibility for a reflected sign would be appropriate.

It can be expected that a person in their own home would be so familiar with their surroundings that they could find their way around even if they could see for only about 10 to 13 ft (3 to 4 m). However, a familiar person in an office building may need to see farther due to the repetitive furnishings common in offices.

The approach that is often used in evaluating visibility is to (1) establish a visibility criterion for the project and (2) calculate the visibility at a number of points. These points need to be chosen to assure that the important locations are included. If all of the calculated visibilities are greater than the criterion, visibility is acceptable. But, the opposite is not necessarily true. It is possible to have a point in a path with smoke exceeding the criterion, but still see through that path. Such paths can be evaluated considering nonuniform smoke, as discussed earlier.

Criteria for visibility have been suggested ranging from 13 to 46 ft (4 to 14 m) (Jin 2008). The factors that should be considered when choosing visibility criteria are (1) familiarity with the building, (2) size of the rooms, (3) size of building, and (4) complexity of building.

Consider a university building consisting of classrooms and corridors with an atrium at the main entrance. It can be expected that most of the occupants would be familiar with the building, and the few not familiar could be expected to move with the rest of the population during evacuation. Because of the size of the building and corridors, the minimum value mentioned above of 13 ft (4 m) may not be enough. A criterion of 25 or 30 ft (7.6 to 9.1 m) might be appropriate for this application. If the spacing of illuminated exit signs is sufficiently close together, visibility can be calculated for illuminated signs.

Consider a museum with a complex design that has an atrium that is five stories high. It can be expected that most of the occupants would be first-time visitors to the building. These people would be unfamiliar with the building, and a criterion of 42 or 46 ft (12.8 to 14 m) might be appropriate for this application. Again, if the spacing of illuminated exit signs is sufficiently close together, visibility can be calculated for illuminated signs.

Toxic Gases Exposure

Exposure to toxic gases can be evaluated by the methods discussed earlier. Alternatively, the approach of calculating the maximum FED possible corresponding to the visibility criterion. This approach is based on considerations of dilution presented by Klote (1999). The products of combustion are considered to be diluted such that the visibility criterion is met, and the FED is calculated for exposure to this smoke for the duration of operation of the smoke control system. This FED is the maximum that is possible for a smoke control system, provided that the visibility criterion is met.

$$FED_{max} = \frac{Kt}{2.303\delta_m S_c LCt_{50}} \tag{6.27}$$

where

FED_{max} = maximum fractional effective dose, dimensionless,

K = proportionality constant (Table 6.3),

t = exposure time, min,

δ_m = mass optical density, ft^2/lb (m^2/g),

S_c = visibility criterion, ft (m),

LCt_{50} = lethal exposure dose from test data, $lb/ft^3 \cdot min$ ($g/m^3 \cdot min$).

This equation is based on dilution of the products of combustion by air neglecting smoke particle aging (agglomeration and deposition) and any reduction of specific gases due to scavenging of molecules at surfaces. These are all conservative assumptions in that they result in higher values of FED_{max}. Further, almost all smoke transport calculations neglect smoke particle aging and reduction of specific gases.

Example 6.9 shows how to calculate the FED_{max} for a visibility criterion of 25 ft (7.6 m) and fuel mixture of 20% flexible polyurethane (PU) foam and 80% cellulosic material. Generally, one component of a mixture would be expected to burn faster than another, but this kind of mixture treatment allows for fuels made of various combinations of materials. Figure 6.9 shows the FED_{max} for combinations of flexible PU foam and cellulosic material.

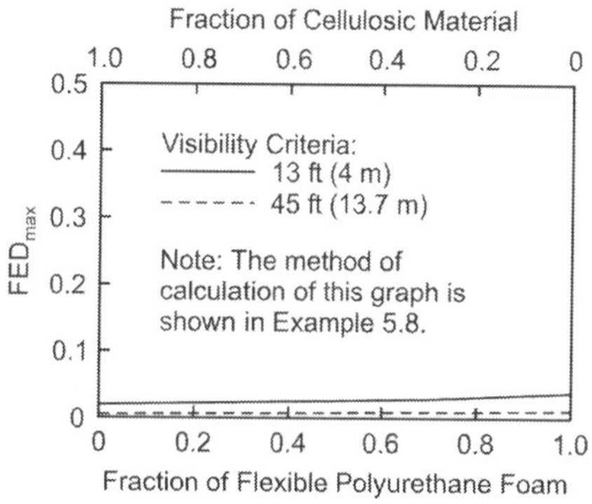

Figure 6.9 *FED_{max} due to mixtures of PU foam and cellulosic material.*

Example 6.9. Maximum FED for a Visibility Criterion

For a design project, the composition of fuel is 20% flexible polyurethane foam and 80% cellulosic material. The visibility criterion of 25 ft (7.6 m) was chosen for illuminated signs. The duration of operation of the smoke control system is 20 min. What is the maximum FED that can result from smoke that meets the visibility criterion?

The parameters are:

t = 20 min	Duration of operation of the smoke control system
S_c = 25 ft	Visibility distance
K = 8	For light emitting signs
δ_m = 1600 ft^2/lb	For smoke from PU foam
δ_m = 1400 ft^2/lb	For smoke from cellulosic material
LCt_{50} = 0.087 $lb/ft^3 \cdot min$.	For smoke from PU foam
LCt_{50} = 0.19 $lb/ft^3 \cdot min$.	For smoke from cellulosic material

$$\delta_m = (0.2)(1600) + (0.8)(1400) = 1440 \text{ ft}^2/\text{lb}$$

$$LCt_{50} = (0.2)(0.087) + (0.8)(0.19) = 0.169 \text{ lb/ft}^3 \cdot \text{min.}$$

$$FED_{max} = \frac{Kt}{2.303\delta_m S_c LCt_{50}} = \frac{8(20)}{2.303(1440)(25)(0.169)} = 0.011$$

This is the FED of a person who would be exposed to smoke meeting the visibility criterion for a duration of 20 minutes. Because FED_{max} is only 0.011, toxic gas exposure is not a concern provided that the visibility criterion is met.

NOMENCLATURE

b = 122,000 for $CO_2 < 5\%$ and –38,600 for $CO_2 > 5\%$

$C_{CO,i}$ = concentration of CO, ppm

$C_{CO2,i}$ = concentration of CO_2, ppm

C_{COHb} = concentration of COHb in the blood, %

$C_{COHb,0}$ = concentration of COHb in the blood at time zero, %

$C_{HBr,i}$ = concentration of HBr, ppm

$C_{HCl,i}$ = concentration of HCl, ppm

$C_{HCN,i}$ = concentration of HCN, ppm

$C_{NO2,i}$ = concentration of NO_2, ppm

$C_{O2,i}$ = concentration of O_2, %s

FED = fractional effective dose, dimensionless

FED_{max} = maximum fractional effective dose, dimensionless

I_o = intensity of light at the beginning of the path length

I_x = intensity of light remaining after it has passed through the path length

K = proportionality constant (Table 6.3)

L = length of path, ft (m)

$LC_{50}(HBr)$ = lethal concentration of HBr, ppm

$LC_{50}(HCl)$ = lethal concentration of HCl, ppm

$LC_{50}(HCN)$ = lethal concentration of HCN, ppm

$LC_{50}(NO_2)$ = lethal concentration of NO_2, ppm

$LC_{50}(O_2)$ = lethal concentration of O_2, %

$\text{LC}t_{50}$ = lethal exposure dose from test data, lb ft^{-3} min (g m^{-3} min)

m = –18 for $CO_2 \leq 5\%$ and 23 for $CO_2 > 5\%$

m_f = mass concentration of fuel burned in units lb/ft^3 (g/m^3)

M_f = mass of fuel burned, lb (g)

$m_{f,i}$ = mass concentration for time interval i, lb/ft^3 (g/m^3)

m_p = mass concentration of particulate lb/ft^3 (g/m^3)

n = number of time intervals

N_{Gas} = N-Gas model indicator, dimensionless

Q = total heat release rate Btu/s (kW)

Q_r = radiant HRR of the fire, Btu/s (kW)

$q''_{r,t}$ = intensity of thermal radiation that can be temporarily tolerated, $Btu/ft^2 \cdot s$ (kW/m^2)

q''_r = intensity of thermal radiation, $Btu/ft^2 \cdot s$ (kW/m^2)

R = minimum distance from the center of the fire to a person, ft (m)

S = visibility, ft (m)

S_{av} = average visibility, ft (m)

S_c = visibility criterion, ft (m)

S_i = visibility at the center of interval i, ft (m)

T = transmittance, dimensionless

t = time from ignition, s; or exposure time, min

t_e = exposure time, min

$t_{r,b}$ = exposure time to blister, s

$t_{r,p}$ = exposure time to pain, s

V = volume of breathed air per minute, L/min

V_c = volume of the space, ft^3 (m^3)

x = distance of light travel, ft (m);

y_p = soot yield (dimensionless)

α = extinction coefficient ft^{-1} (m^{-1})

δ = optical density per unit distance, ft^{-1} (m^{-1})

λ = percent obscuration, dimensionless

ΔH_{ch} = chemical heat of combustion Btu/lb (kJ/kg)

δ_m = mass optical density, ft^2/lb (m^2/g)

α_m = specific extinction coefficient, ft^2/lb (m^2/g)

Δt_i = time interval, min

Δx_i = length of interval i, ft (m)

REFERENCES

Babrauskas, V., et al. 1991. Toxic measurement for fire hazard analysis. NIST Special Publications 827, National Institute of Standards and Technology, Gaithersburg, MD.

Berl, W.C., and B.M. Halpin. 1980. Human fatalities from unwanted fires. *Johns Hopkins APL Technical Digest* 1(2).

Blockley, W.V. 1973. *Biology Data Book*. Bethesda, MD: Federation of American Societies of Experimental Biology.

Bukowski, R.W., et al. 1989. *Technical Reference Guide for HAZARD I Fire Hazard Assessment Method, NIST Handbook 146, vol. II*. National Institute of Standards and Technology, Gaithersburg, MD.

Haber, F., 1924. Funf Vortrange aus den jaren 1920–1923, Verlag von Julius Spanger, Germany.

Harland, W.A., and W.D. Woolley, 1979. Fire fatality study. Borehamwood Information Paper IP 18/79, Building Research Establishment, University of Glasgow.

Hartzell, G.E., A.F. Grand, and W.G. Switzer. 1990. Toxicity of smoke containing hydrogen chloride, fire and polymers—hazards identification and prevention, ed. G.I. Nelson. ASC Symposium Series 425, American Chemical Society, Washington, DC.

ICC. 2012. International Building Code, International Code Council, Country Club Hills, IL.

Jin, T. 1974. Visibility through fire smoke, in main reports on production, movement and control in buildings. *Japanese Association of Fire Science and Engineering*, pp 100–153.

Jin, T. 1975. Visibility thorough fire smoke. *Report of the Fire Research Institute of Japan* 5(42).

Jin, T. 1985. Irritating effects of fire smoke on visibility. *Fire Science and Technology* 5(1).

Jin, T. 2008. *SFPE Handbook of Fire Protection Engineering,* Chapter 2-4, Visibility and Human Behavior in Fire Smoke. Quincy, MA: National Fire Protection Association.

Klote, J.H. 1999. An engineering approach to tenability systems for atrium smoke management. *ASHRAE Transactions* 105(1).

Levin, B.C. 1996. New research avenues in toxicity: 7-Gas N-Gas Model, toxicant suppressants, and genetic toxicology. *Toxicology* 115.(1–3) 89–106.

Levin, B.C. 2000. Personal Communication Between Levin and Klote About the N-Gas Model and LC_{50} Values, March 2000.

Levin, B.C., et al. 1988. Toxicological effects of different time exposures to fire gases: carbon monoxide or hydrogen cyanide or to carbon monoxide combined with hydrogen cyanide or carbon dioxide. 31st Annual Technical/Marketing Conference, Society of Plastics Industry, Polyurethanes 88, October 18–21, Philadelphia, PA.

Levin, B.C., et al. 1989. Synergistic effects of nitrogen dioxide and carbon dioxide following acute exposures in rats. NISTIR 89-4105, National Institute of Standards and Technology, Gaithersburg, MD.

Levin, B.C., et al. 1995. Further development of the N-Gas mathematical model: an approach for predicting the toxic potency of complex combustion mixtures, fire and polymers II: materials and tests for hazard prevention. ACS Symposium Series No. 599, August 21–26, 1994, American Chemical Society, Washington, DC.

Mulholland, G. 2008. *SFPE Handbook of Fire Protection Engineering,* Chapter 2-13, Smoke Production and Properties. Quincy, MA: National Fire Protection Association.

Purser, D.A. 2008. *SFPE Handbook of Fire Protection Engineering,* Chapter 2-6, Assessment of Hazards to Occupants from Smoke. Bethesda, MD: Toxic Gases and Heat. Society of Fire Protection Engineers.

Seader, J., and I. Einhorn. 1976. Some physical, chemical, toxicological, and physiological aspects of fire smokes. NSF Report, Utah University.

Stewart, et al. 1973. Experimental human response to high concentrations of carbon monoxide. *Architectural Environmental Health* 26(1).

Stoll, A.M., and M.A. Chianta. 1969. Method and rating system for evaluation of thermal protection. *Aerospace Medicine* 40:1232–1238.

Tewarson, A. 1995. *SFPE Handbook of Fire Protection Engineering*, 2nd ed., Chapter 3-4, Generation of Heat and Chemical Compounds in Fires. Quincy, MA: National Fire Protection Association.

CHAPTER 7

Air-Moving Systems and Equipment

John H. Klote and Paul G. Turnbull

In the event of a fire, heating, ventilating, and air-conditioning (HVAC) systems are generally shut down to prevent them from transporting smoke, but some HVAC systems go into a zoned smoke control mode in the event of a fire as discussed in Chapter 13. The intent of this chapter is to provide background for the smoke control systems discussed in later chapters, and to provide an indication about the applicability of the different types of systems for smoke control purposes. It is hoped that the information in the chapter will help professionals who do not specialize in HVAC systems to understand this topic. Many publications provide more detailed information about these systems and equipment (for example, ASHRAE 2007, 2011, 2012; Bobenhausen 2005; Bell 2008; Rosaler 2004).

RESIDENTIAL SYSTEMS

Residential heating and cooling is either forced-air or hydronic. A hydronic system is one that uses water as the heat transfer medium. Because hydronic systems do not have the potential to interact with smoke movement or smoke control, they are not discussed further.

Figure 7.1 is a diagram of the kind of forced air system that is commonly used in single family and multifamily residences. This kind of system is also used in some small commercial buildings. In these systems, return air from the conditioned space is pulled in at one location, flows through a filter, fan and coils, and is distributed back to the residence.

Residential systems typically are controlled by a thermostat to maintain the temperature in the living quarters. The most common installations have a two-position control system allowing only on and off operation to maintain temperature conditions. This type of residential forced-air system can be called a constant-volume, single-zone system. A commercial version of the constant-volume, single-zone system that is more complex in a number of respects is discussed later. Residential systems with multiple-speed or variable-speed blower motors are available for installations requiring more precise control or higher efficiency. The multiple-speed or variable-speed residential systems introduce some of the benefits of variable-air-volume (VAV) systems to the residential market. VAV systems are discussed later in this chapter.

PERIMETER AND CORE ZONES

In large buildings, the amount of heating and cooling that is needed often differs considerably from one location to another. Large buildings are commonly

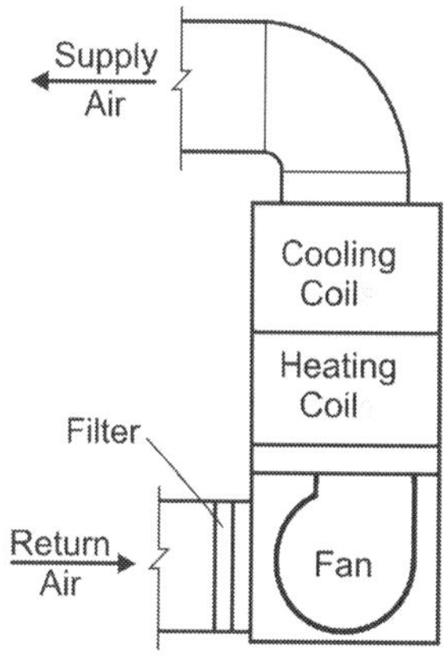

Figure 7.1 *Residential forced-air system.*

divided into perimeter zones and a core zone as is shown in Figure 7.2. The perimeter zones are subjected to heating and cooling requirements due to outdoor temperatures and solar radiation. It is quite common during spring and fall to provide cooling to the side of the building facing the sun, and at the same time, to provide heating to the opposite side of the building. These perimeter zones can be conditioned by a variety of means including fan-coil units, air conditioners, and heat pumps. The core zones often need cooling even in the winter, because of the heat produced by occupants, lighting, and equipment.

Fan-coil units, air conditioners, and heat pumps are generally located along the exterior walls. For smaller spaces, these units may provide the only method of heating or cooling. For larger spaces, these units may be used for supplementary conditioning in combination with a larger forced air system. Fan-coil units often are supplied with hot and cold water to allow the units to be switched between heating and cooling. Ventilation air to fan-coil units and through-the-wall equipment is supplied directly from the outdoors or from a ducted ventilation system.

In large commercial buildings, the perimeter zones may be heated or cooled by ducted forced-air systems. In addition to providing heating or cooling, these forced-air systems provide ventilation air, which is needed to control odors due to cooking, smoking, perspiration, and other processes; to dilute levels of carbon dioxide that would build up in enclosed spaces; and to remove allergens and airborne contaminants. Some types of forced-air systems are capable of satisfying a wide range of needs simultaneously and are used to serve both perimeter and core zones.

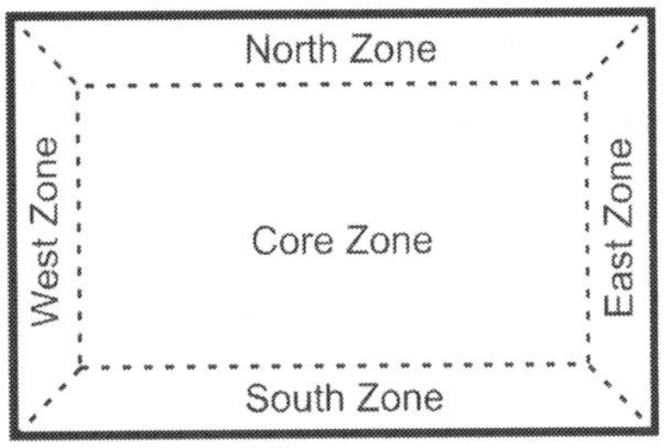

Note: Buildings are often divided in four perimeter zones and a core zone.

Figure 7.2 *Perimeter and core zones.*

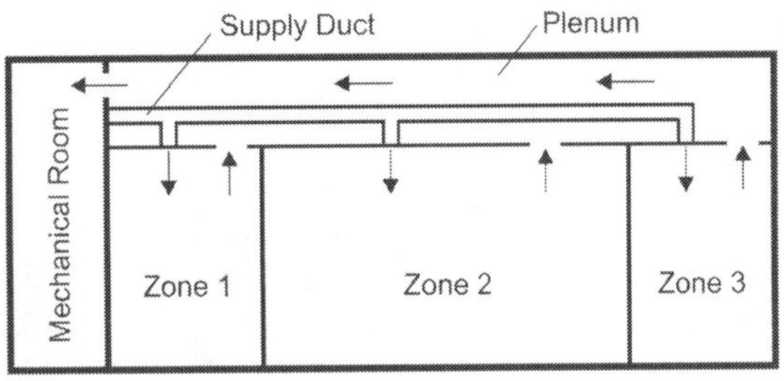

Figure 7.3 *Floor with supply duct and plenum return.*

INDIVIDUAL ROOM UNITS

There is a wide assortment of products used to provide heating and cooling in small spaces, such as hotel rooms, dormitory rooms, or private offices. This category of products includes fan-coil units, unit ventilators, and packaged terminal units. Although they differ in construction, these products share a number of attributes. Typically, these units are not connected to ductwork, but instead are generally installed on an exterior wall so that they can draw in outdoor air directly through the wall. A fan within the unit forces outdoor air across a heating or cooling device, such as a heated- or chilled-water coil, electric heater, or air-conditioning coil. The unit is controlled by a thermostat that is located either on the unit itself or on a wall within the space being served by the unit.

Because these units are designed to heat or cool a relatively small space, they generally have only a small capacity for bringing in outdoor air, and are therefore generally not suitable for performing smoke control functions. When this type of unit is used in conjunction with larger HVAC equipment, such as a central system or VAV system, the individual room units should be shut down during a fire, and smoke control should be provided by the larger system.

FORCED-AIR SYSTEMS

In commercial buildings, air is often distributed through ducts located above a suspended ceiling or through ducts or open space below a raised floor. The return air is pulled through a plenum above the ceiling as shown in Figure 7.3. This figure shows a mechanical room located on the same floor as the zones the system is serving. Often, the equipment in a mechanical room serves more than one floor.

Many buildings have a mechanical room where air is conditioned with return and supply ducts to the other floors located in mechanical shafts. The mechanical room may have one forced-air system capable of serving the needs of many control zones, or it may have multiple forced-air systems. Very large buildings have multiple mechanical floors as shown in Figure 7.4. For complicated buildings such as hospitals and laboratories, the duct systems can be intertwined such that the HVAC drawings need to be studied to understand which systems serve which areas.

Types of Systems

Rather than the simple on/off controls typical of the residential systems, commercial forced-air systems generally use modulating controls that continuously run the equipment at various heating and cooling loads (see Chapter 8). The heat source for heating coils is usually hot water, steam, or electric resistance. Cooling coils are supplied with chilled water or a refrigerant. These forced-air systems can be completely built in a factory, completely built in the field, or may consist of factory-built subsections that are field assembled.

There are many types of commercial systems. The ones discussed here are (1) constant-volume, single-zone systems, (2) constant-volume systems with terminal reheat, (3) variable-air-volume (VAV) systems, and (4) dual duct systems.

Constant-Volume, Single-Zone Systems

Figure 7.5 is a diagram of a commercial constant-volume, single-zone system, and it can be seen that this is much more complex than the residential forced-air system of Figure 7.1. In Figure 7.5, the ducts are represented by lines, and ducts are represented this way in the diagrams of the rest of the commercial forced-air systems. The term *constant volume* is used to mean that the system produces constant or nearly constant volumetric flow rate of air. Temperature control of the building space is provided by varying the temperature of the air supplied to the space. This is done by controlling the outdoor air damper, exhaust damper, and return air damper to achieve the desired temperature. When additional heating or cooling is required, the heating or cooling coils are used. This system is called *single zone* because it serves only one HVAC control zone. Air coming out of all supply grilles or terminal boxes within the zone is supplied at the same temperature.

This system has two fans and is capable of providing ventilation air for odor control. During cold weather, the core zones of many large commercial buildings have so much heat generated by equipment, people, and lighting that cooling is required. Cold outdoor air is often used for this cooling, and the controls and dampers that use outdoor air for cooling are called the economizer. For a system with an economizer, the preheat coil is needed to protect the cooling coils and humidifier from freezing. Often the exhaust fan is sized at 80% to 90% of that of the supply fan so that the building has a slight pressurization to prevent the infiltration of airborne dirt, odors, and pollen from the outdoors.

This type of system may be readily used for smoke control. When used as part of a zoned smoke control system, control signals must be provided to the supply fan, exhaust fan, outdoor air damper, exhaust damper,

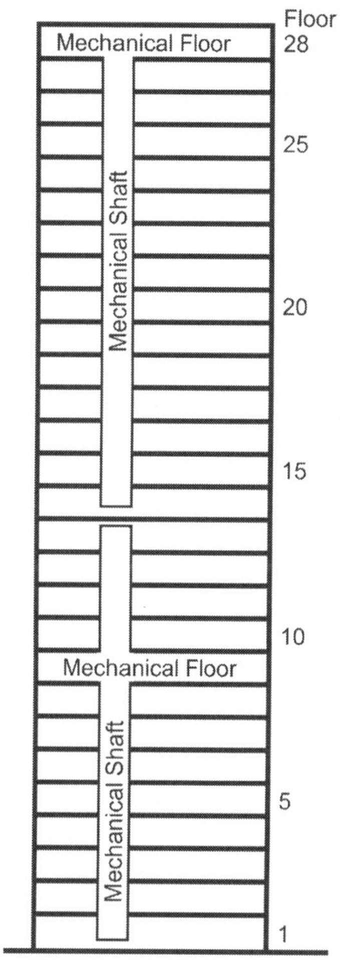

Figure 7.4 *Building with two mechanical floors.*

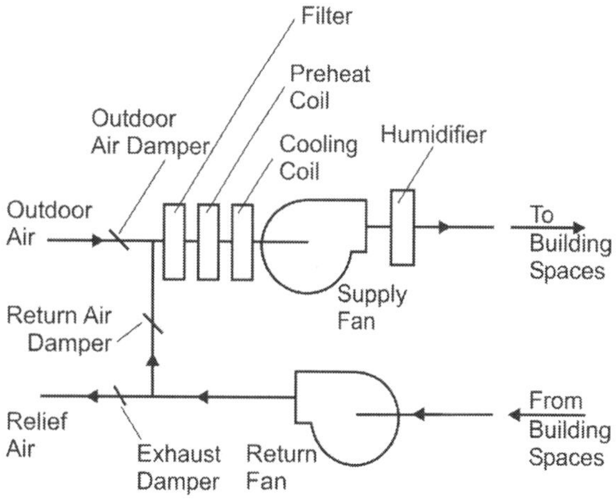

Figure 7.5 *Constant-volume, single-zone system.*

and the return air damper. It should be noted that when a constant-volume system is used only for HVAC purposes, the outdoor air damper, exhaust damper, and return air damper often have mechanical linkages or interlocks, so that a command to the outdoor air damper or exhaust damper also forces the return air damper to move. When a constant-volume system is used for smoke control purposes, mechanical linkages or interlocks cannot be used because it will be necessary to control the return air damper independently from the outdoor air and exhaust dampers.

Constant-Volume Systems with Terminal Reheat

Figure 7.6 shows a diagram of a constant-volume, terminal reheat system, and this system is capable of serving many HVAC zones. As the name implies, the volumetric flow rate of this system is nearly constant. Temperature control of the building spaces is provided by setting the temperature of the air supplied to all zones to the lowest temperature required by any of the connected zones. Terminal boxes in the zone requesting air at this temperature will supply the air without further conditioning, but terminal boxes in zones that require warmer air will activate the reheat coil in the terminal box to raise the temperature of the air before it enters the zone. As with the preceding system, the constant-volume, terminal reheat system can have an economizer.

This type of system may be readily used for smoke control. When used as part of a zoned smoke control system, control signals must be provided to the supply fan, exhaust fan, outdoor air damper, exhaust damper, and the return air damper. As noted, when a constant-volume system is used for smoke control purposes, mechanical linkages or interlocks connecting the outdoor air damper, exhaust damper, and return air damper cannot be used because it will be necessary to control the return air damper independently from the outdoor air and exhaust dampers. Because this system serves multiple zones, it may be necessary to provide additional control commands to isolate zones that are not part of the current smoke control strategy. This will allow the full capacity of the supply and exhaust fans to be focused on the zones being used for smoke control.

Variable-Air-Volume (VAV) Systems

In this system, air is supplied at a constant temperature, and the supply rate of conditioned air is varied based on the system needs. Spaces receive the amount of supply air that they need to satisfy temperature or air quality demands. To satisfy temperature requirements, the damper within the VAV Box is modulated to provide the proper volume of constant-temperature air, which, when mixed with the air in that space, will result in the air in the room being the desired temperature. VAV systems can also be used to comply with air quality requirements; for example, an empty conference room would only require a minimum amount of air, but a conference room filled with people would require much more air. To satisfy air quality needs, it may be necessary to increase or decrease the amount of air supplied to the space, compared to the amount that would be required for temperature control. This will result in the temperature in the room being above or below the desired value. A reheat coil within the VAV Box may be activated to raise the temperature of the air supplied to the space to achieve the desired room temperature. Figure 7.7 shows a VAV system with terminal reheat. The reheat is used to maintain comfort conditions as just discussed. As with the other systems above, the VAV system is often designed to maintain a slight building over pressure.

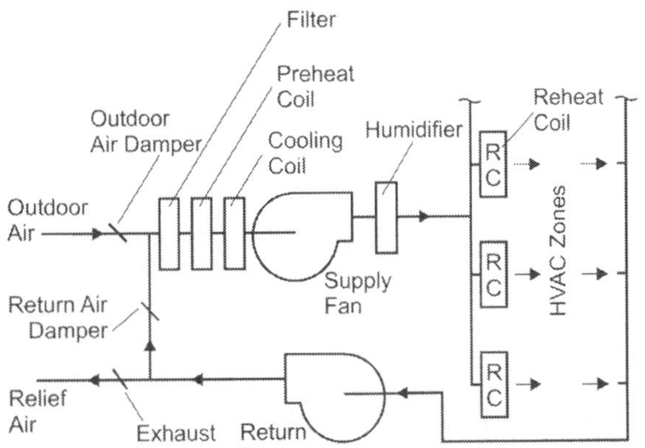

Figure 7.6 *Constant-volume system with terminal reheat.*

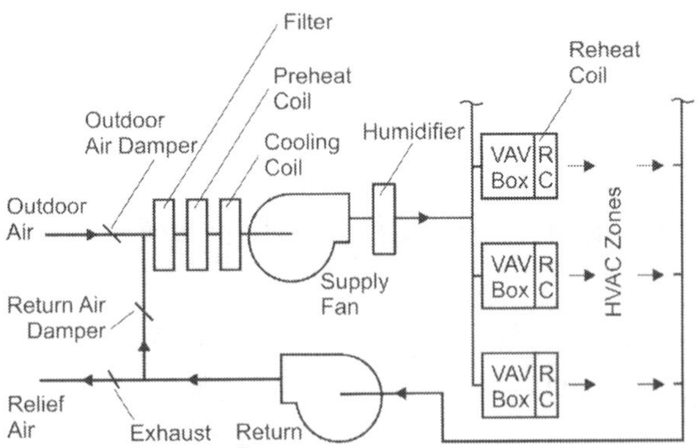

Figure 7.7 *Variable-air-volume (VAV) system with terminal reheat.*

VAV systems may be readily used for smoke control. When used as part of a zoned smoke control system, control signals must be provided to the supply fan, exhaust fan, outdoor air damper, exhaust damper, and the return air damper. Control of the supply fan and exhaust fan should be accomplished by increasing the fan speed to its maximum or decreasing it to its minimum value depending upon the fan's location relative to the fire. When supplying air to pressurize a zone, the VAV Boxes should be commanded to their fully open position. For a more detailed discussion of terminal box control in this application, see the section "Control of Devices that are not Part of the Smoke Control System" in Chapter 8.

As just noted for Constant Volume systems, when a VAV system is used for smoke control purposes, mechanical linkages and interlocks connecting the outdoor air damper, exhaust damper, and return air damper cannot be used because it will be necessary to control the return air damper independently from the outdoor air and exhaust dampers. Because this system serves multiple zones, it may be necessary to provide additional control commands to isolate zones that are not part of the current smoke control strategy. This will allow the full capacity of the supply and exhaust fans to be focused on the zones being used for smoke control.

Dual-Duct Systems

This type of system was widely used from the 1960s through the 1980s, but is not often installed today. It is described here for the sake of completeness. In this system, both heated and cooled air was supplied to all parts of the building from a central location. Two sets of ductwork ran together throughout the building. A mixing box was provided at each point where air was to be supplied to a space within the building. Both the "hot duct" and the "cold duct" were connected to the mixing box, and dampers within the mixing box regulated the amount of heated air and the amount of cooled air that would be supplied to the building space. The heated and cooled air was mixed together in the mixing box to create the desired temperature for the air entering the building space. Figure 7.8 shows a dual-duct system supplying multiple zones.

If a building containing a dual-duct system is being retrofitted, there is no reason that the dual-duct systems cannot be reused, although that may not be desirable for energy-efficiency reasons. Dual-duct systems can be used for smoke control applications. In many cases, the cold duct is larger and carries a larger volume of air than the hot duct. When using a dual-duct system for smoke control, the hot duct is often closed throughout the building and the cold duct is opened and closed in a manner similar to that of a single-duct system.

Fan-Powered Terminal Unit

This type of system is common in applications that require high airflow rates, such as conference rooms, cafeterias, or other spaces where many people gather. The fan-powered terminal unit has a fan that runs constantly during occupied times. The terminal unit is connected to a VAV system, which supplies a variable amount of conditioned air to the terminal unit. Additional air from the building's return plenum is mixed in the terminal unit with the conditioned air to provide the desired temperature for the space. The terminal unit may include an optional reheat coil to raise the temperature of the mixed air. The fan in the unit then introduces this mixed air into the space at a higher velocity than would be provided by the VAV system alone.

During smoke control operations, the VAV system will provide the main control functions. When air is being provided to a building space for pressurization of a smoke control zone, fan-powered boxes should fully open the damper controlling the amount of air from the VAV system. The fan in the terminal unit may continue to run. When the smoke control system requires the air supplied to the zone to be turned off, the fan in the terminal unit should also be turned off.

Other Special-Purpose Systems

Once-Through Systems

Facilities such as laboratories, hospital operating rooms, animal facilities, and other special-purpose occupancies with a high level of concern regarding contamination may employ systems that include high levels of filtration and do not recirculate the air. In these systems, air is drawn from the outdoors, conditioned, introduced into the space, and then exhausted directly to the outdoors. Once-through systems may be constant-volume or variable-volume and operate as just described, except that they do not include a return air damper and do not recirculate any of the air.

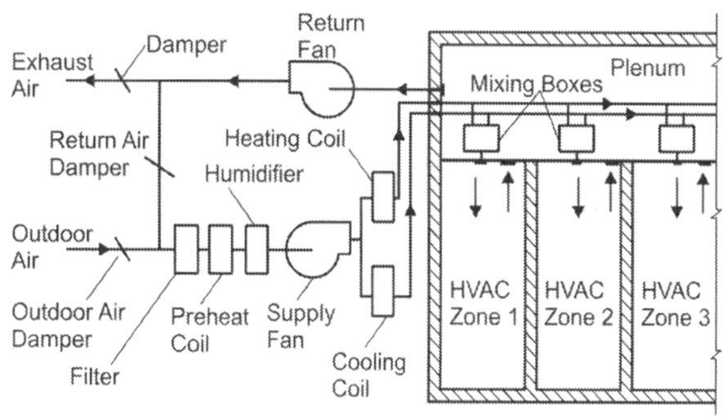

Figure 7.8 *Dual-duct system.*

Once-through systems can be suitable for smoke control applications as previously described for Constant-Volume or VAV systems, except that there is no need to include control signals for a return air damper, because once-through systems do not include this damper.

HVAC Systems with No Outdoor Air

Some systems used for heating or cooling contain no connection to outdoor air. Examples of this type system include radiant heating panels, chilled beams, and self-contained computer room units. Where these systems are used, there will always be another system that provides fresh air to the space to satisfy ventilation requirements.

Because systems with no outdoor air connection can neither supply outdoor air nor exhaust to the outdoors, these systems cannot be used for smoke control purposes. The system that supplies air for ventilation should be used for smoke control purposes, but it should be noted that the ventilation system may not be able to pressurize the space for smoke control purposes if it is designed to provide only the minimum level of ventilation.

FANS

Generally, fans are classified as either centrifugal or axial. Fans used for exhaust in a smoke management system may be subjected to elevated temperatures.

Centrifugal Fans

A characteristic of centrifugal fans is that the air flowing through the fan changes direction as shown in Figure 7.9a. The direction of proper forward rotation of the impeller is shown in Figure 7.9b. The impeller is an assembly that includes a number of blades. When the impeller rotates, air moves between the blades in a radial direction resulting in flow through the fan. If the electrical power to the fan is not wired correctly, the impeller can rotate backward, reducing flow and efficiency, but the airflow will still be in the correct direction.

Centrifugal fans are classified by impeller design as forward curved, backward curved, and airfoil. A forward-curved impeller has many blades (Figure 7.9c), and it rotates at a relatively low speed. Forward-curved fans generally produce high flow rates and low static pressures. Backward-curved impellers (Figure 7.9d) have between 10 and 16 blades, and they rotate at about twice the speed of forward curved impellers. Backward-curved fans are more efficient than forward-curved fans, but the higher rotational speed requires more expensive fan construction. Typically the blades of both forward- and backward-curved impellers are stamped from sheet steel. Airfoil fans are very similar to backward-curved fans, except that airfoil impellers (Figure 7.9e) have blades of varying thickness to improve fan efficiency. Airfoil blades are sometimes made of cast aluminum.

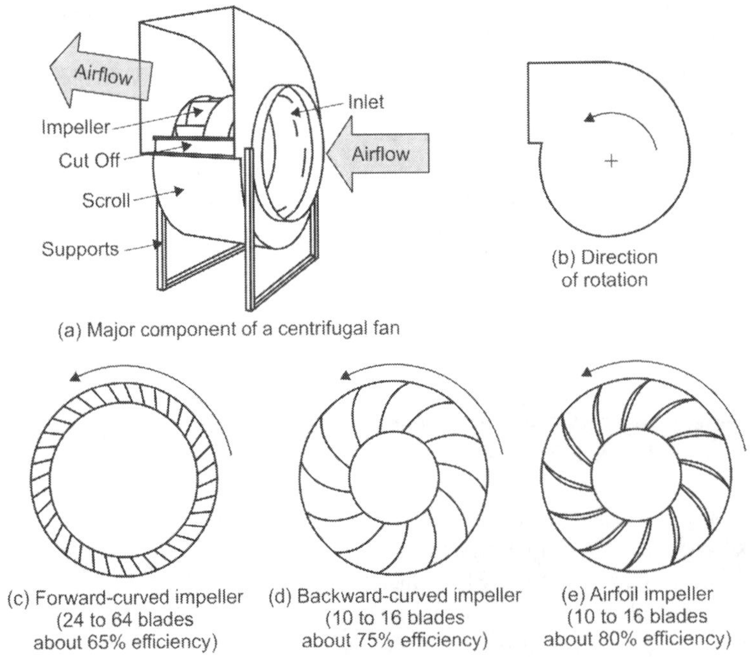

Figure 7.9 *Diagrams of centrifugal fans.*

While the selection of a fan type for a specific project requires considerable knowledge of fan technology, a few generalizations can be made. Forward-curved fans often are used for low-pressure applications including residential furnaces and packaged air-conditioning equipment. Backward-curved fans are used for many general purpose applications, and airfoil fans often are used for large systems where the energy savings are significant.

Figure 7.10 shows a tubular centrifugal fan that is different from the centrifugal fans previously discussed, in that the air flows through the fan without a change in direction. This kind of fan has straightening vanes to direct airflow parallel to the shaft as happens with axial fans. Tubular centrifugal fans are primarily used for low-pressure applications, particularly as return air fans. These fans have significant space savings over other centrifugal fans.

Axial Fans

The flow in axial fans is in the direction of the axis of rotation of the impeller. Propeller fans are frequently mounted on walls and roofs as shown in Figure 7.11. Typically, the impellers of propeller fans are inexpensively made with two or more blades. Propeller fans are used for applications like kitchen exhaust, space ventilation, and stairwell pressurization that require large flows at low pressures.

Figure 7.10d shows a tubeaxial fan. Tubeaxial fans have higher efficiency and higher operating pressures than propeller fans. The blades of tubeaxial fans can be single thickness or airfoil. The blades of all the fans discussed above are fixed to the rest of the impeller, except for vaneaxial fans (Figure 7.11e). The blades of a vaneaxial fan can rotate with respect to the impeller to adjust the pitch of the blade to optimize flow and efficiency. For this reason, vaneaxial fans have even higher efficiencies and operating pressures than tubeaxial fans. The blades of vaneaxial fans are usually airfoil. Like tubular centrifugal fans, tubeaxial and vaneaxial fans have significant space savings over centrifugal fans.

Unlike centrifugal fans, backward rotation of an axial fan normally results in backward airflow. Such backward flow is at a reduced airflow and efficiency. For more about centrifugal and axial fans see Jorgensen (1983) and ASHRAE (2008).

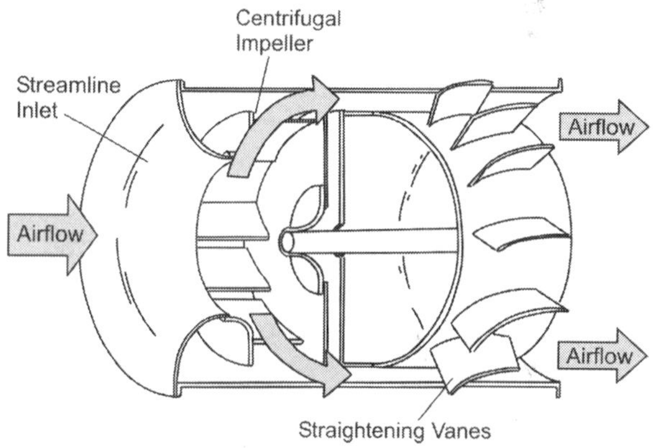

Figure 7.10 *Tubular centrifugal fan.*

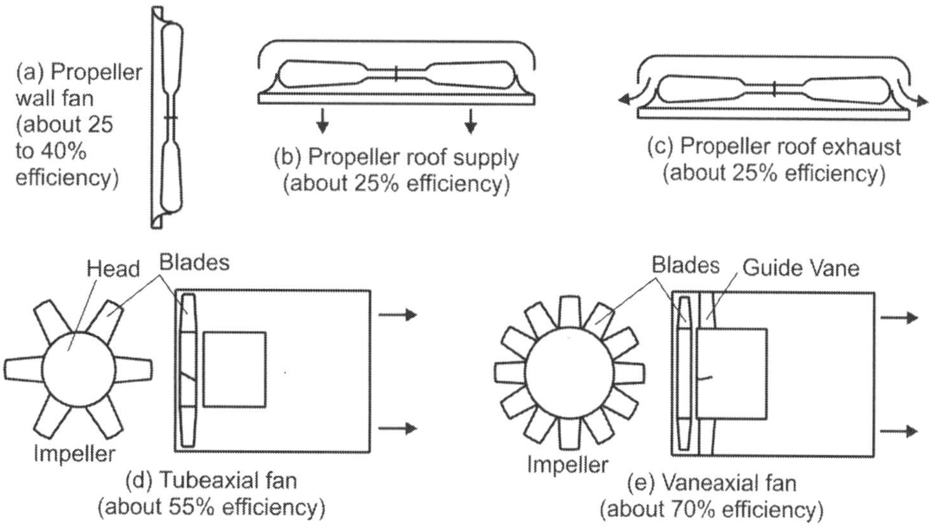

Figure 7.11 *Types of axial fans.*

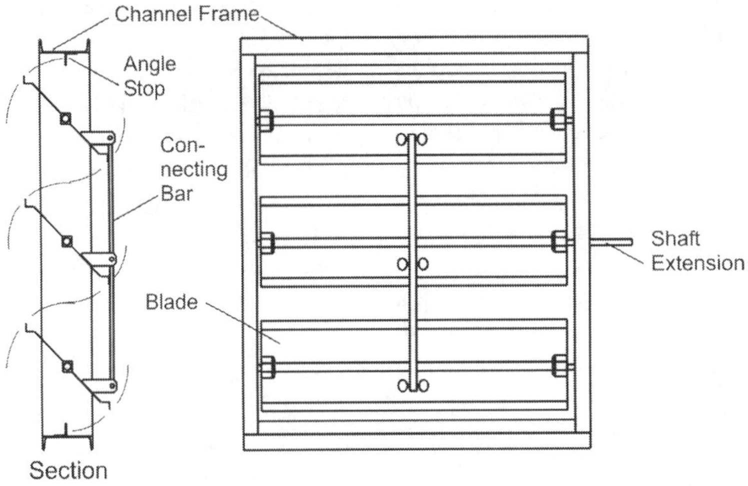

Figure 7.12 *Parallel-blade damper.*

DAMPERS

Dampers are used for one or more of the following purposes: (1) balance flow, (2) control flow, (3) resist the passage of fire, and (4) resist the passage of smoke. Balancing dampers are used to adjust the airflow in the ducts of HVAC systems. Control dampers are used to control the airflow such as the dampers used for HVAC purposes. Dampers intended to resist the passage of fire are called fire dampers, and dampers intended to resist the passage of smoke are called smoke dampers. Dampers that are intended to resist the passage of both fire and smoke are called combination dampers.

Dampers that are operated as part of an engineered smoke control system are generally multiblade dampers. There are two main types of multiblade dampers; parallel-blade and opposed-blade dampers (Figures 7.12 and 7.13). Either type can be used successfully in a smoke control system, but there are differences in the way these dampers affect the air moving through them, which might make one type better than the other in a particular application.

The airflow through a damper vs. the position of the actuator is more linear for a parallel-blade damper than it is for an opposed-blade damper (Figure 7.14). Starting from a fully-closed position and moving to a half-open (45°) position, a parallel-blade damper will pass approximately 25% of the airflow that it will pass in a fully-open position, but an opposed-blade damper will only pass approximately 10% of its fully-open airflow. This means that the parallel-blade damper will begin to pass air more quickly as it opens compared to the opposed-blade damper. Similarly, an opposed-blade damper that starts from a fully-open position and begins to close will be down to 10% of its maximum airflow when it reaches the half-open (45°) position, but a parallel-blade damper would have to close 3/4 of the way (to 22° open) before airflow is reduced to this same level. This suggests that an opposed-blade damper would be better suited for an application that was normally open and only closes for smoke control. The actual airflow through the damper at each position is also affected by the flow curve of the fan being used and the pressure across the damper, but that is a topic for the HVAC system designer; the smoke control system designer is primarily concerned with opening or closing dampers quickly and completely. For more detailed information about dampers, including pressure losses, flow characteristics, actuators, installation, and balancing, see Felker and Felker (2009).

Fire Dampers

Fire dampers are intended to prevent the spread of flames from one part of the building to another through the building's ductwork. It is not expected that a fire damper will prevent airflow between building spaces since gaps of up to 3/8 in. (9.5 mm) are allowed for operating clearances. Fire dampers are rated to indicate the time they can be exposed to flames and still maintain their integrity, with typical ratings of 3 h, 1 1/2 h, 1 h, and less than 1 h. Fire dampers are two-position devices (open or closed), and are usually of either the multiblade design (Figures 7.12 and 7.13) or the curtain design (Figure 7.15). Most multiblade fire dampers are held open by a fusible link and are spring loaded. In a fire situation, hot gases cause this link to come apart so that the spring makes the blades slam shut. Some manufacturers use other heat responsive devices in place of fusible links. Typically, curtain dampers are also held open by a fusible link which also will come apart when it gets hot. Curtain dampers often rely on gravity to make the blades close off

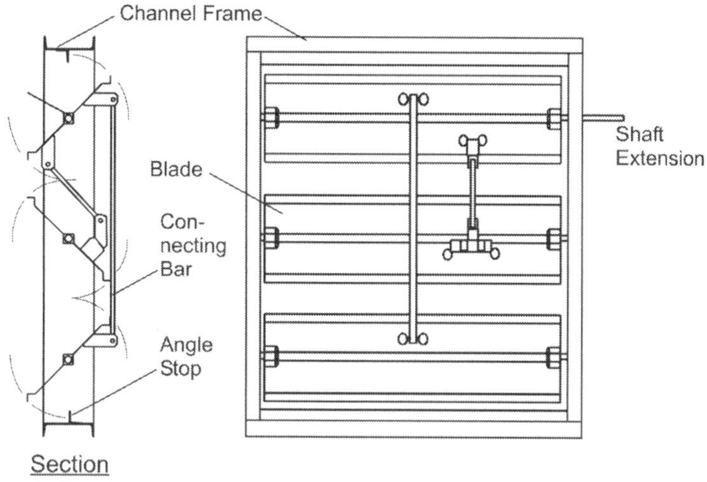

Figure 7.13 *Opposed-blade damper.*

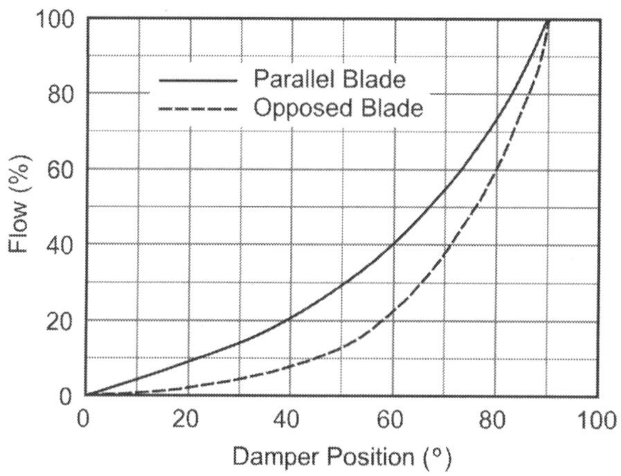

Figure 7.14 *Damper airflow as a function of position.*

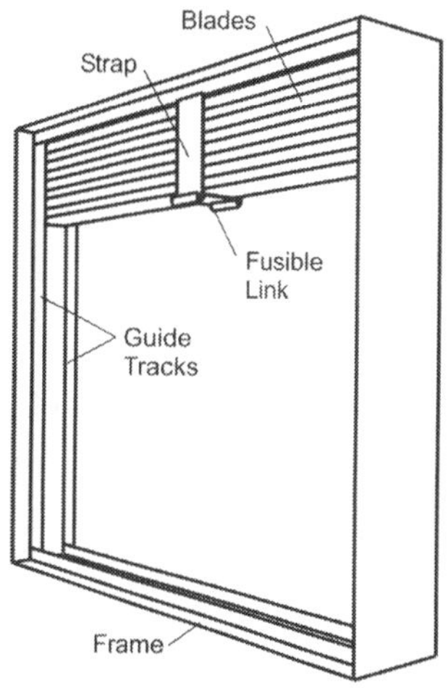

Figure 7.15 *Curtain fire damper.*

the opening, but horizontal (ceiling) type curtain dampers must have spring closure

In the U.S., fire dampers are usually made and labeled in accordance with standard UL 555 (UL 2011a). This standard is for fire dampers that are intended for use (1) where air ducts penetrate or terminate at openings in walls or partitions, (2) in air transfer openings, and (3) where air ducts extend through floors. Fire dampers are subjected to a fire exposure test followed by a hose stream test.

Fire dampers are evaluated for use as (1) static dampers, (2) dynamic dampers, or (3) combination fire and smoke dampers. Static dampers are for applications where the damper will never have to close against an airstream, such as when HVAC systems are automatically shut down upon detection of a fire. Dynamic dampers are for applications where the damper may be required to close against airflow, such as an HVAC system that remains operational for smoke control purposes.

Standard UL 555C (UL 2010) applies to ceiling dampers and ceiling diffusers intended for use in hourly rated fire-resistive floor/ceiling and roof/ceiling assemblies.

Smoke Dampers

Smoke dampers are intended to seal tightly to prevent the spread of smoke from one part of the building to another through the building's ductwork and to allow an engineered smoke control system to build up pressures across zone boundaries. A smoke damper will not withstand high temperature and will not prevent a fire from spreading. Smoke dampers are of the multiblade design (Figures 7.11 and 7.12), and may be either two-

Table 7.1: UL 555S Leakage Classifications for Smoke Dampers

Leakage Class	Maximum Leakage at 4.5 in. water (1.1 kPa)		Maximum Leakage at 8.5 in. water (2.1 kPa)		Maximum Leakage at 12.5 in. water (3.1 kPa)	
	cfm/ft²	m³/s·m²	cfm/ft²	m³/s·m²	cfm/ft²	m³/s·m²
I	8	0.041	11	0.056	14	0.071
II	20	0.102	28	0.142	35	0.178
III	80	0.406	112	0.569	140	0.711

position devices (open and closed), or may be modulated between the open and closed position to serve as both a smoke damper and a control damper.

In the U.S., smoke dampers are usually made and classified for leakage in accordance with standard UL 555S (UL 2011b). This standard includes construction requirements, air leakage tests and the endurance tests of cycling, temperature degradation, salt-spray exposure, and operation under airflow.

Each smoke damper needs to pass testing for (1) reliability, (2) temperature resistance, and (3) air leakage resistance. The operational test consists of confirming proper smoke damper operation after 20,000 cycles, or 100,000 cycles for modulating smoke dampers. The temperature test consists of confirming proper smoke damper operation after 30 min exposure to elevated temperatures. Smoke dampers must meet the requirements at a minimum temperature of 250°F (121°C) and may receive higher temperature ratings in increments of 100°F (56°C).

After the reliability and temperature resistance tests, the air leakage test is conducted. UL defines air leakage classes by the maximum allowable leakage through the closed smoke damper at a minimum pressure difference of 4.5 in. H_2O (1.1 kPa). The smoke damper classes are I, II, and III, and they and the corresponding leakages are listed in Table 7.1.

Designers can use these leakage classes to specify smoke dampers. At a location where very little smoke leakage is acceptable, a class I damper may be needed. At locations where some smoke leakage will not adversely impact smoke control performance, a class II or III damper may be appropriate.

Combination Fire/Smoke Dampers

Where both a fire damper and a smoke damper are required in the same opening, a combination fire/smoke damper may be used. Combination fire and smoke dampers comply with the dynamic fire damper requirements under UL 555 and with the smoke damper requirements under UL 555S.

REFERENCES

ASHRAE. 2000. ANSI/ASHRAE Standard 149, *Laboratory Methods of Testing Fans Used to Exhaust Smoke in Smoke Management Systems*, Atlanta: ASHRAE.

ASHRAE. 2007. *Air Conditioning System Design Manual,* 2nd ed. Atlanta: ASHRAE.

ASHRAE. 2011. *ASHRAE Handbook—Applications*. Atlanta: ASHRAE.

ASHRAE. 2012. *ASHRAE Handbook—HVAC Systems and Equipment*. Atlanta: ASHRAE.

Bell, A. 2008. *HVAC Equations, Data, and Rules of Thumb*, 2nd ed. New York: McGraw-Hill.

Bobenhausen, W. 2005. *Simplified Design of HVAC Systems*. Hoboken, NJ.: John Wiley & Sons.

Felker, L.G., and T.L. Felker. 2009. *Dampers and Airflow Control*. Atlanta: ASHRAE.

Jorgensen, R. 1983. *Fan Engineering*. Buffalo, NY: Buffalo Forge Co.

Rosaler, R. 2004. *The HVAC Handbook*. New York: McGraw-Hill.

UL. 2010. UL 555C, *Standard for Ceiling Dampers*. Northbrook, IL: Underwriters Laboratories, Inc.

UL. 2011a. UL 555, *Standard for Fire Dampers*. Northbrook, IL: Underwriters Laboratories, Inc.

UL. 2011b. UL 555S, *Standard for Smoke Dampers*. Northbrook, IL: Underwriters Laboratories, Inc.

CHAPTER 8

Controls

Paul G. Turnbull

There are many systems present in buildings, including fire alarm, sprinkler, HVAC, and energy management, to name just a few. During a fire, it may be necessary for some of these systems to operate in a mode contrary to their normal mode of operation in order to enhance the life-safety conditions within the building.

To fully understand the importance of coordination between the various building systems, consider the following scenario. When a fire breaks out, the fire alarm system does exactly what it is designed to do—it detects the fire and starts sounding horns and flashing lights to alert the occupants of the presence of a fire. The HVAC system detects that the temperature in the fire zone is above the desired setpoint and does exactly what it is designed to do—it attempts to lower the temperature by blowing lots of cold air into the space. This additional cold air fans the fire, causing the fire to grow. Both systems did exactly what they were designed to do, but unfortunately, the response of the HVAC system was inappropriate during a fire condition. If the HVAC system had been aware that a fire existed in the space, it could have responded in a manner more appropriate to the situation (Turnbull 2005).

This chapter will describe the control systems and strategies that allow the many separate systems in a building to provide a coordinated and appropriate response during a fire. Passive smoke control systems use few, if any, controls, and controls for such passive systems are not included in this chapter.

CONTROL SYSTEMS

Before delving into the details about specific control strategies for smoke control, it should first be understood that many types of systems are capable of performing the functions necessary for smoke control. In some cases, a fire alarm system may be capable of performing these functions. In other cases, the HVAC or building control system may be capable of performing these functions. Occasionally, programmable logic controllers (PLCs) might be used to perform the required functions. Systems might use electric, electronic, or pneumatic signals, or any combination of these signals to accomplish the intended objectives.

Listings

Even though many of the systems installed in a typical building might be capable of monitoring inputs and controlling outputs, the choice of which system to use is often governed by building code requirements. Most building codes require that the system used for smoke control have a specific listing from a nationally recognized testing laboratory. Requiring this listing provides the *authority having jurisdiction* (AHJ) some assurance that a third party has tested and determined that the system is capable of providing the life-safety functions and reliability level associated with that listing, which may not be the case for systems that have other listings or no listing at all. If a specific listing is required, then only systems with that listing may be used for smoke control, even if they are otherwise technically capable of implementing the control strategy.

Building control systems, PLCs, and even fire alarm systems are all technically capable of controlling fans and dampers, but if these devices are not listed as smoke control equipment, they should not be used to initiate smoke control because there are no assurances that they have the reliability and operational features required for life-safety applications. Building control systems and PLCs are generally only listed for electrical safety, and have not been evaluated for their ability to implement the controls hierarchy required of smoke control equipment (described in the Control Priorities section), they have not been evaluated for operation at elevated temperatures and in the presence of high-voltage transient surges, and may not have features that prevent unauthorized changes to the operating program.

There is a common misunderstanding about fire alarm listings that needs to be addressed. Some people believe that fire alarm systems should be used for smoke control because they think of fire alarm systems as being at the top of some hierarchy of building systems. The reality is that fire alarm systems are tested and listed for their ability to detect a fire and alert the building occupants and/or responsible personnel that a fire is present in the protected space. Unlike other systems in the building, fire alarm systems are not tested or listed for any control capabilities unless that specific function is noted as part of the system's listing. Optional control capabilities often assigned to fire alarm systems include fire door or extinguishing agent-releasing capabilities. Operation of relays or other outputs on the fire system that are not associated with releasing service is not evaluated as part of the fire alarm system listing evaluation, so the reliability of these outputs for life-safety applications has not been determined.

Devices that are listed as smoke control equipment, on the other hand, are tested and listed for their ability to control fans and dampers in a prescribed manner in response to specific input signals received, as specified in the standard used for the listing. The smoke control system must also have the capability of prioritizing its response when multiple conflicting inputs are received, as described in the Control Priorities section. Unlike a fire alarm system, smoke control systems are not listed for any detection capabilities.

The preceding discussion should make it clear that the purpose of fire alarm systems and smoke control systems are very different, and therefore fire alarm system listings and smoke control system listings are not interchangeable. It should also be noted that there is no hierarchy, so it should not be said that one system is better than another. Each system is listed for a specific purpose, and it should not be assumed that a system that is listed for one purpose is automatically suitable for another purpose. The system that is listed as a fire alarm system should be used for detection and annunciation, and the system that is listed as a smoke control system should be used to control the fans and dampers in response to specific inputs indicating the presence of a fire. There are systems available today that have been granted both fire alarm and smoke control listings, so these systems may be used for both purposes.

ACTIVATION OF SMOKE CONTROL

Smoke control systems can be activated automatically, manually, or from a *firefighter's smoke control station* (FSCS), which is described later in this section. It is necessary to understand the differences between these activation methods because the smoke control strategy will differ according to how the system is activated. Not every building code requires all three of these methods, so it is important to understand the requirements of the applicable code.

Automatic

The most common way of activating a smoke control system is through automatic activation. In this approach, the smoke control system is activated without manual intervention, in response to one or more fire detection devices. Codes differ in their definition of what constitutes a *fire detection device*. Smoke detectors and sprinkler waterflow switches are generally regarded as *fire detection devices*, but for purposes of activating smoke control strategies, only some codes or standards also include heat detectors in this definition.

Codes and standards also differ in their requirements for smoke detection systems used to activate a smoke control system. Some codes and standards require a total coverage smoke detection system to be used, while others allow use of limited coverage systems where a total coverage system is not otherwise required for fire-protective purposes. When allowed, a limited coverage smoke detection system will provide only the coverage needed to activate the smoke control system, and generally does not satisfy all requirements for a smoke detection system installed for fire-protective purposes. Examples of limited coverage smoke detection systems include detectors installed with spacing in excess of the normal spacing requirements, strategically located beam detectors, and spot detectors installed only in specific locations, such as in ductwork or near return air intakes, but not throughout the occupied areas of the building.

Regardless of the type of devices used to automatically activate the smoke control system, the devices must be arranged so that all devices in a zone of fire-protective devices are contained within one smoke control zone. If this were not the case, and the devices within one zone of fire-protective devices spanned across multiple smoke control zones, the smoke control system would not have enough information to reliably activate the correct smoke control zone. Even though a one-to-one correspondence between fire-protective zones and smoke zones is often used, it is not absolutely necessary. Multiple fire-protective zones could all activate the same smoke control zone without problems if all of the fire-protective zones were contained within the one smoke control zone. For example, if a floor in a building is divided into multiple fire-protective zones to allow pinpointing the location of a fire, but the smoke control zone encompasses the entire floor, there would not be any conflict if all fire-protective zones for that floor were assigned to activate the same smoke control zone.

Devices whose activation is not guaranteed to come from the fire-protective zone containing fire or smoke should only be used to automatically activate smoke control systems that respond identically to activation signals from all fire-protective zones. The stairwell pressurization system is a common example of a system that may respond identically to all activation signals. The most common example of a device whose activation is not guaranteed to come from the fire-protective zone containing fire or smoke is a manual fire alarm pull station. There is a reasonable likelihood that someone could sense a fire in a building and, after fleeing the area, could activate a fire alarm pull station that was physically located in a fire-protective zone other than the zone containing the fire. If smoke control systems that require information about the location of the fire were activated from this manual fire alarm pull station, the wrong smoke control strategy would be activated. By activating only those smoke control systems that respond identically to all activation signals from anywhere in the building, the response will be appropriate regardless of the location of the fire.

Manual

Manual activation of smoke control systems is often misunderstood, possibly because the same term has very different meanings in different codes and standards covering smoke control. Some codes and standards use the term "manual activation" to describe actions performed at the FSCS, but in this text, those actions will be covered in a separate section. This section addresses manual activation signals issued from a location other than from the FSCS.

NFPA 92A-2000 (NFPA 2000) provided a very suitable definition, stating that manual activation "includes all means whereby an authorized person activates one or more smoke control systems by means of controls provided for the purpose." Examples of manual activation of a smoke control system include commands received from an authorized operator at a workstation connected to the smoke control system, or activation signals received from a key-operated switch located within a smoke control zone, where only an authorized operator holds the keys. When using this definition, it should be readily apparent that activation of a manual fire alarm pull station does not constitute manual activation for the purpose of smoke control, because the activation signal does not necessarily come from an authorized user.

The building codes most widely used in North America do not have specific requirements for the manner in which the smoke control system should respond to manual activation signals. However, other standards for smoke control systems and equipment do have requirements for responses to manual activation signals (NFPA 2012), so the topic is included in this text for completeness.

Firefighter's Smoke Control Station (FSCS)

The *firefighter's smoke control station* (FSCS) is a device that provides status indication and manual control of all smoke control system components installed within the building. (Readers should note that this device is referred to as an FSCS by UL and other North American listing organizations, but it is called by other similar names, such as *firefighter's smoke control panel* in some other codes and standards; however, it is the same device described here.) As the name implies, this device is for the use of firefighters or other emergency responders, and is not intended for use by building personnel during normal building operation.

Most codes and standards require that an FSCS be included as part of a required smoke control system. Some codes and standards require that the FSCS be installed in the building's fire command center, but others require only that the FSCS is to be installed in a location acceptable to the AHJ. In the latter case, the FSCS is usually installed on the lowest level of fire department access in a secured location, or secured behind a locked cover in a lobby or central location. The IBC requires that the FSCS be installed in fire command center for high-rise buildings and buildings with smoke-protected assembly seating.

Building codes or standards often contain specific requirements for the design of the FSCS panel. These codes and standards typically specify which smoke control equipment is to be included on the FSCS graphic, the colors and meanings of the indicator lights, and how status indication is to be provided. The FSCS is generally required to display a graphical representation showing the smoke control components, together with the area(s) of the building served by those components. It is not intended that the graphic contain all of the details of a contractor's record drawing, commonly referred to as an as-built drawing, but instead should provide a simplified view that is readily understood by someone who is not familiar with the building.

Some codes and standards contain requirements for *pilot-lamp-type indicators* (ICC 2009; NFPA 2012), while others do not specify any particular type of indicator (UL 2006). In jurisdictions where codes and standards require *pilot-lamp-type indicators*, FSCS panels would need to use lamps or LEDs for indication, and typically also use switches for control functions. In jurisdictions without this requirement, computer-based graphical display panels may be allowed. Of course the AHJ may accept an alternate construction if they feel it satisfies the intent of the code or standard being used.

Depending upon the requirements of the applicable code or standard, control, and status indication of the smoke control equipment may be provided for individual equipment, such as ON/OFF/AUTO or OPEN/CLOSED, or by zone, such as PRESSURIZE/EXHAUST/ISOLATE/AUTO. When not specified by the applicable code or standard, it is recommended that controls be provided by zone rather than by individual equipment, because zone controls can be preprogrammed and pretested so that equipment is operated in the sequence necessary to prevent damage to the system. For example, it may be necessary for a damper to be opened before a fan is started, or for a fan to be shut down before a damper is closed, in order to prevent damage to the ductwork, and preprogrammed zone controls should establish that this occurs. An emergency responder unfamiliar with the building may not know the proper sequencing for individual equipment controls, or may not be able to locate the individual controls in a timely manner. Zone controls also tend to be simpler for emergency responders to understand and operate since they only need to tell the system what they want to occur and do not need to concern themselves with the specific control issues required to make it happen.

Codes and standards may also specify the methods to be used to determine equipment status, such as whether a fan is on or off, or whether a damper is open or closed. Codes and standards in use in North America generally require that equipment status be determined from sensing devices that provide a direct indication of the operational condition, rather than a derived indication. Direct indication of fan status may come from sensing devices that detect a pressure difference across the fan, or devices that indicate the presence of airflow. Other indications, such as measurement of load current, are not considered to be a direct indication of airflow unless they can be calibrated to differentiate between normal operation and fault conditions, and would only be allowed if the code or standard did not require fan status to be indicated directly. Direct indication of damper position is usually accomplished through the use of contact- or proximity-type end switches activated from the blades of the damper. Where the FSCS provides indication of both the OPEN and CLOSED position of the damper, two end switches are generally required on the damper. Other methods of indications, such as damper actuator position, are not considered to be a direct indication of damper position unless they are also able to reliably indicate a fault in the damper linkage, shaft coupling, or blades, and would only be allowed if the code or standard did not require damper position to be indicated directly. See the Sensing Devices section in this chapter for additional discussion about devices that provide direct indication of the desired result.

CONTROL PRIORITIES

With three different ways to activate a smoke control system, it is necessary to have certain rules for prioritizing the system response in the event that multiple inputs are received. This allows the system to respond appropriately even in the presence of conflicting inputs.

Automatic activation of the smoke control system must override normal building control functions for the same equipment. In general, this is accomplished by issuing the commands for smoke control at a higher priority than the commands to the same equipment for normal building operations. The preprogrammed response of an automatically activated smoke control system is generally appropriate only in response to the initial activation signal. As the fire progresses, smoke may spread to other areas of the building, causing additional activation signals. However, in the absence of a responsible person at the site to make decisions, it is generally not appropriate for the smoke control system to automatically respond to these additional input signals, because the worst conditions would normally exist in the space from which the first alarm was received, and because the second response may conflict with the first one.

Some standards (NFPA 2012) do allow the smoke control system to expand its response to include additional input signals, but only if the inputs are received from heat-responsive devices and if the system has been designed with sufficient capacity to pressurize and/or exhaust multiple zones at the same time. Acting only upon signals from heat-responsive devices should confirm that the system is responding to a larger fire scenario, and not simply to a small amount of smoke that has traveled to areas of the building remote from the fire. Requiring that the system be designed with sufficient capacity for multiple zones is intended to establish that the response to the initial signal is not compromised when the system responds to additional zones.

Smoke control may be activated manually by personnel on the site until the firefighters arrive. Manual activation of the smoke control system must override normal building control functions, and must also override any conflicting commands resulting from automatic activation of the smoke control system. As conditions at the site become better known, the operator may determine that a different control strategy is more appropriate than the one initially activated. In order to allow the system to respond to this new information, subsequent manual activations should override any previous manual activation.

Once the fire department or other emergency responders arrive, they will take over control of the smoke control system, using the FSCS. Commands issued from the FSCS have the highest priority of any command to the equipment used for smoke control. Commands from the FSCS must override normal building control functions and any automatically activated smoke control functions. Commands from the FSCS must also override manual smoke control commands issued from any other location in the building. In order to allow the firefighters to modify operation of the smoke control system in response to new information, subsequent activations from the FSCS must override any previous FSCS commands. Once the firefighters assess the situation at the site, they may determine that the best course of action would require overriding the normal protective devices built into the system. For example, the firefighters may determine that it is preferable to override a fan that has shut down due to small amounts of smoke in the duct, in exchange for the benefit of being able to use the fan to pressurize an area of the building.

Smoke control commands, regardless of the type of activation, will need to override the duct smoke detector that would normally shut down the return or exhaust fan. All smoke control activation commands, whether automatic, manual, or from the FSCS, should also override the normal function of equipment protective devices such as freezestats or high-temperature limit cutouts. In an emergency situation where lives are at stake, it is best to run equipment until it fails, rather than protect the equipment while losing the benefit it could provide. Smoke control commands would not, however, override electrical safety devices such as electrical disconnects or electrical overload protection, or heat responsive links on fire-protective equipment, such as fire dampers.

Automatically activated smoke control strategies would not override the duct smoke detectors that are intended to shut down the supply fans, but commands issued manually or from the FSCS may override supply duct smoke detectors; however, it is recommended that this override be performed by a separate command to ensure that the operator intends this action. Regardless of the source of the command, duct static pressure limits that protect the integrity of the ductwork should not be overridden, but programmable duct static pressure limits may be adjusted to higher values that might produce noise or other effects that would be undesirable under normal operation.

CONTROL OF SYSTEM OUTPUTS

In order for a smoke control system to perform its intended function, it must be able to control the air-moving equipment in the building. The smoke control system will typically control fans and dampers, but may also control outputs that bypass certain equipment-protective devices, such as freezestats and high-temperature limit devices. Different combinations of outputs will need to be controlled in response to each separate activation signal.

Activation Schedules

Given the large number of activation signals that can trigger a smoke control system, and the correspondingly large number of separate responses, it is necessary to use some methodical approach to specify in detail what should occur in response to each activation signal. This is often referred to as an *activation schedule*. The specific format of the activation schedule is left to the system designer, but NFPA (NFPA 2012) suggests that the following types of information be included in the activation schedule:

- Fire zone associated with the activation signal
- Type of activation signal (i.e., smoke detector, waterflow, etc.)
- Zones to be exhausted (maximum exhaust and no supply)
- Zones to be pressurized (no exhaust and maximum supply)
- Fans to be ON or FAST
- Fans to be OFF
- Dampers to be OPEN
- Dampers to be CLOSED
- Auxiliary functions (i.e., various overrides or changes to static pressure setpoints)
- Damper position at fan failure

Some standards require an activation schedule to be developed and included with the project documents. Even if an activation schedule is not specifically required, it can be beneficial to develop this information as part of the system design. Having this information helps owners and operators understand the operation of the system, and provides a document from which to test the system for complete and proper operation. An example of an activation schedule is shown in Figure 8.1.

Response Times

Once the smoke control system receives an activation command, it must respond in a timely and deliberate manner. Codes and standards are not in agreement on the specific time interval allowed for the system to respond, so it is important to know which code or standard is in effect for the facility under consideration. Some standards (NFPA 2012; UL 2006) require that the first command in the smoke control strategy must be initiated within 10 s of receipt of the activation signal. Other codes (ICC 2012) require that the smoke control strategy be activated "immediately" upon receipt of an activation signal, although this requirement is subject to interpretation because "immediately" to one person may mean something completely different than to another. With no timescale units presented, this requirement is not specifically measurable.

The other concern regarding system response time has to do with how quickly the smoke control system must fully implement all commands in the smoke control strategy. Again, codes and standards are not in agreement on the specific time interval allowed to achieve full functionality. Standards for some types of systems (NFPA 2012; UL 2006) specify maximum times for each component to achieve the desired state, while standards for other types of systems (ICC 2012, NFPA 2012) require only that the system must be fully operational before conditions in the space exceed the design.

The latter approach is relatively easy to understand —if the system is designed to keep smoke a certain distance above the occupants, then as long as the smoke never drops below that level, the response of the system is fast enough. On the other hand, an approach based on

Activation Signal	Exhaust	Pressurize	Fans Off	Fans On	Dampers Open	Dampers Closed	Override
29th Floor smoke detector	Floor 29	Floor 30 Floor 28	30 Exhaust 29 Supply 28 Exhaust	30 Supply 29 Exhaust 28 Supply	30 Supply 29 Exhaust 28 Supply	30 Exhaust 30 Return 29 Supply 28 Exhaust 28 Return	30 Exhaust duct smoke detector 29 Supply freezestat 29 Supply high temp limit 28 Exhaust duct smoke detector
28th Floor smoke detector	Floor 28	Floor 29 Floor 27	29 Exhaust 28 Supply 27 Exhaust	29 Supply 28 Exhaust 27 Supply	29 Supply 28 Exhaust 27 Supply	29 Exhaust 29 Return 28 Supply 27 Exhaust 27 Return	29 Exhaust duct smoke detector 28 Supply freezestat 28 Supply high temp limit 27 Exhaust duct smoke detector
27th Floor smoke detector	Floor 27	Floor 28 Floor 26	28 Exhaust 27 Supply 26 Exhaust	28 Supply 27 Exhaust 26 Supply	28 Supply 27 Exhaust 26 Supply	28 Exhaust 28 Return 27 Supply 26 Exhaust 26 Return	28 Exhaust duct smoke detector 27 Supply freezestat 27 Supply high temp limit 26 Exhaust duct smoke detector
26th Floor smoke detector	Floor 26	Floor 27 Floor 25	27 Exhaust 26 Supply 25 Exhaust	27 Supply 26 Exhaust 25 Supply	27 Supply 26 Exhaust 25 Supply	27 Exhaust 27 Return 26 Supply 25 Exhaust 25 Return	27 Exhaust duct smoke detector 26 Supply freezestat 26 Supply high temp limit 25 Exhaust duct smoke detector

Figure 8.1 *Example activation schedule.*

specific times for each component appears, on the surface, to be easier to measure, but warrants further explanation. To avoid damage, components of the smoke control system may need to achieve a prerequisite state before the next component is commanded. For example, a damper may need to be opened before a fan is started, or a fan may need to be stopped before a damper is closed. When dealing with codes or standards that have specific maximum timing requirements, it is important to understand when to begin measuring those times. The following example should make this clearer.

Example: When the fire alarm system detects the presence of smoke, it responds by activating the appropriate alarms and sending notification to the smoke control system regarding the location of the fire. Under NFPA 92 (NFPA 2012), which contains specific timing requirements, the smoke control system is allowed up to 10 s to issue the first command in the smoke control strategy. If the strategy is to pressurize a zone, the smoke control system will command the zone's supply damper OPEN. The standard allows up to 75 s to fully open the damper, measured from the time the damper is commanded. Once the damper is open—and it is not necessary to wait the full 75 s if the damper opens in less time—the smoke control system commands the fan ON. The standard allows up to 60 s for the fan to reach full speed, measured from the time the fan is commanded. So, if every component took the maximum allowed time to complete its operation, the fan would reach full speed 145 s (10 + 75 + 60 s) after the smoke control system received the activation command. If any operation completed in less time than it was allowed, then the total strategy would be completed in less time because there is no allowance in the standard for one component to be given more time if another component takes less than its allotted time. Most current systems can reach the fully operational state in much less than the time specified in NFPA 92 (NFPA 2012), but extremely large equipment may require the full allotted time.

INTERFACE TO OTHER BUILDING SYSTEMS

In order to perform their intended function, smoke control systems must receive information about the location of the fire. Because smoke control systems do not detect or annunciate fire conditions, information on the location of the fire must be obtained from the building's fire alarm system. Except for the condition where the fire alarm system is also listed as a smoke control system, information on the location of the fire is transmitted from the fire alarm system to the smoke control system using either point-to-point wiring, or a serial data interface, commonly called a *gateway*. Both of these approaches are discussed in more detail in the following sections. Factors that influence the choice between the two approaches include the availability of one or both methods on the systems being used, hardware and wiring costs for the quantity of information to be passed, and the specific functionality required.

Hardwired

A hardwired interface uses one pair of wires for each unique signal that initiates smoke control. The wires would be run from a relay or contact-closure output located in a fire alarm panel to a dry-contact input in a smoke control panel. While the hardwired approach can be used for any size system, the hardware and wiring costs increase in direct proportion to the number of initiating signals, so hardwiring is generally better suited to smaller systems.

If the fire alarm system does not provide any method to group initiating devices, then one pair of wires would be necessary to pass the activation signal from each initiating device to the smoke control system. The smoke control system logic would then need to be written so that any of the initiating devices in the same smoke control zone would activate the smoke control strategy for that zone. If the fire alarm system allows grouping initiating devices into zones, one pair of wires could be used for each zone, rather than for each detector. Grouping initiating devices at the fire alarm system will require significantly fewer outputs on the fire system, inputs on the smoke control system, and pairs of wire between the two, so it would be preferable if it is available.

When defining zones on the fire alarm system, similar type detection devices within the same zone can be grouped, but different types of detection devices may need to be placed into different groups if the smoke control system needs to respond differently, based on the type of device. For example, all smoke detectors in the same zone could be grouped, regardless of whether they were ionization, photoelectric, or multisensor detectors. However, heat-responsive devices, such as heat detectors and waterflow switches, and manually activated devices, such as pull stations that are located in the same zone, might need to be placed into separate groups if the smoke control system should respond differently to those signals than it would to signals from smoke detectors.

To ensure that the interface is reliable, hardwired interconnections must generally be monitored for integrity in accordance with applicable codes and/or standards. Some codes and standards offer exceptions to the monitoring requirement when the interconnecting wiring is less than a specified distance and also is provided with mechanical protection from injury.

Gateway

A serial data interface, often referred to as a gateway, uses a single pair of wires to communicate data about all points used to initiate smoke control. The wires would be run from a data communications port located on the fire alarm system to a data communications port located on the smoke control system. A communications protocol understood by both the fire alarm and smoke control systems is used to pass data between the systems. While the gateway approach can be used for any size system, the hardware and wiring costs are generally constant regardless of the number of initiating signals, so a gateway is generally better suited to larger systems or systems that may be expanded in the future.

When using a gateway to pass data between the fire alarm and smoke control systems, it is less important whether signals from multiple activation devices are grouped within the fire alarm system or within the smoke control system, than it is when a hardwired interface is used. Grouping signals within the fire alarm system may provide for more intuitive operation than grouping them within the smoke control system, but from a cost and functionality standpoint, there is likely to be little difference.

One functional difference between using a gateway or hardwired inputs is that many gateways can provide more information about each data point than simply whether the point is active or not. For example, with no additional hardware, wiring, or database work, gateways can often display not just whether a fire alarm point or zone is in alarm or not, but also whether the point or zone is in a trouble condition or not. To obtain this additional status information using a hardwired interface, an additional pair of wires and associated hardware would be needed for each additional condition to be monitored. Some gateways can also provide textual information about the fire alarm point or zone, such as its location and/or operator instructions. While trouble conditions and textual information are generally not necessary for automatic operation of the smoke control system, they can be beneficial during manual smoke control operations, and often come at no additional cost.

To ensure that the interface is reliable, serial data interconnections must generally be monitored for integrity in accordance with applicable codes and/or standards. This is typically accomplished through some form of data handshake that provides notification to both systems if communication between them fails. Some codes and standards offer exceptions to the monitoring requirement when the interconnecting wiring is less than a specified distance and also is provided with mechanical protection from injury.

Shared Network Wiring

Most commercial properties now have some form of data network infrastructure installed. These networks are used for multiple purposes, such as connecting computers to each other and to the internet, sharing data between building control panels or industrial automation equipment controllers, and even transmitting telephone voice and data signals within the building. With so much equipment able to connect to the same network infrastructure, building owners expect that other building systems, such as fire alarm systems, security systems, and smoke control systems should also be able to connect to their existing network infrastructure.

While much of this equipment is technically capable of communicating over existing network wiring, codes and standards are just starting to include requirements for life-safety related equipment when it is connected to the same network as non-life-safety equipment. This section is included in this text to discuss the issues that arise when life-safety and non-life-safety systems coexist on the same network. Because the codes, standards, and listing requirements for this application are currently being developed, it is difficult to provide one-size-fits-all guidance in this area. It is expected that many changes will take place as requirements are developed over the next few revision cycles for the relevant codes and standards.

When life-safety equipment is connected to the same network as non-life-safety equipment, the primary concerns are related to system reliability and survivability. Standards for life-safety systems verify that the systems are reliable and continue to function normally in the presence of fault conditions. Life-safety systems are tested at temperatures well beyond expected normal operating temperatures, and they are tested for their ability to survive voltage transients more severe than most systems will ever encounter. Normal data processing equipment that is used for the majority of network infrastructures is not tested beyond normal operating temperatures and voltages, so there is concern that this equipment could fail, leaving the life-safety system unable to pass required signals and therefore unable to perform required operations. There are also concerns that failure of non-life-safety equipment connected to the network may have a detrimental effect on network components that are also used by the life-safety system, causing the life-safety system to be unable to perform required operations, through no fault of its own.

The National Fire Alarm Code (NFPA 2010) describes methods that can be used to increase the integrity level of a network shared by life-safety and non-life-safety equipment. These methods include providing primary and secondary power sources, and continuously monitoring available system bandwidth to confirm that the network is capable of delivering life-safety signals within 10 s. Although these methods are described in the code, few, if any, systems are known to use these methods.

On the other hand, a different approach to ensure the integrity of a network shared by life-safety and non-life-safety equipment has been adopted by many system suppliers. While specifics may differ slightly between systems, the general approach is to use a ruggedized network device to isolate a portion of the network that is used by the life-safety equipment from any non-life-safety equipment that may also be connected to the network. This ruggedized device is subjected to and listed to the same power, environmental, and transient conditions that are required of life-safety equipment. The ruggedized device is additionally subjected to tests to verify that faults such as opens and shorts in the non-life-safety equipment do not affect normal operation of the life-safety equipment. Even though this method does have limitations, such as only being able to connect life-safety equipment to specific network ports instead of any port on the network, this method seems to have become the de facto standard for life-safety equipment that is capable of operating on shared networks.

EXAMPLE CONTROL CIRCUIT DIAGRAMS

The following diagrams are provided to show example control circuits that could be used to implement smoke control strategies. Even though the diagrams look similar, the reader should note the differences between control circuits that use shared ON/OFF controls for both normal building operation and smoke control vs. those that use separate controls for smoke control, and between those for control of dedicated versus nondedicated smoke control equipment.

Nondedicated Fan with Shared ON/OFF Control

Figure 8.2 shows an example control circuit for a nondedicated three-phase fan, where the control contact that turns the fan on and off is shared by both normal operation and smoke control functions. Equipment protective devices (high-static-pressure limit switch, duct smoke detector, high-temperature limit switch, and low-temperature limit switch—also known as a freezestat) are connected in series with the coil of the motor starter, as is common for fans used for normal building operations. These equipment protective devices will open the circuit in the event of a fault condition. Separate control

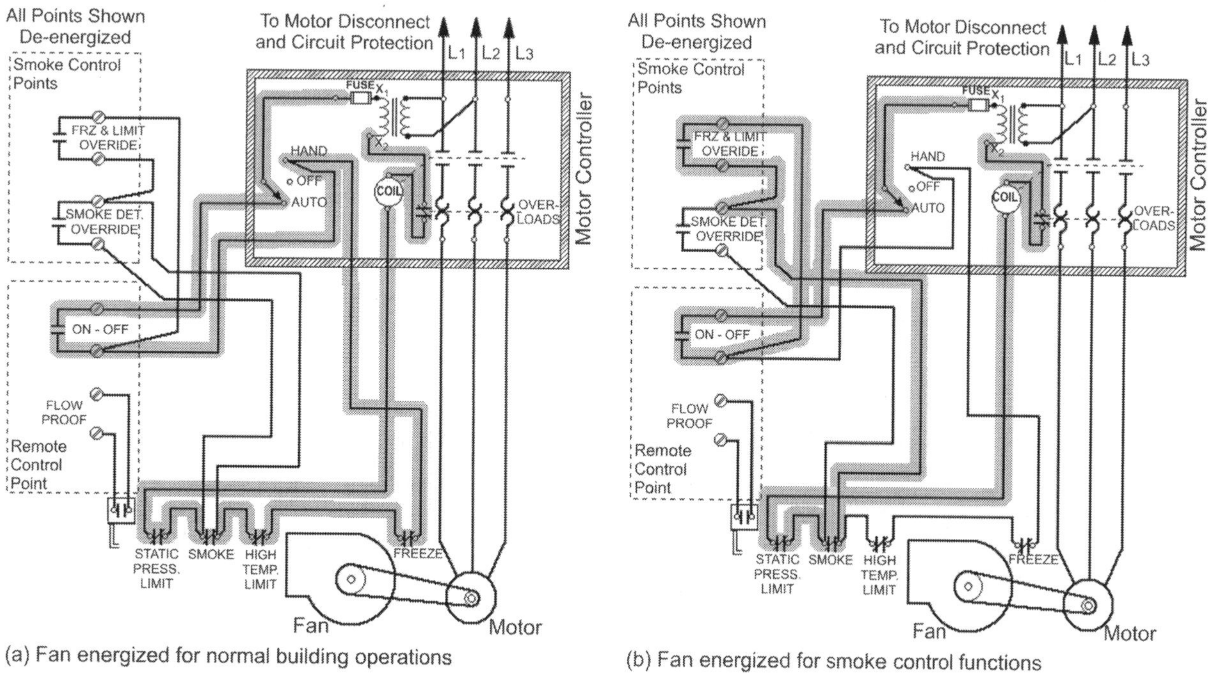

Figure 8.2 *Nondedicated fan with shared ON-OFF controls.*

contacts are provided to override the high- and low-limit switches and the duct smoke detector. The highlighted portion of Figure 8.2a shows the current flow when the fan is energized for normal building operations. Note that the fan may only be controlled when the HAND/OFF/AUTO switch is in the AUTO position. Figure 8.2b shows the current flow when the fan is energized for smoke control functions. Because the control circuit is shared with normal building operations, the fan may only be controlled when the HAND/OFF/AUTO switch is in the AUTO position. This may not be acceptable in all jurisdictions. Note that one of the smoke control contacts is energized to shunt the high- and low-temperature limit switches out of the circuit. Another smoke control contact could be energized to shunt the duct smoke detector out of the circuit to restart the fan if required by emergency response personnel.

Nondedicated Fan with Separate ON/OFF Controls for Smoke Control and Normal Operation

Figure 8.3 shows an example control circuit for a nondedicated three-phase fan, where the control contact that turns the fan on and off for normal building operation is separate from the contacts that turn the fan on and off for smoke control functions. Equipment protective devices (high-static pressure limit switch, duct smoke detector, high-temperature limit switch, and low-temperature limit switch—also known as a freezestat) are connected in series with the coil of the motor starter, as is common for fans used for normal building operations. These equipment protective devices will open the circuit in the event of a fault condition. The highlighted portion of Figure 8.3a shows the current flow when the fan is energized for normal building operations. Note that current flows through a separate, normally-closed smoke control relay which can open the circuit to turn off the fan for smoke control purposes, regardless of the position of the contact used for normal building operations. Figure 8.3b shows the current flow when the fan is energized for smoke control functions. Note that when fan is activated for smoke control, it will activate regardless of the positions of the HAND/OFF/AUTO switch, the relay used for normal building operation, and both the high- and low-temperature limit switches. A separate smoke control contact could be energized to shunt the duct smoke detector out of the circuit to restart the fan if required by emergency response personnel.

Dedicated Stairwell Pressurization Fan

Figure 8.4 shows an example control circuit for a dedicated three-phase fan used for stairwell pressurization. Because this fan is only used for smoke control, the

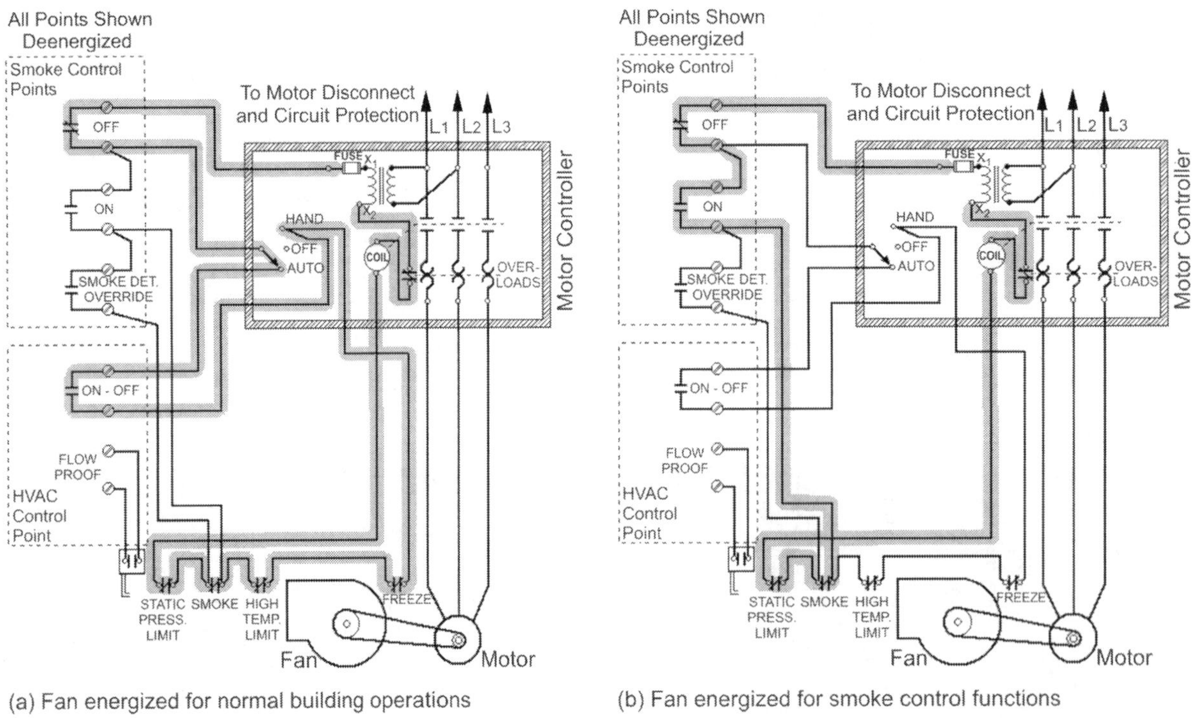

(a) Fan energized for normal building operations

(b) Fan energized for smoke control functions

Figure 8.3 *Nondedicated fan with separate ON/OFF controls for smoke control and normal operation.*

controls are much simpler, without the need to coordinate with shared control circuits or to provide override controls for smoke control operation. Furthermore, there are no high- and low-temperature limit switches included in the circuit because these devices are always overridden for smoke control. A high-static pressure limit switch and duct smoke detector are still provided, and are connected in series with the coil of the motor starter. These equipment protective devices will open the circuit in the event of a fault condition.

The highlighted portion of Figure 8.4 shows the current flow when the fan is energized for smoke control operations. Note that current flows through a separate, normally-closed smoke control relay which can open the circuit to turn off the fan for smoke control purposes, regardless of the position of the manual ON/OFF switch. The relay that turns the fan on is wired in a manner that will energize the fan regardless of the position of the manual ON/OFF switch. A separate smoke control contact could be energized to shunt the duct smoke detector out of the circuit to restart the fan if required by emergency response personnel.

Dedicated Smoke Damper

Figure 8.5 shows an example control circuit for a dedicated smoke damper. This damper is only used for smoke control, so the controls are simple, without the need to coordinate with shared control circuits. Because there are no HAND/OFF/AUTO controls present, there is no need to provide override controls for smoke control operation. One contact is used to open the damper, and another is used to close the damper. If a spring-return damper is used instead of one that is driven in both directions, only one contact would be needed.

SYSTEM RELIABILITY

Smoke control systems are defined in the codes and standards as being life-safety systems. As such, smoke control systems require a higher level of reliability than other systems that might be installed within the building. A variety of methods can be used to provide assurance that the smoke control system is ready to perform its intended function when needed. Reliability of nondedicated smoke control equipment is often verified through use during normal operation, and may also be determined through periodic manual testing. Reliability of dedicated smoke control equipment may be determined through a combination of methods including end-to-end verification, electrical supervision, automatic self-testing, and periodic manual testing. Each of these methods is more fully described in the following sections.

Normal Operation as a Method of Verification

In the codes and standards, requirements are often specified for methods and frequencies of verifying that dedicated smoke control equipment is ready to operate, but requirements for nondedicated equipment are often less rigorous or nonexistent. The reason that nondedicated equipment is not subjected to the same ongoing

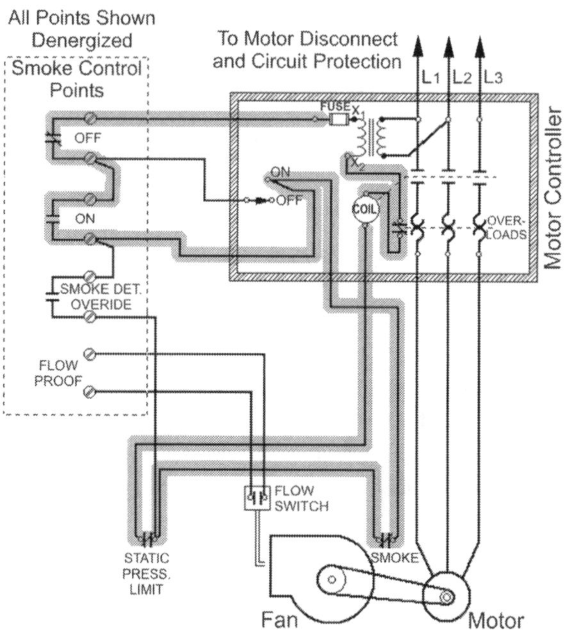

Figure 8.4 *Dedicated stairwell pressurization fan.*

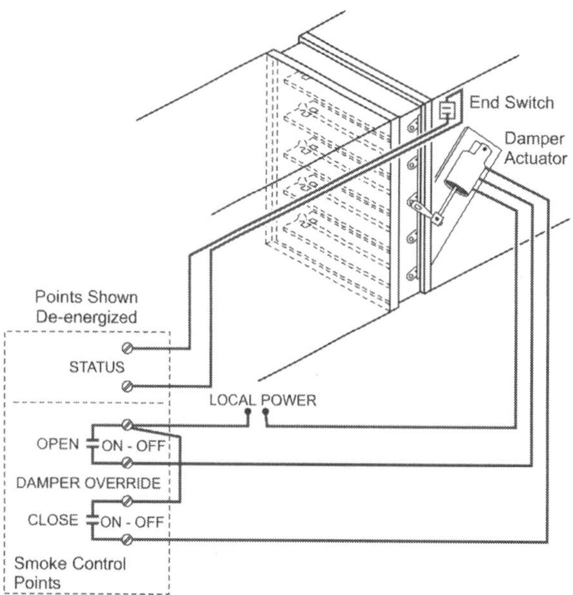

Figure 8.5 *Dedicated smoke damper.*

test requirements as dedicated equipment is that a failure in equipment that is also used for normal building operation will cause the normal building controls to operate incorrectly, resulting in poor airflow and/or temperature control, which will be noticed by building occupants. NFPA 92 (NFPA 2012) echoes this line of thinking with their statement that "impairments to shared equipment required for normal building operation are likely to be corrected promptly." Building occupants are quick to complain when the heating, ventilation, or air-conditioning is not working correctly, and as such, the building occupants function as a form of equipment verification for nondedicated equipment.

Electrical Supervision

Electrical supervision is used to indicate whether control circuit wiring is intact and ready to perform its intended control function. Circuits that are electrically supervised run a trickle current from the control panel, through the circuit wiring, through an end-of-line resistor located at the end of the wiring run, and back to the control panel. If the circuit is cut, then no current flows back to the control panel, indicating a fault condition. If the circuit is shorted, current in excess of the normal value flows back to the control panel, also indicating a fault condition. This method of supervision is familiar to many people, as it is the method used to supervise the wiring for fire alarm systems.

The advantage of electrically supervised circuits is that they continually monitor the condition of the control circuit wiring, and can readily indicate an open or short condition soon after it occurs. In many codes and standards, electrically supervised circuits are required to indicate the presence of a fault condition in 200 s or less.

The weakness of electrically supervised circuits is that they only monitor the integrity of the control circuit; they provide no indication about the integrity of the equipment being controlled. Most codes that regulate electrically supervised circuits require the supervision to extend to within approximately 3 ft (1 m) of the controlled equipment. For all practical purposes, this means that the electrical supervision extends from the control panel to the motor starter or damper actuator. Electrical supervision provides no indication about faults beyond the control wiring, such as broken fan belts, stalled motors, clogged air filters, loose or broken damper actuator linkages, or blocked damper blades.

Because there are so many potential fault conditions that cannot be detected or indicated by electrical supervision, some codes and standards do not view electrical supervision as sufficient to ensure that a smoke control system is ready to operate when called upon. In those cases, the codes and standards require end-to-end verification of smoke control circuits.

End-to-End Verification

End-to-end verification is used to confirm that the electrical and mechanical devices and their controls correctly energize when needed for smoke control. NFPA 92 (NFPA 2012) defines end-to-end verification as "a self-testing method that provides positive confirmation that the desired result (i.e., airflow or damper position) has been achieved when a controlled device has been activated, such as during smoke control, testing, or manual override operations."

End-to-end verification focuses on the desired result, such as the presence of airflow or the position of a damper, rather than just on the control circuit. In order to determine whether the desired result is achieved, sensing devices that respond to the desired result are needed. Commonly used sensing devices that respond to the presence of airflow or damper position include airflow switches, differential pressure switches, and/or damper end-switches. The actual results indicated by these sensing devices are compared to the commanded states of the equipment to determine whether the desired result has been achieved.

The advantage of end-to-end verification is that it verifies proper operation of not only the control system and circuits, but also verifies that the mechanical system (e.g., belts, filters, linkages, pneumatic lines) is functioning correctly, and that power is available to the device being controlled.

The weakness of end-to-end verification is that it can only indicate a fault condition when equipment is supposed to be active, but fails to activate. This is a frequently misunderstood point, which deserves further explanation. If a fault occurs in the control circuit wiring, a fan belt breaks, a damper linkage becomes loose, an air filter becomes clogged, or some other abnormal condition occurs when the device is not supposed to be active, it is not expected that this fault will be indicated until the next attempted activation of the device. Even though this fault may be present when the device is not energized, the "proof" sensors indicate that the actual state of the device matches the desired state of the device, so no fault is indicated. If the fault still exists the next time that the device is supposed to be activated, the device will fail to achieve its desired state, and a fault will be indicated at that time. Because some equipment used for smoke control is activated infrequently, it is necessary to activate the equipment for testing purposes with some regularity in order to detect these faults. The codes and standards that require end-to-end verification also require an automatic self-test of the dedicated smoke control equipment on regular intervals to verify that the electrical and mechanical devices and their controls remain operational. For further information on this topic, refer to the following section on Automatic Testing.

Automatic Testing

Smoke control systems in North America are generally required to be listed to the requirements in UL 864 (UL 2006), under the category of Smoke Control System Equipment. Sometimes, codes or specifications call for a listing under category UUKL, which is UL's Category Control Number for Smoke Control Systems, so these are two different ways of specifying the same listing. It is not within the scope of this text to describe all of the requirements for products to obtain a listing under UL 864 category UUKL, but a section on automatic testing would not be complete without a discussion of the automatic testing requirements that are part of this listing.

UL 864 states "dedicated smoke-control systems shall employ an automatic weekly self-test function. The self-test shall automatically command activation of each associated function. An audible and visual trouble signal shall be annunciated at the FSCS identifying any function that fails to operate within the required time period. The self-test function is not required for nondedicated systems."

The weekly self-test works together with end-to-end verification to confirm that dedicated smoke control equipment remains operational and will function when called upon. As noted above, end-to-end verification confirms that the desired result is achieved when a device is activated, but dedicated smoke control equipment would generally not be activated during normal operation of the building. In order for end-to-end verification to have the opportunity to check for faults, the equipment must be activated periodically. The weekly self-test commands the dedicated smoke control equipment to the state required during smoke control, and the end-to-end verification confirms that the equipment and controls are fully functional. Each test needs to run only long enough for the end-to-end verification to confirm normal operation, after which time the test may be discontinued. Some smoke control equipment, such as dampers, may have multiple states that could be used for smoke control, depending on the location of the fire. In this case, all states that could be used for smoke control must be tested; for example, a damper that may need to be fully open in one smoke control scenario, and fully closed during a different smoke control scenario, must be commanded and verified in both states during the self-test. If a fault is found, the system is required to provide both audible and visual indication to building personnel, indicating the need for corrective action.

While UL 864 requires that products demonstrate they are capable of providing an automatic weekly self-test function as a condition of receiving a listing under category UUKL (Smoke Control System Equipment), it is the building code that actually specifies whether the self-test is required to be run. The International Building Code (IBC) (ICC 2009), for example, requires a "preprogrammed weekly test sequence" to be run, so in jurisdictions covered by this code, the weekly self-test must be run. It should be noted that the local AHJ has the authority to modify any requirement in the building code if, in their opinion, it makes sense to do so. Buildings located in extreme climate areas will find weekly self-testing to be very inefficient from an energy standpoint. For example, testing an atrium exhaust fan in Minneapolis, MN in the winter will exhaust a large volume of heated air and bring in an equal volume of cold outdoor makeup air. Similarly, testing an atrium exhaust fan in Dallas, TX in the summer will exhaust a large volume of cooled air and bring in an equal volume of hot and humid outdoor air. Building owners in areas such as these may want to meet with the AHJ to see whether an alternative testing schedule or method may be used during extreme weather conditions to meet the intent of the code.

Manual Testing

A smoke control system is a life-safety system, just as a fire alarm system is a life-safety system. Fire alarm systems provide electrical supervision of the control wiring but do not test that the input or output devices are operational, so these systems must be manually tested on a periodic basis. Smoke control systems that provide end-to-end verification and a periodic self-test regularly confirm that the devices and their control circuits are operational, so manual testing of these systems may not be necessary. Smoke control systems that do not provide end-to-end verification and a periodic self-test should be manually tested on a periodic basis to confirm that the system is working correctly. The testing frequency may be specified in a code or standard, or may be specified by the AHJ. The testing frequency may be different for dedicated and nondedicated equipment. If so, the dedicated equipment will probably need to be tested more often than nondedicated equipment, which is exercised during normal building operation.

Sensing Devices

As noted in the previous sections, end-to-end verification and self-testing require a determination of whether the desired result has been achieved when a smoke control device has been activated. When a smoke control system commands a fan, the desired result is either the presence or absence of airflow, and not just that the fan motor is running or not. When a damper is commanded, the desired result is that the damper blades are fully open or fully closed, and not just that the damper actuator has moved.

To satisfy the requirements for end-to-end verification, the sensing devices used must respond directly to the desired result in order to detect a failure of either the mechanical or electrical components of the system that could prevent the desired result from being achieved.

When a fan is commanded ON, the desired result is the presence of airflow, so a sensing device that responds directly to airflow must be used. The sensing device must also be able to detect a failure due to loss of fan power, broken belts, clogged air filters, or other conditions that could prevent airflow from occurring. Examples of sensing devices that respond to airflow and can also indicate faults that could prevent airflow include sail-switches located in the airstream and differential pressure switches reading the pressure difference between the intake and the output of the fan.

Similarly, to verify damper operation, the sensing device must be able to reliably indicate when the damper blades have reached their fully open and/or fully closed positions. The sensing device must also be able to detect a failure of the damper blades to reach their intended position due to loss of power or air supply, broken or misadjusted linkage, loose shaft coupling, or other conditions that could prevent the damper blades from fully opening or closing when required. When both the fully open and fully closed positions need to be monitored for smoke control, separate sensing devices are needed to monitor each position. The most common sensing devices that respond to damper blade position and can also indicate faults that could prevent the damper blades from achieving their desired position are damper end switches or proximity switches that are activated directly from the position of damper blades.

Sensing devices that do not respond directly to the desired result should not be used for end-to-end verification unless it can be proven that a direct correlation exists under all conditions. Current monitoring devices that measure motor current can indicate whether the fan motor is running or not, but few can be calibrated precisely enough to indicate the difference between normal fan operation and conditions where the fan motor is running but there is no airflow due to a broken fan belt or a dirty air filter. When multiple fans are configured in parallel within the same duct, a current monitoring device could even provide a normal reading for a fan that failed to activate and was being driven in reverse by airflow from the parallel fan. Switches indicating the end of actuator travel may not directly correlate to the position of the damper blades if the linkage is loose, broken, or misadjusted.

BEST PRACTICES

This section presents ideas that are not necessarily part of any code or standard, but are simply topics that may help the designer avoid common pitfalls.

Use of a Single Control System to Coordinate Smoke Control

One of the biggest pitfalls in the design of smoke control systems is splitting the control functions between multiple systems. If the smoke control response is performed by more than one system, the overall smoke control system will be more costly and more complex than it would be if a single control system were used (Turnbull 2005). While cost is certainly a concern of the building owner, a more important concern is that the additional complexity will cause the overall system to be more difficult to understand, test, and maintain. NFPA 92 (NFPA 2012) addresses this safety concern by stating "a single control system shall coordinate the smoke-control functions provided by the fire alarm system, fire fighters' smoke-control system, and any other related systems with the operation of the building HVAC systems and dedicated smoke-control equipment."

The smoke control functions are most likely to be split between multiple systems when the electrical and mechanical portions of the job specification are not harmonized. A common example of this problem occurs when the electrical specifications call for the fire alarm system to shut down a fan when smoke is detected, and the mechanical specifications call for the building control system to activate the same fan for smoke control purposes. In order for the installing contractors of both systems to comply with their portion of the specifications, the fire alarm installer must include a control circuit to open a series contact to deenergize the fan (Figure 8.6), and the building controls installer must include a control circuit to close a parallel contact to restart the fan (Figure 8.7) for smoke control. These two control circuits are in addition to the control circuit that is provided to control the fan for normal building operation.

If, instead of using the building control system to restart the fan, the specifications called for the fire alarm system to restart the fan for smoke control, the two additional control circuits described in the preceding paragraph would still be required, but both of them would be connected to the fire alarm system (Figure 8.8).

If the control logic for both normal building operation and smoke control were placed into one system, a single control circuit could be provided to start and stop the fan for both purposes. The control logic and priorities would ensure that the state required for smoke control would prevail. This approach avoids the cost and confusion that results from the additional control circuits (Figure 8.9).

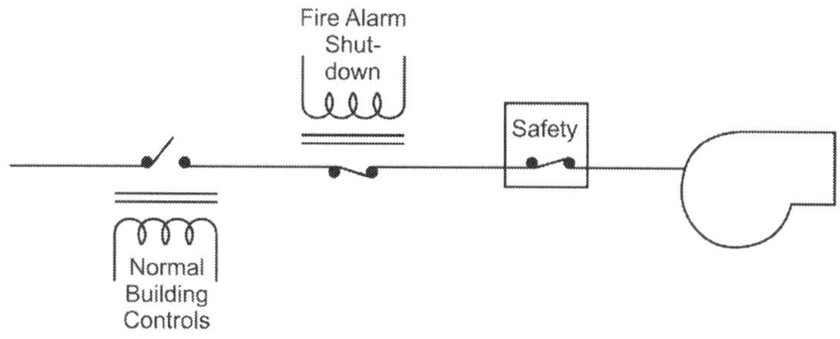

Figure 8.6 *Additional control circuit for fire alarm shutdown.*

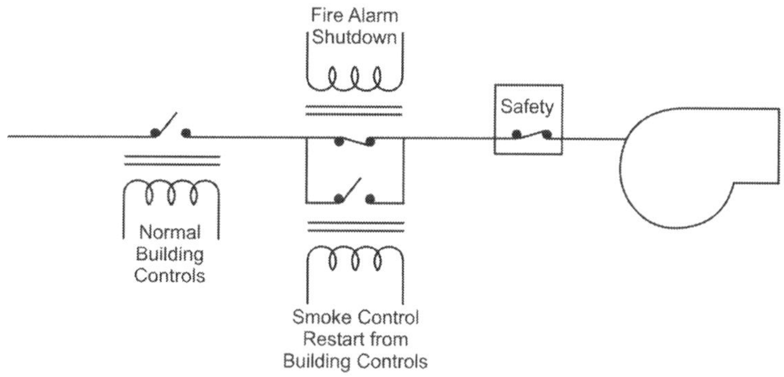

Figure 8.7 *Additional control circuits for fire alarm shutdown and building control restart.*

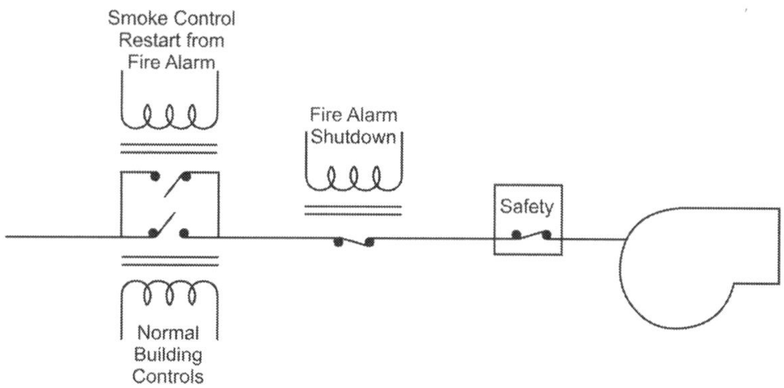

Figure 8.8 *Additional control circuits for fire alarm shutdown and restart.*

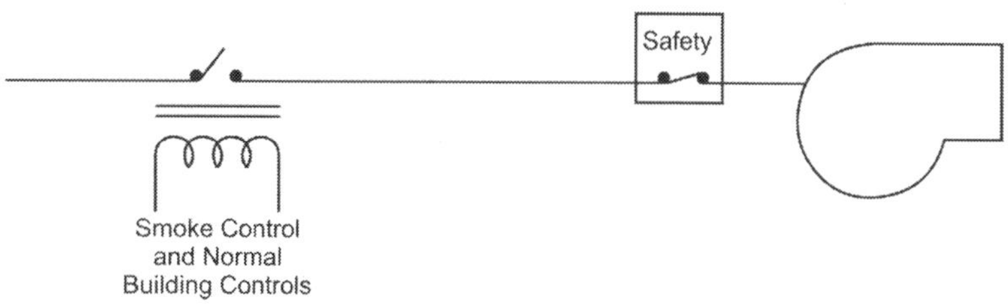

Figure 8.9 *Simplified controls using one control system.*

Control of Devices that are Not Part of the Smoke Control System

In order to perform their intended function, Smoke Control systems control fans used for pressurization or exhaust of a specific area of the building. To direct the airflow from these fans to the correct area of the building, smoke control systems also either open or close smoke dampers that are located within the smoke barrier that forms the boundary of the smoke control zone. These control functions should be described in the activation schedules as part of the overall smoke control strategy.

A point of confusion exists regarding what should be done during smoke control with devices that are part of the air-handling system but are not part of the smoke control system. There is no single answer to this question, but the following discussion should help readers arrive at their own answer applicable to their specific situation. As an example, consider a single zone in a zoned smoke control system. The air supplied to this zone comes through ductwork that feeds many terminal boxes located within the boundary of the zone. These terminal boxes have modulating dampers that are used to adjust the amount of air supplied to the zone to regulate the temperature in the space. The question is what to do with the controls for the terminal boxes when the smoke control system is calling for pressurization of the zone?

To understand what to do with the terminal box controls, first look at whether the smoke control system can achieve its objective without controlling the terminal boxes. If all of the terminal boxes were closed at the time the smoke control system wanted to pressurize this zone, could the zone be pressurized? The answer is "it depends." In many jurisdictions, there are requirements for minimum ventilation to provide specific levels of indoor air quality. If the minimum ventilation requirement ensures that the terminal boxes are never 100% closed, then it is most likely possible to pressurize the zone without controlling the terminal boxes. When attempting to pressurize a zone, the exhaust rate is set to zero, so it takes very little supply air volume to increase the pressure in the zone. Even though it may not be necessary to open the damper in the terminal box any further, commanding the terminal box damper fully open when attempting to pressurize the zone may help to achieve the pressurization more readily, and it would be good engineering practice to do so.

In the opposite scenario, when attempting to exhaust a smoke control zone, the smoke damper in the main supply duct is closed, so it would make no difference at all whether the dampers in the terminal boxes are open or closed.

Why all this concern over whether pressurization can be achieved without controlling the terminal box, when good engineering practice says to open the damper in the terminal box anyway? If pressurization can not be achieved without controlling the terminal box, then the terminal box and the controls associated with it become part of the smoke control system. If the terminal box is part of the smoke control system, then all code requirements for smoke control equipment, such as having appropriate equipment listings, running wiring in conduit, providing redundant power, conducting periodic testing, etc. now apply to the terminal box and its control wiring. If, on the other hand, pressurization can be achieved without controlling the terminal box, then the terminal box becomes an ancillary device not subject to the smoke control requirements. In the latter case, whatever controls the designer chooses to include for the ancillary device are there simply as good engineering practice, and are not mandated by code.

REFERENCES

ICC. 2012. International Building Code®. International Code Council, Washington, DC, Section 909.

NFPA. 2000. NFPA 92A, *Recommended Practice for Smoke-Control Systems*. Quincy, MA: National Fire Protection Association.

NFPA. 2012. NFPA 92, *Standard for Smoke Control Systems*. Quincy, MA: National Fire Protection Association.

NFPA. 2010. NFPA 72, *National Fire Alarm and Signaling Code*. Quincy, MA.: National Fire Protection Association.

Turnbull, P. 2005. Smoke control in integrated buildings. *HPAC Engineering,* Networked Controls Section, October.

UL. 2006. UL 864, *Standard for Control Units and Accessories for Fire Alarm Systems*, Ninth ed. Northbrook, IL: Underwriters Laboratories Inc.

CHAPTER 9

Basics of Passive and Pressurization Systems

John H. Klote

Smoke is commonly recognized as the major killer in building fires, and smoke control systems relying on passive protection and pressurization can provide significant smoke protection. Pressurization smoke control systems are commonly used. Passive systems are sometimes used in conjunction with pressurization smoke control systems. Passive smoke control systems can be used by themselves to provide a tenable environment, and these systems can be analyzed by modern tools. This chapter deals with the basic concepts of these systems.

PASSIVE SMOKE CONTROL

For centuries, compartmentation has been recognized as a way of controlling the spread of fire and smoke. When a person closes the door to a burning room, smoke flow from the room decreases considerably. Also, the amount of air available to the fire drops off. Today, this passive smoke protection is recognized in many building and fire codes even without a design analysis.

To limit the spread of fire, buildings are divided into compartments formed by fire barriers. Fire barriers are not intended to restrict the flow of smoke. These barriers are walls, partitions, and floor-ceiling assemblies that have a level of fire resistance. The traditional approach to evaluate fire resistance is to subject a section of a barrier to a standard fire in a standard furnace. Each building fire is unique in duration and temperature, and it is not surprising the performance of fire barriers in building fires differs to some extent from the performance in standard tests. Historically, the goal of fire resistant construction was property protection, but the goals of current codes focus on protecting human life. For further information about fire resistant construction, readers are referred to Barnett (1991), Boring (1990), Boring et al. (1981), and Bushev et al. (1978).

For purposes of passive smoke control, a passive smoke barrier is a continuous membrane, either vertical or horizontal, such as a wall, floor, or ceiling assembly that is designed and constructed to restrict the movement of smoke. This meaning is consistent with the usage of the term *smoke barrier* in the *International Building Code*® (*IBC*®)(ICC 2012). Such a passive smoke barrier is intended to provide some level of passive smoke protection, but that level of protection is not explicitly defined.

When a passive smoke barrier is part of a passive smoke control system, the construction of the barrier needs to be tight in order to restrict smoke movement. For passive smoke barriers to perform as intended, openings must be properly sealed to limit leakage. In general, this applies to construction cracks, penetrations for ducts and dampers, and other openings. Doors in passive smoke barriers need to close automatically upon smoke detection. A fire rated assembly does not assure the assembly is constructed to restrict smoke migration, but many fire resistance rated separations also act as passive smoke barriers. In Chapter 3 of the IBC publication *A Guide to Smoke Control in the 2006 IBC* (Klote and Evans 2007), smoke barriers for passive smoke protection are discussed.

Depending on the pressure differences across passive smoke barriers, some small amounts of smoke may migrate through them. The intent of smoke barriers in passive smoke control systems is that such smoke migration does not result in untenable conditions on the nonfire side of the barrier for some time after ignition. Small amounts of smoke have the benefit of convincing occupants of the true threat of a fire as no fire alarm can, and this tends to reduce the time that occupants take to get moving in fire situations. This premovement time is discussed in Chapter 4. Engineered smoke control systems can use passive smoke protection by itself or in conjunction with pressurization or other smoke control mechanisms. Engineered smoke control systems that use only passive smoke barriers are a form of tenability systems that can be analyzed by the methods discussed in Chapter 19.

PRESSURIZATION CONCEPT

Many smoke control systems use mechanical fans to control smoke by pressurization. Pressure difference across a barrier can control smoke movement. The idea is that a pressure difference is produced across a barrier such that the smoke on the low-pressure side of the barrier is prevented from migrating to the high-pressure side. Figure 9.1a shows pressurization controlling smoke flow with a relatively small amount of smoke. However, pressurization can control smoke from a large, fully developed fire as shown in Figure 9.1b. In a room with a fully developed fire, every thing that can burn is burning. More information about fully developed fires is provided in Chapter 5.

The common smoke control systems that rely on pressurization are pressurized stairwells, pressurized elevators, and zoned smoke control. These systems are discussed in Chapters 10 through 13.

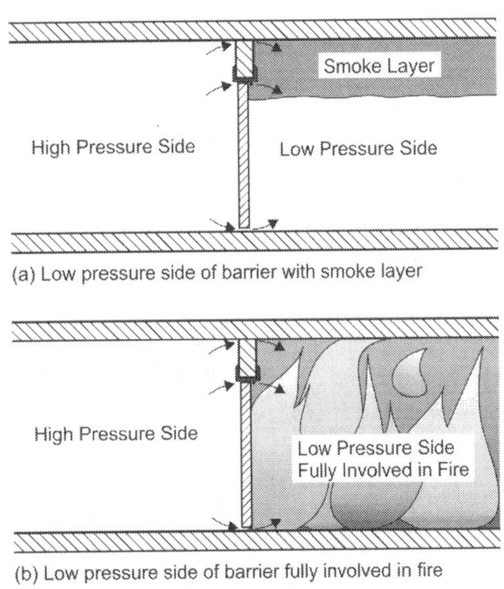

Figure 9.1 *Pressure difference across a barrier can control smoke flow.*

For pressurization smoke control systems, NFPA 92 (NFPA 2012a) uses the term *smoke barrier* in much the same way as the IBC except that the NFPA 92 smoke barrier is to be used in conjunction with a smoke control system. Users of NFPA 92 and the IBC should realize that these documents use the term *smoke barrier* to mean similar, but not the same things.

In this chapter, the term *barrier* is used to mean a barrier of a pressurization smoke control system. As with the passive smoke barriers discussed, this barrier is a continuous membrane, either vertical or horizontal, such as a wall, floor, or ceiling assembly. This barrier is designed and constructed to restrict the movement of smoke in conjunction with a pressurization smoke control system. This barrier needs to be well constructed and free from large openings. Examples of such large openings are open doorways, pass-through wall openings, and open staircases. These barriers are expected to have small openings such as the gaps around closed doors and small construction cracks in walls and floor-ceiling assemblies, but larger construction gaps and cracks are expected to be sealed. As with passive smoke barriers, the doors in barriers of a pressurization smoke control system need to close automatically upon smoke detection. Exterior doors are an exception in that they do not need to close automatically.

Opening and Closing Doors

During a building fire, doors in smoke barriers will be opened and closed as people move through the barriers of a smoke control system. For example, people on the fire floor move through smoke to a stairwell. When they enter the stairwell, some smoke moves with the people, but the stairwell almost certainly will remain tenable. The smoke on the fire floor is probably tenable or people would not be able to move through it, and a small amount of this smoke entering the stairwell cannot make conditions in the stairwell untenable. The failure mode related smoke barrier doors is when these doors are propped open. For stairwell pressurization systems, this failure mode is discussed in Chapter 10.

VALIDATION EXPERIMENTS

The concept of pressurization has been validated by many full-scale fire tests including tests with fully developed fires. The following is a description of some of these tests.

Henry Grady Hotel Tests

In 1973, a series of fire tests of smoke control systems were conducted in the Henry Grady Hotel by the Atlanta City Building Department (Koplon 1973a, 1973b). This hotel was scheduled for demolition. The purpose of the tests was to evaluate the effectiveness of the following smoke control systems: (1) stairwell pressurization without vestibules, (2) stairwell pressurization with vestibules, and (3) elevator hoistway (shaft) pressurization. The stairwell systems were intended to provide "smoke-free" egress, and the elevator system was intended to prevent smoke movement through the hoistway. The design of these smoke control systems was based on the latest papers and ideas about the topic.

Fire tests without vestibules were staged on Floor 3, and fire tests with vestibules were staged on Floor 5 as illustrated in Figure 9.2. The fires consisted of materials that would be expected to be in a hotel room such as a bed with mattress, wood chair, draperies, lamps, and chest of drawers.

Most of the instrumentation was installed and operated by the National Bureau of Standards[1] and Georgia Institute of Technology. Temperature, smoke obscuration, carbon monoxide concentration, and pressure difference were measured and reported for many locations. The impacts of these data on smoke control were discussed. The project demonstrated that pressurization could provide "smoke-free" exits for the fire scenarios and systems tested, and elevator pressurization could keep smoke from flowing through the hoistway.

30 Church Street Tests

At about the same time as the Henry Grady tests, the Brooklyn Polytechnic Institute[2] conducted a series of fire experiments at a 22-story office building at 30 Church Street in New York City (DeCicco 1973). This building was scheduled for demolition. These tests studied the effectiveness of stairwell pressurization systems. Four fires were set to test stairwell pressurization at the locations on the 7th and 10th floors shown on Figure 9.3.

The materials burned were representative of fuels that would be in an office such as desks, chairs, considerable paper on the desks, and boxes of paper. Temperatures and pressure differences were measured. Tests were also conducted to study the effect of open doors on the performance of the pressurized stairwell. As with the Henry Grady tests, this project demonstrated that pressurization could provide "smoke-free" exits for the fire scenarios and systems tested.

This project was different from the Henry Grady project in that a theoretical model of shaft airflow was developed, and airflow and pressure measurements were made in the building stair to evaluate that model. Further, a scale model of the stair shaft at the Church Street building was constructed by Cresci (1973), and a systematic series of experiments were conducted. Cresci describes visualization experiments where stationary vortices were observed at open doorways. These vortices are the reason that the flow coefficient through an open stairwell door is about half of what it would be otherwise.

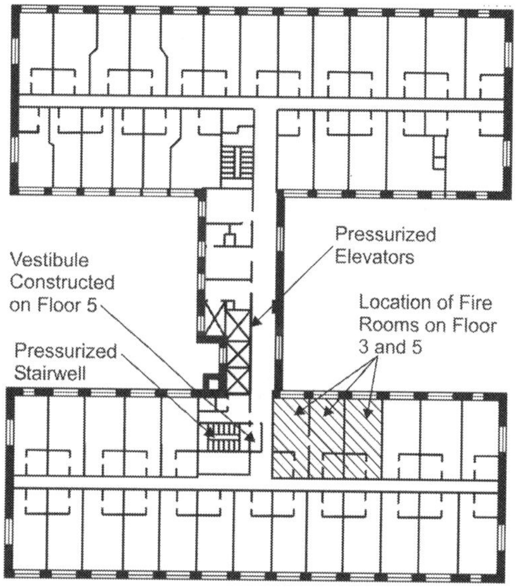

Figure 9.2 *Typical floor plan of the Henry Grady Hotel.*

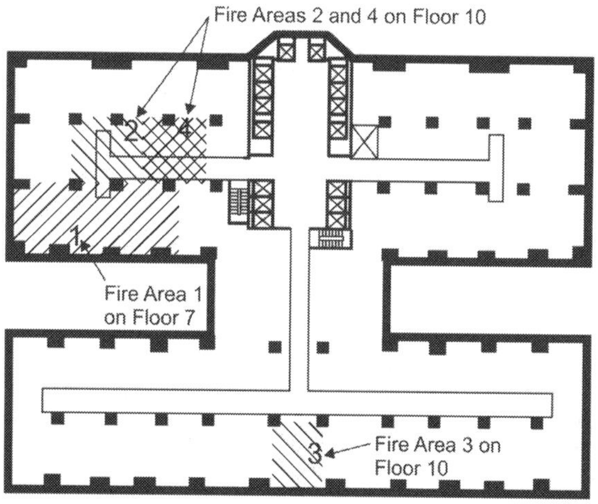

Figure 9.3 *Typical floor plan of the 30 Church Street office building.*

1. The National Bureau of Standards has been renamed the National Institute of Standards and Technology.
2. The Brooklyn Polytechnic Institute has been renamed the Polytechnic Institute of New York University.

Plaza Hotel Tests

In the Spring of 1989, the National Institute of Standards and Technology (NIST) conducted a series of experiments of stairwell pressurization and zoned smoke control at the Plaza Hotel Building in Washington DC (Klote 1990). This seven-story hotel was scheduled for demolition. Fires were set at two locations on the second floor as shown in Figure 9.4. An exhaust fan was installed to depressurize the fire floor, and other fans were installed to pressurize the stairwell and the floors above and below the fire floor. Stairwell pressurization was intended to keep the stairs "smoke-free," and the zoned smoke control was intended to keep the floors above and below the fire "smoke-free." These experiments demonstrated that these smoke control systems can maintain "smoke-free" conditions as intended.

Because the interaction of the smoke control system and the fire was of interest, solid fuels were burned. For each fire, either two or four 150 lb (68 kg) wood cribs were burned (Figure 9.4). A wood crib is a geometrically arranged pile of sticks that is often used in fire research as discussed in Chapter 5. Measurements were made of temperature, smoke obscuration, pressure difference, wind velocity, wind direction, and concentration of carbon monoxide, carbon dioxide, and oxygen.

An analysis based on first principles of engineering was made of the pressure differences produced by the smoke control system during the fires. This analysis supported the use of the hydrostatic equation for both fire modeling and smoke control system analysis.

The NRCC Experimental Fire Tower

In the mid-1980s, the National Research Council of Canada (NRCC) built the 10-story Experimental Fire Tower near Ottawa primarily for smoke control research. Figure 9.5 is a typical floor plan of the tower. The experimental tower is connected to a service tower that has space for researchers to observe experiments in safety. The core of the experimental tower consists of stairs, an elevator shaft, a lobby, and seven air shafts to allow simulation of many different stairwell and elevator smoke control systems. The shafts are connected through a tunnel to a remote building where supply fans are located. Exhaust fans are located on the roof of the experimental tower.

The experimental tower is instrumented for temperature, airflow, smoke obscuration, gas concentrations, and pressure differences at many locations on each floor. Gas burners have been used as the fire for most tests, because they are highly repeatable and easily controlled. To prevent dangers resulting from unburned gas, air is premixed with the gas before it reaches the burners. To prevent unwanted pressurization due to the premixed air, most experiments are conducted with openings in the exterior walls of the fire floor simulating broken fire floor windows.

The Experimental Fire Tower has been used for many smoke control projects. The first project to use this tower was a joint NIST/NRCC project to study the feasibility of elevator smoke control. Some people cannot use stairwells because of physical disabilities, and potential systems that use elevators for evacuation are discussed in Chapter 12. Some of the early full-scale fire projects in the Experimental Fire Tower were conducted by Tamura (1990a, 1990b) and Tamura and Klote (1987a, 1987b, 1988). These projects helped develop much of the foundation of modern smoke control technology, and smoke control research is still being done in this fire tower.

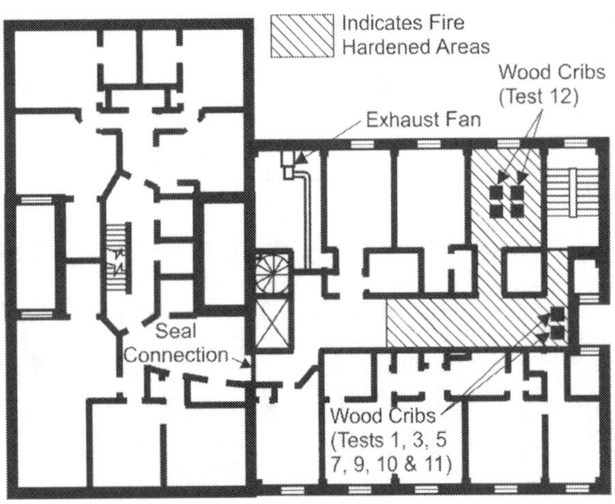

Figure 9.4 *Second floor plan of the Plaza Hotel.*

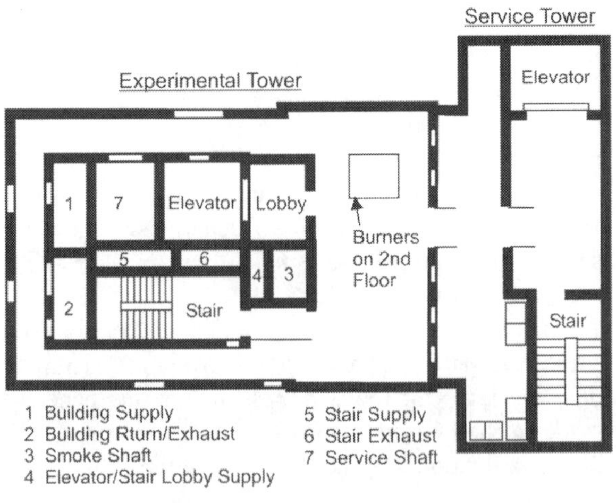

Figure 9.5 *Typical floor plan of the NRCC experimental fire tower.*

SMOKE FEEDBACK

To minimize the potential for smoke feedback from the exhausts or vents into supply air intakes, supply air intakes should be located away from the major openings from which smoke could leave a building. Supply air intakes need to be separated from all building exhausts, heat and smoke vents, open vents of elevator shafts, outlets from a zoned smoke control system, outlets of smoke shafts, and loading docks where large fires can occur inside trucks. For specific applications, smoke might flow from other openings during a building fire. If possible, such other openings should be considered, but it is not possible to anticipate all the possible openings where smoke could leave a building for all possible fires.

The separation between intakes and places where smoke can flow should be as great as practical. Because hot smoke rises, consideration should be given to locating supply air intakes below the openings mentioned. Wind can impact the direction of smoke movement, and this should be taken into account. An approach that sometimes can be used is to have the supply air intakes on one side of the building and the smoke outlets on the other side and on the roof.

WIND

The designers need to address the impact of wind on (1) smoke feedback, (2) fan performance, and (3) stairwell pressure differences. Smoke feedback is previously discussed.

The wind should not adversely impact the performance of a stairwell pressurization system to the extent that acceptable pressurization is not achieved for some wind conditions that can be anticipated. Acceptable pressurization consists of maintaining pressure differences across the stairwell doors that are between the minimum and maximum design values that are discussed below. The usual approach is to design systems that are insensitive to wind effects, but this can be difficult to do for tall, complicated buildings. Analysis of wind effects does not need to be done for such systems that are insensitive to wind effects.

Propeller fans are susceptible to wind forces, and roof-mounted propeller fans should have a top that shields the fan from wind effects. Wind can have a major impact on the flow of propeller fans located on exterior walls, and the wind can even reverse the direction of flow through wall-mounted propeller fans. For this reason, smoke control systems using wall-mounted propeller fans need to have a wind analysis to assure that the system can maintain acceptable pressurization.

Wind speed increases with height above the ground, and wind effects are more significant for tall buildings than they are for short ones. Wind analysis is needed for pressurization smoke control systems in tall buildings. Buildings with operable windows are more susceptible to wind effects than are buildings with windows that do not open. The same is true for buildings with balcony doors that open to building spaces. Wind analysis should be considered for most tall buildings with operable windows or such balconies.

A wind analysis can be done with (1) wind tunnel modeling, (2) network modeling, (3) computational fluid dynamic (CFD) modeling, or a combination of these methods. The methods that are appropriate for a wind analysis depend on the details of the specific application, including the kinds of smoke control systems involved.

DESIGN PRESSURE DIFFERENCES

Pressurization smoke control systems are designed to operate within a pressure difference range. This range is between the minimum design pressure difference and the maximum design pressure difference.

When smoke control systems are tested, both sides of the boundary of the smoke zone are the same temperature, and the pressure difference across the boundary is the same over the height of the boundary. This is shown in Figure 9.6a.

When one side of the boundary is subjected to hot fire gases, the temperature of the smoke has an impact on pressure difference across the boundary. This is illustrated in Figure 9.6b. In this figure, the pressure difference graph looks as it should for acceptable pressurization. If the average pressure difference were much lower, pressurization would be unsuccessful, as shown in Figure 9.6c.

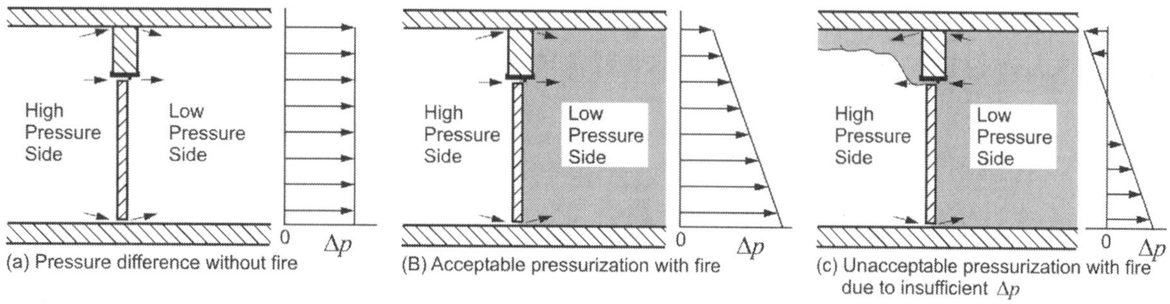

Figure 9.6 *Pressure differences at boundary of a smoke control system.*

Example 9.1. Minimum Design Pressure Difference

Part 1: For a ceiling height of 9 ft (2.74 m), what is the minimum design pressure difference with a fully developed fire?

$\Delta p_{SF} = 0.03$ in. H_2O; $T_o = 70\ °F + 460 = 530\ °R$; $T_F = 1700\ °F + 460 = 2160\ °R$

$$h = \frac{2}{3}(9) = 6 \text{ ft}$$

$$\Delta p_{min} = \Delta p_{SF} + 7.64h\left(\frac{1}{T_o} - \frac{1}{T_F}\right) = 0.03 + 7.64(6)\left(\frac{1}{530} - \frac{1}{2160}\right) = 0.10 \text{ in. } H_2O$$

Part 2: For a sprinklered fire with a smoke layer depth of 0.9 ft (0.27 m) and a floor-to-ceiling height of 9 ft (2.74), calculate the weighted average temperature T_F of the hot gas. The smoke layer temperature is $T_s = 1700$°F, and $T_o = 70$°F.

$$T_F = \frac{T_s d + T_o(H-d)}{H} = \frac{1700(0.9) + 70(9-0.9)}{9} = 233°\text{F}$$

Part 3: For a ceiling height of 9 ft (2.74 m), what is the minimum design pressure difference with a sprinklered fire?

Δp_{SF}, T_o and h are the same as in Part 1. From Part 2, $T_F = 233°F + 460 = 693°R$.

$$\Delta p_{min} = \Delta p_{SF} + 7.64h\left(\frac{1}{T_o} - \frac{1}{T_F}\right) = 0.03 + 7.64(6)\left(\frac{1}{530} - \frac{1}{693}\right) = 0.05 \text{ in. } H_2O$$

Minimum Pressure Difference

A minimum design pressure difference intended to prevent smoke migration across a barrier of a smoke control system is generally stipulated by the code. A smoke control system should be designed to maintain this minimum design pressure difference under likely conditions of stack effect and wind.

The analysis presented here is intended to provide insight into the level of smoke protection that can be anticipated by the values of minimum pressure difference in the code. The minimum design pressure difference can be calculated as a safety factor plus the buoyancy pressure difference of the fire.

$$\Delta p_{min} = \Delta p_{SF} + 7.64h\left(\frac{1}{T_o} - \frac{1}{T_F}\right) \tag{9.1}$$

$$\Delta p_{min} = \Delta p_{SF} + 3460h\left(\frac{1}{T_o} - \frac{1}{T_F}\right) \text{ for SI}$$

where

Δp_{min} = minimum design pressure difference, in. H_2O (Pa),

Δp_{SF} = pressure difference safety factor, in. H_2O (Pa),

h = distance above neutral plane, ft (m),

T_o = absolute temperature of surroundings, °R (K),

T_F = absolute temperature of hot gases, °R (K).

The neutral plane is a horizontal plane between the fire space and surrounding space at which the pressure difference between the fire space and the surrounding space is zero. For a pressurized barrier, there is no neutral plane, but for purposes of calculation a value of the distance above neutral plane, h, is arbitrarily chosen. The safety factor term Δp_{SF} is needed to account for pressure fluctuations due to wind, fan pulsations, and variations in barometric pressure.

For evaluating Equation 9.1, the following conservative values are suggested: (1) h is two thirds the floor to ceiling height, (2) $\Delta p_{SF} = 0.03$ in. H_2O (7.5 Pa), and (3) $T_F = 1700$°F (927°C) for fully developed fires. For a sprinklered fire, temperature of hot gases T_F is a weighted average value of the smoke layer temperature and the lower layer temperature.

$$T_F = \frac{T_s d + T_o(H-d)}{H} \tag{9.2}$$

where

T_F = weighted average temperature of hot gases, °F (°C).

T_s = temperature of the smoke layer, °F (°C),

T_o = temperature of surroundings, °F (°C),

d = depth of the smoke layer, ft (m),

H = floor to height, ft (m).

Example 9.1 illustrates how to calculate minimum pressure differences for sprinklered and unsprinklered buildings. Table 9.1 lists minimum pressure difference calculated from Equations 9.1 and 9.2, and these values are the same as those in NFPA 92.

Table 9.1: Suggested Minimum Pressure Design Difference[1]

Building Type[2]	Ceiling Height ft	m	Design Pressure Difference[3] in. H_2O	Pa
AS	Any	Any	0.05	12.4
NS	9	2.7	0.10	24.9
NS	15	4.6	0.14	34.8
NS	21	6.4	0.18	44.8

[1]For design purposes, a smoke control system should maintain these minimum pressure differences under likely conditions of stack effect or wind.
[2]AS for sprinklered and NS for nonsprinklered.
[3]The pressure difference measured between the smoke zone and adjacent spaces, while the affected areas are in the smoke control mode.

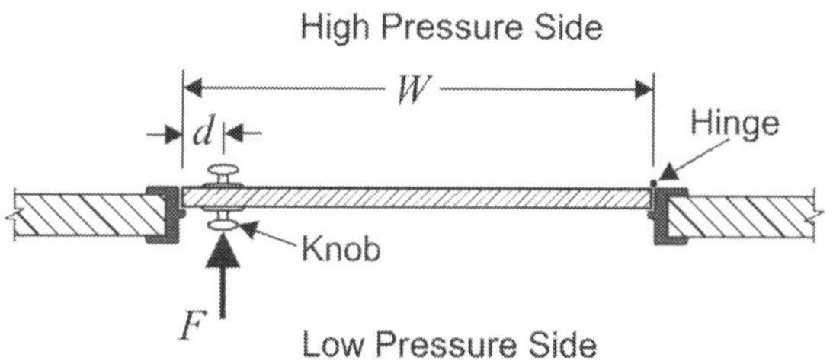

Figure 9.7 *Door-opening force for side-hinged door.*

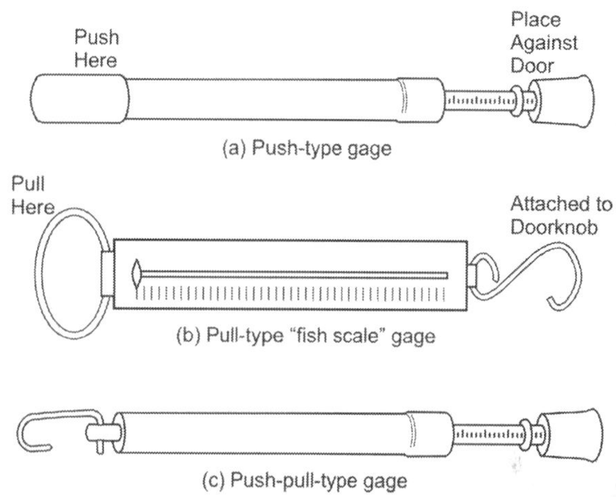

Figure 9.8 *Spring-loaded gages for measuring door-opening force.*

Maximum Pressure Difference

Side-Hinged Swinging Doors

The pressure difference across a barrier must not result in door-opening forces that exceed the maximum values stipulated in codes. For example, in NFPA 101 (NFPA 2012b) this maximum force is 30 lb (133 N).

The force to open a side-hinged swinging door is shown in Figure 9.7. The force required to open such a door when the smoke control system is operating is

$$F = F_{dc} + \frac{5.2 WA\Delta p}{2(W-d)}$$
$$F = F_{dc} + \frac{WA\Delta p}{2(W-d)} \text{ for SI} \tag{9.3}$$

and this equation is written for the pressure difference as

$$\Delta p = \frac{2(W-d)(F-F_{dc})}{5.2WA}$$
$$\Delta p = \frac{2(W-d)(F-F_{dc})}{WA} \text{ for SI} \tag{9.4}$$

where:

A = door area, ft^2 (m^2),
d = distance from doorknob to knob side of door, ft (m),
F = total door-opening force, lb (N),
F_{dc} = door closer force, lb (N),
W = door width, ft (m),
Δp = pressure difference, in. H_2O (Pa).

Equations 9.3 and 9.4 are applicable for doors that do not stick to the frame and have properly lubricated hinges. The total door-opening force, F, can be measured with a door-opening force gage. Typically these gages are spring loaded, and they can be the push-type, pull-type, or the push-pull-type as shown in Figure 9.8. Regardless of the type of gage used, the door latch needs to be disengaged before measuring the door-opening force.

The door-closer force is the force required to overcome the closing device. The door-closer force is not constant over the swing of the door, and it can increase as the door opens. When a door of a smoke control system is opened, the pressure drops when the door is just slightly cracked open. For this reason, the door-closer force F_{dc} in Equations 9.3 and 9.4 is the force needed to initially get the door just cracked open.

Table 9.2: Maximum Pressure Difference (in. H_2O) across Doors with 30 lb Door-Opening Force

Door Closer Force (lb)	Door Width				
	32 in.	36 in.	40 in.	44 in.	48 in.
6	0.45	0.40	0.37	0.34	0.31
7	0.43	0.39	0.35	0.32	0.30
8	0.41	0.37	0.34	0.31	0.28
9	0.39	0.35	0.32	0.29	0.27
10	0.37	0.34	0.30	0.28	0.26
11	0.35	0.32	0.29	0.27	0.24
12	0.34	0.30	0.27	0.25	0.23
13	0.32	0.29	0.26	0.24	0.22
14	0.30	0.27	0.24	0.22	0.21

Note: The door height is 7 ft, and the distance from the doorknob to the knob side of the door is 3 in.

An easy way to determine the force F_{dc} is to use a spring-loaded gage measure the total door-opening force when there is no pressure difference across the door. From Equation 9.3, it can be seen that when the pressure difference across the door is zero, the total door-opening force F is the same as the force F_{dc} required to overcome the closing device.

For a door-opening force of 30 lb (133 N), Tables 9.2 and 9.3 list the maximum pressure differences calculated from Equation 9.4 for a range of door-closer forces. Example 9.2 calculates the door-opening force for a door with a pressure difference across it.

Elevator Doors

For pressurized elevator systems, the maximum pressure difference across elevator doors is based on concern about elevator doors jamming in the closed position. While not supported by research, the following discussion supports the idea that jammed doors may require only modest force to open.

John Klote has conducted considerable research with elevator smoke control systems, including research on elevators with pressurized shafts. In this research, Klote encountered no elevators with doors that jammed shut. Before this research, Klote encountered elevator doors jammed shut on an elevator in normal service. The elevator car had smooth metal center-opening doors.Placing the palms of his hands flat on the doors, relying only on the friction of his hands, Klote easily opened the doors.

To prevent injury to the passengers of elevators with automatic doors, the Elevator Code (ASME 2010) restricts door-closing forces and speed. It is customary for elevator mechanics to adjust the elevator door mechanisms with the seasons of the year so that elevator doors will open and close without jamming when subjected to different pressure differences caused by stack effect. In fire situations, the elevators are put into Phase II operation, which is sometimes called firefighter's service. During Phase II, the elevators are only used by firefighters who are equipped with various tools and are more than capable of opening a door that has been jammed shut.

There has been no research about the maximum design pressure difference for elevator pressurization systems, but the 2012 IBC prescribes a maximum pressure difference of 0.25 in. H_2O (62.2 Pa). For doors that are only to be used by firefighters, this maximum pressure difference is probably conservative.

ANALYSIS APPROACH FOR PRESSURIZATION SYSTEMS

The purposes of analysis of a pressurization smoke control system are to: (1) determine if a particular smoke control system in a particular building is capable of being balanced such that it will perform as intended, and (2) size the fans for the system. For some simple systems, some designers may know from experience that they will work as intended, and they can be sized by simple calculations.

Many smoke control systems need analysis for the first purpose above. This is especially true for buildings that have a number of smoke control systems. For such complex applications, analysis needs to be done with a network model such as CONTAM, which is discussed in Chapter 14.

Analysis of pressurization smoke control systems must take into account all of the smoke control systems in the building operating together as they would during a building fire. This is because the pressurization smoke control systems in a building interact with each other. Air flows from a pressurization system into the building, where it encounters the air flowing from the other pressurization systems, and all of this air has to flow through various flow paths in the building to the outdoors.

Table 9.3: Maximum Pressure Difference (Pa) across Doors with 133 N Door-Opening Force

Door Closer Force (N)	Door Width				
	0.81 m	0.91 m	1.02 m	1.12 m	1.22 m
25	113	102	92	84	78
30	108	97	88	80	74
35	103	93	83	77	71
40	98	88	79	73	67
45	92	83	75	69	64
50	87	78	71	65	60
55	82	74	66	61	56
60	77	69	62	57	53
65	71	64	58	53	49

Note: The door height is 2.13 m, and the distance from the doorknob to the knob side of the door is 0.76 mm.

Example 9.2. Door-Opening Force

What is the door-opening force for a side hinged swinging door 3 ft (9.1 m) wide by 7 ft (2.13 m) high with a door closer that requires 9 lb (40 N) of force and a pressure difference across it of 0.35 in. H_2O (87 Pa)? The knob is 3 in. (0.25 ft) from the edge of the door.

$$W = 3 \text{ ft};\ F_{dc} = 9 \text{ lb};\ A = 3 \times 7 = 21 \text{ ft}^2;\ d = 0.25 \text{ ft};\ \Delta p = 0.35 \text{ in. } H_2O$$

The door-opening force is: $F_{dc} + \dfrac{5.2WA\Delta p}{2(W-d)} = 9 + \dfrac{5.2(3)(21)(0.35)}{2(3-0.25)} = 30 \text{ lb } (133 \text{ N})$

NOMENCLATURE

A = door area, ft^2 (m^2)

d = depth of the smoke layer, or distance from doorknob to knob side of door, ft (m)

F = total door-opening force, lb (N)

F_{dc} = door closer force, lb (N)

H = floor to height, ft (m)

h = distance above neutral plane, ft (m)

T_F = weighted average temperature of hot gases, °F (°C); or absolute temperature of hot gases, °R (K)

T_o = temperature of surroundings, °F (°C); or absolute temperature of surroundings, °R (K)

T_s = temperature of the smoke layer, °F (°C)

W = door width, ft (m)

Δp = pressure difference, in. H_2O (Pa)

Δp_{min} = minimum design pressure difference, in. H_2O (Pa)

Δp_{SF} = pressure difference safety factor, in. H_2O (Pa)

REFERENCES

ASME. 2010. ASME A17.1, *Safety Code for Elevators and Escalators*. New York: American Society of Mechanical Engineers.

Barnett, J.R. 1991. New design approach for steel structures exposed to fires. *Journal of Fire Protection Engineering* 3(1).

Boring, D.F. 1990. Primer/fireproofing compatibility. *Building Standards* 59(5).

Boring, D.F., J.C. Spence, W.G. Wells. 1981. *Fire Protection Through Modern Building Codes*, 5th ed. New York: American Iron and Steel Institute.

Bushev, et al. 1978. *Fire Resistance of Buildings*, 2nd ed, revised and supplemented, translated from Russian. New Delhi: Amerind Publishing Company Pvt. Ltd.

Cresci, R.J. 1973. Smoke and fire control in high-rise office buildings—Part II: analysis of stair pressurization systems. Symposium on Experience and Applications on Smoke and Fire Control, ASHRAE Annual Meeting, June, Louisville.

DeCicco, P.R. 1973. Smoke and fire control in high-rise office buildings—Part I: full-scale tests for establishing standards. Symposium on Experience and Applications on Smoke and Fire Control, ASHRAE Annual Meeting, June, Louisville.

ICC. 2012. *International Building Code*® (*IBC*®). International Code Council, Country Club Hills, IL.

Klote, J.H. 1990. Fire experiments of zoned smoke control at the Plaza Hotel in Washington, DC. *ASHRAE Transactions* 96(2).

Klote, J.H., and D.H. Evans. 2007. *A Guide to Smoke Control in the 2006 IBC*. Country Club Hills, IL: International Code Council.

Koplon, N.A. 1973a. Report of the Henry Grady fire tests. City of Atlanta Building Department, Atlanta.

Koplon, N.A. 1973b. A partial report of the Henry Grady fire tests (Atlanta GA— July 1972). Symposium on Experience and Applications on Smoke and Fire Control, ASHRAE Annual Meeting, June, Louisville.

NFPA. 2012a. NFPA 92, *Standard for Smoke Control Systems*. Quincy, MA: National Fire Protection Association.

NFPA. 2012b. NFPA 101, *Life Safety Code*. Quincy, MA: National Fire Protection Association.

Tamura, G.T. 1990a. Fire tower tests of stair pressurization systems with overpressure relief. *ASHRAE Transactions* 96(2).

Tamura, G.T. 1990b. Fire tower tests of stair pressurization systems with mechanical venting of the fire floor. *ASHRAE Transactions* 96(2).

Tamura, G.T., and J.H. Klote. 1987a. Experimental fire tower studies on elevator pressurization systems for smoke control. *ASHRAE Transactions* 93(2).

Tamura, G.T., and J.H. Klote. 1987b. Experimental fire tower studies on mechanical pressurization to control smoke movement caused by fire pressures. Proceedings of the 2nd International Symposium on Fire Safety Science, Tokyo, Japan.

Tamura, G.T., and J.H. Klote. 1988. Experimental fire tower studies on adverse pressures caused by stack and wind action: studies on smoke movement and control. ASTM International Symposium on Characterization and Toxicity of Smoke, December 5, Phoenix, AZ.

CHAPTER 10

Pressurized Stairwells

John H. Klote

Analysis of pressurized stairwell systems can be done with algebraic equations or with a network model such as CONTAM. CONTAM is so extensively used for analyses of pressurization smoke control systems that it has become the de facto standard. In this chapter when CONTAM is discussed, much of that discussion could apply to other network models. For more information about network modeling and CONTAM, see Chapter 14.

DESIGN AND ANALYSIS

The factors involved with design and analysis of stairwell pressurization systems are (1) building height, (2) stairwell height, (3) floor plans, (4) flow areas of building components, (5) minimum design pressure difference, (6) maximum design pressure difference, (7) atmospheric pressure, (8) building temperature, (9) outdoor temperature, (10) stairwell temperature, (11) type of stairwell pressurization system, (12) wind effects, and (13) smoke feedback. There is some general information about minimum and maximum design pressure differences, wind effects, and smoke feedback in Chapter 9. For summer and winter outdoor design temperatures, atmospheric pressure, and design wind speed, see Chapter 2.

Purposes of CONTAM: The purposes of CONTAM analysis of pressurized stairwells are (1) to determine if the kind of stairwell pressurization system in a particular building is capable of being balanced to perform as intended and (2) to size the fans for the system.

For buildings with pressurized stairwells and other pressurization smoke control systems, analysis of these systems should be done considering all of the pressurization systems operating together as they would during a fire. This is because the pressurization smoke control systems in a building interact with each other. Air flows from a pressurization system into the building where it encounters the air flowing from the other pressurization systems, and all of this air has to flow through various flow paths in the building to the outdoors. These flows can be very complex. For buildings with multiple pressurization smoke control systems, analysis with CONTAM is recommended.

Simple Systems in Simple Buildings

For simple systems in simple buildings, some designers may know from experience that the pressurized stairwell will work as intended, and the fans can be sized by simple calculations. A simple stairwell pressurization system is one that (1) has air supplied to at a constant (or nearly so) volumetric flow rate, (2) is intended to maintain acceptable pressurization with all the doors closed, and (3) has no features to prevent loss of pressure when stair doors are opened. Acceptable pressurization consists of maintaining pressure differences across the stairwell doors that are between the minimum and maximum design values as discussed in Chapter 9. As discussed later, a compensated stairwell system has features intended to prevent loss when stair doors are opened, and compensated stairwell systems are not simple.

Figure 10.1 is an example of a simple building. The algebraic equations in this chapter can be used to size the supply fans. Some engineers have developed their own rules of thumb that are appropriate for certain kinds of stairwell pressurization systems in some buildings.

Systems in Complicated Buildings

For buildings that are relatively complicated, CONTAM analysis of the pressurized stairwells is often needed to determine if the stairwell systems are capable of being balanced to perform as intended. For stairwells pressurized with untreated air, building complexity often is more important than stack effect. Even buildings that are not especially tall are in this category when the floors plans vary significantly from floor to floor.

The building of Figure 10.2a consists of underground parking levels, condominium floors, and a penthouse. It can be difficult to maintain acceptable pressurization of stairwells that extend from the parking levels to the penthouse, because the plans of these floors are so different. Wind effects for this building are complicated, because the condominium units have operable windows and balconies open to the outdoors. For this building, analysis of the pressurized stairwell systems with CONTAM is needed.

The building of Figure 10.2b is another complicated building. Wind effects may not be so important for this building, because of lack of operable windows and balconies open to the outdoors. However, the floors with the general hotel spaces add considerable complexity to this building. For the building of Figure 10.2b, analysis of the pressurized stairwell systems with CONTAM is suggested.

When a CONTAM analysis shows that a stairwell pressurization system in a particular building cannot be balanced to perform as intended, a new approach is needed. There are two categories of new approaches: (1) use an alternate stairwell pressurization system, and (2) modify the building. Various stairwell pressurization systems are discussed later. Modifications to the building can consist of a number of simple things. The new approach needs a CONTAM analysis to determine if it is capable of being balanced to perform as intended.

For example, during windy conditions, the pressure difference across one stairwell door could be too high, and a new approach that could be considered is use of an automatically opening door that has the high pressure difference. Another new approach could be an automatically opening vent at that location. Any such vent in a fire-rated wall should have a fire damper.

Another example is a stairwell pressurization system in a complicated building such that acceptable pressurization is not possible for a portion of the stairwell. Possible systems that could be used to deal with this problem are (1) stairwell pressurization with fire floor exhaust, and (2) stairwell pressurization with stairwell compartmentation. These systems are discussed later. Separating the stairs into two sections by using a transfer corridor might be sufficient.

Stairwells in tall and supertall buildings can be especially challenging, and CONTAM analysis is necessary. For information about tall and supertall buildings, see Webb (2011) and Tubbs et al. (2011).

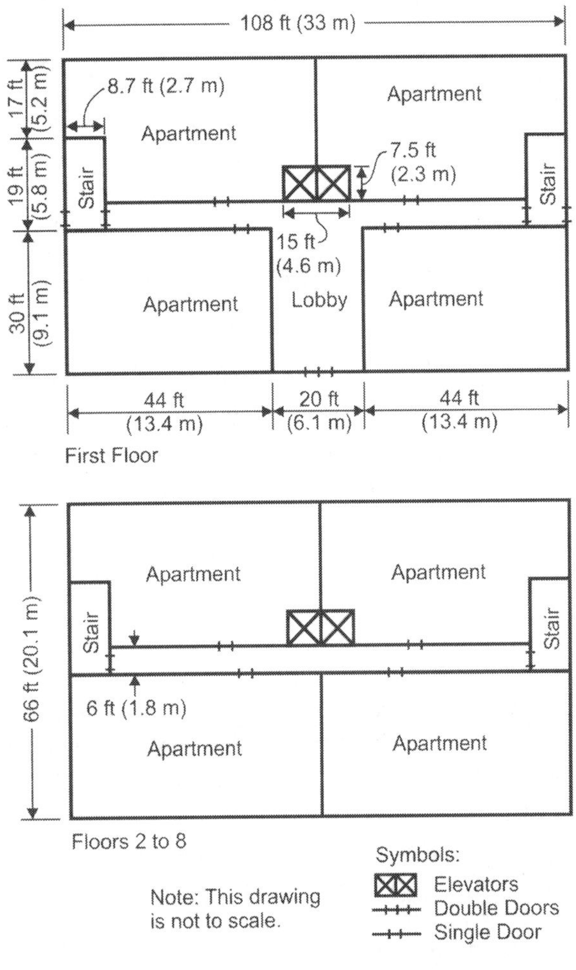

Figure 10.1 *Simple eight-story apartment building.*

PRESSURIZATION SYSTEMS

The number of possible stairwell pressurization systems is very large, and this section is a discussion of some of these systems. Stairwell ventilation is a form of stairwell smoke control that does not rely on pressurization, and it has the potential to provide smoke protection for very tall stairwells in complex buildings. Computational fluid dynamics (CFD) simulations of stairwell ventilation systems are discussed in Chapter 20.

Single and Multiple Injection

A single injection system has pressurization air supplied at one location. Air can be supplied at the top of the stairwell, the bottom, or at a location in between. Figures 10.3(a) and (b) are top injection systems, and Figure 10.3c is a bottom injection system. As explained in Chapter 9, roof-mounted propeller fans should have tops that shield the fan from wind effects, and wall-mounted propeller fans should not be used because they can be adversely impacted by the wind unless a wind analysis indicates otherwise.

With a bottom injection system (Figure 10.3c), some of the supply air can short circuit the system by flowing directly out the opened exterior doorway reducing system effectiveness. Simulation of such detailed fluid flow is beyond the capability of CONTAM, but it can be simulated with CFD. It is recommended that bottom injection systems be analyzed using CFD to determine the extent to which supply air flows out of an open exterior door. Alternatively, the air can be introduced into the stairwell at least one floor above or below the exterior doors.

For tall stairwells, single injection systems can fail when a few doors near the air injection point are open. Much of the pressurization air can be lost through these open doors, and the system will then fail to maintain positive pressures across doors further from the injection

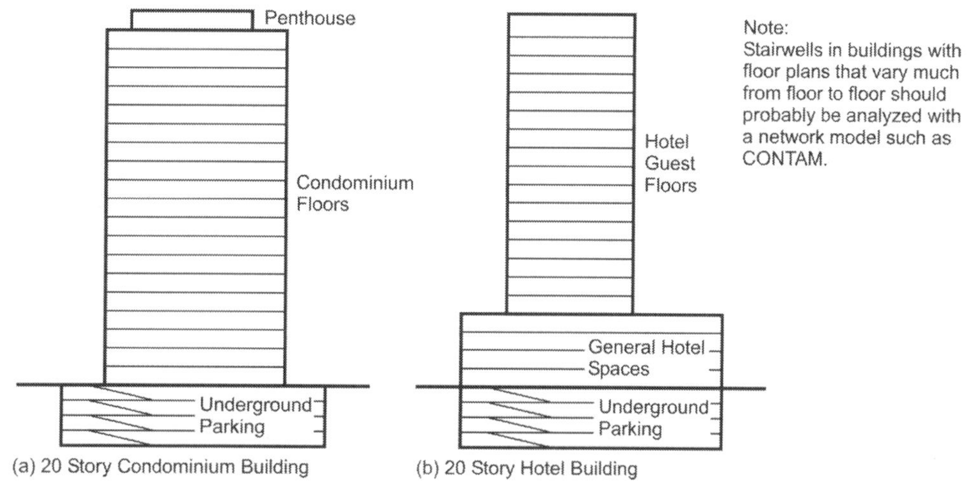

Figure 10.2 *Examples of buildings with floors that vary significantly from floor to floor.*

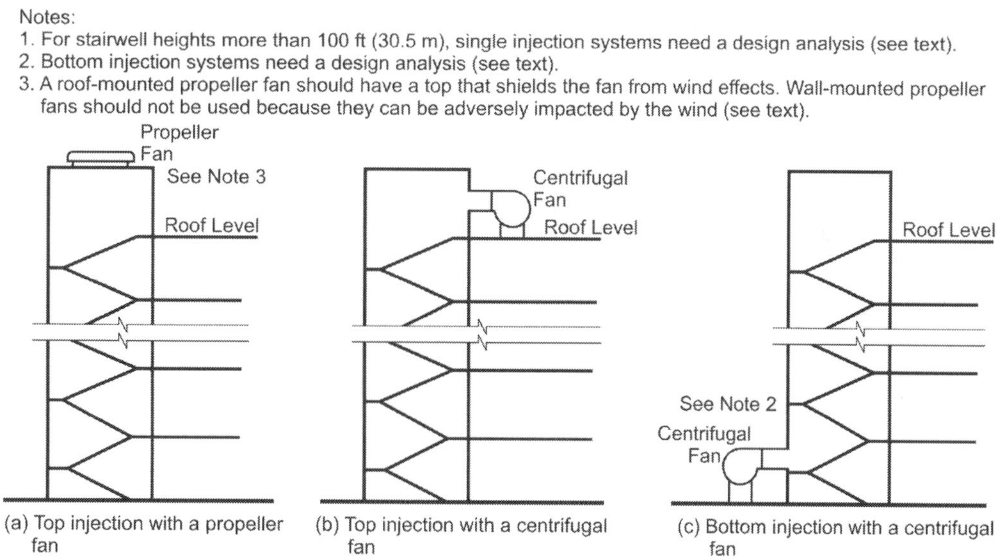

Figure 10.3 *Some arrangements of single-injection systems.*

point. To reduce the potential for such failure, multiple injection systems are to be used. Multiple injection systems can consist of one fan supplying air through a duct located in a shaft, and Figures 10.4a and 10.4b are examples of such ducted systems. Multiple injection systems can eliminate the shaft and much of the duct by using more than one fan as shown in Figure 10.4c.

There has been no research on this subject, but the consensus is that single injection systems for stairwell heights more than 100 ft (30.5 m) need a design analysis using CONTAM. For multiple injection systems supplying air through a duct in a shaft, injection points are usually one to three floors apart. Multiple injection systems that have a separate fan at each injection point can have injection points much further apart. For systems with two injection points, one at the top and another near the bottom, a CONTAM analysis is recommended for stairwell heights more than 200 ft (61 m).

Compartmentation

Stairwell compartmentation is not often used, but it is included here for completeness. This system consists of dividing a stairwell into a number of sections as shown in Figure 10.5, and each compartment has at least one supply air injection point. The compartments are separated by walls with normally closed doors. The main advantage of compartmentation is that it allows acceptable pressurization of stairwells that are otherwise too tall for acceptable pressurization. A disadvantage is the increase in floor area needed for the walls and doors that separate the stairwell sections.

When the doors between compartments are open, the effect of compartmentation is lost. For this reason, compartmentation is inappropriate for densely populated buildings, where total building evacuation by the stairwell is planned in the event of a fire. Stairwell compartmentation may be practical for (1) buildings that are not densely populated or (2) buildings where total building evacuation is not planned during fire situations. It is suggested that an analysis of people movement be performed for a stairwell compartmentation system to determine the extent to which doors between compartments would be open during a fire situation, and that the system be designed accordingly.

Vestibules

Pressurized stairwells with vestibules are occasionally used. The vestibules can be (1) unpressurized, (2) pressurized, (3) ventilated, or (4) a combination of pressurized and ventilated. Vestibules provide an additional barrier around a stairwell, and vestibules have the potential to reduce the probability of an open-door connection existing between the stairwell and the building.

An evacuation analysis can be performed to determine the extent to which both vestibule doors are likely to be opened simultaneously. For densely populated buildings, it is expected that on many floors both vestibule doors would be opened simultaneously. Therefore, vestibules may provide little benefit of an extra barrier for densely populated buildings.

The algebraic equation method of analysis that is discussed later can be extended to a pressurized stairwell with an unpressurized vestibule as discussed later. The pressure differences and flows of stairwell systems with any kind of vestibules can be analyzed by CONTAM. It is possible to evaluate the benefits of ventilated vestibules using tenability analysis. For information about tenability analysis, see Chapters 19 and 20.

System with Fire Floor Exhaust

This system can achieve acceptable pressurization of tall stairwells in very complex buildings. A relatively

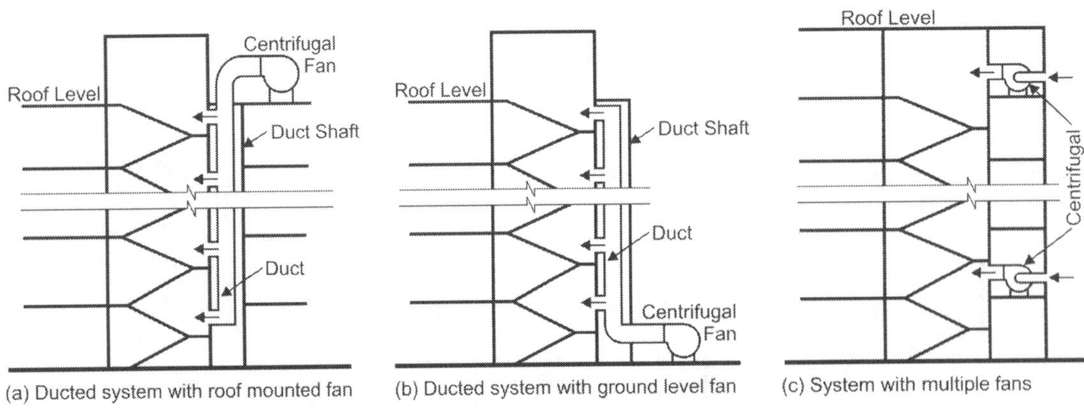

Figure 10.4 *Some arrangements of multiple injection systems.*

small amount of air is supplied to the stairs, and the fire floor is exhausted such that acceptable pressurization is maintained on the fire floor where it is needed. It is common to also exhaust one or two floors above and below the fire floor. Fire floor exhaust is a form of zoned smoke control, and stairwell pressurization with fire floor exhaust is discussed further in Chapter 13.

STAIRWELL TEMPERATURE

The temperature of the supply air to stairwells has an impact on the performance of pressurized stairwells. Much that has been written about stairwell pressurization was based on the tacit assumption that the supply air would be treated to nearly the same conditions as the building air. In the early days of smoke control, this idea was based on concern about cold air blowing on building occupants using the stairwell. Today, the supply air for most stairwells in North America is not treated so that pressurized stairwells are hot in the summer and cold in the winter.

In many applications, the use of untreated supply air can be justified for the following reasons: (1) fire drills are usually held in the spring or fall when the outdoor temperature usually is mild, and (2) during a fire emergency, being exposed to nearly outdoor temperatures seems reasonable considering occupants are often traveling to the safety of the outdoors.

When pressurization air is untreated in cold climates, there is a concern about the water freezing in sprinkler standpipes in stairwells. To prevent such freezing, the standpipe can be heat traced, and both the standpipe and the heat trace should be insulated. The heat trace can be electric resistance, hot-water pipe, or steam pipe. Alternately, pressurization air can be treated to a minimum temperature in the range of 45°F to 50°F (7°C to 10°C) to prevent such freezing. Using the heat trace and untreated pressurization air has the advantage of minimizing the impact of stack effect, but using air treated to a minimum temperature has the advantage of minimizing the potential of freezing water on the stairwell floor during firefighting.

Untreated Pressurization Air

As stated, the use of untreated pressurization air has the benefit on minimizing the adverse impact of stack effect on stairwell pressurization. For purposes of this chapter, stack effect is taken to mean any impact resulting from the temperature difference between the stairwell or another shaft and the outdoors.

The stairwell temperature can be expressed as

$$T_S = T_O + \eta(T_B - T_O) \qquad (10.1)$$

where

T_S = temperature in the stairwell, °F (°C),
T_O = temperature outdoors, °F (°C),
T_B = temperature in the building, °F (°C),
η = heat transfer factor (dimensionless).

There has been little research on the heat transfer factor. For supply air that is conditioned to the building temperature, the heat transfer factor is one.

For untreated supply air, it is believed that the heat transfer factor is in the range of 0.05 to 0.15. In the absence of better data for a specific application, a heat transfer factor of 0.15 is suggested as being conservative regarding the impact of stack effect.

For untreated supply air, it takes a few minutes for the temperature in the stairwell to stabilize near that of the outdoors. During this stabilization, excessive pressure differences could be produced. To prevent this, supply air can gradually be increased so that when the stairwell temperature is near that of the building there is insufficient flow to cause excessive pressurization. If needed, the temperature stabilization can be evaluated by a heat transfer analysis.

ANALYSIS BY ALGEBRAIC EQUATIONS

The algebraic equation method of analysis is for an idealized building, and this method is appropriate for sizing pressurization fans and to gain conservative information about the feasibility of achieving acceptable pressurization. The idealized building (1) has no vertical leakage through the floors and shafts, and (2) has leakage that is the same from floor to floor. The algebraic equation method applies to stairwells when all the stair doors are closed. Stairwells with open doors are discussed later in this chapter.

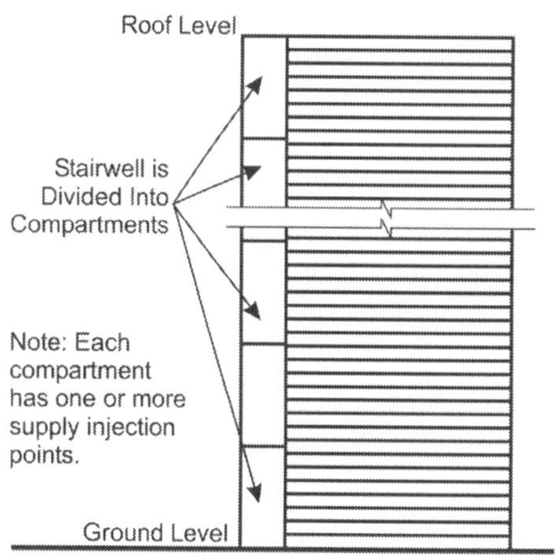

Figure 10.5 *Compartmentation of a stairwell.*

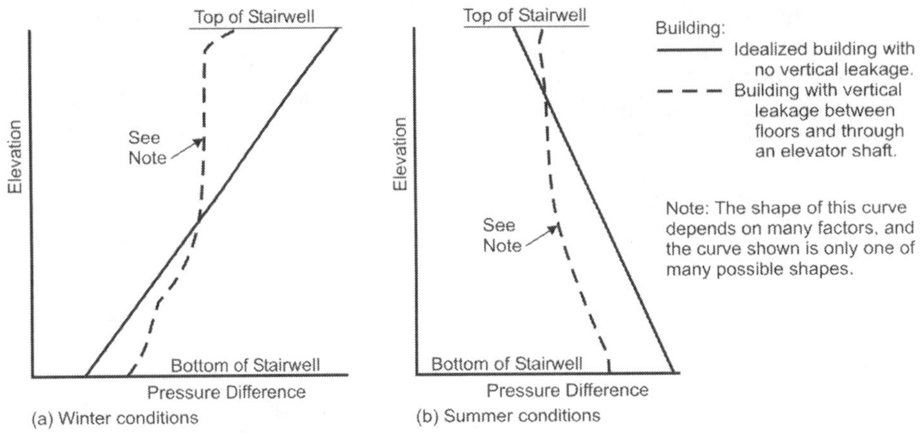

Figure 10.6 *Pressure profiles for pressurized stairwells in buildings with different leakage characteristics.*

Pressure differences across a stairwell tend to vary over the height of the stairwell. Figure 10.6 shows pressure profiles for pressurized stairwells in an idealized building and in a more realistic building with vertical leakage through floors and an elevator shaft. When it is cold outdoors, the pressure differences tend to be smaller at the bottom of the stairwell than at the top, as can be seen in Figure 10.6a. When it is hot outdoors, the trend is the opposite, as can be seen in Figure 10.6b. For both winter and summer conditions, it can be seen on this figure that the pressure profile for the idealized building is a straight line.

The pressure profiles of stairs in real buildings depends on many factors including: (1) the leakage values of the building components, (2) the building floor plans, (3) the size of the elevator shaft or shafts and the number of elevator doors, (4) the presence or absence of elevator vents, and (5) the leakage through other shafts. There are many possible shapes for such pressure profiles in real buildings.

For a building with vertical leakage, the flows through the floors and shafts act to even out to some extent the highest and lowest pressure differences across the stairwell. The profile for a building with vertical leakage is bounded by the extremes of the pressure profile of the idealized building. This means that other things being equal, the smallest pressure difference of the idealized analysis will be less than that of the realistic building, and that the largest pressure difference of the idealized analysis will be more than that of the realistic building. This is the reason that the algebraic equations discussed next are conservative.

Strictly speaking, the method applies to a building with only one stairwell, but it is extended to any number of similar stairwells by the concept of symmetry. This symmetry is accounted for by dividing the flow area from the building to the outdoors by the number of stairwells as discussed later.

Pressure Differences

The pressure difference at elevation y from the stair to the building is

$$\Delta p_{SBy} = \Delta p_{SBb} + \frac{yF_T}{F_R} \tag{10.2}$$

where

Δp_{SBy} = pressure difference from stairwell to building at y, in. H_2O (Pa),

Δp_{SBb} = pressure difference from stairwell to building at bottom, in. H_2O (Pa),

y = elevation above bottom of stairs, ft (m),

F_T = temperature factor, in. H_2O/ft (Pa/m),

F_R = flow area factor, dimensionless (see Equation 10.6.

The top pressure difference, Δp_{SBt}, from the stair to the building is

$$\Delta p_{SBt} = \Delta p_{SBb} + \frac{HF_T}{F_R} \tag{10.3}$$

where

H = stairwell height, ft (m).

The stairwell height, H, is the distance from the bottom to the top of the stairwell. The temperature factor is

$$F_T = \frac{0.798 g P_{atm}}{R}\left(\frac{1}{T_O} - \frac{1}{T_S}\right)$$

$$F_T = \frac{g P_{atm}}{R}\left(\frac{1}{T_O} - \frac{1}{T_S}\right) \text{ for SI} \tag{10.4}$$

where

g = acceleration of gravity, ft/s^2 (m/s^2),

p_{atm} = atmospheric pressure, psi (Pa),

Example 10.1. Flow Area A_{SB} with a Vestibule

For a stairwell with an unpressurized vestibule, calculate A_{SB}. The building is shown in Figure 10.7a. Use the flow areas listed in Table 10.1 with the high leakage values of single doors and stairwell walls. The floor-to-floor height is 10 ft (3.05 m). This calculation uses effective flow areas, which are discussed in Chapter 3.

The areas A_1, A_2, and A_3 are shown on Figure 10.7b. For this building, $A_{SB} = N(A_1 + A_{23e})$, where N is the number of floors, and A_{23e} is the effective flow area of A_2 and A_3. For this building, there are 14 floors. These areas are calculated below.

Area	Wall Area, ft^2	Wall Leakage, ft^2	Number of Doors	Door Leakage, ft^2	Total Flow Area, ft^2
A_1	10(19)	$190(3.5\times10^{-4})$	0	0	0.067
A_2	10(8.7)	$87(3.5\times10^{-4})$	1	0.24	0.270
A_3	10(8.7+7.9)	$166(3.5\times10^{-4})$	1	0.24	0.298

Areas A_2 and A_3 are in series, and the effective flow area is calculated as

$$A_{23e} = \left(\frac{1}{A_2^2} + \frac{1}{A_3^2}\right)^{-1/2} = \left(\frac{1}{0.27^2} + \frac{1}{0.298^2}\right)^{-1/2} = 0.20 \text{ ft}^2$$

$$A_{SB} = N(A_1 + A_{23e}) = 14(0.067 + 0.20) = 3.74 \text{ ft}^2 \ (0.347 \text{ m}^2)$$

R = gas constant, 53.34 ft·lbf/lbm·°R (287 J/kg·K).

T_O = absolute temperature outdoor, °R (K),

T_S = absolute temperature in stairwell, °R (K).

For sea level, Equation 10.4 becomes

$$F_T = 7.08\left(\frac{1}{T_O} - \frac{1}{T_S}\right) \qquad (10.5)$$

$$F_T = 3200\left(\frac{1}{T_O} - \frac{1}{T_S}\right) \text{ for SI.}$$

The flow area factor is

$$F_R = 1 + \frac{A_{SB}^2 T_B}{A_{BO}^2 T_S} \qquad (10.6)$$

where

A_{SB} = flow area between the stairwell and the building, ft^2 (m^2),

A_{BO} = flow area per stairwell between the building and the outdoors, ft^2 (m^2),

T_S = absolute temperature in stairwell, °R (K),

T_B = absolute temperature in building, °R (K).

The area, A_{SB}, is the total flow area between the stairwell and the building which would typically include the gaps around all the closed doors and the leakage paths in the walls. For a stairwell with an unpressurized vestibule, A_{SB} is the sum of the effective flow areas for all floors from the stairwell to the building. Example 10.1 is of the flow area A_{SB} with a vestibule as shown in Figure 10.7.

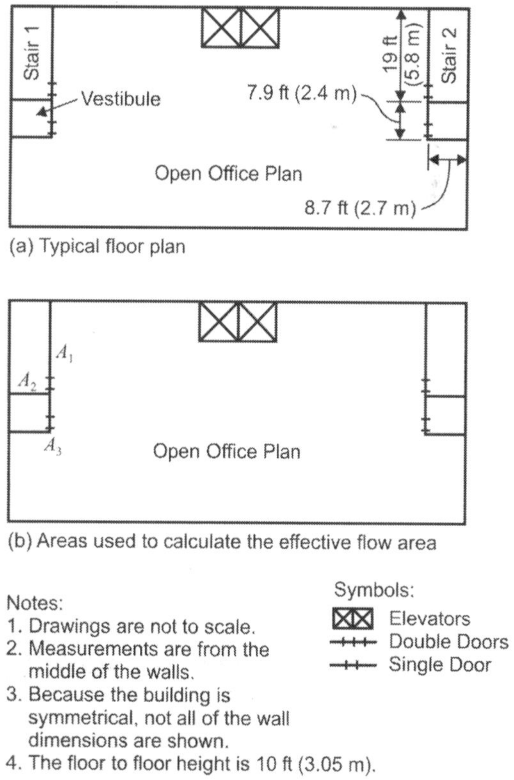

Figure 10.7 *Building of Example 10.1.*

As already mentioned, the area, A_{BO}, is on a per stairwell basis because of symmetry considerations. For a building with an open floor plan, A_{BO} consists of the total leakage area of the exterior walls divided by the number of stairwells. For more complex floor plans, an effective flow area needs to be used for A_{BO} as is done in Example 10.2 for the building of Figure 10.1 with flow areas as shown in Figure 10.8.

The algebraic equation method does not explicitly include the leakages of the building roof, toilet exhausts, and the HVAC system, but the leakage value used for A_{BO} can include an allowance for these leakages. This is done in later examples by using a high leakage value for the building walls (see Table 10.1).

The pressure difference from the stairwell to the outdoors at the bottom of the stair is

$$\Delta p_{SOb} = F_R \Delta p_{SBb} \tag{10.7}$$

where

Δp_{SOb} = pressure difference from the stairwell to the outdoors at stair bottom, in. H_2O (Pa),

Δp_{SBb} = pressure difference from the stairwell to the building at stair bottom, in. H_2O (Pa),

F_R = flow area factor, dimensionless.

The pressure difference from the stairwell to the outdoors at the top of the stair is

$$\Delta p_{SOt} = F_R \Delta p_{SBt} \tag{10.8}$$

where

Δp_{SOt} = pressure difference from stairwell to outdoors at top of stairwell, in. H_2O (Pa),

Δp_{SBt} = pressure difference from stairwell to building at top of stairwell, in. H_2O (Pa),

F_R = flow area factor, dimensionless.

Average Pressure Differences

The average pressure difference is defined as the pressure difference that will result in the same total flow as the pressure profile that varies with elevation. The average pressure difference from a stairwell to the building is

$$\Delta p_{SBav} = \frac{4}{9}\left(\frac{\Delta p_{SBt}^{3/2} - \Delta p_{SBb}^{3/2}}{\Delta p_{SBt} - \Delta p_{SBb}}\right)^2 \tag{10.9}$$

where

Δp_{SBav} = average pressure difference from stairwell to building, in. H_2O (Pa),

Δp_{SBb} = bottom pressure difference from stairwell to building, in. H_2O (Pa),

Δp_{SBt} = top pressure difference from stairwell to building, in. H_2O (Pa).

The average pressure difference from a stairwell to the outdoors is

$$\Delta p_{SOav} = \frac{4}{9}\left(\frac{\Delta p_{SOt}^{3/2} - \Delta p_{SOb}^{3/2}}{\Delta p_{SOt} - \Delta p_{SOb}}\right)^2 \tag{10.10}$$

where

Δp_{SOav} = average pressure difference from stairwell to outdoors, in. H_2O (Pa),

Δp_{SOb} = bottom pressure difference from stairwell to outdoors, in. H_2O (Pa),

Δp_{SOt} = top pressure difference from stairwell to outdoors, in. H_2O (Pa).

For most calculations, Equations 10.9 and 10.10 can be approximated as

$$\Delta p_{SBav} = \frac{\Delta p_{SBb} + \Delta p_{SBt}}{2} \tag{10.11}$$

and

$$\Delta p_{SOav} = \frac{\Delta p_{SOb} + \Delta p_{SOt}}{2} \tag{10.12}$$

Figure 10.9 shows the error of Equation 10.11. In this figure the error is $(\Delta p_{SBav,\,Ap} - \Delta p_{SBav,\,Ex})/\Delta p_{SBav,\,Ex}$ where the subscripts *Ap* and *Ex* are for approximate and exact. It can be seen from this figure that the error of the Equation 10.4 is relatively small (less than 3%), and this error is conservative in that Equation 10.4 overpredicts the average pressure difference. The error situation of Equation 10.12 is similar.

Stairwell Supply Air

The flow of supply air to the stairwell equals the sum of the flow from the stairwell. Part of the flow from the stairwell goes to the building, and the rest goes directly outdoors. The following mass flow equations include flow through uniform paths that are the same over the height of the stairs. The flow rate of supply air to the stairwell can be expressed as

$$m_T = m_{SB} + m_{SOu} + \sum_{i=1}^{N} m_{SOi} \tag{10.13}$$

where

Example 10.2. Flow Area A_{BO} for an Apartment Building

For the apartment building of Figure 10.1, calculate A_{BO}. The relevant areas for this calculation are shown in Figure 10.8, and the calculation is for one side of the axis of symmetry shown on this figure. The floor to floor height is 10 ft (3.05 m). Use the flow areas listed in Table 10.1 with the high leakage of single doors and average leakage of stairwell walls. For this building, there are 8 floors.

As with Example 10.1, this calculation uses effective flow areas. For these calculations, the relevant flows are from the corridor to the outdoors. Because the calculations are based on the idealized building without vertical leakage, the elevator shaft is not included in the calculations. Strictly speaking, it is not correct to use effective flow areas for this evaluation, because there is some flow from the stairs to the apartments. If this flow is much less than that from the stairwell to the corridor, the effective flow area calculations are meaningful. Otherwise, CONTAM analysis should be used.

These areas are calculated below.

Area	Wall Area, ft^2	Wall Leakage, ft^2	Number of Doors	Door Leakage, ft^2	Total Flow Area, ft^2
A_{11}	10(17 + 54)	$710(1.1\times10^{-4})$	0	0	0.078
A_{12}	10(54 – 8.7)	$453(1.1\times10^{-4})$	1	0.24	0.290
A_{13}	10(44 – 8.7 + 30)	$653(1.1\times10^{-4})$	1	0.24	0.279
A_{14}	10(30 + 44)	$740(1.1\times10^{-4})$	0	0	0.081
A_{15}	10(10)	$100(1.1\times10^{-4})$	0.5	0.17	0.181
A_{21}	Same as A_{11}	Same as A_{11}	Same as A_{11}	Same as A_{11}	0.078
A_{22}	Same as A_{12}	Same as A_{12}	Same as A_{12}	Same as A_{12}	0.290
A_{23}	Same as A_{12}	Same as A_{12}	Same as A_{12}	Same as A_{12}	0.290
A_{24}	10(30 + 54)	$840(1.1\times10^{-4})$	0	0	0.092

$$A_{11,12e} = \left(\frac{1}{A_{11}^2} + \frac{1}{A_{12}^2}\right)^{-1/2} = \left(\frac{1}{0.078^2} + \frac{1}{0.290^2}\right)^{-1/2} = 0.075 \text{ ft}^2$$

$$A_{13,14e} = \left(\frac{1}{A_{13}^2} + \frac{1}{A_{14}^2}\right)^{-1/2} = \left(\frac{1}{0.279^2} + \frac{1}{0.081^2}\right)^{-1/2} = 0.078 \text{ ft}^2$$

$$A_{boe1} = A_{11,12} + A_{13,14} + A_{15} = 0.075 + 0.078 + 0.181 = 0.334 \text{ ft}^2$$

$$A_{21,22e} = \left(\frac{1}{A_{21}^2} + \frac{1}{A_{22}^2}\right)^{-1/2} = \left(\frac{1}{0.078^2} + \frac{1}{0.290^2}\right)^{-1/2} = 0.075 \text{ ft}^2$$

$$A_{23,24e} = \left(\frac{1}{A_{23}^2} + \frac{1}{A_{24}^2}\right)^{-1/2} = \left(\frac{1}{0.290^2} + \frac{1}{0.092^2}\right)^{-1/2} = 0.088 \text{ ft}^2$$

$$A_{boe2} = A_{21,22e} + A_{23,24e} = 0.075 + 0.088 = 0.163 \text{ ft}^2$$

$$A_{BO} = A_{boe1} + (\text{N} - 1)A_{boe2} = 0.334 + (8 - 1)0.163 = 1.48 \text{ ft}^2\ (0.137 \text{ m}^2)$$

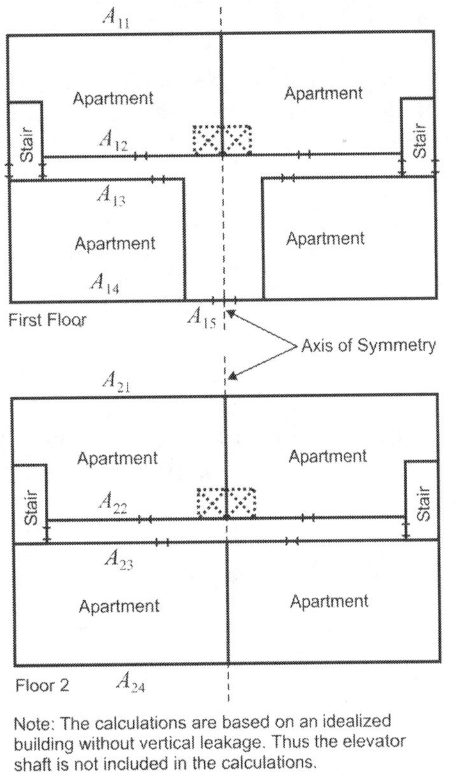

Figure 10.8 *Building of Example 10.2.*

m_T = mass flow of supply air to the stairwell, lb/s (kg/s),

m_{SB} = mass flow from the stairwell to the building, lb/s (kg/s),

m_{SOu} = flow through uniform paths from the stairwell to the outdoors, lb/s (kg/s),

m_{SOi} = flow through specific path i at location y_i, lb/s (kg/s),

N = number of specific paths, dimensionless.

The flow rate from the stairwell to the building is

$$m_{SB} = 12.9CA_{SB}\sqrt{2\rho_s \Delta p_{SBav}}$$
$$m_{SB} = CA_{SB}\sqrt{2\rho_s \Delta p_{SBav}} \text{ for SI} \qquad (10.14)$$

where

C = flow coefficient, dimensionless,

ρ_S = density of air in stairwell, lb/ft^3 (kg/m^3).

The values of C depend on many factors, and theoretically each flow equation would have a different value. For wall leakage and gaps around closed doors, a general value of $C = 0.65$ can be used. To simplify calculations, it is considered that all the flow coefficients have the same value of 0.65. The flow rate through uniform paths from the stairwell to the outdoors is

Table 10.1: Flow Areas* for Examples 10.1, 10.2, and 10.3

Flow Path	ft^2	m^2
Single door, closed		
Low	0.08	0.0074
Average	0.16	0.015
High	0.24	0.022
Double door, closed	0.34	0.032
Stairwell Walls	**Per Unit Area**	
Low	1.4×10^{-5}	
Average	1.1×10^{-4}	
High	3.5×10^{-4}	
Building Walls**	3.5×10^{-4}	
Roof	5.2×10^{-5}	

*A flow coefficient of 0.65 is used with the flow areas of this table.
**The calculations using the equation method do not explicitly include the leakages of the building roof, toilet exhausts, and the HVAC system. A high leakage flow area for the building was chosen as a partial allowance for such leakages.

$$m_{SOu} = 12.9CA_{SOu}\sqrt{2\rho_s \Delta p_{SOav}}$$
$$m_{SOu} = CA_{SOu}\sqrt{2\rho_s \Delta p_{SOav}} \text{ for SI} \qquad (10.15)$$

where

A_{SOu} = flow area of uniform paths between the stairwell to the outdoors, ft^2 (m^2).

As mentioned, the uniform paths are the same over the height of the stairs. The flow through a specific path is

$$m_{SOi} = 12.9CA_{SOi}\sqrt{2\rho_s \Delta p_{SOyi}}$$
$$m_{SOi} = CA_{SOi}\sqrt{2\rho_s \Delta p_{SOyi}} \text{ for SI} \qquad (10.16)$$

where

A_{SOi} = flow area of specific path i between the stairwell to the outdoors, ft^2 (m^2),

Δp_{SOyi} = pressure difference through specific path i, in. H_2O (Pa).

In general, the pressure difference, Δp_{SOyi}, can be calculated from Equation 10.2. At the top of the stairwell, the pressure difference, Δp_{SOyi}, can be calculated from Equation 10.3. Examples of specific flow paths are the exterior door, roof door, and leakage through a roof. For a standard temperature of 70°F (21°C) and standard atmospheric pressure, the flow equations become

$$m_{SB} = 4.99 C A_{SB} \sqrt{\Delta p_{SBav}}$$
$$m_{SB} = 1.41 C A_{SB} \sqrt{\Delta p_{SBav}} \text{ for SI}, \quad (10.17)$$

and

$$m_{SOu} = 4.99 C A_{SOu} \sqrt{\Delta p_{SOav}}$$
$$m_{SOu} = 1.41 C A_{SOu} \sqrt{\Delta p_{SOav}} \text{ for SI}, \quad (10.18)$$

and

$$m_{SOi} = 4.99 C A_{SOi} \sqrt{\Delta p_{SOyi}}$$
$$m_{SOi} = 1.41 C A_{SOi} \sqrt{\Delta p_{SOyi}} \text{ for SI}. \quad (10.19)$$

The density of air in the stairwell is

$$\rho_S = \frac{144 p_{atm}}{R T_S}$$
$$\rho_S = \frac{p_{atm}}{R T_S} \text{ for SI} \quad (10.20)$$

where

ρ_S = density of the air in stairwell, lb/ft^3 (kg/m^3),

p_{atm} = atmospheric pressure, psi (Pa),

R = gas constant, 53.34 ft·lbf/lbm·°R (287 J/kg·K).

The density of outdoor air is

$$\rho_O = \frac{144 p_{atm}}{R T_O}$$
$$\rho_O = \frac{p_{atm}}{R T_O} \text{ for SI} \quad (10.21)$$

where

ρ_O = density of outdoor air, lb/ft^3 (kg/m^3).

The volumetric flow of supply air to the stairwell is

$$V_T = \frac{60 m_T}{\rho_O}$$
$$V_T = \frac{m_T}{\rho_O} \text{ for SI} \quad (10.22)$$

where

V_T = volumetric flow of supply air to the stairwell, cfm (m^3/s).

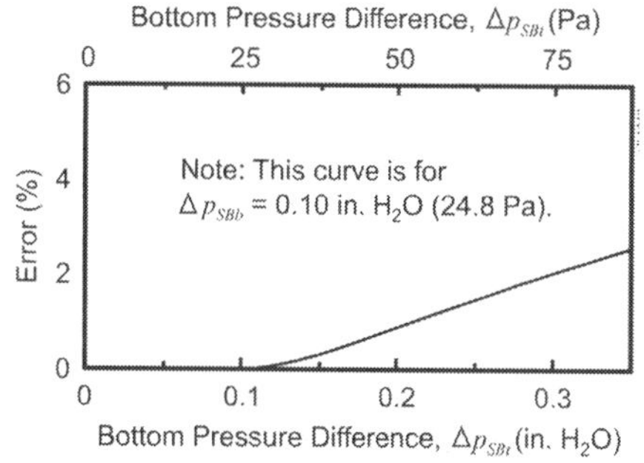

Figure 10.9 *Error of the approximate pressure difference equation for* Δp_{SBav}.

Height Limit

For some tall stairwells, acceptable pressurization may not be possible because of the impact of the indoor to outdoor temperature differences. This is more likely with systems with treated supply air than those with untreated supply air.

The height limit is the height above which acceptable pressurization is not possible for an idealized building. The height limit is

$$H_m = \frac{1.13 R F_R (\Delta p_{max} - \Delta p_{min})}{g p_{atm} \left| \frac{1}{T_O} - \frac{1}{T_S} \right|}$$
$$H_m = \frac{R F_R (\Delta p_{max} - \Delta p_{min})}{g p_{atm} \left| \frac{1}{T_O} - \frac{1}{T_S} \right|} \text{ for SI} \quad (10.23)$$

For standard atmospheric pressure at sea level, Equation 10.19 becomes

$$H_m = 0.131 \frac{F_R (\Delta p_{max} - \Delta p_{min})}{\left| \frac{1}{T_O} - \frac{1}{T_S} \right|}$$
$$H_m = 2.89 \times 10^{-4} \frac{F_R (\Delta p_{max} - \Delta p_{min})}{\left| \frac{1}{T_O} - \frac{1}{T_S} \right|} \text{ for SI} \quad (10.24)$$

where

H_m = height limit, ft (m),

Δp_{max} = maximum design pressure difference, in. H_2O (Pa),

Δp_{min} = minimum design pressure difference, in. H_2O (Pa).

If the height limit is greater than the height of a stairwell, acceptable pressurization is possible. However, it does not follow that acceptable pressurization is not possible when the height limit is less than the height of a stairwell. The height limit apples to the idealized building described above. For a real building, acceptable pressurization may be possible for some stairwells that are taller than the height limit. In such a situation, analysis using CONTAM is recommended.

Example Calculations

Example 10.3 consists of the calculations of the flow areas of the 16-story building of Figure 10.10. The building of this figure is of a simple floor plan that is the same for all floors with the exception of the ground floor exterior doors.

Examples 10.4 and 10.5 consist of calculations of the amount of supply air needed for acceptable pressurization of the same building in winter and summer. In these examples, the supply air is treated to the same temperature as the building. The calculations show that acceptable pressurization of the stairwells in this building is possible.

When using the algebraic equation method, the calculations will show if acceptable pressurization is possible. For simple buildings such as those of Figures 10.1, 10.7, and 10.10, the algebraic equation method is appropriate. When using a rule of thumb to size pressurization fans, the height limit can be used to get some information about acceptable pressurization (see Example 10.6).

Rule of Thumb

As mentioned earlier, some designers size fans for pressurized stairwells using their own rules-of-thumb, which are generally in the range of 300 to 550 cfm (0.14 to 0.26 m^3/s) per floor. Such rules-of-thumb can be

Example 10.3. Flow Areas for a 16-Story Building

For the building of Figure 10.10, calculate the flow areas that would be used in an algebraic equation analysis of Stairwell 1. The floor-to-floor height (and floor-to-roof) is 10 ft (3.05 m). The building and stairwells all have a height of 160 ft (48.8 m). Use the flow areas listed in Table 10.1 with the high leakage values of single doors and stairwell walls. The stairwells have a roof access hatch, and the flow area of a closed single door is to be used for the hatch.

	ft²	
Calculate A_{SB}		
Walls: 16 × 10 × (19 + 8.7) = 4432 ft² at 3.5 × 10^{-4} ft² per ft² of wall =	1.55	
Gaps around closed doors: 16 doors at 0.24 ft² =	3.84	
A_{SB} =	5.39	(0.494 m^2)
Calculate A_{BO}		
Walls: 16 × 10 × (90 + 42) – 2 × 16 × 10 × (19 + 8.7) = 12,260 at 3.5 × 10^{-4} ft² per ft² of wall =	4.29	
Door: 1 double door at 0.34 ft² =	0.34	
Subtotal	4.36	
Because there are 2 stairwells, A_{BO} is half the subtotal. A_{BO} =	2.32	(0.216 m^2)
Calculate A_{SOu}		
Walls: 16 × 10 × (19 + 8.7) = 4432 ft² at 3.5 × 10^{-4} ft² per ft² of wall =	1.55	
A_{SOu} =	1.55	(0.144 m^2)
Calculate A_{SO1}		
Door at stair bottom: 1 door at 0.24 ft² =	0.24	
A_{SO1} =	0.24	(0.0223 m^2)
Calculate A_{SO2}		
Roof: 19 × 8.7 = 165 ft² at 5.2 × 10^{-5} ft² per ft² of roof =	0.01	
Roof access hatch: 1 hatch at 0.24 ft² =	0.24	
A_{SO2} =	0.25	(0.0232 m^2)

appropriate for simple buildings like those discussed above. For untreated supply air, the primary factors are the flow areas and the outdoor temperature.

Table 10.2 lists the amount of untreated supply air per floor needed for acceptable pressurization of the building of Figure 10.10. When using untreated air, stack effect is minimized, and the summer design temperature is the most important temperature because of the impact it has on the volumetric flow of supply air.

It can be seen from Table 10.2 that the stairwell leakage is most important in determining that amount of supply air. The flow areas used for this leakage are listed in Table 10.1. From Table 10.2 it can be seen that a low leakage stairwell would need less than 90 cfm (42 m^3/s) per floor, and an average leakage stairwell would need less than 240 cfm (0.113 m^3/s) per floor. A high leakage stairwell would need as much as 560 cfm (0.264 m^3/s) per floor. The leakage areas of buildings including the stairwells depends more on construction quality than on materials. Buildings made of the same materials by different workmen can have very different leakages.

Engineers determining a rule of thumb for stairwell pressurization should take into account the building specifications and the anticipated quality of construction. Of course, experienced engineers develop rules-of-thumb including an allowance to avoid the expense of replacing fans, motors, and electrical wiring in the event that the stairwell would be somewhat more leaky than anticipated.

SYSTEMS WITH OPEN DOORS

As discussed, when any stair door is opened in a simple stairwell pressurization system, the pressure difference drops significantly. When all doors are closed suddenly in such a simple system, the pressure difference increases significantly. A compensated stairwell pressurization system is one that adjusts for changing conditions either by modulating supply airflow or by relieving excess pressure. The intent of a compensated system is to maintain acceptable pressurization when doors are opening and closing.

Doors Propped Open

For safety reasons, it has long been recognized that stairwell doors must not be propped open, and most building codes require most stairwells to be enclosed with self-closing or automatic closing and self-latching doors. Klote (2004) conducted a study of the effect of stairwell doors being improperly propped open. This open stairwell door study was a tenability analysis using the computer models CONTAM and CFAST. For information about tenability systems and CONTAM, see Chapter 19. Klote analyzed 80 scenarios considering many factors including weather, stair geometry, building leakage, fire type, fire location and number, and location of open doors. The buildings were 7 stories and 21 stories. Pressure compensated systems were not included in this study.

The open stairwell door study confirmed the conventional wisdom that the stairwell door on the fire floor needs to be closed for proper operation of a pressurized stairwell. Further, if there is an open stairwell door on the fire floor, the stairwell will fill with smoke regardless of whether the stairwell is pressurized or nonpressurized. This mode of stairwell failure is one reason that buildings need to have more than one stairwell.

For reasons discussed in Chapter 9, people opening and closing the stair door on the fire floor during evacuation is not much of a concern.

Need for Compensated Systems

In the U.S., most codes do not require pressurized stairwells to be compensated, and such stairwells are designed to maintain pressurization only when all the stair doors are closed. Traditionally, some engineers felt that pressurized stairwells need to be compensated, but an incidental finding of open stairwell door study by Klote mentioned above casts doubt on this opinion. For two simulations in this study with a closed stair door on the fire floor and some other stair doors open, the stairwell

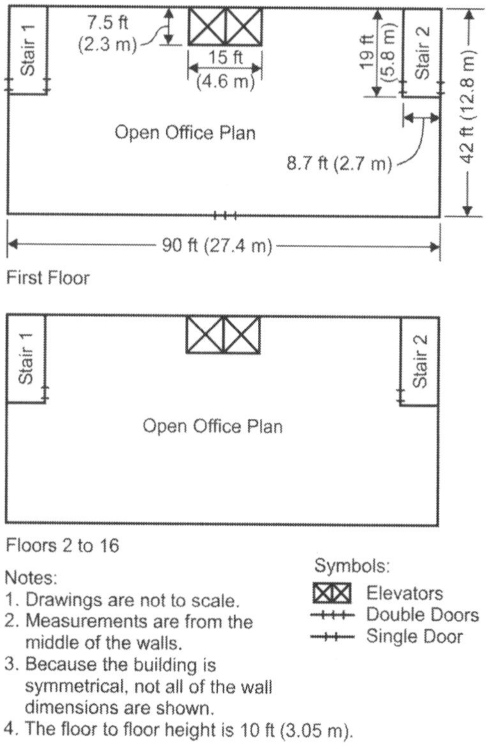

Figure 10.10 *Building of Examples 10.3 to 10.5.*

Example 10.4. Untreated Supply Air During Winter

Calculate the supply air needed to pressurize Stairwell 1 shown in Figure 10.10. Use the flow areas from Example 10.3. The minimum and maximum design pressure differences are 0.10 and 0.35 in. H_2O (24.9 and 87 Pa).

The parameters are: P_{atm} = 14.7 psi; T_O = 10°F; T_S = 70°F; T_B = 70°F; g = 32.2 ft/s²; R = 53.34 lbf/lbm°R; H = 160 ft; C = 0.65. The absolute temperatures are: T_O = 10 + 460 = 470°R; T_S = 70 + 460 = 530°R; T_B = 530°R.

For untreated supply air, and use $T_S = T_O$, and the temperature factor is

$$F_t = 7.08\left(\frac{1}{T_O} - \frac{1}{T_S}\right) = 7.08\left(\frac{1}{470} - \frac{1}{530}\right) = 0.001705 \text{ in. H}_20/\text{ft}$$

The flow area factor is

$$F_R = 1 + \frac{A_{SB}^2}{A_{BO}^2} = 1 + \frac{(5.39)^2}{(2.32)^2} = 6.40$$

For the idealized building in winter, the pressure difference is lowest at the bottom. For this reason, Δp_{SBb} is chosen as Δp_{SBb} = 0.10 in. H_2O. Next the pressure differences are calculated.

$$\Delta p_{SBt} = \Delta p_{SBb} + \frac{HF_T}{F_R} = 0.10 + \frac{160(0.001705)}{6.4} = 0.143 \text{ in. H}_20$$

These values of Δp_{SBb} and Δp_{SBt} show that acceptable pressurization is possible.

$$\Delta p_{SBav} = \frac{\Delta p_{SBb} + \Delta p_{SBt}}{2} = \frac{0.1 + 0.143}{2} = 0.122 \text{ in. H}_20$$

$$\Delta p_{SOb} = F_R \Delta p_{SBb} = 6.4(0.10) = 0.64 \text{ in. H}_20$$ (This high value is OK since the door here swings out.)

$$\Delta p_{SOt} = F_R \Delta p_{SBt} = 6.4(0.143) = 0.915 \text{ in. H}_20$$ (This high value is OK since there is no door here.)

$$\Delta p_{SOav} = \frac{\Delta p_{SOb} + \Delta p_{SOt}}{2} = \frac{0.64 + 0.915}{2} = 0.778 \text{ in. H}_20$$

$$m_{SB} = 4.99CA_{SB}\sqrt{\Delta p_{SBav}} = 4.99(0.65)(5.39)\sqrt{0.122} = 6.11 \text{ lb/s}$$

$$m_{SOu} = 4.99CA_{SOu}\sqrt{\Delta p_{SOav}} = 4.99(0.65)(1.55)\sqrt{0.778} = 4.43 \text{ lb/s}$$

Because A_{SO1} is the exterior door leakage, $\Delta p_{SOy1} = \Delta p_{SOb}$.

$$m_{SO1} = 4.99CA_{SO1}\sqrt{\Delta p_{SOb}} = 4.99(0.65)(0.24)\sqrt{0.64} = 0.62 \text{ lb/s}$$

Because A_{SO2} is at the roof, $\Delta p_{SOy2} = \Delta p_{SOt}$.

$$m_{SO2} = 4.99CA_{SO2}\sqrt{\Delta p_{SOt}} = 4.99(0.65)(0.25)\sqrt{0.915} = 0.78 \text{ lb/s}$$

$$m_T = m_{SB} + m_{SOu} + \sum_{i=1}^{N} m_{SOi} = m_{SB} + m_{SOu} + m_{SO1} + m_{SO2} = 6.11 + 4.43 + 0.62 + 0.78 = 11.94 \text{ lb/s}$$

$$\rho_O = \frac{144p_{atm}}{RT_O} = \frac{144(14.7)}{53.34(470)} = 0.0844 \text{ lb/ft}^3$$

$$V_T = \frac{60m_T}{\rho_O} = \frac{60(11.94)}{0.0844} = 8490 \text{ cfm } (4.01 \text{ m}^3/\text{s})$$

Example 10.5. Untreated Supply Air During Summer

Calculate the supply air needed to pressurize Stairwell 1 shown in Figure 10.9. This example is the same as Example 10.4 except that the outdoor temperature is 96°F (36°C).

$$T_O = 96 + 460 = 556°\text{R}$$

The temperature factor is

$$F_T = 7.08\left(\frac{1}{T_O} - \frac{1}{T_S}\right) = 7.08\left(\frac{1}{556} - \frac{1}{530}\right) = -0.000625 \text{ in. } H_20/\text{ft}$$

The flow factor is the same as Example 10.2: $F_R = 6.40$

For the idealized building in summer, the pressure difference is lowest at the top. For this reason, Δp_{SBt} is chosen as $\Delta p_{SBt} = 0.10$ in. H_2O. Next the pressure differences are calculated.

$$\Delta p_{SBb} = \Delta p_{SBt} - \frac{HF_T}{F_R} = 0.10 - \frac{160(-0.000625)}{6.4} = 0.116 \text{ in. } H_20$$

These values of Δp_{SBb} and Δp_{SBt} show that acceptable pressurization is possible.

$$\Delta p_{SBav} = \frac{\Delta p_{SBb} + \Delta p_{SBt}}{2} = \frac{0.116 + 0.10}{2} = 0.108 \text{ in. } H_20$$

$$\Delta p_{SOb} = F_R \Delta p_{SBb} = 6.4(0.116) = 0.742 \text{ in. } H_20$$

$$\Delta p_{SOt} = F_R \Delta p_{SBt} = 6.4(0.10) = 0.64 \text{ in. } H_20$$

$$\Delta p_{SOav} = \frac{\Delta p_{SOb} + \Delta p_{SOt}}{2} = \frac{0.742 + 0.64}{2} = 0.691 \text{ in. } H_20$$

$$m_{SB} = 4.99CA_{SB}\sqrt{\Delta p_{SBav}} = 4.99(0.65)(5.39)\sqrt{0.108} = 5.75 \text{ lb/s}$$

$$m_{SOu} = 4.99CA_{SOu}\sqrt{\Delta p_{SOav}} = 4.99(0.65)(1.55)\sqrt{0.691} = 4.18 \text{ lb/s}$$

$$m_{SO1} = 4.99CA_{SO1}\sqrt{\Delta p_{SOb}} = 4.99(0.65)(0.24)\sqrt{0.742} = 0.67 \text{ lb/s}$$

Because A_{SO2} is at the roof, $\Delta p_{SOy2} = \Delta p_{SOt}$.

$$m_{SO2} = 4.99CA_{SO2}\sqrt{\Delta p_{SOt}} = 4.99(0.65)(0.27)\sqrt{0.64} = 0.70 \text{ lb/s}$$

$$m_T = m_{SB} + m_{SOu} + \sum_{i=1}^{N} m_{SOi} = m_{SB} + m_{SOu} + m_{SO1} + m_{SO2} = 5.75 + 4.18 + 0.67 + 0.70 = 11.3 \text{ lb/s}$$

$$\rho_O = \frac{144 p_{atm}}{RT_O} = \frac{144(14.7)}{53.34(556)} = 0.0714 \text{ lb/ft}^3$$

$$V_T = \frac{60 m_T}{\rho_O} = \frac{60(11.3)}{0.0714} = 9500 \text{ cfm} \quad (4.48 \text{ m}^3/\text{s})$$

The flow is greater than that of Example 10.2, and the fan would be sized at 9500 cfm (4.48 m^3/s).

Example 10.6. Height Limit

For the building of the previous examples, calculate the height limit. From the other examples, the parameters are: Δp_{min} = 0.10 in. H_2O; Δp_{max} = 0.35 in. H_2O; F_R = 6.4; T_O = 470°R; T_S = 530°R; H = 160 ft.

$$H_m = 0.131\frac{F_R(\Delta p_{max} - \Delta p_{min})}{\left|\frac{1}{T_O} - \frac{1}{T_S}\right|} = 0.131\frac{6.4(0.35 - 0.10)}{\left|\frac{1}{470} - \frac{1}{530}\right|} = 870 \text{ ft}(265 \text{ m})$$

H_m is greater than H. Therefore, acceptable pressurization of this stairwell is possible.

Table 10.2: Untreated Supply Air Needed Per Floor to Pressurize Building in Figure 10.10*

Outdoor Design Temperature, T_O		Low Leakage Stairwell**		Average Leakage Stairwell**		High Leakage Stairwell**	
°F	°C	cfm	m^3/s	cfm	m^3/s	cfm	m^3/s
80	27	84.6	0.0399	230	0.108	549	0.259
90	32	86.3	0.0407	233	0.110	554	0.262
100	38	88.0	0.0415	236	0.111	560	0.264

*The flows were calculated by the algebraic equation method with the following parameters: P_{atm} = 14.7 psi (101 kPa); H = 160 ft (48.8 m); C = 0.65, and T_B = 70°F (21°C). The stairwell temperature T_S, was calculated with a heat transfer factor of 0.15. The flow areas are listed in Table 10.1.

**Low leakage means that the low values of flow area were used for the single doors and stairwell walls. Average leakage means that average values of flow area were used for the single doors and stairwell walls. High leakage means that high values of flow area were used for the single doors and stairwell walls.

remained tenable. The reason the stairwell remained tenable was that the smoke that leaked into the stairwell was diluted by the large amount of air supplied to the stairwell. In light of this finding, ASHRAE is sponsoring a research project to study the need for compensated stair systems.

Compensated and the Wind

Wind can have a serious impact on compensated stair systems. During design analysis of some compensated stair systems, some engineers have encountered very high pressure differences during some wind conditions. For example, when an exterior door is opened during the design wind speed, a compensated stair system may supply so much air that the pressure difference across some stair doors may exceed the maximum design value. It is possible to exceed this design value by as much as 100%. During such an occurrence, it would be impossible or extremely difficult for occupants to enter the stairwell. For this reason, it is recommended that design analysis of compensated stairwell pressurization systems include CONTAM simulations under wind conditions.

Compensated Systems

The following are types of compensated systems: (1) open exterior door system, (2) outdoor overpressure relief system, (3) building barometric damper system, (4) bypass system, and (5) variable-air-volume (VAV) system. The two general categories of compensated systems are overpressure relief systems and modulating systems. Systems 1 and 2 are overpressure relief systems, and Systems 4 and 5 are modulating systems. System 3 has the features of both categories in that excess pressure is relieved through a vent, but the barometric damper acts to modulate the extent to which the vent is open.

The types of compensated systems are shown in Figure 10.11. This figure is of compensated stairwell pressurization systems that are single injection systems with fans at the top, but the fans could be located elsewhere. Compensated stairwell systems also can be multiple injection systems. Compensated systems are designed for a design number of open doors. The number can be prescribed by code or based on an evacuation analysis. Because of advances in VAV fan technology, most compensated systems are of the VAV type.

Open Exterior Door System

The open exterior door system has constant-supply airflow, and an exterior stairwell door that opens automatically upon system activation (Figure 10.11a). This system is sometimes called the Canadian system because it originated in Canada, and it has been used extensively there. The supply air rate is not actually constant, but it varies to some extent with the pressure across the fan. For centrifugal fans, this variation in flow is generally small. However, the term *constant-supply* is used to differentiate

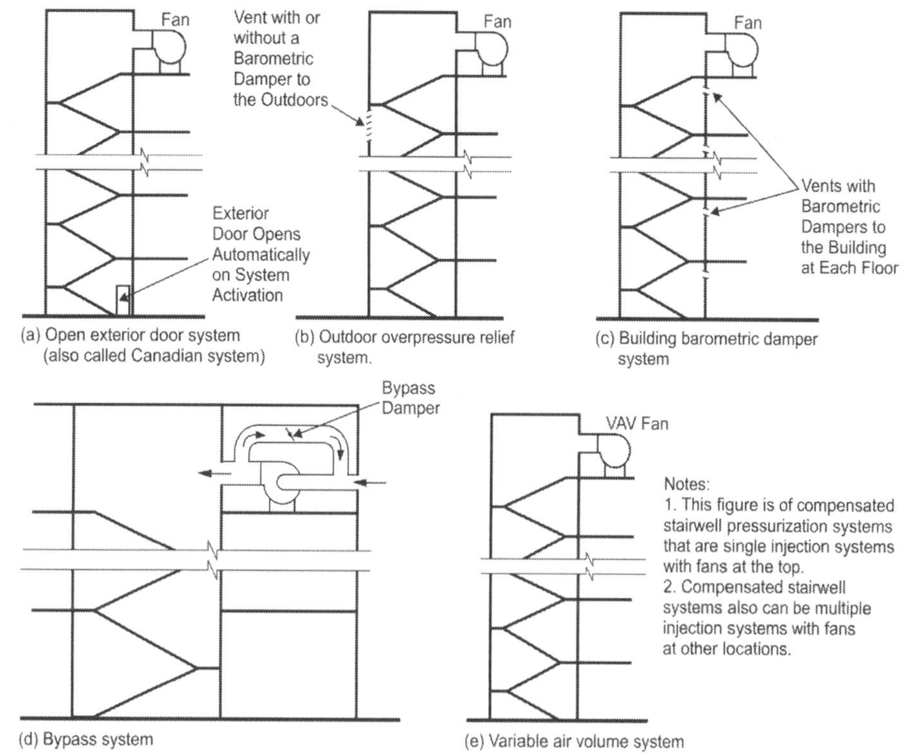

Figure 10.11 *Pressure compensating systems.*

this system from the systems where the supply air intentionally changes.

By eliminating opening and closing of the exterior stairwell door during system operation, the Canadian system eliminates the major source of pressure fluctuations. This system is simple and relatively inexpensive, but there are many locations where opening exterior doors automatically raises issues of building security. For complex buildings, it is recommended that this system has a CONTAM analysis to assure that it operates as intended.

Outdoor Overpressure Relief

Figure 10.11b is an outdoor overpressure relief system with single injection. This system has constant-supply airflow, and air is vented to the outdoor through either an open vent or a vent with a barometric damper. The version of this system with the open vent operates much like the open exterior door system except that (1) the vent has to be designed for the exterior door being opened and closed, and (2) the security issue of an automatically opening exterior door is eliminated. The flow through exterior vents can be adversely impacted by the wind, and these systems need a CONTAM analysis to assure that it operates as intended.

For the version of this system with a barometric damper, the amount of air vented depends on the pressure in the stairwell. When all the stair doors are closed, part of the supply air is relieved through the vent with an open barometric damper, which prevents excessive pressurization. When doors open, the pressure in the stairwell drops and the barometric damper closes, which helps keep the pressure difference above the minimum level.

Barometric dampers have a tendency to chatter when the wind is gusting. This chatter can be so annoying that there is a tendency for building operators to wire these dampers shut to stop the noise, but the operators unwittingly eliminate the benefit of the dampers during building fires. A tight smoke damper in series with the barometric damper can reduce the potential for annoying damper chatter. Such a smoke damper should be closed when the stairwells are not pressurized and open when the pressurization system is activated. For complex buildings, it is recommended that these systems have a CONTAM analysis to assure that they operate as intended.

Building Barometric Dampers

Figure 10.11c is a building barometric damper system with single injection. This system has constant-supply airflow, and air is vented to the building through a vent with a barometric damper. Venting to the building has the advantage that the barometric dampers maintain the pressure difference of interest which is from the

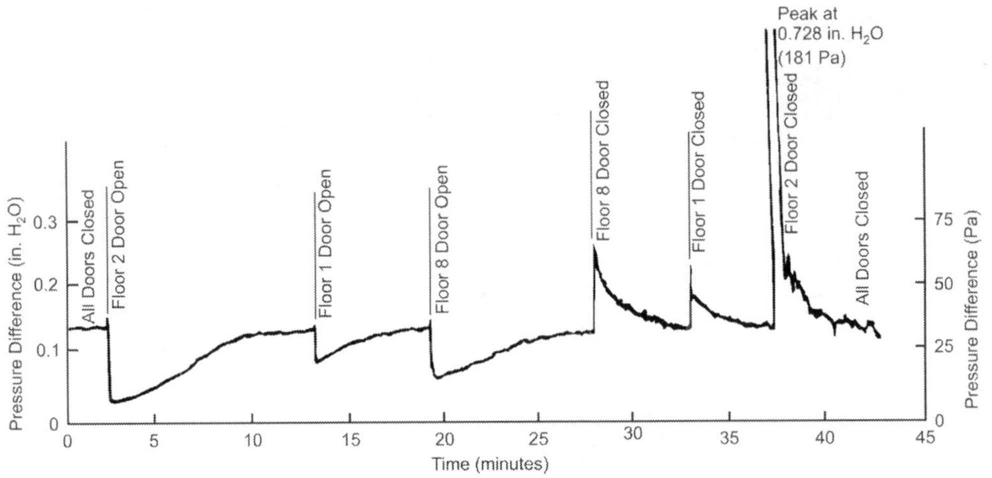

Figure 10.12 *Response time of a variable-air-volume stairwell compensated system.*

stairwell to the building. This system has the disadvantage that the air vented can supply oxygen to the fire. With this system, the vents typically have a fire damper in series with the barometric damper.

As mentioned, barometric dampers have a tendency to chatter, and a tight smoke damper in series with the barometric damper can reduce this tendency. A combination fire/smoke damper can be used. For complex buildings, it is recommended that this system has a CONTAM analysis to assure that it operates as intended.

Bypass and VAV Systems

With the bypass system and the VAV system, the flow rate of supply air to the stairwell is adjusted to account for opening and closing of doors. These systems are shown in Figures 10.11d and 10.11e. For complex buildings, it is recommended that these systems have a CONTAM analysis to assure that it operates as intended.

With both of these systems, the flow of supply air to the stairwell is controlled by one or more static pressure sensors that sense the pressure difference from the stairwell to the building. When doors are opened, the stairwell pressure drops and the flow rate of supply air is increased to achieve at least the minimum design pressurization. When all the doors are closed, the stair pressure increases, and the supply air is reduced to prevent excessive pressure differences.

A bypass system has a single-speed fan, and the fan is sized to provide the volume of air required under worst-case conditions of temperature, wind, and the design number of doors open in the stairwell. Ductwork is constructed connecting the intake and outlet of the fan, as shown in Fig 10.11d. A damper in this bypass ductwork is fully closed when the fan should deliver maximum airflow to the space, and is opened varying amounts to allow some of the air to short-circuit back to the fan intake, thereby reducing airflow to the space when less than the maximum airflow is required.

A VAV system will increase or decrease the air volume being introduced into the stairwell by increasing or decreasing fan speed and/or adjusting the pitch of the fan blades in response to decreases or increases in pressure in the stairwell caused by opening or closing of doors.

Tamura (1990) conducted research on VAV systems and bypass damper systems at the National Research Council of Canada. Time response curves for these systems are shown in Figures 10.12 and 10.13. It can be seen from these figures that the pressure drops after a door open, but in about three to seven minutes the pressure recovers to the initial value.

From Figures 10.12 and 10.13 it can also be seen that there is a pressure spike when a door is closed. When there is only one open door and it is closed, the resulting pressure spike can be very large. From Tamura's research, these spikes had peaks of 0.728 in. H_2O (181 Pa) for the VAV system and 1.47 in. H_2O (365 Pa) for the bypass damper system. These spikes only lasted about 30 or 40 s, but the peaks were much more than any reasonable maximum design pressure difference. These peaks are a concern. A person encountering such a peak would probably not be able to open the stair door, but they could open it a minute or so later provided they knew enough to try. It is possible that a person encountering such a peak would think the stair door was locked, and he or she might not try to open it again.

Because of the concerns about the wind problems mentioned, these systems need to have a CONTAM analysis performed to determine if the system is capable of being balanced to perform as intended.

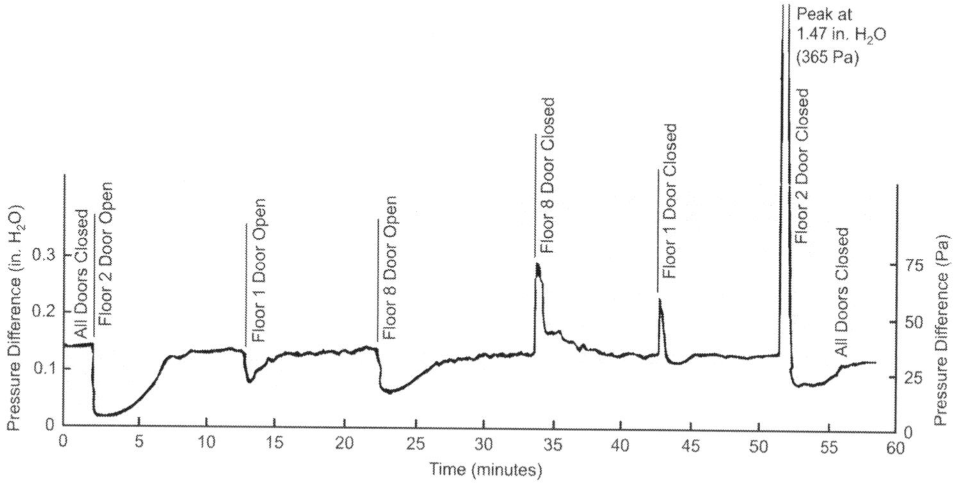

Figure 10.13 *Response time of a bypass damper stairwell compensated system.*

NOMENCLATURE

A = flow area, ft^2 (m^2)
C = flow coefficient (dimensionless)
F_R = flow area factor (dimensionless)
F_T = temperature factor, in. H_2O/ft (Pa/m)
g = acceleration of gravity, ft/s^2 (m/s^2)
H = stairwell height, ft (m)
H_m = height limit, ft (m)
m = mass flow, lb/s (kg/s)
N = number of specific paths (dimensionless)
p_{atm} = atmospheric pressure, psi (Pa)
R = gas constant, 53.34 ft·lbf/lbm·°R (287 J/kg·K)
T = temperature, °F (°C); or absolute temperature, °R (K)
y = elevation above the bottom of the stairs, ft (m)
η = heat transfer factor (dimensionless)
ρ = density, lb/ft^3 (kg/m^3).
Δp = pressure difference, in. H_2O (Pa)

Subscripts:

av = average
b = bottom
B = building
i = specific path i
max = maximum
min = minimum
S = stairwell
t = top
T = total

REFERENCES

Klote, J.H. 2004. Tenability and open doors in pressurized stairwells. *ASHRAE Transactions* 110(1).

Tamura, G.T. 1990. Fire tower tests of stair pressurization systems with overpressure relief. *ASHRAE Transactions* 96(2).

Tubbs, J.S., M. Johann, and A. Neviackas. 2011. Smoke control for tall buildings—an integrated approach to life safety. *ASHRAE Transactions* 117(1).

Webb, W.A. 2011. Super tall buildings—special smoke control requirements? *ASHRAE Transactions* 117(1).

CHAPTER 11

Pressurized Elevators

John H. Klote

The elevator pressurization systems discussed in this chapter are intended to prevent smoke from flowing from the fire floor through an elevator shaft and threatening life on floors away from the fire floor. The material in this chapter is based on design experience and the treatment of pressurized elevators in SFPE smoke control seminars (Klote and Turnbull 2010; Klote and Ferreira 2011).

Analysis of pressurized elevators can be done with a network model such as CONTAM. CONTAM is so extensively used for analyses of pressurization smoke control systems that it has become the de facto standard. In this chapter, when it is stated that analysis be done with CONTAM, it should be recognized that analysis with another network model is possible. For more information about network modeling and CONTAM, see Chapter 14.

Usually, pressurized elevators are in buildings that have pressurized stairwells, and the focus of this chapter is on both of these pressurization systems operating together. In the rare situation where pressurized elevators are the only pressurization smoke control system in a building, the information in this chapter should be useful. Readers of this chapter should be familiar with stairwell pressurization (Chapter 10).

DESIGN AND ANALYSIS

Design of pressurized elevators is much more complicated than design of pressurized stairwells, but there are a number of systems that can deal with this complexity. The reasons for this complexity are (1) often the building envelope is not capable of effectively handling the large airflow resulting from pressurization, and (2) open exterior doors on the ground floor can cause high pressure differences across the elevator shaft at the ground floor. The CONTAM simulations that are presented later were chosen to help explain these reasons. Systems that eliminate the first reason also tend to eliminate the second reason, but the opposite does not follow.

Elevators need much more pressurization air than stairwells, and much of this air flows from the shafts through the building to the outdoors. If the building envelope cannot release this flow to the outdoors in a desired manner, excessive pressurization can result.

Usually, a number of exterior doors on the ground floor are open during a building fire. During a fire, the fire service opens these doors or keeps these them open while fighting the fire. Occupants also open these doors during evacuation. The shaft pressurization system needs to operate as intended with these exterior doors open, and the CONTAM simulations discussed later address these open doors. Large airflows through these open doorways can cause high pressure differences across the ground floor doors of pressurized elevators.

Generally, a CONTAM analysis is needed to determine if pressurized elevators and pressurized stairwells in a particular building are capable of being balanced to perform as intended. While it may be theoretically possible to use only a rule of thumb to design these systems, a CONTAM analysis is strongly recommended.

When a CONTAM analysis shows that the elevator and stairwell systems in a particular building cannot be balanced to perform as intended, a new approach is needed. The categories of new approaches are (1) use an alternate elevator pressurization system, (2) use an alternate stairwell pressurization system, and (3) modify the building. A number of elevator pressurization systems are discussed later, and various stairwell pressurization systems are discussed in Chapter 10. The new approach needs a CONTAM analysis to determine if it capable of being balanced to perform as intended.

Design Pressure Differences

Pressurization smoke control systems are designed to operate within ranges of pressure difference across the stair doors and elevator doors. The minimum pressure difference is intended to prevent smoke from entering the elevators and the stairs. For stair doors, the intent of the maximum pressure difference is to prevent excessive door-opening forces. For pressurized elevators, the maximum pressure difference across elevator doors is based on concern about elevator doors jamming shut in the closed position. The term *successful pressurization* of a shaft means that the pressure differences across that shaft are within the design minimum and maximum pressure differences.

For example simulations presented in this chapter, the design pressure differences listed in Table 11.1 are used, and these values are consistent with the *International Building Code®* (*IBC®*) (ICC 2012). For reasons discussed in Chapter 9, the maximum pressure difference listed in Table 11.1 for elevators is probably conservative when the elevators are only used by firefighters.

Shaft Temperature

Elevator equipment has a typical range of operating temperature. There is usually no effort to maintain this temperature for passenger elevators during building fires, because the elevators are taken out of service. Supply air to elevator shafts is usually untreated such that the temperature in pressurized shafts is close to the outdoor temperature. As with pressurized stairwells, the use of untreated air has the benefit on minimizing the adverse impact of stack effect. The shaft temperature can be expressed as $T_S = T_O + \eta\ (T_B - T_O)$ where T_S is the temperature in the shaft in °F (°C), T_O is the temperature outdoors in °F (°C), T_B is the temperature in the building in °F (°C), and η is a dimensionless heat transfer factor.

As with pressurized stairwells, there has been little research conducted on the heat transfer factor for elevator shafts. For supply air that is conditioned to the building temperature, the heat transfer factor is one. For untreated supply air, the temperature within the shaft depends on the same factors as that for stairwells. As with stairwells, a heat transfer factor of 0.15 is suggested as being conservative regarding the impact of stack effect.

Elevator Top Vent

The requirements for vents at the top of the elevator shafts have been in codes for so many decades that the original intent of the vents is uncertain. The most common reasons that people have given for these vents are that they (1) vent odorous gases, (2) vent smoke during a building fire, (3) and prevent excessive pressures at the top of the elevator shaft due to a rising elevator car.

An historical perspective is needed to understand the idea of venting odorous gases. In 1853, Elisha Otis invented a safety device to prevent elevator cars from falling. By the 1880s, elevators were extensively used in many large cities. In the 19th and early 20th centuries, the standards of sanitation were not advanced, and it is likely that open elevator hoistways were used as trash chutes by some people. Further, it is possible that vents were needed at the top of elevator hoistways to relieve some of the malodorous gases emanating from rotting food waste and other garbage at the bottom of the hoistway.

The idea that the vents are needed to prevent excessive pressures is not likely for two reasons. First, vents would also be needed at the bottom of the shafts if the pressures from moving elevator cars needed to be relieved. Second, the pressures produced by moving elevator cars are relatively small as described in the section about elevator piston effect in Chapter 3.

The idea that vents can significantly improve smoke conditions during building fire has gained wide acceptance even in the absence of supporting research. The idea is that the buoyancy of hot smoke would make it flow out the vents, but buoyancy can also make the smoke flow from the elevator shaft to the building, especially on the upper floors of buildings.

Table 11.1: Pressure Differences Criteria for Example Simulations[1]

System	Minimum		Maximum	
	in. H_2O	Pa	in. H_2O	Pa
Pressurized elevators	0.10	25	0.25	62
Pressurized stairwells	0.10	25	0.35	87

[1]Criteria are for the simulations discussed in this chapter, and some projects may have different criteria depending on code requirements and requirements of specific applications.

For the elevator pressurization systems discussed in this chapter, there is either no top vent or the top vent is considered to be closed. For energy conservation, these top vents are often normally closed. Such normally closed vents should remain closed during elevator pressurization unless the open vent is part of the pressurization system design. The capability for remote operation of top vents may be desired by the fire service.

Piston Effect

Elevator car motion results in increased air pressure in the direction of car motion. There is a concern that piston effect could reduce the effectiveness of pressurization smoke control systems. In Chapter 3, there is information about calculating the upper limit of the pressure difference across elevator lobby doors due to piston effect. The piston effect produces a pressure spike when a car passes a particular floor, and this happens for only a few seconds during the run of an elevator. The upper limit of the pressure difference is the maximum value of this pressure spike. For elevators in multiple car shafts with car velocities less than 1000 fpm (5 m/s), piston effect should not adversely impact the performance of elevator pressurization. For elevators in single car shafts with car velocities less than 500 fpm (2.5 m/s), piston effect should not adversely impact the performance of elevator pressurization.

Volumetric Flow

Flow of pressurization air and exhaust air are entered in CONTAM as mass flows, and the corresponding volumetric flows can be calculated as $V = 60\ \rho\ /m$ where V is volumetric flow in cfm, ρ is density in lb/ft^3, and m is mass flow in lb/s. In SI units, this equation is $V = \rho\ /m$ where V is volumetric flow in m^3/s, ρ is density in kg/m^3, and m is mass flow in kg/s.

The density can be calculated from the ideal gas law as $\rho = 144p_{atm}/(RT)$ where p_{atm} is atmospheric pressure in psi, R is the gas constant which is 53.34, and T is absolute temperature of air in °R. The SI units, this equation is as $\rho = p_{atm}/(RT)$ where p_{atm} is atmospheric pressure in Pa, R is the gas constant which is 287, and T is absolute temperature of air in K.

For a shaft pressurized with untreated air, the temperature used to calculate density is the outdoor air temperature. For treated outdoor air, T is the outdoor air temperature when the fan is upstream of the heating and cooling coils, and T is the temperature of the treated air when the fan is downstream of the heating and cooling coils. For exhaust air, T is the temperature of the space being exhausted.

PRESSURIZATION SYSTEMS

The systems discussed here are (1) the basic system, (2) the exterior vent system, (3) the floor exhaust system, and (4) the ground floor lobby system. As mentioned, these systems are for use in buildings with pressurized stairwells. A concern with pressurized elevators in a building without pressurized stairwells is that is that air from the elevators could force smoke into the stairs.

Basic System

In the basic system, each stairwell and elevator shaft has one or more dedicated fans that supply pressurization air. In most buildings, the basic system does not result in successful pressurization, and the other systems to be discussed consist of the basic system plus features to improve performance. Figure 11.1 shows a basic system in a building with two stairwells and one elevator shaft. In this figure, the stair pressurization and elevator pressurization are single injection subsystems, but they can be multiple injection subsystems. The stair subsystems are not compensated systems.

The basic system is not appropriate for most buildings, but the example CONTAM simulations of the

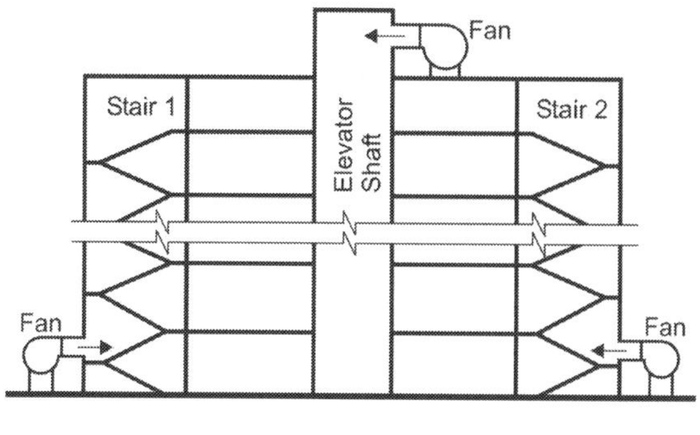

Notes:
1. In the basic system, each stairwell and elevator shaft has one or more dedicated fans that supply pressurization air.
2. In most buildings the basic system does not perform well, but there are other systems that have features to improve performance.
3. In this figure, the stair pressurization and elevator pressurization are single injection subsystems, but they can be multiple injection subsystems.
4. The stair subsystems are not compensated systems.

Figure 11.1 *Basic elevator pressurization system.*

basic system that follow demonstrate the importance of building envelope tightness on system performance.

Example Building

The example building shown in Figure 11.2 is used for CONTAM simulations of the basic system. This building also is used for CONTAM simulations of the other systems to be discussed. This building is completely symmetrical except for the double doors in the north wall. This exception was done to show that a building does not need to be geometrically symmetrical to be symmetrical with respect to flow. This means that the pressure differences across Elevator Shaft 1 are the same as those across Elevator Shaft 2, provided that both shafts have the same amount of pressurization air. This symmetry of flow simplifies the discussion of the results of the example simulations because the results for one elevator shaft are the same as for other elevator shaft. Also, this is true for the stairwells.

The flow areas and flow coefficients used for these simulations are listed in Tables 11.2 and 11.3. The amount of pressurization air is very dependent on the flow areas for the closed doors. The values of closed doors listed in Table 11.2 were chosen to be relatively large to aid in sizing the pressurization fans. For some buildings, the leakage of these elements may be even larger, and greater fan capacity could be required.

Because the leakage of the building envelope is a significant factor for the systems discussed here, Table 11.3 includes exterior wall leakage classifications ranging from tight to very loose. Because of the emphasis on energy conservation, some people may have unrealistic expectations about the tightness of building envelopes. The building envelope can be looser than many people realize (Persily 1999). The leakages of toilet exhausts and the HVAC system are not explicitly included CONTAM simulations discussed in this chapter, but it is recognized that they are part of the leakage of the building envelope. For some kinds of systems, the floor leakage and the curtain wall gap may be important, and Table 11.3 includes a number of these values. For more information about leakage in buildings, see Chapter 3.

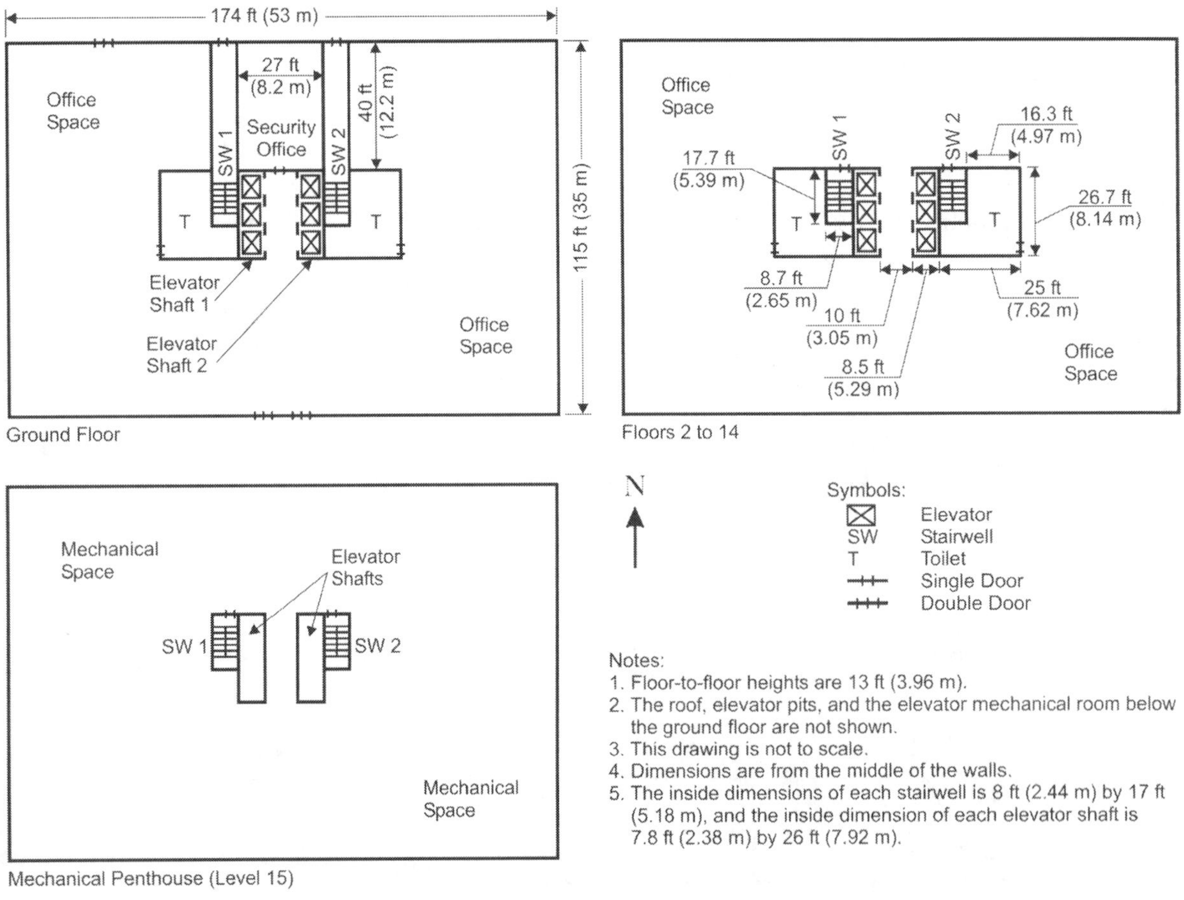

Figure 11.2 *Floor plans of example 14-story open plan office building.*

The temperatures used for the simulations are listed in Table 11.4. Areas and lengths of curtain wall gaps of the Example Building are shown in Figure 11.3. Figure 11.4 shows the CONTAM representation of the Example Building.

CONTAM Simulations

The simulations of the basic system for the example building are summarized in Table 11.5. As explained next, these simulations show that the basic system can result in successful pressurization for buildings with very loose exterior walls, but successful pressurization cannot be achieved with exterior walls of average or loose leakage. While simulations were not made with tight exterior walls, successful pressurization with this wall leakage also in not possible.

The first five runs (runs BA01 to BA05) were made with very loose exterior walls, and the others with loose and average walls. The very loose exterior walls were chosen to see how well they could help accommodate the large airflows involved with this system. The mass flows of supply air listed in Table 11.5 were determined by using CONTAM in a trial and error way to get the

Table 11.2: Flow Areas and Flow Coefficients of Doors Used for Examples[1]

Flow Path	Path Name[2]	Flow Coefficient	Flow Area	
			ft^2	m^2
Single door (closed)	DOOR-SC	0.65	0.25	0.023
Single door (opened)	DOOR-SO	0.35	21	2.0
Double door (closed)	DOOR-DC	0.65	0.48	0.045
Double door (opened)	DOOR-DO	0.35	42	3.9
Elevator door (closed)	DOOR-EC	0.65	0.65	0.06
Elevator door (opened)	DOOR-EO	0.65	6	0.56

[1]The values in this table were chosen for the example simulations of this chapter. The flow areas and flow coefficients appropriate for a design analysis of a specific building may be different. For more information about flow areas and flow coefficients, see Chapter 3.
[2]The path name is an identifier used in the CONTAM simulations.

Table 11.3: Flow Areas and Flow Coefficients of Leakages Used for Examples

Flow Path	Leakage Classification	Path Name	Flow Coefficient	Flow Area	
				ft^2 per ft^2 of wall	m^2 per m^2 of wall
Exterior walls	Tight	WALL-EXT	0.65	0.50×10^{-4}	0.50×10^{-4}
	Average			0.17×10^{-3}	0.17×10^{-3}
	Loose			0.35×10^{-3}	0.35×10^{-3}
	Very Loose			0.12×10^{-2}	0.12×10^{-2}
Interior walls	Loose	WALL	0.65	0.35×10^{-3}	0.35×10^{-3}
Floor (or roof)	Tight	FLOOR	0.65	0.66×10^{-5}	0.66×10^{-5}
	Average			0.52×10^{-4}	0.52×10^{-4}
	Loose			0.17×10^{-3}	0.17×10^{-3}
				ft^2 per ft of wall	m^2 per m of wall
Curtain wall gap	Tight	FLOORW	0.65	0.002	0.00061
	Loose			0.02	0.0061

See notes on Table 11.1.

Table 11.4: Temperatures Used for Examples

Outdoors in winter	10°F	–12°C
Outdoors in summer	96°F	36°C
Building spaces	73°F	23°C
Pressurized shafts[1]:		
Winter	19°F	–7°C
Summer	93°F	34°C

[1]These temperatures were calculated with a heat transfer factor of 0.15.

minimum pressure difference across the stairwell doors and elevator doors to be about 0.10 or 0.11 in. H_2O (25 or 27 Pa). These flows are converted to volumetric flows in Table 11.6.

The CONTAM generated pressure differences for the basic system are summarized in Table 11.7. For the first five runs, the pressure difference across the elevator shafts ranged from 0.11 to 0.15 in. H_2O (27 to 37 Pa), and the pressure difference across the stairwell ranged from 0.11 to 0.18 in. H_2O (27 to 45 Pa). From Table 11.1, the criteria for elevators is a minimum of 0.10 in. H_2O (25 Pa) and a maximum of 0.25 in. H_2O (62 Pa), and the criteria for stairwells is 0.10 in. H_2O (25 Pa) and a maximum of 0.35 in. H_2O (87 Pa). Thus,

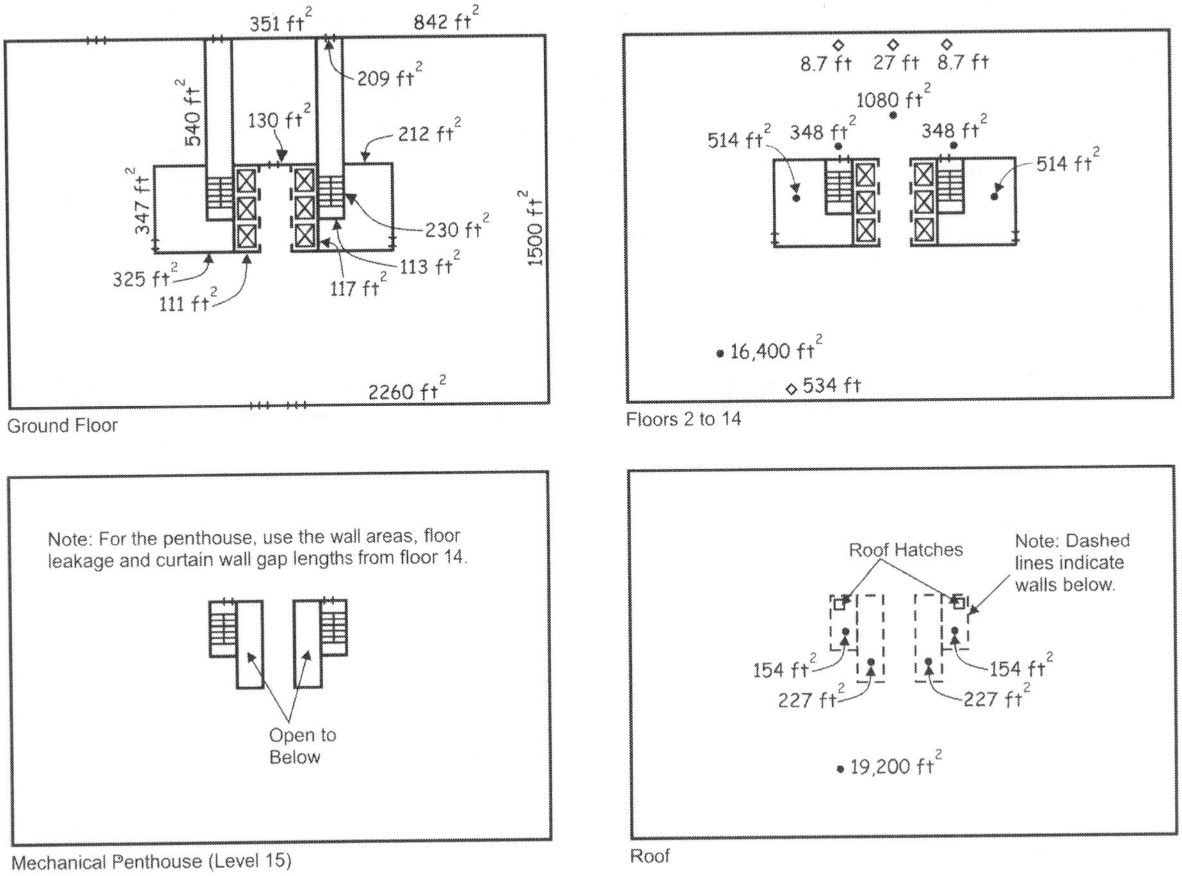

Notes:

1. The convention used here is that the wall areas are the product of the wall length times the floor-to-floor height (or the floor-to-roof height).
2. The cross-sectional area of each stairwell is 136 ft^2 (12.6 m^2).
3. Each elevator shaft has a perimeter of 67.6 ft (20.6 m) and a cross-sectional area of 203 ft^2 (18.9 m^2).
4. Not all of the areas are shown in this figure. Because of symmetry the other areas can be determined from those supplied.
5. Unit conversions are: 1 ft = 0.3048 m and 1 ft^2 = 0.0929 m^2.

Symbols:

- • Floor or roof leakage
- ◇ curtain wall gap length

Figure 11.3 *Areas and lengths of curtain wall gaps of example 14-story office building.*

Table 11.5: CONTAM Simulations of Basic System

Run	Exterior Doors Open	Season	Exterior Wall Leakage	Floor Leakage	Curtain Wall Gap	Elevator Supply Air[1]		Stairwell Supply Air[1]	
						lb/s	kg/s	lb/s	kg/s
BA01	3	Winter	Very Loose	Average	Tight	36.6	16.6	8.50	3.85
BA02	1	Winter	Very Loose	Average	Tight	36.6	16.6	8.50	3.85
BA03	0	Winter	Very Loose	Average	Tight	36.6	16.6	8.50	3.85
BA04	3	Summer	Very Loose	Average	Tight	33.0	15.0	7.80	3.54
BA05	3	Summer	Very Loose	Average	Loose	33.0	15.0	7.80	3.54
BA06	3	Winter	Loose	Average	Tight	40.0	18.1	9.00	4.08
BA07	3	Winter	Average	Average	Tight	56.0	25.4	11.5	5.22

[1]These flow rates were determined by using CONTAM in a trial and error way to get the minimum pressure difference across the stairwell doors and elevator doors to be about 0.10 or 0.11 in. H_2O (25 or 27 Pa).

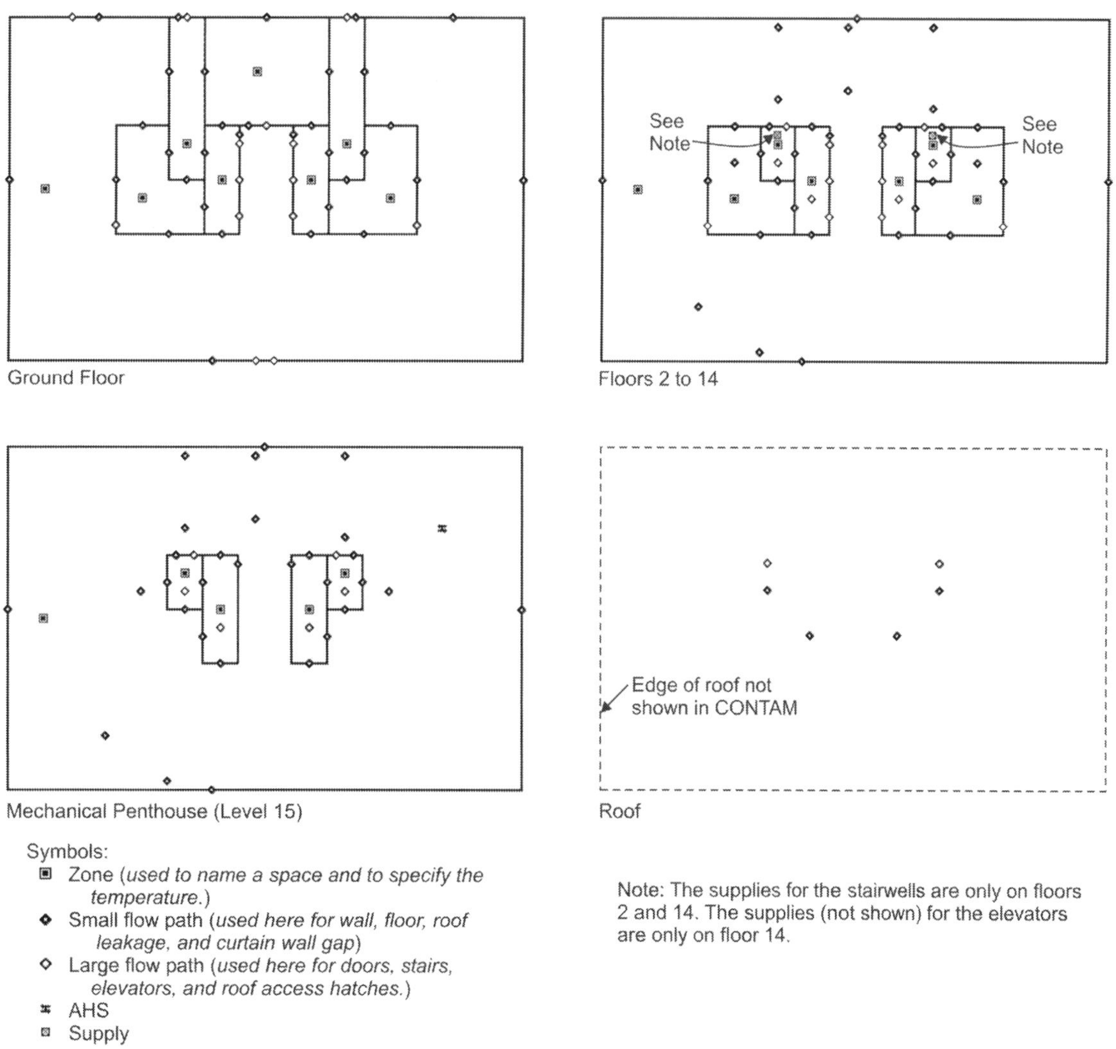

Figure 11.4 *CONTAM representation of example 14-story office building.*

for runs with very loose exterior walls (runs BA01 to BA05), both the elevators and stairwells meet the pressure difference criteria.

For the example building with very loose walls, runs BA01 to BA05 show that elevator and stairwell pressurization are feasible for (1) winter and summer temperatures, (2) for any combination of open exterior doors, and (3) a curtain wall gap that is tight or loose.

Run BA06 is with loose wall leakage, and run BA07 is with average wall leakage. For runs BA06 and BA07, the stairwell pressure differences range from 0.10 to 0.34 in. H_2O (25 to 86 Pa), which meet the criteria. From Table 11.7, it can be seen that maximum pressure difference across the elevator door for runs BA06 and BA07 are 0.53 and 1.9 in. H_2O (130 and 470 Pa), respectively. These large pressure differences are shown in Figure 11.5.

The very large pressure differences of runs BA06 and BA07 are due to the combination of less leaky exterior walls and open exterior doors. The open exterior doors result in much more flow at the ground floor, and this flow results in high pressure differences across the elevator doors at the ground floor. The open exterior doors on the ground floor cause high pressure differences across the elevator shaft at the ground floor.

For the few buildings that have very leaky building envelopes, the basic system can be a simple way to pressurize elevators and stairwells. For less leaky buildings, the following discussed systems can be considered.

Exterior Vent (EV) System

The idea of this system is to use vents in the exterior walls to increase the leakiness of the building envelope such that successful pressurization can be achieved. The vents are usually closed, but they open when the pressurization system is activated. The vents should be located in a manner to minimize adverse wind effects, and the supply intakes need to be located away from the vents to minimize the potential for smoke feedback into the supply air. These vents may need fire dampers depending on code requirements.

Table 11.6: Volumetric Supply Flows Used in Basic System Simulations[1]

Run	Elevator Supply Air		Stairwell Supply Air	
	cfm	m^3/s	cfm	m^3/s
BA01	26,000	12.3	6,040	2.85
BA02	26,000	12.3	6,040	2.85
BA03	26,000	12.3	6,040	2.85
BA04	27,700	13.1	6,560	3.09
BA05	27,700	13.1	6,560	3.09
BA06	28,400	13.4	6,400	3.02
BA07	39,800	18.8	8,170	3.86

[1]These volumetric flows were calculated from the mass flow in Table 11.5 using the density calculated from ideal gas law at the outdoor temperature and at 14.7 psi (101 kPa).

Table 11.7: Pressure Differences from CONTAM Simulations of Basic System

Run[1]	Elevator Minimum			Elevator Maximum			Stairs Minimum			Stairs Maximum		
	in. H_2O	Pa	Floor	in. H_2O	Pa	Floor	in. H_2O	Pa	Floor	in. H_2O	Pa	Floor
BA01	0.11	27	2–6	0.14	35	14	0.11	27	2-5	0.18	45	MP
BA02	0.11	27	2–6	0.14	35	14	0.11	27	2-5	0.18	45	MP
BA03	0.11	27	G-5	0.14	35	14	0.11	27	2-5	0.18	45	MP
BA04	0.11	27	3-14	0.15	37	G	0.11	27	6-13	0.14	35	MP
BA05	0.11	27	4-13	0.15	37	G	0.11	27	4-13	0.13	32	MP
BA06	0.11	27	5-12	0.53	130	G	0.10	25	6-10	0.27	67	MP
BA07	0.10	25	7-11	1.9	470	G	0.10	25	7-11	0.34	86	2

[1]The pressure differences are acceptable for runs BA01 to BA04 For runs BA05 and BA06, the pressure differences on the ground floor are more than the allowable minimum value, but this can be prevented by not using the basic system with the example building except with very loose exterior walls.

Figure 11.6 shows a typical floor of the example building with vents in the exterior walls. For the example building, the vents can be sized to assure the design criteria are met. The example building has open office plan. For buildings with corridors, the simple EV approach of Figure 11.6 is not appropriate. The flow resistance of corridor walls and other walls has a negative impact on system performance when the vents are located in the exterior walls, but this can be overcome by use of ducted vents.

Figure 11.7 shows a ducted EV system that can be used for an office building with permanent corridors. The ducted EV system can be used for other occupancies such as hotels and condominiums. Duct penetrations of a fire rated wall may have fire resistance requirements depending on code requirements. For a building where the floors can be either open office plan or divided by tenant installed partitions, an EV system can be achieved by wall vents above a suspended ceiling

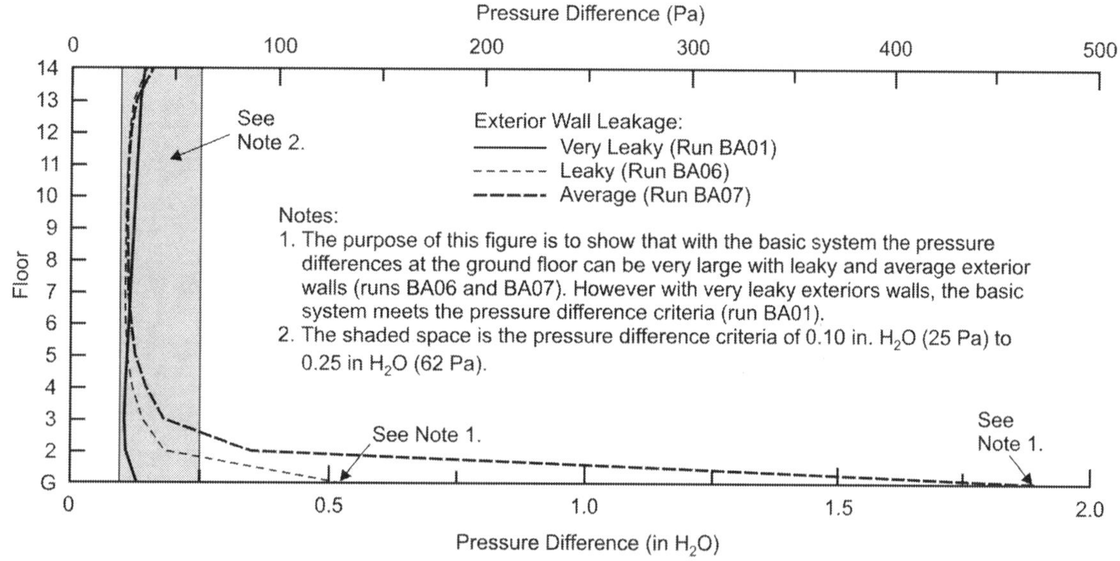

Figure 11.5 *Elevator pressure differences for basic system runs BA01, BA06, and BA07.*

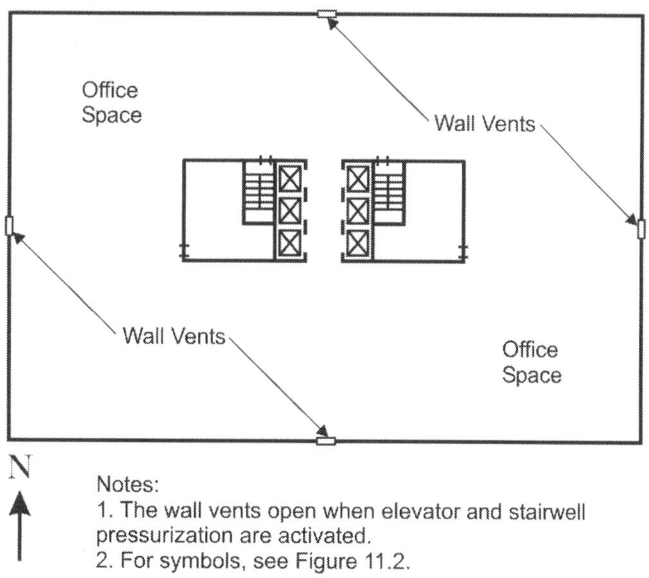

Figure 11.6 *Typical floor of the example building with the EV system.*

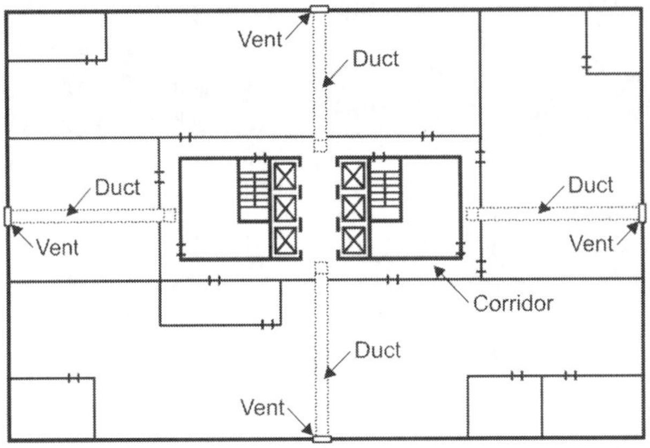

Figure 11.7 *Floor plan of ducted vents used for EV system in building with permanent corridors.*

and one or more air transfer grills in the ceiling of the elevator lobby.

In Figures 11.6 and 11.7, the vents are in all four exterior walls with the intent of minimizing any adverse impact of the wind. It is suggested that the vent area be proportional to the area of the exterior walls as was done in the simulations discussed below. If fewer vents are used, it is suggested that wind effects be incorporated in the CONTAM analysis.

CONTAM Simulations

CONTAM simulations were made for an EV system for the example building shown in Figure 11.6. These simulations are summarized in Table 11.8, and the volumetric flows are listed in Table 11.9. As about to be explained, these simulations show that the EV system can result in successful pressurization for buildings with tight exterior walls. While simulations were not made with other wall leakages, the EV system can achieve successful pressurization with any kind of exterior wall leakage in the example building.

For the simulations of the EV system (Table 11.8), the exterior wall leakage was tight. On Floors 2 to 14 and the mechanical penthouse, vents of 2.60 ft^2 (0.242 m^2) are in the north and south walls, and vents of 1.73 ft^2 (0.161 m^2) are in the east and west walls. The areas of the vents were chosen so that the leakage of the tight walls plus the vents was about the same as that of the very loose walls of the basic system simulations (runs BA01 to BA05). Because of the large flow area of the open doors on the ground floor, no vents were used on that floor.

The CONTAM-generated pressure differences for the EV system are summarized in Table 11.10. Runs EV01, EV02 and EV04 had one or three exterior doors open (Table 11.8). For these runs, the pressure differences across shafts ranged from 0.11 to 0.20 in. H_2O (27 to 50 Pa), which is within the acceptable range listed in Table 11.1.

For run EV03, the elevator pressure difference on the ground floor is 0.05 in. H_2O (12 Pa) which is less than the allowable minimum of 0.10 in H_2O (25 Pa). For the example building, this low pressure can be prevented by (1) using more supply air, (2) adding exterior vents on the ground floor, or (3) using one or more exterior doors that open automatically on activation of the shaft pressurization systems.

Floor Exhaust (FE) System

The FE system deals with the building envelope issue by reducing the amount of supply air used. In the FE system, a relatively small amount of air is supplied to the elevator shafts and the stairwells, and the fire floor is exhausted such that acceptable pressurization is maintained on the fire floor where it is needed. It is common to also exhaust one or two floors above and below the fire floor.

The FE system is a kind of zoned smoke control. As discussed in Chapter 13, exhausting air from the fire floor and some floors above and below the fire floor has

Table 11.8: CONTAM Simulations of EV System[1]

Run	Exterior Doors Open	Season	Exterior Wall Leakage	Floor Leakage	Curtain Wall Gap	Elevator Supply Air[1]		Stairwell Supply Air[2]	
						lb/s	kg/s	lb/s	kg/s
EV01	3	Winter	Tight	Average	Tight	36.6	16.6	8.50	3.85
EV02	1	Winter	Tight	Average	Tight	36.6	16.6	8.50	3.85
EV03	0	Winter	Tight	Average	Tight	36.6	16.6	8.50	3.85
EV04	3	Summer	Tight	Average	Tight	33.0	15.0	7.80	3.54

[1]On Floors 2–14 and the mechanical penthouse, vents of 2.60 ft^2 (0.242 m^2) are in the north and south walls, and vents of 1.73 ft^2 (0.161 m^2) are in the east and west walls.

[2]Except for run EV03, these flow rates were determined by using CONTAM in a trial and error way to get the minimum pressure difference across the stairwell doors and elevator doors to be about 0.10 or 0.11 in. H_2O (25 or 27 Pa).

Table 11.9: Volumetric Supply Flows Used in EV System Simulations[1]

Run	Elevator Supply Air		Stairwell Supply Air	
	cfm	m^3/s	cfm	m^3/s
EV01	26,000	12.3	6040	2.85
EV02	26,000	12.3	6040	2.85
EV03	26,000	12.3	6040	2.85
EV04	27,700	13.1	6560	3.09

[1]These volumetric flows were calculated from the mass flow in Table 11.8 using the density calculated from ideal gas law at the outdoor temperature and at 14.7 psi (101 kPa).

Table 11.10: Pressure Differences from CONTAM Simulations of EV System

Run[1]	Elevator Minimum			Elevator Maximum			Stairs Minimum			Stairs Maximum		
	in. H_2O	Pa	Floor	in. H_2O	Pa	Floor	in. H_2O	Pa	Floor	in. H_2O	Pa	Floor
EV01	0.11	27	2–5	0.14	35	14	0.11	27	3–4	0.18	45	MP
EV02	0.11	27	2–5	0.14	35	14	0.11	27	2–4	0.18	45	MP
EV03	0.05	12	G	0.15	37	14	0.11	27	2	0.19	47	MP
EV04	0.11	27	4–13	0.15	37	G	0.11	27	11–12	0.14	35	MP

[1]The pressure differences are acceptable for runs EV01, EV02, and EV04. For run EV03, the pressure difference on the ground floor is less than the allowable minimum value, but this can be prevented in a number of ways as discussed in the text.

a beneficial impact on shaft pressurization. Often, this system can achieve successful pressurization in tall and very complex buildings.

Typically, the exhaust is through a shaft with a fan located in a mechanical floor or on the roof, and dampers between the shaft and the floors are closed on all floors when the system is not operating. On system activation, the dampers open on the floors to be exhausted. The outlet of the exhaust fan needs to be located away from the inlets the supply fans to minimize the potential for smoke feedback into supply air.

For the example building, the FE system is shown in Figure 11.8. For a building with many interior partitions, the exhaust can be from the corridor that the elevators and stairwells open onto, and this is shown in Figure 11.9.

CONTAM Simulations

CONTAM simulations were made for an FE system for the example building shown in Figure 11.8. The simulations with the FE system are summarized in Table 11.11. As to be discussed, these simulations show that the FE system can result in successful pressurization for buildings with average exterior walls. While not shown included in the simulations, this EV

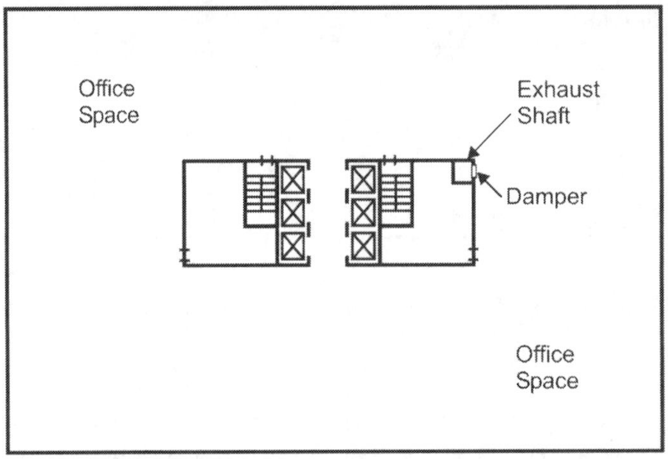

Notes:
1. The exhaust shaft has a fan (not shown) located in the mechanical penthouse, and the dampers are closed on all floors when the system is not operating. On system activation, the dampers open on the floors to be exhausted.
2. For symbols, see Figure 11.2.

Figure 11.8 *Typical floor of example building with FE system.*

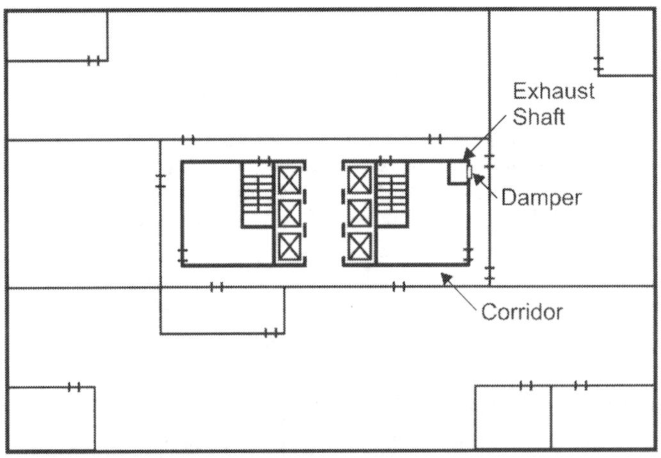

Note: See notes on Figure 11.7.

Figure 11.9 *Floor plan of FE system in building with tenant spaces.*

system can achieve successful pressurization with other exterior wall leakage.

The first four runs (FE01 to FE04) are for a fire on Floor 10, and the rest of the runs (FE05 to FE08) are with a fire on the ground floor or Floor 2. This system exhausts three floors with the middle floor being the fire floor. The exceptions to this are when the when the fire floor is the top or bottom floor of the building. Thus, when a fire is on Floor 10 (FE01 to FE04), Floors 9 to 11 are exhausted. When a fire is on the ground floor or Floor 2 (FE05 to FE08), the ground floor and Floors 2 and 3 are exhausted.

The mass flows of supply air are listed in Table 11.11, and the exhaust flows are listed in the notes at the bottom of this table. These flows were chosen so that on floors being exhausted, the minimum pressure difference across shafts would be in the range of 0.10 to 0.13 in. H_2O (25 to 32 Pa). The volumetric flows of supply air for these simulations are listed in Table 11.12.

The results of the FE simulations are summarized in Table 11.13 by floor types. These types are floors being exhausted and floors not being exhausted. For the floor being exhausted, the pressure differences for the elevators ranged from 0.10 to 0.20 in. H_2O (25 to 50 Pa), and the pressure differences for the stairwells ranged from 0.11 to 0.16 in. H_2O (27 to 40 Pa). These pressure differences meet the criteria (Table 11.1). This demonstrated that for the example building, the FE system can be balanced to meet the pressure difference criteria in summer and winter with the exterior doors open and closed.

Ground Floor Lobby (GFL) System

This system has an enclosed elevator lobby on the ground floor to reduce the tendency of open exterior doors to cause high pressure differences across the elevator shaft at the ground floor. The GFL system often has a vent between the enclosed lobby and the building with the intent of preventing excessive pressure differences across the lobby doors. The lobby doors are the doors between the enclosed lobby and the building.

The pressure difference across the lobby door and the elevator door depend on the area of the vent. There is no established criterion for the maximum pressure difference across the lobby doors, but the pressure should not be so high as to prevent the doors from remaining closed. This value depends on the specific doors and hardware. For discussion here, a maximum pressure difference for the lobby doors was chosen as 0.35 in. H_2O (87 Pa), but this value can be much different for specific applications. The vent should have a fire damper and a control damper in series. The control damper can be used to adjust the flow area of the vent so it can be balanced during commissioning. Figure 11.10 shows the ground floor of the example building with a GFL system.

As previously stated, the intent of the elevator pressurization systems discussed in this chapter is to prevent smoke from flowing from the fire floor through an elevator shaft and threatening life on floors away from the fire floor. In the GFL system, the enclosed lobby on the ground floor protects the elevator from smoke from a fire on the ground floor. For this reason, the minimum elevator

Table 11.11: CONTAM Simulations of FE System[1]

Run	Fire Floor	Exterior Doors Open	Season	Exterior Wall Leakage	Floor Leakage	Curtain Wall Gap	Elevator Supply Air[2]		Stairwell Supply Air[2]	
							lb/s	kg/s	lb/s	kg/s
FE01[2]	10	3	Winter	Average	Average	Tight	18.0	8.16	4.50	2.04
FE02[2]	10	1	Winter	Average	Average	Tight	18.0	8.16	4.50	2.04
FE03[2]	10	0	Winter	Average	Average	Tight	18.0	8.16	4.50	2.04
FE04[2]	10	3	Summer	Average	Average	Tight	18.0	8.16	4.50	2.04
FE05[3]	G or 2	3	Winter	Average	Average	Tight	18.0	8.16	4.50	2.04
FE06[3]	G or 2	1	Winter	Average	Average	Tight	18.0	8.16	4.50	2.04
FE07[3]	G or 2	0	Winter	Average	Average	Tight	18.0	8.16	4.50	2.04
FE08[3]	G or 2	3	Summer	Average	Average	Tight	18.0	8.16	4.50	2.04

[1]This system is intended to maintain acceptable pressure differences across the elevator shafts and stairwells on the floor below the fire floor, on the fire floor and on the floor above the fire floor. The floor below the fire floor, the fire floor, and the floor above the fire floor were exhausted.

[2]Floors 9, 10, and 11 were exhausted at 6.70 lb/s (3.04 kg/s) each. This flow is 5400 cfm (2.55 m^3/s).

[3]Floors G, 2, and 3 were exhausted at 6.00 lb/s (2.72 kg/s) each. Floor G is the ground floor. This flow is 4800 cfm (2.28 m^3/s).

Table 11.12: Volumetric Supply Flows Used in FE System Simulations[1]

Run	Elevator Supply Air		Stairwell Supply Air	
	cfm	m^3/s	cfm	m^3/s
FE01	12,800	6.04	3200	1.51
FE02	12,800	6.04	3200	1.51
FE03	12,800	6.04	3200	1.51
FE04	15,100	7.14	3800	1.79
FE05	12,800	6.04	3200	1.51
FE06	12,800	6.04	3200	1.51
FE07	12,800	6.04	3200	1.51
FE08	15,100	7.14	3800	1.79

[1]These volumetric flows were calculated from the mass flow in Table 11.11 using the density calculated from ideal gas law at the outdoor temperature and at 14.7 psi (101 kPa).

Table 11.13: Pressure Differences from CONTAM Simulations of FE System

Run	Floor Types[1]	Elevator Minimum		Elevator Maximum		Stairs Minimum		Stairs Maximum	
		in. H_2O	Pa	in. H_2O	Pa	in. H_2O	Pa	in. H_2O	Pa
FE01	Ex	0.10	25	0.12	30	0.11	27	0.13	32
	Non-Ex	0.01	2	0.09	22	0.01	2	0.04	10
FE02	Ex	0.10	25	0.12	30	0.11	27	0.13	32
	Non-Ex	0.01	2	0.09	22	0.01	2	0.04	10
FE03	Ex	0.11	27	0.14	35	0.12	30	0.14	35
	Non-Ex	0.01	2	0.04	10	0.01	2	0.04	10
FE04	Ex	0.11	27	0.12	30	0.11	27	0.12	30
	Non-Ex	0.01	2	0.13	32	0.01	2	0.03	7
FE05	Ex	0.11	27	0.17	42	0.11	27	0.14	35
	Non-Ex	0.01	2	0.04	10	0.01	2	0.04	10
FE06	Ex	0.11	27	0.17	42	0.11	27	0.14	35
	Non-Ex	0.01	2	0.04	10	0.01	2	0.04	10
FE07	Ex	0.11	27	0.16	40	0.11	27	0.14	35
	Non-Ex	0.01	2	0.04	10	0.01	2	0.04	10
FE08	Ex	0.13	32	0.20	50	0.14	35	0.16	40
	Non-Ex	0.01	2	0.03	7	0.01	2	0.04	10

[1]"Ex" indicates floors being exhausted, and "Non-Ex" indicated floors not being exhausted. For runs FE01 to FE04, the floors being exhausted are Floors 9, 10, and 11. For runs FE05 to FE08, the floors being exhausted are floors the ground floor and Floors 2 and 3.

pressure difference criterion of Table 11.1 does not apply to the ground floor for a GFL system. The other criteria of Table 11.1 apply. Table 11.14 lists the criteria that are used for the GFL system simulations. For the GFL system of the simulations discussed next, successful pressurization consists of meeting the criteria listed in Table 11.14.

For fires in high-rise buildings, the fire service frequently uses the elevators for rescue and for mobilization of firefighting equipment. When ground floor lobby doors are opened, the pressure difference may exceed the maximum pressure difference. If this can happen for a particular design, the fire service should be contacted to determine if this is acceptable to them.

The floor-to-floor leakage can have a significant impact on the performance of a GFL system. This leakage consists of the leakage of the floor and that of the curtain wall gap (Table 11.3).

CONTAM Simulations

CONTAM simulations were made for a GFL system for the example building with the lobby as shown in Figure 11.10. These simulations are summarized in

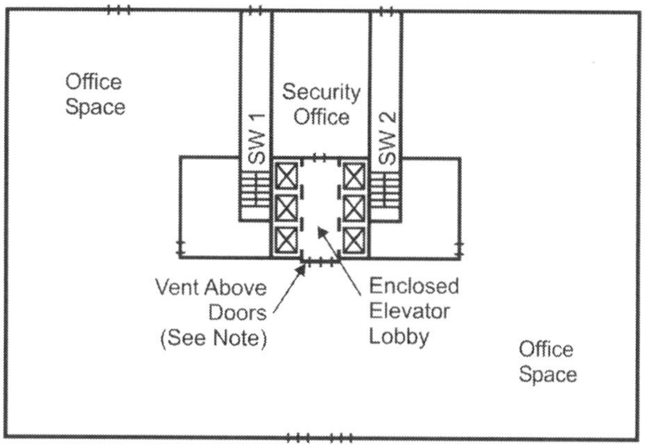

Figure 11.10 *Ground floor of example building with GFL system.*

Table 11.14: Pressure Differences Criteria for GFL Simulations[1]

Location	Criteria Number	Minimum in. H_2O	Minimum Pa	Maximum in. H_2O	Maximum Pa
Pressurized elevators on ground floor	1	NA	NA	0.25	62
Pressurized elevators on other floors	2	0.10	25	0.25	62
Pressurized stairwells on all floors	3	0.10	25	0.35	87
Ground floor elevator lobby door	4	NA	NA	0.35	87

[1]These pressure differences are with stairwell doors closed, the elevator doors closed, and the ground floor lobby door closed. The above criteria are for the GFL simulations discussed in this chapter, and some projects may have different criteria depending on code requirements and requirements of specific applications.

Table 11.15, and the supply flows for these simulations are listed in Table 11.16. The variables in the simulations were status of the ground floor lobby door, status of the exterior doors, flow area of the vent, season, exterior wall leakage, floor leakage, and curtain wall gap.

As explained later, the simulations show that the vent area can be adjusted during commissioning to provide successful pressurization in the example building for the following conditions: (1) loose exterior walls, (2) a wide range of floor leakage, (3) tight curtain wall gap, (4) closed elevator lobby door on the ground floor, and (5) any number of open exterior doors. For the example building, loose curtain wall gaps need to be sealed in order to be able to balance the GFL system. For an example building with exterior walls of average leakage, it would be difficult or impossible to balance the GFL system to provide acceptable pressurization. Simulations were not needed to know that it would be impossible to balance the GFL system to provide acceptable pressurization in the example building with exterior walls of tight leakage. The simulations show that there can be excessive pressure differences when the ground floor elevator lobby doors are open, and this needs to be taken into account by the fire service.

The results of the simulations are summarized in Tables 11.17 and 11.18 for I-P and SI units. The last column in these tables indicates system performance, and this column has an OK for simulations that result in successful pressurization. Numbers in this column are criteria numbers that are not met for that simulation. The criteria numbers are listed in Table 11.14. For example, run GFL01 has a system performance of 4 on Tables 11.17 and 11.18, which means that for run GFL01, the criteria number 4 on Table 11.14 is not met. It can be seen that the pressure differences of run GFL01 meet the rest of the criteria (Table 11.17 or Table 11.18).

Runs GFL01 to GF03 (Table 11.15) are for loose exterior walls, tight floor leakage and tight curtain walls. These runs have various vent areas, but the only one that

Table 11.15: CONTAM Simulations of GFL System

Run	Ground Floor Lobby Door	Exterior Doors Open	Flow Area of Vent[1] ft²	Flow Area of Vent[1] m²	Season	Exterior Wall Leakage	Floor Leakage	Curtain Wall Gap
GFL01	C	3	2.0	0.19	Winter	Loose	Tight	Tight
GFL02	C	3	3.0	0.28	Winter	Loose	Tight	Tight
GFL03	C	3	4.0	0.37	Winter	Loose	Tight	Tight
GFL04	C	3	3.0	0.28	Summer	Loose	Tight	Tight
GFL05	C	3	2.0	0.19	Winter	Loose	Loose	Tight
GFL06	C	3	4.0	0.37	Winter	Loose	Loose	Tight
GFL07	C	3	8.0	0.74	Winter	Loose	Loose	Tight
GFL08	C	3	4.0	0.37	Summer	Loose	Loose	Tight
GFL09	C	3	2.0	0.19	Winter	Loose	Tight	Loose
GFL10	C	3	5.0	0.46	Winter	Loose	Tight	Loose
GFL11	C	3	8.0	0.74	Winter	Loose	Tight	Loose
GFL12	C	3	5.0	0.46	Winter	Average	Tight	Tight
GFL13	C	3	5.0	0.46	Winter	Average	Loose	Tight
GFL14	O	3	3.0	0.28	Winter	Loose	Tight	Tight
GFL15	O	3	4.0	0.37	Winter	Loose	Loose	Tight
GFL16	C	1	4.0	0.37	Winter	Loose	Loose	Tight
GFL17	C	0	4.0	0.37	Winter	Loose	Loose	Tight

[1]The vent is between the enclosed ground floor lobby and the building.

Table 11.16: Supply Flows Used in GFL System Simulations

	Mass Flow of Supply Air[1]				Volumetric Flow of Supply Air[2]			
	Elevator		Stairwell		Elevator		Stairwell	
Run	lb/s	kg/s	lb/s	kg/s	cfm	m³/s	cfm	m³/s
GFL01	38.0	17.2	9.0	4.08	27,000	12.7	6400	3.02
GFL02	38.0	17.2	9.0	4.08	27,000	12.7	6400	3.02
GFL03	38.0	17.2	9.0	4.08	27,000	12.7	6400	3.02
GFL04	34.0	15.4	8.6	3.90	28,600	13.5	7230	3.41
GFL05	38.0	17.2	9.0	4.08	27,000	12.7	6400	3.02
GFL06	38.0	17.2	9.0	4.08	27,000	12.7	6400	3.02
GFL07	38.0	17.2	9.0	4.08	27,000	12.7	6400	3.02
GFL08	35.0	15.9	9.0	4.08	29,400	13.9	7570	3.57
GFL09	42.2	19.1	10.5	4.76	30,000	14.2	7460	3.52
GFL10	42.2	19.1	10.5	4.76	30,000	14.2	7460	3.52
GFL11	42.2	19.1	10.5	4.76	30,000	14.2	7460	3.52
GFL12	42.2	19.1	11.4	5.17	30,000	14.2	8100	3.82
GFL13	42.2	19.1	11.4	5.17	30,000	14.2	8100	3.82
GFL14	38.0	17.2	9.0	4.08	27,000	12.7	6400	3.02
GFL15	40.0	18.1	9.0	4.08	28,400	13.4	6400	3.02
GFL16	38.0	17.2	9.0	4.08	27,000	12.7	6400	3.02
GFL17	38.0	17.2	9.0	4.08	27,000	12.7	6400	3.02

[1]These flow rates were determined by using CONTAM in a trial and error way to get the minimum pressure difference across the stairwell doors and elevator doors to be about 0.10 or 0.11 in. H_2O (25 or 27 Pa).

[2]The volumetric flows were calculated from the mass flows using the density calculated from ideal gas law at the outdoor temperature and at 14.7 psi (101 kPa).

Table 11.17: Pressure Differences in I-P Units from CONTAM Simulations of GFL System

Run	Ground Floor Elevator Door in H_2O	Elevator[1] Minimum in H_2O	Elevator[1] Maximum in H_2O	Stairwell Minimum in H_2O	Stairwell Maximum in H_2O	Elevator Lobby Door in H_2O	System Performance[2]
GFL01	0.14	0.11	0.17	0.11	0.35	0.43	4
GFL02	0.21	0.11	0.16	0.10	0.34	0.34	OK
GFL03	0.27	0.11	0.16	0.10	0.33	0.27	1
GFL04	0.20	0.10	0.15	0.10	0.29	0.30	OK
GFL05	0.13	0.11	0.22	0.10	0.22	0.38	4
GFL06	0.24	0.10	0.21	0.10	0.24	0.24	OK
GFL07	0.36	0.10	0.20	0.10	0.20	0.10	1
GFL08	0.25	0.10	0.20	0.11	0.19	0.22	OK
GFL09	0.15	0.11	0.33	0.11	0.33	0.41	2,4
GFL10	0.32	0.10	0.31	0.10	0.31	0.21	1,2
GFL11	0.40	0.10	0.30	0.10	0.31	0.11	1,2
GFL12	1.12	0.10	0.25	0.10	0.43	0.78	1,3,4
GFL13	1.10	0.10	0.26	0.10	0.40	0.76	1,2,3,4
GFL14	0.48	0.10	0.15	0.10	0.31	0.02	1
GFL15	0.48	0.10	0.22	0.10	0.21	0.01	1
GFL16	0.23	0.10	0.21	0.10	0.21	0.23	OK
GFL17	0.11	0.12	0.17	0.11	0.19	0.10	OK

[1]This is for elevator on Floors 2 to 14.

[2]For system performance, "OK" indicates all criteria met, and the numbers are the criteria numbers from Table 11.14 that are not met.

Table 11.18: Pressure Differences in SI Units from CONTAM Simulations of GFL System

Run	Ground Floor Elevator Door Pa	Elevator[1] Minimum Pa	Elevator[1] Maximum Pa	Stairwell Minimum Pa	Stairwell Maximum Pa	Elevator Lobby Door Pa	System Performance[2]
GFL01	35	27	42	27	87	107	4
GFL02	52	27	40	25	85	85	OK
GFL03	67	27	40	25	82	67	1
GFL04	50	25	37	25	72	75	OK
GFL05	32	27	55	25	55	95	4
GFL06	60	25	52	25	60	60	OK
GFL07	90	25	50	25	50	25	1
GFL08	62	25	50	27	47	55	OK
GFL09	37	27	82	27	82	102	2,4
GFL10	80	25	77	25	77	52	1,2
GFL11	100	25	75	25	77	27	1,2
GFL12	279	25	62	25	107	194	1,3,4
GFL13	274	25	65	25	100	189	1,2,3,4
GFL14	119	25	37	25	77	5	1
GFL15	119	25	55	25	52	2	1
GFL16	57	25	52	25	52	57	OK
GFL17	27	30	42	27	47	25	OK

[1]This is for elevator on Floors 2 to 14.

[2]For system performance, "OK" indicates all criteria met, and the numbers are the criteria numbers from Table 11.14 that are not met.

meets all the criteria is run GFL02 (Table 11.17 or 11.18). This run has a vent area of 3.0 ft^2 (0.28 m^2). Run GFL04 is the same as GFL02 except that GFL04 is in summer, and GFL02 is in winter. It can be seen that GFL04 also meets all the criteria. This means that in both winter and summer, the example building with loose exterior walls, tight floor leakage and tight curtain walls can be balanced to meet all the criteria.

Runs GFL05 to GFL08 are similar to runs GFL01 to GF04 except for the vent areas and the floor leakage. Runs GFL05 to GFL08 have loose floor leakage. It can be seen from Table 11.17 or Table 11.18 that runs GFL06 and GFL08 meet all the criteria. Both runs have a vent area of 4.0 ft^2 (0.37 m^2). Run GFL06 is for winter, and run GFL08 is for winter. This means that in both winter and summer, the example building with loose exterior walls, loose floor leakage and tight curtain walls can be balanced to meet all the criteria.

From runs GFL01 to GFL08, it can be concluded that the example building can be balanced to achieve successful pressurization over a wide range of floor leakage for both winter and summer provided that (1) the lobby doors are closed, the exterior walls have loose leakage, and (3) tight curtain wall gap is tight.

Runs GFL09 to GFL011 have loose curtain wall gaps with a range of vent areas, and none of these runs meet the criteria. For the example building, this indicates that the curtain wall gaps need to be sealed in order to be able to balance the GFL system.

Runs GFL12 and GFL13 have average exterior wall leakage, and these runs do not result in successful pressurization. This indicates that it is difficult or impossible to balance the GFL system in a building with average exterior walls.

Runs GFL14 and GFL15 have open ground floor lobby doors. These runs fail to meet criterion 1 which is the pressure difference across the ground floor elevator door. In both runs, the pressure difference across the lobby door is 0.48 in. H_2O (119 Pa). This is much larger than maximum criterion of 0.25 in. H_2O (62 Pa), which is intended to minimize the potential of elevator doors jamming closed. To deal with this, the fire service could use these elevators with the lobby doors closed, or they could be prepared to open these doors in the event of door jamming.

All of the runs discussed are with three exterior doors open. Runs GFL16 and GFL17 have one and zero exterior doors open. Except for the numbers of open exterior doors, these runs are the same as rum GFL06. All these runs meet all the performance criteria. For the example building, run GFL16 and GFL17 show that GFL system can operate over a wide range of open and closed exterior doors.

REFERENCES

ICC. 2012. *International Building Code® (IBC®)*. International Code Council, Country Club Hills, IL.

Klote, J.H., and M.J. Ferreira. 2011. Seminar: Smoke Control Session I—Fundamentals and Pressurization Systems, Society of Fire Protection Engineers, October 27, Bethesda, MD.

Klote, J.H. and P.G. Turnbull. 2010. Seminar: Smoke Control Session I—Fundamentals and Pressurization Systems. Society of Fire Protection Engineers, October 27, Bethesda, MD.

Persily, A.K. 1999. Myths about building envelopes. *ASHRAE Journal* 41(3).

CHAPTER 12

Elevator Evacuation Systems

John H. Klote

Throughout most of the world, warning signs next to elevators indicate that they should not be used in fire situations (Figure 12.1). These elevators are not intended as a means of fire egress, and they should not be used for fire evacuation. In many municipalities, the fire service generally uses elevators during fires for mobilization and rescue. The idea of using elevators to speed up emergency evacuation has gained considerable attention, especially for very tall buildings (Bukowski 2009; Tubbs and Meacham 2009; Kuligowski and Bukowski 2004). For people who cannot use stairwells because of physical disabilities, fire evacuation is recognized as a serious problem, and elevator evacuation has the potential to be used as a means of emergency egress for people with such disabilities.

This chapter addresses the basic concepts of elevator evacuation and smoke protection for elevator evacuation systems. These concepts include system protection, elevator smoke control, human considerations, and calculation of elevator evacuation time.

ELEVATOR EVACUATION CONCEPT

An emergency elevator evacuation system (EEES) includes the elevator equipment, hoistway, machine room, elevator pits, elevator lobbies, and other equipment and controls needed for safe operation of the elevator during the evacuation process (Figure 12.2). The elevator lobbies are included because people must be protected from fire and smoke while they wait for an elevator. Such protected elevator lobbies also help prevent the fire from activating elevator buttons so that cars are prevented from being called by the fire to the fire floor. Even buttons that are not heat sensitive can short out when subjected to the elevated temperatures of a fire.

An EEES must have protection from heat, smoke, water, overheating of elevator room equipment, and loss of electrical power. In addition, an EEES must have a control approach that provides protection of people traveling in the elevator. In areas of high seismic activity, attention must be paid to earthquake design. Further, the development of an EEES needs to take into account human behavior so that building occupants will be willing and capable to use the system in an emergency. The following sections address these issues.

Availability

When an elevator in an EEES is out of service for scheduled or unscheduled maintenance, it cannot be used for evacuation. If there are many elevators in a building, the number of elevators used for evacuation

Figure 12.1 *Typical sign indicating elevators are not to be used during fires.*

can be selected to allow for a percentage that may be out of service.

In buildings with only one elevator, the redundancy approach to assuring availability is not possible. Scheduled maintenance can be done during off hours when the building is shut down or in a low state of activity. Maintenance contracts can put a premium on fast repair for unscheduled maintenance. When an elevator is out of service, a sign should clearly indicate this so that valuable evacuation time is not wasted waiting for an elevator that cannot come.

Elevator Control

Readers are familiar with automatic elevator controls that are common in passenger elevators in most of the world. There are two other modes of operation: elevator recall and firefighters service. Upon receipt of an alarm from a smoke detector in an elevator lobby, the elevator goes into recall mode in which cars are moved to the exit landing and removed from service. In the event of a fire on the exit floor, the cars are recalled to an alternate floor. This recall is called *Phase I*. After recall, firefighters can operate elevators, and such operation is under the control of the firefighter inside the elevator car. This firefighter's operation is referred to as *Phase II*.

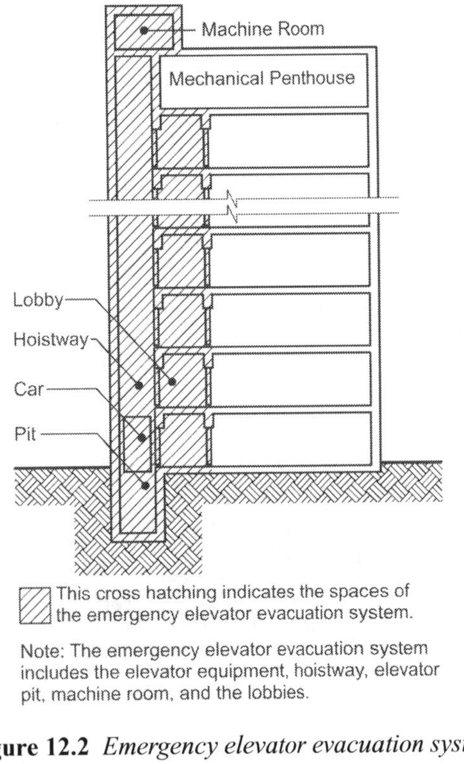

Figure 12.2 *Emergency elevator evacuation system (EEES).*

Control approaches that might be used to control elevators during a fire emergency evacuation are (1) Phase II and (2) another mode of control. Probably for most or all applications, elevator evacuation using normal control is impractical because occupants might travel to the fire floor and be exposed to fire.

The fire service, using Phase II, could also use elevators for evacuation of small numbers of people such as those with mobility limitations. Further, it is possible that building personnel could operate elevators under Phase II for emergency evacuation before the fire service arrives. Use of Phase II by nonfire service people would require that the elevator operators be trained and that the general control approach not adversely affect fire service operations. Phase II service must only be provided by people who are aware of the location and extent of the fire and of its potential for endangering people using elevators.

A new control mode could be developed specifically for emergency operation. Features that could be considered for this emergency mode include the ability to prioritize evacuation and the ability to communicate with people waiting for elevators. Such capabilities may be needed for evacuation of large numbers of people. For more information about elevator controls with considerations of fire emergencies, see Bukowski et al. (2006).

Human Considerations

Designing and building elevators that can be safely used during a fire emergency is only half the job. The building occupants must be willing and able to use the EEES. Groner (2009) developed an approach to determining the information needed by people to make critical decisions when elevators are used during fire emergencies.

Studies have been conducted regarding acceptance of elevator evacuation during fire (Heyes and Spearpoint 2009; Levin and Groner 1994). These studies consisted of interviews and surveys. It was not surprising that most people would prefer to use stairs, but most people indicated willingness to use elevators if necessary but with some concern. Occupants of higher floors tended to be more willing to accept elevators than those of lower floors.

To develop confidence in EEES, education is needed to describe the safety features. Such education should address general aspects of fire evacuation. This education can be by a combination of formal training classes, viewing videos, performing drills, reading an emergency plan, and group discussions. Levin and Groner provide more information about such education.

EEES PROTECTION

Heat and Flame

Compartmentation is one of the oldest methods of fire protection and has been extensively used to limit the spread of fire. Fire barriers are intended to resist heat and flame, but they are not intended to restrict the flow of smoke. These barriers are walls, partitions and floor-ceiling assemblies that have a level of fire resistance. The traditional approach to evaluate fire resistance is to subject a section of a barrier to a standard fire in a standard furnace. This technology is well established.

Smoke

As mentioned, the EEES needs protection from smoke. Because smoke is the major killer in fire situations, the people waiting in elevator lobbies especially need protection from smoke. Elevator smoke control is discussed later.

Water

During a building fire, water from sprinklers and fire hoses has the potential to damage electronic, electrical, and mechanical components. Klote and Braun (1996) conducted experiments of water flow around elevator doors at a specially built facility at the U.S. National Institute of Standards Technology (NIST). Water leakage of elevator doors was measured for conditions of (1) a ceiling mounted sprinkler, (2) a sidewall sprinkler, (3) standing water in the lobby, and (4) a fire hose stream aimed at the elevator doors. For the sprinklers, the leakage ranged from 2.1 to 3.3 gpm (0.13 to 0.22 L/s). For standing water of 0.5 in. (13 mm), the leakage was 13 gpm (0.84 L/s). The hose stream resulted in leakage of 210 gpm (13.5 L/s).

For fires outside the EEES, the locations of major concern about water damage are the machine room and the hoistway. Potential approaches to minimize water damage are (1) use elevator components that can function in a wet environment, and (2) prevent water from entering the hoistway or machine room.

Some methods that might be used to minimize or prevent water from entering a hoistway are use of sloping floors, floor drains, and doors with seals. Other methods might include exterior elevators or elevators located in their own towers and separated from the building by a section of exterior walkway or an exterior lobby.

Currently, no elevators have been developed with water resistant components for operation during fire evacuation. However, many elevators operate outdoors on exterior walls of buildings with many system components exposed to rain, wind, and extreme conditions. These outdoor conditions are believed to be much more severe than most of the conditions associated with water flow inside a hoistway due to a building fire. Without routine testing for water exposure, components that degrade from years of use or were accidentally damaged would go undetected and unrepaired. For this reason, routine testing of these components would be needed.

Overheating of Elevator Room Equipment

Loss of cooling can result in loss of elevator service due to overheating of electrical equipment, and precautions need to be taken to minimize the likelihood of such overheating. The maximum operating temperatures of most elevator equipment are in the range of 86°F to 95°F (30°C to 35°C). There are several approaches to providing the necessary machine room cooling, but dedicated air-conditioning equipment has significant advantages. Dedicated equipment located in the machine room or outside the building eliminates the possibility of damage to this equipment from fire outside the machine room to the extent that the fire resistive construction withstands the fire. Further, dedicated equipment uses less electrical power than nondedicated equipment with resulting advantages concerning reliability of electric power.

Electrical Power

Reliability of electric power consists of providing a source of power and providing continued distribution of power to where it is used. Some components that can be used for reliable power are fire protected distribution, redundant feeds, power from multiple substations outside the building, and emergency generator sets. Because elevator evacuation can tolerate short duration power loss, uninterrupted power supplies may not be necessary. Any consideration of reliability of electric power should consider potential causes of power failure and the consequences of that failure.

Earthquakes

The concern with earthquakes is that the counterweight could become dislodged from its rails resulting in a collision between the elevator car and the counterweight. Such a collision could result in injury or fatality to elevator passengers. In areas of high seismic activity, some elevators have strengthened rails and a seismic switch to sense significant acceleration. The strengthened rails allow safe elevator operation up to a specific level of earthquake-induced acceleration. If the seismic switch senses acceleration greater than this specific level, the elevators are put into an emergency mode to prevent collision and then taken out of service. Such an approach can be applied to EEESs that are in areas of high seismic activity.

Fire Inside the EEES

For fires in the hoistway, elevator lobbies or machine room, the most appropriate action seems to be to take the elevators out of service. Fires in the hoistway or elevator lobbies can easily result in untenable conditions within the EEES. Further, an elevator cannot be expected to operate when there is a fire in the machine room because of elevator equipment exposure to elevated temperatures. If there is a fire in the hoistway, elevator lobbies, or machine room, the EEES should be shut down. Because of limited fuel load, relatively small compartment size and the fire resistance of construction, fires in the hoistway, elevator lobbies, or machine room are not believed to have as high a potential for hazard as fires in many other locations. If evacuation is needed, other vertical paths (other elevators and stairs) can be used.

ELEVATOR SMOKE CONTROL

This chapter addresses smoke control by pressurization for EEESs. It is also possible to use a tenability system for elevator smoke control. Tenability systems are discussed in Chapters 18 and 19.

Design Pressure Differences

The minimum design pressure difference and maximum design pressure difference are generally stipulated by the code. For a pressurization smoke control system for an EEES, these pressure differences are across the elevator lobby doors. Acceptable pressurization consists of maintaining pressure differences across doors in barriers of smoke control systems between the minimum and maximum design values. There is some general information about minimum and maximum design pressure differences in Chapter 9.

Analysis

In Chapter 10, the reasons for using CONTAM for design analysis of stairwell pressurization systems are discussed. Most of that discussion also applies to smoke control by pressurization for EEESs except that an algebraic method of analysis of the systems for elevators is not presented. While it is theoretically possible to develop such an algebraic method of analysis for elevators, more realistic analysis with CONTAM is needed for practically all applications. This is especially so for buildings with pressurized elevators and other pressurization smoke control systems. CONTAM (Chapter 14) has become the de facto standard network model for analysis of pressurization smoke control systems, but it is possible to use another network model.

When a CONTAM analysis shows that an elevator smoke control system in a particular building cannot be balanced to perform as intended, a new approach is needed. There are two categories of new approaches: (1) use an alternate EEES pressurization system and (2) modify the building. Because there have been few smoke control systems for ESSS, there is limited experience with such systems. Shaft pressurization used with floor exhaust is expected to work well for most complicated buildings. The systems discussed in Chapter 11 may also be useful.

Piston Effect

Elevator car motion results in increased air pressure in the direction of car motion. There is a concern that this piston effect could reduce the effectiveness of pressurization smoke control systems. In Chapter 3, there is information about calculating the upper limit of the pressure difference across elevator lobby doors due to piston effect. Piston effect induces pressure spikes as a car passes a particular floor, and this happens for only a few seconds during the run of an elevator. The upper limit of the pressure difference is the maximum value of this pressure spike. For elevators in multiple car shafts with car velocities less than 1000 fpm (5 m/s), piston effect should not adversely impact the performance of elevator pressurization. For elevators in single car shafts with car velocities less than 500 fpm (2.5 m/s), piston effect should not adversely impact the performance of elevator pressurization.

Top Vent

For the elevator smoke control systems discussed in this chapter, there is either no top vent or the top vent is closed. For energy conservation, these top vents are often normally closed. Such normally closed vents should remain closed during elevator pressurization unless the open vent is part of the pressurization system design. The capability of remote operation of top vents may be desired by the fire service. For more information about elevator top vents, see Chapter 11.

Pressurization Systems

Much of the information in this section is based a joint project of NIST and the National Research Council of Canada (NRCC) to evaluate the feasibility of using elevators for evacuation during fires (Klote and Tamura 1986, 1991a, 1991b; Tamura and Klote 1987, 1989, 1990). Before this joint project, Klote (1983, 1984) studied elevator evacuation and elevator smoke control.

The elevator smoke control systems must provide smoke protection for the elevator lobbies, hoistways, and machinery rooms. Figure 12.3 shows some system arrangements that can be used, and the elevator cars are shown recalled to the ground floor. Figures 12.3a and 12.3b have the pressurization air supplied directly to the elevator shaft, and the lobbies are pressurized indirectly by air flowing from the shaft through the gaps around the elevator doors into the lobby.

Shaft pressurization systems can have the air supplied at the bottom (Figure 12.3a), at the top (Figure 12.3b), or somewhere in between. Some shaft pressurization systems may need pressurization air supplied at more than one location. The pressure losses of flow in elevator shafts due to friction are much lower than those of stairwells. For this reason, fewer pressurization supply points are needed for shaft pressurization systems than are needed for pressurized stairwells. Elevators 20 or 30 stories may need only one supply location, but it is suggested that the number of supply locations be evaluated by CONTAM analysis.

Figure 12.3c shows a lobby pressurization system with a bottom fan, and Figure 12.3d shows a similar system with a roof-mounted fan. Lobby pressurization systems need dampers at each floor that can be used to balance the airflow to each lobby with the intent of obtaining acceptable pressurization.

With the systems discussed, some smoke can move into the lobby on the fire floor when occupants enter the lobby. Before system activation, there can be smoke in the lobby on the fire floor. Fire tests have shown that lobby pressurization systems tend to dilute smoke in the fire floor lobby more quickly than shaft pressurization systems do. Considering that conditions on the fire floor are probably tenable in order for occupants to move to the lobby, the smoke dilution benefit of lobby pressurization systems probably is not significant.

ELEVATOR EVACUATION TIME

Bazjanac (1974, 1977), Klote et al. (1992), and Kinsey, et al. (2009) used people movement calculations to show how much elevators could be used to speed up evacuation time. Horizontal motion and movement on stairs can be calculated using the methods described in Chapter 4.

Klote and Alvord (1992) developed the computer program, ELVAC, for calculating elevator evacuation time based on accepted principles of elevator engineering. For information about these principles, see Strakosch and Caporale (2010). ELVAC calculates the evacuation time for a group of elevators. Input consists of floor-to-floor heights, number of people on floors, number of elevators in the group, elevator speed, elevator acceleration, elevator capacity, elevator door type and width, and various inefficiency factors. The output is a table of elevator travel time, round trip time, people moved, and number of round trips for each floor plus the total evacuation time. Reneke, Peacock, and Hoskins (2012) developed the Egress Estimator model that is capable of calculating evacuation time by stairs and elevators. For practical purposes, elevator evacuation time calculated by Egress Estimator is the same as that calculated by ELVAC. The following sections describe the method used in ELVAC.

Evacuation Time

Analysis of people movement during elevator evacuation must take into account the number and arrangement of elevators in a building. Generally, elevators are located in groups of up to eight elevators. Elevators in a group are located near each other and are controlled together to efficiently move people. Arrangements of elevator groups are

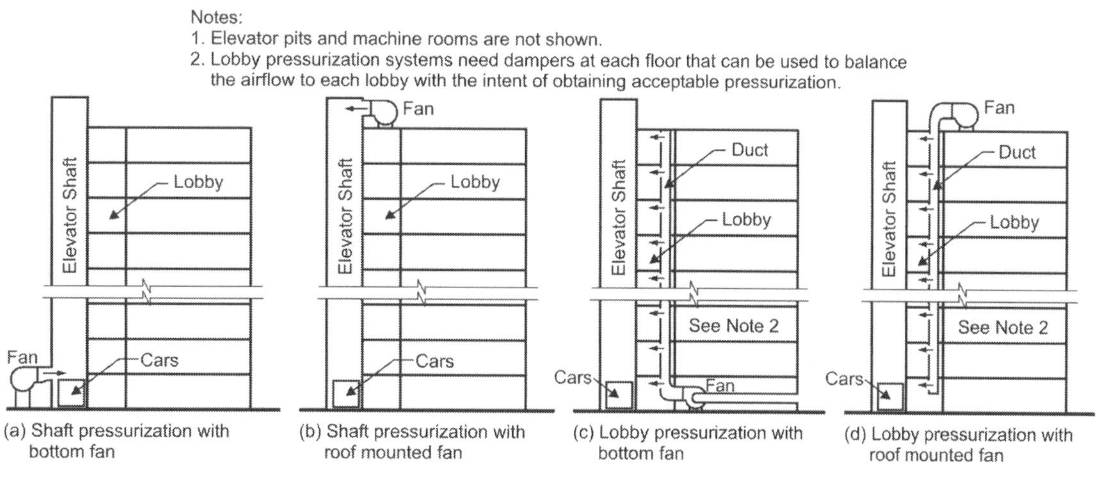

Figure 12.3 *Some arrangements of elevator pressurization systems.*

discussed later. The method of analysis discussed here is for the calculation of the evacuation time for one group of elevators. For buildings with multiple groups of elevators, the approach presented here can be applied separately to each group of elevators.

Ideally, the time to evacuate a number of people using one group of elevators consists of the sum of all the round trip times divided by the number of elevators, plus the time needed to start up the elevator evacuation and the travel time from the elevator lobby to the outdoors (or to another safe location). Accounting for inefficiencies of elevator operation, this evacuation time can be expressed as

$$t_e = t_a + t_o + \left(\frac{1+\eta}{J}\right)\sum_{j=1}^{m} t_{r,j} \qquad (12.1)$$

where

t_e = total evacuation time, s,

t_a = elevator evacuation start up time, s,

t_o = travel time from the elevator lobby to a safe location, s,

$t_{r,j}$ = time for round trip j, s,

m = number of round trips, dimensionless,

J = number of elevators, dimensionless,

η = trip inefficiency, dimensionless.

In many applications, the safe location is outside the building, but it can be a safe location inside the building. The round trip time depends on the travel time of the elevator and on the number of people carried by the elevator as discussed later. The travel time from the elevator lobby to a safe location can be evaluated by conventional methods of people movement (Chapter 4). The trip inefficiency accounts for trips to empty floors and trips to pick only a few stragglers. The elevator evacuation start-up time is discussed in the next section.

The number of elevators J used in Equation 12.1, may be less than the number of elevators in the group to account for out-of-service elevators. The probability of elevators being out-of-service depends on a number of factors including the age of the elevators and the quality of maintenance. Because the out-of-service condition can significantly increase elevator evacuation time, any analysis of elevator evacuation should take this condition into account.

Start-Up Time

The elevator evacuation start-up time is the time from activation to the start of the round trips that evacuate people. For automatic elevator operation during evacuation, a simple approach is to start elevator evacuation after all of the elevators have been moved to the discharge floor. For this approach, the start-up time t_a consists of the time for elevators to go to the discharge floor plus the time for the passengers to leave the elevators. This can be expressed as

$$t_a = t_T + (t_u + t_d)(1 + \mu) \qquad (12.2)$$

where

t_T = travel time for elevator car from farthest floor to discharge floor, s,

t_u = time for passengers to leave the elevator, s,

t_d = time for doors to open and close once, s,

μ = the total transfer inefficiency, dimensionless.

The terms in Equation 12.2 are discussed in detail later. An alternative to the simple approach discussed consists of starting the evacuation operation individually for each elevator when it reaches the discharge floor. This alternative could result in slightly reduced evacuation time. This alternative is not discussed further here, because of its limited benefit and added complexity.

For manual elevator operation, the time for elevator operators to be alerted and then get to the elevators must be included in the estimate of start-up time. This additional time may be considerably greater than that calculated from Equation 12.2.

Elevator Round Trip Time

The round trip starts at the discharge floor and consists of the following sequence: elevator doors close, car travels to another floor, elevator doors open, passengers enter the car, doors close, car travels to discharge floor, doors open, and passengers leave the car. The round trip time t_r is can be written as

$$t_r = 2t_T + t_s \qquad (12.3)$$

where

t_s = standing time, s,

t_T = travel time for one way of the round trip, s.

This equation is based on the elevator only stopping at one floor to pick up passengers. It is expected that most elevators will fill up on one floor and proceed to the discharge floor. What constitutes a full elevator is discussed later. If an elevator stops to pick up passengers at more than one floor during a round trip, Equation 12.3 can be modified accordingly. However, the trip inefficiency accounts for such multiple stops.

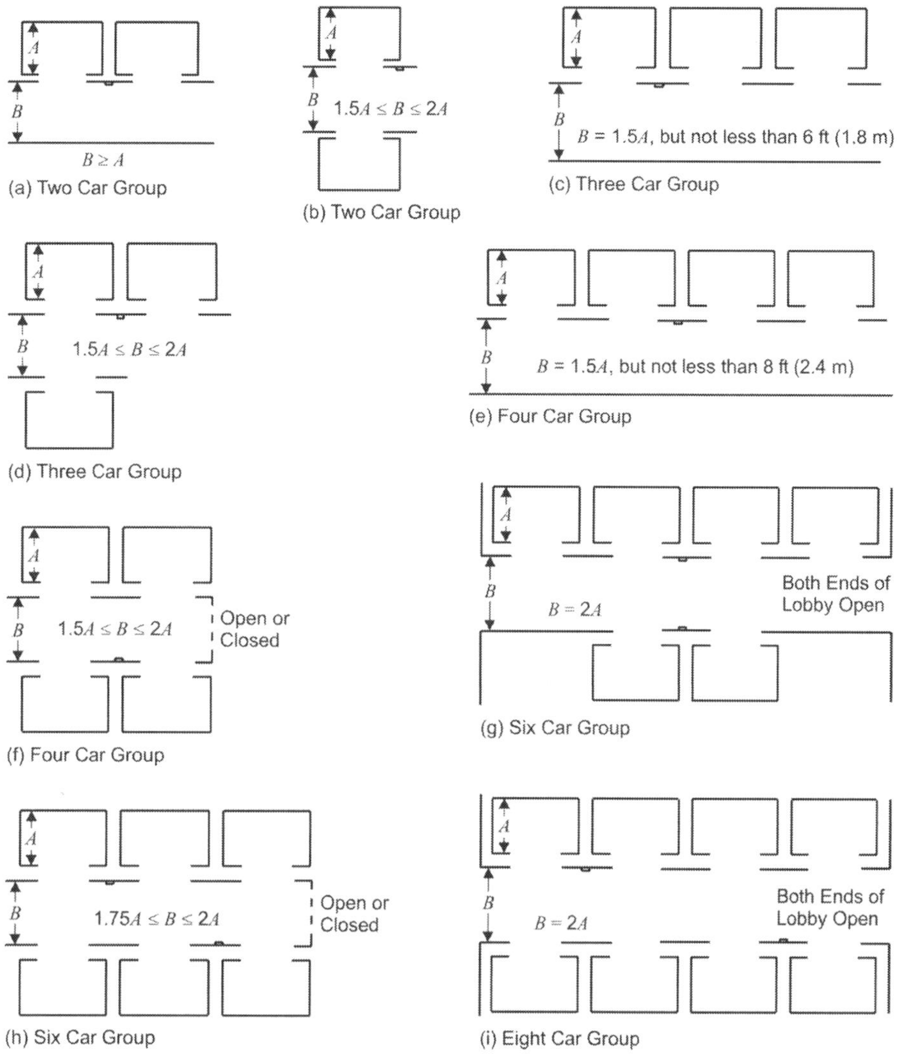

Figure 12.4 *Commonly accepted elevator arrangements.*

Standing Time

The standing time is the sum of the time to open and close the elevator doors twice, the time for people to enter the elevator, and the time for people to leave the elevator. Considering transfer inefficiencies, the standing time for a round trip can be expressed as

$$t_s = (t_i + t_u + 2t_d)(1 + \mu) \qquad (12.4)$$

where

t_i = time for people to enter the car, s,

t_u = time for people to leave the car, s,

μ = total transfer inefficiency ($\mu = \alpha + \varepsilon + \gamma$),

α = basic transfer inefficiency,

ε = door transfer inefficiency,

γ = other transfer inefficiency.

The basic transfer inefficiency α allows for rounding off of probable stops, door operating time, door starting and stopping time, and the unpredictability of people. Typically, a value of 0.10 is used for the basic transfer inefficiency for commonly accepted arrangements of elevator groups as illustrated in Figure 12.4. For each of these arrangements, the configuration of the elevator lobby is such that passengers can recognize which elevator has arrived and get on the elevator without excessive delay. Further, these lobbies have sufficient space so that people exiting one elevator will have a minimal impact on the flow of people leaving another elevator.

Arrangements of elevator groups other than those commonly accepted can be less efficient and require an increased value of the basic transfer inefficiency. These unusual arrangements include cars separated, too many cars in a line, cornered arrangement, and angular

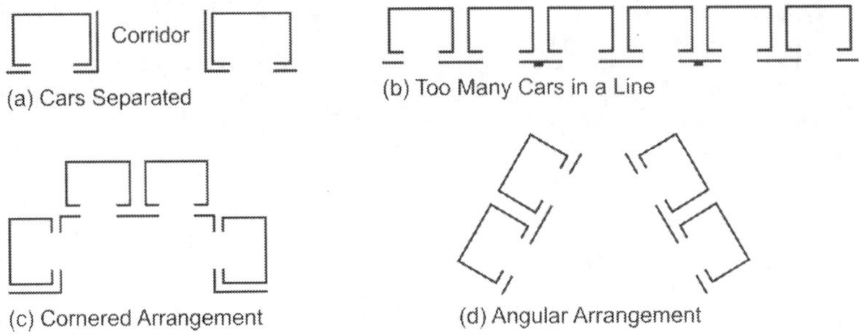

Figure 12.5 *Unusual elevator arrangements resulting in inefficient people movement.*

Table 12.1: Door Operating Time and Transfer Inefficiency

Door Type	Width		Time[1] to Open and Close	Door Transfer Inefficiency
	in.	mm	t_d (s)	ε
Single-slide	36	900	6.6	0.10
Two-speed	36	900	5.9	0.10
Center-opening[2]	36	900	4.1	0.08
Single-slide	42	1100	7.0	0.07
Two-speed	42	1100	6.6	0.07
Center-opening[2]	42	1100	4.6	0.05
Two-speed	48	1200	7.7	0.02
Center-opening[2]	48	1200	5.3	0
Two-speed	54	1400	8.8	0.02
Center-opening[2]	54	1400	6.0	0
Two-speed	60	1600	9.9	0.02
Center-opening[2]	60	1600	6.5	0
Two-speed, center-opening[2]	60	1600	6.0	0

[1]Time to open and close doors includes 0.5 s for car to start.
[2]When preopening can be used, the time to open and close these doors can be reduced by 1 s.

arrangement (Figure 12.5). Separation of elevators results in increased boarding time for people waiting by one elevator who have to walk to another when it arrives. If the separation is too large, some passengers choose to let elevators go by without boarding. Use of too many elevators in a line has similar inefficiencies (Figure 12.5b). In the cornered arrangement (Figure 12.5c), passengers entering or leaving corner cars tend to interfere with each other. With the angular arrangement (Figure 12.5d), cars at the narrow end tend to be too close together while cars at the wide end tend to be too far apart.

The door inefficiency ε is used to adjust for any increase in transfer time over that of a 48 in. (1200 mm) wide center opening door. Values of ε are listed in Table 12.1. The inefficiency γ is used to account for any other inefficiencies in people transfer into or out of elevators, such as increased movement times within an elevator car due to an unusual elevator car shape or limited physical capability of passengers. For example, γ often is chosen to be 0.05 for hospital elevators. Generally for office buildings, γ is taken as zero.

The time t_d for the doors to open and close depends on the width and type of the doors as listed in Table 12.1. The kinetic energy of closing doors is limited by elevator safety codes and is usually not more than 7 ft ·pdl (foot-poundal[1]) (0.29 J). This is why doors from different manufacturers take about the same time to open and close.

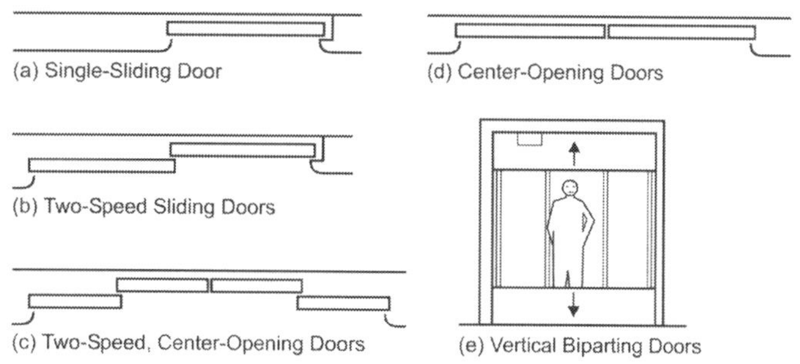

Figure 12.6 *Types of elevator doors.*

Table 12.2: Car Size and Observed Loading in I-P Units

Capacity, lb	Car Inside		Area, ft^2	Observed Loading[1], people
	Width, in.	Depth, in.		
2000	68	51	24.1	8
2500	82	51	29.0	10
3000	82	57	32.5	12
3500	82	66	37.6	16
3500 (alt.)	92	57	36.4	16
4000	82	73	41.6	19
4000 (alt.)	92	66	42.2	19
4500	92	72	46.0	21
5000	92	77	49.2	23
6000	92	90	57.5	27

[1]This loading is the value for which people will not board an elevator and choose to wait for the next one.

Types of elevator doors are shown in Figure 12.6. Door operating time is important because of the many times that doors open and close during an evacuation.

Further, an elevator cannot leave a floor before the doors are closed and locked, and passengers cannot leave an elevator until the doors are fully opened or nearly fully opened. Generally, elevator doors do not open until the car has stopped and is level with the floor. However, some center opening doors start opening while the car is leveling, and the times listed in Table 12.1 should be reduced by one second for these preopening doors.

The time, t_i, for people to enter an elevator depends on the number N of people entering and on the door operation. As previously stated, it is expected that most elevators will fill up on one floor and proceed to the discharge floor. However, elevators will be less than full when there are not enough people waiting in the lobby to fill an elevator or elevators. Thus, the analysis must include partially filled elevators.

The maximum loading for a car in ASME A17.1 (2010) is 1.5 ft^2 (0.14 m^2) per person, but this high density is not achieved in normal practice. The commonly accepted observed loading is 2.3 ft^2 (0.22 m^2) of floor space in the elevator car per person, and this loading is used to determine the number of persons in a full elevator car listed in Tables 12.2 and 12.3.

When elevator doors open, the doors remain open for at least fixed time referred to as the dwell-time, t_{dw}. The time that the door is open can be extended beyond the dwell-time by blocking the light beam across the door opening or by pushing the door safety edge. The time t_i for N people to enter an elevator car can be expressed as

1. The poundal is the unit of force in the pound mass-poundal system of units, and one poundal equals 0.0311 pounds force.

$$t_i = \begin{cases} t_{dw} & \text{for } N \le 2 \\ t_{dw} + t_{io}(N - N_{dw}) & \text{for } N > 2 \end{cases} \qquad (12.5)$$

where the N_{dw} is the number of people entering the elevator during the dwell time, and t_{io} is the average time for one person to enter the elevator. The number of people entering the elevator during the dwell time is the term (t_{dw}/t_{io}) rounded down to the nearest integer. The time for N people to leave an elevator can be expressed in a similar manner.

$$t_u = \begin{cases} t_{dw} & \text{for } N \le 2 \\ t_{dw} + t_{uo}(N - N_{dw}) & \text{for } N > 2 \,. \end{cases} \qquad (12.6)$$

For the computer program of this paper, the dwell-time is taken to be 4 s, the average time for one passenger to enter an elevator is taken to be 1 s, and the average time for one passenger to leave an elevator is taken to be 0.6 s.

Travel Time

Elevator motion is depicted in Figure 12.7a for most trips. Motion starts with constant acceleration, followed by transitional acceleration, and constant velocity motion. Constant acceleration ends when the elevator reaches a pre-determined velocity which is typically about 60% of the normal operating velocity ($V_1 = 0.6V_m$). For office buildings, the normal operating velocity is generally from 200 to 1800 fpm (1 to 9 m/s), and acceleration is from 2 to 8 ft/s^2 (0.6 to 2.4 m/s^2). Deceleration has the same magnitude as the acceleration, and the total acceleration time equals the total deceleration time $t_2 = t_5 - t_3$. The method of analysis that follows takes advantage of this symmetry.

Analysis of elevator motion that reaches the normal operating velocity is presented next. For short trips, elevators do not always reach the normal operating velocity, and methods of analysis for these short trips are presented later.

Motion Reaching Normal Operating Velocity

The time to complete constant acceleration motion (going to Point 1 on Figure 12.7a) is

$$t_1 = \frac{V_1}{a} \qquad (12.7)$$

where

V_1 = velocity at point 1, ft/s (m/s),

a = constant acceleration, ft/s^2 (m/s^2).

The distance traveled during constant acceleration is

$$S_1 = \frac{V_1^2}{2a} \qquad (12.8)$$

where S_1 is the distance traveled during constant acceleration in ft (m). Transitional acceleration is approximated by considering the product of velocity and acceleration to be a constant. The time to reach the end of transitional acceleration (Point 2 of Figure 12.7a) is

$$t_2 = t_1 + \frac{V_m^2 - V_1^2}{2aV_1} \qquad (12.9)$$

where

t_2 = time to the end of transitional acceleration, s,

V_m = normal operating velocity, ft/s (m/s).

The distance traveled by the end of transitional acceleration is

$$S_2 = S_1 + \frac{1}{3a}\left(\frac{V_m^3}{V_1} - V_1^2\right) \qquad (12.10)$$

Table 12.3: Car Size and Observed Loading in SI Units

Capacity, kg	Car Inside: Width, mm	Car Inside: Depth, mm	Area, m^2	Observed Loading[1], people
1200	2100	1300	2.73	10
1400	2100	1450	3.05	12
1600	2100	1650	3.47	16
1600 (alt.)	2350	1450	3.41	16
1800	2100	1800	3.78	18
1800 (alt.)	2350	1650	3.88	18
2000	2350	1800	4.23	20
2250	2350	1950	4.58	22
2700	2350	2150	5.05	25

[1]This loading is the value for which people will not board an elevator and choose to wait for the next one.

where S_2 is the distance traveled at the end of transitional acceleration in ft (m). The one way travel time is

$$t_5 = 2t_2 + \frac{S_T - 2S_2}{V_m} \tag{12.11}$$

where t_5 is the time traveled at the end of constant deceleration in seconds. The leveling time must be added to the above time to get the total travel time for a one way trip.

$$t_T = t_5 + t_h \tag{12.12}$$

where

t_T = time to end of car motion, s,

t_h = leveling time, s.

Usually, elevators do not stop exactly at the desired floor at the end deceleration, so the elevator must be moved slowly up or down to get it nearly level with the floor. Unless there is better data, a leveling time t_h of 0.5 s is suggested.

Motion Reaching Transitional Acceleration

If the trip is too short for the elevator to reach the normal operating velocity, but it reaches transitional acceleration, the velocity is represented by Figure 12.7b. The time t_1 and distance S_1 traveled during constant acceleration are given by Equations 7 and 8. The velocity at the end of transitional acceleration is

$$V_2 = \left[V_1^3 + 3aV_1\left(\frac{S_T}{2} - S_1\right)\right]^{1/3}. \tag{12.13}$$

The time at the end of transitional acceleration is

$$t_2 = t_1 + \frac{V_2^2 - V_1^2}{2aV_1}. \tag{12.14}$$

The one way travel time is

$$t_T = 2t_2 + t_h. \tag{12.15}$$

Motion Not Reaching Transitional Acceleration

When the trip does not go beyond constant acceleration, the motion is illustrated in Figure 12.7c. The one way travel time is

$$t_T = t_h + 2\sqrt{\frac{S_T}{a}}. \tag{12.16}$$

Example 12.1 shows how to calculate the time it takes for round trip from the ground floor to pick up a full load of passengers at Floor 21.

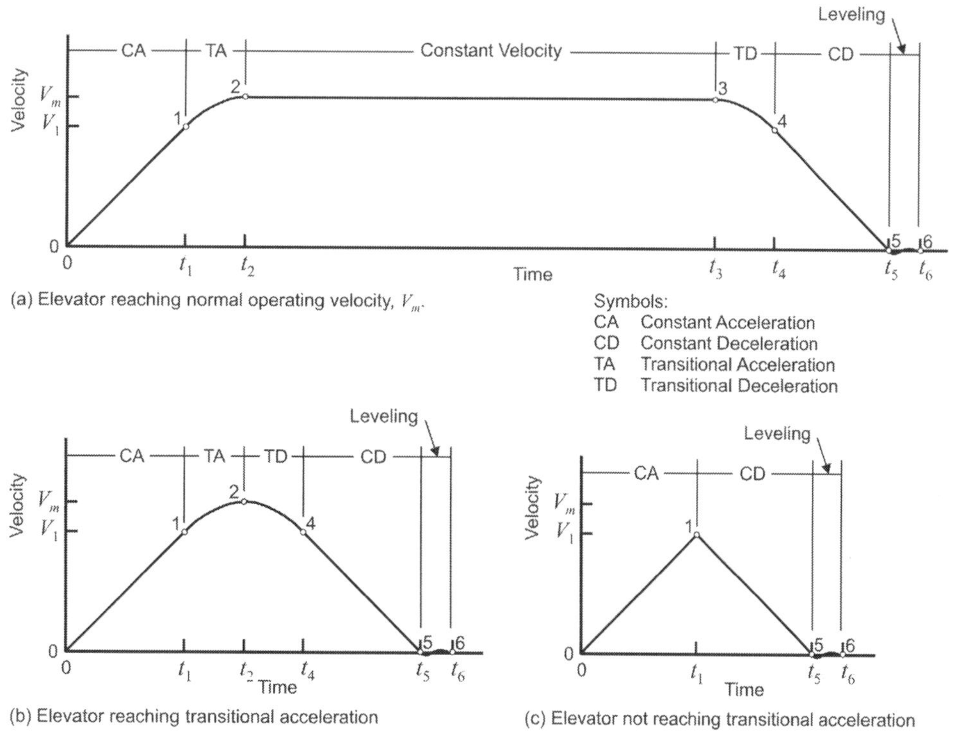

Figure 12.7 *Velocity of elevator car for different conditions of top speed.*

Example 12.1. Round Trip Time in I-P Units

A 3500 lb elevator in an office building makes a round trip from the ground floor to pick up a full load of passengers from the 21st floor and return them to the ground floor. The operating velocity is 600 fpm with an acceleration of 4 ft/s^2, and the elevator door is 48 in. wide center-opening. The distance between floors is 10.5 ft, and the total travel distance, S_T, is 210 ft.

From Table 12.2, the number of people in the full elevator is approximated at 16. From Table 12.1, t_d is 5.3 s, and ε is 0. The elevator shape is not unusual and the passenger capability is normal, so γ is 0. The total transfer inefficiency is

$$\mu = \alpha + \varepsilon + \gamma = 0.10 + 0 + 0 = 0.10 .$$

From Equation 12.5, the time for 16 people to enter the elevator is $t_i = N = 16\text{ s}$.

From Equation 12.6, the time for 16 people to leave the elevator is $t_u = 4 + 0.6(N-6) = 4 + 0.6(16-6) = 10\text{ s}$.

From Equation 12.4, the standing time is $t_s = (t_i + t_u + 2t_d)(1+\mu) = [16 + 10 + 2(5.3)](1 + 0.1) = 40.26\text{ s}$.

The normal operating velocity is $V_m = 600 \dfrac{\text{ft}}{\text{min}} \dfrac{1\text{ min}}{60\text{ s}} = 10\text{ ft/s}$.

Consider V_1 is 60% of V_m, then $V_1 = 0.6V_m = 0.6(10) = 6\text{ ft/s}$.

From Equation 12.7, the time at the end of constant acceleration is $t_1 = V_1/a = 6/4 = 1.5\text{ s}$.

From Equation 12.8, the distance traveled during constant acceleration is $S_1 = \dfrac{V_1^2}{2a} = \dfrac{(6)^2}{2(4)} = 4.5\text{ ft}$.

From Equation 12.9, the time at the end of transitional acceleration is $t_2 = t_1 + \dfrac{V_m^2 - V_1^2}{2aV_1} = 1.5 + \dfrac{(10)^2-(6)^2}{2(4)(6)} = 2.83\text{ s}$.

From Equation 12.10, the distance traveled by the end of transitional acceleration is

$$S_2 = S_1 + \frac{1}{3a}\left(\frac{V_m^3}{V_1} - V_1^2\right) = 4.5 + \frac{1}{3(4)}\left(\frac{10^3}{6} - 6^2\right) = 15.4\text{ ft}.$$

The one way travel time is calculated from Equation 12.11 $t_5 = 2t_2 + \dfrac{S_T - 2S_2}{V_m} = 2(2.83) + \dfrac{210 - 2(15.4)}{10} + 236.6\text{ s}$.

The total travel time is calculated from Equation 12.12 $t_T = t_5 + t_h = 23.6 + 0.5 = 24.1\text{ s}$.

The round trip time is calculated from Equation 12.3 $t_r = 2t_T + t_s = 2(24.1) + 40.3 = 88.5\text{ s}$.

NOMENCLATURE

a = acceleration, ft/s^2 (m/s^2)

J = number of elevators

m = number of round trips

N = number of people entering or leaving the elevator

N_{dw} = number of people entering or leaving the elevator during the dwell time

S = distance, ft (m)

S_T = total travel distance for trip, ft (m)

t = time, s,

t_a = elevator evacuation start up time, s,

t_d = time for elevator doors to open and close, s,

t_{dw} = dwell time for elevator doors, s,

t_e = total evacuation time, s,

t_h = time for leveling of elevator car, s,

t_i = time for N people to enter elevator car, s,

t_{io} = time for one person to enter elevator car, s,

t_o = travel time from elevator lobby to outdoors or another safe location, s,

t_r = time for elevator car to make a round trip, s,

t_s = standing time, s,

t_u = time for N people to leave elevator car, s,

t_{uo} = time for one person to leave elevator car, s,

V = velocity, ft/s (m/s)
V_m = normal operating velocity, ft/s (m/s)
α = basic transfer inefficiency
μ = total transfer inefficiency, $\mu = \alpha + \varepsilon + \gamma$
ε = door transfer inefficiency
γ = other transfer inefficiency
η = trip inefficiency

Subscripts

T = end of leveling car motion (also end of travel)
1 = end of constant acceleration motion
2 = end of transitional acceleration motion
3 = end of constant velocity motion
4 = end of transitional deceleration motion
5 = end of constant deceleration motion

REFERENCES

ASME. 2010. Safety Code for Elevators and Escalators. American Society of Mechanical Engineers, New York.

Bazjanac, V. 1974. Another Way Out? *Progressive Architecture*, April.

Bazjanac, V. 1977. Simulation of elevator performance in high-rise buildings under conditions of emergency. *Human Response to Tall Buildings*, ed. by D.J. Conway. Stroudsburg, PA: Dowden, Hutchinson & Ross.

Bukowski, R.W. 2009. Emergency egress from buildings: Part 1: history and current regulations for egress systems design and Part 2: new thinking on egress from buildings. NIST TN 1623, Gaithersburg, MD.

Bukowski, R.W., et al. 2006. Elevator controls. *NFPA Journal* 100(2).

Groner, N. 2009. A situation awareness requirements analysis for the use of elevators during fire emergencies. 4th International Symposium on Human Behaviour in Fire, July 13–15, Robinson College, Cambridge, UK.

Heyes, E., and M. Spearpoint. 2009. Lifts for evacuation—human behaviour considerations. 4th International Symposium on Human Behaviour in Fire, July 13–15, Robinson College, Cambridge, UK.

Kinsey, M., et al. 2009. Investigating the use of elevators for high-rise building evacuation through computer simulation. 4th International Symposium on Human Behaviour in Fire, July 13–15, Robinson College, Cambridge, UK.

Klote, J.H. 1983. Elevators as a means of fire escape. *ASHRAE Transactions* 89(1B).

Klote, J.H. 1984. Smoke control for elevators. *ASHRAE Journal* 26(4).

Klote, J.H., and D.M. Alvord. 1992. Routine for analysis of the people movement time for elevator evacuation. NISTIR 4730, National Institute of Standards and Technology, Gaithersburg, MD.

Klote, J.H., and E. Braun. 1996. Water leakage of elevator doors with application to building fire suppression. NISTIR 5925, National Institute of Standards and Technology, Gaithersburg, MD.

Klote, J.H., and G. Tamura. 1986. Smoke control and fire evacuation by elevators. *ASHRAE Transactions*, 92(1A).

Klote, J.H., et al. 1992. Feasibility and design considerations of emergency evacuation by elevators. NISTIR 4870, National Institute of Standards and Technology, Gaithersburg, MD.

Klote, J.H., and G.T. Tamura. 1986. Smoke control and fire evacuation by elevators. *ASHRAE Transactions*, 92(1A).

Klote, J.H., and G.T. Tamura. 1991a. Design of elevator smoke control systems for fire evacuation. *ASHRAE Transactions* 97(2).

Klote, J.H., and G.T. Tamura. 1991b. Smoke Control Systems for Elevator Fire Evacuation. Elevators and Fire, Council of American Building Officials and National Fire Protection Association. February 19–20, Baltimore, MD.

Kuligowski, E., and R. Bukowski. 2004. Design of occupant egress systems for tall buildings. CIB World Building Congress 2004 Proceedings. CIB HTB T3S1 Design for Fire Safety, May 1–7, Toronto, Canada.

Levin, B.M., and N.E. Groner. 1994. Human factors considerations for the potential use of elevators for fire evacuation of FAA air traffic control towers. NIST GCR 94-656, National Institute of Standards and Technology, Gaithersburg, MD.

Reneke, P.A., R.D. Peacock, and B.L. Hoskins. 2012. Simple estimates of combined stairwell/elevator egress in buildings. NIST Technical Note 1722, National Institute of Standards Technology, Gaithersburg, MD.

Strakosch, G.R., and R.S. Caporale. 2010. *The Vertical Transportation Handbook*, 4th ed. Hoboken, NJ: Wiley & Sons.

Tamura, G.T., and J.H. Klote. 1987. Experimental fire tower studies of elevator pressurization systems for smoke control. *ASHRAE Transactions* 93(2).

Tamura, G.T., and J.H. Klote. 1989. Experimental fire tower studies on mechanical pressurization to control smoke movement caused by fire pressures. Proceedings of International Association for Fire Safety Science. Fire Safety Science, 2nd International Symposium. June 13–17, 1988, Tokyo, Japan.

Tamura, G.T., and J.H. Klote. 1990. Experimental fire tower studies on controlling smoke movement caused by stack and wind action. Proceedings of International Association for Fire Safety Science. Fire Safety Science, 2nd International Symposium. June 13–17, 1988, Tokyo, Japan.

Tubbs, J., and B. Meacham. 2009. Selecting appropriate evacuation strategies for super tall buildings: current challenges and needs. 4th International Symposium on Human Behaviour in Fire, July 13–15, Robinson College, Cambridge, UK.

CHAPTER 13

Zoned Smoke Control

John H. Klote

The traditional approach for HVAC systems is to shut them down during building fires, but HVAC system can be operated in a smoke control mode during building fires. Zoned smoke control consists of exhausting the zone of the fire and possibly pressurizing the surrounding zones. For reasons discussed later, pressurizing the surrounding zones is not recommended for zoned smoke control systems in tall buildings. For zoned smoke control systems that rely on smoke exhaust only, the zoned smoke control can complement the performance of stairwell pressurization in tall and complex buildings. In addition to using the HVAC system, dedicated equipment can be used for zoned smoke control.

ZONED SMOKE CONTROL CONCEPT

In zoned smoke control, a building is divided into a number of zones, each separated from the others by barriers. In the event of a fire, the zone with the fire is called the smoke zone, and the others are called the nonsmoke zones. The zones that border on the smoke zone are called the surrounding zones. Passive smoke protection or pressurization smoke protection is used to limit the extent of smoke spread beyond the smoke zone. It is beyond the capability of smoke control to make conditions tenable in the smoke zone, and it is intended that occupants evacuate the smoke zone as soon as possible.

Often, the smoke zone is one floor of the building as shown in Figure 13.1a. In this figure, the smoke zone is indicated by a minus sign and the surrounding zones are indicated by a plus sign. The smoke zone can consist of a number of floors. A common approach is to make the smoke zone be the fire floor plus the floor directly above and below the fire floor, as shown in Figure 13.1b. In a relatively low sprawling building made of a number of wings, the smoke zone can be part of a floor as in Figure 13.1c.

A surrounding zone can be one floor as in Figures 13.1a and b, and it can be part of a floor as in Figure 13.1c. A surrounding zone can also be a number of floors as shown in Figure 13.1d.

The methods that can be used to treat the smoke zone are: (1) fan-powered exhaust, (2), passive smoke control using smoke barriers (3) exterior wall vents, or (4) smoke shafts. Fan-powered smoke exhaust is the most common method, and passive smoke control using smoke barriers may be satisfactory when fan powered exhaust is not practical. Exterior wall vents and smoke shafts are not commonly used, but they are discussed later.

The methods that can be used for the zones surrounding the smoke zone are (1) fan-powered pressurization or (2) passive smoke control using smoke barriers. Fan-powered pressurization of the surrounding zones has a negative impact on stairwell pressurization as discussed in the next section. For the rest of this chapter, fan-powered pressurization will be called pressurization, and fan-powered exhaust will be called exhaust.

Considering wide variations in buildings and the treatments that are possible for zones, very many kinds of zoned smoke control systems are possible. In this chapter some of these systems will be discussed.

Smoke Zone Size and Arrangement

Traditionally, the smoke zone is large enough so that any hot gases from the fire space are mixed with

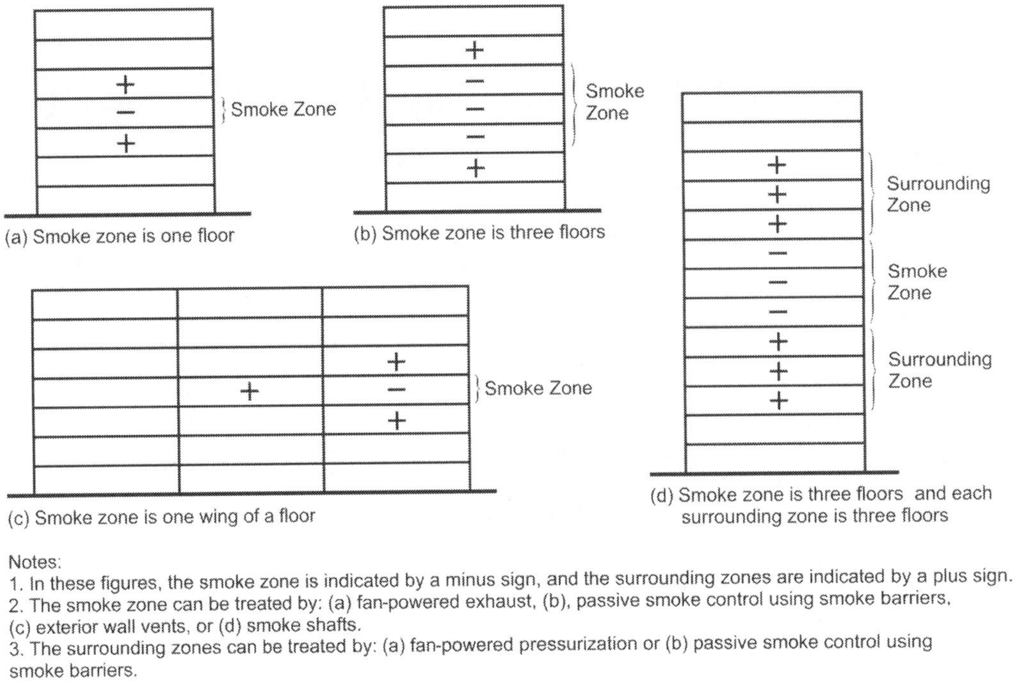

Figure 13.1 *Smoke arrangements of smoke control zones.*

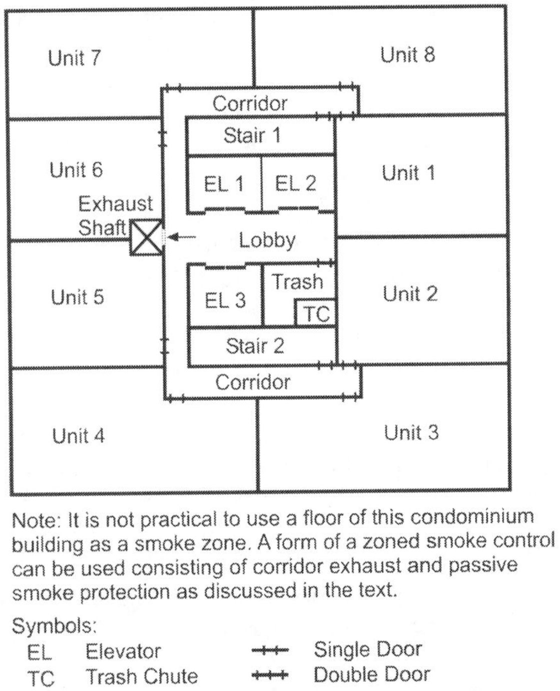

Figure 13.2 *Example of corridor exhaust and passive smoke control.*

air from other spaces, and this reduces the temperature of the gases in the exhaust fan. Calculation of this temperature is addressed later. Also, the traditional approach to zoned smoke control was intended for spaces that are open so that exhaust and pressurization can be achieved throughout the space. For open plan office building, making the smoke zone one or more floors is practical. The presence of a few private offices in an open office plan floor will not prevent the floor from being used as a smoke control zone or a floor of a smoke control zone.

When the floors of a building are divided into many rooms with normally closed doors, these floors do not lend themselves to the traditional concept of zoned smoke control. This can also be said of wings of a building that are divided into many rooms with normally closed doors. For such applications, a form of zoned smoke control can be used that relies on a combination of corridor exhaust and passive smoke control using smoke barriers. Figure 13.2 shows a floor of a condominium building that can be considered a form of a smoke zone. The floor has corridor exhaust, and the other spaces rely on passive smoke protection of the corridor walls and ceiling floor assembly of the other spaces. This passive protection tends to minimize smoke flow through the ceiling floor assembly during a building fire.

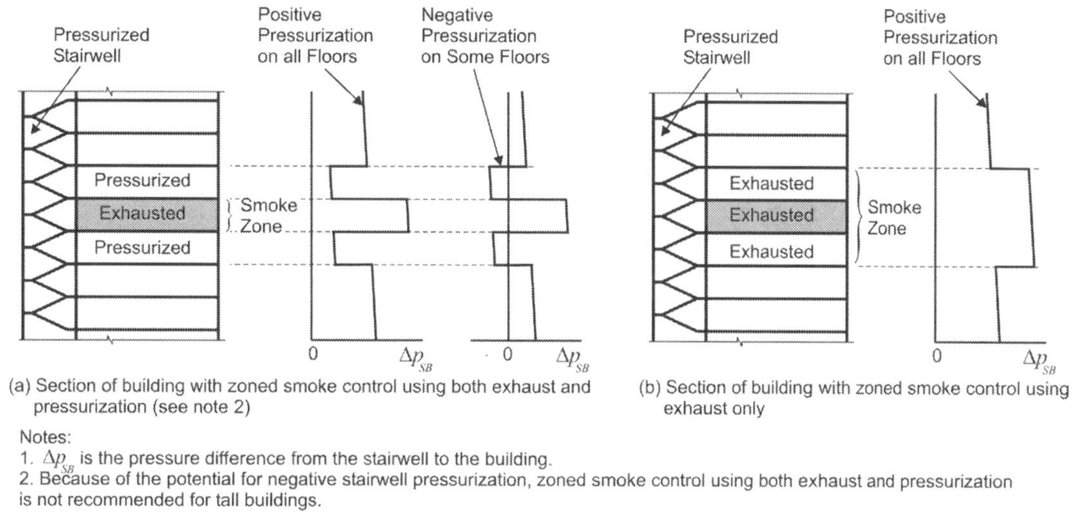

Figure 13.3 *Interaction between zoned smoke control and pressurized stairwells.*

Interaction with Pressurized Stairs

The interaction of zoned smoke control with pressurized stairwells can have a significant impact on the pressure differences across the stairwell doors. The following discussion is about smoke zones that are one floor and surrounding zones consisting of one floor above and one floor below. However, the same kind of interactions can happen with smoke zones and surrounding zones that are more than one floor.

For zoned smoke control using both exhaust and pressurization, the pressure differences, Δp_{SB}, from the stairwell to the building are shown in Figure 13.3a. Exhaust of the smoke zone increases the pressure difference across pressurized stairwell doors on the floor or floors of the smoke zone. Pressurization of the surrounding zones decreases the pressure difference across pressurized stairwell doors on these floors.

It can be seen on Figure 13.3a that Δp_{SB} can be positive on all the floors or on some floors Δp_{SB} can be negative. Negative pressurization can happen on the floors that are pressurized, and this negative pressurization has the potential for the significant failure mode discussed below.

In Figure 13.3a, smoke should be prevented from reaching the floor above the smoke zone, and negative stairwell pressurization should not compromise tenability of the stairwell. The effectiveness of this depends on proper identification of the fire floor. Properly maintained fire alarm systems are very good at identifying the location of a fire, but no system is perfect. In some fires, the first smoke detector to activate is a floor or so above the fire floor. This can be attributed to any of the following: (1) smoke flowing through a complex route to a floor above the fire, (2) smoke detectors not working properly on the fire floor, and (3) signals from smoke detectors being misidentified.

Regardless of the reason, when a fire floor is incorrectly identified, the smoke zone is incorrectly chosen. In this situation, the failure mode is that inadvertent pressurization of the fire floor can push smoke into the stairwells. An additional concern is that if this failure mode happens, it will probably happen to all the stairwells serving the fire floor. This failure mode is more of a concern for tall buildings because: (1) acceptable pressurization is more difficult in tall buildings than in short ones and (2) stairwell smoke protection is more important in tall buildings than in short ones. Occupant density is another factor regarding the importance of stairwell smoke protection. In this context, a tall building might be thought of as one having a minimum of about 10 stories. Because of this failure mode, it is recommended that zoned smoke control using systems employing both exhaust and pressurization not be used for tall buildings where protection of the stairwells is especially important. Alternatively, an analysis of this failure mode could be performed that includes factors such as evacuation time, emergency response time, and probability using the FSCS for corrective action.

The zoned smoke control shown in Figure 13.3b does not have this failure mode. The zoned smoke control system of this figure consists of a three-story smoke zone that is exhausted and the surrounding zones rely on passive smoke protection. The exhaust acts to increase Δp_{SB} for the three floors of the smoke zone. Because this system does not have pressurization of surrounding zones, Δp_{SB} is not reduced for surrounding zones, and this eliminates the failure mode discussed.

In Figure 13.3b, the fire floor is shaded, and the smoke zone consists of the fire floor and the floors directly above and below. It is expected that there will be some smoke flow to the floor above the fire floor, and there may be some smoke flow to the floor below the fire floor. This smoke flow is restricted by the floor-ceiling assembly. A floor-ceiling assembly is a passive smoke barrier like those discussed in Chapter 9, and it has significant resistance to smoke flow. Even a floor-ceiling assembly not constructed as a passive smoke barrier has considerable resistance to smoke flow provided that the only openings in it are construction cracks and small cracks around penetrations. This means that there will be some amount of time for occupants of the floors directly above and below the fire floor to evacuate those floors. Further, the small amount of smoke on these floors should act to convince occupants of the serious nature of the fire such that premovement time will be significantly reduced. For tall buildings that are to have zoned smoke control, the kind of system shown in Figure 13.3b is suggested in place of the one in Figure 13.3a.

Analysis

In Chapter 10, the reasons for using CONTAM for design analysis of stairwell pressurization systems are discussed. Much of that discussion applies to zoned smoke control systems, except that an algebraic method of analysis of zoned smoke control systems is not presented. It may be possible to use some algebraic equations or rules of thumb for simple zoned smoke systems in simple buildings, but more realistic analysis with CONTAM is needed for practically all applications. This is especially so for buildings that have zoned smoke control and other kinds of smoke control systems. CONTAM (Chapter 14) has become the de facto standard network model for analysis of pressurization smoke control systems, but it is possible to use another network model.

When a CONTAM analysis shows that a specific zoned smoke control system in a building cannot be balanced to perform as intended, a new approach is needed. There are two categories of new approaches (1) use an alternative zoned smoke control system, and (2) modify the building. Various zoned smoke control systems are discussed later. The new approach needs a CONTAM analysis to determine if it capable of being balanced to perform as intended.

For example, during wind conditions, the pressure difference across one stairwell door could be too high, and a new approach that could be considered is use of an automatic door opener on the door at that location. Another new approach could be an automatically opening vent at that location. Any such vent in a fire-rated wall needs a fire damper.

USE OF HVAC SYSTEM

The HVAC system discussed here is a variable-air-volume (VAV) system with an economizer. For reasons of energy conservation, this system is extensively used. The economizer allows outdoor air to be used for cooling when conditions permit. Because of the large amounts of heat generated by people and equipment, office buildings often need cooling even during cold weather. Figure 13.4a shows this system in normal HVAC operating mode. This mode includes the return damper, exhaust damper, and outdoor air damper modulating to adjust the amount of outdoor air supplied to the building. The VAV system has VAV fans for supply and return that are used to adjust the flow rates of the HVAC system as needed. The system serves a number of HVAC zones, and each zone has a terminal box that controls the amount of air supplied to that space. For more information about this and other HVAC systems, see Chapter 7. For information about the control of terminal boxes in a smoke control mode, see "Control of Devices that are not Part of the Smoke Control System" in Chapter 8.

The mode of operation used for an HVAC system depends on the kind of zoned smoke control and on the arrangement of HVAC zones in the building. The arrangements of the HVAC zones discussed are (1) separate HVAC systems for each floor and (2) one HVAC system for many floors of a building. There are other HVAC arrangements, and the following ideas can be adapted to those systems.

Separate HVAC Systems for Each Floor

Zoned Smoke Control by Pressurization and Exhaust

For buildings that have separate HVAC systems for each floor, zoned smoke control can consist of putting the HVAC systems in the pressurization mode or the smoke exhaust mode as appropriate. HVAC systems serving other floors either may be shut down or allowed to operate in the normal node.

For an HVAC system in the pressurization mode (1) the return damper and exhaust damper are closed, (2) the outdoor air damper is opened, (3) the return fan is shut down, and (4) the supply fan is set to a flow rate determined during balancing of the smoke control system. This pressurization mode is shown in Figure 13.4b.

For an HVAC system in the exhaust mode: (1) the return damper and outdoor air damper are closed, (2) the

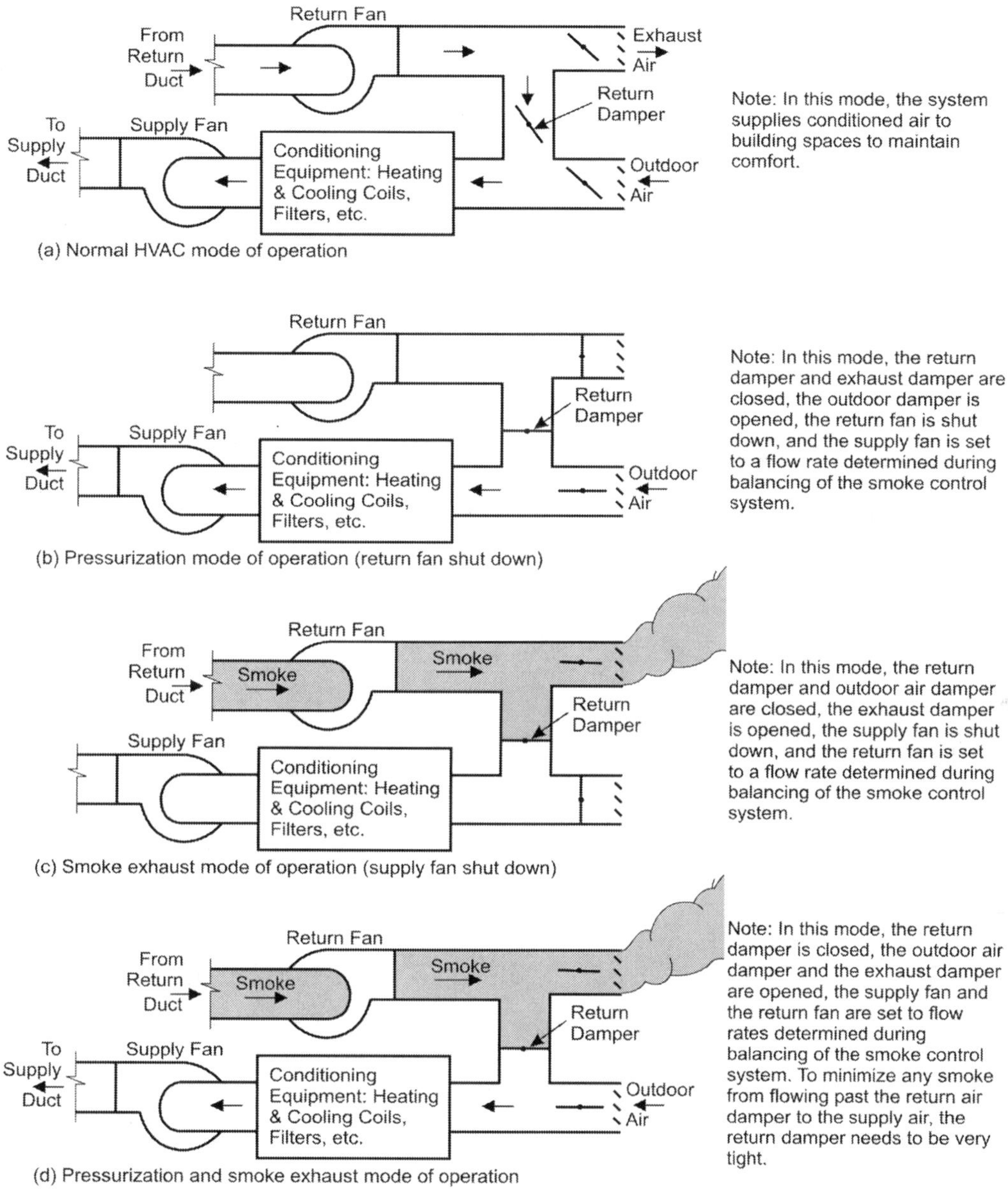

Figure 13.4 *VAV system with economizer can be used for zoned smoke control.*

exhaust damper is opened, (3) the supply fan is shut down, and (4) the return fan is set to a flow rate determined during balancing of the smoke control system. This exhaust mode is shown in Figure 13.4c.

Zoned Smoke Control by Exhaust Only

For buildings that have separate HVAC systems for each floor, this system consists of putting the HVAC systems in the smoke exhaust mode (previously discussed) as appropriate.

HVAC System for Many Floors

An HVAC system can serve many floors or all the floors of a building. The discussion that follows focuses on an HVAC system that serves all the floors, but the concepts also apply to HVAC systems that serve many floors. Figure 13.5a shows an HVAC system in normal operating mode serving all the floors of a building. For this building, the stairwells are pressurized in the event of a fire. For simplicity, the distribution ducts on each floor and equipment in the penthouse are not shown on Figure 13.5.

Zoned Smoke Control by Pressurization and Exhaust

Figure 13.5b shows zoned smoke control consisting of fire floor exhaust with pressurization floors directly above and below. In the event of a fire with this system: (1) the smoke damper in the return duct is open on the fire floor, (2) the rest of the smoke dampers in the return duct are closed, (3) the smoke dampers in the supply duct on the floors directly above and below the fire floor are open, and (4) the rest of the smoke dampers in the supply duct are closed. These damper settings are intended to (1) concentrate exhaust on the fire floor, (2) concentrate pressurization on the floors above and below the fire floor, and (3) prevent smoke from migrating to other zones through the return ductwork.

For this zoned smoke control system, the mode of operation of the equipment in the mechanical penthouse is shown in Figure 13.4d. In this mode (1) the return damper is closed, (2) the outdoor air damper and the exhaust damper are opened, and (3) the supply fan and the return fan are set to flow rates determined during balancing of the smoke control system. To minimize any smoke from flowing past the return air damper to the supply air, the return damper needs to be very tight. It is suggested that this return damper be a Class I smoke damper.

Zoned Smoke Control by Exhaust Only

Figure 13.5c shows zoned smoke control consisting of fire floor exhaust and exhaust of the floors directly above and below the fire floor. In the event of a fire, (1)

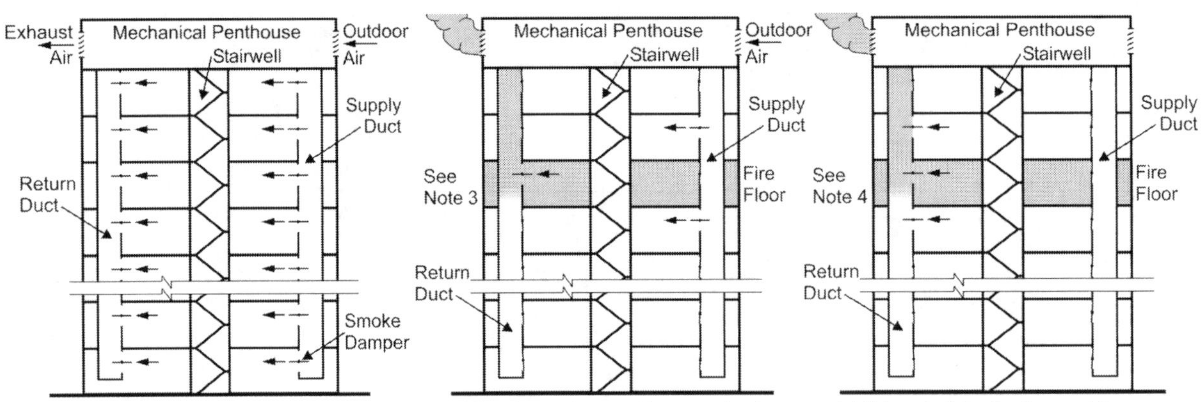

(a) Normal HVAC Operation

(b) Fire floor exhaust with pressurized floors directly above and below.

(c) Exhaust of the fire floor and the floors directly above and below.

Notes:

1. For the systems shown above, the stairwells are pressurized in the event of a fire.
2. For simplicity, the distribution ducts on each floor and equipment in the penthouse are not shown.
3. In the event of a fire with system (b) above: (1) the smoke damper in the return duct is open on the fire floor, (2) the rest of the smoke dampers in the return duct are closed, (3) the smoke dampers in the supply duct on the floors directly above and below the fire floor are open, (4) the rest of the smoke dampers in the supply duct are closed, (5) the HVAC equipment in the mechanical penthouse is in the mode of operation shown in Figure 13.3(d). Because of a failure mode, this system is not recommended for tall buildings. For more information about this failure mode, see the text.
4. In the event of a fire with system (c) above: (1) the smoke dampers in the return duct are open on the fire floor and the floors directly above and below the fire floor, (2) the rest of the smoke dampers in the return duct are closed, (3) the smoke dampers in the supply duct are closed, (4) the HVAC equipment in the mechanical penthouse is in the mode of operation shown in Figure 13.3(c).
5. The supply and return ducts that penetrate floors in the systems above are in fire-rated shafts.

Figure 13.5 *Zoned smoke control with HVAC system that serves all floors of building.*

the smoke dampers in the return duct are open on the fire floor and the floors directly above and below the fire floor, (2) the rest of the smoke dampers in the return duct are closed, and (3) all smoke dampers in the supply duct are closed.

For this zoned smoke control system, the mode of operation of the equipment in the mechanical penthouse is shown in Figure 13.4c. In this mode (1) the return damper and outdoor air damper are closed, the exhaust damper is opened, (2) the supply fan is shut down, and (3) the return fan is set to a flow rate determined during balancing of the smoke control system.

DEDICATED EQUIPMENT

The preceding discussion focused on use of the HVAC system for smoke control, but smoke control equipment can be dedicated completely to smoke control use. This dedicated equipment can consist of fans, ducts, and dampers. A major advantage of using dedicated equipment over using HVAC equipment is that dedicated equipment is not likely to be inadvertently modified during modifications to the HVAC system. Dedicated equipment can be used when the HVAC system does not lend itself to zoned smoke control applications.

Zoned Smoke Control by Pressurization and Exhaust

This system has the following dedicated equipment: supply fan, supply duct, exhaust fan, and exhaust duct. The ducts are fire rated shafts. The supply and exhaust shafts each have a damper between the duct and the building space on every floor. When it is used with stairwell pressurization, the stairwell has its own dedicated supply fan. This system is shown in Figure 13.6a. When the system is not operating, the fans are off and the dampers are closed.

In the event of a fire with this zoned smoke control system (1) the smoke damper in the exhaust duct on the fire floor is opened, (2) the rest of the smoke dampers in the exhaust duct remain closed, (3) the smoke dampers in the supply duct on the floors directly above and below the fire floor are opened, (4) the rest of the smoke dampers in the supply duct remain closed, and (5) the supply and exhaust fans are set to flow rates determined during balancing of the smoke control system. To prevent unwanted interaction with the zoned smoke control system, the HVAC system needs to be shut down either to (1) the floor being exhausted and the floors being pressurized or (2) the entire building. Because of the potential for negative stairwell pressurization, this system is not recommended for tall buildings for the reasons discussed.

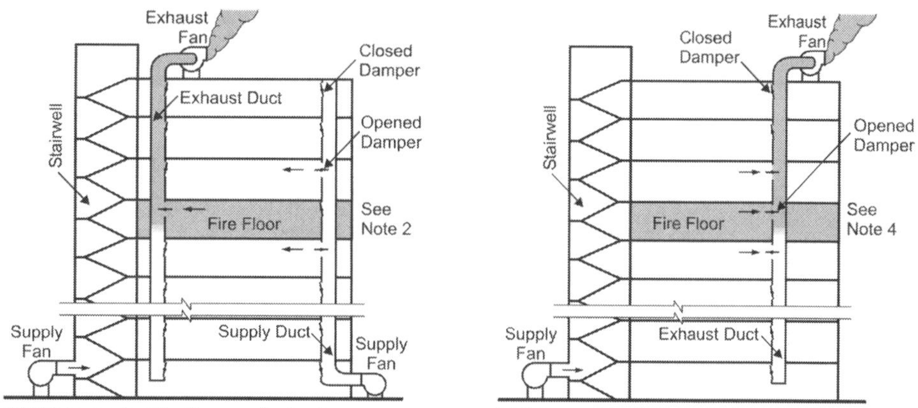

(a) Fire floor exhaust with pressurization floors directly above and below (see note 3)

(b) Exhaust of the fire floor and the floors directly above and below.

Notes:
1. For the systems shown above, the stairwells are pressurized in the event of a fire.
2. In the event of a fire with system (a) above: (1) the smoke damper in the exhaust duct on the fire floor is opened, (2) the rest of the smoke dampers in the exhaust duct remain closed, (3) the smoke dampers in the supply duct on the floors directly above and below the fire floor are opened, (4) the rest of the smoke dampers in the supply duct remain closed, and (5) the supply and exhaust fans are set to flow rates determined during balancing of the smoke control system.
3. Because of a failure mode, the system (a) is not recommended for tall buildings. For more information about this failure mode, see the text.
4. In the event of a fire with system (b) above: (1) the smoke dampers in the exhaust duct are opened on the fire floor and the floors directly above and below the fire floor, (2) the rest of the smoke dampers in the return duct remain closed, and (3) the exhaust fan is set to a flow rate determined during balancing of the smoke control system.
5. In system (a) above, the supply duct and exhaust duct are in fire rated shafts (not shown). In system (b) above, the exhaust duct is located in a fire rated shaft (not shown).

Figure 13.6 *Zoned smoke control with stairwell pressurization using dedicated equipment.*

Zoned Smoke Control by Exhaust Only

This system has a dedicated exhaust fan and a dedicated exhaust shaft. The exhaust duct is in a fire rated shaft. As with the previous system, there is a second dedicated supply fan when there is stairwell pressurization. This system is shown in Figure 13.6b. When the system is not operating, the fans are off and the dampers are closed. In the event of a fire with this zoned smoke control system, (1) the smoke dampers in the exhaust duct are opened on the fire floor and the floors directly above and below the fire floor, (2) the rest of the smoke dampers in the return duct remain closed, and (3) the exhaust fan is set to a flow rate determined during balancing of the smoke control system. To prevent unwanted interaction with the zoned smoke control system, the HVAC system needs to be shut down either to (1) the floors being exhausted or (2) the entire building.

EXHAUST FAN TEMPERATURE

When an exhaust fan moves hot smoke, the operating temperature should be determined so that an appropriate fan can be specified. Fans that operate at elevated temperatures need to be rated for at least the calculated fan operating temperature, T_{fan}.

The mass flow through an exhaust fan is

$$m_e = 0.0167\rho_{fan}V_{fan}$$
$$m_e = \rho_{fan}V_{fan} \text{ for SI} \tag{13.1}$$

where

m_e = mass flow rate of exhaust fan, lb/s (kg/s),
ρ_{fan} = density of gases in exhaust fan, lb/ft^3 (kg/m^3),
V_{fan} = volumetric flow rate of exhaust fan, ft^3/min (m^3/s).

Using the ideal gas equation, the mass flow rate through the fan is a function of the absolute temperature of the gases in the fan.

$$m_e = \frac{1.067V_{fan}}{RT_{fan}}$$
$$m_e = \frac{V_{fan}}{RT_{fan}} \text{ for SI} \tag{13.2}$$

where

T_{fan} = absolute temperature of gases in fan, °R (K),
R = gas constant, 53.34 ft lbf/lbm·°R (287 J/kg·K).

Increased fan temperature decreases the mass flow rate of the exhaust fan resulting in a reduction in pressure differences produced by the smoke control system. The temperature of the gases in the fan depends on (1) the temperature of gases entering the exhaust duct system, (2) the mass flow of gases entering the exhaust duct system, and (3) the heat transfer from the exhaust duct system to the surroundings. Analysis of this heat transfer is somewhat cumbersome because (1) the gas temperature changes through the duct, (2) the convection coefficient from the hot gases to the duct change as the gas flows through the duct, and (3) the convection coefficient from the hot duct to the surroundings change as the gas flows through the duct. For steady flow when the mass flows through the duct system are known, the duct can be divided into sections, and basic equations of heat transfer can be used to calculate the temperature at each section of duct. These basic equations are in a number of heat transfer texts such as Holman (2002) and Incropera and DeWitt (2002). The temperature at the last section of duct is the temperature of the gases entering the exhaust fan.

An upper limit of the exhaust fan temperature can be calculated by neglecting heat transfer which results in a conservatively high temperature. The following equations use this approach. Considering constant specific heat, the fan temperature can be expressed as

$$T_{fan} = \frac{\sum_{j=1}^{n} \rho_j V_j T_j}{\sum_{j=1}^{n} \rho_j V_j} \tag{13.3}$$

where

T_{fan} = temperature of gases in exhaust fan, °F (°C),
ρ_j = density of gases in space j, lb/ft^3 (kg/m^3),
V_j = volumetric flow rate of exhaust from space j, cfm (m^3/s),
T_j = temperature of gases in space j, °F (°C),
n = number of spaces.

The temperatures T_j in the above equation can obtained from a fire simulation by a zone fire model or a computational fluid dynamic (CFD) model. Fire test data could also be used, provided that the test conditions are similar to those anticipated for the fire in question. It is suggested that the fire be a fully developed fire, but for a minimum, a shielded fire can be considered. For information about these fires see Chapter 5, and for information about zone fire modeling see Chapter 18. For information about CFD modeling see Chapter 20.

Example 13.1. Exhaust Fan Temperature

For the smoke zone in Figure 13.7, calculate the exhaust fan temperature neglecting heat transfer from the exhaust duct. This smoke zone is divided into types of spaces, and the exhaust temperatures of these spaces were calculated with a zone fire model with a fully developed fire. The temperatures, densities, and exhaust flow rates of these spaces are listed below.

	Temperature		**Density**		**Exhaust Flow**	
Type of Space	**°F**	**°C**	**lb/ft³**	**kg/m³**	**cfm**	**m³/s**
Fire space	1700	927	0.0184	0.294	600	0.283
Communicating space	800	427	0.0315	0.504	600	0.283
Removed spaces	400	204	0.0462	0.739	3300	1.56

$$T_{fan} = \frac{\sum_{j=1}^{n} \rho_j V_j T_j}{\sum_{j=1}^{n} \rho_j V_j} = \frac{0.184(600)(1700) + 0.0315(600)(800) + 0.0462(3300)(400)}{0.184(600) + 0.0315(600) + 0.0462(3300)} = 520°\text{F } (270°\text{C})$$

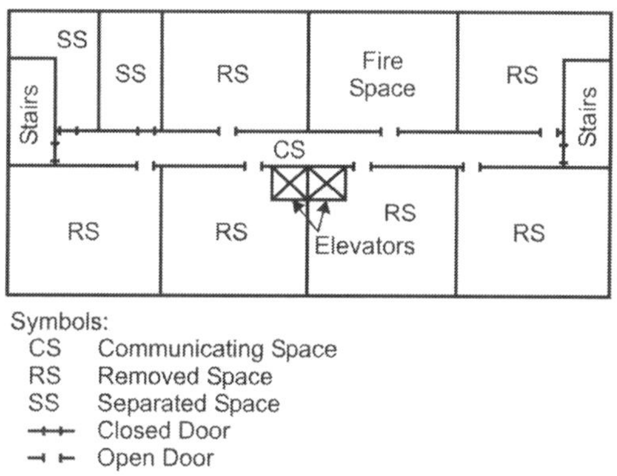

Figure 13.7 *Spaces in smoke zone of Example 13.1.*

Example 13.1 shows calculation of fan temperature for the smoke zone of Figure 13.7

The fan temperature also can be calculated as

$$T_{fan} = \frac{T_r}{(1 - \phi)} \tag{13.4}$$

where

T_{fan} = absolute temperature of gases in exhaust fan, °R (K),

T_r = absolute temperature of gases in fan under normal conditions, °R (K),

ϕ = fraction reduction in mass flow rate through fan.

EXTERIOR WALL VENTS

Exterior wall vents are not normally used for zoned smoke control, but they are included for completeness. Possibly this system could be used in or adapted to a sustainable building. This system is shown in Figure 13.8a, and in the fire mode (1) wall vents open on the fire floor, (2) the floors directly above and below the fire floor are pressurized, and (3) the HVAC system on the fire floor is shut down.

The National Research Council of Canada conducted a study of exterior wall vents, which included experiments and theoretical analysis (Tamura 1978; Tamura and Shaw 1978). They developed a relationship between pressure differences and flow areas:

$$\frac{\Delta p_{BF}}{\Delta p_{BO}} = \frac{(A_v / A_e)^2}{1 + (A_v / A_e)^2} \tag{13.5}$$

where

Δp_{BF} = pressure difference from surrounding zones to smoke zone, in. H_2O (Pa),

Δp_{BO} = pressure difference from surrounding zones to outdoor, in. H_2O (Pa),

A_v = flow area of exterior vent, ft² (m²),

A_e = effective flow area of enclosure of smoke zone to surrounding zones, ft² (m²).

Equation 13.5 is shown graphically in Figure 13.8b. A shortcoming of this equation is that it is not appropriate for buildings with pressurized stairwells. The exterior vents can consist of windows or panels that open automatically when the smoke control system is activated. For information about effective flow areas see Chapter 3. It is recommended that design analysis of wall vents be done with a network model like CONTAM, and such an analysis needs to include stack effect, reverse stack effect, and wind.

SMOKE SHAFTS

Smoke shafts are not normally used today, but they were used in Europe at one time. Possibly, smoke shafts could be used in or adapted to sustainable buildings. The original system did not take advantage of dampers, but an approach using dampers is discussed here for completeness. A smoke shaft is a vertical shaft intended to be a path for smoke movement from the fire floor to above the level of the roof.

This system is shown in Figure 13.9, and there are dampers at every floor and at the roof. In the event of a fire (1) the damper opens on the fire floor and (2) the dampers at the top of the smoke shaft open. This system relies on the buoyancy of hot smoke to force smoke through the shaft and out at the top of the building. This flow can be aided by normal stack effect, but reverse stack effect can have a negative impact on system performance.

Tamura and Shaw (1973) developed an analysis of this kind of smoke shaft, but their analysis does not include wind effects or reverse stack effect. It is recommended that design analysis of smoke shafts be done with a network model like CONTAM, and such an analysis needs to include stack effect, reverse stack effect, and wind.

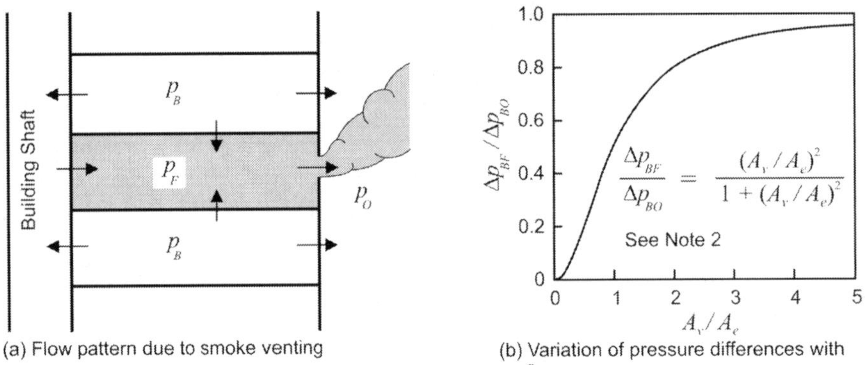

Notes:
1. Exterior wall vents are not normally used, but they are included for completeness.
2. This system does not include pressurized stairwells.
3. In the event of a fire with system above: (1) wall vents open on the fire floor, (2) the floors directly above and below the fire floor are pressurized, and (3) the HVAC system on the fire floor is shut down.
4. For the meaning of the variables, see the text.

Figure 13.8 *Exterior wall vents with pressurization of floors directly above and below fire floor.*

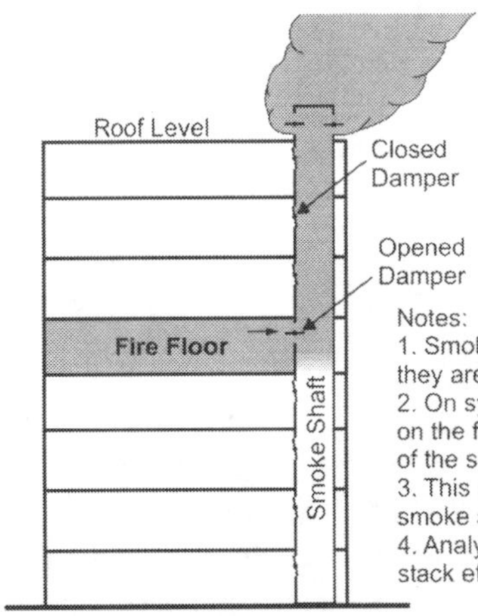

Notes:
1. Smoke shafts are not normally used, but they are included for completeness.
2. On system activation, the damper opens on the fire floor, and the dampers at the top of the smoke shaft open.
3. This system relies on buoyancy of hot smoke and stack effect.
4. Analysis of this system needs to include stack effect, reverse stack effect and wind.

Figure 13.9 *Smoke venting through smoke shaft.*

NOMENCLATURE

A_e = effective flow area of enclosure of the smoke zone to surrounding zones, ft^2 (m^2)

A_v = flow area of exterior vent, ft^2 (m^2)

m_e = mass flow rate of exhaust fan, lb/s (kg/s)

n = number of spaces

R = gas constant, 53.34 ft lbf/lbm·°R (287 J/kg·K)

T_{fan} = absolute temperature of gases in fan, °R (K)

T_j = temperature of gases in space j, °F (°C)

T_r = absolute temperature of gases in fan under normal conditions, °R (K)

V_{fan} = volumetric flow rate of exhaust fan, ft^3/min (m^3/s)

V_j = volumetric flow rate of exhaust from space j, cfm (m^3/s)

ϕ = allowable fraction reduction in mass flow rate through fan

ρ_{fan} = density of gases in exhaust fan, lb/ft^3 (kg/m^3)

ρ_j = density of gases in space j, lb/ft^3 (kg/m^3)

Δp_{BF} = pressure difference from surrounding zones to smoke zone, in. H_2O (Pa)

Δp_{BO} = pressure difference from surrounding zones to outdoors, in. H_2O (Pa)

REFERENCES

Holman, J.P. 2002. *Heat Transfer*, 10th ed. New York: McGraw-Hill.

Incropera, F.P., and DeWitt, D.P. 2002. *Fundamentals of Heat and Mass Transfer*, 5th ed. Hoboken, NJ: Wiley.

Tamura, G.T. 1978. Exterior wall venting for smoke control in tall office buildings. *ASHRAE Journal* 20(8).

Tamura, G.T., and C.Y. Shaw. 1973. Basis for the design of smoke shafts. *Fire Technology* 9(3).

Tamura, G.T., and C.Y. Shaw. 1978. Experimental studies of mechanical venting for smoke control in tall office buildings. *ASHRAE Transactions* 86(1).

CHAPTER 14

Network Modeling and CONTAM

John H. Klote

Network models are a class of software that can simulate the flow of air or water through a complex system of paths which is called the network. This chapter discusses network modeling in a general way to help readers understand this type of modeling, and this chapter discusses some of the features of CONTAM (Walton and Dols 2005), including user information, specifically for applications of pressurization smoke control systems.

CONTAM was developed for indoor air quality applications, but it is probably the most used computer software in the world for analysis of smoke control systems that rely on pressurization. This software can analyze airflow in buildings, and also has the ability to simulate the flow of contaminants. CONTAM is a product of the U.S. National Institute of Standards and Technology (NIST), and it can be downloaded from the NIST Web site at no cost.

The equations of this chapter are intended to explain the concepts of network modeling and the models that perform the calculations. Therefore, units are not given for the variables of this chapter, but these equations are valid for SI units as discussed in Chapter 1.

PURPOSE OF NETWORK MODELING

The primary purpose of network simulations is to determine if a particular smoke control system in a particular building is capable of being balanced such that it will perform as intended. Network models are capable of simulating the pressures and flows throughout very large and complex building networks with high accuracy. Figure 14.1 shows CONTAM representations of some floors of projects that illustrate the complexity of buildings that can be analyzed with CONTAM. The symbols on this figure are discussed later.

There are many flow paths in buildings including gaps around closed doors, open doors, and construction cracks in walls and floors. These flow paths can only be approximated for a design analysis. For this reason, the results of a network model simulation are only approximations, and the actual pressures and flows may be somewhat different. However, these approximate results can be useful in identifying problems with specific smoke control systems. If such problems are identified, the smoke control system can be modified appropriately. A secondary purpose of these simulations is to provide information to help size the system components such as supply fans, exhaust fans, and vents.

EARLY NETWORK MODELS

In the rest of this chapter, the term network model will be used to mean a model that simulates the flow of air in buildings, and many of these models are also capable of simulating contaminant flow. The National Research Council of Canada (NRCC) developed airflow programs (Sander 1974; Sander and Tamura 1973). While ASCOS (Klote 1982) was extensively used for smoke control design for much of the 1980s and 1990s, it was only intended to be a research tool. Yoshida et al. (1979), Butcher et al. (1969), Barrett and Locklin (1969), Evers and Waterhouse (1978), and Wakamatsu (1977) developed programs that also simulate smoke concentrations.

The early network models suffered from convergence problems. An ASHRAE-funded research project (Wray and Yuill 1993) evaluated numerical routines that could be used for analysis of smoke control systems.

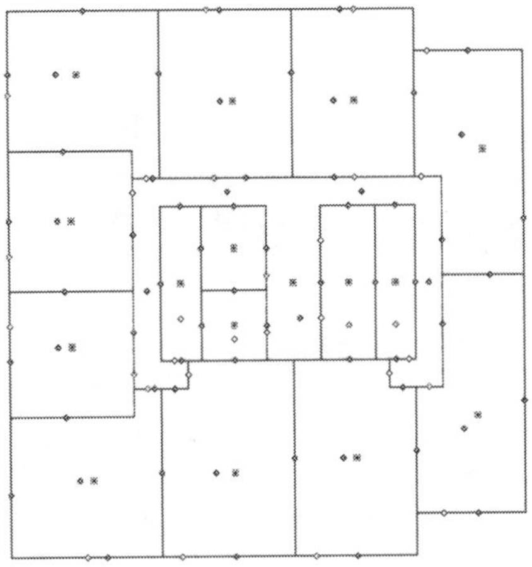

(a) Level 8 of a condominium building

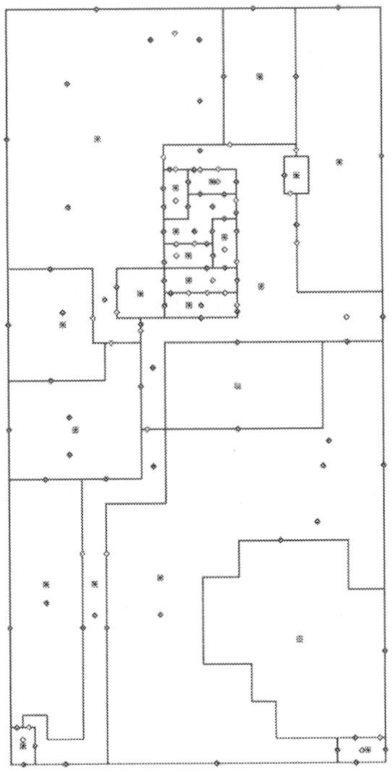

(b) Ground level of a hotel

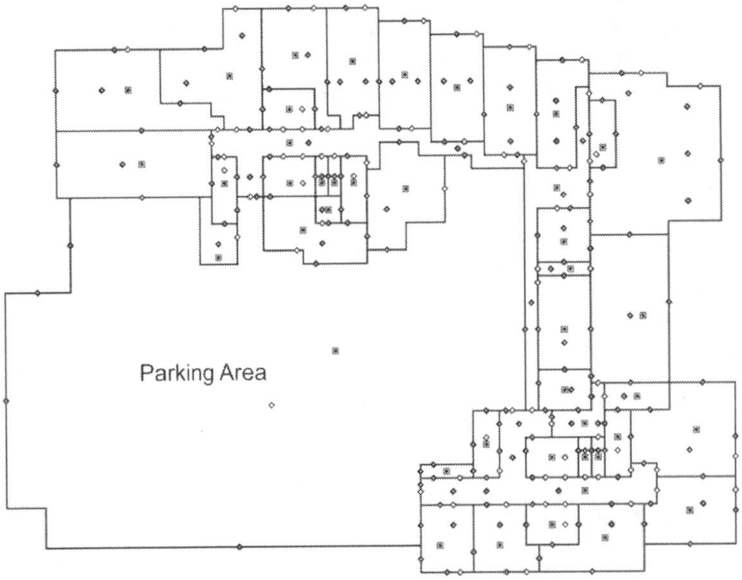

(c) Level 5 of a shopping mall/condominimum complex

Figure 14.1 *CONTAM representations illustrating complex buildings that can be analyzed.*

This study showed that the AIRNET routine developed by Walton (1989) was the best algorithm based on convergence, computational speed, and use of computer memory. CONTAM was developed with an improved version of the algorithm from AIRNET. None of the routines of this study take advantage of the repetitive nature of building flow networks, so data entry for these routines is difficult and time consuming. CONTAM has a sophisticated graphic interface that eases data entry and helps reduce data input errors.

NETWORK MODEL

These network models represent a building by a network of spaces or nodes, each at a specific pressure and temperature. The stairwells and other shafts can be modeled by a vertical series of spaces, one for each floor. Air flows through leakage paths from regions of high pressure to regions of low pressure. These leakage paths are doors and windows that may be opened or closed. Leakage can also occur through partitions, floors, and exterior walls and roofs. The airflow through a leakage path is a function of the pressure difference across the leakage path.

In this model, air from outside the building can be introduced by a pressurization system into any level of a shaft or other building spaces. This allows simulation of stairwell pressurization, elevator shaft pressurization, stairwell vestibule pressurization, and pressurization of any other building space. Also, building spaces can be exhausted. This allows analysis of zoned smoke control systems and other systems that include fire floor exhaust.

The pressures throughout the building and steady flow rates through all the flow paths are obtained by solving the airflow network, including the driving forces such as wind, the pressurization system, and indoor-to-outdoor temperature difference.

Mass Flow Equations

The following is a generalized treatment of network models. This overview only considers one flow path between any two nodes, but many network models allow a number of flow paths between the same two points. The mass flow in a path between two nodes can be represented as

$$m_{ij} = f_{ij}(\Delta p_{ij}) \tag{14.1}$$

where

m_{ij} = mass flow from node i to node j,

f_{ij} = denotes functional relationships appropriate for path between nodes i and j,

Δp_{ij} = pressure difference from node i to node j.

A number of functional relationships for flow usually includes the orifice equation, and some models are capable of using many functional relationships for various types of flows and flow elements. A function can also be used to represent the flow of a fan, which is an exception in that fan flow is from a node of lower pressure to a node of higher pressure.

The pressure difference can be expressed as

$$\Delta p_{ij} = p_i - p_j + \rho_i g(z_i - z_j) \tag{14.2}$$

where

p_i = pressure at node i,

p_j = pressure at node j,

ρ_i = gas density at node i,

z_i = elevation of node i,

z_j = elevation of node j,

g = acceleration of gravity.

For steady flow, conservation of mass at node i can be stated as the sum of the mass flows leaving node i is zero. In equation form, this is

$$\sum_{j=1}^{M} f_{ij}(\Delta p_{ij}) = 0 \tag{14.3}$$

where M is the number of flow paths between node i and other spaces. The mass flows entering node i have negative values. Writing the conservation of mass equations for each node in the building results in

$$\begin{aligned} f_{11}(\Delta p_{11}) + f_{12}(\Delta p_{12}) + \ldots f_{1N}(\Delta p_{1N}) &= 0, \\ f_{21}(\Delta p_{21}) + f_{22}(\Delta p_{22}) + \ldots f_{2N}(\Delta p_{2N}) &= 0, \\ &\vdots \\ f_{N1}(\Delta p_{N1}) + f_{N2}(\Delta p_{N2}) + \ldots f_{NN}(\Delta p_{1N}) &= 0. \end{aligned} \tag{14.4}$$

Substituting Equation 14.2 into this set of equations yields

$$\begin{aligned} F_1(p_1, p_2, \ldots p_N) &= 0, \\ F_2(p_1, p_2, \ldots p_N) &= 0, \\ &\vdots \\ F_N(p_1, p_2, \ldots p_N) &= 0, \end{aligned} \tag{14.5}$$

where F_i is the functional relationship for flows into node i.

The solution to Equations 14.5 are the pressures (p_1, p_2, ... p_N) for which all of the right hand side of each of these equations is zero. From these pressures, all of the pressure differences and flows throughout the building can be calculated. The convergence problems associated with the early network models were because Equation 14.5 is nonlinear.

Contaminant Flow

At time, $t = k\Delta t$, the concentration of a contaminant at node i can be expressed as

$$C_{i,k+1} = C_{i,k} + \frac{1}{\rho V_i} \left[g_{i,k}\Delta t + \sum_j C_{j,k} m_{ji} + \sum_j C_{i,k} m_{ij} \right] \quad (14.6)$$

where

$C_{i,k}$ = concentration at node i at time step k,

$C_{j,k}$ = concentration at node j at time step k,

$C_{i,k+1}$ = concentration at node i at time step $k + 1$,

$g_{i,k}$ = contaminant generation rate in node i at time step k,

m_{ij} = mass flow from node i to node j,

m_{ji} = mass flow from node j to node i,

V_i = volume at node i,

t = time,

Δt = time interval,

k = integer (k =1, 2, …).

The fire space has a contaminant generation rate, and for nonfire spaces, $g_{i,k} = 0$. Equation 14.6 applies for concentrations that are much smaller than one which is appropriate for smoke control applications. Once the steady mass concentrations have been calculated by solution of Equations 14.5, Equation 14.6 can be used to calculate the concentrations at all the nodes for one time step after another. The use of Equation 14.6 is an explicit method that has the drawback of needing relatively small time steps. Wakamatsu (1977) developed a more complicated implicit method that does not have this drawback. Also, CONTAM does not have this drawback.

CONTAM FEATURES

Like other network models, CONTAM simulates the flows and pressures throughout building networks. Contaminant flows can be simulated, which allows tenability calculations. CONTAM is unique in that it can be used in conjunction with a computational fluid dynamic (CFD) model, and this is discussed in Chapter 18. CONTAM refers to the nodes as zones, and pressures in zones in a more general way than other network models.

Zone Pressures

The pressure in a zone can be described by the equation

$$p_i = p_{io} - \rho_i g z \quad (14.7)$$

where

p_i = pressure in zone i at elevation z,

p_{io} = pressure at floor ($z = 0$) of zone i,

g = acceleration of gravity,

ρ_i = density of air in zone i,

z = elevation above floor of zone i.

The CONTAM documentation refers to the model as a multizone model where the zones would represent rooms, plenum, or floors of a shaft. Like other network models, CONTAM does not include an energy equation, and so the temperature of zones needs to be designated by the user.

CONTAM documentation calls the flow paths flow elements, which can include exponential flow, orifice flow, stairwell flow, shaft flow, and bidirectional (two-way) flow. For smoke control applications, the orifice flow element is used for flow through construction cracks, gaps around closed doors, open doors, and other large openings. The stairwell flow element accounts for pressure losses due to friction in stairwells. The duct flow element is used for calculation of the pressure losses in ducts and elevator shafts due to friction.

Bidirectional flow happens between zones when they are at different temperatures, as discussed in Chapter 3. This is not relevant for smoke control systems that rely on pressurization, but it can be significant in tenability applications where zones near the fire have higher temperatures than those farther from the fire.

CONTAM allows users to enter the temperature of each zone, the ambient temperature, and the ambient pressure. All calculations in CONTAM are in SI units, but the user can choose either SI or I-P units as the default for input. When entering data, the user can choose from a number of units as shown in the pop-up menu of Figure 14.2.

Wind

CONTAM is capable of simulating the effects of wind on building pressures and airflow. When wind is simulated, users need to enter the wind speed and the wind direction. Additionally, users can calculate the wind speed modifier from equations in Chapter 3, or CONTAM can calculate this modifier. When CONTAM calculates the wind speed modifier, the terrain parameters listed in Table 14.1 should be used. These parameters are different from those listed in Chapter 3. The terrain parameters listed in Chapter 3 may be somewhat more accurate than those used in CONTAM, but the ones in CONTAM are sufficient for the purposes of network simulations of smoke control systems.

When CONTAM calculates the modifier, users need to enter the roof height, local terrain constant, and velocity profile exponent. For some applications, the relative north also needs to be entered. Figure 14.3 shows the azimuth angles used in CONTAM for standard north. Standard north is when the north arrow points upward. For standard north, the relative north azimuth angle is 0° as can be seen from Figure 14.3.

The convention for wind is that it is flowing from the direction associated from it. For example, a north wind is blowing from the north. The convention for walls is that the wall is associated with the direction that it faces. This means that a north wind blows directly into a north wall. Figure 14.4 is an example of the curves that can be used in CONTAM for wind coefficients.

At flow paths in exterior walls, the user can assign wind pressure coefficients. The average pressure coefficients provided in Chapter 3 can be used. When a wind tunnel study has been conducted for a building, it is recommended that the pressure coefficients in a CONTAM analysis be based on that study. Alternatively, the wind can be analyzed by a CFD model. CONTAM 3.0 has the ability to perform CFD simulations on the building exterior to develop a set of wind pressure coefficient profiles for each airflow path on the outside of a building. This method is explained in a tutorial available on the CONTAM Web site.

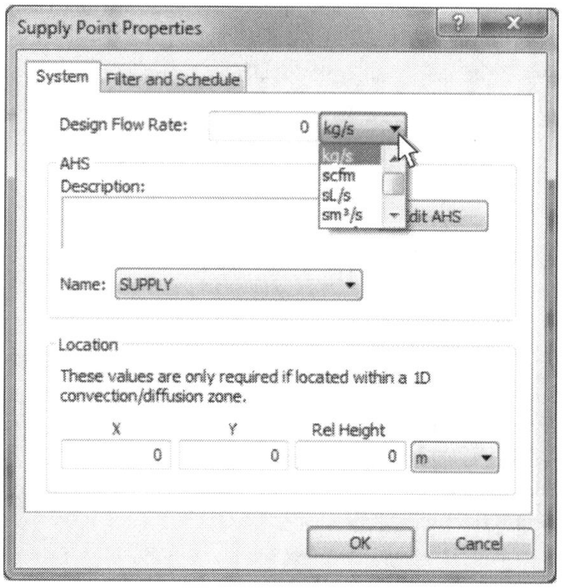

Figure 14.2 *Point source window showing some units that can be chosen for data.*

CONTAM Output

After a simulation, the results can be immediately viewed on the sketchpad in what is called results mode. When in this mode, the user puts the caret at a flow path, and the pressure difference and flow at that flow path are displayed on the status bar. The results mode is discussed later in the examples. When in the results mode, many of the tools cannot be used. To be able to use these tools, the user needs to go back to the normal mode.

Results also can be examined by generating a shaft report. For example, a shaft report can be made of a pressurized stairwell showing the pressured differences and flows at two flow paths from the stairwell at every floor.

CONTAM USER INFORMATION

People using CONTAM for the first time may be confused by its extensive capabilities, many of which are not usually used for smoke control analysis. For example, the supply point window (Figure 14.2) has input for the supply point location, but this feature is not needed for analysis of pressurization smoke control systems. CONTAM can also be used as a part of a tenability analysis, but this is not addressed here except for a few minor comments.

The user information provided here is intended to help get people started using the software for analysis of smoke control systems that rely on pressurization.

Table 14.1: Wind Terrain Parameters used in CONTAM.

Terrain Type	Local Terrain Constant	Velocity Profile Exponent
Urban	0.35	0.40
Suburban	0.60	0.28
Airport	1.00	0.15

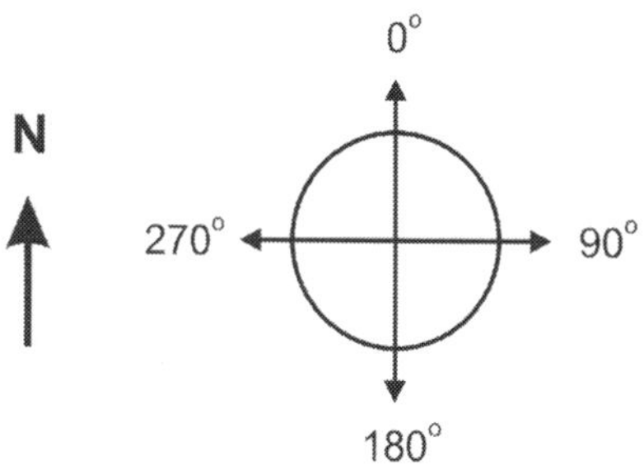

Figure 14.3 *Azimuth angles for standard north.*

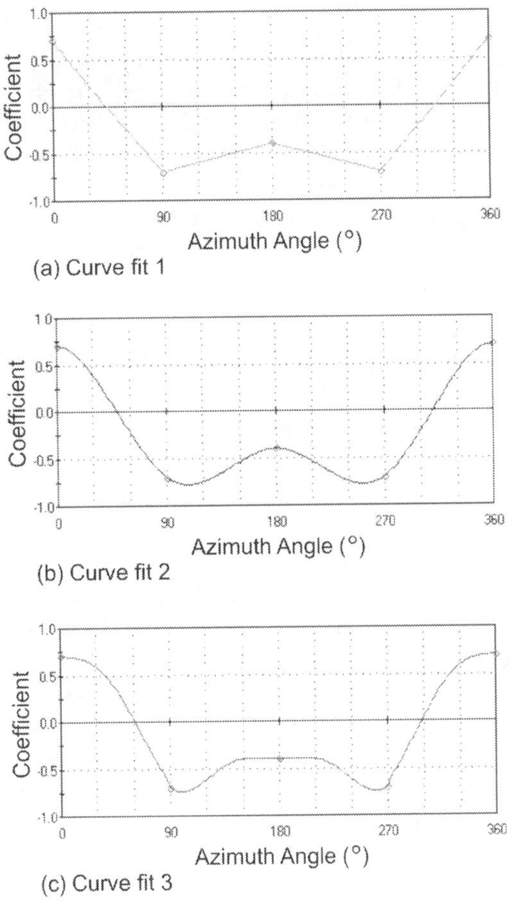

Figure 14.4 *CONTAM wind coefficient options.*

Explaining how to use CONTAM is best done face-to-face using the software itself on a computer to illustrate what is said. When this is not possible, it is hoped that the treatment provided here will be useful. The discussion that follows is for readers familiar with Windows programs.

It is suggested that while reading this material, people have CONTAM running, and that they do things on their own computers as they are discussed in the text. When working the examples, some readers may need to go back in the text for explanations. CONTAM conforms to the usual conventions of software written for Microsoft Windows. For this reason, it is expected that readers who are experienced with Windows programs will be familiar with many things about CONTAM. As with many other Windows programs, the approach to learning by "playing around with the software" works with CONTAM.

The following material was written for CONTAM version 3.0. Because the developers of CONTAM have kept the look and feel of software similar between versions, it is expected that the following information will be useful for subsequent versions. A limited number of features are addressed. For information about features not addressed here, readers should see the latest version of the CONTAM documentation.

Check Units: When entering data, users always need to check the units. For example, a user wants to enter mass flow in scfm, but the input space is set for kg/s. The user needs to change the units to scfm as shown in Figure 14.2.

Save Often: In spite of the fact that CONTAM is reliable and well written, it occasionally crashes. Users should save their work often.

CONTAM Representation of a Floor

CONTAM can be used to simulate flows in very large and complex buildings. Figure 14.5 shows the floor plan and the CONTAM representation of Floor 15 of a condominium building. As can be seen from this figure, the CONTAM representation does not have to model the building to scale. What is important is that the network modeled by CONTAM is representative of the building.

Each space in CONTAM is referred to as a zone, and the zone icon, ■, is used to name the space and specify the temperature of the space. The corridors and elevator lobby are open to each other, so they are defined as one zone as shown on Figure 14.5b. Figure 14.5 does not include the interior wall of the condominium units, because (1) the interior doors of the unit are often open and (2) typically this level of detail is not necessary for design of smoke management systems.

The CONTAM representation needs to include sufficient detail such that the results of the simulations will be sufficient for the purpose of the simulation. To illustrate this, the trash chute was not included in representation of Figure 14.5b. Provided that the leakage paths to this chute are very small, this simplification is justified. Such very small leakage can be obtained by the absence of significant leakage paths including tight wall construction and tight well-fitting gasketed chute doors. With experience, it is possible for CONTAM users to make such simplifications and have simulations appropriate for their purpose.

In CONTAM, flow paths are represented by a small flow icon, ◆, and a large flow icon, ◇. These icons are selected by the user when defining a flow path. For Figure 14.5b, the small icons were used for the leakages of the walls and floors, and the large icon was used for the

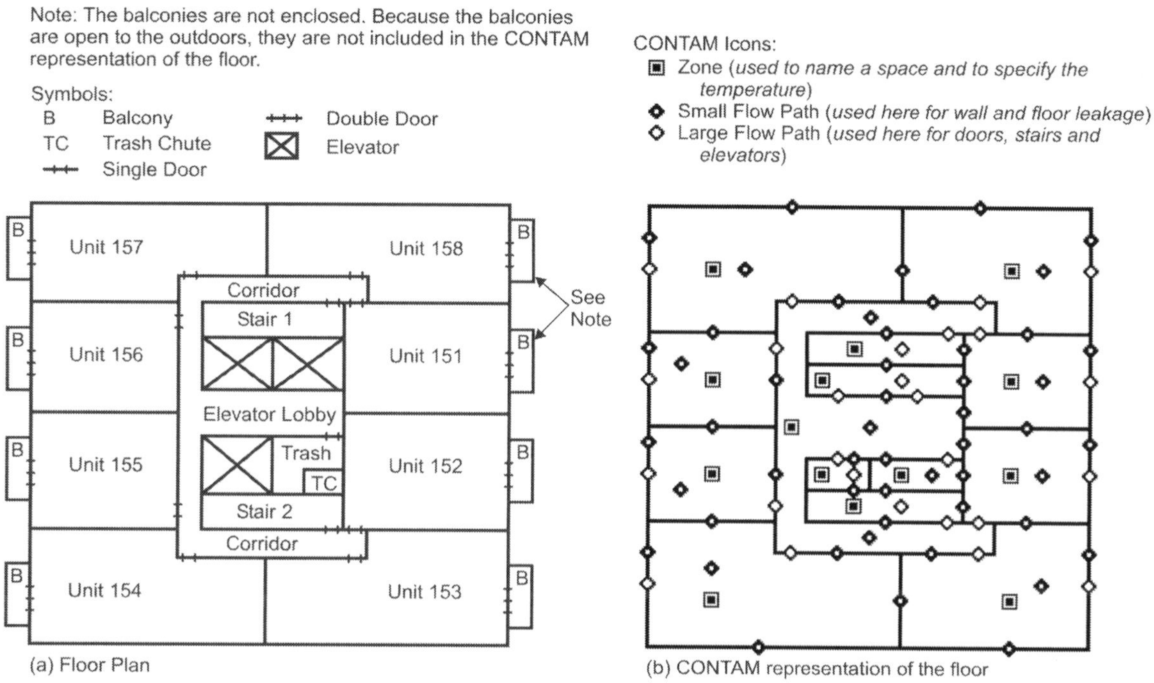

Figure 14.5 *Floor plan and CONTAM representation of Floor 15 of condominium building.*

flow paths of the doors, stairs, and elevator shaft. The flow paths of the stairs and elevator shafts account for the pressure losses due to flow in these shafts. Defining zones and flow paths are discussed later.

CONTAM Window

Figure 14.6 shows the CONTAM window much as it looks when the program is first opened. The largest part of this window is the sketchpad where the user draws a representation of a floor or other level. In addition to the sketchpad, the window has the title bar, menu, toolbar, and status bar. The basic idea of CONTAM input is to draw a schematic floor plan of a building on the sketchpad including flow paths and other elements to be discussed. This schematic floor plan does not need to be to scale. A number of floors can be defined so that an entire building is represented in CONTAM.

On Figure 14.6 there is a cursor which can be moved around by the user by moving the mouse. When CONTAM is first opened, the caret is located as shown in Figure 14.6. The caret can be relocated on the sketchpad by moving the cursor to that location and clicking the left button of the mouse. When it is said that a CONTAM user clicks on an item or a location, it means that the user moves the cursor there and clicks the left button of the mouse. To double click means to click twice quickly.

Title Bar

The project title is listed on the title bar after the project has been saved. For an unsaved project, the title bar indicates that it is untitled as in Figure 14.6.

Menu

The menu is a line of pull down menus that should look familiar to Windows users. The File menu is similar to that in most Windows programs in that it lets users start, open, and save projects. Also, the Edit menu is similar to the ones in other Windows programs in that it can be used to cut, copy, paste, delete, find, and replace. Cut, copy, and paste work with items on the sketchpad such as flow paths, air-handling systems (AHS), supply points and return points. Unlike most Window programs, CONTAM has no undo feature.

The View menu allows the user to change various features of display on the sketchpad. In the View menu, the user can choose the normal mode or the results mode that were just discussed. The user can also choose the wind pressure mode that can be useful in verifying wind-related input data. An important feature under the View menu is the Options pop-up window which allows the user to select defaults for (1) the unit system (SI or I-P), (2) the units of mass flow (kg/s, lb/s, scfm, etc.), and (3) the temperature of the zones or rooms. This Options window also allows selection of the size and appearance of the sketchpad.

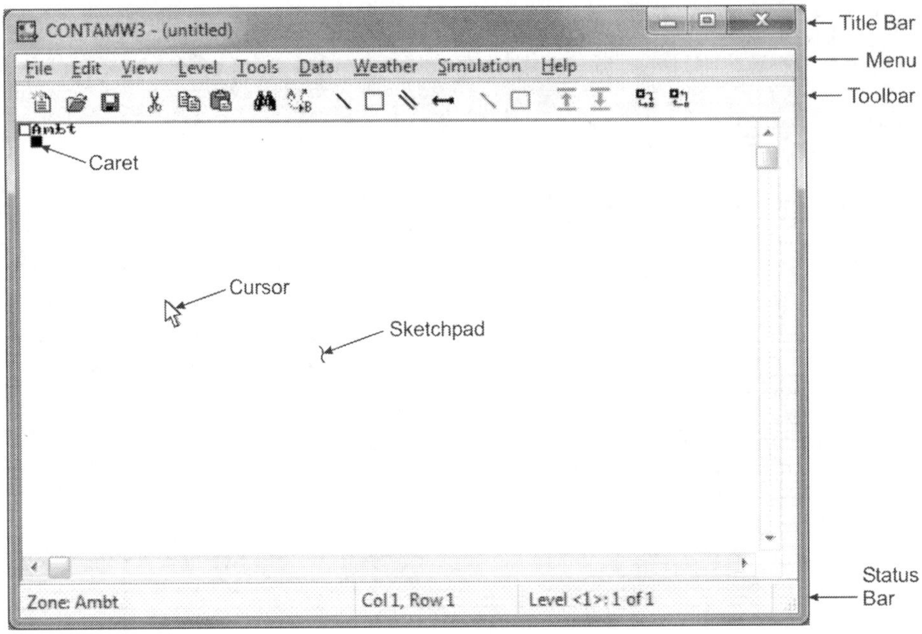

Figure 14.6 *The CONTAM Window.*

In CONTAM floors are referred to as levels, and the Level menu allows levels to be edited, copied, pasted, and deleted. Levels cannot be copied and pasted from the Edit menu; it must be done from the Level menu. The Level menu also allows users to change the level that is displayed on the sketchpad, but this can be done more efficiently by using tools on the toolbar as discussed below.

The Tools menu lets the user draw lines, boxes, ducts, as well as specify flow paths, AHS, supply points, and return points, but much of this can be done faster by using tools on the toolbar as mentioned in the following section. The Data menu is not needed for most smoke control applications, and it is not discussed here.

The Weather menu produces the Weather and Wind pop-up window that allows specification of many related properties including the ambient temperature, absolute pressure, wind speed, wind direction, and other wind parameters.

The Simulation menu allows the user to run CONTAM to obtain flows and pressure differences throughout the building network. Other features of the Simulation menu are not discussed in here. As with many other Windows programs, the Help menu provides added explanation and steps for program features.

Toolbar

Figure 14.7 shows the tools that are located on the Toolbar, and these tools are discussed here. The tools called new project, save project and open project (, , and) can be used in that same way as they can in most other Windows programs.

The cut, copy, and paste tools (, , and) work much like they do in other Windows programs, except that they cannot be used on levels. The find tool ( ) and the find and replace tool ( ) require that the user know the item numbers assigned by CONTAM. These tools are not used often.

The draw lines tool ( ) is used to draw walls on the sketchpad. After clicking on this tool, the mouse is used to draw walls. When finished drawing walls, this tool needs to be deactivated by clicking on it again.

The draw boxes tool ( ) is used to draw walls that form rectangles on the sketchpad, and this tool is used with the mouse in a manner similar to the draw lines tool. The draw lines and boxes tools are black, and they are different from the area delete and box delete tools which are red.

As the name implies, the draw ducts tool ( ) is used to draw ductwork. For most smoke control simulations, this tool is not necessary. In some applications simulation of ducts is needed, such as a vent at the top of an elevator shaft that is ducted through rooms. In most applications, pressurization and exhaust air can be defined without ducts by using the AHS, supply points, and return points as discussed later.

The draw controls tool ( ) is not necessary for simulations of pressurization smoke control systems. The approach usually used is to simulate a smoke

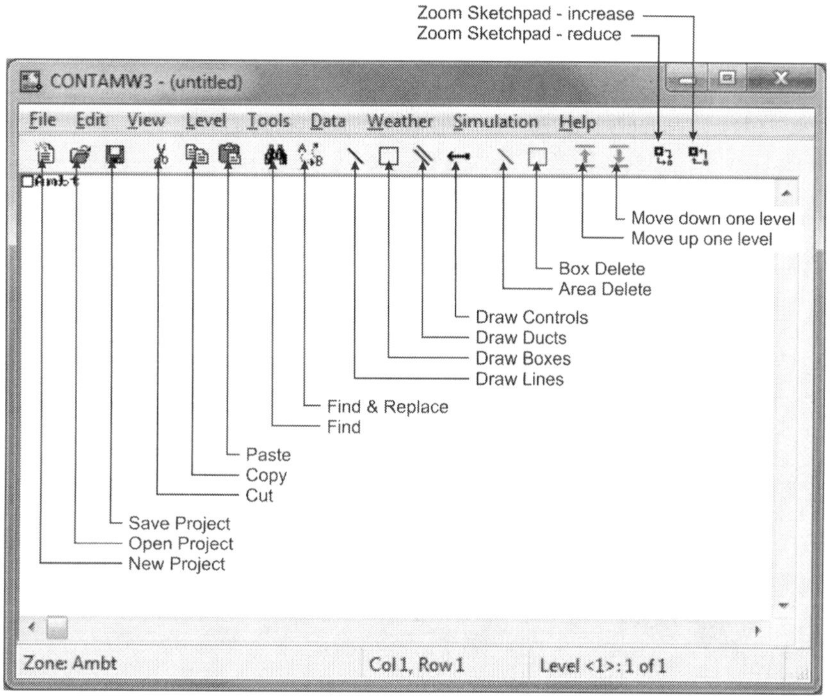

Figure 14.7 *Items on the CONTAM Toolbar.*

control system performance after activation and full operation has been achieved.

As already stated, the area delete and the box delete tools look like the draw lines tools and the draw boxes tool except that the delete tools are red. These tools are a little tricky, but can be useful when deleting many things from the sketchpad. The user clicks on one of these tools, draws an area or a box, and presses the delete key. Then a window appears where the user can select which icons are to be deleted. Another way to delete an item is to click on the item to be removed and press the delete key.

The move up one level tool (⭱) is used to change the level displayed in the sketchpad to the level above, and the move down one level tool (⭳) does the same thing for the level below. The page up and page down keys do the same thing, and these keys can be held down to work much faster when moving many levels.

The zoom sketchpad tools (⧉) and (⧉) are used to decrease and increase the size of the sketchpad on the monitor. This only changes the appearance of the sketchpad on the monitor. If it is desired to change the sketchpad to accommodate a very large or complicated building, the number of cells in the sketch pad can be increased from the options feature mentioned.

Status Bar

The status bar at the bottom of the CONTAM window provides information about the location of the caret (Figure 14.6) and the current level. The examples will show how the status bar works.

Pop-Up Menu

Figure 14.8 shows a menu that pops up by right clicking inside the sketchpad. This menu is a quick way to do a number of things on a floor plan. The cut, copy, and paste items on this pop-up menu perform the cut, copy, and paste tools discussed. The other items on this menu are discussed next.

Zone

Spaces on the floor plans are most often defined as zones. Clicking Zone produces a red zone icon (■) at the caret. Double click on this icon, and the zone properties window appears. For our applications, the only thing on this window that needs to be done is to enter a zone name. After the zone properties window is closed, the zone icon becomes black.

The zone properties window also allows the user to define the following zone properties: volumes, floor area, temperature, and pressure. The volume and floor area are only needed when contaminate flow is simulated, so these are not needed for our applications. The default temperature

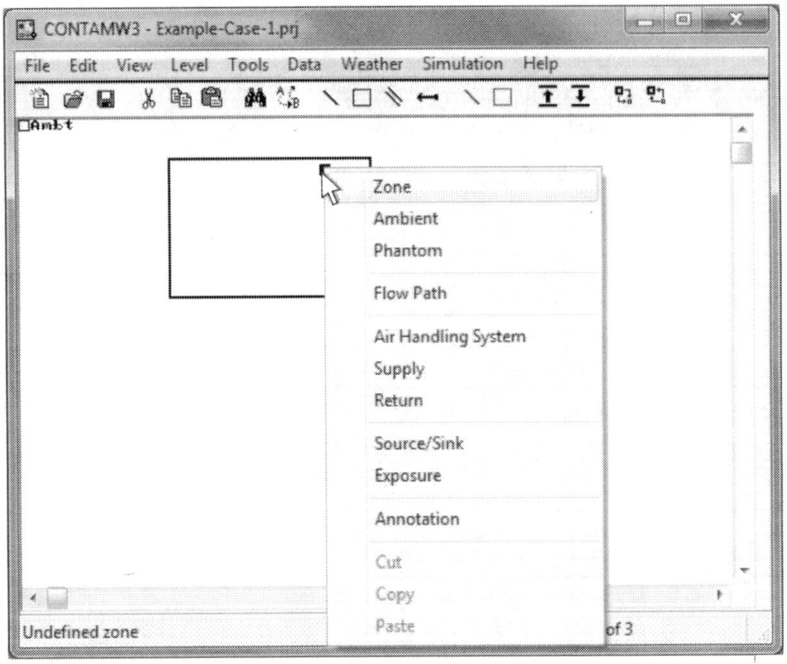

Figure 14.8 *Pop-up menu produced by a right click.*

is assigned to newly defined zones, and this can be changed as needed. Also, a scheduled temperature can be useful for shafts pressurized with untreated outdoor air. The pressure can be defined as either variable or constant. Variable pressure is the default, and it always needs to be kept that way for the applications of this chapter.

Keep Pressure Variable: When defining a zone, users need to keep the pressure variable. Entering a constant pressure can result in significant error for the applications discussed here.

Ambient

By clicking ambient on the pop-up menu, the space where the caret is located becomes an ambient space identified on the sketchpad by the ambient icon (□). For example, a courtyard enclosed on all sides can be defined as an ambient space.

Phantom Zone

A space that is open to the space below can be defined as a phantom zone (▣). This kind of zone can be useful for the second level of a two-story lobby, and it can be used for the upper levels of arenas and auditoriums.

Flow Path

When flow path is clicked on the pop-up menu, a red version of the small flow icon (◆) appears. Double click on this icon, and the airflow path properties window appears (Figure 14.9). An airflow element needs to be selected or a new one defined for this flow path.

When defining a new element, the user needs to select from many element models, and the most commonly used are the orifice, stairwell, and shaft. The orifice model is used for flow paths such as open doorways, gaps around doors, construction cracks, and other connections between zones. When using the orifice model there are many fields that can be filled out, but users should only enter (1) name, (2) cross-sectional area and (3) discharge coefficient. The discharge coefficient is the same as the flow coefficient discussed in Chapter 3, and the flow areas and flow coefficients in Chapter 3 can be used in CONTAM when defining an orifice flow element.

When defining a flow element, users may want to enter a description of the flow element. Users can also select the large flow icon (◇). Some users select the small flow icon for building leakage in walls and floor due to construction cracks, and use the large icon for other flow elements.

Air-Handling System, Supply and Return

Systems using pressurization and exhaust can be simulated by a combination of AHS, supply points, and return points. The AHS icon is ⌶. The supply point icon is ⊠, and the exhaust point icon is ⊡. The AHS needs to be defined before the supply and exhaust points. For most smoke control applications, it is suggested that users define one AHS for supply air and another AHS for exhaust air. Otherwise, care must be taken to understand the flow balance algorithm of the AHS. Many users locate the AHS icon () where the fan or fans are located in the building, but location of this icon has no impact on the results of CONTAM simulations.

SPEEDING UP DATA INPUT

Of course the cut, copy, and paste can be used to speed up data input. The following are some less obvious ways to save time.

Check for Missing Items

Before copying and pasting levels, the floor plan on the sketchpad should be checked carefully for missing items. It seems that every CONTAM user has stories about copying a large number of levels, only to find out that a flow path or other essential item is missing on all the levels. To fix this the user needs to either (1) correct each of these levels or (2) delete the copied levels, correct the original level, and copy and paste levels again. This amounts to a lot of wasted time that can be avoided by checking before copying and pasting levels.

Paste Groups of Levels Quickly

When the user copies a level, CONTAM seems to "forget" the level as soon as anything else is done other than pasting that level. For example, you copy a level, paste it, and name it. Because you named the level, CONTAM "forgot" that level. If you want to paste it again, you need to copy it again. The quick way to copy and paste a large group of levels is to copy it once and paste it over and over. The levels can all be named afterword.

Use the Multiplier with Leakages

Time can be saved by using a feature called a multiplier so that only a few flow elements need to be defined for walls and floors. To understand the advantage of this, a few things need to be explained about flow paths and flow elements. A flow path consists of many things including the flow element, path elevation, multiplier, and possibly some information about the wind. The kinds of flow elements used for most smoke control applications are orifice paths, stairwells, and shafts. Orifice paths are used for the leakage of walls and floors. For example, a flow element can be defined as the leakage per square foot of wall, and the multiplier can be the wall area in ft^2. The same flow element can be used for many or all the walls in a building by using different multipliers. This saves considerable time when defining

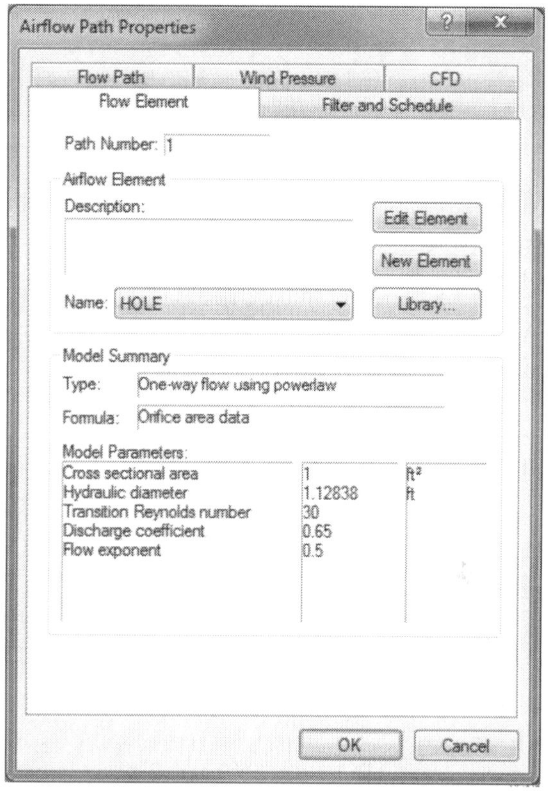

Figure 14.9 *Airflow path properties window.*

flow paths. It has the added advantage of allowing all these leakages to be changed by editing only one flow element.

Use Dummy Wind Data

Often users model the building leakage without wind first, and they add the wind effects later. When it comes time to add the wind, the user has to laboriously modify each flow path in the exterior walls to include the wind. A lot of work can be eliminated by using a dummy wind data. The dummy wind data consists of a zero wind speed, wind profile, and other parameters. When it is time for simulations including wind, the appropriate values for wind speed, wind profile, and other parameters can be entered.

Use Temperature Schedule

In a CONTAM simulation, most zones are at a default temperature, but common exceptions to this are unheated garages and shafts pressurized with untreated outdoor air. For example, a building has a temperature of 72°F (22°C) for most spaces, and the shafts are at 20°F (–7°C) in the winter, 72°F (22°C) in the between season, and 90°F (32°C) in the summer. The default temperature should be set at 72°F (22°C) so that the zones will be at this temperature unless specifically changed. The time-consuming way to deal with the shafts temperatures is to simply assign them to each zone of each shaft. For a large building, this can be a lot of unnecessary work, not to mention the potential for error. An example of the efficient way to deal with this is to define a temperature profile at the winter shaft temperature and assign it to all the shaft zones. When it comes time to make the summer simulations, the temperature of this profile can be changed to the summer shaft temperature. This will change all the temperatures in the shaft zones to the summer shaft temperature.

CONTAM EXAMPLES

The following examples consider that readers are familiar with other Windows programs. Table 14.2 lists a number of CONTAM operations with explanations of how to perform them. These operations can be used as examples of how to do the steps in the examples.

Example 14.1 is a very simple example of stack effect, and it shows how to draw walls, define zones, define flow elements, assign temperatures, run simulation, and read CONTAM results. There are equations for this kind of stack effect (Chapter 3), and it would not normally be analyzed by CONTAM. This example is here for purposes of illustration.

Wind data input is complicated, but Example 14.2 shows how to input wind data, as well as how to do a number of other things. Wind data input consists of entering wind speed, wind direction, terrain data, and wind pressure coefficients. The only driving force of air movement in this example is the wind, thus this example shows the results of wind by itself. The wind pressure mode of the View menu can be used to display wind pressures.

Example 14.3 consists of pressurized stairwells in an eight-story condominium building during winter. This example shows how to pressurize stairs with a simple AHS, and it also includes copying levels, pasting levels, and editing levels. Because CONTAM calculates the pressures and flows in the building, the user needs to make a guess about the amount of supply air to a stairwell and use CONTAM to calculate the pressure differences. Usually more than one guess is needed to arrive at acceptable pressure difference. This is illustrated in Example 14.3. This building is so simple and short that a design analysis using CONTAM would not be expected, but it is done here to illustrate the method.

Example 14.4 is like Example 14.3 except that it is in summer, and it shows how simple it is to use the results of one example to make a new one.

Table 14.2: Example CONTAM Operations

Item	Example Operation	Steps
Air-handling system	Define an AHS named SUPPLY on the roof of a building.	Put the caret on the desired location, right click, and click Air-Handling System. Double click on the red blinking AHS icon, . For the system name enter SUPPLY. Click OK.
Ambient	A space is enclosed on all sides but open to outdoors at top. Define this space as an ambient zone.	Put the caret in the space, right click, and click on Ambient.
Copy/paste	Copy the flow path for a door and paste it to another location.	Put the caret on the flow path of the door, and click the copy tool, . Move the caret to the desired location, and click the paste tool, . (For copying and pasting levels see the item Level.)
Cut/paste	Cut the flow path for a door and paste it to another location.	Put the caret on a single door flow path, and click the cut tool, . Move the caret to the desired location, and click the paste tool, .
Default units	Set the default units to I-P with mass flow in standard cfm, and a default zone temperature of 70°F.	Click the View menu, and click Options. Select default units I-P, and select flow units of scfm. Enter a Default Zone and Junction Temperature of 70°F. If CONTAM asks about resetting existing zones, enter Yes. Otherwise click OK.
Draw	Draw floor plan walls.	Click and use the mouse to draw the walls. After finishing the walls, click to deactivate the draw walls tool. (The walls can also be drawn with the draw boxes tool, .)
Exhaust (see Return)		

Table 14.2: Example CONTAM Operations ***(Continued)***

Item	Example Operation	Steps
Flow path	Define a flow path named in a wall with an area of 1 ft^2 with a flow coefficient of 0.65 and at 3.5 ft above the floor. Name this path OPENING.	Move the caret to the desired location on the wall, and right click. Click Flow Path. Double click on the red flow path icon that appears. Click New Element. Select Orifice Area Data, and click OK. Enter the name OPENING, enter the Cross-Sectional Area of 1 ft^2, and enter the Discharge coefficient of 0.65. Click on Flow Path, enter the relative elevation of 3.5 ft. Under icon, click on Large Opening ◇. Click OK.
Flow path	Define the wall leakage per ft^2 of wall area (WALL) with an area of 0.35×10^{-3} ft^2 in a 270 ft^2 wall. Use a flow coefficient of 0.65.	Move the caret to the desired wall location, and right click. Click Flow Path. Double click on the red flow path icon that appears. Click New Element. Select Orifice Area Data, and click OK. Enter the name WALL, enter the Cross-Sectional Area of 0.35E-3 ft^2, and enter the Discharge coefficient of 0.65. Click OK. Click on Flow Path, enter a Multiplier of 270. Click OK.
Flow path	Define the floor leakage per ft^2 of floor area (FLOOR) with an area of 0.17×10^{-3} ft^2 in 120 ft^2 of floor. Use a flow coefficient of 0.65.	Move the caret to the desired floor location, and right click. Click Flow Path. Double click on the red flow path icon that appears. Click New Element. Select Orifice Area Data, and click OK. Enter the name FLOOR, enter the Cross-Sectional Area of 0.17E-3 ft^2, and enter the Discharge coefficient of 0.65. Click OK. Click on Flow Path, enter a Multiplier of 120. Click OK.
Flow path	Change the relative elevation of a flow path in a wall to 3.5 ft.	Double click on the flow path. Click on Flow Path, and for the relative location enter 3.5 ft. Click OK.
Flow path—elevator	Define a flow path for the elevator shaft that has an area of 102 ft^2, a perimeter of 43 ft, and a roughness of 0.33 ft.	Move the caret to a location inside the elevator shaft, and right click. Click Flow Path. Double click on the red flow path icon that appears. Click New Element. Select Shaft, and click OK. Enter the name ELEVATOR, enter the Cross-Sectional Area of 102 ft^2, enter a Perimeter of 43 ft, and enter a Roughness of 0.33 ft. Click OK twice.
Flow path—stairwell	Define the flow path for a stairwell that has an area of 150 ft^2, a people density of zero, and closed tread.	Move the caret to a location inside the stairwell, and right click. Click Flow Path. Double click on the red flow path icon that appears. Click New Element. Select Stairwell, and click OK. Enter the name STAIRS, enter the Cross-Sectional Area of 150 ft^2, enter a people density of 0, and choose closed stair treads. Click OK twice.
Level	Change the height of level 1 to 12 ft.	With level 1 displayed on the sketchpad, Click the Level menu, click Edit Level Data, for Distance to level above enter 12 ft. Click Replace, and answer the question about adjusting volumes Yes. Click Go To Level.
Level	Name current level G.	Click the Level menu, click Edit Level Data, enter the name G. Click Replace, and click Go To Level.
Level	Copy level 1 and paste it above.	With level 1 displayed on the sketchpad, Click the Level menu, and click Copy Level. Click the Level menu, click Paste Level, and click Above current level. (Note: After this operation, the new level is displayed on the sketchpad.)
Level	Reveal the level below the current one.	Click the Level menu, click Reveal Level Below. (Note: This makes the level below show up in grey. To make the level below to disappear, repeat the steps above.)

Table 14.2: Example CONTAM Operations ***(Continued)***

Item	Example Operation	Steps
Level	Check the current level.	Click the Level menu, click Check Current Level. (Note: This checks for missing items such as zone definitions, but it does not check for missing flow paths.)
Phantom zone	A hotel lobby is two stories tall. Use a phantom zone for the top level of this lobby.	On the top level of the lobby, do not define a zone. In this part of the lobby, right click, and click Phantom.
Return	In the corridor of the third floor, use a return to define an exhaust of 300 scfm. Use the AHS named EXHAUST.	Right click on the desired location, click Return. Double click on the red blinking return icon, ⧄. For the design flow rate, enter 300 scfm. Under AHS name, select EXHAUST. (Note: The AHS named EXHAUST needs to be defined before the return can be defined.)
Run	Run a simulation.	Click Simulation menu, click Run Simulation, click Start Simulation, and click Close. (Note: Before a simulation is run, CONTAM automatically saves the project file. If the project has not yet been saved, the user will be prompted to provide a name and location for the project file.)
Save	Save a new project and name it Hotel-01.	Click, enter the project name Hotel-01, and press Save.
Shaft report	After a simulation, make a shaft report for stairwell, SW1.	Click Simulation menu, click Generate Shaft Report. A window appears explaining to click on the shaft icon, click on the first path, and click on the second path. Follow these steps, and the shaft report appears. (Note: For this report to work, the zone icon and the path icons need to be in the same location on each level.)
Supply	On the second floor of a stairwell, define a supply of 1100 scfm. Use the AHS named SUPPLY.	Right click on the desired location, click Supply. Double click on the red blinking supply icon, ⊠. For the design flow rate, enter 1100 scfm. Under AHS name, select SUPPLY. (Note: The AHS named SUPPLY needs to be defined before the supply can be defined.)
Temperature	Enter the default temperature for an entire project of 72°F.	Click the View menu, click Options. Enter 72°F in the box identified as the Default Zone and Junction Temperatures. If CONTAM asks about resetting temperatures for existing zones, enter Yes. Otherwise click OK.
Temperature	Enter an outdoor temperature of 10°F.	Click Weather menu, and click Edit Weather Data. Enter 10°F for the Ambient Temperature. Click OK.
Temperature	Enter a temperature of 15°F in a room.	Double click on the zone icon in the room, and enter a temperature of 15°F. Click OK.
Temperature	For Stair-1, use the temperature schedule Shaft.	Double click on the zone icon of Stair-1, under Temperature choose Scheduled, for Temperature Schedule Name select the schedule named Shaft. Click OK. (Note: If the schedule Shaft does not exist, one needs to be made as described below.)
Temperature	Make a new temperature schedule named Shaft at 15°F.	With the zone properties window open and scheduled temperature selected, click on New Schedule. Enter name Shaft for Week Schedule. Click on New Day Schedule. Enter name D-Shaft for the name of the day schedule. Enter 15°F and press insert for time 00.00.00. Click on time 24:00:00, enter 15°F and press insert. Click OK. For each day and number on the week schedule, click Replace to assign the day schedule named D-Shaft. When all the day schedules are assigned, click OK.

Table 14.2: Example CONTAM Operations *(Continued)*

Item	Example Operation	Steps
View mode	In results mode, convert to normal mode.	Click View menu, and click Normal Mode.
View mode	In normal mode, convert to results mode.	Click View menu, and click Results Mode.
Wind	Enter the following parameters for wind simulations: wind speed of 23 mph, relative north of 0°, roof height of 108 ft, local terrain constant of 0.60, and a velocity profile exponent of 0.28.	Click Weather menu, and click Edit Weather Data. Enter a wind speed of 23 mph. Click Wind and enter the following: relative north of 0°, roof height of 108 ft, local terrain constant of 0.60, and a velocity profile exponent of 0.28. Click OK. Answer the question about revising wind pressure modifiers Yes. (Note: When the above steps are done, CONTAM calculates a wind speed modifier of about 0.702 which is shown in the window. This modifier is used to calculate wind pressures for flow paths to the outdoors.)
Wind	For a flow path in an exterior wall, make the wind pressure variable using the following pairs of angle and coefficient: 0, 0.7; 90, –0.7; 180, –0.4; 270, –0.7. Name the wind profile WIND1.	Double click on the flow path. Click on Wind Pressure. For wind pressure option, choose Variable. Click New Profile. Enter the profile name WIND1. Enter the following pairs of angle in degrees and coefficient: 0, 0.7; 90, –0.7; 180, –0.4; 270, –0.7. Click Redraw, and see if the graph of the coefficients looks as expected. If it looks right, click OK, and click OK again. (Note: The above coefficients are plotted in Figure 14.4.)
Zone	Define the zone for stairwell 1, and name it SW1.	Move the caret inside the walls of stairwell 1, and double click. The zone properties window will appear. Enter the zone name, SW1. Click OK.
Zoom	Make the cell size on the sketchpad larger.	Click the Zoom Sketchpad Increase tool, . (Note: the Zoom Sketchpad Reduce tool, , will make the cell size smaller. Cell size can also be changed from the cell/Icon Size window which is reached by clicking on the View menu followed by clicking on Options.)

Example 14.1. Stack Effect

Use CONTAM to calculate the flows and pressure differences at the holes of the shaft shown in Figure 14.10. The temperatures and dimensions are on this figure. Open CONTAM and do the following steps. The example operations of Table 14.1 may be helpful for these steps.

- Choose default I-P units and flow in scfm; plus enter a default zone temperature of 70°F.
- Enter an outdoor temperature of 0°F.
- Save the project and name it Shaft.
- Set the floor-to-floor height to 37 ft.
- Draw a rectangle for the shaft walls.
- Define the zone inside the rectangle, and name it SHAFT.
- Define the lower leakage hole (name the flow element HOLE) of 1 ft^2 and flow coefficient 0.65 and at 2 ft above the floor.
- Copy the flow path for the above hole and paste in higher on the same wall.
- Change the relative elevation of this second hole to be 35 ft above the floor.
- Run simulation.

Figure 14.11 is the CONTAM window in results mode after the simulation. The caret is on the flow path of the upper hole, and the pressure difference and mass flow at this location are displayed on the status bar. The flow is 511 scfm (0.290 kg/s) at a pressure difference of about 0.039 in H_2O (9.7 Pa). When the caret is moved to the lower hole, it can be seen that the flow at the lower hole is the same, but the pressure difference across it is about 0.034 in. H_2O (8.5 Pa).

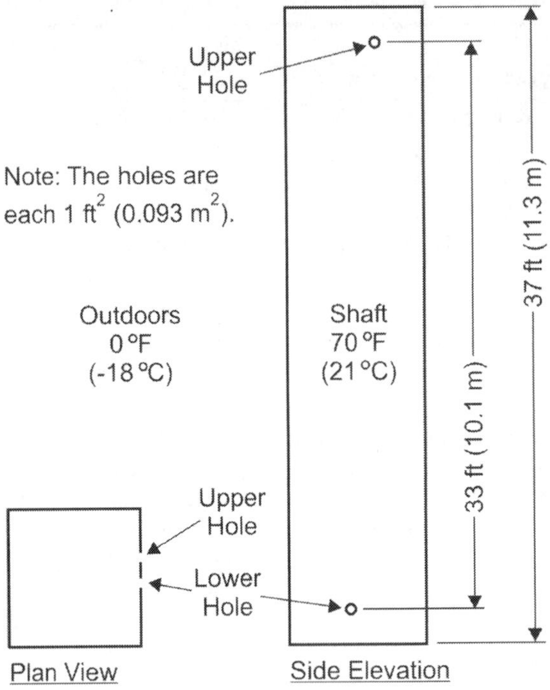

Figure 14.10 *CONTAM window in results mode for Example 14.1.*

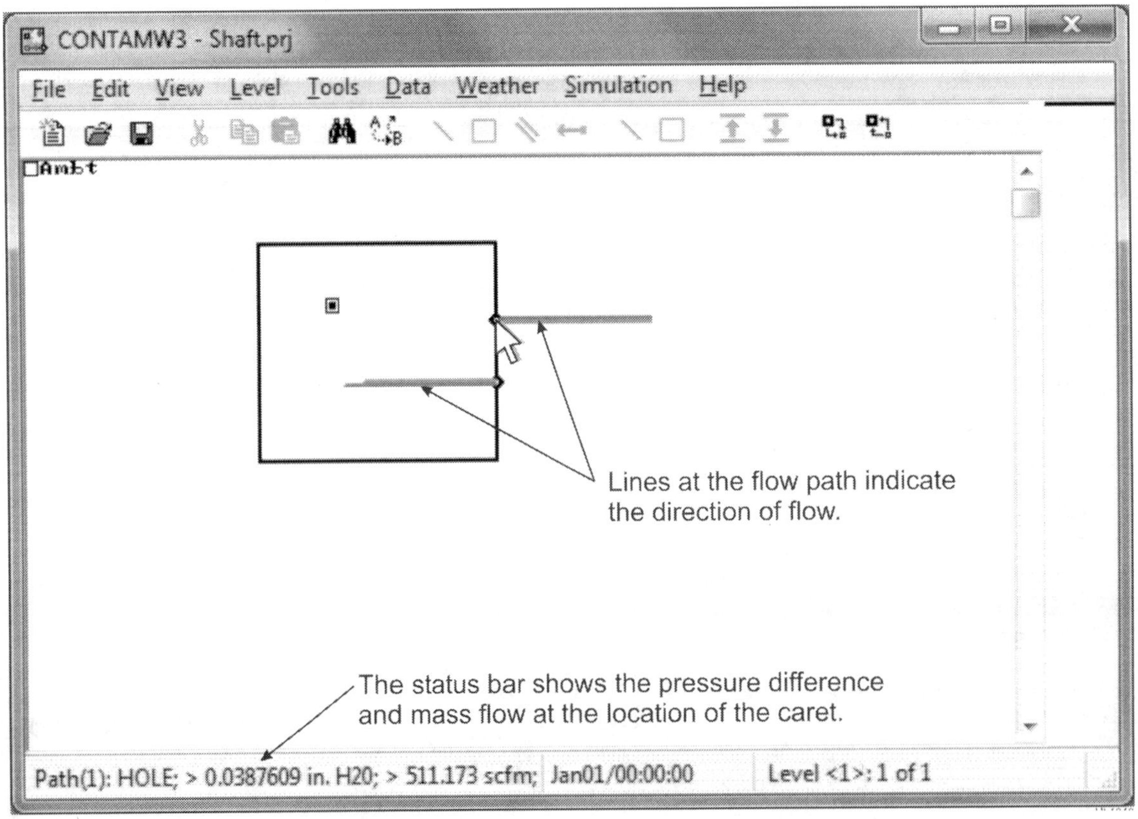

Figure 14.11 *Shaft with two holes of Example 14.1.*

Example 14.2. Wind

Use CONTAM to calculate the pressures produced by the wind on the building of Figure 14.12. The building dimensions, temperatures, wind speed, and leakages are on this figure. For this building use the following wind parameters for suburban terrain: local terrain constant of 0.60 and velocity profile exponent of 0.28 (Table 14.1).

$\overline{C}_w$ for this building are listed in Table 14.3. In CONTAM, these coefficients are entered in terms of the azimuth angle. This angle is zero at the north, and it goes clockwise around the building. For the building of this example, the north wind has an azimuth angle of $\alpha = 0$. The wind coefficient profile for this example is listed in Table 14.4.

Open CONTAM and do the following steps. The example operations of Table 14.2 may be helpful.

- Choose default I-P units and flow in scfm; plus enter a default zone temperature of 70°F.
- Enter an outdoor temperature of 70°F.
- Save the project and name it Wind.
- Change the floor height to 12 ft.
- Draw a rectangle for the shaft walls.
- Define the zone inside the rectangle, and name it ROOM.
- Enter the following parameters for wind simulations: wind speed of 23 mph, relative north of 0°, roof height of 108 ft, local terrain constant of 0.60, and a velocity profile exponent of 0.28.
- Define the leakage for one of the short exterior walls (name the flow element WALL) at 0.5 ft^2 and flow coefficient 0.65 and at 6 ft above the floor.
- For the above flow path, make the wind pressure variable using the following pairs of angle and coefficient: 0, 0.7; 90, –0.7; 180, –0.4; 270, –0.7. Name the wind pressure profile WIND1.
- Copy this flow path, and paste it to the other short wall.
- Paste this path to one of the long exterior walls, and use a multiplier of 3 for this path. (The multiplier 3 is used because the long wall has three times the area of the short wall.)
- Copy this flow path, and paste it to the other long exterior wall.
- Copy the level, and paste it above.
- In the floor of this new level, define a leakage named FLOOR with an area of 0.3 ft^2 and a flow coefficient of 0.65.
- Copy the level, and paste it above over and over until the building has 9 levels.
- Run simulation.

Figure 14.13 is the CONTAM window in results mode after this simulation. The caret is on the flow path in the north wall, and the pressure difference and mass flow at this location are displayed on the status bar. The flow is 1550 scfm (0.881 kg/s) at a pressure difference of about 0.158 in. H_2O (39 Pa). When the caret is moved to the other paths in the south wall, their properties are displayed on the status bar. It can be seen that the flows and pressure differences are the same at each floor. This is because (1) the indoor and outdoor temperatures are the same, and (2) the only driving force for this example is wind with only one wind pressure profile.

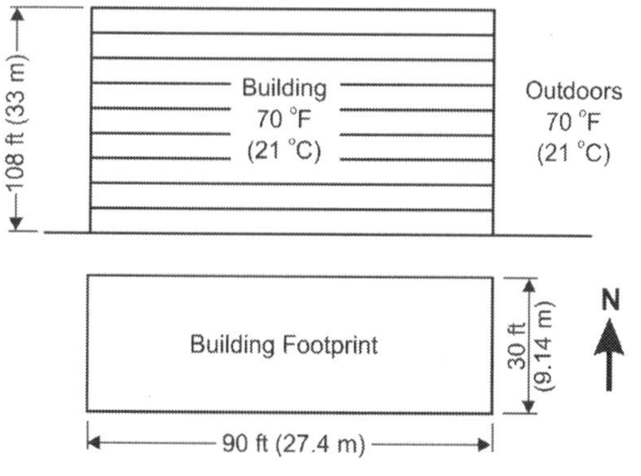

Notes:
1. The wind speed is 23 mph (10.2 m/s) from the north.
2. The leakage between floors is 0.3 ft^2 (0.028 m^2).
3. The leakage of the east and west walls are both 0.5 ft^2 (0.046 m^2).
4. The leakage of the north and south walls are both 1.5 ft^2 (0.14 m^2).

Figure 14.12 *Simple building of Example 14.2.*

Table 14.3: Average Wind Pressure Coefficients for Example 14.2

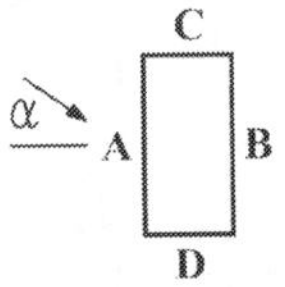

	Wind Angle α	$\bar{C}_w$ for Surface A	B	C	D
	0°	+0.7	−0.4	−0.7	−0.7
	90°	−0.5	−0.5	+0.8	−0.1

Note: This data is for the building of this example. For other buildings, see the figure of average pressure coefficients in Chapter 3.

Table 14.4: Pressure Coefficients in Terms of Azimuth Angle for North Wind of Example 14.2

Wall Facing	Wall Azimuth Angle, °	Coefficient $\bar{C}_w$
North	0	0.7
East	90	–0.7
South	180	–0.4
West	270	–0.7

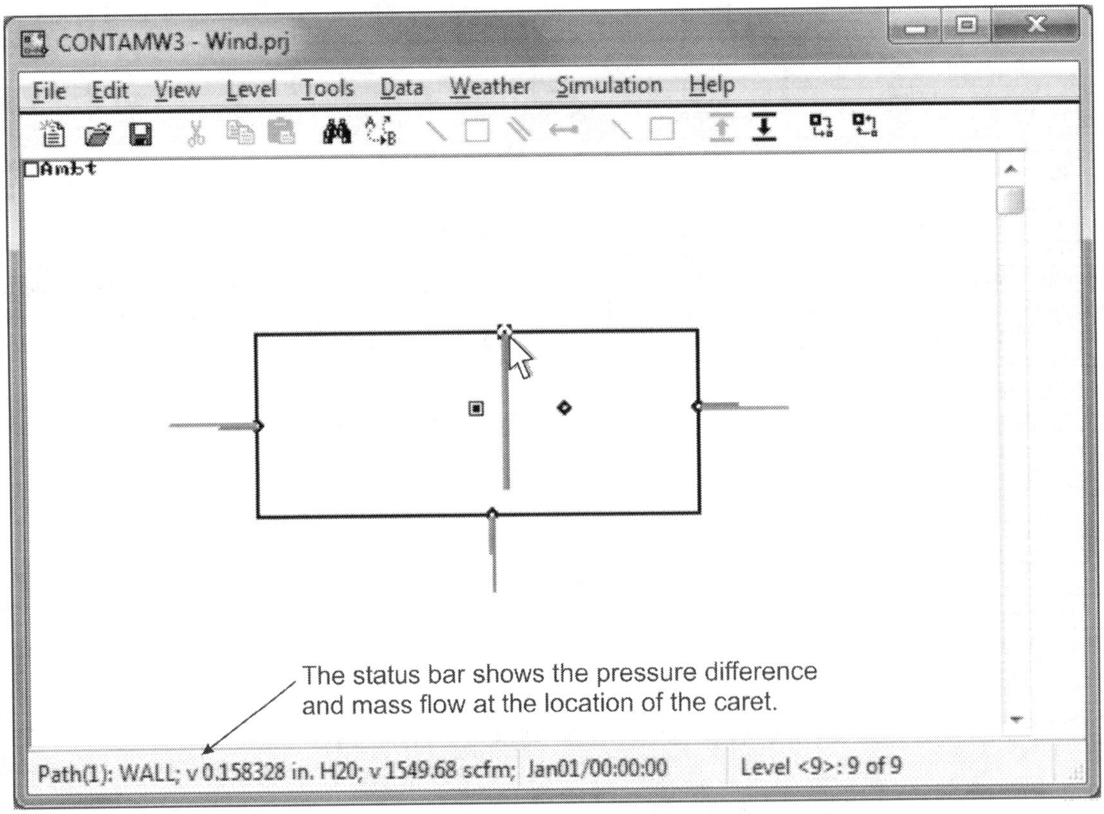

Figure 14.13 *CONTAM window in results mode for level 9 of Example 14.2.*

Example 14.3. Eight-Story Condominium Building

Use CONTAM to estimate the amount of pressurization air needed for stairs of the building of Figure 14.14 with all the doors closed in winter. This figure has building dimensions, design pressure differences, and barometric pressure. The temperatures for this example are listed in Table 14.5. The flow areas and flow coefficients are listed in Table 14.6. The supply air to the stairs needs to be determined such that the minimum pressure difference across the stairs is the same as or slightly more than the minimum design pressure of 0.10 in. H_2O.

Open CONTAM and do the following steps. The example operations of Table 14.2 may be helpful.

- Choose default I-P units and flow in scfm; plus enter a default zone temperature of 73°F.
- Enter an outdoor temperature of –4 °F, and an absolute pressure of 14.3 psi.
- Save the project and name it Condo-01. (Save the project occasionally during the following steps.)
- Change the floor height to 9 ft, and name the level G.
- Draw the walls of the ground floor.
- Outside the building, enter an AHS named SUPPLY.
- Define the zones, and for the zones of the stairs use a temperature schedule set at 8°F.
- Enter the doors using the names, flow coefficients and flow areas from Table 14.6. The relative elevation of these doors is 3.5 ft.
- Enter the wall leakage flow paths using the name, flow coefficient and flow area per ft^2 of wall area from Table 14.6, plus use the wall areas in ft^2 from Figure 14.15a as the multiplier for the flow paths. At this point, the CONTAM sketchpad should look like Figure 14.16a.
- Copy the level, paste it above, edit this level like Figure 14.14b, revise door flow paths, and revise wall flow paths.
- Enter the floor leakage flow paths using the name, flow coefficient and flow area per ft^2 of floor area from Table 14.6, plus use the floor areas in ft^2 from Figure 14.15b as the multiplier for the flow paths.
- Define the flow path for stair 1 that has an area of 150 ft^2, a people density of zero, and closed tread.
- Copy this path to stair 2.
- Define the flow path for the elevator shaft that has an area of 102 ft^2, a perimeter of 43 ft, and a roughness of 0.33 ft. At this point, the CONTAM sketchpad should look like Figure 14.16b except that the supplies have not been added to the stairs.
- Check level 2 that all the items above have been done.
- Copy level 2, and paste it over and over until the building has 7 levels.
- Rename these levels from 2 to 6 and Roof.
- Move to the top level (Roof), delete the zones and walls except for the elevator shaft, delete elevator doors, and delete the shaft leakage of the stairs. Reveal the level below. Add roof leakages for the stairs, and add the flow paths for the roof access hatches using flow element DOOR-SC for the hatches. At this point, the CONTAM sketchpad should look like Figure 14.16c.
- Insert a blank level above, name it Roof2, and enter a flow path for the leakage of the elevator shaft roof using a multiplier of 113 for the area of this roof.
- On level 2 of both stairwells, enter a supply of 2500 scfm using the AHS named SUPPLY.
- Run simulation.
- By examining the pressure differences across the stair doors on each level, it can be seen that the pressure differences range from about 0.04 to 0.13 in. H_2O as shown in Table 14.7. This is not acceptable because the minimum design pressure is 0.10 in. H_2O. On level 2, the supply air to both stairs are changed and simulations are run a number of times until at 3300 scfm the desired pressure differences are reached (Table 14.7).

Because the building is symmetrical, the flows and pressure differences should be the same for both stairs.

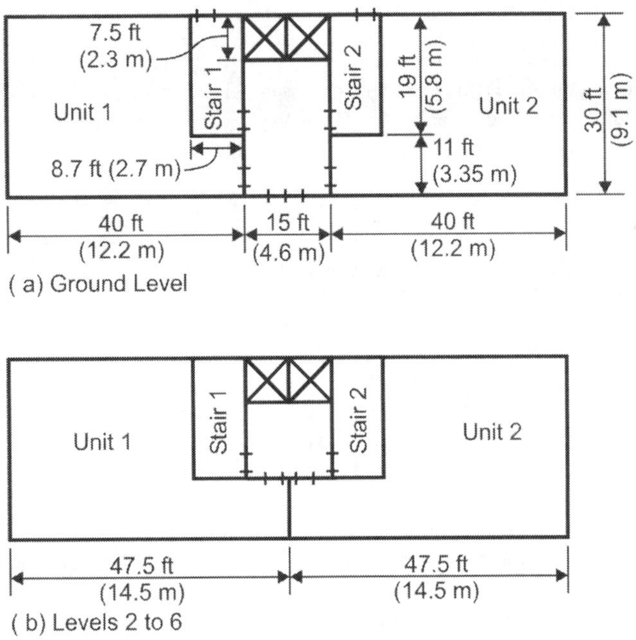

Notes:
1. Drawings are not to scale.
2. Measurements are from the middle of the walls.
3. Because the building is symmetrical, not all of the wall dimensions are shown.
4. The interior dimensions of the stairwells are 8.1 ft by 18.5 ft (2.47 m by 5.64m).
5. The interior dimensions of the hoistway are 7 ft by 14.5 ft (2.13 m by 4.42 m).
6. The stairwells are to be pressurized.
7. The minimum and maximum design pressure differences are 0.10 and 0.35 in H_2O (25 to 87 Pa).
8. The standard barometric pressure at this location is 14.3 psi (98.6 kPa).
9. The floor-to-floor height is 9 ft (2.74 m).

Symbols:
Single Door
Double Door
Elevator

Figure 14.14 *Condominium building of Example 14.3.*

Table 14.5: Temperatures Used for Example 14.3

Winter Conditions:		
	Outdoor	–4°F
	Building	73°F
	Stairwell	8°F
Summer Conditions:		
	Outdoor	92°F
	Building	73°F
	Stairwell	89°F

Note: The stairwell temperatures were calculated from $T_S = T_O + \eta\,(T_B - T_O)$, where T_S is the shaft temperature, T_O is the outdoor temperature, T_B is the building temperature, and η is the shaft temperature factor estimated at 0.15.

Table 14.6: Flow Areas and Flow Coefficients Used for Example 14.3

Path	Flow Element Name	Flow Coefficient	Flow Areas, ft^2	Flow Areas, m^2
Single door (closed)	DOOR-SC	0.65	0.17	0.0158
Double door (closed)	DOOR-DC	0.65	0.34	0.0316
Elevator door (closed)	DOOR-EL	0.65	0.60	0.0557
Wall per ft^2 (or m^2)	WALL	0.65	0.35×10^{-3}	0.35×10^{-3}
Floor or roof per ft^2 (or m^2)	FLOOR	0.65	0.17×10^{-3}	0.17×10^{-3}

Note: The flow areas for a loose building were chosen for this table for the purpose of conservatively sizing the fans.

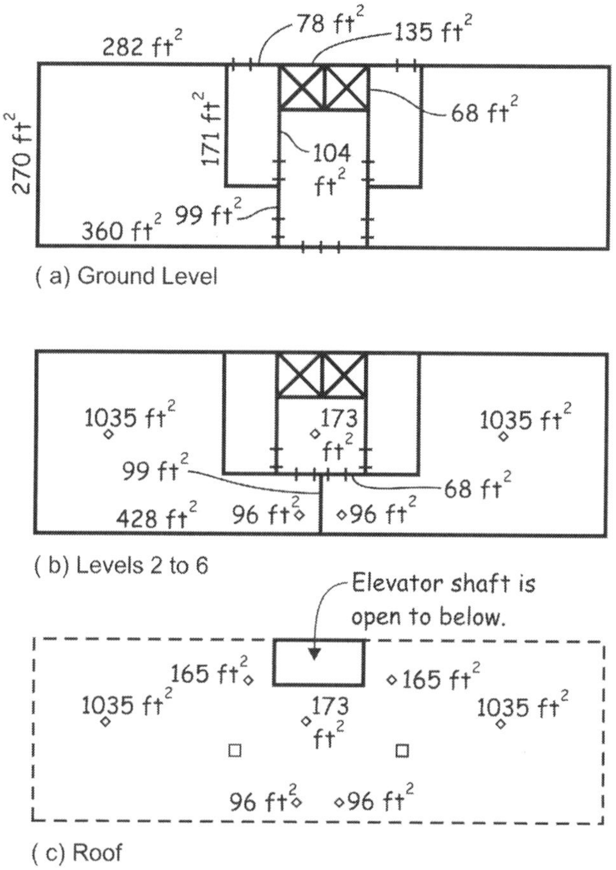

Notes:

1. The convention used here is that wall areas are the product of the wall length and the floor-to-floor height or floor to roof height.
2. Because the building is symmetrical, not all of the wall areas are shown.
3. The roof of the elevator shaft has an area of 113 ft² (not shown).

Symbols:

◇ Floor or Roof Leakage

□ Roof Access Hatch

Figure 14.15 *Areas of building for Example 14.3.*

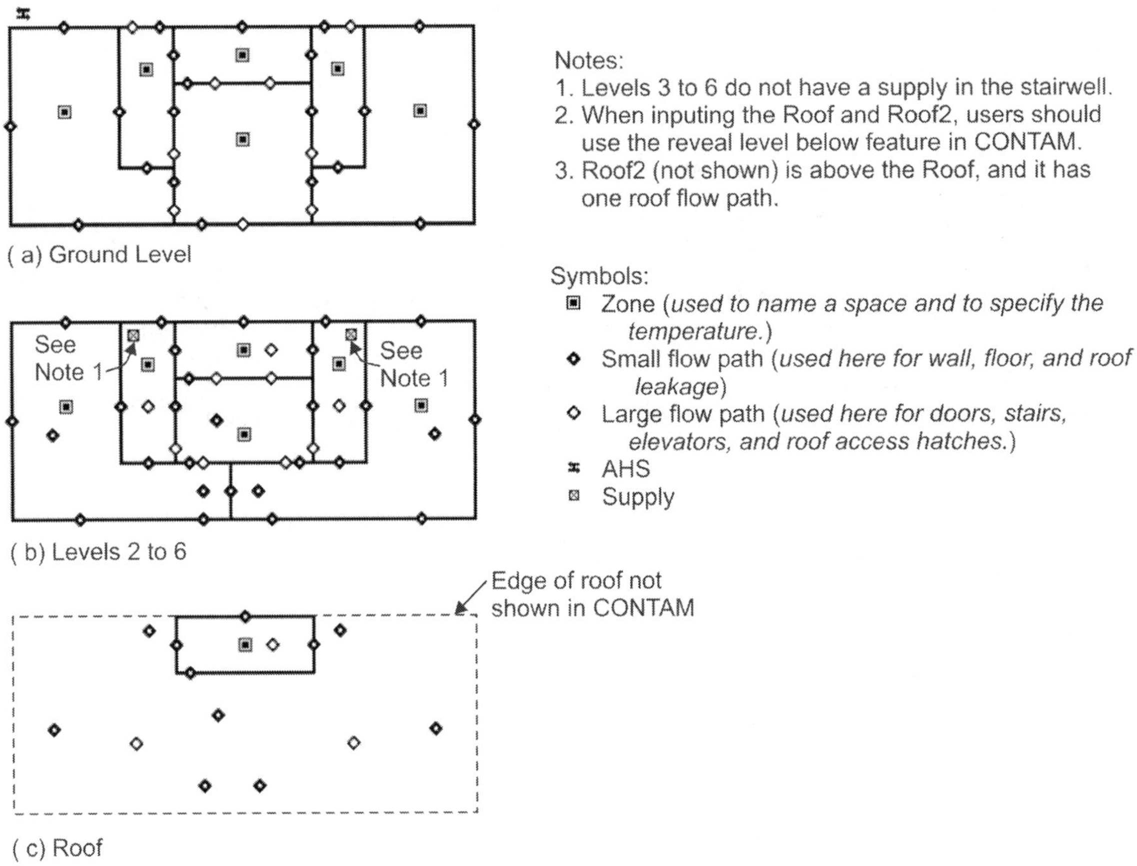

Figure 14.16 *CONTAM representation of the building of Example 14.3.*

Table 14.7: Pressure Difference across Stairs from Example 14.3 for Different Values of Supply Air

Level	Supply Air of 2500 scfm		Supply Air of 3000 scfm		Supply Air of 3500 scfm		Supply Air of 3300 scfm	
	in. H_2O	Pa	in. H_2O	Pa	in. H_2O	Pa	in. H_2O	Pa
6	0.04	10	0.08	20	0.13	32	0.11	27
5	0.06	15	0.10	25	0.14	35	0.12	30
4	0.08	20	0.12	30	0.16	40	0.14	35
3	0.09	22	0.13	32	0.18	45	0.16	40
2	0.11	27	0.15	37	0.19	47	0.18	45
G	0.13	32	0.17	42	0.22	55	0.20	50

Example 14.4. Summer Temperatures

Determine the pressure differences across the building of Example 14.3 in the summer with supply air of 3300 scfm on level 2 of both stairs. The summer temperatures are listed in Table 14.5.

Open CONTAM and do the following steps.

- Open the project Condo-01, and save the project as Condo-02.
- Enter an outdoor temperature of 92°F.
- Edit the temperature schedule for the stairs to 89°F.
- Run simulation.
- Generate a shaft report for stair 2.

The pressure differences across the stair 2 range from about 0.18 to 0.19 in. H_2O, which is acceptable.

NOMENCLATURE

$\bar{C}_w$ = average wind pressure coefficient
$C_{i,k}$ = concentration at node *i* at time step *k*
$C_{i,k+1}$ = concentration at node *i* at time step $k + 1$
$C_{j,k}$ = concentration at node *j* at time step *k*
f_{ij} = denotes functional relationships appropriate for path between nodes *i* and *j*
g = acceleration of gravity
$g_{i,k}$ = contaminant generation rate in node *i* at time step *k*
m_{ij} = mass flow from node *i* to node *j*
m_{ji} = mass flow from node *j* to node *i*
p_i = pressure at node *i*
p_j = pressure at node *j*
t = time
V_i = volume at node *i*
z_i = elevation of node *i*
z_j = elevation of node *j*
Δp_{ij} = pressure difference from node *i* to node *j*
Δt = time interval
ρ_i = gas density at node *i*

REFERENCES

Barrett, R.E., and D.W. Locklin. 1969. A computer technique for predicting smoke movement in tall buildings. *Symposium on Movement of Smoke on Escape Routes in Buildings, Watford College of Technology, Watford, Herts, U.K.*, pp. 78–87.

Butcher, E.G., P.J. Fardell, and P.J. Jackman. 1969. Prediction of the behavior of smoke in a building using a computer. *Symposium on Movement of Smoke in Escape Routes in Buildings, Watford College of Technology, Watford, Herts, England,* pp. 70–75.

Evers, E., and A. Waterhouse. 1978. *A Computer Model for Analyzing Smoke Movement in Buildings*. Borehamwood, Herts, U.K.: Building Research Est.

Klote, J.H. 1982. A computer program for analysis of smoke control systems. NBSIR 82-2512, National Bureau of Standards, Gaithersburg, MD.

Sander, D.M. 1974. *FORTRAN IV Program to Calculate Air Infiltration in Buildings*, DBR Computer Program No. 37. National Research Council Canada, Ottawa, Canada.

Sander, D.M., and G.T. Tamura. 1973. *FORTRAN IV Program to Simulate Air Movement in Multistory Buildings*, DBR Computer Program No. 35. National Research Council Canada, Ottawa, Canada.

Wakamatsu, T. 1977. Calculation methods for predicting smoke movement in building fires and designing smoke control systems. *Fire Standards and Safety*, ASTM STP-614, A.F. Robertson, ed., Philadelphia, PA, American Society for Testing and Materials.

Walton, G.N. 1989. *AIRNET—A Computer Program for Building Airflow Network Modeling*, National Institute of Standards and Technology, Gaithersburg, MD.

Walton, G.N., and W.S. Dols. 2005. CONTAM 2.4 user guide and program documentation. NISTIR 7251, revised 2010, National Institute of Standards and Technology, Gaithersburg, MD.

Wray, C.P. and G.K. Yuill. 1993. An evaluation of algorithms for analyzing smoke control systems. *ASHRAE Transactions* 99(1).

Yoshida, H., et al. 1979. *A FORTRAN IV Program to Calculate Smoke Concentrations in a Multistory Building*. National Research Council Canada, Ottawa, Canada.

CHAPTER 15

Basics of Atrium Smoke Control

John H. Klote

Smoke is commonly recognized as the major killer in building fires. Smoke control in large-volume spaces is based on a long history of experience and research going back to the 1881 Ring Theater fire in Vienna, which killed 449 people. There had already been many theater fires with high fatalities, but this time the Austrian Society of Engineers conducted reduced scale fire tests that showed how roof vents over the stage would have protected the audience from smoke. As a result, many theaters had vents installed over the stage, but it took a long time to get the vents to work properly. It was not until the Palace Theater fire in Edinburgh in 1911 that these vents worked as intended.

In addition to such natural smoke venting, today there are a number of design approaches to deal with smoke in large-volume spaces. A large-volume space is a space that is at least two stories high such as an atrium, a sports arena, or an airplane hangar. In this handbook, the term *atrium* is used in a generic sense to mean any large-volume space.

DESIGN SCENARIOS

A design scenario is the outline of events and conditions that are critical to determining the outcome of alternate situations or designs. In addition to the fire location and heat release rate (HRR), a design scenario may include many other conditions such as the materials being burned, the weather, the status of the HVAC system, and doors that are opened and closed. A design analysis should include a number of design scenarios to provide a level of assurance that the smoke control system will operate as intended.

Design fires need to be realistically selected as discussed in Chapter 5. In general, a design analysis needs to include design fires located in the atrium and in communicating spaces. A communicating space is one that has an open pathway to an atrium such that smoke from a fire either in the atrium or the communicating space can move from one to the other without restriction. Figure 15.1a illustrates these spaces.

A separated space is one that is isolated from the atrium by smoke barriers (Figure 15.1a). For this handbook, a smoke barrier is a continuous membrane, either vertical or horizontal, that is designed and constructed to restrict the movement of smoke in conjunction with a smoke control system. Smoke movement at these smoke barriers can be controlled by pressurization or by compartmentation alone.

Figure 15.1b shows a fire in the atrium with smoke rising above the fire to form a smoke layer under the ceiling of the atrium. The most widely used approach to atrium smoke control is smoke exhaust, but other approaches can also be used. Regardless of the smoke control approach, there is a distance around the fire where occupants cannot go because of the intensity of the fire. To determine the minimum distance that a person can be from a fire for a few minutes without unbearable pain, see Chapter 6.

For a scenario with the fire in the atrium, the design fire does not normally take into account any benefit of sprinklers. In spaces with high ceilings, the temperature of the smoke plume can drop so much that sprinklers may not activate or activation may be so delayed that the spray may evaporate before it reaches the fire. For information, see the section "Smoke Layer with Sprinkler Action" and Chapter 5.

Smoke from a fire in a communicating space can flow into the atrium and form a balcony spill plume as shown in Figure 15.1c. This figure shows smoke blocking

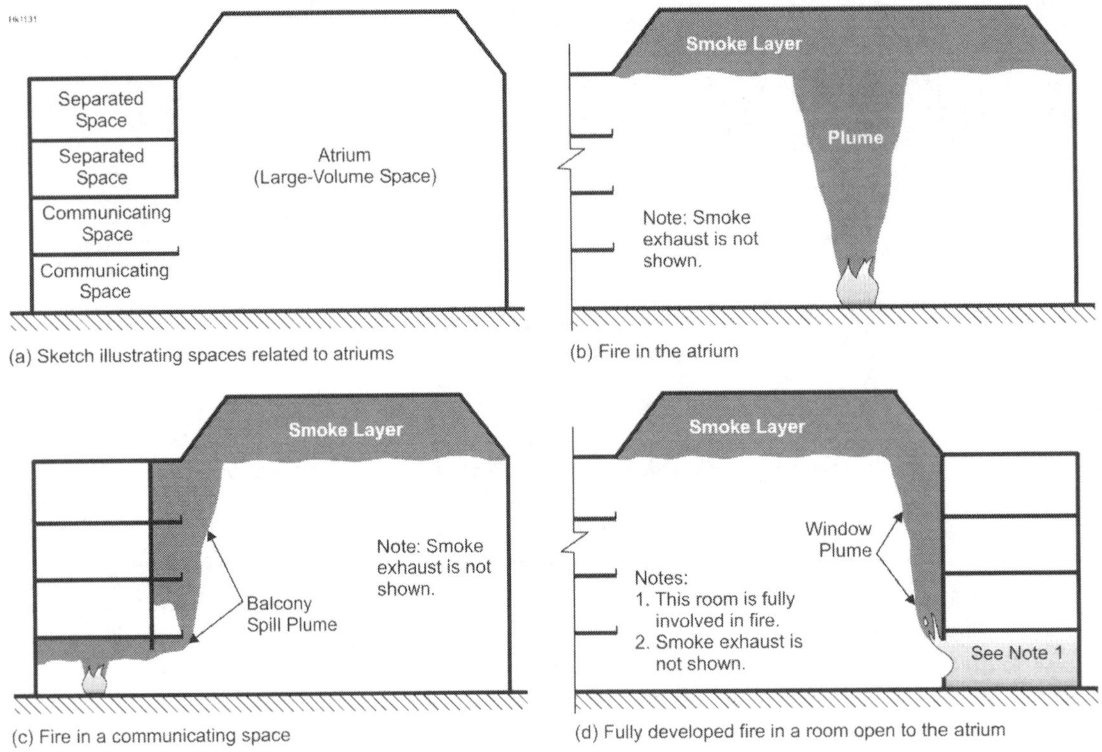

Figure 15.1 *Fire locations for atrium smoke control analysis.*

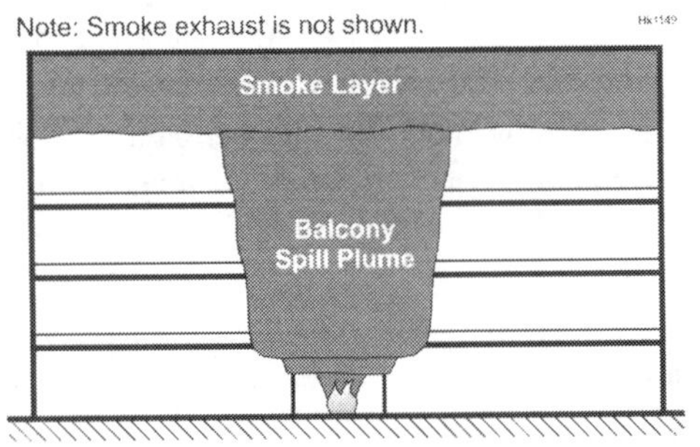

Figure 15.2 *Front view of balcony spill plume.*

of parts of balconies above the fire. It is beyond the capability of smoke control technology to prevent such smoke blocking, but the balcony is not blocked away from the balcony spill plume (Figure 15.2). The comments regarding the minimum distance that a person can be from a fire also apply here. For a scenario with the fire in a communicating space, the growth of the design fire generally stops upon sprinkler activation.

Figure 15.1d shows a fully developed fire and smoke forming a window plume. A fully developed fire would not happen when a sprinkler system is operating properly. Because most new commercial buildings in the U.S. are fully sprinklered, design fire scenarios that include a fully developed fire are uncommon there. In countries where fully sprinklered buildings are uncommon, design fire scenarios may include fully developed fires. It is also possible that some building owners or building managers may want the very high level of protection associated with a smoke control system that can handle even a fully developed fire.

DESIGN APPROACHES

Design approaches that have been used for atrium smoke control are (1) natural smoke filling, (2) steady mechanical smoke exhaust, (3) unsteady mechanical smoke exhaust, (4) steady natural smoke venting, and (5) unsteady natural smoke venting. These approaches are discussed later. Airflow can also be used to control smoke flow in conjunction with these approaches, but care needs to be exercised because airflow has the potential to provide combustion air to the fire.

Smoke Exhaust through a Plenum with a Suspended Ceiling Not Recommended: The pressures produced by the exhaust flow through a plenum with a suspended ceiling can be high enough to lift ceiling tiles out of their frames. Such relocation of ceiling tiles could have an adverse impact on the performance of the smoke exhaust system. The effort involved with periodic testing of such a smoke exhaust systems can be significantly increased due to the need for repair of suspended ceilings after testing.

Many design approaches are intended to prevent occupants from coming into contact with smoke. The idea is to control smoke so that it descends only to a predetermined height during the operation of the smoke control system. In many locations, there are code requirements for the predetermined height. This height is often in the range from 6 to 10 ft (1.83 to 3.05 m) above the highest walking surface that forms a portion of a required egress in the atrium.

Other design approaches are intended to maintain a tenable environment when people come into contact with smoke. When the products of combustion are sufficiently diluted, the resulting diluted smoke can be tenable, and tenability analyses routinely deal with reduced visibility and exposure to toxic gases, heat, and thermal radiation. See Chapter 6 for more information about tenability.

The following discussion of design approaches addresses systems that are intended to prevent occupant contact with smoke, but these systems can be modified to ones that address tenability.

Natural Smoke Filling

This approach consists of allowing smoke to fill the atrium without any smoke exhaust or other smoke removal. For some spaces, the smoke filling time with the design fire is more than sufficient for evacuation. The smoke filling time is the time from ignition until the smoke descends to the predetermined height. Applications that are appropriate for natural smoke filling are not common, because there needs to a very large space above the highest occupied level of the atrium. Any of the following methods of analysis can be used for this system. It is essential that calculations of evacuation time include the times needed for recognition, validation and premovement as discussed in Chapter 4.

Steady Mechanical Smoke Exhaust

This is the most commonly used approach in North America. This system consists of mechanical smoke exhaust sized to keep the bottom of the smoke layer at the predetermined height for the design fire.

Unsteady Mechanical Smoke Exhaust

This approach also uses mechanical smoke exhaust, but the flow rate of the exhaust is less than steady mechanical exhaust such that the exhaust only slows the rate of smoke layer descent for a time that allows occupants to safely egress from the space. This method needs to maintain at least the predetermined height previously mentioned for the time it takes the occupants to safely evacuate. The considerations about calculation evacuation time for natural smoke filling systems also apply here.

Steady Natural Venting

As already mentioned, this kind of venting has a history going back to the Ring Theater fire of 1881. This approach is not common in the U.S., but it is common in Europe, Australia, New Zealand, and Japan. Rather than using exhaust fans, this approach uses nonpowered smoke vents at or near the top of the atrium. Often this kind of venting is called *gravity venting* because the smoke is vented due to buoyancy.

The flow rate of the smoke through the vents needs to be such that the bottom of the smoke layer is kept at the predetermined height for an indefinite time. The previous comments regarding the predetermined height also apply here. There is an equation for the steady mass flow rate through a natural vent, and this is discussed later. It is recommended that steady natural venting systems be analyzed with the aid of a computational fluid dynamics (CFD) model, and this is discussed in Chapter 20.

Unsteady Natural Venting

This approach is like the previous one, except that the smoke venting rate is such that it only slows the rate of smoke layer descent for a time that allows occupants to safely egress from the space. This method needs to maintain at least the predetermined height previously mentioned for the time it takes the occupants to safely evacuate. It also is recommended that unsteady natural venting systems be analyzed with the aid of a CFD model. The considerations about calculation evacuation time for natural smoke filling systems also apply here.

METHODS OF ANALYSIS

The methods that can be used for analysis of atrium smoke control systems are algebraic equations, zone fire modeling, CFD modeling, and scale modeling.

Algebraic Equations

Atrium smoke control makes use of many algebraic equations. Some of these are based on the fundamental principles of engineering, and others are empirical correlations based on experimental data. Equations for smoke filling, natural venting, and the airflow velocity to prevent smoke backflow are discussed later in this chapter.

Chapter 16 addresses the algebraic equations for steady mechanical smoke exhaust, and these equations are based on the zone fire model concepts discussed in the next section. In the following section on zone fire modeling, the discussion about smoke exposure in the transition zone also applies to systems designed with the algebraic equations of Chapter 16.

When another method of analysis is used, algebraic equations are often used to determine starting points for the analysis.

Zone Fire Modeling

In an atrium fire, smoke flows upward in a plume that entrains air as it rises. When the plume reaches the ceiling, it turns and becomes a ceiling jet that flows under the ceiling. This is illustrated in Figure 15.3a, and Figure 15.3b shows an idealized zone model representation of an atrium fire.

Zone fire models are simple models that consider a fire compartment to be divided into two zones: (1) a smoke layer and (2) a lower layer that is free or nearly free of combustion products. The smoke layer can change in size based on the mass flowing into and out of it. In a real fire, the temperature and concentration of contaminants vary throughout the smoke layer with the highest values tending to be near the top of the smoke layer. In real fires, there is also a gradual transition zone between the smoke layer and the lower layer as shown in Figure 15.3a.

In a zone fire model, the smoke layer has a uniform temperature and uniform concentrations. This means that the temperature at any place in the smoke layer is the same as everywhere else in the smoke layer, and the same can be said about the concentration of each contaminant.

Zone fire models do not simulate the transition zone, but the bottom of the smoke layer is simulated as a horizontal plane called the *smoke layer interface* as shown in Figure 15.3b. The zone model considers the air a fraction of an inch (or centimeter) below the smoke layer interface to be as free of smoke as the rest of the lower layer. Occupants in the lower layer near the smoke layer interface will actually be in the transition zone exposed to some smoke. Unfortunately, neither zone fire models nor the algebraic equations of Chapter 16 can be used to evaluate this smoke exposure. It is believed that in many situations, conditions in the transition zone may be tenable. CFD modeling can be used to evaluate tenability at this location.

Zone fire models do not simulate the time it takes for the plume to reach the ceiling, which is small in a normal size room but larger in an atrium. Empirical equations for this lag time are discussed later in this Chapter. Zone fire models do not simulate plume flow, but they use empirical equations to calculate plume temperature and the mass flow.

Even with these limitations, zone fire models have proven to be very useful tools for many applications, but they need to be used with care. Chapter 18 has more detailed information about zone fire models.

CFD Modeling

CFD consists of dividing a space of interest such as an atrium into a large number of cells and using a computer program to solve the governing equations for each cell. CFD is capable of highly realistic simulations. The plume, ceiling jet, smoke layer, and the transition zone are all simulated by the CFD model. CFD models are capable of simulating plugholing, and they can simulate any adverse effects of makeup air velocity on plume formation. Plugholing is discussed later.

CFD modeling requires a level of knowledge and experience beyond that of zone fire modeling, and CFD simulations typically require hours, and sometimes days, of computer time. For more information about this kind of modeling, see Chapter 20.

Scale Modeling

Scale modeling is capable of highly realistic simulations. This kind of modeling consists of conducting fire tests in a small model of the atrium or other facility, and converting the data from those tests to the full scale facility. Scale modeling is addressed in Chapter 21.

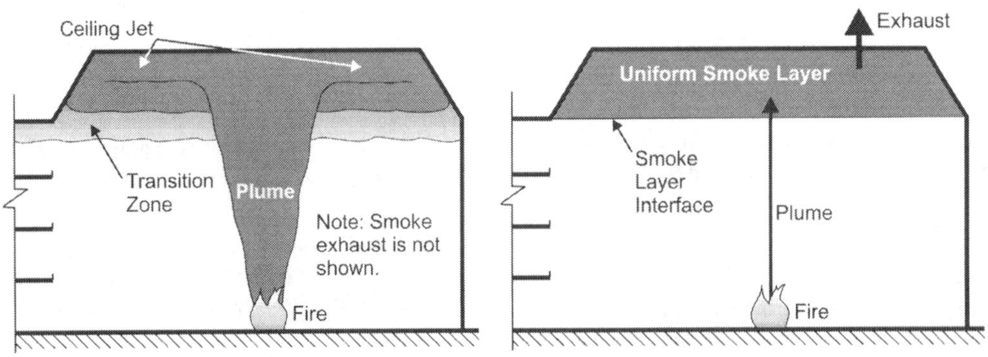

(a) Sketch of an atrium fire (b) Zone model Idealization of an atrium fire

Figure 15.3 *Sketch of idealized zone model representation of atrium fire.*

ATRIUM TEMPERATURE

For systems that rely on mechanical smoke exhaust, the temperature of the air below the smoke layer quickly approaches the outdoor temperature. This is because of the very large amounts of makeup air that enter the atrium. For design analysis of systems using mechanical smoke exhaust, the outdoor design temperature should be used for the ambient temperature of the atrium.

As the gas temperatures increase, the density of the gas decreases, and the volumetric flow rate needed to maintain a constant mass flow increases. Atrium exhaust fans need to be sized for the maximum volumetric flow needed to control smoke for the design conditions. This maximum volumetric flow will happen when the summer outdoor design temperature is used for the ambient temperature of the atrium. For this reason, smoke exhaust fans need to be sized with an ambient temperature of the atrium equal to the summer outdoor design temperature.

MINIMUM SMOKE LAYER DEPTH

The minimum smoke layer depth must be 20% of the floor-to-ceiling height except when an engineering analysis using full scale data, scale modeling, or CFD modeling indicates otherwise. The formation of the minimum smoke layer depth is shown in Figure 15.4. When a smoke plume reaches the ceiling, the smoke flows away from the point of impact in a radial direction forming a ceiling jet. When the ceiling jet reaches a wall, the smoke flow turns down and flows back under the ceiling jet. The ceiling jet has a depth of about 10% of the floor-to-ceiling height, and the smoke flow under the ceiling jet is also about 10% of the floor-to-ceiling height. This means that the smoke layer depth is about 20% of the floor-to-ceiling height.

MAKEUP AIR

Makeup air is outdoor air either supplied by openings to the outdoors or by mechanical fans. For systems that have fan-powered smoke exhaust, makeup air must be provided by mechanical fans or by openings to the outdoors. Makeup air has to be provided so that the exhaust fans can remove the design quantities of smoke and that the door-opening force requirements are not exceeded. Makeup air must be supplied far enough below the smoke layer interface so it does not disrupt the smoke layer.

When providing makeup air through openings to the outdoors, some air also flows by way of leakage paths. The large openings (such as vents, doors, and windows) need to open automatically on system activation. The leakage paths consist of construction cracks, gaps around closed doors, gaps around closed windows, and other similar small paths. The large openings should be sized to provide about 85% to 95% of the makeup air with the rest coming through the leakage paths.

When makeup air is provided by mechanical fans, the makeup air should be less than the mass flow rate of the mechanical smoke exhaust. It is recommended that makeup air for fan-powered smoke exhaust systems be designed at 85% to 95% of the exhaust. The idea is that the remaining air (5% to 15%) will enter the large-volume space through leakage paths, preventing positive pressurization of the atrium.

The makeup air must not exceed 200 fpm (1.02 m/s), where the makeup air could come into contact with the plume unless a higher makeup air velocity is supported by an engineering analysis. The primary reason for this 200 fpm (1.02 m/s) limit is to prevent significant deflection of the plume and disruption of the smoke layer (Hadjisophocleous and Zhou 2008). Deflection of the plume results in increased air entrainment, which can cause smoke control system failure. A secondary reason for this velocity restriction is that it reduces the potential for fire growth and spread due to airflow.

The 200 fpm (1.02 m/s) limitation is not relevant in communicating spaces one story high that are sprinklered. At these locations, successfully sprinklered fires do not form plumes as they would in the atrium, and successfully sprinklered fires limit fire growth. However, air introduced in a communicating space must slow down to meet the 200 fpm (1.02 m/s) limitation when it reaches the atrium. For example, consider makeup air supplied to the communicating space on the ground floor of Figure 15.1a. The makeup air enters the communicating space at a velocity above the limitation. A jet of supply air forms as it would from an HVAC diffuser. The velocity of this jet must drop to 200 fpm (1.02 m/s) or less when it reaches the atrium space. The

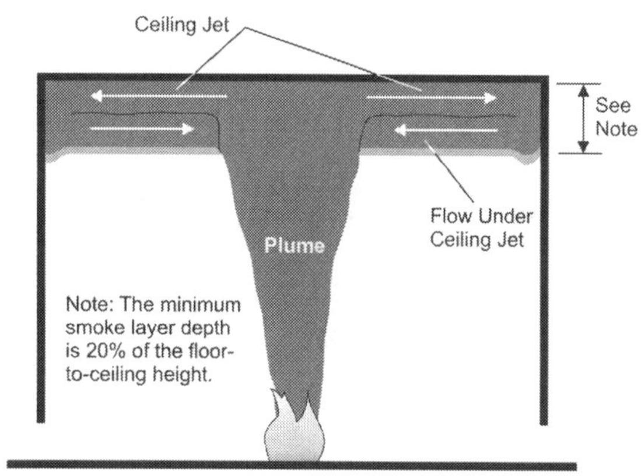

Figure 15.4 *Minimum smoke layer depth.*

design calculations need to include velocity calculations of this makeup air jet at the point where it reaches the atrium.

When makeup air is provided by openings to the outdoors, the design analysis of the system needs to address wind effects as discussed in the next section.

WIND

Atrium smoke control systems should be designed to minimize the potential for wind to result in (1) velocities greater than 200 fpm (1.02 m/s) where the makeup air could come into contact with the plume, and (2) smoke feedback from the smoke exhaust (or smoke vents) into the makeup air.

When makeup air openings face in different directions, wind forces can result in velocities exceeding 200 fpm (1.02 m/s) inside the atrium. The wind can "blow" into openings facing one direction and out the other openings. A simple approach for minimizing wind effects inside an atrium is to have all the makeup air openings face in the same direction. Another simple approach is using mechanical fans for both smoke exhaust and makeup air such that the impact of the wind is minimized. When such simple approaches are not feasible, a detailed analysis is needed that takes into account the prevailing wind directions. Such an analysis can be done with a network model or a CFD model.

Smoke can be carried by the wind from the smoke exhaust or from smoke vents to makeup air openings or inlets. The simple approach to minimize the potential for this is to locate the smoke exhaust (or vents) and makeup air openings (or inlets) (1) far away from each other, and (2) such that the prevailing wind directions carry the smoke away from makeup air openings (or inlets). When this simple approach is not feasible, CFD analysis or wind tunnel analysis is needed to evaluate the potential for smoke feedback into the atrium.

PLUGHOLING

Plugholing is a phenomenon in which air from below the smoke layer is pulled through the smoke layer into the smoke exhaust. Plugholing can cause system failure, but it can be easily prevented. Plugholing reduces the exhaust from the smoke layer which tends to lower the smoke layer and expose occupants to smoke.

The following discussion of plugholing applies to steady mechanical smoke exhaust systems. Figure 15.5a shows a smoke layer that is at the intended design height, but the layer is still descending due to plugholing. As the smoke layer depth increases, the buoyancy forces of the smoke layer increase, and the amount of plugholing decreases. Eventually, the smoke layer becomes deep enough that a state of equilibrium is achieved with a constant smoke layer height as shown in Figure 15.5b. Plugholing has resulted in a smoke layer below what was intended.

The important forces for plugholing are the kinetic forces of the smoke exhaust and the buoyancy forces of the smoke layer. When kinetic forces dominate, there will be plugholing. When the buoyancy forces dominate, there will be no plugholing. The kinetic forces depend on the flow rate of the smoke exhaust, and the buoyancy forces depend on the temperature and depth of the smoke layer. When these forces are balanced at an exhaust inlet, the flow at that inlet is the maximum that can be achieved without plugholing. Plugholing can be prevented by using a number of smoke exhaust inlets

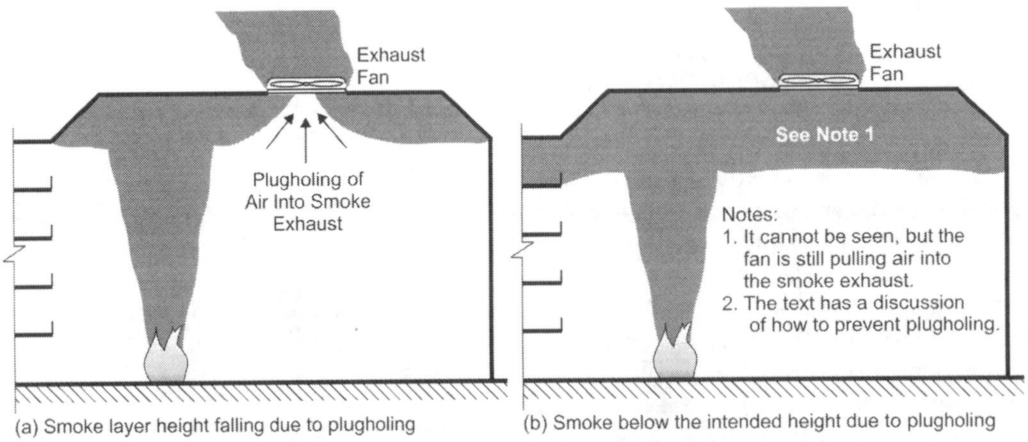

Figure 15.5 *Plugholing causing smoke layer to fall below intended height.*

such that the flow rate at each inlet is at or below this maximum value.

There is an empirical equation in Chapter 16 for the maximum volumetric flow rate that can happen at an exhaust inlet without plugholing. This equation and the previous discussion also apply to systems that use natural venting.

Scale modeling and CFD modeling can simulate plugholing without the use for the empirical maximum flow rate equation of Chapter 16. This empirical equation can be conservative, and it is possible that an analysis using scale modeling or CFD modeling would result in a lower number of exhaust inlets than an analysis using the empirical equation.

CONTROL AND OPERATION

Atrium smoke control systems need to be activated automatically to quickly provide smoke protection for the occupants. For atria where smoke stratification can happen, one of the detection schemes described later needs to be used. The smoke control system needs to reach full operation before conditions in the atrium reach the design conditions. Determination of the time for the system to become operational needs to take into account (1) the time for detection of the fire and (2) the HVAC system activation time including shutdown and start-up of air-handling equipment, opening and closing dampers, and opening and closing natural ventilation devices.

A means of manually starting and stopping the smoke control system needs to be provided at a location acceptable to the fire department. These manual controls need to be able to override the automatic controls. For general information about controls of smoke control systems see Chapter 8.

STRATIFICATION

A hot layer of air can form under the ceiling of an atrium due to solar radiation on the atrium roof. The temperature of such a layer can be 120°F (50°C) or more. When the average temperature of the plume is less than that of the hot-air layer, a stratified smoke layer can form under the hot-air layer preventing smoke from reaching ceiling-mounted smoke detectors.

When smoke stratification can occur, one of the following detection schemes of projected beam smoke detectors should be used: (1) upward-angled beam to detect the smoke layer, (2) horizontal beams to detect the smoke layer at various levels, and (3) horizontal beams to detect the smoke plume. These schemes are shown in Figure 15.6.

Upward-Angled Beam to Detect the Smoke Layer: The upward-angled beams are intended to quickly detect the development of the smoke layer at whatever temperature and height it occurs. More than one upward-angled beam detector is suggested. The manufacturer's recommendation should be reviewed when using beam smoke detectors for this application because some beam detectors are not recommended for upward angled installation.

Horizontal Beams to Detect the Smoke Layer at Various Levels: This scheme consists of using horizontal beams with the intent of quickly detecting the development of the smoke layer at whatever height it occurs. One or more beam detectors are located at the ceiling and additional beam detectors are located at lower levels. The exact positioning of the beams depends on the specific design but should include beams at the bottom of any unconditioned (dead-air) space and at or near the design level of the bottom of the smoke layer.

Figure 15.6b shows suggested beam spacing for a simple atrium arrangement. The rationale behind this recommended spacing is that the smoke layer is expected to be about 20% of the floor-to-ceiling height, and placing detectors every 10% of the height is almost guaranteed to have at least one detector in the smoke layer.

Horizontal Beams to Detect the Smoke Plume: This scheme uses horizontal beams with the intent of quickly detecting the development of the smoke plume rather than the smoke layer. For this scheme, the beams need to be located below the lowest expected level of smoke stratification, and the spacing between the beams needs to be based on the narrowest plume width at the level of detection.

Figure 15.6c shows suggested beam spacing for this scheme. The rationale behind this recommendation is that the width of the smoke plume expected to be about half of the height, and spacing detectors at 25% of the height, is almost guaranteed to have at least one detector within the smoke plume.

SMOKE FILLING EQUATIONS

In addition to the equations of this section, smoke filling can be analyzed by zone fire modeling and CFD modeling. The following empirical filling equations are based on smoke filling tests (Heskestad and Delichatsios 1977; Nowler 1987; Mulholland et al. 1981; Cooper et al. 1981; Hagglund, Jansson, and Nireus 1985). In these tests, the smoke layer was visually determined as the first indication of smoke. This first indication of smoke is different from the smoke layer interface predicted by zone fire models.

In actual fires, there is a gradual transition zone between the lower cool layer and upper hot layer. While there is no correlation between the first indication of

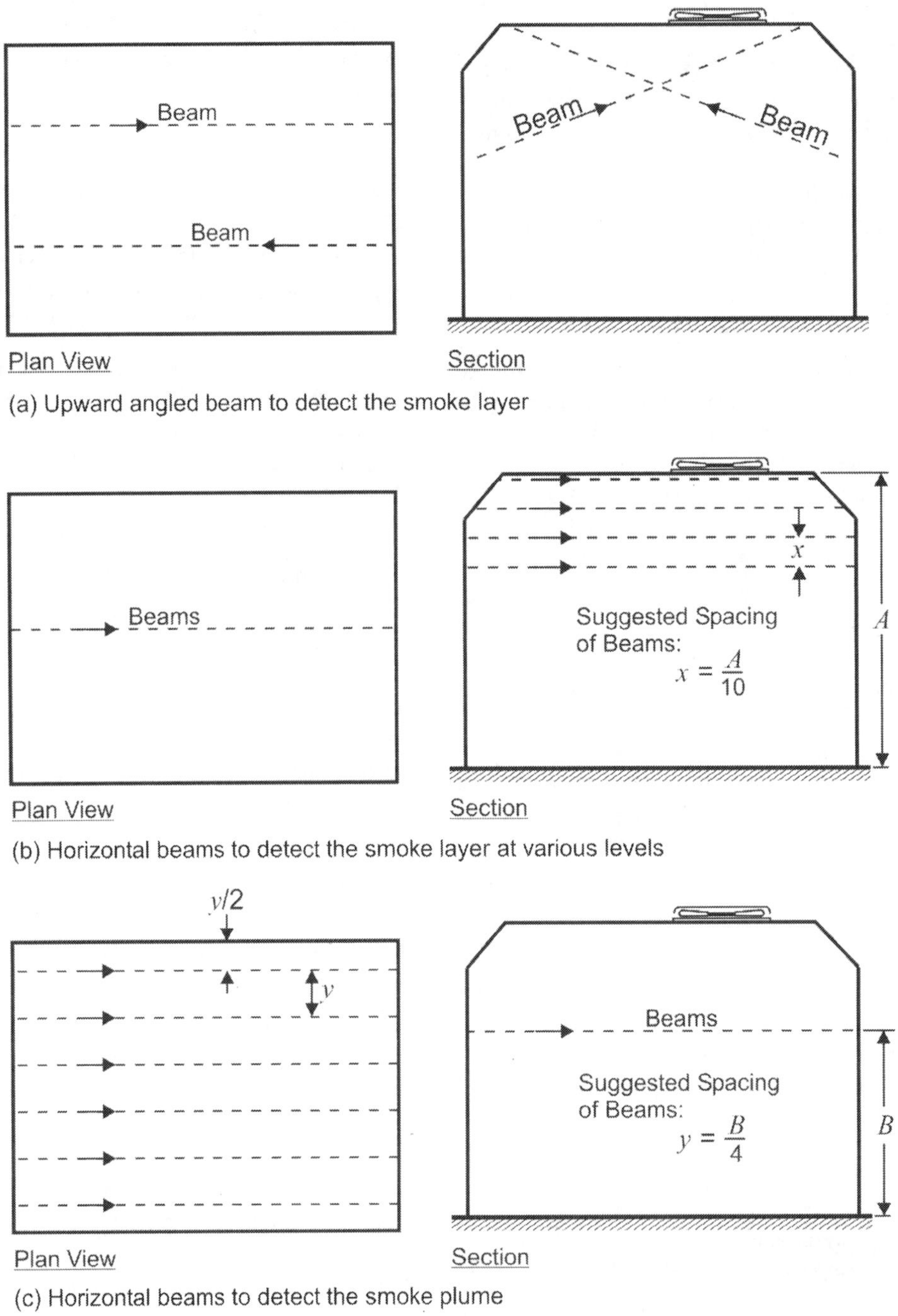

Figure 15.6 *Arrangements of beam smoke detectors for smoke stratification.*

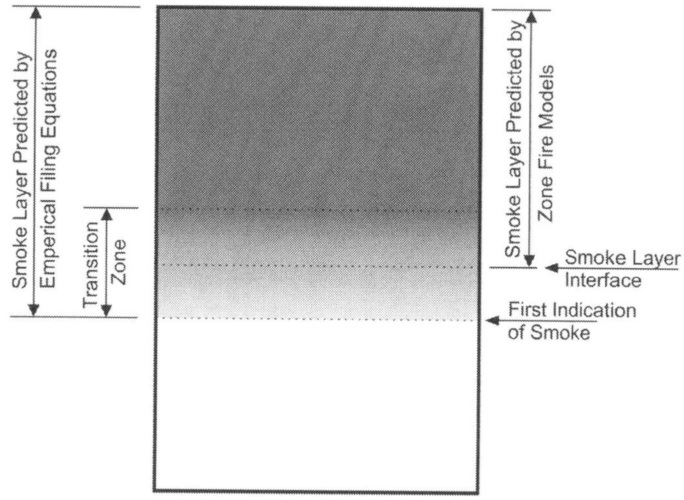

Figure 15.7 *Illustration of smoke layer interface and first indication of smoke.*

Example 15.1. Steady Smoke Filling

An atrium has a floor-to-ceiling height of 65 ft (19.8 m) and a constant cross-sectional area of 20,000 ft^2 (1860 m^2).

Part 1: For a steady 2000 Btu/s (2100 kW) fire, how long will it take for the smoke layer to descend to 13 ft above the floor of the atrium?

$A/H^2 = 20{,}000/(65)^2 = 4.73$, and $z/H = 13/65 = 0.20$. These values are in the range of applicability of Equation 15.2.
From this equation, the time for the smoke layer to descend is 526 s (8.8 min).

Part 2: If the above fire were 5000 Btu/s (5300 kW), what would the filling time be?
From Equation 15.2, the filling time would be 388 s (6.5 min).

smoke and the smoke layer interface, Figure 15.7 shows a graphical representation of both. As might be expected, predictions using the equations of this section differ from predictions using a zone fire model.

Steady Filling

For a steady fire, the smoke filling can be approximated as

$$\frac{z}{H} = 0.67 - 0.28\ln\left(\frac{t\,Q^{1/3}H^{-4/3}}{A/H^2}\right)$$
$$\frac{z}{H} = 1.11 - 0.28\ln\left(\frac{t\,Q^{1/3}H^{-4/3}}{A/H^2}\right) \text{ for SI} \quad (15.1)$$

and solving for time t, this becomes

$$t = \frac{A\,H^{4/3}}{H^2 Q^{1/3}}\exp\left[\frac{1}{0.28}\left(0.67 - \frac{z}{H}\right)\right]$$
$$t = \frac{A\,H^{4/3}}{H^2 Q^{1/3}}\exp\left[\frac{1}{0.28}\left(1.11 - \frac{z}{H}\right)\right] \text{ for SI} \quad (15.2)$$

provided that A is a constant with respect to z, $0.2 \le \frac{z}{H} \le 1.0$, and $0.9 \le \frac{A}{H^2} \le 14$,

where

z = distance above base of fire to first indication of smoke, ft (m),
H = ceiling height above base of the fire, ft (m),
t = time, s,
Q = heat release rate, Btu/s (kW),
A = cross-sectional area of atrium, ft^2 (m^2).

These equations are based on a plume that has no contact with the walls. Because wall contact reduces entrainment of air, this condition is conservative. Equations 15.1 and 15.2 are also conservative because they estimate the height of the first indication of smoke. These filling equations are for a constant cross-sectional area, A. The equations are appropriate for A/H^2 from 0.9 to 14 and for values of z greater than or equal to 20% of H. A value of z/H greater than one from Equation 15.1 means that the smoke layer under the ceiling has not yet begun to descend. Example 15.1 illustrates steady smoke filling calculations.

Unsteady Filling

For the following *unsteady filling equation*, the fire continues to grow throughout the filling process.

$$\frac{z}{H} = 0.23\left[t t_g^{-2/5} H^{-4/5}\left(\frac{A}{H^2}\right)^{-3/5}\right]^{-1.45}$$

$$\frac{z}{H} = 0.91\left[t t_g^{-2/5} H^{-4/5}\left(\frac{A}{H^2}\right)^{-3/5}\right]^{-1.45} \text{ for SI} \quad (15.3)$$

and solving for time t, this becomes

$$t = 0.363 t_g^{2/5} H^{4/5}\left(\frac{A}{H^2}\right)^{3/5}\left(\frac{z}{H}\right)^{-0.69}$$

$$t = 0.937 t_g^{2/5} H^{4/5}\left(\frac{A}{H^2}\right)^{3/5}\left(\frac{z}{H}\right)^{-0.69} \text{ for SI} \quad (15.4)$$

provided that A is a constant with respect to H, $0.2 \le \frac{z}{H} \le 1.0$, $0.9 \le \frac{A}{H^2} \le 14$,

where

z = distance above base of fire to first indication of smoke, ft (m),

H = ceiling height above base of fire, ft (m),

t = time, s,

t_g = growth time, s,

A = cross-sectional area of the atrium, ft^2 (m^2).

As with the steady filling equations, Equations 15.3 and 15.4 are conservative in that they estimate the height of the first indication of smoke, and they are for a plume that has no wall contact. Equations 15.3 and 15.4 are also for a constant cross-sectional area with respect to height. These equations are appropriate for A/H^2 from 0.9 to 23 and for values of z greater than or equal to 20% of H. A value of z/H greater than one from Equation 15.3 means that the smoke layer under the ceiling has not yet begun to descend.

These unsteady filling equations are based on a fire that grows with the square of time from ignition which is called a t-squared fire. Because of this fire growth, the HRR of the fire can become extremely large. When using Equations 15.3 or 15.4, it is recommended that the fire size be calculated at the end of the filling process as is done in Example 15.2. See Chapter 5 for information about the t-squared fire and values of growth time t_g.

IRREGULAR GEOMETRY

The smoke filling equations (Equations 15.1 to 15.4) and zone fire models are for atria with cross-sectional areas that are constant over the height of the atrium. For irregular shaped atria, CFD modeling and scale modeling can be used to accurately analyze smoke filling. Alternately, the following approximate methods can be used.

The zone fire model AZONE was capable of simulating smoke filling with irregular geometries (Klote and Milke 2002), but this model is no longer supported because its approach to plugholing is out of date and CFD models are capable of much more realistic flows regarding all aspects atrium of smoke control, including smoke filling.

Slightly Irregular Ceilings

When the difference in elevation between the highest and lowest parts of an atrium ceiling are less than 10% of the maximum floor-to-ceiling height, a weighted average ceiling height can be used with a zone model or smoke filling equations. The weighted average ceiling height is

$$H_{av} = \frac{1}{A}\sum_{i=1}^{n} H_i A_i \quad (15.5)$$

where:

H_{av} = weighted average ceiling height, ft (m),

H_i = ceiling height i, ft (m),

A = total cross-sectional area of the atrium, ft^2 (m^2),

A_i = area at H_i, ft^2 (m^2),

n = number of ceiling heights.

Example 15.2. Unsteady Smoke Filling

The atrium of Example 15.1 has a t-squared fire with a growth time of t_g = 150 s.

Part 1: How long does it take for the smoke layer to descend to 13 ft above the floor of the atrium?

From the values of A/H^2 and z/H of Example 15.1, it can be seen that Equation 15.4 is applicable.

From this equation, the time for the smoke layer to descend is 586 s (9.8 min).

Part 2: When the smoke layer reaches 13 ft above the atrium floor, how big is the fire?

$Q = 1000(t/t_g)^2 = 1000(586/150)^2$ = 15,300 Btu/s (16,100 kW). This fire is extremely large.

Sensitivity Analysis

A sensitivity analysis considers the extent to which variations in model inputs influence model output. This kind of analysis can be used to provide information to aid engineering judgment regarding the smoke filling equations and zone fire models. The idea is to choose a number of regular shapes related in some way to the atrium of concern, and to calculate the time it takes to fill these spaces with smoke down to a predetermined height. The predetermined height is also called minimum smoke layer height. The analysis can provide insight about smoke filling, and it can put bounds on the filling time.

The related spaces consist of a minimum space, a maximum space, and volume-equivalent spaces. The minimum and maximum spaces are chosen such that the smoke filling time of the space would be less and more than that of the atrium. When the atrium and a volume-equivalent space are filled with smoke to the predetermined height, they have the same volume of smoke. This volume of smoke is

$$S_{vt} = \sum_{i=1}^{n} S_{v,i} \tag{15.6}$$

where

S_{vt} = total smoke volume, ft^3 (m^3),

$S_{v,i}$ = smoke volume i, ft^3 (m^3),

n = number of smoke volumes.

For a volume-equivalent space, the height, width, and length are related as

$$W = \frac{S_{v,t}}{L(H-x)} \tag{15.7}$$

and

$$H = x + \frac{S_{v,t}}{LW} \tag{15.8}$$

where

L = length of volume-equivalent atrium, ft (m),

W = width of volume-equivalent atrium, ft (m),

H = ceiling height above fire, ft (m),

x = minimum smoke layer height, ft (m),

S_{vt} = total smoke volume, ft^3 (m^3).

The cross-sectional area of the volume-equivalent atrium is the length times the width ($A = LW$). Figure 15.8 shows an atrium that has five smoke volumes, and Example 15.3 is a sensitivity analysis of this atrium.

NATURAL VENTING EQUATION

Extensive research in natural venting has been conducted at the Fire Research Station in the UK (Thomas et al. 1963; Hansell and Morgan 1985, 1990, 1994; Morgan 1979, 1998; Morgan and Hansell 1987).

When the vents open for the smoke and makeup air, the atrium quickly fills with outdoor air, and the ambient temperature in the atrium becomes the outdoor temperature. Based on this research, the equation for the steady mass flow rate through a natural vent is

$$m_v = \frac{C A_v \rho_o [2 g d_b (T_s - T_o)(T_o - T_s)]^{1/2}}{[T_s + (A_v/A_i)^2 T_o]^{1/2}} \tag{15.9}$$

where

m_v = mass flow rate through the vent, lb/s (kg/s),

C = flow coefficient, dimensionless,

A_v = smoke vent area, ft^2 (m^2),

A_i = inlet air opening area, ft^2 (m^2),

ρ_o = outdoor air density, lb/ft^3 (kg/m^3),

g = acceleration of gravity, ft/s^2 (m/ s^2),

d_b = depth of smoke layer below the smoke vent, ft (m),

T_o = absolute temperature of outdoor air, °R (K),

T_s = absolute temperature of smoke, °R (K).

For general information about flow coefficients, see Chapter 3. Natural venting is shown in Figure 15.9, and Example 15.4 illustrates calculations using Equation 15.9. This equation does not include wind effects which can be very important with natural venting, and it is suggested that it be used in conjunction with other methods of analysis that include wind effects. Both the smoke vent area and the inlet air opening area are important. Equations in Chapter 16 can be used with Equation 15.9 to calculate the mass flow of the plume and the smoke layer temperature. Calculations like those of Example 15.4 can provide a starting point for CFD analysis of natural venting systems.

Because buoyancy of hot smoke is the driving force of natural venting, the mass flow rate through the vent increases with increasing smoke temperature. As the HRR increases, the mass flow rate of the plume into the upper layer increases, and the temperature of the smoke layer increases. For a fire larger than the design fire, the smoke temperature goes above the design value, and the mass flow rate through the vent increases above the design value. This benefit is unique to natural venting, and it helps offset the greater amount of smoke produced by fires that might exceed the design fire.

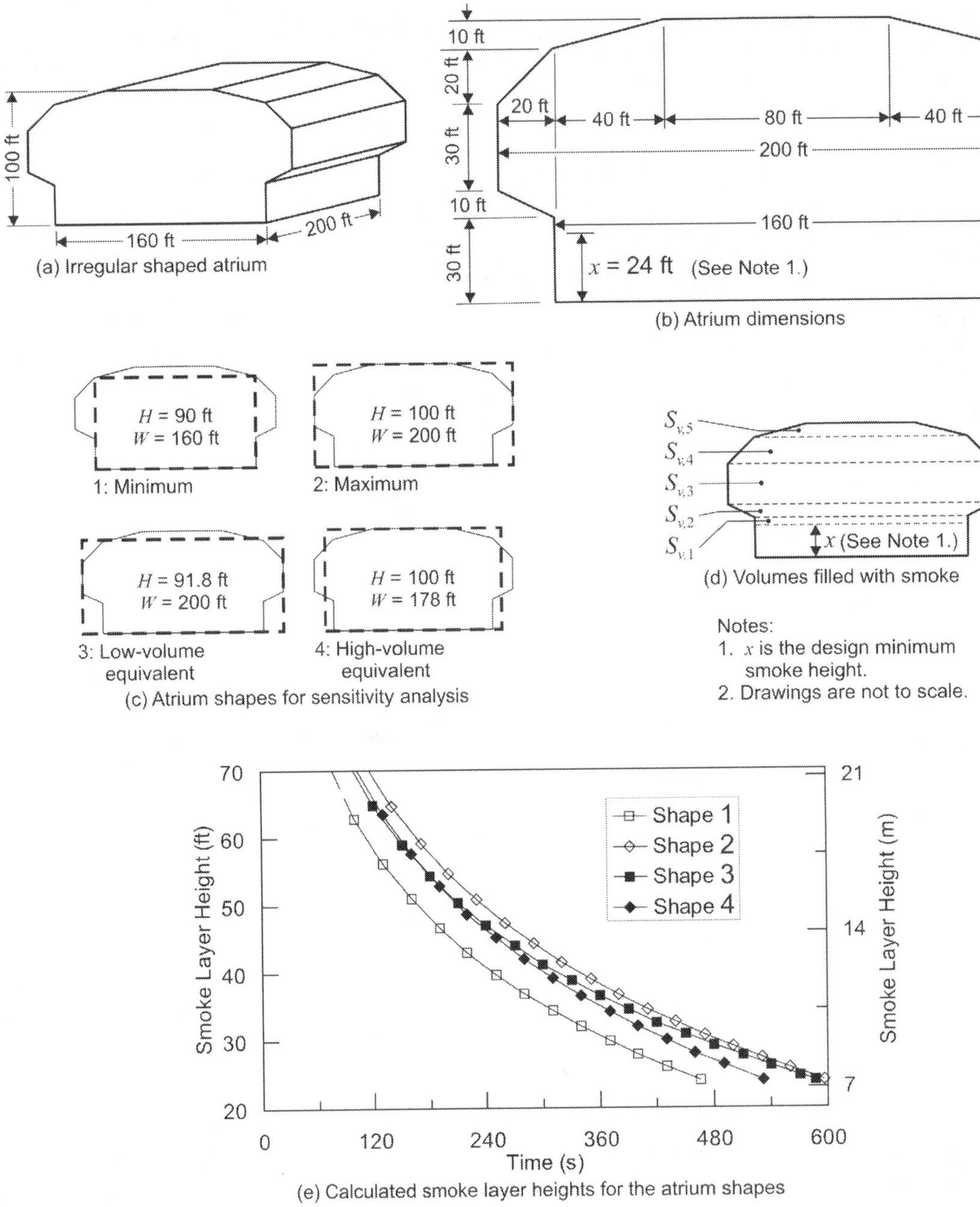

Figure 15.8 *Irregular shaped atrium of Example 15.3.*

Example 15.3. Smoke Filling Sensitivity Analysis

Estimate the time it takes to fill the atrium of Figure 15.8 with smoke to 24 ft (7.3 m) above the floor for a 3000 Btu/s (3170 kW) fire.

Using Figure 15.8, the smoke volumes are: $S_{v,1} = (30 - 24)(160)(200) = 192{,}000$ ft^3, $S_{v,2} = 10(20/2 + 40 + 80 + 40 + 20/2)(200) = 360{,}000$ ft^3, $S_{v,3} = 30(200)(200) = 1{,}200{,}000$ ft^3, $S_{v,4} = 20(20/2 + 40 + 80 + 40 + 20/2)(200) = 720{,}000$ ft^3, and $S_{v,5} = 10(40/2 + 80 + 40/2)(200) = 240{,}000$ ft^3. The total smoke volume is $S_{vt} = S_{v,1} + S_{v,2} + S_{v,3} + S_{v,4} + S_{v,5} = 2{,}712{,}000$ ft^3.

The sensitivity analysis is conducted with regular shaped spaces 1, 2, 3 and 4 shown in Figure 15.8c. All of these spaces have a length of 200 ft. For the volume-equivalent atria, H and W were evaluated from Equations 15.7 and 15.8. The atrium shapes are shown in Figure 15.8c. The minimum shape (1) was chosen with a relatively small H and W. The maximum shape (2) has the largest H and W. The low volume-equivalent shape (3) was selected with a width of 200 ft. The high volume-equivalent shape (4) was selected with a height of 100 ft.

This ratio, A/H^2, needs to be in the range of 0.9 to 14 for Equation 15.1. For shapes 1, 2, 3, and 4, A/H^2 is 3.95, 4.0, 4.75, and 3.56, respectively; thus the A/H^2 values are acceptable. The predetermined height, x, cannot be less than 20% of H for Equation 15.1. For shapes 1, 2, 3, and 4, 20% of H ranges from 0.18 to 0.20 ft, and $x = 24$ ft. Thus the value of x is acceptable. Equation 15.1 is used to calculate the height of the first indication of smoke for each of the sizes above, and these are graphed in Figure 15.8e.

For shapes 1, 2, 3, and 4, the times for the smoke layer to descend to 24 ft are 466 s, 597 s, 587 s, and 532 s. The actual time to descend is between that of shape 3 and shape 4. The time to descend is estimated conservatively at the value for shape 4 which is 532 s.

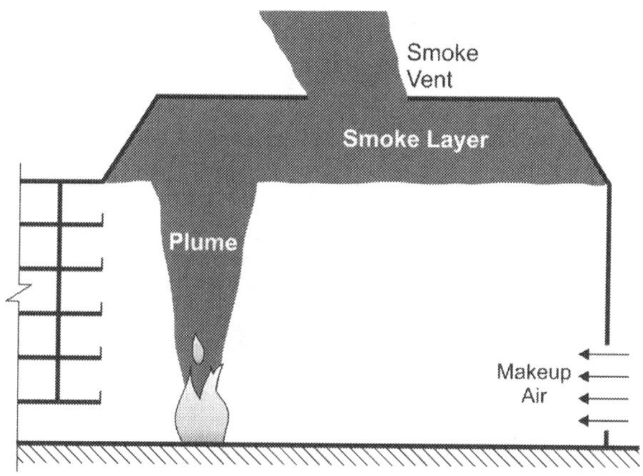

Figure 15.9 *Natural venting.*

AIRFLOW EQUATIONS

The airflow approach can be used to control smoke flow between the atrium and communicating spaces. The amount of airflow needed to control smoke is usually large, and this approach is only occasionally used. The airflow approach is capable of maintaining tenable conditions in the spaces to be protected, but small amounts of smoke infiltrate to the protected space. There could be low levels of products of combustion in spaces being protected by the airflow approach.

This section presents empirical equations for airflow that are based on research by Heskestad (1989). This flow can also be analyzed by CFD modeling. The limiting average velocity is the minimum value that can prevent smoke flow for a particular application. For a fire in a communicating space below the smoke layer interface (Figure 15.10), the limiting velocity can be calculated from

$$v_e = 38\left(gH\frac{T_f - T_o}{T_f}\right)^{1/2}$$

$$v_e = 0.64\left(gH\frac{T_f - T_o}{T_f}\right)^{1/2} \text{ for SI} \qquad (15.10)$$

where

v_e = limiting average air velocity, fpm (m/s),

g = acceleration of gravity, ft/s^2 (m/s^2),

H = height of opening, ft (m),

T_f = absolute temperature of smoke, °R (K),

T_o = absolute temperature of ambient air, °R (K).

Equation 15.10 also can be used to calculate the velocity to keep smoke out of a communicating space that is above the smoke layer interface (Figure 15.11), and the equations of Chapter 16 can be used to calculate T_f for this situation.

To keep smoke from a plume from entering a communicating space as shown in Figure 15.12, the limiting velocity can be calculated from

$$v_e = 17\left(\frac{Q}{z}\right)^{1/3}$$

$$v_e = 0.057\left(\frac{Q}{z}\right)^{1/3} \text{ for SI} \qquad (15.11)$$

Example 15.4. Natural Venting

An atrium has a total smoke vent area of 350 ft² (32.5 m²) and a total inlet air opening of 300 ft² (27.9 m²). The winter design temperature is 10°F (–12°C). For a smoke layer depth of 12 ft (3.66 m) and temperature of 140°F (60°C), what is volumetric flow through the smoke vent?

The parameters are: $C = 0.65$, $A_v = 350$ ft², $A_i = 300$ ft², $g = 32.2$ ft/s², $d_b = 12$ ft, $T_o = 10 + 460 = 470$ °R, $T_s = 140 + 460 = 600$°R.

The density is $\rho_o = \frac{144P_{atm}}{RT_o}$ where P_{atm} is atmospheric pressure of 14.7 psi, R is the gas constant 53.34.

$$\rho_o = \frac{144P_{atm}}{RT_o} = \frac{144(14.7)}{53.34(470)} = 0.0844 \text{ lb/ft}^3$$

$$m_v = \frac{CA_v\rho_o[2gd_b(T_s - T_o)(T_o/T_s)]^{1/2}}{[T_s + (A_v/A_i)^2T_o]^{1/2}}$$

$$m_v = \frac{0.65(350)(0.844)[2(32.2)(12)(600-470)(470/600)]^{1/2}}{[600+(350/300)^2 470]^{1/2}} = 153 \text{ lb/s}$$

The volumetric flow is $V = 60\frac{m_v}{\rho_o} = 60\frac{153}{0.0844} = 108,800 \text{ cfm}(51.3 \text{ m}^3/\text{s})$

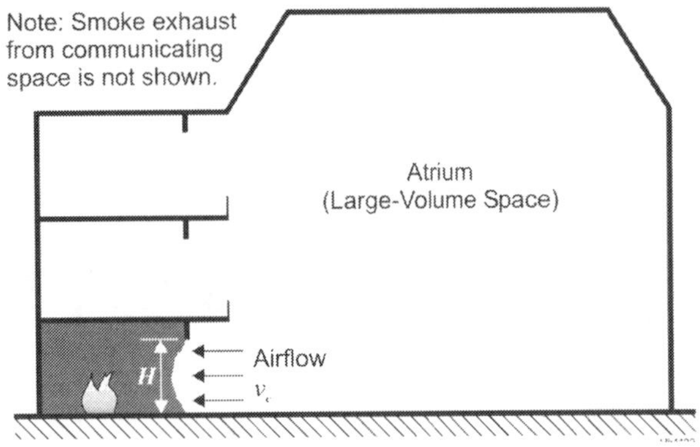

Figure 15.10 *Airflow to control smoke flow from communicating space.*

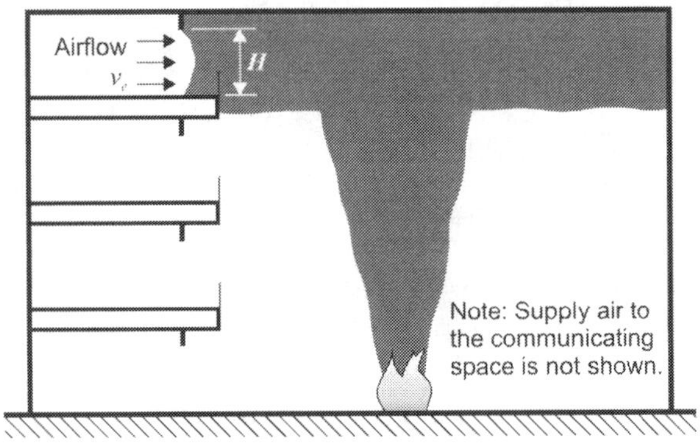

Figure 15.11 *Airflow to control smoke flow from smoke layer.*

where

v_e = limiting average air velocity, fpm (m/s),

Q = heat release rate, Btu/s (kW),

z = distance from base of fire to bottom of opening, ft (m).

Equation 15.11 is not applicable when z is less than 10 ft (3 m).

The velocity limitation of 200 fpm (1.02 m/s) for makeup air also applies here. This means that when the value of v_e, calculated from Equations 15.10 or 15.11, is greater than 200 fpm (1.02 m/s), airflow must not be used to control smoke flow. Examples 15.5 and 15.6 illustrate airflow calculations.

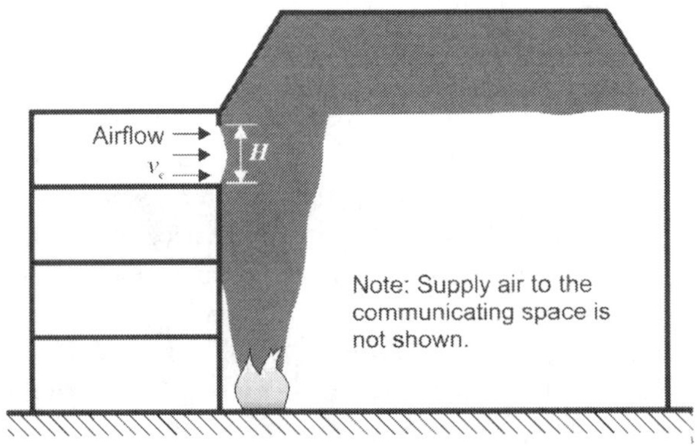

Figure 15.12 *Airflow to control smoke flow from plume.*

Example 15.5. Airflow Approach for Smoke from the Smoke Layer

Part 1: Calculate the velocity needed to prevent smoke flow into a communicating space from the smoke layer as shown in Figure 15.10. The height of the opening is 9 ft (2.74 m). The ambient temperature is 70°F (21°C), and the smoke temperature is 151°F (66°C). (Note: This smoke temperature is that of the smoke layer which can be calculated from equations in Chapter 16.)

The parameters are: $g = 32.2$ ft/s^2, $H = 9$ ft, $T_f = 151 + 460 = 611$°R, $T_o = 70 + 460 = 530$°R.

$$v_e = 38\left(gH\frac{T_f - T_o}{T_f}\right)^{1/2} = 38\left(32.2(9)\frac{611-560}{611}\right)^{1/2} = 236 \text{ fpm } (1.20 \text{ m/s})$$

This velocity would prevent smoke from entering the communicating space, but it is greater than 200 fpm (1.02 m/s), so the airflow approach cannot be used for this application.

Part 2: In the above calculation, if the smoke temperature were 120°F (49°C), what velocity would have been needed?

$$T_f = 120 + 460 = 580\text{°R}$$

$$v_e = 38\left(gH\frac{T_f - T_o}{T_f}\right)^{1/2} = 38\left(32.2(9)\frac{580-560}{580}\right)^{1/2} = 190 \text{ fpm } (0.96 \text{ m/s})$$

This velocity is less than 200 fpm (1.02 m/s), so the airflow approach can be used.

Example 15.6. Airflow Approach for from a Plume

For a 3500 Btu/s (3700 kW) fire, what is the limiting average velocity to prevent a smoke plume from entering a communicating space 25 ft (7.6 m) above the bottom of the fire (Figure 15.11)?

$$v_e = 17\left(\frac{Q}{z}\right)^{1/3} = 17\left(\frac{3500}{25}\right)^{1/3} = 88 \text{ fpm} (0.45 \text{ m/s})$$

This is much less than 200 fpm, so the airflow approach can be used for this application.

Airflow Can Supply Oxygen to the Fire: The airflow approach can supply oxygen to the fire, which can result in catastrophic failure. Even full sprinkler protection does not completely eliminate this potential. For any application that uses the airflow approach, this failure mode needs to be addressed in the design analysis.

TIME LAG

Figure 15.13 illustrates plume and ceiling jet development. Zone fire models neglect the time it takes the plume to rise to the ceiling and the ceiling jet to form. For normal size rooms like living rooms and bedrooms, the errors from such simplifications are insignificant. In an atrium, these time lags can be much larger. Newman (1988) and Mowrer (1990) developed relationships for the time lag of plumes from steady and t-squared fires.

The total time lag is the sum of that for the plume and the ceiling jet to form ($t_t = t_{pl} + t_{cj}$ where t_t is the total lag time, t_{pl} is time lag of plume, and t_{cj} is the time lag of ceiling jet).

Steady Fires

The development time of a plume from a steady fire is

$$t_{pl} = 0.135\frac{H^{4/3}}{Q^{1/3}}$$

$$t_{pl} = 0.67\frac{H^{4/3}}{Q^{1/3}} \text{ for SI} \tag{15.12}$$

and the development time for a ceiling jet from a steady fire is

$$t_{cj} = 0.168\frac{r^{11/6}}{Q^{1/3}H^{1/2}}$$

$$t_{cj} = 0.833\frac{r^{11/6}}{Q^{1/3}H^{1/2}} \text{ for SI} \tag{15.13}$$

where

t_{pl} = transport time lag of plume, s,

t_{cj} = transport time lag of ceiling jet, s,

H = ceiling height above top of fuel, ft (m),

Q = heat release rate, Btu/s (kW),

r = radius or horizontal distance from centerline of plume, ft (m).

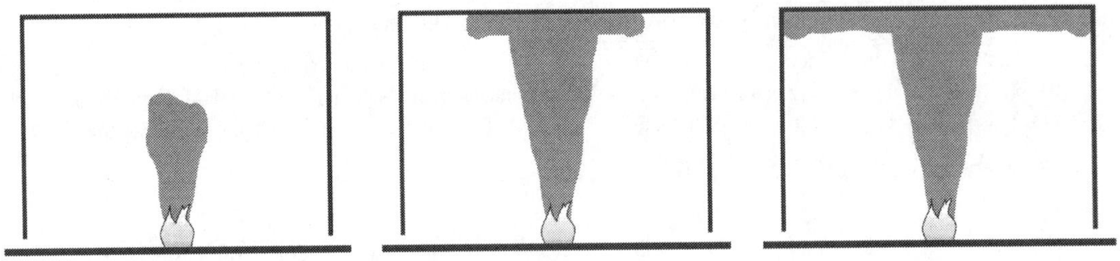

Figure 15.13 *Development of plume and ceiling jet.*

Example 15.7. Time Lag for a Steady Fire

What is the total time lag to form the plume and the ceiling jet from a 2000 Btu/s (2110 kW) fire in the center of an atrium 80 by 100 by 50 ft (24.4 by 30.5 by15.2 m) high?

It is considered that the fuel is at the floor, and $H = 50$ ft.

The development time of the plume is $t_{pl} = 0.135\frac{H^{4/3}}{Q^{1/3}} = 0.135\frac{50^{4/3}}{2000^{1/3}} = 2.0\text{ s}$.

The ceiling jet is completely formed when it reaches the corner of the room. Because the fire is at the center of the atrium, the distance to the corner is $r = \sqrt{40^2 + 50^2} = 64\text{ ft}$.

The development time for the ceiling jet is $t_{cj} = 0.168\frac{r^{11/6}}{Q^{1/3}H^{1/2}} = 0.168\frac{64^{11/6}}{2000^{1/3}50^{1/2}} = 3.9\text{ s}$.

The total time lag is 2.0 + 3.9 = 5.9 s.

Example 15.8. Time Lag for a T-squared Fire

What would be the time lag if the fire in Example 15.6 were a slow t-squared fire (t_g = 600s)?

The development time of the plume is $t_{pl} = 0.039H^{4/5}t_g^{2/5} = 0.039(50^{4/5})(600^{2/5}) = 11.5\text{ s}$.

The development time for the ceiling jet is $t_{cj} = 0.278\frac{rt_g^{2/5}}{H^{1/5}} = 0.278\frac{64(600^{2/5})}{50^{1/5}} = 105.5\text{ s}$.

The total time lag is 11.5 + 105.5 = 117 s.

T-Squared Fires

The development time of a plume from a t-squared fire is

$$t_{pl} = 0.039H^{4/5}t_g^{2/5}$$
$$t_{pl} = 0.1H^{4/5}t_g^{2/5} \text{ for SI} \tag{15.14}$$

and the development time for a ceiling jet from a t-squared fire is

$$t_{cj} = 0.278\frac{rt_g^{2/5}}{H^{1/5}}$$
$$t_{cj} = 0.72\frac{rt_g^{2/5}}{H^{1/5}} \text{ for SI} \tag{15.15}$$

where

t_{pl} = transport time lag of plume, s,
t_{cj} = transport time lag of ceiling jet, s,
t_g = growth time, s,
H = ceiling height above top of fuel, ft (m),
r = radius or horizontal distance from centerline of plume, ft (m).

As with the steady fire, the total time lag is the sum of that for the plume and the ceiling jet. The time lag equations for the t-squared fires are more useful than those for the steady fire, because fires tend to start out small and grow larger. For a comparison between time lags for steady and t-squared fires, see Examples 15.7 and 15.8.

SMOKE LAYER WITH SPRINKLER ACTION

As already stated, the temperature of a smoke plume can drop so much that sprinklers may not activate or activation may be so delayed that the spray may evaporate before it reaches the fire. The probability of sprinkler activation is less for fires in spaces with high ceilings than it is in spaces with low ceilings. A fire in an atrium space with a 30 ft (9.1 m) ceiling has a high probability of activating sprinklers. There is little chance that a fire in an atrium space with a 50 ft (15.2 m) ceiling would activate any sprinklers.

There is a mistaken belief that sprinkler action in an atrium will always drive the smoke layer down to the bottom of the atrium. In corridors and relatively small rooms, sprinkler action does tend to mix the smoke throughout the space. When sprinklers activate in an atrium, such smoke mixing may not happen. Full-scale fire tests with sprinklers conducted at the BHP Laboratory in Australia showed that sprinklered fires in a communicating space can produce buoyant smoke that flows out of the communicating space and upward to the ceiling of the large volume space (Bennetts et al. 1997).

The impact of sprinklers should be incorporated in the determination of design fires for atrium smoke control systems. Traditionally, the impact of sprinklers is not incorporated in the analysis of plumes and the smoke layer. This traditional approach is used for most atrium smoke transport calculations, including the equation approach discussed in Chapter 16, but it is possible to simulate the impact of sprinklers to some extent with some CFD models.

NOMENCLATURE

A = cross-sectional area of the atrium, ft^2 (m^2)

A_i = area at H_i, ft^2 (m^2); or inlet air opening area, ft^2 (m^2)

A_v = smoke vent area, ft^2 (m^2)

C = discharge coefficient, dimensionless

d_b = depth of smoke layer below smoke vent, ft (m)

g = acceleration of gravity, ft/s^2 (m/s^2),

H = ceiling height above base of fire; or ceiling height above top of fuel; or height of opening, ft (m),

H_{av} = weighted average ceiling height, ft (m),

H_i = ceiling height i, ft (m),

L = length of volume-equivalent atrium, ft (m)

m_v = mass flow rate through vent, lb/s (kg/s)

n = number of ceiling heights; or number of smoke volumes

Q = heat release rate, Btu/s (kW),

r = radius or horizontal distance from centerline of plume, ft (m)

$S_{v,i}$ = smoke volume i, ft^3 (m^3)

S_{vt} = total smoke volume, ft^3 (m^3)

t = time, s

t_{cj} = transport time lag of ceiling jet, s

T_f = absolute temperature of smoke, °R (K)

t_g = growth time, s

T_o = absolute temperature of ambient air; or absolute temperature of outdoor air, °R (K)

t_{pl} = transport time lag of plume, s

T_s = absolute temperature of smoke, °R (K)

v_e = limiting average air velocity, fpm (m/s)

W = width of volume-equivalent atrium, ft (m)

x = minimum smoke layer height, ft (m)

z = distance from base of fire to bottom of opening, ft (m); or distance above base of fire to first indication of smoke, ft (m)

ρ_o = outdoor air density, lb/ft^3 (kg/m^3)

REFERENCES

Bennetts, I.D., et al. 1997. Fire safety in shopping centres. Broken Hill Proprietary Company Limited, Mulgrave, Australia.

Cooper, L.Y. et al. 1981. An experimental study of upper hot layer stratification in full scale multiroom fire scenarios. Paper 81-HT-9. New York: American Society of Mechanical Engineers.

Hadjisophocleous, G., and J. Zhou. 2008. Evaluation of atrium smoke exhaust makeup air velocity. *ASHRAE Transactions*, 114(1): 147–155.

Hagglund, B., R. Jansson, and K. Nireus. 1985. Smoke filling experiments in a 6 × 6 × 6 meter enclosure. FOA Report C 20585-D6, National Defense Research Institute of Sweden, Stockholm.

Hansell, G.O., and H.P. Morgan. 1985. Fire sizes in hotel bedroom—implications for smoke control design. *Fire Safety Journal* 8(3).

Hansell, G.O., and H.P. Morgan. 1990. Smoke control in atrium buildings using depressurization Part 2: Considerations affecting practical design. *Fire Science and Technology* 10(1).

Hansell, G.O., and H.P. Morgan. 1994. Design approaches for smoke control in atrium buildings. BR 258, Fire Research Station, Borehamwood, Herts, UK.

Heskestad, G. 1989. Inflow of air required at wall and ceiling apertures to prevent escape of fire smoke. FMRC No. 01836.20, Factory Mutual Research Corp., Norwood, MA.

Heskestad, G., and M.A. Delichatsios. 1977. Environments of fire detectors Phase I: Effect of fire size, ceiling height and materials. Volume I and II Measurements (NBS GCR 77 86; NBS GCR 77 95), National Bureau of Standards, Gaithersburg, MD.

Klote J.H., and J.A. Milke. 2002. *Principles of Smoke Management*. Atlanta: ASHRAE.

Morgan, H.P. 1979. Smoke control methods in enclosed shopping complexes of one or more stores: A design summary. Building Research Establishment.

Morgan, H.P. 1998. Fire safety—smoke control: developments in European standards. CIBSE-ASHRAE Seminar on Fire Safety—Smoke Control: Standards and Practice, March 23, London, UK, Institution of Mechanical Engineers, London, UK.

Morgan, H.P., and G.O. Hansell. 1987. Atrium buildings: Calculating smoke flows in atria for smoke-control design. *Fire Safety Journal* 12(1):9–35.

Mowrer, F.W. 1990. Lag times associated with fire detection and suppression. *Fire Technology* 26(3).

Mulholland, G., et al. 1981. Smoke filling in an enclosure. Paper 81-HT-8. New York: American Society of Mechanical Engineers.

Newman, J.S. 1988. Principles of fire detection. *Fire Technology* 24(2).

Nowler, S.P. 1987. Enclosure environment characterization testing for the baseline validation of computer fire simulation codes. Report, Sandia National Laboratories, Albuquerque, New Mexico.

Thomas, P.H., et al. 1963. Investigation into the flow of hot gases in roof venting. Tech. Paper No. 7, Fire Research Station, Boreham Woods, Herts, UK.

CHAPTER 16

Equations for Steady Atrium Smoke Exhaust

John H. Klote

This chapter addresses the algebraic equations used for analysis of steady mechanical smoke exhaust, which is the most common design approach in North America. In this approach, the atrium exhaust is sized to maintain a steady smoke layer height for a fire with a steady heat release rate (HRR). As stated in the last chapter, the term atrium is used in this handbook in a generic sense to mean any large-volume space.

At the end of this chapter there is a case study that includes three examples illustrating analysis of steady mechanical smoke exhaust. There are also a number of other examples.

SMOKE PRODUCTION

Smoke production depends on the heat release rate of the fire and the kind of smoke plume that rises above the fire. This section addresses the axisymmetric plume, balcony spill plume, wall plume, corner plume, and the window plume. The equations for these plumes are for strongly buoyant plumes, and they are not to be used when the temperature rise of the plume above ambient is less than 4°F (2.2°C). This temperature rise is the average plume temperature minus the ambient temperature ($T_p - T_o$ where T_p is the average plume temperature and T_o is the ambient temperature).

The convective portion of the heat release rate is

$$Q_c = \chi_c Q \qquad (16.1)$$

where

Q_c = convective heat release rate of fire, Btu/s (kW),

Q = heat release rate of fire, Btu/s (kW),

χ_c = convective fraction of heat release, dimensionless.

The convective fraction of heat release varies from about 0.4 to 0.9, but $\chi_c = 0.7$ is commonly used for most design applications.

Axisymmetric Plume

The smoke plumes that are of concern for fire protection in buildings are by nature unsteady and turbulent processes. The empirical equations for plumes are based on time-averaged flow that considers the plume coming from a point called the virtual origin.

Figure 16.1a is a sketch of an axisymmetric plume. The idealized plume model is cone shaped, and Figure 16.1b shows the virtual origin of this idealized plume above the base of the fire. Along the edges of the plume, air is entrained from the surroundings into the plume. This entrained air is proportional to the velocity of the plume at that height. Temperature and velocity distributions are shown in Figure 16.1b. The maximum for both of these distributions is at the centerline of the plume. The section of the idealized axisymmetric plume is round as shown in Figure 16.1c.

The basic concepts of plume analysis were developed by Morton, Taylor, and Turner (1956). Empirical plume equations were developed at the California Institute of Technology (Cetegan, Zukoski, and Kubota 1982), National Institute of Standards and Technology (McCaffrey 1983), and Factory Mutual Research Corporation (Heskestad 1983, 1984). The following equations are primarily based on the work at Factory Mutual Research Corporation.

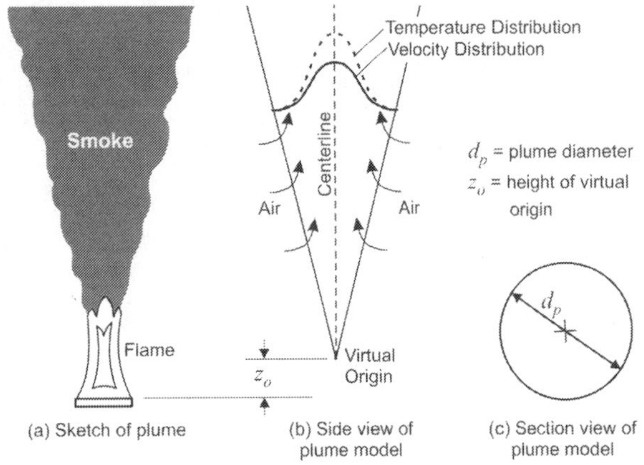

Figure 16.1 *Sketch and idealized structure of an axisymmetric plume.*

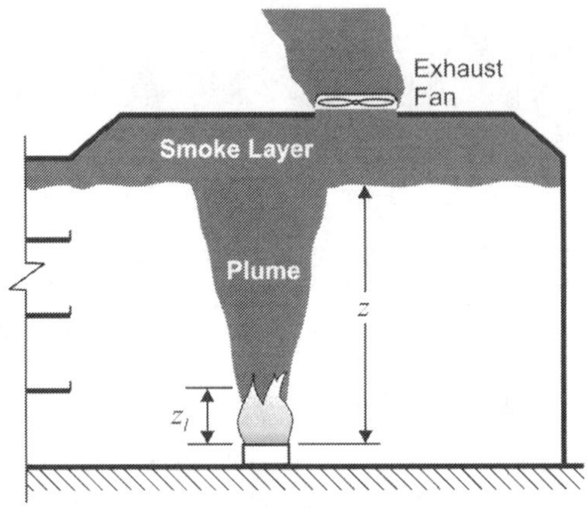

Figure 16.2 *Fire in an atrium with an axisymmetric plume.*

Mass Flow

For z greater than or equal to z_l ($z \geq z_l$), the mass flow is

$$m = 0.022Q_c^{1/3}(z-z_o)^{5/3}[1+0.19Q_c^{2/3}(z-z_o)^{-5/3}]$$

$$m = 0.071Q_c^{1/3}(z-z_o)^{5/3}[1+0.027Q_c^{2/3}(z-z_o)^{-5/3}] \text{ for SI} \tag{16.2}$$

and for z less than z_l ($z < z_l$), the mass flow is

$$m = \frac{0.013Q_c z}{z_l}$$

$$m = \frac{0.0056Q_c z}{z_l} \text{ for SI} \tag{16.3}$$

where

m = mass flow of an axisymmetric plume at height z, lb/s (kg/s),

Q_c = convective heat release rate of fire, Btu/s (kW),

z = distance above base of fire, ft (m),

z_o = distance to virtual origin, ft (m),

z_l = limiting elevation, ft (m).

Equations 16.2 and 16.3 are the axisymmetric plume equations used in NFPA 204 (NFPA 2012a). Because smoke includes the air that is entrained with the products of combustion, all of the mass flow in the axisymmetric plume is smoke. It follows that these equations can be thought of as equations for the production of smoke from a fire. Simplified versions of these equations are discussed later, and the same comments also apply to them. Figure 16.2 shows a fire in an atrium with an axisymmetric plume.

There is a jump in the mass flow calculated from Equations 16.2 and 16.3 at $z = z_l$, which is called a discontinuity. This discontinuity does not impact the utility of these equations for the purposes of this chapter. Because discontinuities cause convergence failures in computer programs, direct use of these equations is unsuitable for zone fire models.

Virtual Origin

The distance to the virtual origin is

$$z_o = 0.278Q^{2/5} - 1.02D_f$$

$$z_o = 0.083Q^{2/5} - 1.02D_f \text{ for SI} \tag{16.4}$$

where

z_o = distance to virtual origin, ft (m),

Q = heat release of the fire, Btu/s (kW),

D_f = diameter of fire, ft (m).

The distance to the virtual origin can also be expressed in terms of heat release density as

$$z_o = 0.278Q^{2/5} - 1.15(Q/q)^{1/2}$$

$$z_o = 0.083Q^{2/5} - 1.15(Q/q)^{1/2} \text{ for SI} \tag{16.5}$$

where

z_o = distance to virtual origin, ft (m),

Q = heat release of fire, Btu/s (kW),

q = heat release density, Btu/s ft^2 (kW/m^2).

Figure 16.3 shows the distance to virtual origin z_o over a wide range of values of Q and q. From Equation 16.2, it can be seen that as z_o decreases, the mass flow of the plume increases. The opposite is also true. Because the fan exhaust rate is related to the mass flow of the plume, negative values of z_o are especially significant.

Strictly speaking, the fire for an axisymmetric plume has to be circular in shape, but other fire shapes can produce plumes that become similar to the axisymmetric plume at some distance above the base of the fire. For other shaped fires, the effective diameter can be used as an approximation of the fire diameter. The effective diameter is

$$D_e = \frac{4A}{P} \quad (16.6)$$

where

De = effective diameter, ft (m),

A = cross-sectional area of fire, ft^2 (m^2),

P = perimeter of fire, ft (m).

For a rectangular fire, the effective diameter is

$$D_e = \frac{4ab}{2(a+b)} \quad (16.7)$$

where

D_e = effective diameter, ft (m),

a = width of fire, ft (m),

b = length of fire, ft (m).

Equations for the distance to virtual origin and the limiting elevation are addressed later, and Example 16.1 shows how to calculate the mass flow of an axisymmetric plume including the virtual origin.

Limiting Elevation

The limiting elevation can be considered the mean flame height where the maximum plume temperature is 900°R (500 K). This elevation depends on the fire geometry, the ambient conditions, the heat of combustion and the stoichiometric ratio. A general relationship for flame height is

$$z_l = 0.788Q^{2/5} - 1.02D_f$$
$$z_l = 0.235Q^{2/5} - 1.02D_f \quad \text{for SI} \quad (16.8)$$

where

z_l = limiting elevation, ft (m),

Q = heat release of fire, Btu/s (kW),

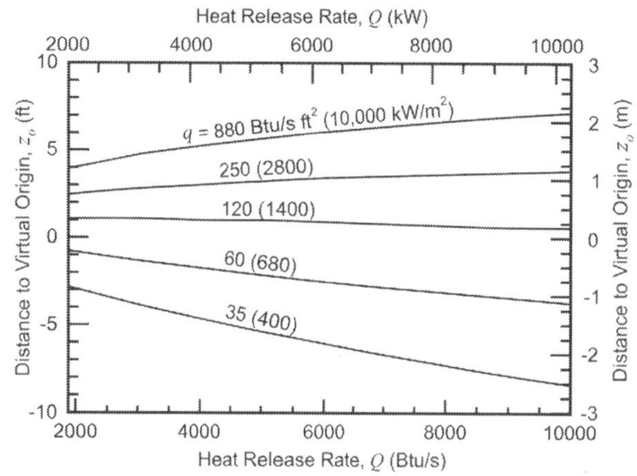

Figure 16.3 *Distance to the virtual origin in terms of heat release density, q.*

D_f = fire diameter, ft (m).

The limiting elevation also can be expressed in terms of heat release density as

$$z_l = 0.788Q^{2/5} - 1.15(Q/q)^{1/2}$$
$$z_l = 0.235Q^{2/5} - 1.15(Q/q)^{1/2} \quad \text{for SI} \quad (16.9)$$

where

z_l = limiting elevation, ft (m),

Q = heat release of fire, Btu/s (kW),

q = heat release density, Btu/s ft^2 (kW/m^2).

Centerline Temperature

As previously mentioned, the temperature of an axisymmetric plume is greatest at the centerline. This centerline temperature is of interest for fire reconstruction or in the unusual situations when an atrium is tested by a real fire. The centerline temperature is

$$T_{cp} = T_o + 9.1\left(\frac{T_o}{gc_p^2\rho_o^2}\right)^{1/3} \frac{Q_c^{2/3}}{(z-z_o)^{5/3}} \quad (16.10)$$

where

T_{cp} = absolute centerline axisymmetric plume temperature at elevation z, °R (K),

T_o = absolute ambient temperature, °R (K),

c_p = specific heat of plume gases, 0.24 Btu/lb·°F (1.0 kJ/kg·K),

ρ_o = density of ambient air, lb/ft^3 (kg/m^3),

g = acceleration of gravity, ft/s^2 (m/s^2),

Q_c = convective heat release rate of fire, Btu/s (kW),

Example 16.1. Plume with Virtual Origin

Part 1: An 8 ft (2.44 m) diameter fire has a heat release rate of 1900 Btu/s (2000 kW). At 40 ft (12.2 m) above the base of the fire, what is the mass flow of the plume? Use a convective fraction of 0.7.

The parameters are: χ_c, = 0.7, D_f = 8 ft, Q = 1900 Btu/s, z = 40 ft.

The convective heat release rate is $Q_c = \chi_c\ Q = 0.7(1900) = 1330$ Btu/s.

The distance to the virtual origin is

$$z_o = 0.278Q^{2/5} - 1.02D_f = 0.278(1900)^{2/5} - 1.02(8) = -2.46 \text{ ft}$$

Because z_o is negative, we know that the virtual origin is below the fire.

The limiting elevation is

$$z_l = 0.788Q^{2/5} - 1.02D_f = 0.788(1900)^{2/5} - 1.02(8) = -7.98 \text{ ft}$$

Because z is greater than z_l, the mass flow of the plume is calculated as follows:

$$m = 0.022Q_c^{1/3}(z - z_o)^{5/3}[1 + 0.19Q_c^{2/3}(z - z_o)^{-5/3}]$$

$$m = 0.022(1330^{1/3})[40 - (-2.46)]^{5/3}[1 + 0.19(1330^{2/3})(40 - (-2.46))^{-5/3}] = 131 \text{ lb/s (59 kg/s)}$$

Part 2: If the diameter of the fire above were 4 ft (1.22 m), what would the mass flow be?

$$z_o = 0.278Q^{2/5} - 1.02D_f = 0.278(1900)^{2/5} - 1.02(4) = 1.62 \text{ ft}$$

$$z_l = 0.788Q^{2/5} - 1.02D_f = 0.788(1900)^{2/5} - 1.02(4) = 11.9 \text{ ft}$$

$$m = 0.022(1330^{1/3})(40 - 1.62)^{5/3}[1 + 0.19(1330)^{2/3}(40 - 1.62)^{-5/3}] = 111 \text{ lb/s (50 kg/s)}$$

This is about 15% less than Part 1.

z = distance above base of the fire, ft (m),

z_o = distance to virtual origin, ft (m).

Example 16.2 shows how to calculate the centerline temperature of an axisymmetric plume.

Simplified Axisymmetric Plume

The simplified axisymmetric plume equations do not include the distance to the virtual origin z_o. These simplified equations are often used for atrium smoke control analysis with the following justifications. It is not possible to determine z_o with confidence in many applications because the fuel is not known for a fire that has yet to happen. Also, the impact of z_o on the mass flow rate of the plume can be relatively small, because z usually is much larger than z_o. When the fuel can be determined with some confidence, analysis that incorporates the virtual origin can be used. The simplified axisymmetric plume equations are listed here.

For z greater than or equal to z_l ($z \geq z_l$), the mass flow is

$$m = 0.022Q_c^{1/3}z^{5/3} + 0.0042Q_c$$
$$m = 0.071Q_c^{1/3}z^{5/3} + 0.0018Q_c \text{ for SI} \qquad (16.11)$$

and for z less than z_l ($z < z_l$), the mass flow is

$$m = 0.0208Q_c^{3/5}z$$
$$m = 0.032Q_c^{3/5}z \text{ for SI} \qquad (16.12)$$

where

m = mass flow in axisymmetric plume at height z, lb/s (kg/s),

Q_c = convective heat release rate of fire, Btu/s (kW),

z = distance above base of fire, ft (m),

z_l = limiting elevation, ft (m).

As with Equations 16.2 and 16.3, Equations 16.11 and 16.12 are not to be used when the temperature rise of the plume above ambient is less than 4°F (2.2°C). Equations 16.11 and 16.12 also have a discontinuity which makes direct use of the equations unsuitable for zone fire models. Chapter 18 explains how zone fire models deal with this discontinuity.

Example 16.2. Plume Centerline Temperature

What is the centerline temperature at 40 ft (12.2 m) above the base of the fire for the plume of Part 1 of Example 16.1? The ambient temperature is 70°F (21°C).

The parameters are: $T_o = 70 + 460 = 530$°R, $g = 32.2$ ft/s^2, $C_p = 0.24$ Btu/lb°R, $z = 40$ ft, $z_o = -2.46$ ft, $Q_c = 1330$ Btu/s

The density of ambient air is $\rho_o = \frac{144 p_{atm}}{RT} = \frac{144(14.7)}{53.34(530)} = 0.0749 \text{ lb/ft}^3$

The centerline temperature is

$$T_{cp} = T_o + 9.1\left(\frac{T_o}{gC_p^2\rho_o^2}\right)^{1/3} \frac{Q_c^{2/3}}{(z - z_o)^{5/3}}$$

$$T_{cp} = 530 + 9.1\left(\frac{530}{32.2[0.24]^2[0.07488]^2}\right) \frac{(1330)^{2/3}}{(40 - [2.46])^{5/3}} = 609°\text{R}$$

This temperature is 609 – 460 = 149°F (65°C).

Equations 16.11 and 16.12 are used in NFPA 92 (NFPA 2012b). For these equations, the limiting elevation is

$$z_l = 0.533 Q_c^{2/5}$$
$$z_l = 0.166 Q_c^{2/5} \text{ for SI} \qquad (16.13)$$

where

z_l = limiting elevation, ft (m).

Q_c = convective heat release rate, Btu/s (kW).

Figure 16.4 is a graph of the volumetric flow rate of smoke exhaust for a fire in an atrium with a simplified axisymmetric plume. It can be seen from this figure that the smoke exhaust increases with the HRR, Q, and the distance above the base of the fire, z. The curves in this figure were calculated from Equations 16.1 and 16.11 and equations discussed later for the smoke layer temperature, density, and volumetric flow. The case study at the end of the chapter has an example that illustrates how to calculate the volumetric flow using a simplified axisymmetric plume.

Plume Diameter

As a plume rises, it entrains air and widens. The diameter of an axisymmetric plume is calculated as

$$d_p = K_d z \qquad (16.14)$$

where

d_p = plume diameter, ft (m).

z = distance above base of fire, ft (m).

K_d = diameter constant (dimensionless).

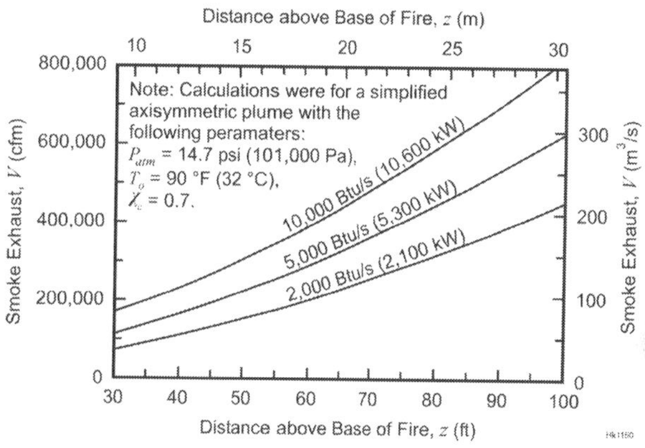

Figure 16.4 *Smoke exhaust required for fires in atrium.*

The plume diameter varies considerably such that K_d ranges from about 0.25 to 0.5. To achieve conservative results, the following values should be used: (1) K_d = 0.5 for plume contact with walls, and (2) K_d = 0.25 for beam detection of the smoke plume.

When the calculated plume diameter indicates that the plume is in contact with all the walls of an atrium, the point of contact can be considered the smoke layer interface.

Wall and Corner Plumes

For a fire located next to a wall, the plume comes into contact with the wall, resulting in less air entrained into the plume. Provided that z is large enough, the fire and the plume may be considered half that of the idealized axisymmetric plume (Figure 16.5). Thus the mass flow rate of a wall plume is half that of an axisymmetric plume, but the analogous fire for the axisymmetric plume is twice that of the "real" fire of the wall plume. The mass flow rate of a wall plume can be estimated as (1) let Q be twice that of the wall plume, (2) apply a plume equation to calculate the mass flow m of an axisymmetric plume, and (3) the mass flow of the wall plume is half of the calculated value. This approach can be used with any equation for the mass flow of an axisymmetric plume.

This reasoning can be extended to the configurations of fire contact with corners. For an axisymmetric plume including the virtual origin, the mass flow for wall and corner plumes is

$$m = 0.022n^{2/3}Q_c^{1/3}(z-z_o)^{5/3}\left[1+0.19(Q_c/n)^{2/3}(z-z_o)^{-5/3}\right] \quad (16.15)$$

$$m = 0.071n^{2/3}Q_c^{1/3}(z-z_o)^{5/3}\left[1+0.027(Q_c/n)^{2/3}(z-z_o)^{-5/3}\right] \text{ for SI}$$

For a simplified axisymmetric plume, the mass flow is

$$m = n[0.022(Q_c/n)^{1/3}z^{5/3} + 0.0042(Q_c/n)]$$

$$m = n[0.071(Q_c/n)^{1/3}z^{5/3} + 0.0018(Q_c/n)] \text{ for SI} \quad (16.16)$$

where

m = mass flow in axisymmetric plume at height z, lb/s (kg/s),
n = fire location factor, dimensionless,
Q_c = convective heat release rate of fire, Btu/s (kW),
z = height above fuel, ft (m),
z_l = limiting elevation, ft (m).

The fire location factor is the ratio of the angle of the corner to 360° ($n = \theta/360$ where θ is the angle of the corner as shown in Figure 16.6a). Figure 16.6 also shows some arrangements of walls and corners including values of n. When $n = 1$, Equation 16.15 is the same as Equation 16.2, and Equation 16.16 is the same as Equation 16.11.

While there is no consensus about how large z must be for the wall and corner plumes equations to be applicable, it is suggested that z be at least three times the limiting elevation z_l provided that the fire is in contact with the wall. For Equation 16.15, the limiting elevation is calculated from Equations 16.8 or 16.9. For Equation 16.16, it is calculated from Equation 16.13.

Example 16.3 shows that a corner plume produces less smoke than one that is away from walls. For this reason, corner plumes are usually not used for calculations that would size atrium smoke exhaust fans.

Balcony Spill Plume

Figure 16.7a shows a balcony spill plume from a fire in a room open to an atrium. Smoke flows out of room, under the balcony, around the balcony edge, and up to form a smoke layer under the ceiling of the atrium. This figure shows smoke blocking parts of balconies, which sometimes happens. It is beyond the capability of smoke control technology to assure that such smoke blocking will not occur, but balconies away from the spill plume will not be blocked.

The following balcony spill plume equations are based on extensive research including scale model fire experiments, full scale fire experiments, and analytical studies (Law 1986; Morgan and Marshall 1979; Lougheed and McCartney 2008a, 2008b; Lougheed, McCartney, and Gibbs 2007; McCartney, Lougheed, and Weckman 2008; Ko, Hadjisophocleous, and Lougheed 2008).

There are three regions of mass flow for the balcony spill plume, and factors defining these regions are (1) the height z_b of the plume above the balcony edge and (2) width W of the plume as it spills over the balcony edge. This width is also called the *length of the*

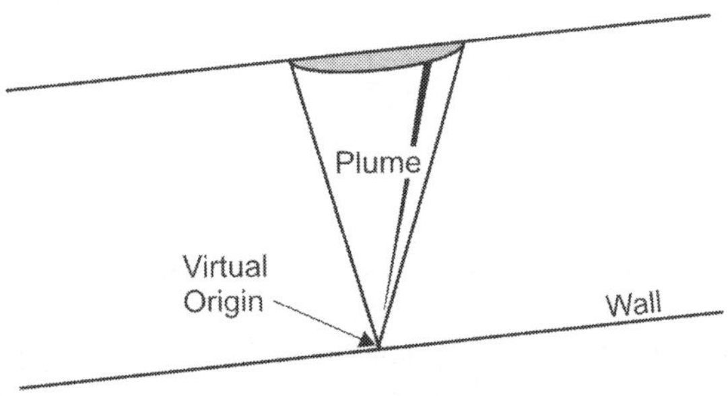

Figure 16.5 *Idealized wall plume model.*

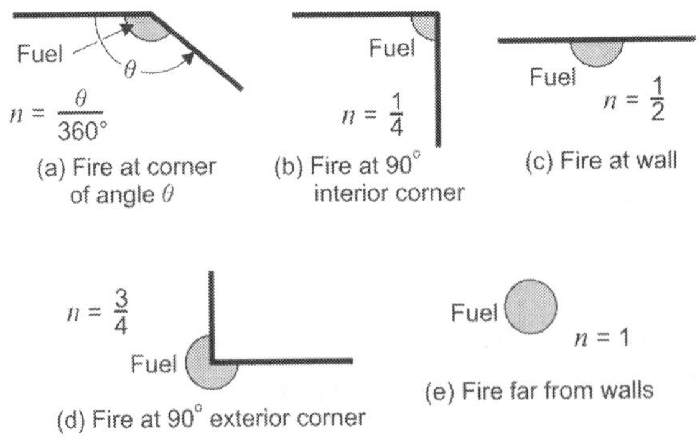

Figure 16.6 *Fire and wall configurations.*

spill. Some applications have draft curtains perpendicular to the opening to channel smoke, and the length of the spill is the distance between draft curtains as shown in Figure 16.7b. The depth of these draft curtains needs to be at least 10% of the floor-to-ceiling height of the balcony.

In the absence of draft curtains, the smoke spilling over the balcony edge becomes thin at the ends of the spill, and determination of the length of the spill is subjective (Figure 16.7c). The equivalent length of the spill is

$$W = w + b \tag{16.17}$$

where

W = length of spill, ft (m),

w = width of opening from area of origin, ft (m),

b = distance from opening to balcony edge, ft (m).

The length of the spill calculated from Equation 16.17 may not be the same as the actual length of the spill, but it is appropriate for calculation of the plume mass flow in the following equations.

Example 16.3. Corner Plume

Part 1: For a 90° interior corner fire as shown in Figure 16.6b, the heat release rate is 3500 Btu/s (3700 kW). What is the mass flow 40 ft (12.2 m) above the base of the fire? Consider the plume to be a simplified axisymmetric plume, and use a convective fraction of 0.7.

The parameters are: $n = 0.25$, $\chi_c, = 0.7$, $Q = 3500$ Btu/s, $z = 40$ ft

The convective heat release rate is $Q_c = \chi_c\ Q = 0.7(3500) = 2450$ Btu/s

The limiting elevation is $z_l = 0.533 Q_c^{2/5} = 0.533(2450)^{2/5} = 12.1$ ft

Because z is more than 3 times z_l, the mass flow of the plume is calculated as follows:

$$m = n[0.022(Q_c/n)^{1/3} z^{5/3} + 0.0042(Q_c/n)]$$

$$m = 0.25[0.022(2450/0.25)^{1/3}(40)^{5/3} + 0.0042(2450/0.25)] = 65.4 \text{ lb/s (29.6 kg/s)}$$

Part 2: If the fire in Part 1 were for a fire away from any walls, what would the mass flow be?

The parameters are the same as above, except that $n = 1$.

$$m = [0.022(2450)^{1/3}(40)^{5/3} + 0.0042(2450)] = 149 \text{ lb/s (67.6 kg/s)}$$

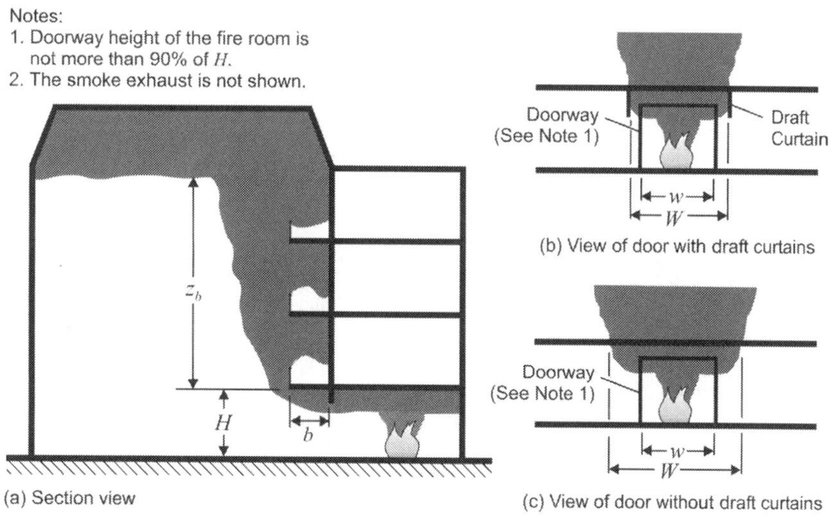

Figure 16.7 *Balcony spill plume.*

Region 1 is for a height of the plume above the balcony edge less than 50 ft (15 m). Region 2 is for a height of the plume above the balcony edge greater than or equal to 50 ft (15 m), and the length of the spill less than 32.8 ft (10 m). Region 3 is for a height of the plume above the balcony edge greater than or equal to 50 ft (15 m) and the length of the spill is between 32.8 ft (10 m) and 45.9 ft (14 m). For each region, there is an equation that is used to calculate the mass flow of the plume into the smoke layer.

In mathematical terms, these regions are[1]:

Region 1: $z_b < 50$ ft ($z_b < 15$ m),

Region 2: $z_b \geq 50$ ft and $W < 32.8$ ft ($z_b \geq 15$ m and $W < 10$ m),

Region 3: $z_b \geq 50$ ft and $32.8 \text{ ft} \geq W \geq 45.9$ ft

($z_b \geq 15$ m and $10 \text{ m} \geq W \geq 14$ m).

For Region 1, the mass flow of the plume is

$$m = 0.12(QW^2)^{1/3}(z_b + 0.25H)$$
$$m = 0.36(QW^2)^{1/3}(z_b + 0.25H) \text{ for SI} \quad (16.18)$$

For Region 2, the mass flow of the plume is

$$m = 0.32Q_c^{1/3}W^{1/5}(z_b + 0.098W^{7/15}H + 19.5W^{7/15} - 49.2)$$
$$m = 0.59Q_c^{1/3}W^{1/5}(z_b + 0.17W^{7/15}H + 10.35W^{7/15} - 15) \text{ for SI} \quad (16.19)$$

For Region 3, the mass flow of the plume is

$$m = 0.062(Q_cW^2)^{1/3}(z_b + 0.51H + 52)$$
$$m = 0.2(Q_cW^2)^{1/3}(z_b + 0.51H + 15.75) \text{ for SI} \quad (16.20)$$

where

m = mass flow rate in plume, lb/s (kg/s),

Q = heat release rate of fire, Btu/s (kW),

Q_c = convective heat release rate of fire, Btu/s (kW),

W = length of spill, ft (m),

z_b = height of plume above balcony edge, ft (m),

H = height of balcony above fuel, ft (m).

The case study at the end of the chapter has an example that illustrates how to calculate the volumetric flow using a balcony spill plume. Equations 16.18, 16.19, and 16.20 do not include the effect of sprinklers, which will reduce plume buoyancy to some extent. It is believed that these equations can be useful for designs that include sprinklered fires.

For balcony spill plumes that have either geometry different from that of Figure 16.7 or do not fit in one of the three regions, scale modeling or computational fluid dynamic (CFD) modeling should to be used. It is suggested that sprinklers be included in such modeling.

Window Plume

A window plume comes from a room that is completely involved in fire. Because such a fully developed fire is not expected with a properly functioning sprinkler system, window plumes are appropriate only for unusual applications.

A window plume can flow from a window or other opening as shown in Figure 16.8. As discussed in Chapter 5, the HRR of a fully developed fire in a room with only one opening is

$$Q = 61.2A_wH_w^{1/2}$$
$$Q = 1260A_wH_w^{1/2} \text{ for SI} \quad (16.21)$$

where

Q = heat release rate of fire, Btu/s (kW),

A_w = area of ventilation opening, ft² (m²),

H_w = height of ventilation opening, ft (m).

The mass entrainment for window plumes is given as

$$m = [0.077(A_wH_w^{1/2})^{1/3}(z_w + a)^{5/3}] + 0.18A_wH_w^{1/2}$$
$$m = [0.68(A_wH_w^{1/2})^{1/3}(z_w + a)^{5/3}] + 0.159A_wH_w^{1/2} \text{ for SI} \quad (16.22)$$

where

m = mass flow rate in plume, lb/s (kg/s),

A_w = area of ventilation opening, ft² (m²),

H_w = height of ventilation opening, ft (m),

z_w = distance from the smoke layer interface to top of the window, ft (m),

a = $(2.40A_w^{2/5}H_w^{1/5}) - 2.1H_w$, ft (m).

1. The regions and mass flow equations listed here have been corrected. In the 2012 version of NFPA 92, there is an error in one of the bounds for region 2 and errors in the I-P versions of Equations 16.19 and 16.20. NFPA has issued an errata sheet correcting the equations. A correction of the bounds of region 2 is expected in the future.

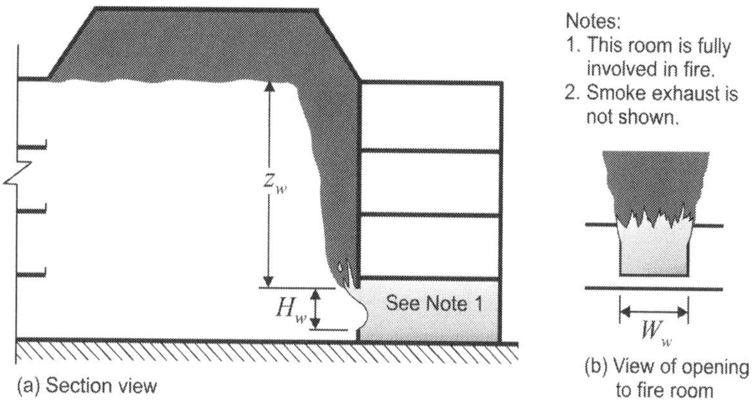

Figure 16.8 *Window plume.*

Example 16.4. Window Plume

A window plume is coming from a room that has a 6 by 8 ft high (1.83 by 2.44 m high) opening to an atrium. The distance from the smoke layer interface to the top of the window is 30 ft (9.14 m). What is the heat release rate of the fire and the mass flow of the plume into the smoke layer?

The parameters are: $z_w = 30$ ft, $H_w = 8$ ft, $W_w = 6$ ft.

The area of the window is $A_w = 8(6) = 48$ ft^2.

The heat release rate of the fire is $Q = 61.2 A_w H_w^{1/2} = 61.2(48)(8^{1/2}) = 8310$ Btu/s.

$$a = (2.40 A_w^{2/5} H_w^{1/5}) - 2.1 H_w = [2.4(48^{2/5})(8^{0.5})] - 2.1(8) = 0.31 \text{ ft.}$$

$$m = [0.077(A_w H_w^{1/2})^{1/3}(z_w + a)^{5/3}] + 0.18 A_w H_w^{1/2}.$$

$$m = [0.077(48(8^{1/2}))^{1/3}(30 + 0.31)^{5/3}] + 0.18(48)(8^{1/2}) = 141 \text{ lb/s (63.9 kg/s).}$$

Example 16.4 calculates the mass flow for a window plume coming from an opening that is 6 by 8 ft high (1.83 by 2.44 m high).

AVERAGE PLUME TEMPERATURE

The average temperature of the plume is

$$T_p = T_o + \frac{Q_c}{m c_p} \tag{16.23}$$

where

T_p = average plume temperature at elevation z, °F (°C),

T_o = ambient temperature, °F (°C),

Q_c = convective heat release rate, Btu/s (kW),

c_p = specific heat of plume gases, 0.24 Btu/lb·°F (1.0 kJ/kg·K),

m = mass flow rate of the plume where it enters smoke layer, lb/s (kg/s).

This equation is based on the first law of thermodynamics, and it applies to any plume. Figure 16.9 shows the average plume temperatures of simplified axisymmetric plumes. It can be seen that the plume temperature decreases with elevation. For 40 ft (12 m) above a 2000 Btu/s (2100 kW) fire, the average plume temperature is about 140°F (60°C). The curves in Figure 16.9 were calculated from Equations 16.1 and 16.11 plus the equation for density that is discussed later. Example 16.5 shows how to calculate the average plume temperature.

SMOKE LAYER TEMPERATURE

The smoke layer temperature is calculated from

$$T_p = T_o + \frac{K Q_c}{m c_p} \tag{16.24}$$

where

T_p = average plume temperature at elevation z, °F (°C),

T_o = ambient temperature, °F (°C),

K = fraction of convective heat release contained in smoke layer, dimensionless,

Q_c = convective heat release rate, Btu/s (kW),

c_p = specific heat of plume gases, 0.24 Btu/lb·°F (1.0 kJ/kg·K),

m = mass flow rate of the plume where it enters smoke layer, lb/s (kg/s).

For atrium smoke control systems, it is believed that K varies from 0.5 to 1.0. For calculating the volumetric flow rate of smoke exhaust, a value of $K = 1.0$ should be used because it results in the highest smoke exhaust which is conservative. For plugholing calculations, a value of $K = 0.5$ should be used because it results in the largest number of exhaust inlets, which is also conservative. Other values of K may be used for these applications, provided that they are supported by test data or an engineering analysis.

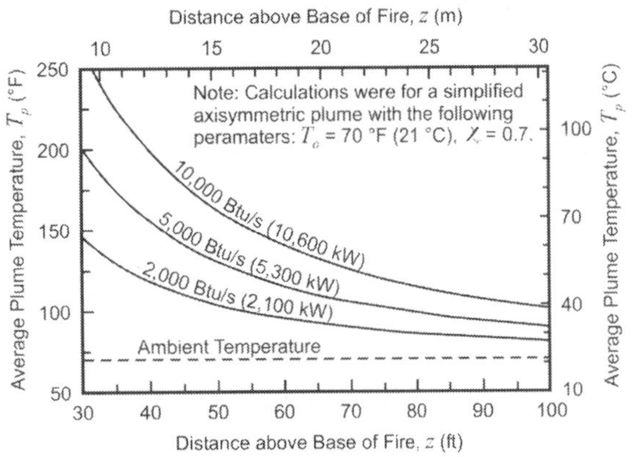

Figure 16.9 *Average temperature of simplified axisymmetric plume.*

PLUGHOLING

Plugholing is the condition where air from below the smoke layer is pulled through the smoke layer due to a high exhaust rate. The following method is based on research with round exhaust inlets, and it can be used for rectangular inlets. However, the method of this section is not appropriate for slot inlets where the length is more than ten times the width.

The number of exhaust inlets needs to be chosen so that the flow rate at exhaust inlets does not result in plugholing. The maximum volumetric flow rate that can be exhausted by a single exhaust inlet without plugholing is

$$V_{max} = 452\gamma d^{5/2}\left(\frac{T_s - T_o}{T_o}\right)^{1/2}$$

$$V_{max} = 4.16\gamma d^{5/2}\left(\frac{T_s - T_o}{T_o}\right)^{1/2} \quad \text{for SI} \tag{16.25}$$

where

V_{max} = maximum volumetric flow rate without plugholing at T_s, ft^3/min (m^3/s),

T_s = absolute temperature of smoke layer, °R (K),

T_o = absolute ambient temperature, °R (K),

d = depth of smoke layer below lowest point of exhaust inlet, ft (m),

γ = exhaust location factor, dimensionless.

Example 16.5. Average Plume Temperature

For a 4000 Btu/s (4200 kW) fire, what is the average plume temperature 38 ft (11.6 m) above the base of the fire?

Use the parameters: $Q = 4000$ Btu/s, $z = 38$ ft, $T_o = 70$°F, $\chi_c = 0.7$.

The convective HRR is $Q_c = \chi_c Q = 0.7(4000) = 2800$ Btu/s.

The limiting elevation is $z_l = 0.533 Q_c^{2/5} = 0.533(2800)^{2/5} = 12.8$ ft.

Because z is greater than z_l, the mass flow of the plume is calculated with the following equation:

$$m = 0.022 Q_c^{1/3} z^{5/3} + 0.0042 Q_c$$

$$m = 0.022(2800)^{1/3}(38)^{5/3} + 0.0042(2800) = 145 \text{ lb/s}$$

The average plume temperature is $T_p = T_o + \dfrac{Q_c}{mc_p} = 70 + \dfrac{2800}{145(0.24)} = 150°\text{F } (66°\text{F}).$

This equation is based on research at the Fire Research Station in the UK (Spratt and Heselden 1974). An exhaust location factor of one should be used for a ceiling-mounted inlet that is centered at least two diameters from the nearest wall as shown in Figure 16.10a. This can also be stated $\gamma = 1$ for $L \geq 2D_i$ where D_i is the diameter of the inlet and L is the distance from the center of the inlet to the nearest wall.

An exhaust location factor of one half should be used for a ceiling-mounted inlet that is centered less than two diameters from the nearest wall as shown in Figure 16.10b. This can be stated as stated $\gamma = ½$ for $L < 2D_i$.

For an exhaust inlet located in a wall, an exhaust location factor γ of one half should be used as shown in Figure 16.10c. This can be stated $\gamma = ½$ for wall inlets. The ratio d/D_i needs to be greater than two, where D_i is the diameter of the inlet, and d is the smoke layer depth below the lowest point of the exhaust inlet.

For rectangular exhaust inlets, the effective value of the inlet diameter should be used

$$D_i = \frac{2ab}{a+b} \tag{16.26}$$

where

D_i = effective diameter of inlet,

a = length of inlet,

b = width of inlet.

If inlets are located too close together, they act as one inlet with respect to plugholing. In order to be considered separate inlets for plugholing purposes, the minimum separation between exhaust inlets is

$$S_{\min} = 0.065 V_e^{1/2}$$
$$S_{\min} = 0.9 V_e^{1/2} \text{ for SI} \tag{16.27}$$

where

$S_{\min}$ = minimum edge-to-edge separation between inlets, ft (m),

V_e = volumetric flow rate of one exhaust inlet, ft³/min (m³/s).

VOLUMETRIC FLOW RATE

The volumetric rate of smoke exhaust is

$$V = 60\frac{m}{\rho}$$
$$V = \frac{m}{\rho} \text{ for SI} \tag{16.28}$$

where

V = volumetric flow rate of smoke exhaust, ft³/min (m³/s).

m = mass flow rate of smoke exhaust, lb/s, (kg/s).

ρ = density of smoke, lb/ft³ (kg/m³).

The volumetric flow rate is determined using the smoke density ρ at the smoke layer temperature.

DENSITY OF SMOKE

The density of smoke can be calculated from

$$\rho = \frac{144 p_{atm}}{R(T_s + 460)}$$
$$\rho = \frac{p_{atm}}{R(T_s + 273)} \text{ for SI} \tag{16.29}$$

where

ρ = density of smoke, lb/ft³ (kg/m³),

p_{atm} = atmospheric pressure, lb/in² (Pa),

R = gas constant, 53.34 (287),

T = temperature of smoke, °F (°C).

The standard atmospheric pressure p_{atm} for many locations is provided in Chapter 2.

CASE STUDY

This case study is an engineering analysis for the three-story atrium shown in Figure 16.11. This atrium is the entrance to the headquarters of a corporation making soft drinks, and tours of the facility start on the ground floor. There is a gift shop for the visitors. The second

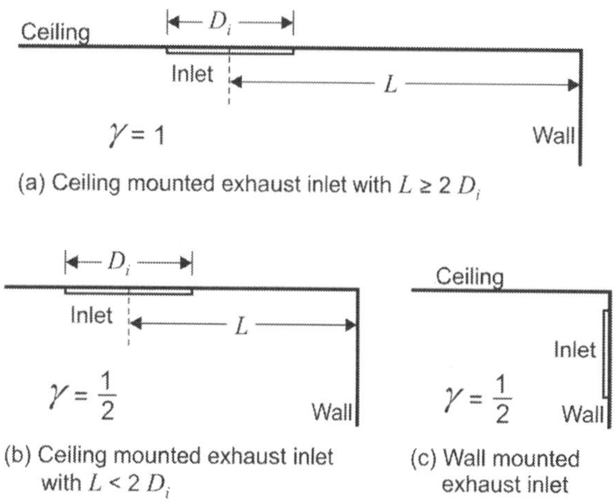

Figure 16.10 *Location factors γ for exhaust inlets.*

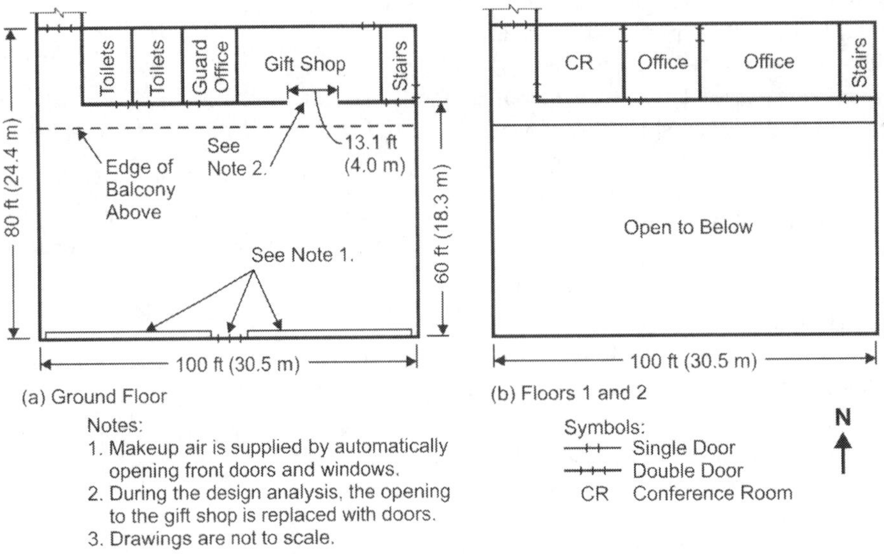

Figure 16.11 *Floor plans for case study.*

Table 16.1: Climatological Data for Case Study

Winter design temperature:	10°F (–12°C)
Summer design temperature:	92°F (33°C)
Design wind speed:	23 mph (37 km/h)
Prevailing winds:	west and southwest
Atmospheric pressure:	14.7 lb/in.^{2} (101300 Pa)

and third floors have executive offices and conference rooms. The engineering analysis addresses wind effect, stack effect, minimum smoke layer depth, system activation, design scenarios, smoke exhaust flow rate, makeup air, and plugholing. In this context, stack effect is used to mean the impact of the indoor-to-outdoor temperature difference.

The weather data for this project is listed in Table 16.1, and wind effect is evaluated in Example 16.6. The minimum smoke layer depth is evaluated in Example 16.7. For much of the year, a layer of hot air under the atrium ceiling is anticipated. Accordingly, system activation is evaluated in Example 16.8. Design scenarios consist of fires in the atrium and the gift shop, and these fires are shown in Figures 16.12 and 16.13. The design fires are evaluated in Example 16.9. The atrium design fire is 2000 Btu/s (2100 kW), and the gift shop design fire is 1000 Btu/s (1050 kW).

The initial plan for the atrium considers the doors to the offices, conference rooms, guard office, and other parts of the building will automatically close in the event of a fire. The opening to the gift shop has a roll down security screen. Example 16.10 evaluates the fire in the atrium which has an axisymmetric plume. This example shows that a smoke exhaust of 92,600 cfm (43.7 m^3/s) is needed for the fire in the atrium.

Example 16.11 evaluates the gift shop fire that has a balcony spill plume shown in Figure 16.13. This example shows that a smoke exhaust of 209,000 cfm (98.6 m^3/s) is needed for the gift shop fire. This exhaust is so large that the design team decides to put doors on the gift shop that close automatically in the event of a fire to prevent this kind of balcony spill plume. For this reason, the design will be based only on a fire in the atrium.

The velocity through the makeup air vents is

$$U_{mu} = \frac{V_{mu}}{A_{fv}} \tag{16.30}$$

where

U_{mu} = velocity of makeup air, fpm (m/s),

V_{mu} = volumetric flow of makeup air, cfm (m^3/s),

A_{fv} = total free area of vents, ft^2 (m^2).

This equation is used in Example 16.12, which evaluates the makeup air of the case study, and it shows that the planned openings are sufficient. Example 16.13 evaluates plugholing and determines that four exhaust inlets are needed in the atrium ceiling, and these inlets need to be separated end-to-end by at least 9.9 ft (3.02 m). The average velocity at the exhaust inlet is 1500 fpm (7.6 m/s). A square inlet with sides of at least 5.55 ft (1.69 m) is used. The nearest edge of an inlet must not be less than 11.1 ft (3.38 m) from any wall. The sequence of operation for the smoke control system is listed Table 16.2.

Example 16.6. Case Study: Wind Effect

Atrium smoke control systems should be designed to minimize the potential for wind to result in (1) velocities greater than 200 fpm (1.02 m/s) where the makeup air could come into contact with the plume, and (2) smoke feedback from the smoke exhaust into the makeup air.

This design was developed to be relatively insensitive to wind effects. The smoke exhaust fan is a centrifugal fan, and so the exhaust flow rate is only slightly affected by the wind. The makeup air is by way of doors and windows in the front of the atrium that automatically open in the event of a fire in the atrium or communicating spaces. Because all the makeup air openings are on one side of the atrium and the other doors to the atrium are closed, the probability of the wind producing a velocity in the atrium in excess of 200 fpm (1.02 m/s) is minimized.

The smoke exhaust outlet is located at the east end of the north wall of the building. The wind direction is the direction from which the wind is coming. This means that the prevailing west and southwest winds will tend to blow the smoke exhaust away from the makeup air vents in the front of the building. No detailed wind analysis is needed.

Example 16.7. Case Study: Minimum Smoke Layer Depth

As discussed in Chapter 15, the minimum smoke layer depth must be 20% of the floor-to-ceiling height of the atrium. The local codes in effect at this building require that the smoke layer be 6 ft (1.83 m) above the highest walking surface that forms a portion of a required egress in the atrium. The highest walking surface is the third floor, which is 31.2 ft (9.5 m) above the floor of the atrium.

The smoke layer interface is at 31.2 + 6 = 37.2 ft. The smoke layer depth is 46.75 – 37.2 = 9.55 ft.

This is 100(9.55)/46.75 = 20.4% of the floor-to-ceiling height.

The smoke layer depth is acceptable, because it is greater than 20% of the floor-to-ceiling height of the atrium.

Example 16.8. Case Study: System Activation

The roof of the atrium is exposed to the sun, and a stratified layer of hot air can be expected at times. To ensure quick activation, system activation should be by projected beam smoke detectors in one of the schemes as discussed in the section about stratification in Chapter 15. Signals from other detectors or from sprinkler water flow can also be used in addition to these beam detectors.

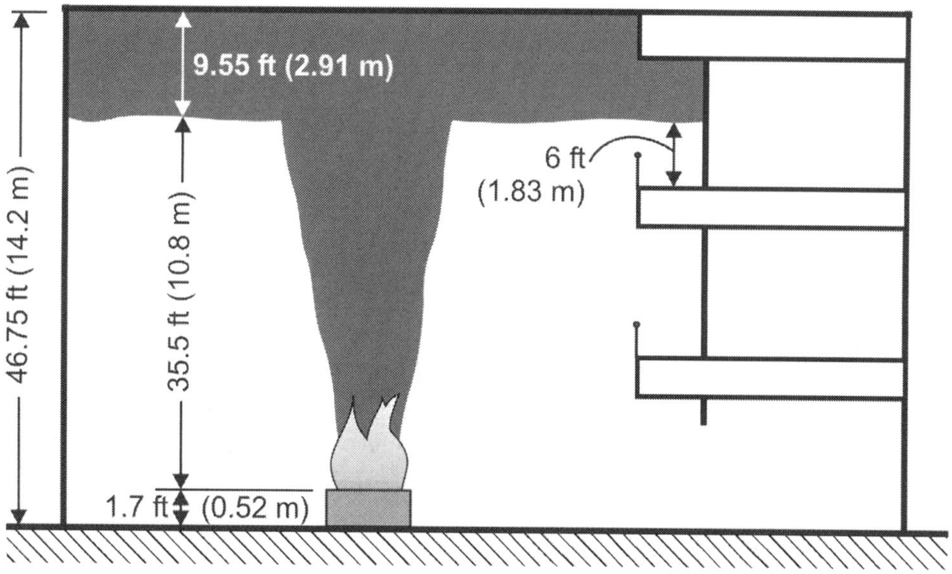

Figure 16.12 *Fire in atrium of case study.*

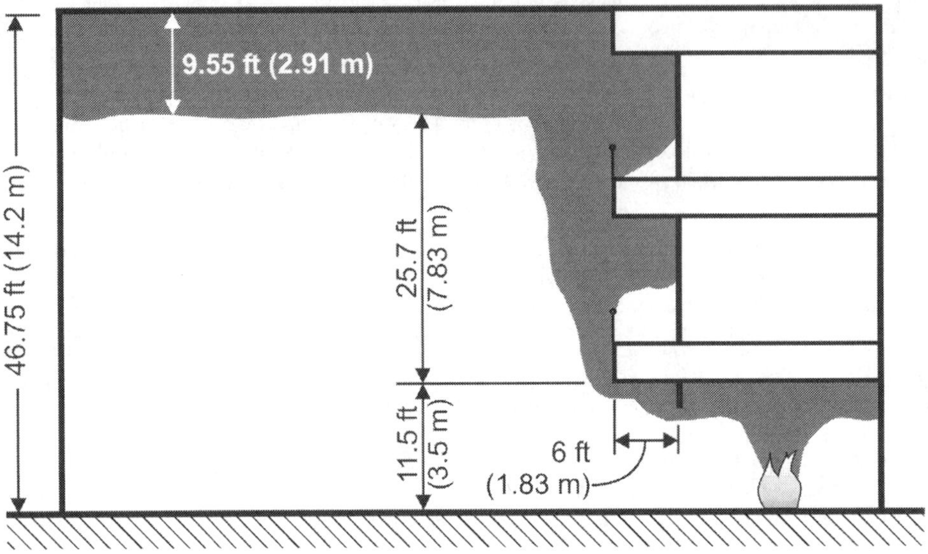

Figure 16.13 *Fire in the gift shop of the case study.*

Example 16.9. Case Study: Design Scenarios

Fire in Atrium: The following information about the project has been provided by the architect. The building materials are concrete, brick, terrazzo, and glass. The ground floor of the atrium has an information desk, potted plants, a rock garden with water falls, and some park benches. The information desk and the park benches are noncombustible, and only small amounts of paper will be kept at the information desk.

The materials normally in the atrium make up only a small fuel load, and the design fire needs to be sized for transient materials as discussed in Chapter 5. Materials are delivered to the building at a loading dock located far from the atrium, but it is possible that some newly delivered materials may be put in the atrium en route to their ultimate destination. A design fire of 2000 Btu/s (2100 kW) with the base of the fire 1.7 ft (0.52 m) above the atrium floor is chosen.

Fire in Gift Shop: Figure 16.13 shows the fire in the gift shop and the resulting balcony spill plume. The shop will be selling an array of items including stuffed toy animals and clothing. It is possible to have a fire of cardboard boxes of merchandise under the checkout counter. At times there would be temporary display tables, and it is possible to have a fire of cardboard boxes of merchandise under the table. For both of these possible design fires, the fire would be shielded from the sprinkler spray.

Chapter 5 discusses such shielded fires, and the following points are relevant. The smoke from the fire does not rise like an axisymmetric plume, but it flows under the counter or table and spreads out. This means that calculation of sprinkler activation time using a zone fire model is not appropriate for shielded fires. Two design fires have been developed for shielded fires in offices: (1) one with a peak of 474 Btu/s (500 kW) for an office that has only modest amounts of fuels, and (2) another with a peak of 948 Btu/s (1000 kW) for an office with a very large amount of material that is shielded. These fires were primarily of paper in corrugated cardboard boxes with a small amount of polymers.

Considering the larger shielded fire, the design fire is conservatively chosen at 1000 Btu/s (1050 kW).

Example 16.10. Case Study: Fire in the Atrium

Calculate the smoke exhaust needed for the fire in the atrium of the case study (Figure 16.12).

Quickly after system activation, the ambient temperature in the atrium is essentially the same as the outdoor temperature. The maximum exhaust happens at the summer outdoor design temperature, which for this application is 92°F.

The parameters are: $Q = 2000$ Btu/s, $z = 35.5$ ft, $T_o = 92$°F, $p_{atm} = 14.7$ psi, $\chi_c = 0.7$.

The convective HRR is $Q_c = \chi_c\, Q = 0.7(2000) = 1400$ Btu/s.

The limiting elevation is $z_l = 0.533 Q_c^{2/5} = 0.533(1400)^{2/5} = 9.7$ ft.

Because z is greater than z_l, the mass flow of the plume is calculated with the following equation:

$$m = 0.022 Q_c^{1/3} z^{5/3} + 0.0042 Q_c$$

$$m = 0.022(1400)^{1/3}(35.5)^{5/3} + 0.0042(1400) = 100.3 \text{ lb/s}.$$

The smoke layer temperature is $T_s = T_o + \dfrac{KQ_c}{mc_p}$. For calculation of smoke exhaust, $K = 1$ is used.

$$T_s = 92 + \frac{(1)(1400)}{100.3(0.24)} = 150.2°\text{F}$$

The density of the smoke is $\rho = \dfrac{144 p_{atm}}{R(T+460)} = \dfrac{144(14.7)}{53.34(150.2+460)} = 0.0650 \text{ lb/ft}^3$.

The smoke exhaust is $V = 60\dfrac{m}{\rho} = 60\dfrac{100.3}{0.0650} = 92,600 \text{ cfm } (43.7 \text{ m}^3/\text{s})$.

Example 16.11. Case Study: Fire in Gift Shop

Calculate the smoke exhaust needed for the fire in the gift shop of the case study (Figure 16.13).

For the same reason as in Example 16.10, the ambient temperature is the same as the summer outdoor design temperature, which for this application is 92°F.

The parameters are: $Q = 1000$ Btu/s, $z_b = 25.7$ ft, $H = 11.5$ ft, $b = 6$ ft, $w = 13.1$ ft, $T_o = 92$°F, $p_{atm} = 14.7$ psi, $\chi_c = 0.7$.

The length of the spill is $W = w + b = 13.1 + 6 = 19.1$ ft.

The convective HRR is $Q_c = \chi_c\, Q = 0.7(1000) = 700$ Btu/s.

Because of the value of z_b, the balcony spill plume is in Region 1, and the mass flow is calculated as

$$m = 0.12(QW^2)^{1/3}(z_b + 0.25H) = 0.12[1000(19.1)^2]^{1/3}[25.7 + 0.25(11.5)]$$

$$m = 245 \text{ lb/s}.$$

The smoke layer is at $T_s = T_o + \dfrac{KQ_c}{mc_p}$ where $K = 1$. $T_s = 92 + \dfrac{(1)(700)}{245(0.24)} = 104°\text{F}$.

The density of the smoke is $\rho = \dfrac{144 p_{atm}}{R(T+460)} = \dfrac{144(14.7)}{53.34(104+460)} = 0.0704 \text{ lb/ft}^3$.

The smoke exhaust is $V = 60\dfrac{m}{\rho} = 60\dfrac{245}{0.0704} = 209,000 \text{ cfm } (98.6 \text{ m}^3/\text{s})$.

Example 16.12. Case Study: Makeup Air

The makeup air velocity must not exceed 200 fpm (1.02 m/s). The makeup air is by way of doors and windows in the front of the atrium that automatically open in the event of a fire in the atrium or communicating spaces. The front doors are 6 by 7 ft high (1.83 by 2.13 m high). The window panels open at a 30° angle. There are two windows each 40 by 10 ft (12.2 by 3.05 m) high.

The free area of the open doors is 6 (7) = 42 ft^2.

The total area of the two windows is 2 (40) (10) = 800 ft^2. Because the free area of these windows is about 50% of the total area, the free area of the windows is 0.5 (800) = 400 ft^2.

The total free area of vents is 42 plus 400 = 442 ft^2.

For this project, makeup air is set at 95% of the smoke exhaust, which is 92,600 cfm (0.95) = 88,000 cfm.

The velocity of the makeup air is $U_{mu} = \frac{V_{mu}}{A_{fv}} = \frac{88,000}{42} = 199$ ft/min (1.01 m/s).

Because this velocity is less than 200 fpm (1.02 m/s), the vents are large enough.

Example 16.13. Case Study: Plugholing Evaluation

This example makes calculations regarding plugholing for the case study.

After making the calculations of Example 16.7, it was decided that the gift shop would be separated from the atrium by doors that would automatically close in the event of a fire in either the atrium or the gift shop. For this reason, plugholing calculations are based on the fire in the atrium (Example 16.10). The smoke layer depth is 9.55 ft as shown in Figure 16.12. Plugholing calculations will be made for an ambient temperature 70°F. Because the exhaust inlets are to be in the ceiling away from the walls, γ is one.

The parameters are: $d = 9.55$ ft, $T_o = 70$°F (530°R), $\gamma = 1$, $Q_c = 1400$ Btu/s, $m = 100.3$ lb/s, $c_p = 0.24$ Btu/lb-°F.

The smoke layer temperature is $T_s = T_o + \frac{KQ_c}{mc_p}$. For plugholing calculations, $K = 0.5$ needs to be used.

$$T_s = T_o + \frac{KQ_c}{mc_p} = 70 + \frac{(0.5)(1400)}{100.3(0.24)} = 99°\text{F (559°R)}$$

The maximum flow that can be used without plugholing is

$$V_{max} = 452\gamma d^{5/2}\left(\frac{T_s - T_o}{T_o}\right) = 452(1)(9.55)^{5/2}\left(\frac{559-530}{530}\right)^{1/2} = 29,800 \text{ cfm.}$$

From Example 16.10, the total smoke exhaust is 92,600 cfm. Dividing this exhaust by V_{max} is 92,600/29,800 = 3.11. This means that four exhaust inlets are needed. Each inlet will have a flow of V_e = 92,600/4 = 23,150 cfm (11.9 m^3/s).

The edge-to-edge separation between inlets must be at least $S_{min} = 0.065\, V_e^{1/2} = 0.065\,(23{,}150)^{1/2} = 9.89$ ft (3.01 m).

The average velocity at the exhaust inlet is chosen as 1500 fpm. The free area needed for the exhaust inlet is 23,150 cfm / 1500 fpm = 15.4 ft^2. The free area is about 50% of the total area of the exhaust inlet. The area of the exhaust inlet is 15.4/0.5 = 30.8 ft^2. For a square exhaust inlet, the side needs to be at least (30.8)$^{1/2}$ = 5.55 ft. The ceiling exhaust needs to be at least two diameters from the nearest wall. This means that the nearest edge of an inlet must not be less than 11.1 ft (3.38 m) from any wall.

Table 16.2: Sequence of Operation for Case Study

Item	Operation
Makeup air vents (doors and windows)	open
Smoke exhaust fan(s)	turn on
Double doors on Levels 1, 2, and 3 to rest of building	close
Office doors	close
Conference room doors	close
Gift shop doors	close
Guard office door	close

Note: These operations are done automatically.

NOMENCLATURE

A = cross-sectional area of fire, ft^2 (m^2),

a = length of inlet; or width of fire, ft (m); or $a = (2.40A_w^{2/5}\,H_w^{1/5}) - 2.1H_w$, ft (m)

A_{fv} = total free area of vents, ft^2 (m^2).

A_w = area of ventilation opening, ft^2 (m^2),

b = width of inlet; or length of fire, ft (m); or distance from opening to the balcony edge, ft (m).

c_p = specific heat of plume gases, 0.24 Btu/lb·°F (1.0 kJ/kg·K)

d = depth of smoke layer below the lowest point of exhaust inlet, ft (m)
D_e = effective diameter, ft (m)
D_f = diameter of fire, ft (m)
D_i = diameter of the inlet
d_p = plume diameter, ft (m)
g = acceleration of gravity, ft/s^2 (m/s^2)
H = height of balcony above fuel, ft (m)
H_w = height of ventilation opening, ft (m)
H_w = height of ventilation opening, ft (m)
K = fraction of convective heat release contained in smoke layer, dimensionless
K_d = diameter constant (dimensionless)
m = mass flow rate, lb/s (kg/s)
n = fire location factor, dimensionless
P = perimeter of the fire, ft (m)
p_{atm} = atmospheric pressure, lb/in^2 (Pa)
Q = heat release rate of the fire, Btu/s (kW)
q = heat release density, Btu/s ft^2 (kW/m^2)
Q_c = convective heat release rate, Btu/s (kW)
R = gas constant, 53.34 (287)
S_{min} = minimum edge-to-edge separation between inlets, ft (m)
T_{cp} = absolute centerline axisymmetric plume temperature at elevation z, °R (K)
T_o = absolute ambient temperature, °R (K); or ambient temperature, °F (°C)
T_p = average plume temperature at elevation z, °F (°C)
T_s = absolute temperature of the smoke layer, °R (K)
U_{mu} = velocity of makeup air, fpm (m/s)
V = volumetric flow rate of smoke exhaust, ft^3/min (m^3/s)
V_e = volumetric flow rate of one exhaust inlet, ft^3/min (m^3/s)
V_{max} = maximum volumetric flow rate without plugholing at T_s, ft^3/min (m^3/s)
V_{mu} = volumetric flow of makeup air, cfm (m^3/s),
W = length of the spill, ft (m)
w = width of the opening from the area of origin, ft (m)
z = distance above base of fire, ft (m); or height above fuel, ft (m)
z_b = height of plume above balcony edge, ft (m)
z_l = limiting elevation, ft (m)
z_o = distance to virtual origin, ft (m)
z_w = distance from smoke layer interface to top of window, ft (m)
γ = exhaust location factor, dimensionless
χ_c = convective fraction of heat release dimensionless
ρ = density of smoke, lb/ft^3 (kg/m^3)
ρ_o = density of ambient air, lb/ft^3 (kg/m^3)

REFERENCES

Cetegan, B.M., E.E. Zukoski, and T. Kubota. 1982. Entrainment and flame geometry of fire plumes. PhD Thesis of Cetegan, California Institute of Technology, Pasadena.

Heskestad, G. 1983. Virtual origins of fire plumes. *Fire Safety Journal* 5(2).

Heskestad, G. 1984. Engineering relations for fire plumes. *Fire Safety Journal* 7(1).

Ko, Y., G. Hadjisophocleous, G.D. Lougheed. 2008. CFD study of the air entrainment of balcony spill plumes at the balcony edge. *ASHRAE Transactions* 114(1).

Law, M. 1986. A note on smoke plumes from fires in multilevel shopping malls. *Fire Safety Journal* 10(3).

Lougheed, G.D., C.J. McCartney, and E. Gibbs. 2007. Balcony spill plumes. RP-1247, ASHRAE, Atlanta.

Lougheed, G.D., C. McCartney. 2008a. Balcony spill plumes: Full-scale experiments, Part 1. *ASHRAE Transactions* 114(1).

Lougheed, G.D., C. McCartney. 2008b. Balcony spill plumes: Full-scale experiments, Part 2. *ASHRAE Transactions* 114(1).

McCaffrey, B.J. 1983. Momentum implications for buoyant diffusion flames. *Combustion and Flame* 52(2).

McCartney, C., G.D. Lougheed, E.J. Weckman 2008. CFD investigation of balcony spill plumes in atria. *ASHRAE Transactions* 114(1).

Morgan, H.P., and N.R. Marshall. 1979. Smoke control measures in covered two-story shopping malls having balconies and pedestrian walkways. BRE CP 11/79, Borehamwood, UK.

Morton, B.R., G. Taylor, and J.S. Turner. 1956. Turbulent gravitational convection from maintained and instantaneous sources. *Proceedings of the Royal Society of London,* Vol. 234, pp 1–23.

NFPA. 2012a. NFPA 204, *Standard for Smoke and Heat Venting*. Quincy, MA: National Fire Protection Association.

NFPA. 2012b. NFPA 92, *Standard for Smoke Control Systems.* Quincy, MA: National Fire Protection Association.

Spratt, D. and A.J.M. Heselden. 1974. Efficient extraction of smoke from a thin layer under a ceiling. Fire Research Note 1001, Fire Research Station, Building Research Establishment, Garston, UK.

CHAPTER 17

Fire and Smoke Control in Transport Tunnels

Ahmed Kashef

A transport tunnel is an enclosed facility that carries different types of traffic including vehicles, trucks, buses, rolling stock, etc. A tunnel can run underwater, through mountains, or be an urban type. Tunnels may also be created by the development of air-right structures (structures other than a skywalk bridge that are built over roadways using the roadway's air rights).

Tunnels can be classified according to mode of transport (road, railway, both, bi/unidirectional), length (roughly 1000 ft to 27 mi [300 m to 50 km]), traffic density (urban, rural), cross-section (rectangular, round, arched, horse-shoe), possible fire hazards (hazardous freight, vehicles, tunnel itself), and ventilation systems (longitudinal, transverse, hybrid).

FIRE SAFETY ISSUES IN TUNNELS

In general, fires in tunnels are rare events, so the statistical significance of the rates of fires is limited; the rates can change considerably by only one fire event. According to the statistics available, urban tunnels tend to have a higher fire rate than other tunnels (PIARC 2007). Fires are mainly generated by the traffic (95%) passing through the tunnel (collisions, electrical defects, brake overheating, or other defects leading to the self-ignition of a vehicle) and not by tunnel equipment or maintenance work (PIARC 2007). As such, the likelihood of tunnel fires is mainly related to items like tunnel length, traffic density, type of traffic and combustible load, speed control, and slope of the road.

Fires in tunnels pose major safety issues and challenges to the designer, especially with the increase in the number of tunnels, their length, and number of people using them. Life can be threatened in a number of ways: the inhalation of combustion products such as carbon monoxide and carbon dioxide, and the exposure to high temperatures and heat fluxes. Furthermore, evacuation can be significantly hindered by poor visibility, power failure, blocked exits due to traffic jams or crashed vehicles, or obstruction resulting from a collapse or explosion in the tunnel. For safe evacuation, tolerable temperatures, acceptable visibility, and adequate air quality must be maintained. The main fire safety issues include safeguard of tunnel users, safe rescue operations, minimal effects on the environment due to the release of combustion gases, and a minimal loss of property.

In the event of an incident or accident, the first ten to fifteen minutes are crucial when it comes to people saving themselves and limiting damage. If the fire attains high levels of energy release rates (50,000 Btu/s [50 MW] or more), it becomes difficult to approach it. The reported major fire events reveal the difficulty of extinguishing the fire at this stage either due to the density of smoke or the intensity of radiation (Temperatures up to 2500°F [1350°C], [PIARC 2007; Lacroix 1998]) and heat fluxes in excess of 26 Btu/s·ft^2 [300 kW/m^2] [Lacroix 1998] preventing the fire service approaching the fire source. The prevention of critical events or early intervention are therefore the number-one priority, which means that the most important measures to be taken may have to be of a preventive nature.

The Standard for Road Tunnels, Bridges, and Other Limited Access Highways—NFPA 502 (NFPA 2011) requires a tenable environment to be maintained in the tunnel and dictates that motorists should not be exposed to air temperatures that exceed 140°F (60°C) during emergencies and radiant heat of 0.22 Btu/s·ft^2 (2.5 kW/m^2) for more than 30 min. Furthermore, it states that smoke obscuration levels should be continuously kept below the point at which a sign internally illuminated at 7.43 footcandle (80 lx) is discernible at 98 ft (30 m) and doors and walls are discernible at 33 ft (10 m).

The World Road Association (PIARC 1999) recommends a tenable tunnel environment during the self-evacuation phase in which: air temperature should not be higher than 176°F (80°C) for longer than 15 min, radiation does not exceed a level of 0.22 Btu/s·ft^2 (2.5 kW/m^2) for users and 0.44 Btu/s·ft^2 (5 kW/m^2) for fire fighters, and a minimum visibility of 23 ft (7 m) (for walking through smoke) and 49 ft (15 m) (for reading signs) is maintained. The guideline dictates that people must, in any case, be able to reach a safe place in a reasonably short time and covering a reasonably short distance. Therefore, facilities such as emergency exits or fireproof shelters should be provided whenever necessary.

FIRE PROTECTION MATRIX

A fire incident involves a sequence of events or phases occurring, which include ignition, fire development, self-evacuation and assisted-evacuation, and firefighting after the arrival of the emergency services. Prior to the activation of emergency equipment, the fire must be detected and the detection be confirmed.

Emergency equipment is necessary to enhance safety in the event of a fire. This equipment is an integral part of modern tunnels (Figure 17.1). The type and amount of equipment is usually based on tunnel type and length, traffic, etc. The equipment covers all facilities necessary in case of fire or some other accident such as detection and lighting systems, ventilation systems, fixed firefighting systems, emergency telephone, fire extinguishers, equipment too close the tunnel, radio transmissions, loudspeakers, variable text signs, and emergency exits. Many countries use a tunnel classification system that categorizes tunnels into groups and specifies the necessary emergency equipment for each group (NFPA 2011; PIARC 1999).

FIRE DEVELOPMENT IN TUNNELS

Fires within confines of underground systems are among the most difficult to extinguish due to restricted space of operation and limited access points for fire fighters, limited emergency exits locations, restricted availability of water supply, and ventilation capabilities.

From the beginning of a fire incident, the airflow in the tunnel is considerably modified and becomes highly transient. The modifications are not only due to the fire itself, but also to the emergency ventilation operation, and the change in the traffic in the tunnel. The smoke progress and its degree of stratification depend on airflow, heat release rate, natural ventilation, tunnel geometry and slope and traffic. With no air current in the fire zone, the smoke progresses in a symmetrical way on both sides of the fire. Underneath the smoke layer, fresh air is drawn in towards the fire source in an opposite direction to the spreading smoke. This separation

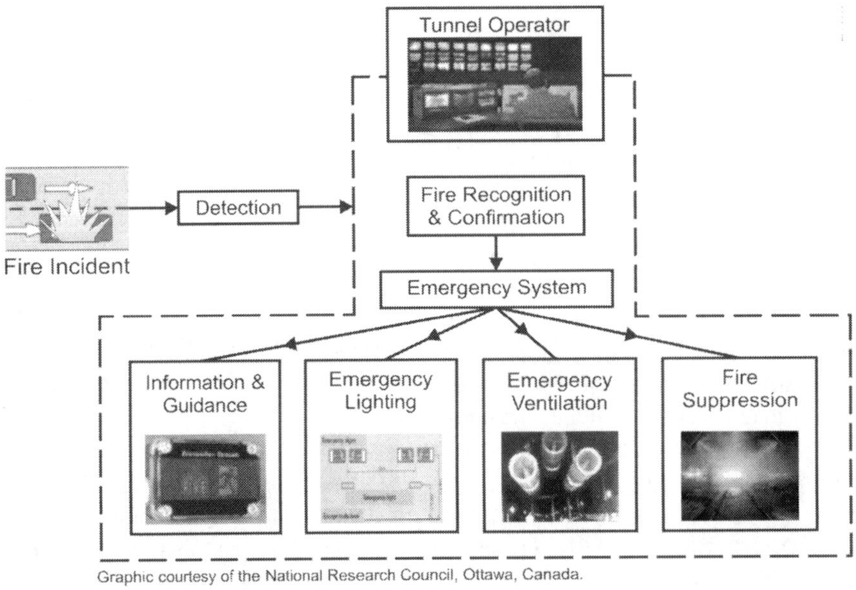

Figure 17.1 *Fire safety matrix in road tunnel.*

between the hot upper layers and cooler lower layers is termed *stratification*. The combined effects of convective heat exchange with the tunnel walls and the mixing between the smoke and the fresh air layer causes the smoke to cool down and lose its stratification. After a period of typically 5 to 10 min (Heselden 1976), both upstream and downstream sections of the tunnel can be completely filled with smoke (Figure 17.2). The tunnel slope could induce an acceleration of the smoke towards the ascending direction (stack effect).

The tunnel ceiling restricts impinging flames from a substantial fire to travel horizontally forming a hot layer at the ceiling. For example, a flame in the open that is 23 ft (7 m) high could give flames 33 to 66 ft (10 to 20 m) long when placed in a tunnel 16 ft (5 m) high (Heselden 1976). This elongation arises due to the much slower entrainment of air into the horizontally traveling flame compared to the case in which the flame is traveling vertically. In this case, the turbulent eddies are opposed by buoyancy forces and the rate of air mixing is reduced. In addition, the mixing with the horizontal flame can only occur on one side of the flame. Consequently, a hot smoky flow of combustion products under the ceiling will travel for long distances with limited mixing of air into the layer. This will inhibit dilution of the smoke, resulting in high concentrations of smoke and combustion products (for most fuels, the gases should be diluted with approximately 100–1000 times (Heselden 1976) their weight with air before they can be brought to a condition which can be tolerated by human beings). This will seldom be possible in tunnels, complicating egress and firefighting processes.

In a tunnel environment, the type of fuel and the resultant fire load can vary enormously due to the different types of materials being transported. Also, multiple fires can occur as the incident escalates. The fire, for example, may undergo an incubation period (slow burning) of 3 to 8 min (Kashef et al. 2011) due to the ignition of a poor combustible material. On the other hand, an explosion can be expected in the event of ignition of large loads of highly flammable hydrocarbons. The burning rates for most fires in tunnels will not be controlled by the rate of air supply to the fire. Rather, the burning rate and hence the heat release rate will depend on the amount of fuel and its characteristics. In fact, higher burning rates may be more possible in tunnels than in open air due to the intense heat fed back from the heated walls and ceiling and from the hot gas layer flowing away from the fire.

During the initial stages, the traffic in the tunnel is modified. First, the fire creates an obstacle to the traffic. In the second stage, barriers and/or traffic lights will be operated to stop the traffic upstream of the fire and at tunnel portals. After vehicles present in the tunnel upstream of the fire are stopped, they create an obstacle to the airflow in the tunnel and thus reduce the performance of the ventilation system. These changes in the traffic flow also modify the piston effect induced by the circulation of the vehicles in the tunnel.

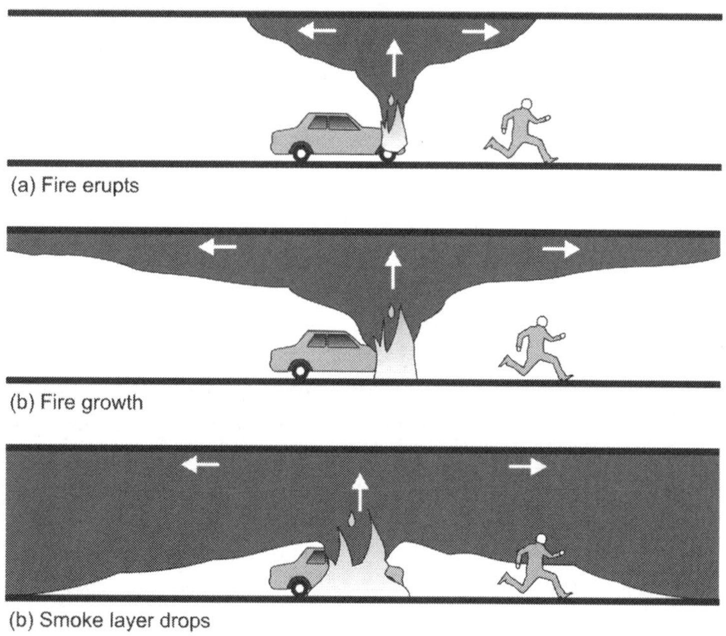

Figure 17.2 *Fire development in road tunnels (no ventilation).*

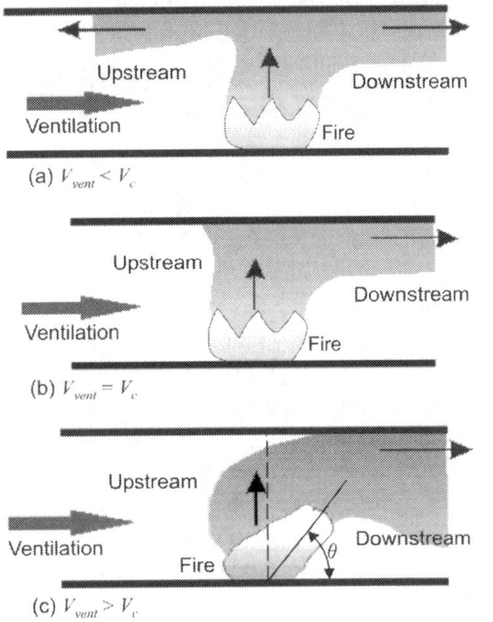

Graphic courtesy of the National Research Council, Ottawa, Canada.

Figure 17.3 *Influence of longitudinal air velocity on smoke progress.*

Backlayering

The backlayering phenomenon is defined as the situation in which the hot smoke layer moves against the provided ventilation upstream of the fire (Figure 17.3a), creating an environment that poses a danger to both tunnel users and emergency responders. If the airflow has a high enough velocity V_{vent}, the longitudinal airflow pushes the smoke to move to the downstream of the fire, while at the upstream side the progression of the smoke layer is slowed down. Further increase of the longitudinal airflow velocity will at a certain value, called critical velocity V_c, completely stop the backlayering and reduce the concentration of smoke (Figure 17.3b). In this case, smoke stratification is unlikely. With higher ventilation velocities the smoke stratification will be completely disrupted (Figure 17.3c).

The occurrence of backlayering depends on many factors including the intensity of fire, the grade and geometry of the tunnel, and the velocity of the ventilating air approaching the fire.

Smoke Layer Speed and Depth

The smoke layer initial moving velocity u_{so} and initial thickness d_{so} for the situation where no longitudinal ventilation is provided can be estimated from (Heselden 1976):

$$u_{so} = C_s\left(\frac{g t_F Q}{T_\infty^2 \rho_\infty C_p W}\right)^{1/3}$$

$$d_{so} = \frac{\dot{m}_s}{\rho_s u_{so} W} \tag{17.1}$$

where:

u_{so} = initial smoke layer moving velocity, ft/s (m/s),
d_{so} = initial smoke layer thickness, ft (m),
g = acceleration due to gravity, ft/s^2 (m/s^2),
W = tunnel width, ft (m),
Q = fire heat release rate, Btu/s (kW),
c_p = specific heat of air, Btu/lb·°R (kJ/kg·K),
ρ_∞ = density of ambient air, lb/ft^3 (kg/m^3),
ρ_s = density of smoke, (lb/ft^3) (kg/m^3),
T_∞ = temperature of ambient air, °R (K),
T_F = average temperature of fire site gases and smoke, °R (K),
$\dot{m}_s$ = smoke production rate, lb/s (kg/s),
C_s = experimental constant, 0.80.

Tables 17.1 and 17.2 show some information on the smoke layer speed and thickness calculated by Heselden (1976) for a hypothetical tunnel for different fire sizes. The speed of movement of the smoke layer dictates to a great degree the minimum distance for the smoke to be extracted through the ceiling vents.

Methods of Smoke Management

In a tunnel environment, various smoke management strategies can be employed to achieve the goals of offsetting buoyancy of the smoke and external atmospheric conditions and to prevent backlayering. These strategies include extraction, transport, control of the direction of movement, and dilution. Smoke can be extracted using localized extraction points, a transverse, or a semi-transverse ventilation system.

Dilution is usually an efficient method for normal operation, in which case the objective is to maintain air quality and visibility. It can increase tenability by reducing concentrations of toxic gases. During emergency operation, smoke management is ideally achieved by removal of vitiated air, for example, by the extraction of air and smoke. Hence, vitiated air is replaced by clean or smoke-free air, which is either supplied mechanically or drawn in through the portals.

The smoke layer can be transported by creating a longitudinal airflow along the length of the tunnel by introducing or removing air from the tunnel at a limited number of points. This approach is the most effective

Table 17.1: Smoke Layer Characteristics in Hypothetical Tunnel in I-P Units (Heselden 1976)

Fire Size (Btu/s)	2800	9500	19,000	47,000	95,000
$\dot{m}_s$ (lb/s)	37	53	77	106	209
u_{so} (ft/min)	260	430	600	1040	1300
d_{so} (ft)	2.3	3.0	3.9	5.6	8.9

Table 17.2: Smoke Layer Characteristics in Hypothetical Tunnel in SI Units (Heselden 1976)

Fire Size (MW)	3	10	20	50	100
$\dot{m}_s$ (kg/s)	17	24	35	48	95
u_{so} (m/s)	1.3	2.2	3.0	5.3	6.7
d_{so} (m)	0.7	0.9	1.2	1.7	2.7

method of smoke control in tunnels with unidirectional traffic, in which case airflow is in the direction of the traffic and therefore smoke is pushed downstream of the fire.

Visibility

In a tunnel environment, visibility tends to be the most restrictive criterion for tenability. Evacuation can be significantly hindered by poor visibility. For acceptable visibility and therefore safe evacuation, reliable and robust control of airflow velocity is essential at all times.

Smoke stratification in tunnels is a transient phenomenon that typically lasts 5 to 10 min. A stratified smoke layer allows evacuees adequate visibility in the region under the smoke layer. Thus, maintaining the stratification of the smoke for the longest period of time is essential for tunnel users to rescue themselves (self-evacuation phase).

Visibility can be estimated based on the smoke optical density δ. The parameter δ indicates level of smoke obscuration. The higher the value of δ, the higher the smoke obscuration and the lower the visibility. The visibility S may be calculated from δ as follows:

$$S = \frac{K}{\alpha} \qquad S = \frac{K}{2.303\delta} \tag{17.2}$$

where:

α = extinction coefficient, ft^{-1} (m^{-1}),
 = 2.303 δ,
K = proportionality constant, dimensionless (8 for illuminated signs, 3 for reflected signs and building components in reflected light),
S = visibility, ft (m).

The extinction coefficient can be expressed in terms of the specific extinction coefficient, α_m ft^2/lb (m^2/g) and mass concentration of particulate, m_p, lb/ft^3 (g/m^3):

$$\alpha = \alpha_m m_p \tag{17.3}$$

where:

$$m_p = \frac{K_f\, y_p Q t}{\Delta H_c V_s} \tag{17.4}$$

where:

y_p = particulate yield, dimensionless,
t = time from ignition, s,
V_s = volume of smoke in the space, ft^3 (m^3),
ΔH_c = heat of combustion, Btu/lb (kJ/kg),
K_f = 1 (1000).

Thus, Equation 17.2 can be written as:

$$S = \frac{K \Delta H_c V_s}{K_f\, \alpha_m y_p Q t} \tag{17.5}$$

The NFPA 502 (NFPA 2011) defines the smoke obscuration levels that should be considered to maintain a tenable environment for periods of short duration. Smoke obscuration levels should be continuously maintained below the point at which a sign illuminated at 7.43 footcandle (80 lx) or equivalent brightness for luminated signs is discernible at 99 ft (30 m), and doors and walls that are discernible at 33 ft (10 m). PIARC (PIARC 1999) recommends maintaining a minimum visibility of 23 to 49 ft (7 to 15 m) for evacuation and firefighting operations. For more information about visibility, see Chapter 6.

EXITS AND OTHER SAFETY FACILITIES

Road Tunnels

For tunnels with unidirectional traffic (two tubes or more), evacuees could escape the fire through tunnel portals on foot, cross-passage between tunnel tubes, direct communications to the open, or through separate escape corridors. The escape corridors should be lighted and have a special ventilation system.

The most common escape route in two-tube tunnels is through cross-passages between the two tubes. The distance between cross-passages should depend on traffic density and emergency rescue scenarios; for instance, 330 to 660 ft (100 to 200 m) in cities. This distance should be designed so that people can walk to the nearest exit before smoke reduces visibility. When such cross-connections are used, the tunnel operator must consider that people will walk into the second tube. As a consequence, the traffic in this other tube must be stopped immediately. All cross-connections have to be closed by doors in order to prevent the circulation of smoke to the unharmed tube.

It is very important to sign all emergency exit possibilities with internationally standardized signs. The signs should have the international exit symbol used in buildings and show direction as well as distance to the nearest escape point. The signs should be internally lighted and connected to an UPS (uninterrupted power supply) or have a battery backup. In high-traffic tunnels, there should be a minimum safety lighting connected to a UPS. In low-traffic tunnels with no UPS, one of every three or four tunnel lights should be fitted with a battery backup. In tunnels with heavy traffic, there should also be a separate system of evacuation lights (marker lights). These lights should be placed as low as possible on the sidewall (1 m or lower) and the distance between the lights should be 25 m or less. All evacuation systems should be kept always lighted to educate drivers and show that the systems are in working order.

Rail and Subway Tunnels

In subway systems, the train on fire should, if possible, be sent to the closest station to facilitate the evacuation from the train directly to the station platform. Other nearby trains should be stopped to eliminate the piston effect of moving trains and shift smoke control entirely to the ventilation systems. If the evacuation is taking place away from the station, traction power should be deenergized on the involved track so evacuees and first responders are not at risk. Furthermore, adjacent tunnels should be kept available for a rescue train or a fire department access train in the event that the train is stranded in the tunnel. The rescue train can expedite evacuation of passengers and transport of first responders to the scene.

NFPA 130 (NFPA 2010) requires that each vehicle shall be provided with a minimum of two means of emergency egress on the sides or at the ends and that emergency exits shall be provided so that the maximum travel distance is less than or equal to 1250 ft (381 m).

SMOKE MANAGEMENT SYSTEMS IN TUNNELS

All tunnels require ventilation to maintain acceptable levels of contaminants produced by vehicle engines during normal and congested traffic operation, and to remove and control smoke and hot gases during a fire emergency (emergency ventilation). The ultimate goals of smoke management systems are to:

- provide an environment sufficiently clear of smoke and hot gases to permit safe evacuation, and
- allow relatively safe access for rescue services as a function of the fire scenario.

In designing the smoke systems, one should differentiate between phases of emergency operation. The first phase, called *self-evacuation phase*, occurs immediately after the fire incident is detected and in which the tunnel users commence their evacuation to the nearest exit or safe shelter. The self-evacuation phase could last between 4 to 15 min depending on the fire severity, tunnel environment, users' experience with these situations, and availability of exits. The second phase, called *assisted-evacuation and firefighting phase*, occurs upon the arrival of emergency services to the fire scene. The strategies of smoke control may be completely different during these two phases.

Establishing airflow requirements in the tunnel, and consequently the capacity of the ventilation system, is challenging due to the difficulty of controlling many variables (Kashef and Benichou 2008; Kashef et al. 2009). Among those variables are the possibility of occurrence of many vehicle combinations, combustible loads, and traffic situations during the lifetime of the facility. Smoke management in tunnels can be achieved using either natural or mechanical systems.

Natural Ventilation Systems

The consideration of natural smoke venting in the design of new tunnels is gaining more importance with the continued drive toward environmentally sustainable infrastructures to reduce energy consumption and save costs.

Natural systems count on the pressure differential created by the piston effect of moving vehicles, the meteorological conditions (external wind, temperature differentials between the portals), and differences in elevation to produce airflow through the tunnel. The created pressure must be large enough to overcome tunnel resistance and is therefore a function of tunnel length, cross-sectional geometry, wall roughness, number of vehicles in the tunnel, and air density. Any change in meteorological and operating conditions significantly influence the performance of natural ventilation systems.

Natural and traffic-induced ventilation is adequate for relatively short tunnels and tunnels with low traffic volume (or density). The use of natural ventilation for long and heavily traveled tunnels should be carefully investigated to assure its reliability. Otherwise, mechanical ventilation systems should be used.

Mechanical Ventilation Systems

Mechanical ventilation systems use series of fans to produce airflow and ducts and dampers to distribute this airflow. Regardless of mechanical ventilation equipment, the natural effects previously mentioned are present in all tunnels to a varying extent. A mechanical ventilation system is generally classified based on the direction of airflow it creates in the tunnel. There are two basic types of ventilation airflow systems applied in transport tunnels: longitudinal and transverse. Longitudinal ventilation systems produce longitudinal airflow in the direction of the tunnel axis; whereas transverse ventilation systems produce airflow that is perpendicular to the tunnel axis in the plane of a cross-section. The choice of what ventilation system to use depends on several parameters that include tunnel length, cross-section, and grade; meteorological conditions (prevailing wind and temperature); traffic type, direction, and volume; design fire size; and construction cost.

Longitudinal Ventilation

The longitudinal ventilation system (Figure 17.4) creates a longitudinal flow along the roadway tunnel by introducing or removing air from the tunnel at a limited number of points. The ventilation is provided either by injection (Figure 17.4a), by jet fans (Figure 17.4), or by a combination of injection or extraction at intermediate points in the tunnel (Figure 17.4c). The longitudinal form of ventilation is the most effective method of smoke control in a transport tunnel with unidirectional traffic. In a number of situations, if a longitudinal ventilation strategy is adopted, it is necessary to provide massive extractions points of smoke.

In the event of a fire in a unidirectional tunnel (Figure 17.5d), it is usually assumed that the traffic ahead of the fire will proceed to the exit portal and the traffic behind the fire will come to a stop (Figure 17.5e). The ventilation system should therefore be operated to force the smoke and hot gases in the direction of the empty tunnel to provide a clear and safe environment behind the fire for evacuees and fire fighters. In this case, the main measure of the adequacy of the system for smoke management is its ability to prevent backlayering (Figure 17.5a and b).

The challenges occur in a unidirectional tunnel with congested traffic (Figure 17.5c and f) or in a bidirectional tunnel (Figure 17.6), and so the simple philosophy previously described would not be achievable. As a consequence, users may be on either side of the fire limiting the ability of a longitudinal system to properly manage the smoke. In this case, it is crucial to initially limit the longitudinal air current in order to promote smoke stratification and allow users to evacuate underneath the smoke layer in the self-evacuation phase. Subsequently, the strategy of pushing the smoke in one direction can be applied to allow for the intervention of rescue and firefighting services.

As the length of the tunnel increases, excessive air velocities could be expected in one or a few locations across the tunnel. Moreover, in the event of a fire, smoke could be drawn throughout the entire length of the tunnel downstream of the fire.

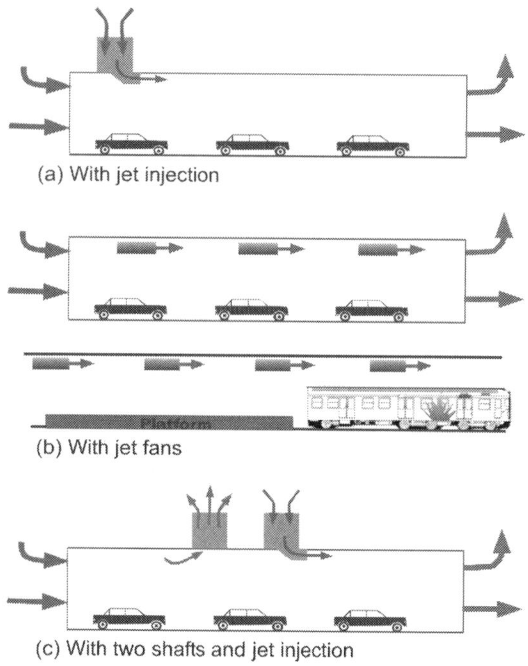

Graphic courtesy of the National Research Council, Ottawa, Canada.

Figure 17.4 *Longitudinal ventilation.*

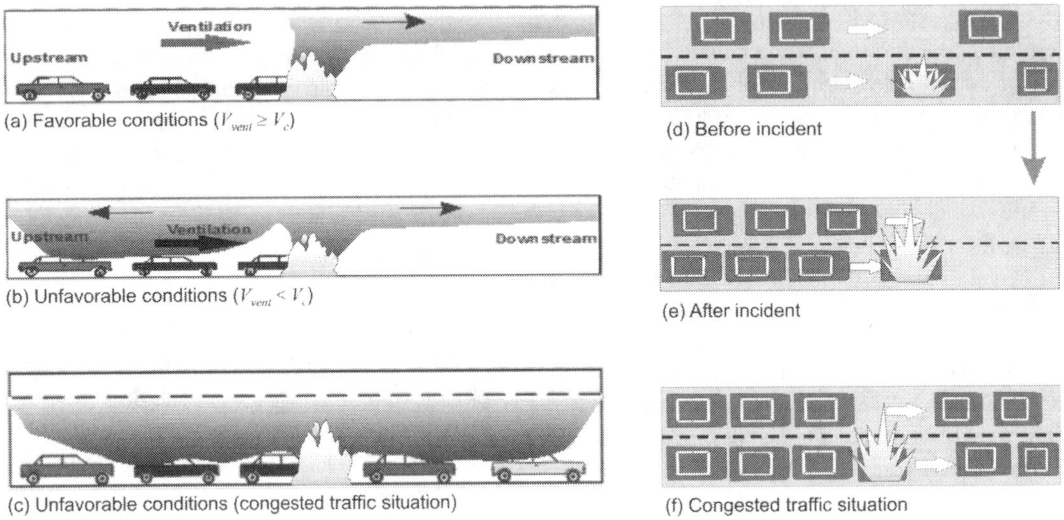

Graphic courtesy of the National Research Council, Ottawa, Canada.

Figure 17.5 *Backlayering—unidirectional traffic.*

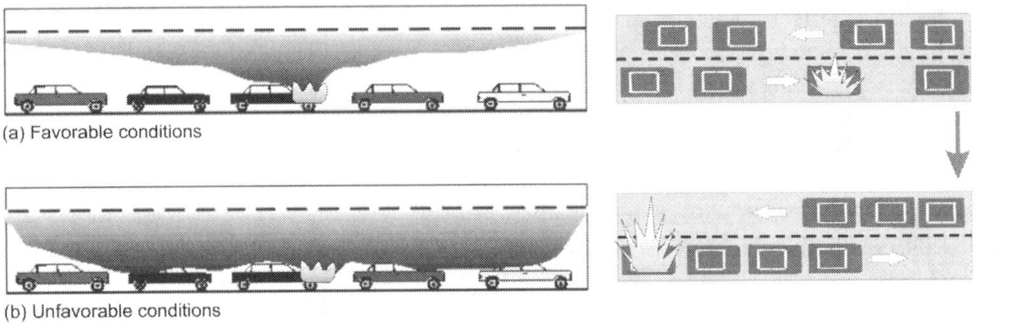

Graphic courtesy of the National Research Council, Ottawa, Canada.

Figure 17.6 *Backlayering—bidirectional traffic.*

Critical Velocity

The ventilation system must generate sufficient longitudinal air velocity to prevent backlayering of smoke. The air velocity necessary to prevent backlayering of smoke over the stalled motor vehicles is the minimum steady-state velocity needed for smoke control in a longitudinal ventilation system and is known as the critical velocity. The critical velocity depends on the fire heat release rate (fire size), the slope, and the tunnel section geometry. The critical velocity is used to define the fan requirements for smoke control from possible fires and can be determined by the simultaneous solution of Equations 17.6 and 17.7 (Kennedy et al. 1996):

$$V_c = K_1 K_2 \left(\frac{gHQ}{\rho_\infty C_p A T_F} \right)^{1/3} \quad (17.6)$$

$$T_F = \left(\frac{Q}{\rho_\infty c_p A V_c} \right) + T_\infty \quad (17.7)$$

where:

V_c	=	the critical velocity, ft/s (m/s),
g	=	acceleration of gravity, ft/s^2 (m/s^2),
H	=	tunnel height, ft (m) ,
Q	=	fire heat release rate, Btu/s (kW),
A	=	tunnel cross-sectional area, ft^2 (m^2),
c_p	=	specific heat of air, Btu/lb·°R (kJ/kg·K),
ρ_∞	=	density of ambient air, lb/ft^3 (kg/m^3),
T_∞	=	temperature of ambient air, °R (K),
T_F	=	average temperature of fire site gases, °R (K),
K_1	=	constant ($K_1 = F_r^{-1/3}$),

Example 17.1

For a tunnel with cross-section dimensions of 5 m in height and 22 m in width (~5 lanes) and grade of 3%, estimate the value of the critical speeds for a design fire of 100 MW. Assume ambient air temperature 20°C.

Solution:

$$A = 5 \cdot 22 = 110 \text{ m}^2$$

$$T_\infty = 20 + 273.15 = 293.15\text{K}$$

$$\rho_\infty = \frac{P}{RT_\infty} = \frac{101325}{287 \cdot 293.15} = 1.204 \text{ kg/m}^3$$

$$K_2 = 1 + 0.0374\,(grade\,\%)^{0.80} = 1.09$$

Assume $V_c = 1.5$ m/s, then

$$T_F = \left(\frac{10^8}{1.204 \cdot 1006 \cdot 1.5 \cdot 110}\right) + 293.15 = 793\text{K}$$

The new value of V_c will be:

$$V_e = 0.606 \cdot 1.09\left[\frac{9.81 \cdot 5 \cdot 10^8}{1.204 \cdot 1006 \cdot 110 \cdot 793}\right]^{1/3} = 2.4 \text{ m/s}$$

Using a value of V_c of 2.4 m/s, the new T_F will be 609°K. After few iterations of solving the two equations 17.6 and 17.7, the final values of the two parameters will be:

$$V_c = 2.6 \text{ m/s}$$
$$T_F = 578°\text{K}$$

F_r = Froude Number for a Flow ventilating a fire ($F_r = 4.5$),

K_2 = grade factor ($K_2 = 1 + 0.0374[grade\%]^{0.80}$).

The grade is positive if it is a descending grade in the traffic direction. If the longitudinal air velocity is much greater than the critical velocity, the high flow rates may have the advantage of reducing temperature and decreasing toxicity in the tunnel. However, they will completely destroy the smoke stratification and may cause the fire to grow faster to higher heat release rates. Furthermore, excessive longitudinal air velocity can lead to a faster fire spread among vehicles trapped in the tunnel. Example 17.1 illustrates calculation of the critical velocity using the two equations 17.6 and 17.7.

As shown in Figure 17.7, the critical velocity increases rapidly with the fire size up to about 28,000 Btu/s (30 MW) and then only increases slightly with increased heat release rate. The same trend is true for different tunnel grades with higher values of the critical velocity corresponding to higher grades for the same fire size (e.g., for a 95,000 Btu/s [100 MW] fire and grade of 3%, V_c = 520 ft/min [2.64 m/s] versus 467 ft/min (2.38 m/s) at 0% grade).

The value of the critical velocity is influenced by the tunnel cross-section dimensions. Reducing the width of the tunnel or increasing its height will increase the value of the critical velocity (Figure 17.8).

While evaluating the required longitudinal ventilation system capacity in case of fire, it must be assumed that a certain number of vehicles can be trapped in the tunnel and their presence reduces the performance of the ventilation system. The number of vehicles trapped can be assessed according to the design mix of traffic (% of passenger cars and heavy vehicles) for the specific tunnels. PIARC guidelines (PIARC 1999) recommended a design airflow velocity of 600 fpm (3 m/s) for all fires which do not involve a heavy goods vehicle carrying very flammable dangerous goods.

Smoke Stratification Versus Longitudinal Airflow

Figure 17.3 shows that the airflow in the tunnel affects not only the backlayering phenomenon, but also the degree of smoke stratification downstream of the fire.

If the airflow has a lower velocity V_{vent} than the critical velocity V_c the smoke layer will progress upstream of the fire causing the backlayering phenomenon to occur.

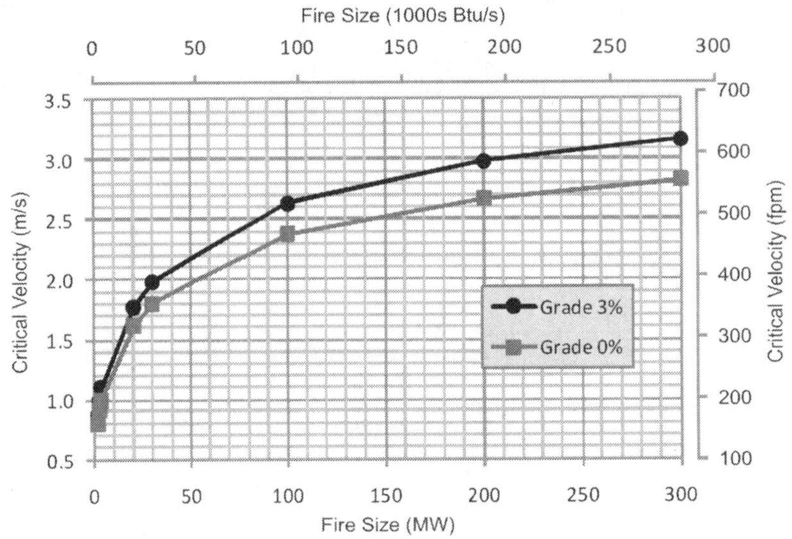

Graphic courtesy of the National Research Council, Ottawa, Canada.

Figure 17.7 *Effect of fire size on critical velocity.*

The length of the backlayering distance B_L can be estimated from (Beard and Carvel 2005):

$$B_L = K_B\left(\frac{gh^2Q}{\rho_\infty c_p V_{vent}^3 T_\infty}\right)^{1/3} \qquad (17.8)$$

where K_B is the proportionality constant in the range of 0.60–2.2, (Beard and Carvel 2005) and V_{vent} is the airflow velocity in the tunnel, ft/s (m/s). Equation 17.8 does not consider other geometrical dimensions (e.g., cross-section shape, tunnel width, or tunnel slope) than the height of the tunnel.

If the approaching airflow has a relatively higher velocity V_{vent} than the critical velocity, the longitudinal airflow will greatly tilt the fire plume, pushing the smoke to move to the downstream of the fire where the smoke stratification will be completely disrupted (Figure 17.3c). The degree of deflection of the plume depends on both the approaching airflow velocity and the plume upward velocity. Raj (Raj et al. 1979) suggested the following relationship to calculate the plume deflection of LNG pool fires:

$$\sin\theta = \left[\left(\frac{V_{vent}}{w}\right)\left(\frac{2c_pT_\infty\rho_\infty}{\pi\rho_f\Delta H_c}\right)^{-1/3}\right]^{-1/2} \qquad (17.9)$$

where ρ_∞ and ρ_f are densities of ambient air and fuel vapor, lb/ft^3 (kg/m^3), respectively, and w is the characteristic plume velocity, ft/s (m/s), and may be estimated from (McCaffrey 1976):

$$w = C_\theta Q_c^{1/5} \qquad (17.10)$$

where Q_c is the convective portion of the total heat release, Btu/s (kW) and C_θ is the velocity constant in the intermittent region of the fire plume, which is 6.301 (1.9). For example, for a propane pool fire size of 19,000 Btu/s (20 MW), $Q_c \approx$ 13,300 Btu/s (14 MW), a ventilation airflow speed of 600 fpm (3.0 m/s) will bend the plume by $\theta \approx 70°$.

The deflection of the plume and the disturbance of the smoke layer should be carefully considered as they can create hazardous conditions downstream of the fire. It could greatly impede the evacuation and promote the fire spread to nearby objects due to direct flame impingement or elevated levels of irradiance. Depending on the level of irradiance, the minimum separation distance R between vehicles to minimize fire spread can be calculated from NFPA 92b (NFPA 2005):

$$R = \sqrt{\frac{Q}{12\pi q''}} \qquad (17.11)$$

where q'' is the incident radiant heat flux required for piloted ignition, Btu/s ft^2 (kW/m^2).

Transverse Ventilation

A tunnel that is relatively long, has a heavy traffic flow, or is bidirectional requires different schemes of ventilation than the longitudinal method to properly manage the smoke. Options include semitransverse ventilation and full transverse ventilation. The main strategy adopted for these systems to manage the smoke is by maintaining smoke stratification near the

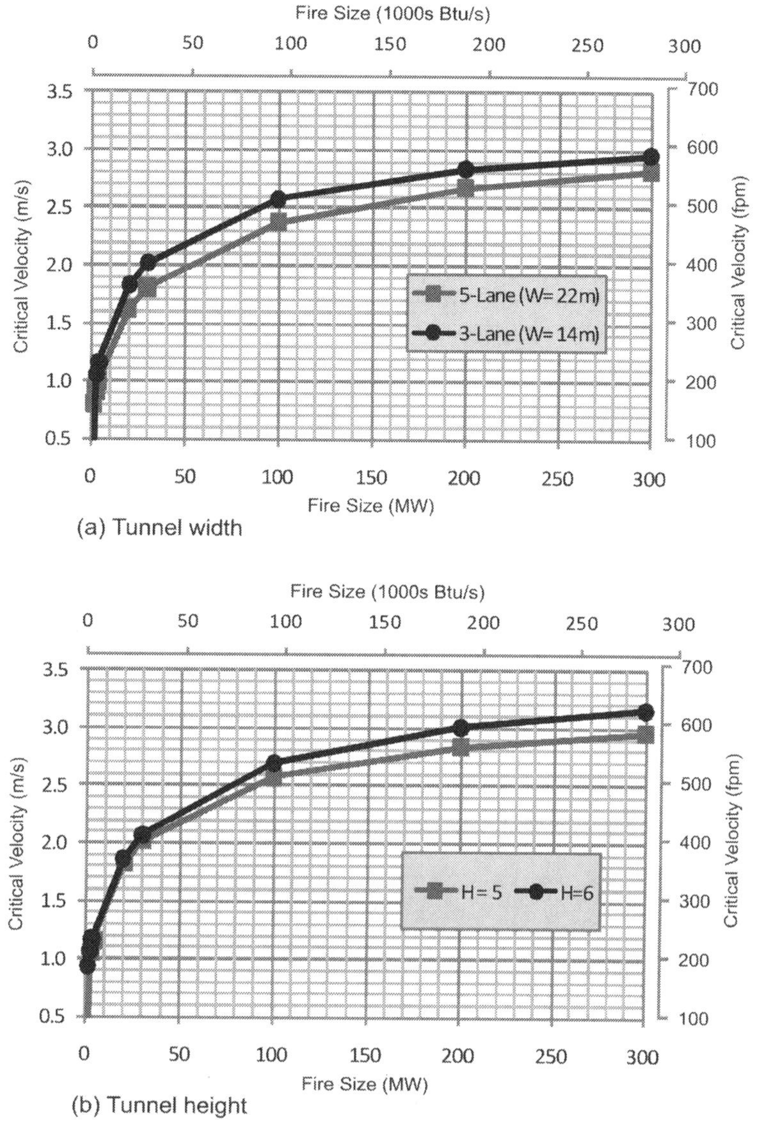

Graphic courtesy of the National Research Council, Ottawa, Canada.

Figure 17.8 *Effect of tunnel dimensions on critical velocity.*

fire and extracting it from the ceiling or from high openings along the tunnel wall(s). In this approach, longitudinal airflow velocities must to be kept at low magnitudes.

In designing these systems, means to avoid the occurrence of the plugholing phenomenon should be carefully considered. Plugholing refers to a situation where an exhaust fan has such a large capacity that it creates a hole in the smoke layer, exhausting clean air from below the smoke layer as well as smoke and therefore reducing the efficiency of smoke extraction. To sustain smoke layer stratification, the smoke exhaust rate, E_S, ft^3/min (m^3/s) should be maintained lower than

$$E_S = K_S\left(\frac{g d_S^5 T_\infty \Delta T_F}{T_F^2}\right)^{1/2} \tag{17.12}$$

(Heselden 1976) where d_S is the thickness of the smoke layer in ft (m) and K_S is constant, 102 (1.7).

Semitransverse Systems

A semitransverse ventilation system can be either a supply or an exhaust system (Figure 17.9a and b). This type of ventilation system induces (supply system Figure 17.9a) or collects (exhaust system Figure 17.9b) air uniformly throughout the length of a road tunnel in a duct fitted with supply outlets or exhaust inlets spaced at predetermined distances. Semitransverse ventilation is normally used in tunnels up to about 6600 ft (2000 m). Beyond that length, the tunnel air speed near the portals becomes excessive. This type of ventilation has the advantage of being less affected by atmospheric conditions since the tunnel airflow is fan generated.

For semitransverse ventilation in the normal mode of operation, fresh air is best introduced at the vehicle exhaust pipe level to dilute the exhaust gases immediately. An adequate pressure differential between the duct and the roadway must be generated and maintained to counteract the piston effect induced by the traffic flow and adverse atmospheric winds.

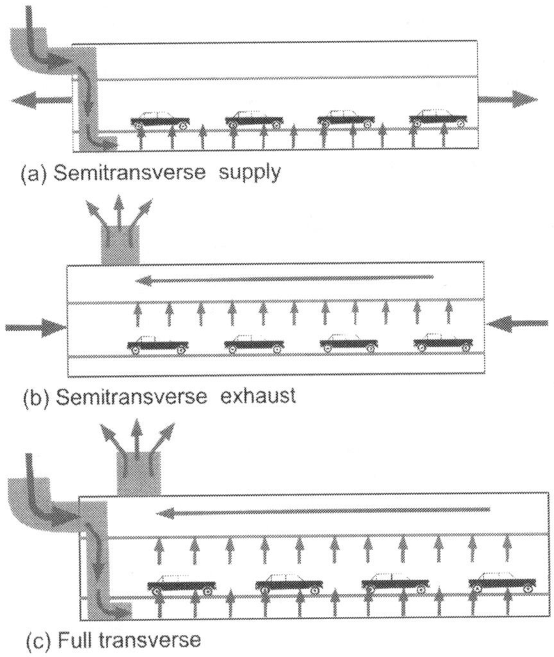

Graphic courtesy of the National Research Council, Ottawa, Canada.

Figure 17.9 *Transverse ventilation.*

In an emergency situation (e.g., if a fire occurs in the tunnel) initially the supply air dilutes the smoke. Upon the arrival of fire fighters to the tunnel, the system should be operated in the emergency mode such that fresh air enters the tunnel through the portals and smoke exits through the ceiling, hence creating a tenable environment during the assisted evacuation and firefighting phases. To implement this kind of smoke management strategy, this system should have ceiling or at-the-top-of-the-walls exhaust vents and reversible fans. The reversible fans are essential to be able to operate in the supply mode during normal operations and in the exhaust mode during emergencies.

Full Transverse Ventilation

Full transverse ventilation (Figure 17.9c) is used in extremely long tunnels (more than 6600 ft [2000 m] long) and in tunnels with heavy traffic volume. This ventilation system is usually a balanced system and comprises both supply and exhaust ducts to achieve uniform distribution of supply air and uniform collection of vitiated air throughout the tunnel length. Consequently, this configuration produces a uniform pressure along the tunnel length with no longitudinal airflow being generated except that created by the traffic piston effect.

An adequate pressure differential between the ducts and the roadway should be maintained to ensure proper air distribution under all ventilation conditions. In the event of a fire, the exhaust fan should attain its maximum available capacity while the supply should be maintained at a relatively low capacity. This scheme ensures that the stratified smoke at the ceiling remains at that higher elevation and is extracted by the exhaust ducts without mixing with the lower fresh air. In this way, a tenable environment could be maintained in the tunnel for firefighting and emergency egress.

Smoke management strategies should be configured such that smoke is extracted in the fire zone. The objective is to limit the smoke spread within the extraction zone and enable safe evacuation on both sides of the

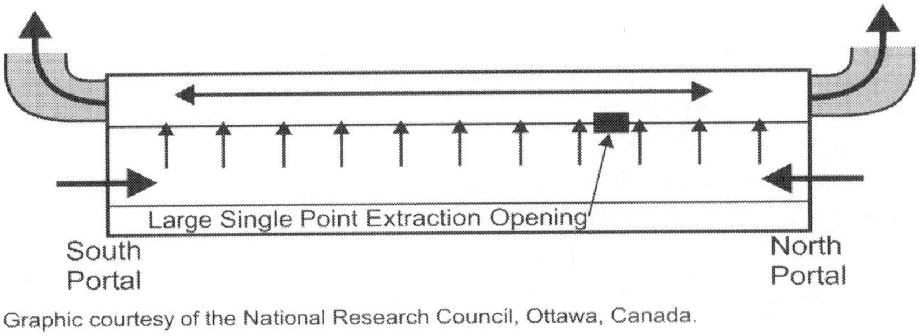

Graphic courtesy of the National Research Council, Ottawa, Canada.

Figure 17.10 *Single-point extraction.*

fire. The extraction zone should be as short as practicable while still covering the fire area, to maximize airflow to the affected location. The extraction rate is sometimes estimated as 150% of the smoke-production rate of the design fire at a distance of 330 ft (100 m) from the fire (RABT 2006). Alternatively, the smoke extraction rate, in ft^3/min (m^3/s), can be determined as 600 ft/min (3 m/s) to 800 ft/min (4 m/s) times the tunnel cross section (ASTRA 2006). Sufficient air needs to be provided in the direction of the extraction point (both upstream and downstream) to control the smoke flow and maximize the extraction efficiency.

This method is essential in case of rescue from a fire in a tunnel with bidirectional traffic or traffic congestion. Some systems employ remotely controlled dampers that enable the point extraction of smoke around the fire. It should be noted that the construction and maintenance costs for these systems are higher than for longitudinal systems. Moreover, a larger design fire requires larger duct sizes impacting the resulting investment costs. Careful consideration should be given for the maintenance of dampers in areas of extreme weather (e.g., cold climates).

Other Mechanical Ventilation Systems

The ventilation systems described earlier could be combined into a hybrid system to address a specific problem in a tunnel. A few modifications can be made to improve the performance of these systems. Among these are single-point extraction and oversized exhaust ports.

Single-Point Extraction

A single-point extraction could be added to a transverse system with large openings to the exhaust duct (Figure 17.10). These openings include devices that can be operated during a fire emergency to extract a large volume of smoke as close to the fire source as possible. Tests conducted as a part of the Memorial Tunnel Fire Ventilation Test Program "MTFVTP" (FHWA 1999) concluded that this concept is effective in reducing the temperature and smoke in the tunnel. The size of openings tested ranged from 100 to 300 ft^2 (9.3 to 27.9 m^2).

Oversized Exhaust Ports

These (Figure 17.11) are simply an expansion of the standard exhaust port installed in the exhaust duct of a transverse or semitransverse ventilation system. Two methods can be used to create such a configuration. One is to install on each port expansion a damper with a fusible link; the other uses a material that when heated to a specific temperature melts and opens the airway. Several tests of such meltable material were conducted as part of the Memorial Tunnel Fire Ventilation Test Program but with limited success.

In addition, other modifications could be implemented for the transverse ventilation system to allow for single-zone or two-zone ventilation. In the two-zone configuration, as opposed to the single-zone ventilation, the tunnel is divided into two ventilation zones such that one half of the tunnel acts as a supply zone and the other half as an exhaust zone (Figure 17.12).

Ventilation Shafts

In many railway and subway tunnels, ventilation shafts are used, among other functions, to manage the smoke during a fire event. They could be used to supply fresh air or exhaust smoke out the tunnel. Traditionally, two main ventilation strategies could be used, depending on shaft design and location, fire location, the *push-pull* and *pull-pull* schemes. In the push-pull strategy (Figure 17.13a), the fresh air is supplied through the shafts upstream of the fire and smoke is pulled through shafts downstream of the fire. As such a longitudinal airflow is created across the fire to direct smoke towards outside the tunnel. In the other ventilation scheme, the pull-pull scheme (Figure 17.13b), the smoke is extracted from both sides of the fire.

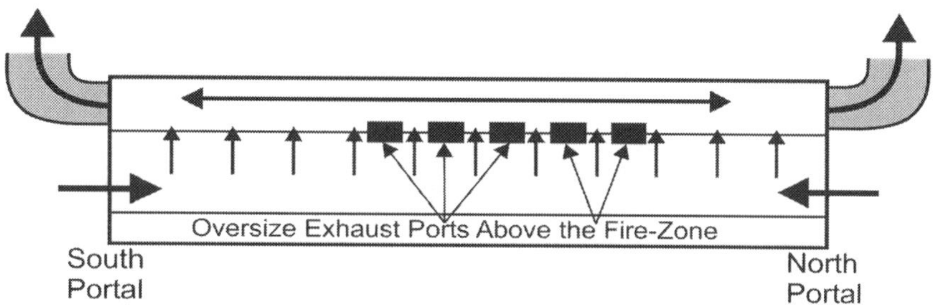

Graphic courtesy of the National Research Council, Ottawa, Canada.

Figure 17.11 *Oversized exhaust ports.*

On-Site Evaluation of Ventilation Systems Performance

Fire tests can be carried out on site to verify the adequacy of the ventilation systems and to define more precisely the operating instructions to meet the safety objectives (Kashef et al. 2003). Test configurations (fire location, initial air velocity in the tunnel, and fire heat release rate) must be designed to take into account as many situations as possible. If only a few configurations can be tested, it is advisable to choose those that are a priority and most unfavorable with regards to user safety.

On-site fire tests can prove to be valuable as a commissioning tool. A number of operational defects can be identified in the fire tests that could not otherwise be identified in the normal commissioning process. In particular, these can include the communication or lack of communication between systems such as smoke detection systems, smoke exhaust fans, smoke dampers and emergency warning systems. In addition, obstructions left in air ducts or faulty wiring of fans and dampers can be recognized.

On-site fire tests are delicate operations because they must be carried out before the opening of the tunnel or during its operation, with tight time schedules. It is thus necessary that these tests be led correctly to minimize disruptions and maximize safety.

Procedures to protect people must be established before the tests. Protection also relates to both the equipment that can be isolated and the structure of the tunnel, which should not be exposed to excessive temperatures. In other words, security management must be considered in the design of the tests. The fire heat release rate is inevitably limited and the scenarios must take into account the protection requirements. In the case of a fire test during operation in a tube where traffic is maintained in the other tube, it is necessary to ensure that smoke is not recycled in the tube open to traffic.

Measurements

Measurements have two objectives:

- Monitoring the temperatures during each test. This information can be used to manage safety during the test.

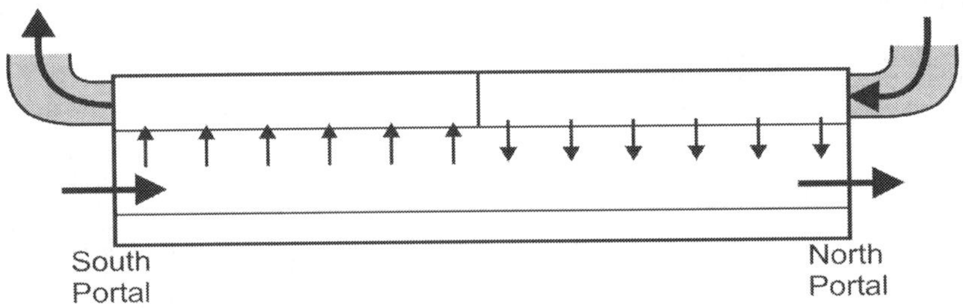

Graphic courtesy of the National Research Council, Ottawa, Canada.

Figure 17.12 *Two-zone configuration.*

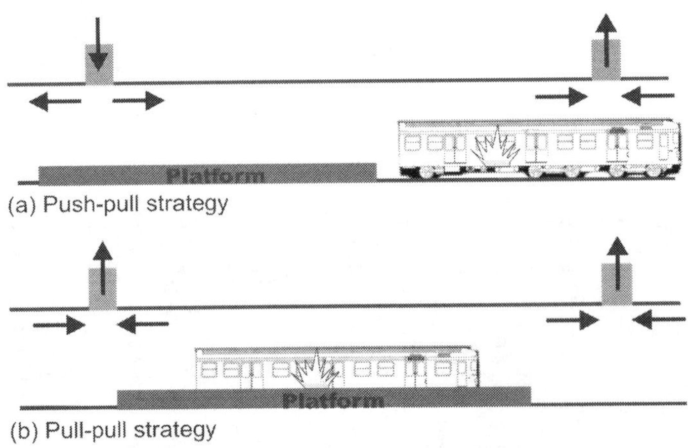

Graphic courtesy of the National Research Council, Ottawa, Canada.

Figure 17.13 *Ventilation shafts.*

- Monitoring airflow velocity, smoke stratification, visibility, and backlayering.

The first objective requires the installation of thermocouples on exposed places in the fire zone (equipment and structure of the tunnel). The second objective requires a methodical approach in which it is necessary, before the tests, to identify the phenomena to be characterized. At the end of this analysis, it is necessary to determine the nature, location, and number of sensors to be installed in the tunnel.

Fire Source

Different sources of smoke can be used to represent fires in tunnels, such as cold smoke, pool fires, and real fires. The use of cold-smoke-producing products is not representative of a fire. The production of heat by the fire is not taken into account. This limits the representation of the fire phenomena in terms of critical velocity and natural smoke stratification. This approach is not recommended for establishing operating instructions, because the phenomena related to the presence of a real fire are not reproduced.

Realistic fires generally use wrecks of road vehicles. The heat release rate developed by this type of fire is well known. Second order variations, such as turbulence or the chaotic emissions of puffs of smoke, result in a smoke behavior that is much more difficult to characterize and introduce substantial differences compared to calibrated fires tests.

Pool fires and hot-smoke tests generally involve hydrocarbon pool fires (heptane or fuel oil). These fires are well known. The advantage of these fires is their stability and therefore leads to well-characterized situations that emphasize the effect of ventilation on smoke behavior (Kashef and Benichou 2008). With hydrocarbon fires, it is generally possible to reach several steady-state situations during the same fire test, and thus to test various aerodynamic configurations.

DESIGN FIRE

Design fires are an intrinsic part in designing tunnels to withstand fires. They provide, quantitatively, the fire characteristics that are used to establish the sizing of equipment in tunnels and the scenarios to consider when developing emergency response plans. They are also used, indirectly, when considering the impact of fires on the structure. As such, design fires form the base input for emergency ventilation, evacuation, and structural design analyses.

A design fire is an idealization of a real fire that might occur in a tunnel and is generally defined in terms of heat release rate and species output as functions of time. It is a set of data that provides the actual fire characteristics such as ignition sources, nature and configuration of the fuel, fire growth, peak heat release, production rates of combustion products (smoke CO, CO_2, etc.), and extinction. For the design purposes, it is necessary to choose a typical design fire corresponding to the traffic type and pattern in the tunnel and whether hazardous transports are permitted.

A prescriptive approach has traditionally been adopted in which a specific fire size, usually the peak heat release rates depending on the type of vehicle (passenger cars, buses, heavy goods vehicles, pool fires, etc.), is chosen as a basis for the tunnel fire life-safety design (Tables 17.3 and 17.4). The adequacy of the design fire sizes used for the design of fire protection systems used in road tunnels was seriously questioned following the occurrence of major fire catastrophes in late 1990s. This has promoted the shift from prescriptive- to performance-based regulations. Performance-based designs are usually based upon explicitly stated objectives that allow the freedom to develop innovative designs satisfying these objectives. Such innovative designs often lead to lower fire protection costs.

The performance-based design approach makes possible the evaluation of the tunnel fire safety as a whole. An important step in the performance-based design is the establishment of possible fire scenarios. Different fire scenarios are created to instigate the design analyses of emergency ventilation, egress, structural, and fire safety tunnel equipment (e.g., detection and fixed firefighting systems). A design fire scenario qualitatively describes the key time events following the ignition of a fire, such as: quantity and characteristics of combustible materials, material arrangement and location, tunnel geometry, environment, fire protection systems, etc. The design fires are the cornerstone in developing such fire scenarios. As such, design fires are the underpinning in conducting a performance-based design.

Table 17.3: Fire Data for Typical Vehicles (NFPA 502 Table A.10.5.1 [NFPA 2011])

Vehicles	Peak Fire Heat Release Rates, MW
Passenger car	5–10
Multiple passenger cars (2–4 vehicles)	10–20
Bus	20–30
Heavy goods truck	70–200
Tanker	200–300

Note: 1 MW = 948 Btu/s.

Table 17.4: Fire Sizes Adopted in Different Countries (PIARC 2011)

Country	Fire Size, MW	Notes
Australia	50	With FFFS (deluge system), for ventilation only
Austria	30	High risk category: 50 MW
France	30–200	200 MW when transports of dangerous goods allowed, but only applied for longitudinal ventilation
Germany	30–100	Depending on length and HGV in tunnel
Greece	100	Longitudinal ventilation
Italy	20–200	
Japan	30	
Netherlands	100–200	100 MW if tankers are not allowed, otherwise 200 MW for ventilation system
Norway	20–100	Depending on risk class, always longitudinal ventilation
Portugal	10–100	Based on traffic type
Russia	50–100	
Singapore	30–200	Depends on vehicle types allowed
Spain	≥ 30	
Sweden	100	Longitudinal ventilation
Switzerland	30	Smoke extraction equals 3.3–4 m/s times cross section
UK	30–100	
USA	30–300	300 MW if dangerous goods allowed

Note: 1 MW = 948 Btu/s.

Different aspects of a design fire are more important to certain types of analysis than others. For example, the peak heat release (PHRR) and burning duration are important to evaluating structural response to fire. The HRR at the end of evacuation and the PHRR are considerations in evaluating tunnel ventilation equipment and is of concern for the life safety of the fire service during the firefighting phase. The objective during this phase is to provide tenable conditions for safe firefighting activities.

The early transient stages of fire development during the growth phase affect the conditions in the tunnel during the *self-rescue* phase and are therefore important to life-safety analyses. An understanding of how fast a fire might grow, and the subsequent spread of smoke and hot gases, is a factor in the design of ventilation, suppression, and detection systems as well as the determination of evacuation strategies.

Design Fire Scenarios

To achieve optimum fire prevention strategy for tunnels, a number of fire scenarios should be considered during the design stage. These should include

- egress analysis,
- thermal action on structures,
- ventilation systems design and assessment, and
- the safety of tunnel fire equipment (e.g., detection and fixed firefighting systems)

Each of these scenarios must be well described prior to the design process. Following are some guidelines for their selection and description:

- Description of the aim of the scenario
- Thorough definition of the fire parameters:
    - Heat release versus time
    - Number of vehicles involved: incidents with one vehicle (car, bus, truck, fuel, tanker) or collision incidents of two to three vehicles
- Natural ventilation of the tunnel
- Effective escape and rescue possibilities:
    - Availability of firefighting equipment (e.g., fire extinguishers)
    - Availability of detection systems
    - Time of arrival of the fire brigade
    - Availability of emergency exits
    - Ability to control smoke and visibility
    - Possibility of traffic control
- Traffic situation encountered when dealing with questions about tunnel ventilation and operation

(e.g., stop-and-go situation, congested traffic, mode of traffic flow)

- Specifications to be fulfilled by material, equipment, and structure with regard to fire prevention strategies (e.g., temperature at concrete reinforcement should not exceed 572°F [300°C])

Worst cases should not necessarily be considered for design when their probability is very low. For instance, very few fires result from a collision while this case leads to the highest heat release rates and temperatures. If the consequences may be catastrophic (e.g., collapse of an immersed tunnel), such very severe scenarios should be taken into account for design.

NUMERICAL MODELING

The applicable NFPA standards for tunnels (e.g., NFPA 502) require engineering analysis for tunnels greater than a certain length to assist in evaluating whether the smoke and heat layer is properly managed. Traditionally, engineers and designers have shown compliance with the codes and standards requirements by using one-, two- or three-dimensional numerical models.

One-Dimensional Models (1D)

One-dimensional models provide simple design tools for the transient calculation of networks. The fundamental equations of fluid thermodynamics are solved, but only one dimension is considered. That means all the conditions are homogeneous in the cross-section. As they cannot take the layering phenomena into account, they cannot be applied in the fire vicinity. Nevertheless, these principles seem sufficient for studying the conditions far from the source in an underground road network or a very long tunnel, and for providing boundary conditions to a CFD code in the case where the whole tunnel is not modeled by this latter model.

The appropriateness of such tools for special applications, particularly when the tunnel is wide or high with respect to the physical size of the fire, should be carefully validated. In these situations, the studies in the MTFVTP (FHWA 1999) indicated that these design tools need improvement to better predict the critical longitudinal air velocity required to prevent backlayering and allow for the control of smoke and hot gas spread in a tunnel. Examples of this family of tools include:

Subway Environment Simulation (SES): The predominant worldwide tool for analyzing the aero-thermodynamic environment of rapid transit rail tunnels is the SES computer program (DOT 1997). SES is a one-dimensional network model that is used to evaluate longitudinal airflow in tunnels. The model predicts airflow rates, velocities, and temperatures in the subway environment due to train movement or fans, as well as the station cooling loads required to maintain the public areas of the station to predetermined design conditions throughout the year. This program contains a fire model that can simulate longitudinal airflow required to overcome backlayering and control smoke movement in a tunnel. The SES program is in the public domain, available from the Volpe National Transportation Systems Center in Cambridge, MA.

TUNVEN: This program solves coupled one-dimensional, steady-state tunnel aerodynamic and advection equations. It can predict quasi-steady-state longitudinal air velocities and concentrations of CO, NO_x, and total hydrocarbons along a road tunnel for a wide range of tunnel designs, traffic loads, and external ambient conditions. The program can also be used to model all common road tunnel ventilation systems (i.e., natural, longitudinal, semitransverse, and transverse). The user needs to update emissions data for the calendar year of interest. The program is available from the National Technical Information Service (NTIS 1980).

Zone Models (2D Models)

Zone models are seldom used to study the spread of smoke and temperature in tunnels, but they are commonly used in buildings. They generally describe a room or a corridor as a homogeneous zone where a fresh air layer lies under a smoke layer, each of them having constant characteristics—including their thickness—on the whole zone. The fire and the exchanges between the layers and between the neighboring zones are governed by partially empirical equations.

Such models are relatively flexible, and they can be investigated on a desktop computer and are well adapted to investigate the smoke and heat propagation in a complex system of communicating rooms. Unfortunately they are not well adapted to studying fires in tunnels, where the main problem is to predict the evolution of the smoke plume inside a large zone, moreover submitted to a longitudinal airflow, whose influence is determining. Few validations have been performed with such models and their success still seems uncertain for tunnel fires.

Computational Fluid Dynamics (CFD) (3D)

CFD modeling techniques are sophisticated and computationally intensive design tools. They can model actual conditions in tunnels and predict three-dimensional patterns of airflow, temperature, and other flow variables, including concentration of species, as function of time and space. In areas of geometrical complexity, CFD is the appropriate tool.

SOLVENT: SOLVENT was developed for tunnel ventilation simulation in Phase IV of the MTFVTP (Parsons Brinckerhoff 2000). The model is applicable to different ventilation modes, including longitudinal ventilation using jet fans, transverse ventilation, and natural ventilation. SOLVENT is geared towards individuals (analysts, designers, and tunnel operators) concerned with fire/life safety in tunnels. The primary objective established for the model was the ability to simulate the interactive effects of a tunnel fire and the ventilation system to determine the unsafe regions of the tunnel. That is, the regions where the hazardous effects of the fire (smoke and high temperatures) are confined, and how these regions are affected by the ventilation system configuration, capacity, and operation. The program results have been validated against data from the MTFVTP.

Fire Dynamics Simulator (FDS): FDS (McGrattan et al. 2011) is a CFD model of buoyancy-driven fluid flow from a fire. FDS consists of two computer programs, the first called simply FDS, a Fortran 90 computer program that solves the governing equations. The second, called Smokeview, is an OpenGL graphics program that allows one to visualize the results. All of the input parameters required by FDS to describe the particular scenario of interest are conveyed via one or two text files created by the user. FDS can model pollutant levels outside the portals and around the exhaust stacks of tunnels. Both of these public domain programs are under active development and can be obtained from National Institute of Standards and Technology (NIST).

Other CFD programs, both commercially available and in the public domain, have been used to model fire scenarios in road and rapid transit tunnels and stations, the list of which is too numerous to include here. The strengths and weaknesses of each program should be investigated beforehand, and validation of the results against experimental data or another equivalent program is encouraged.

DETECTION

The dynamics of smoke movement in tunnels is such that evacuation or rescue operations should start within 5 to 15 min from the start of fire incident. As such, a fire detection system is a critical element of the fire prevention and protection matrix in tunnels. Quick and reliable fire detection allows for timely activation of other elements of the fire protection matrix which may include one or all of the following systems: emergency

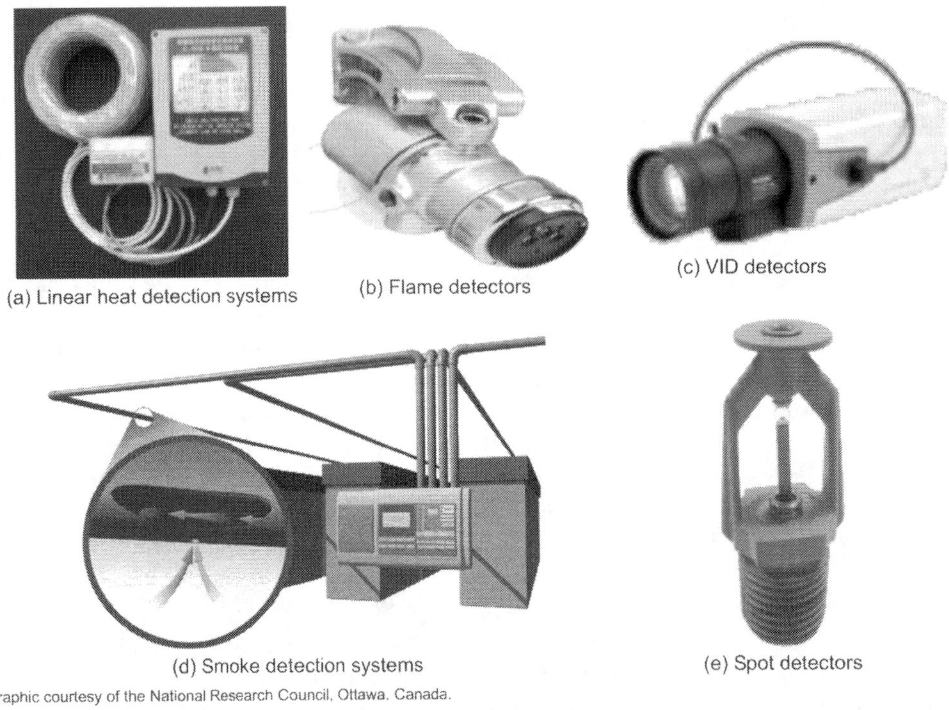

Figure 17.14 *Fire detection technologies (Kashef et al 2008).*

lighting, public address, emergency ventilation, and fixed firefighting systems (Figure 17.1). Detection can make the difference between a manageable fire and one that gets out of control. On other hand, false or nuisance alarms are not only costly but also can promote a lack of confidence in the reliability of detection systems. Automatic fire detection has been used in tunnels for several years.

Detection is conducted based on exceeding threshold values for a prescribed duration (Kashef et al. 2008). It is useful to include the rates of change of the measurements in the evaluation. In this context, it is important to divide the tunnel into well-defined sections to enable accurate information regarding the location of an incident to the operator. Particularly when using smoke extraction, the location of the fire needs to be detected in order to incorporate the correct response with respect to ventilation control.

Normally, smoke detection is less accurate in determining the location of the fire than is a high-temperature alarm using a linear heat detector. Moreover, the reaction due to several independent fire detectors by one or more systems has to be considered. This concerns the detection of moving fire sources (moving trains on fire) as the location of the initial detection of the fire might not be the same as the location where the vehicle comes to a standstill (in particular, information retrieved from VID and smoke detection).

Performance Criteria

Many factors affect the performance of detection systems in the harsh environment of tunnels. Pollution, wind speed, tunnel geometry, traffic congestions, fire type, size, and location are a few examples. Various types of detection systems are affected to a different degree by these factors. Performance of fire detection systems is usually evaluated based on the requirements for tunnel protection (Kashef et al. 2008):

- Response capability to tunnel fire incidents involving fire size, type, growing rate, and location (measuring parameter: time [min])
- Locating capability to the fire position in the tunnel (measuring parameter: distance [m])
- Monitoring capability of some systems to a fire incident (i.e., fire growing and developing direction in the tunnel)
- As well as their reliability in harsh tunnel environments, including their nuisance alarm immunity and requirements for maintenance and operating costs

Available Detection Technologies

There are a range of methods available to detect fire and smoke within tunnels. Each system is designed to detect a certain fire-related signature. There are five types of currently available technologies: linear heat detection (Figure 17.14a), flame detectors (Figure 17.14b), video image detectors (VID) (Figure 17.14c), smoke (Figure 17.14d) and heat detectors, and spot heat detectors (Figure 17.14 e). Fire detection systems should be selected to support the fire safety goals and objectives and the overall fire safety program, which can include notifying occupants to allow for safe evacuation, modifying tunnel ventilation or operations, and notifying emergency responders.

Table 17.5 lists the main five types of detection technologies, their principal method of detection, along with general assessment of their performance in road tunnels. All these systems are required to have a guaranteed backup of operational elements (redundancy), for both detection devices and the system as a whole. It is a prerequisite that the system ensures execution of predefined tasks if a total breakdown situation comes up.

NOMENCLATURE

A = tunnel cross-sectional area, ft (m^2)
C_θ = velocity constant in intermittent region of fire plume
C_p = specific heat of air, (Btu/lb·°R) (kJ/kg·K)
C_s = experimental constant
d_{so} = smoke layer initial thickness, ft (m)
d_s = smoke layer thickness, ft (m)
E_s = smoke exhaust rate, ft^3/min (m^3/s)
g = acceleration due to gravity, ft/s^2 (m/s^2)
H = tunnel height, ft (m)
K = proportionality constant
K_1 = constant
K_2 = grade factor
K_B = proportionality constant
m_p = mass concentration of particulate, lb/ft^3 (g/m^3)
$\dot{m}_s$ = smoke production rate, lb/s (kg/s)
Q = fire heat release rate, Btu/s (kW)
Q_c = convective portion of fire heat release rate, Btu/s (kW)
q'' = incident radiant heat flux required for piloted ignition, kW/m^2 (Btu/s m^2)
t = time from ignition, s (s)
T_∞ = temperature of ambient air, °R (K)
T_F = average temperature of fire site gases, °R (K)

Table 17.5: Status of Fire Detection Technologies (Kashef et al. 2008)

	Linear Heat Detection Systems (Figure 17.14a)	Flame Detectors (Figure 17.14b)	VID detectors (Figure 17.14c)	Smoke Detection Systems (Figure 17.14d)	Spot Detectors (Figure 17.14e)
Detecting principle	Heat	Radiation	Image/smoke	Smoke	Heat, smoke, gas, etc.
Detecting capability	Response Locate & monitor fire	Fast response Locate fire	Fast response Locate & monitor fire	Fast response Locate fire	Moderate response Locate fire
Reliability	High	Moderate to high	Moderate to high	Low	Moderate to high
Availability	High	Moderate	Moderate	Moderate	Moderate to high
Applications	Europe	Japan	Unknown	None	Sprinkler head

w = characteristics plume velocity, m/s (ft/s)
W = tunnel width, m (ft)
V = airflow velocity in tunnel, ft/s (m/s)
S = visibility, ft (m)
V_s = volume of smoke in the space, ft^3 (m^3)
V_c = the critical velocity, m/s (ft/s)
y_p = particulate yield, dimensionless
u_{so} = smoke layer initial moving velocity, ft/s (m/s)
α = extinction coefficient, ft^{-1} (m^{-1})
ΔH_c = heat of combustion, Btu/lb (kJ/kg)
ρ_∞ = density of ambient air, lb/ft^3 (kg/m^3)
ρ_s = density of smoke, lb/ft^3 (kg/m^3)
α_m = specific extinction coefficient, ft^2/lb (m^2/g)
ρ_f = density of fuel vapor, lb/ft^3 (kg/m^3)

REFERENCES

ASHRAE. 2011. *ASHRAE Handbook—HVAC Applications*, Chapter 17. Atlanta: ASHRAE.

ASTRA. 2006. Lüftung der Straßentunnel - Systemwahl, Dimensionierung und Ausstattung. Herausgeber: Bundesamt für Straßen, Bern.

Beard, A. and R. Carvel. 2005. *The Handbook of Tunnel Fire Safety*. London: Thomas Telford Ltd.

DOT. 1997. *Subway Environment Simulation (SES), Version 4: User's Manual and Programmer's Manual.* Pub. No. FTA-MA-26- 7022-97-1. U.S. Department of Transportation, Washington, DC.

Heselden, A.J.M. 1976. Studies of fire and smoke behaviour relevant to tunnels. *Proceedings of Second International Symposium on the Aerodynamics and Ventilation of Vehicle Tunnels, BHRA, Cambridge, UK,* paper J1.

Innovative Research, Inc./Parsons Brinckerhoff, Inc. 2000. *SOLVENT,* Version 1.0.

McGrattan, K.B., H.R. Baum, R.G. Rehm, G.P. Forney, J.E. Floyd, and S. Hostikka. 2011. Fire dynamics simulator (Version 5), technical reference guide. Technical Report NISTIR 6783, National Institute of Standards and Technology, Gaithersburg, Maryland.

Kashef, A., and N. Bénichou. 2008. Investigation of the performance of emergency ventilation strategies in the event of fires in a road tunnel—a case study. *Journal of Fire Protection Engineering* 18(3).

Kashef, A., N. Bénichou, G.D. Lougheed. 2003. Numerical modelling of movement and behaviour of smoke produced from fires in the Ville-Marie and L.-H.-La Fontaine Tunnels: Literature review. Research Report, NRC Institute for Research in Construction, 141 (RR-141).

Kashef, A., Z. Liu, G.D. Lougheed, G.P. Crampton, K. Yoon, G.V. Hadjisophocleous, and K. Almand. 2008. Findings of the international road tunnel fire detection research project, *Fire Technology Journal* 45(2).

Kashef, A., H.H. Saber, L. Gao. 2009. Optimization of emergency ventilation strategies in a roadway tunnel. *Fire Technology Journal* 45(4).

Kashef, A., J. Viegas, A. Mos, and N. Harvey. 2011. Proposed idealized design fire curves for road tunnels. 14th International Symposium on Aerodynamics and Ventilation of Tunnels, May 5, Dundee, Scotland.

Kennedy, W.D., J.A. Gonzalez, and J.G. Sanchez. 1996. Derivation and application of the SES critical velocity equations. *ASHRAE Transactions* 102(2).

Lacroix, D. 1998. The new PIARC report on fire and smoke control in road tunnels. *Third International Conference on Safety in Road and Rail Tunnels, Nice, France*, pp. 185–97.

FHWA. 1999. Memorial Tunnel fire test ventilation program. Phase IV Report, Commonwealth of Massachusetts, Massachusetts Highway Department/ Federal Highway Administration, Central Artery/ Tunnel Project.

McCaffrey, B.J. 1976. Purely buoyant diffusion flames: Some experimental results. NBSIR 79-1910, National Bureau of Standards, Gaithersburg, MD.

NFPA. 2005. NFPA 92B, *Standard for Smoke Management Systems in Malls, Atria, and Large Areas*. Quincy, MA.: National Fire Protection Association.

NFPA. 2010. NFPA 130, *Standard for Fixed Guideway Transit and Passenger Rail Systems*. Quincy, MA: National Fire Protection Association.

NFPA. 2011. NFPA 502, *Standard for Road Tunnels, Bridges, and Other Limited Access Highways*. Quincy, MA: National Fire Protection Association.

NTIS. 1980. User's guide for the TUNVEN and DUCT programs. Publication PB80141575, National Technical Information Service, Springfield, VA.

PIARC. 1999. Fire and smoke control in road tunnels. Technical Committee on Road Tunnels, reference 20.05.B, the World Road Association (PIARC).

PIARC. 2007. Systems and equipment for fire and smoke control in road tunnels. Technical Committee on Road Tunnels, the World Road Association (PIARC).

PIARC. 2011. Design fire characteristics for road tunnels. Working Group 4, Technical Committee on Road Tunnels, C4, Committee on Tunnel Operations, the World Road Association (PIARC).

RABT. 2006. Richtlinie für die Ausstattung und den Betrieb von Straßentunneln, Herausgeber: Forschungsgesellschaft für Strassen- und Verkehrswesen, Deutschland. ISBN 3-937356-87-8.

Raj, P., A. Moussa, and K. Aravamudan. 1979. Experiments involving pool and vapour fires from spills of liquefied natural gas on water. Report No. CG-D-55-79, U.S. Coast Guard.

CHAPTER 18

Zone Fire Modeling

John H. Klote

Zone fire modeling is a simple approach to simulating smoke transport. Zone fire modeling cannot simulate smoke flow with the complexity and level of detail of computational fluid dynamics (CFD) as described in Chapter 19, but zone fire models can be useful tools provided that they are used with care. The equations in this chapter are used for explanation, and units are not given for the variables. These equations are valid for the SI system (Chapter 1).

ZONE MODEL CONCEPT

The idea of the zone fire model came from observations of early room fire experiments. Figure 18.1a is a sketch of a room fire with a smoke plume rising above the fire, and there is a smoke layer under the ceiling. There is a gradual transition zone at the bottom of the smoke layer. The plume continues into the smoke layer and forms a ceiling jet under the ceiling. The temperature of the plume decreases with height above the fire, and the mass flow of the plume increases with height above the fire. Smoke forms a doorjet flowing though the doorway into the next room.

A zone fire model considers a fire compartment to consist of an upper smoke layer and a lower nonsmoke layer as shown in Figure 18.1b. In the zone model, the mass flows of the smoke plume and the doorjet are calculated from empirical equations. For the zone model idealization, the temperature and concentrations of constituents are considered to be constant throughout each layer. These properties change only as a function of time.

Most zone models consider that ceilings are flat and that rooms have uniform cross-sectional areas. The height of the discontinuity between these layers is considered to be the same everywhere. This discontinuity is called the smoke layer interface. In the idealized model

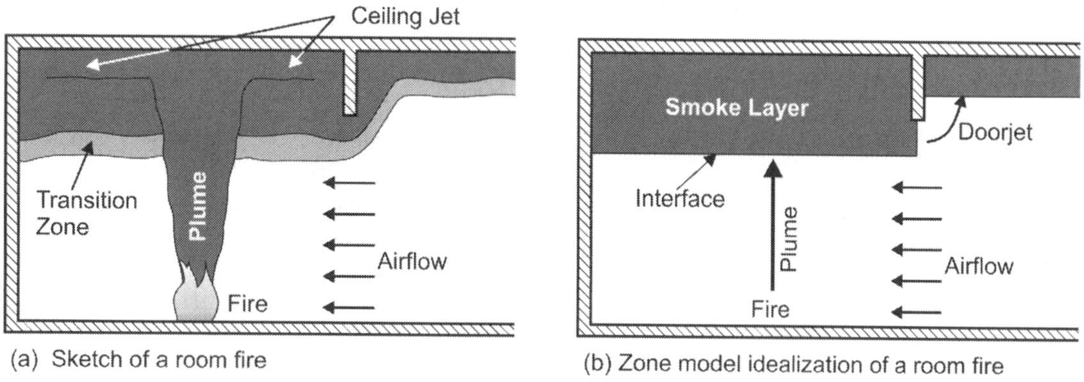

Figure 18.1 *Sketch of room fire and an idealized zone model representation.*

at an infinitesimal distance above the interface, the temperature and contaminant concentrations are those of the smoke layer. At an infinitesimal distance below the interface, the temperature and contaminant concentrations are those of the lower layer. Even with these simplifications, zone fire models have proven to be very useful tools for many applications, but they need to be used with care.

The following zone fire models were developed at the National Institute of Standards (NIST): ASET (Cooper 1985), ASET-B (Walton 1985), CCFM (Cooper and Forney 1990), CFAST (Jones et al. 2009; Peacock, R. D. et al. 2008a), LAVENT (Davis and Cooper 1989), and JET (Davis 1999). The Harvard Code (Mitler and Emmons 1981) was initially developed at Harvard with later work done at NIST. These models are available from NIST at no cost. NIST's flagship zone fire model is CFAST, which is a multiroom model with many features including heat transfer to walls and forced ventilation. The Tanaka model was developed at the Building Research Institute in Japan (Tanaka 1983). MAGIC was developed in France specifically for nuclear power plant applications (Gay 2005).

Early zone fire models were quasi-steady models based on systems of algebraic equations, and later models were based on a differential equation approach. Both of these approaches are discussed later.

Sprinkler Actuation

Actuation of thermal devices such as sprinklers and heat detectors depends on: (1) the response time index (RTI) of the device, (2) the gas temperature, and (3) the gas velocity. The RTI is discussed in Chapter 5. These devices are located in the ceiling jet, which is hotter than the temperature of the smoke layer calculated by a zone fire model. Empirical equations can be used to calculate the gas temperature and velocity of the ceiling jet. Some zone fire models that are capable of predicting actuation times are CFAST, LAVENT, and JET. DETACT-QS and DETACT-T2 are computer programs specifically developed to predict actuation times of thermal devices like sprinklers (Evans and Stroup 1986; Evans, Stroup and Martin 1986).

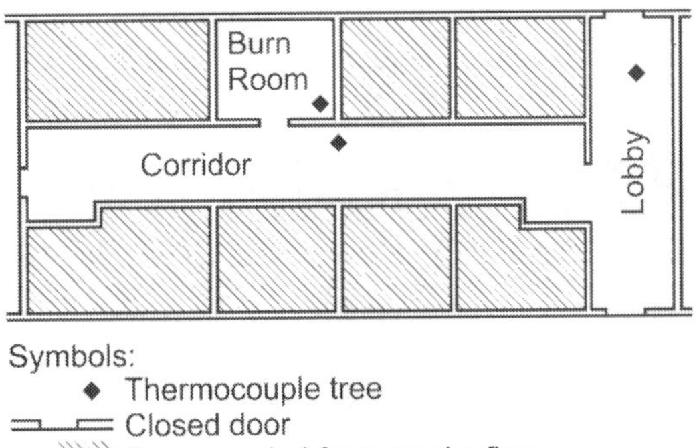

Figure 18.2 *Facility for NIST zone model validation experiments.*

Model Evaluation

Full scale fire experiments and validation studies of CFAST and the Harvard code were conducted at NIST (Mitler and Rockett 1986; Peacock et al. 1988, 1991, 1993). CFAST, MAGIC and the computational fluid dynamic model, FDS, were verified in a study of fire modeling for nuclear power plant applications conducted by the Nuclear Regulatory Commission and the Electric Power Research Institute (NRC 2007).

Figure 18.2 is a plan of the test facility used by at NIST to verify the Harvard Code (Rocket et al. 1987). This facility consisted of a burn room, corridor, and lobby, and each space had a thermocouple tree which is a vertical group of thermocouples. For a 100 kW fire, Figure 18.3 shows the measured and calculated temperature rise at 200 s after ignition. From this figure, it can be seen that in the burn room, the measured temperature rise has an appearance similar to that calculated by the zone model. This is not so for the corridor and the lobby. This kind of agreement is typical of zone fire models.

ALGEBRAIC EQUATION APPROACH

Many zone fire models such as the Harvard Code and the Tanaka model use a sophisticated formulation that sometimes fails to converge to a solution. An explicit method is described here that is easier to understand and less prone to convergence failure.

This explicit method uses the calculations of the previous time step to make calculations for the current time step. If the time steps are too large, the simulation will not be accurate. The size of the time interval needed depends on the application, and it can be evaluated by a sensitivity analysis. For atrium smoke filling applications, a time interval of 0.05 s results in accurate simulations over a wide range of atrium sizes.

The method described in this section is only for one room, and it can be used for atrium smoke control simulations. The change in mass of the smoke layer can be expressed as

$$\Delta m = (\dot{m}_p - \dot{m}_e)\Delta t \tag{18.1}$$

where

Δm = change in mass of smoke layer during time interval,

Δt = time interval,

$\dot{m}_p$ = mass flow of plume into smoke layer,

$\dot{m}_e$ = mass flow of smoke exhaust.

The mass flow of the plume is discussed later. Typically, the volumetric flow of smoke exhaust is constant. The exhaust is expressed as $m_e = \rho \dot{V}$ where ρ is density and $\dot{V}$ is the volumetric flow. The density can be calculated from the ideal gas law.

The mass of smoke in the layer at the end of the interval can be expressed as

$$m_2 = m_1 + \Delta m \quad (18.2)$$

where

m_1 = mass of smoke layer at beginning of interval,

m_2 = mass of smoke layer at end of interval.

The change in energy of the smoke layer can be expressed as

$$\Delta E = c_p[\dot{m}_p T_p - \dot{m}_e T_{s1} - \eta \dot{m}_p (T_p - T_o)]\Delta t \quad (18.3)$$

where

ΔE = change in energy of smoke layer,

c_p = specific heat of smoke,

η = wall heat transfer fraction,

T_p = absolute temperature of plume gases entering smoke layer,

T_{s1} = absolute temperature of smoke layer gases at beginning of time interval,

T_o = absolute ambient temperature.

The energy in the smoke layer is

$$E_2 = E_1 + \Delta E \quad (18.4)$$

where

E_1 = energy of smoke layer at beginning of interval,

E_2 = energy of the smoke layer at end of interval.

The smoke temperature, T_{s2}, at the end of the time interval is

$$T_{s2} = \frac{E_2}{c_p m_2} . \quad (18.5)$$

The density of the smoke layer at the end of the time interval is

$$\rho_{s2} = \frac{p_o}{RT_{s2}} \quad (18.6)$$

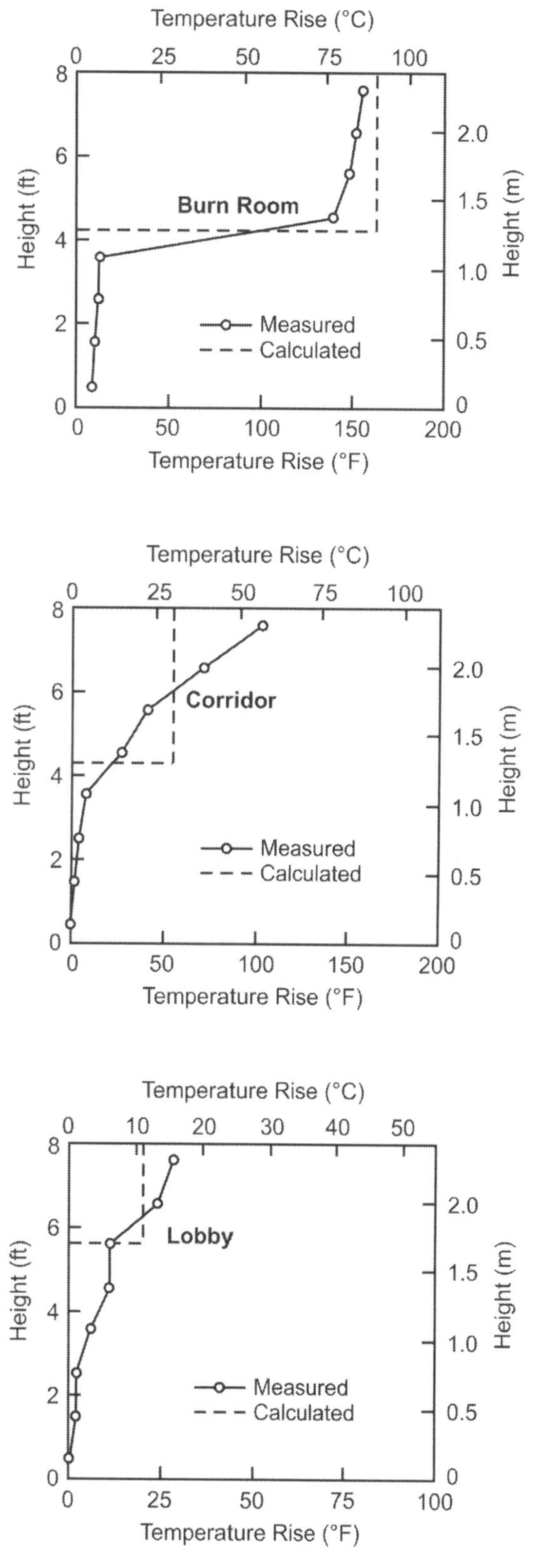

Figure 18.3 *Temperature profiles for 100 kW fire at 200 s after ignition from NIST zone model validation study.*

where

ρ_{s2} = smoke density at end of interval,

p_o = ambient pressure,

R = gas constant.

The volume V_2 of the smoke layer at the end of the time interval is

$$V_2 = \frac{m_2}{\rho_{s2}}. \quad (18.7)$$

The height of the smoke layer at the end of the time interval is

$$z_2 = H - H_{\text{fuel}} - \frac{V_2}{A} \quad (18.8)$$

where

z_2 = height of smoke layer above base of fire,

H = height of atrium,

H_{fuel} = height of base of fire,

A = cross-section area of atrium.

Plume Flow

The mass flow of the plume can be calculated for an axisymmetric plume, a balcony spill plume, or a window plume. In addition to the plume equations in Chapter 16, a number of other plume equations can be used (Beyler 1986). Many plume equations have discontinuities that can cause convergence failures in computer programs. Sometimes computer programs can be used for a while before such failures happen. A common approach to preventing such failures is to slightly adjust a coefficient to eliminate the discontinuity without adversely impacting the engineering usefulness of mass flow predictions. This approach is used in CFAST, and it is described here for a simplified axisymmetric plume called the Heskestad plume. To be consistent with the rest of this chapter, the notation used here is different from that used in Chapter 16, but the plume mass flow equations are the same.

The convective portion of the heat release rate is

$$\dot{Q}_c = \chi_c \dot{Q} \quad (18.9)$$

where

$\dot{Q}_c$ = convective heat release rate of fire,

$\dot{Q}$ = heat release rate of the fire,

χ_c = convective fraction of heat release.

The program can use a specific value for the convective fraction of heat release, or the commonly used value of 0.7 can be used.

The limiting elevation is

$$z_1 = 0.166\dot{Q}_c^{2/5} \quad (18.10)$$

where

z_l = limiting elevation,

$\dot{Q}_c$ = convective heat release rate.

For $z \geq z_l$, the mass flow is

$$\dot{m}_p = 0.071\dot{Q}_c^{1/3}z^{5/3} + 0.0018\dot{Q}_c \quad (18.11)$$

and for $z < z_l$, the mass flow is

$$\dot{m}_p = C\dot{Q}_c^{3/5}z \quad (18.12)$$

where

$\dot{m}_p$ = mass flow in axisymmetric plume at height z,

$\dot{Q}_c$ = convective heat release rate of fire,

z = distance above base of fire,

C = coefficient to be adjusted.

To eliminate the discontinuity, Equations 18.11 and 18.12 need to predict the same mass flow at $z = z_l$. The limiting elevation z_l is put into Equations 18.11 and 18.12. The equations are equated and rearranged to yield

$$C = \frac{0.071\dot{Q}_c^{1/3}z_1^{5/3} + 0.0018\dot{Q}_c}{\dot{Q}_c^{3/5}z_1} \quad (18.13)$$

The limiting elevation z_l and coefficient C need to be recalculated every time there is a change in the value of $\dot{Q}_c$.

DIFFERENTIAL EQUATION APPROACH

The differential equation approach is theoretically more exact than the algebraic equation method, and it is used in many later zone fire models including CCFM, CFAST, LAVENT, and JET. The following is a general mathematical development for a zone fire model, and specific zone models differ in some respects.

The upper and lower layers of a room in a zone fire model are considered control volumes as shown in Figure 18.4. The approach used with zone modeling is to write the conservation of mass and energy equations for the upper and lower layers. Conservation of mass for the upper layer is

$$\frac{dm_u}{dt} = \dot{m}_{u,in}\dot{m}_{u,out} \quad (18.14)$$

where

m_u = mass in upper layer,

$\dot{m}_{u,in}$ = mass flow rate into upper layer,

$\dot{m}_{u,out}$ = mass flow rate out of upper layer.

The mass flow rates in this equation depend on the specific computer model. For complex multiroom zone models, $\dot{m}_{u,in}$ is the sum of all mass flows into the upper layer (plume, doorjet from another room, HVAC flow, etc.), and $\dot{m}_{u,out}$ is the sum of all mass flows out of the upper layer (doorjet from another room, HVAC flow, etc.).

The conservation of energy equation is the first law of thermodynamics. Because changes in potential energy and kinetic energy are negligible, the conservation of energy equation for the upper layer is

$$\dot{Q}_u + \sum \dot{m}_{u,in} h_{u,in} = \dot{W} + \sum \dot{m}_{u,out} h_{u,out} + \frac{d(m_u e_u)}{dt} \quad (18.15)$$

where

$\dot{Q}$ = heat transferred to upper layer,

$\dot{W}$ = work done by smoke layer on surroundings,

$h_{u,in}$ = enthalpy of mass flow into upper layer,

$h_{u,out}$ = enthalpy of mass flow out of upper layer,

e_u = internal energy of upper layer.

The heat transfer term $\dot{Q}_u$ should not be confused with the heat release rate of a fire. In thermodynamics, the heat transfer term is the thermal energy that flows into the upper layer due to a temperature difference. The upper layer is relatively hot so that the heat transfer generally is from the upper layer such has heat conduction and thermal radiation from hot smoke to the walls. This means that the term $\dot{Q}_u$ is generally negative.

The thermodynamic definition of work is the product of a force F acting through a displacement l (in differential form work is $dW = F\,dl$). The displacement for the upper layer is the moving smoke interface which is also the surface of the control volume. The force is the product of absolute pressure at the interface and the area of this surface ($F = pA_u$). The work term is

$$\dot{W} = \frac{dW}{dt} = pA\frac{dl}{dt} = p\frac{dV_u}{dt} \quad (18.16)$$

Most zone fire models consider the gases as ideal gases, which behave as

$$pV = mRT \quad (18.17)$$

where

p = absolute pressure,

V = volume of gas,

m = mass of gas,

R = gas constant,

T = absolute temperature of gas.

For an ideal gas, the gas constant is

$$R = c_p - c_v \quad (18.18)$$

where

c_p = constant-pressure specific heat,

c_v = constant-volume specific heat.

The enthalpy of an ideal gas is

$$h = c_p T \quad (18.19)$$

The internal energy of an ideal gas is

$$e = c_v T \quad (18.20)$$

The ratio of specific heats γ is

$$\gamma = \frac{c_p}{c_v} \quad (18.21)$$

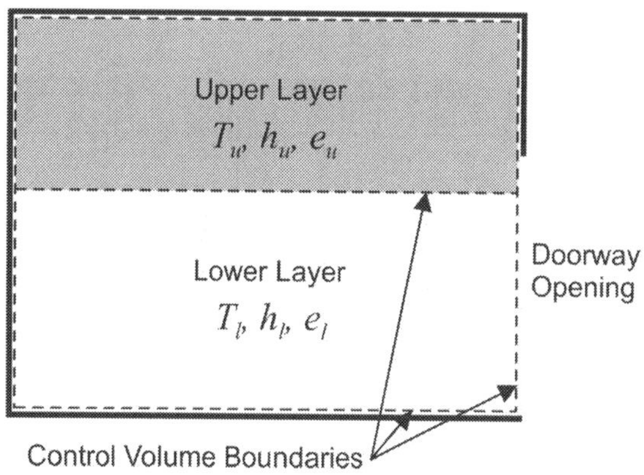

Figure 18.4 *Control volumes for upper and lower layers of zone fire model.*

For an ideal gas, c_p, c_v, R, and γ are constants. The time derivative of Equation 18.17 is

$$p\frac{dV}{dt} + V\frac{dp}{dt} = RT\frac{dm}{dt} + mR\frac{dT}{dt} \tag{18.22}$$

Substituting this into Equation 18.16 results in

$$\dot{W} = RT_u\frac{dm_u}{dt} + m_uR\frac{dT_u}{dt} - V_u\frac{dp}{dt}. \tag{18.23}$$

Combining Equations 18.15, 18.18, 18.19, 18.20, and 18.23 yields

$$T_u\frac{dm_u}{dt} + m_u\frac{dT_u}{dt} - \frac{V_u}{c_p}\frac{dp}{dt} = \frac{\dot{Q}_u}{c_p} + \dot{m}_{u,in}T_{u,in} - \dot{m}_{u,out}T_{u,out}. \tag{18.24}$$

This is a form of the conservation of energy equation for the upper layer. The following conservation of mass and energy equations for the lower layer can be developed in a similar manner:

$$\frac{dm_l}{dt} = \dot{m}_{l,in} - \dot{m}_{l,out} \tag{18.25}$$

and

$$T_l\frac{dm_l}{dt} + m_l\frac{dT_l}{dt} - \frac{V_l}{c_p}\frac{dp}{dt} = \frac{\dot{Q}_l}{c_p} + \dot{m}_{l,in}T_{l,in} - \dot{m}_{l,out}T_{l,out}. \tag{18.26}$$

The conservation equations can be rearranged as

$$\frac{dp}{dt} = \frac{\dot{s}}{(\beta - 1)V}, \tag{18.27}$$

$$\frac{dT_u}{dt} = \frac{1}{\beta}\left(\frac{T_u}{pV_u}\right)\left[\dot{E}_u + \frac{V_u}{(\beta-1)V}\dot{s}\right] \tag{18.28}$$

$$\frac{dT_l}{dt} = \frac{1}{\beta}\left(\frac{T_l}{pV_l}\right)\left[\dot{E}_l + \frac{V_l}{(\beta-1)V}\dot{s}\right] \tag{18.29}$$

$$\frac{dV_u}{dt} = \frac{1}{p\beta}\left(c_p\dot{m}_uT_u + \dot{E}_u - \frac{V_u}{V}\dot{s}\right) \tag{18.30}$$

where

$\dot{s}$ = $c_p\dot{m}_uT_u + c_p\dot{m}_lT_l + \dot{E}_u + \dot{E}_l$,

β = $c_p/R = \gamma/(\gamma - 1)$,

$\dot{E}_u$ = net energy release rate for upper layer,

$\dot{E}_l$ = net energy release rate for lower layer,

V = room volume ($V = V_u + V_l$).

Equations 18.27 to 18.30 were developed by Jones et al. (1984), and he provides a detailed description of the net energy release rate terms. To facilitate numerical solution, the equations solved in CFAST are a rearranged version of these equations.

CFAST

As already mentioned, CFAST is a multiroom model with many features. CFAST can be used to simulate many kinds of atrium smoke control systems. Because CFD models, especially Fire Dynamics Simulator (FDS), produce much more detailed and realistic simulations, it is suggested that CFD modeling be considered for atrium smoke control simulations. For information about CFD modeling, see Chapter 20. CFAST can be used in the development of design fires, and it is particularly useful in the estimation of sprinkler activation time.

The default plume equation in CFAST is the McCaffrey axisymmetric plume (McCaffrey 1984). Users can select the Heskestad axisymmetric plume, which is used in the algebraic equation method for analysis of steady atrium smoke exhaust (Chapter 16).

Because CFAST has many features that are not needed for smoke control, learning the model can be time consuming. This section consists of limited instructions in using CFAST to get users started with the program. Complete user information is in the CFAST User's Guide (Peacock 2008b).

The information here is for version 6 of this model, but this information should also be useful for later versions. CFAST can be downloaded from NIST at no cost. When CFAST is installed, Smokeview and CEdit are also installed. Smokeview can produce graphic output of CFAST simulations. CEdit is a Microsoft Windows program that can be used to generate text data files for CFAST and make CFAST simulations. The instructions here are for users who have some experience with Windows programs.

For projects that require large numbers of CFAST simulations, a text editor can be used to generate CFAST data files, and the simulations can be made efficiently in batch mode. New users should focus on using CFAST with CEdit, and only use the batch approach after gaining experience with CFAST. It is suggested that a new user start with an existing data file, run it as is, and then make the appropriate changes to the input file for the desired scenario.

Example Input File

New users are encouraged to look over the sample input file, "standard," before attempting to create an input file. This example simulates a gas burner igniting a wood wall in a large room with a doorway open to the outdoors. After CFAST installation, the input file, "standard," is located in the folder "Samples," which is in the folder "CFastData." To keep from accidentally modifying this input file, users should copy it to another folder such as the one named CFAST-runs before examining it.

Double clicking on the "standard" file in this folder should open the CEdit in the Simulation Environment, which should look like Figure 18.5. Clicking on the Run button at the bottom of the window will make CFAST run the example simulation, and the output window will appear. This window should look the same as the output window at the end of Chapter 2 of the User Manual. The example, "standard," uses many features of the model, and to this extent it verifies proper installation of CFAST.

The Simulation Environment (Figure 18.5) of this example shows the simulation times, ambient conditions, and exterior conditions. If CFAST has just been installed, the units in the Simulation Environment will be the default SI units. It can be seen that the Simulation Time is 1800 s, and this is the length of time that will be simulated. Another important time in the Simulation Environment is the spreadsheet output interval.

By clicking on the Compartment Geometry tab, the room dimensions can be seen. By clicking on the Horizontal Flow Vents tab, the dimensions of the open doorway can be seen. Clicking on the Vertical Flow Vents tab shows that there are not openings that would allow flow into or out of the room in the vertical direction. Clicking on the Mechanical Flow Vents tab shows that there is no forced airflow into or out of the room. Clicking on the Fire tab shows information about the fires. Clicking on the Detection/Suppression tab shows that

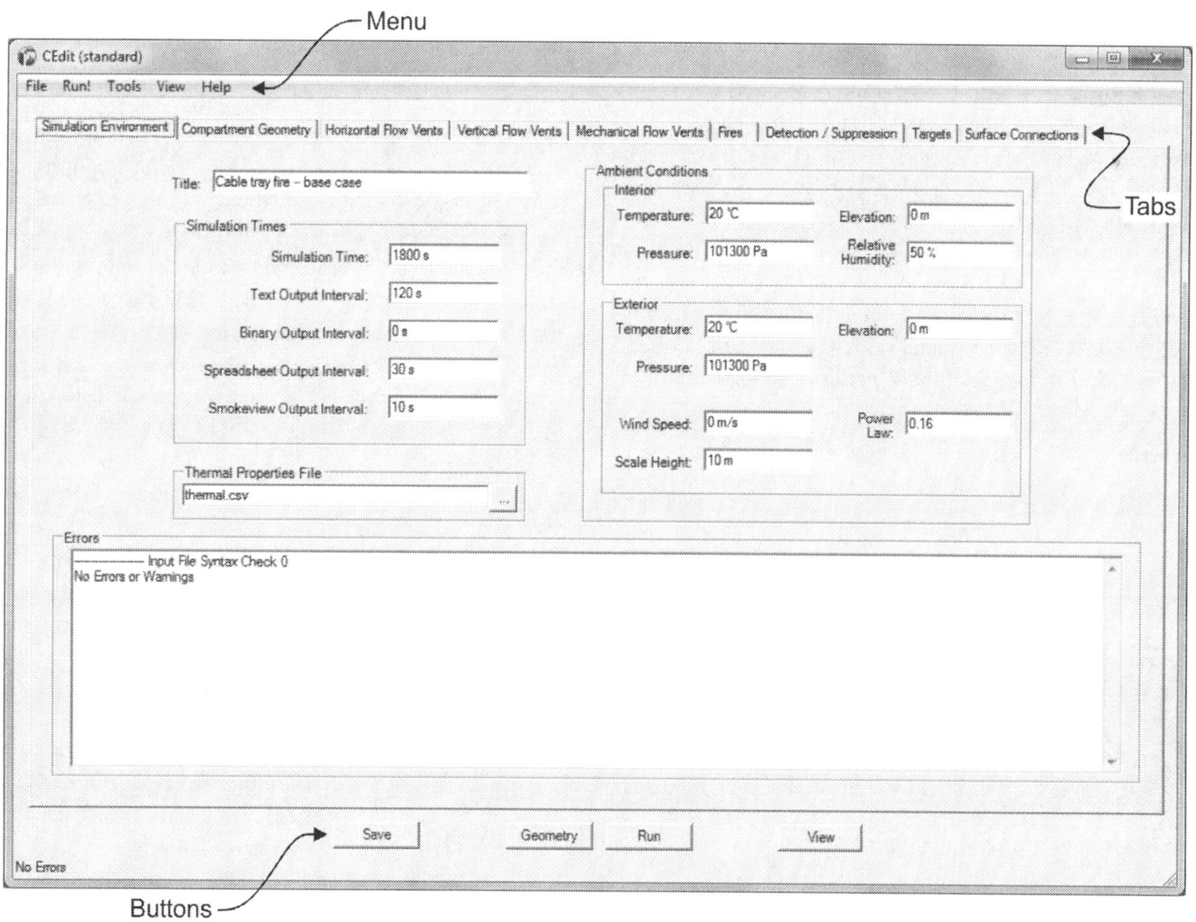

Figure 18.5 *CEdit window for sample input file,* standard, *open in Simulation Environment.*

the simulation does not include detectors or sprinklers. Clicking on the Targets tab shows information about a target for which the radiant flux will be included in the output. Clicking on the Surface Connections tab shows that this feature was not used.

Output Files

In the CFAST user manual, the term *filename* stands for any character string that helps to identify the simulation. The sample input file above the filename is "standard," and the output files use this file name. Examination of the folder from which the sample input file, "standard," was run shows many files with the prefix "standard."

For this example, there are four spreadsheet output files: "standard.n," "standard.w," "standard.s," and "standard.f." The primary spreadsheet is "standard.n." The file "standard.w" has data about temperatures and fluxes at walls and other surfaces. The file "standard.s" has data about gas (O_2, CO_2, CO, etc.) concentrations and other tenability variables. The file "standard.f" has data about flows through vents and between layers in the compartment.

Opening the spreadsheet shows computer generated data for time from 0 to 1800 s at intervals of 30 s. The simulation time and the spreadsheet output interval were defined in the Simulation Window (Figure 18.5). The spreadsheet files capture a snapshot of the modeling data at an instant of time. The data in this spreadsheet is about layer temperatures, layer height, pressure, pyrolysis, and fire size. Pyrolysis is the time rate of mass loss of a fire, and the fire size is the heat release rate of the fire. The units for this data are not included in the spreadsheet. For commonly used variables, the units are listed in Table 18.1.

Menus

There is a line of pulldown menus that should look familiar to Windows users. The File menu is similar to most Windows programs in that it lets users start, open and save projects. The Run! menu allows users to create a geometry file, model a CFAST simulation, make a Smokeview visualization, and select an option for the output spreadsheet files. As already mentioned, a CFAST simulation can also be made by clicking on the Run button at the bottom of the CEdit window.

The Tools menu allows users to edit thermal properties, edit fire objects, select engineering units, and set a maximum simulation time. Selection of engineering units is useful to users who want input data to be in IP units such as feet, pounds, degrees Fahrenheit, and inches of water. These engineering units are only for CEdit, and they do not change the units of the output spreadsheets. The View menu allows users to see three CFAST files, and this may be useful to advanced users. As with many other Windows programs, the Help menu provides help features.

Table 18.1: CFAST Spreadsheet Units

Variable	Units
Time	s
Temperature	°C
Layer height	m
Pressure	Pa
Pyrolysis	kg/s
Fire size (HRR)	W

Note: For variables not listed here, see the CFAST User Manual.

Fires

Figure 18.6 shows the CEdit window with the Fire tab open. It is possible to have a number of fires in various compartments. At the top of this window there is a table that lists the current fires. A fire can be removed from the list by clicking the Remove button, and a fire can be duplicated by clicking the Duplicate button.

A fire can be added by clicking the Add button, and a fire can be selected from already defined fires from a list of Fire Objects. CFAST comes with a number of these predefined fires including a kiosk, sofa, and upholstered chair. Clicking the nearby Edit button will open the Fire Objects window (Figure 18.7).

From the Fire Objects window, existing fires can be removed and edited, but it is recommended not to edit the fires that come installed with CFAST. T-squared fires are frequently used, and these can be added by clicking the Add t^2 button which makes the Define New Fire window appear (Figure 18.8).

In the Define New Fire, the user can choose fire growth rates of custom, slow, medium, fast, and ultra fast. As can be seen from Figure 18.8, these fires consist of a growth stage, a constant HRR stage, and a decay stage.

Do Not Use CFAST to Calculate Sprinkler Activation Times of Shielded Fires: CFAST uses axi-symmetric plumes, but the plumes of shielded fires are very different from what CFAST simulates. For more information, see Chapter 5.

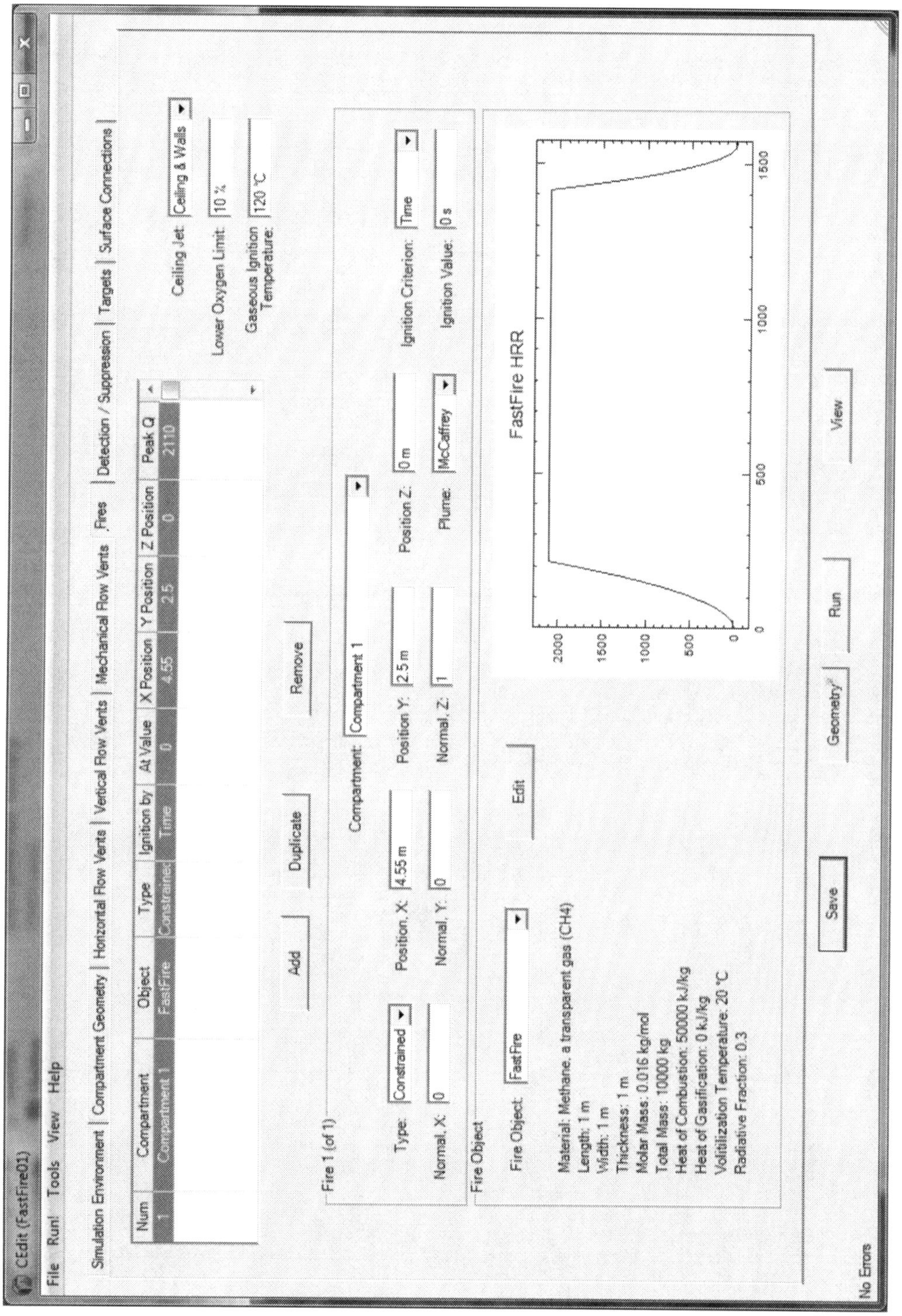

Figure 18.6 *CEdit window with Fire tab open.*

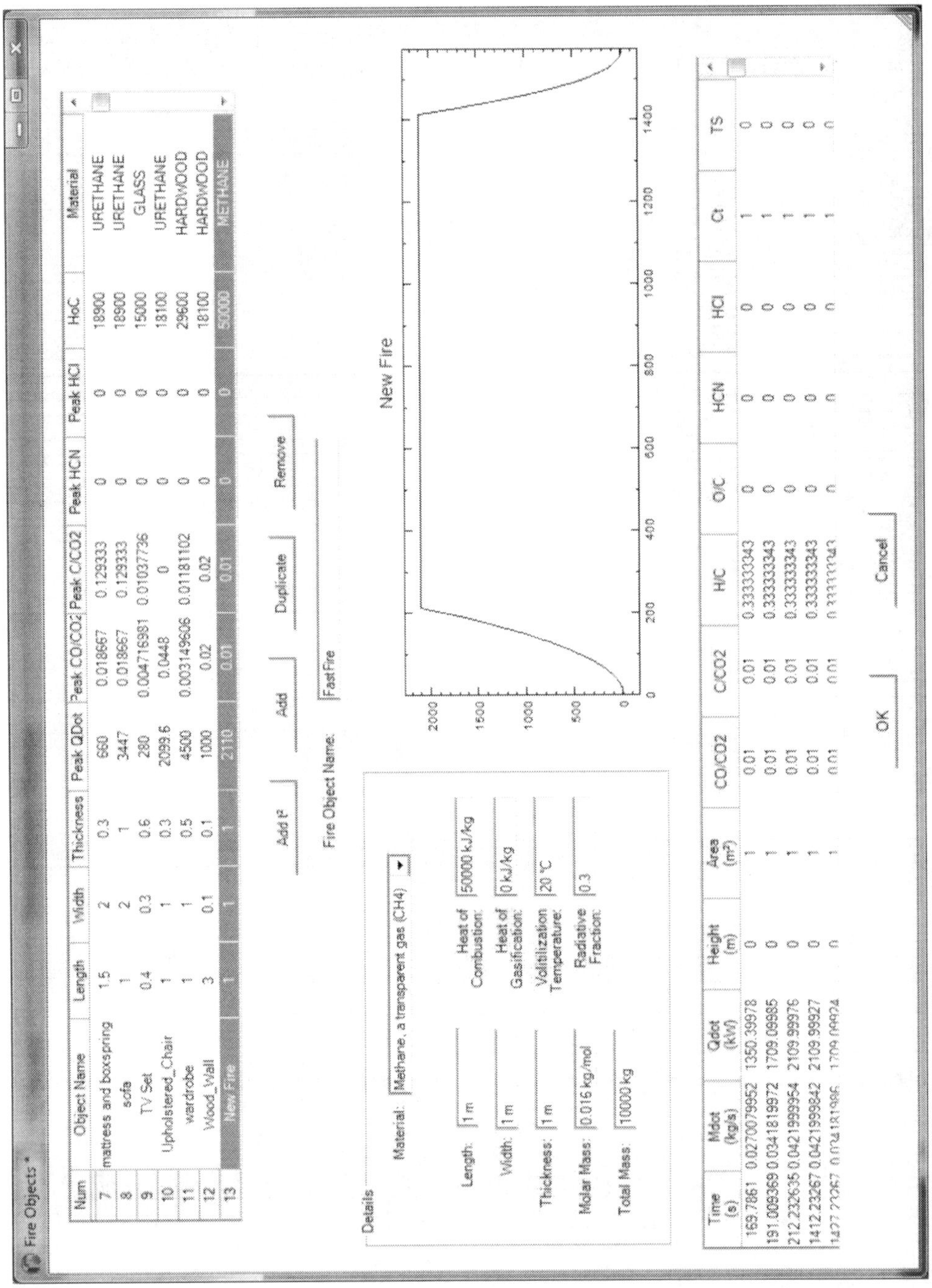

Figure 18.7 *Fire Objects Window for Example 18.1.*

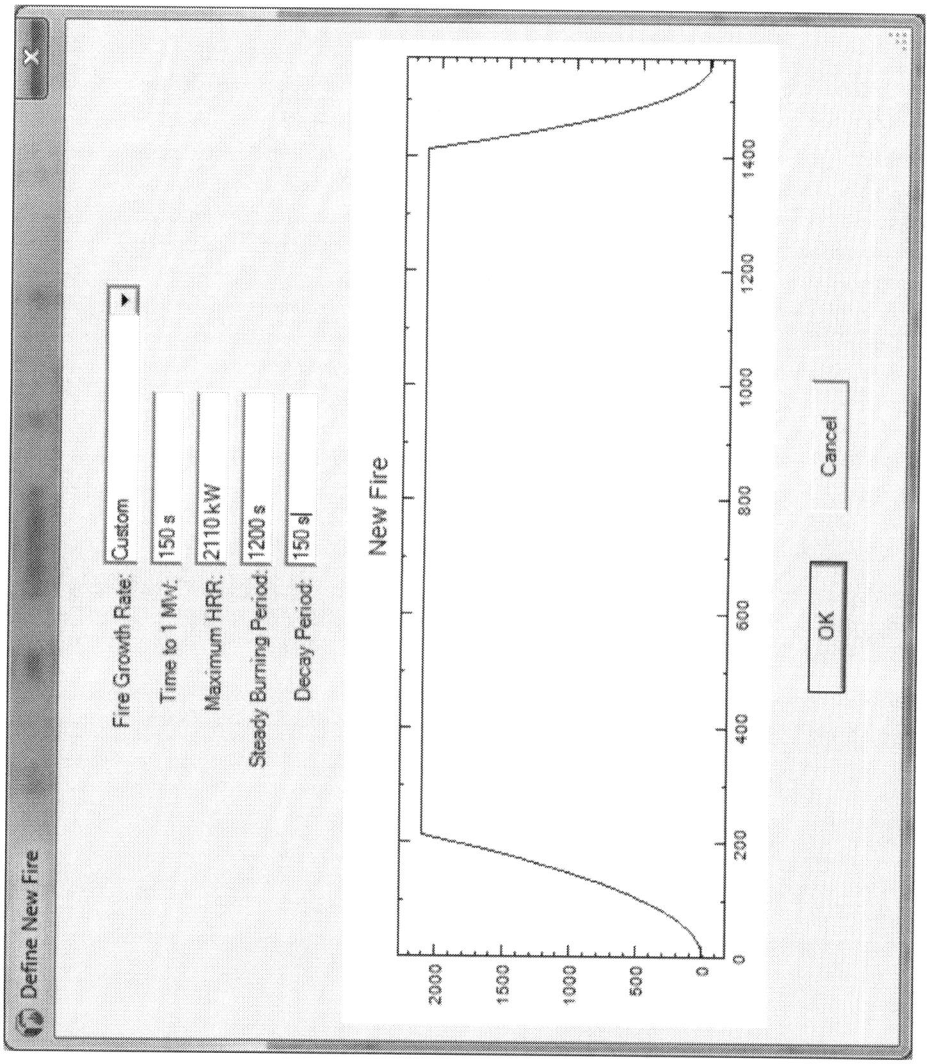

Figure 18.8 *Define New Fire window in CEdit.*

Examples

For those who want to learn to use CFAST, it is recommended that they read the following and do the examples with their computer. Example 18.1 shows how to define a t-squared fire, and it compares the impact of using the McCaffrey and Heskestad plumes. Example 18.2 shows how to use sprinklers. This example shows that the McCaffrey and Heskestad plumes have little impact on sprinkler activation time.

Example 18.1. Fast T-Squared Fire

Part 1: Make a CFAST simulation with a fire with a fast t-squared growth rate up to 2000 Btu/s (2110 kW), and which remains constant for 1200 s. The fire is located at the floor in the center of the room, and the plume is the McCaffrey axisymmetric plume, which is the CFAST default plume.

Open the example input file, "standard", and save it as FastFire01. Enter Fast Fire as the title. Click on the Fire tab, and remove the two existing fires. Click on Add to add another fire. Click on Edit, and the Fire Objects window pops up. Click the button Add t^2, and the Define New Fire window pops up. For fire growth rate, select Fast. For maximum HRR, enter 2110 kW. For steady burning period, enter 1200 s. The Define New Fire window should look like Figure 18.8. Click OK.

For fire object name, enter FastFire. The fire objects window should look like Figure 18.7. It can be seen that at about 212 s and 1412 s, the HRR is about 2110 kW. Click OK. Under fire Object, select FastFire. Select Compartment 1. Locate the fire in the center of the room as follows. For position X, enter 4.55 m. For Position Y, enter 2.5 m. For position Z, enter 0 m. The window should look like Figure 18.6. Click the Run button at the bottom of the window.

Part 2: Make the same simulation as above, except with the Heskestad plume.

Open FastFire01, and save as FastFire02. Click the Fire tab, and for the plume select Heskestad. Click the Run button at the bottom of the window.

Part 3: Compare the results of Parts 1 and 2 above.

Much of the simulated data from Parts 1 and 2 are in spreadsheets, FastFire01.n and FastFire02.n. By opening these spreadsheets, the data can be observed and plotted. It should be noted that the time interval is 30 s, and this means that the HRRs at 212 s and 1412 s are not in the spreadsheets. If we wanted these values included in the spreadsheets, smaller intervals would be needed.

Figure 18.9a shows the layer heights, and it can be seen that Heskestad plume results in a lower smoke layer than the McCaffrey plume. The maximum difference between curves is about 1.1 ft (0.32 m) which is about 30% of the smoke layer depth. This means that the Heskestad plume tends to be more conservative for evaluation of smoke layer heights.

Figure 18.9b shows the upper layer (smoke layer) temperatures, and it can be seen that these temperatures are relatively close together. The maximum difference between these curves in 22°F (12°C), which is about 3% of the maximum temperature rise.

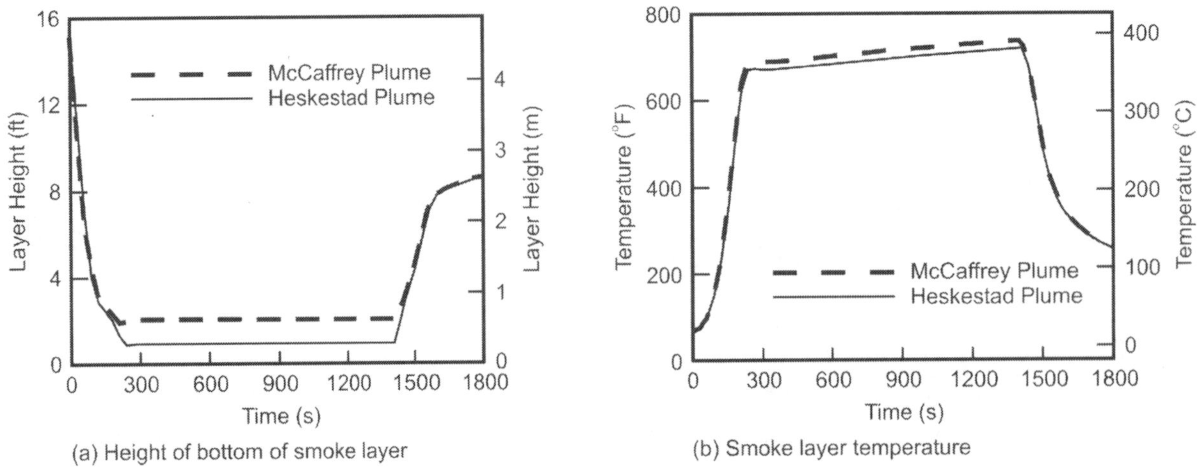

Figure 18.9 *Comparison of plume types from Example 18.1.*

Example 18.2. Sprinklered Fire

For the room of the example input file, use a fast t^2 fire, and add a sprinkler. The sprinkler is located at the ceiling 7 ft (2.13 m) horizontally from the fire. This means that the sprinkler has an X position of 21.9 ft (6.68 m), a Y position of 8.2 ft (2.5 m), and a Z position of 15.1 ft (4.6 m).

The sprinkler has an activation temperature of 165°F (73.9°C), an RTI of 145 $ft^{1/2}\ s^{1/2}$ (80 $m^{1/2}\ s^{1/2}$), and a spray density of 2.29×10^{-4} ft/s (7×10^{-5} m/s).

Part 1: Use the McCaffrey plume. Open FastFire01, and save it as Sprink01. Change the title to Fast Fire Sprinklered. Change the simulation time to 300 s. Change the text output interval to 30 s. Change the spreadsheet output interval to 1 s. Click the Detection/ Suppression tab. Click the Add button. For type select Sprinkler. For compartment select Compartment 1. For activation temperature, enter 73.9°C. For RTI enter 80 $m^{1/2}\ s^{1/2}$. For spray density, enter $7\text{x}10^{-5}$ m/s. Enter an X position of 21.9 ft (6.68 m). Enter a Y position of 8.2 ft (2.5 m). Enter Z position of 15.1 ft (4.6 m). Click on Save. Click on the run button.

Open spreadsheet Sprink01.w, and look at the column named "Sensor 1 is a Heat detector activated." Initially all the values are zero, and they change to ones at 205 s. This means that the sprinkler activation time is 205 s or 3 min and 25 s.

Part 2: Make the same simulation as above except with the Heskestad plume.

Open Sprink01, and save as Sprink02. Click the Fire tab, and for the plume select Heskestad. Click the Run button at the bottom of the window.

Open spreadsheet Sprink02.w, and as above look for the sprinkler activation time, which is 206 s or 3 min and 26 s. This shows that the choice of McCaffrey or Heskestad plume has little impact on sprinkler activation time.

NOMENCLATURE

A = cross-section area of atrium
C = coefficient to be adjusted
C_p = constant-pressure specific heat, or specific heat of smoke
C_v = constant-volume specific heat
E_1 = energy of smoke layer at beginning of interval
E_2 = energy of smoke layer at end of interval
e_u = internal energy of upper layer
$\dot{E}_l$ = net energy release rate for lower layer
$\dot{E}_u$ = net energy release rate for upper layer
H = height of atrium
H_{fuel} = height of the base of fire
$h_{u,in}$ = enthalpy of mass flow into upper layer
$h_{u,out}$ = enthalpy of mass flow out of upper layer
m = mass of gas
m_1 = mass of smoke layer at beginning of interval
m_2 = mass of smoke layer at end of interval
$\dot{m}_p$ = mass flow of plume
$\dot{m}_e$ = mass flow of smoke exhaust
m_u = mass in upper layer
$\dot{m}_{u,\,in}$ = mass flow rate into upper layer
$\dot{m}_{u,\,out}$ = mass flow rate out of upper layer
p = absolute pressure
p_o = ambient pressure
$\dot{Q}$ = heat release rate of fire
$\dot{Q}_c$ = convective heat release rate
$\dot{Q}_u$ = heat transferred to upper layer
R = gas constant
T = absolute temperature of gas
T_o = absolute ambient temperature
T_p = absolute temperature of plume gases entering smoke layer
T_{s1} = absolute temperature of smoke layer gases at beginning of time interval
V = volume
$\dot{W}$ = work done by smoke layer on surroundings
z = distance above base of fire
z_2 = height of smoke layer above base of fire
z_l = limiting elevation
η = wall heat transfer fraction

ΔE= change in energy of smoke layer

Δm = change in mass of smoke layer during time interval
ρ_{o2} = smoke density at end of interval
Δt = time interval
χ_c = convective fraction of heat release

REFERENCES

Beyler, C.L. 1986. Fire plumes and ceiling jets. *Fire Safety Journal* 11:53–75.

Cooper, L.Y. 1985. ASET—A computer program for calculating available safe egress time. *Fire Safety Journal* 9.

Cooper, L.Y. and G.P Forney.1990. The Consolidated Compartment Fire Model (CCFM) Computer Code Application CCFM. VENTS—Part I: Physical Basis. NISTIR 90-4342, National Institute of Standards and Technology, Gaithersburg, MD.

Davis, W.D. 1999. The zone model JET: A model for the prediction of detector activation and gas temperature in the presence of a smoke layer. NISTIR 6324, National Institute of Standards and Technology, Gaithersburg, MD.

Davis, W.D. and L.Y. Cooper. 1989. Estimating the environment and the sprinkler links in compartment fires with draft curtains and fusible-link-actuated ceiling vents—Part II: User guide for the computer code LAVENT. NISTIR 89-4122, National Institute of Standards and Technology, Gaithersburg, MD.

Evans, D.D., and D.W. Stroup. 1986. Methods to calculate the response time of heat and smoke detectors installed below large unobstructed ceilings. *Fire Technology* 22(1).

Evans, D.D., D.W. Stroup, and P. Martin. 1986. Evaluating thermal fire detection systems (SI Units), NBSSP 713, National Bureau of Standards, Gaithersburg, MD.

Gay, L. 2005. *User Guide of the MAGIC Software V4.1.1*, EdF HI82/04/23/A, Electricité de France, France.

Jones, W.W., et al. 1984. CFAST—Consolidated model of fire growth and smoke transport, technical reference guide. NIST Special Publication 1030, National Institute of Standards and Technology, Gaithersburg, MD.

Jones, W.W., et al. 2009. CFAST—Consolidated model of fire growth and smoke transport (version 6), technical reference guide. NIST Special Publication 1026, National Institute of Standards and Technology, Gaithersburg, MD.

McCaffrey, B.J. 1984. Fire plume dynamics. Conference on Large-Scale Fire Phenomenology, September 10–13, Gaithersburg, MD.

Mitler, H.E., and H.W. Emmons. 1981. Documentation for CFC V, the fifth Harvard computer code. Home Fire Project Tech. Rep. #45, Harvard University, Cambridge, MA.

Mitler, H.E., and J.A. Rockett. 1986. How accurate is mathematical fire modeling? NBSIR 86-3459, National Bureau of Standards, Gaithersburg, MD.

NRC. 2007. Verification and validation of selected fire models for nuclear power plant applications. Vols. 1–7, NUREG-1824, U.S. Nuclear Regulatory Commission, Washington DC.

Peacock, R.D., et al. 1988. Experimental data set for the accuracy assessment of room fire models. NBSIR 88-3752, National Bureau of Standards, Gaithersburg, MD.

Peacock, R.D., et al. 1991. Data for room fire model comparisons. *Journal of Research of the National Institute of Standards and Technology* 96(4).

Peacock, R.D., et al. 1993. Verification of a model of fire and smoke transport. *Fire Safety Journal* 21(2).

Peacock, R.D., et al. 2008a. CFAST—Consolidated model of fire growth and smoke transport (version 6)—software development and model evaluation guide. NIST Special Publication 1086, National Institute of Standards and Technology, Gaithersburg, MD.

Peacock, R.D., et al. 2008b. CFAST—Consolidated model of fire growth and smoke transport (version 6)—user's guide. NIST Special Publication 1041, National Institute of Standards and Technology, Gaithersburg, MD.

Rocket, et al. 1987. Comparison of NBS/Harvard VI simulations and full scale, multiroom fire test data. NBSIR 87-3567, National Bureau of Standards, Gaithersburg, MD.

Tanaka, T. 1983. A model of multiroom fire spread. NBSIR 83-2718, National Bureau of Standards, Gaithersburg, MD.

Walton, W.D., 1985. ASET-B: A room fire program for personal computers. NBSIR 85-3144-1, National Bureau of Standards, Gaithersburg, MD.

CHAPTER 19

Tenability Analysis and CONTAM

John H. Klote

Smoke is the major killer in building fires. Smoke can flow far from the fire, endangering life and damaging property. This chapter discusses the use of the model CONTAM to simulate smoke movement in multistory buildings and the resulting impact on tenability. CONTAM is especially good for simulating smoke flow far from the fire. The model is not very good at simulating conditions near the fire, and this limitation and a method of dealing with it are discussed later.

CONTAM is a network model that is extensively used for analyses of pressurization smoke control systems as discussed in Chapter 14. Users of the material in this chapter need to familiar with modeling of airflow in buildings with CONTAM and modeling fires in rooms with CFAST (Chapter 18). Users should also be familiar with a spreadsheet program such as Microsoft® Excel.

NEAR FIRE LIMITATION

In network models including CONTAM, spaces such as rooms and corridors are represented as zones where the temperature and contaminants are uniform. This means that the smoke layer in the fire space cannot be simulated by CONTAM. Also, the smoke layers in spaces open to the fire space cannot be simulated by CONTAM.

In an effort to overcome this limitation, research has been conducted on hybrid models that are combinations of network models and zone fire models (Floyd et al. 2005; Hadjisophocleous et al. 2011; Kashef et al. 2011). Zone fire models are discussed later. This research has not yet resulted in a generally accepted model appropriate for engineering applications.

As mentioned in Chapter 14, Version 3 of CONTAM is unique in that it is coupled with a computational fluid dynamic (CFD) model (Wang 2007; Wang, Dols, and Chen 2010). This model is called CFD0, and it solves the governing equations for incompressible flow using the Boussinesq approximation for buoyancy. For fire applications, the governing equations for compressible flow are generally used to appropriately deal with fire related buoyancy. As of the date of this publication, CFD0 has not been verified for any fire applications. CONTAM with CFD0 may be an indication of future trends in analysis of tenability systems; possibly CONTAM could be coupled with a CFD model more appropriate for fire applications such as FDS. For additional information about CFD, see Chapter 20.

As smoke moves away from the fire room, the smoke temperature drops due to heat transfer and mixing with ambient air. At some distance from the fire, the mixing can become such that considering a space to have a uniform temperatures and contaminants is appropriate. This is a basic idea of the approach discussed next.

THE TWO FIELD APPROACH

The two field approach is not a sophisticated modeling technique, but the rough results of this approach can be very useful. This approach uses CONTAM in combination with a zone fire model. The National Institute of Standards and Technology (NIST) conducted one of the first applications of this approach for a study of the concept of staging areas (Klote et al. 1992). This approach has been used for many design analyses of smoke control systems for large buildings, and examples of these are reported by Ferreira (1998, 2002).

A study funded by NIST used this approach to evaluate the hazards due to smoke movement through elevator shafts in office buildings (Klote 2004a). The smoke flow and the resulting hazard were analyzed for 27 scenarios in

5 buildings ranging from 5 to 58 stories. An ASHRAE-funded study also used this approach to evaluate the hazards due to stairwell doors inadvertently propped open during fires (Klote 2004b). In this study, the consequences of smoke transport were evaluated for 80 scenarios in two apartment buildings. One was 7 stories, and the other was 21 stories. A study at the National Research Council of Canada compared the results of smoke movement experiments to calculations using the two field approach (Hadjisophocleous, Fu, and Lougheed 2002).

In the two field approach, building fires are considered to consist of: (1) a near fire field where smoke buoyancy dominates and there is a well-defined smoke layer, and (2) a far field where mixing dominates and there not is a well-defined smoke layer. At the fire tests in the Plaza Hotel Building, videos of the floor above the fire floor showed that while there was much smoke flow through small holes in the floor, no well-defined smoke layer resulted (Klote 1990).

The two field approach consists of the following steps: (1) modeling the near field with a zone fire model, (2) adapting zone fire model results for CONTAM, and (3) modeling the near and far fields with CONTAM. These steps are discussed in the following text.

Zone Fire Modeling of the Near Field

In a zone fire model, rooms and other spaces are considered to be made up of an upper smoke layer and a lower nonsmoke layer. In a room fire, a smoke plume rises above the fire, and a smoke layer forms under the ceiling. The depth of the smoke layer changes with time. Zone fire models can simulate smoke flowing from the fire room to other spaces where other smoke layers form. For more information about zone fire models, see Chapter 18.

CFAST is a multiroom zone fire model with many features including heat transfer to walls and forced ventilation. CONTAM is a product of NIST, and it can be downloaded from the NIST Web site at no cost. There may be some applications for which a CFD model may be more appropriate for simulation of the near field conditions. However, for this chapter, it is considered that the near field conditions will be modeled by CFAST.

The spaces to be modeled with CFAST need to be chosen with care. The general rule is that the near field should consist of the fire room and spaces directly or indirectly open to the fire room. A space that is directly open to the fire room is a space that is separated from the burn room by a barrier with at least one large opening. Some examples of large openings are an open doorway, a large open archway, an open pass through, and a stair opening between floors. A space that is indirectly open to the fire room has other spaces between it and the burn room, and these other spaces can be directly or indirectly open to the burn room.

Figure 19.1 illustrates some near field situations where the only large openings are open doorways. Figure 19.1a is a floor of a condominium building

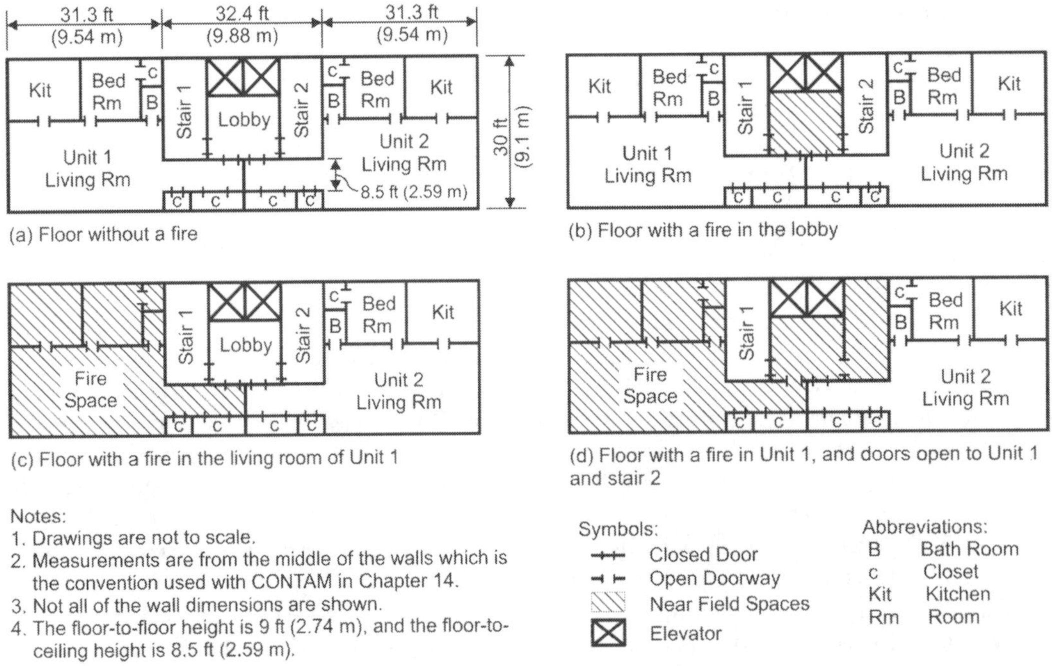

Figure 19.1 *Examples of near-field spaces.*

without a fire. In Figure 19.1b there is a fire in the lobby. Because all the lobby doors are closed, the near field can be taken as only the lobby. Elevator doors are typically leakier than other doors, and the elevator shaft could also be included in the near field spaces. If the elevator shaft is not included in the near field spaces, the elevator shaft temperature could be adjusted based on engineering judgment.

In Figure 19.1c there is a fire in the living room of unit 1. The kitchen, bed room, and bath room are directly open to the living room. The closet off the bed room is indirectly open to the fire room. The near field spaces consist of all of the direct and indirect spaces as shown in Figure 19.1c. Typically this near field space would be modeled with CFAST as one compartment. Alternatively, two compartments could be modeled with CFAST: (1) the living room and (2) all the other near field spaces lumped together.

Figure 19.1d is the same as Figure 19.1c except the doors to unit 1 and stair 2 are open. These doors would normally be closed, but sometimes doors are inadvertently propped open. The near field spaces consist of unit 1, the lobby, and stair 2. The near field spaces would be modeled with CFAST using separate compartments for unit 1, the lobby, and stair 2. Stair 2 could be modeled in CFAST as a number of vertical compartments connected together.

When very long corridors are part of the near field space, the temperature in the corridor decreases with distance from the fire, and this temperature decrease needs to be accounted for. Because the smoke layer and lower layer in CFAST have uniform temperatures, very long corridors should not be modeled in CFAST as a single compartment. One approach is to model the corridor as more than one compartment, and another approach is to use CFD modeling.

Another issue is when the general rule results in a very large number of rooms in the near field, some of which have more than one room between them and the fire room. The temperatures of rooms far away from the fire room may be low enough that they do not need to be modeled with CFAST. This needs to be evaluated individually using engineering judgment.

Example 19.1 describes CFAST simulations for a lobby fire with the near field shown in Figure 19.1b and a condominium fire with the near field shown in Figure 19.1c. Both of these simulations had a prescribed fire that grew as fast t-squared fires until they reached 2000 Btu/s (2110 kW). For the lobby fire, there was insufficient oxygen to support the prescribed

Example 19.1. CFAST Simulation of Near-Field Spaces

Part 1—Lobby Fire: For a lobby fire, the near field spaces are shown in Figure 19.1b. Make a CFAST simulation with a fast t-squared growth rate up to 2000 Btu/s (2110 kW) of Example 18.1. The inside dimensions of the lobby are 14.6 by 11.1 by 8.5 ft high (4.45 by 3.38 by 2.59 m high). Before the fire, the building temperature is 73°F (23°C). With the doors closed, the lobby has a leakage area of 2.11 ft^2 (0.196 m^2) consisting of construction cracks and gaps around the doors. This leakage will be accounted for by an opening of 8 ft high by 0.264 ft wide (2.44 m by 0.0805 m wide).

Because CFAST was run in SI units, the following steps are in SI units.

From Example 18.1, open project FastFire01, and save it as NearFld01. Change both the interior and exterior temperatures to 23°C. Relocate the fire inside the lobby at X = 1.30 m, Y = 1.71 m, and Z = 0 m. Edit the compartment geometry to width X = 4.45 m, depth Y = 3.38 m, and height Z = 2.59 m. Change the materials to a ceiling and floor of normal weight concrete and walls of gypsum wallboard. Change the dimensions of the horizontal flow connection to the following. Sill: 0 m, soffit: 2.44 m, and width: 0.0805 m. Press the Save button, and press the Run button.

The results of this simulation of interest for this application are in the spreadsheet file NearFld01.n. In this file, the upper and lower layer temperatures are in °C, the layer height is in m, and the fire size is the HRR in W. This simulated data was converted to I-P units and graphed as shown in Figure 19.2.

Part 2—Condominium Unit Fire: The near field spaces for this fire are shown in Figure 19.1c. Make a CFAST simulation with the fire used in Part 1 above. The duration of the simulation is 20 min. A CFAST compartment with the following dimensions has the same volume as the near field spaces: 30 by 34 by 8.5 ft (9.14 by 10.4 by 2.59 m high). Unit 1 has an open window 4.2 ft (1.28 m) wide by 4 ft (1.22 m) high. The outdoor temperature is –4°F (–20°C).

As with Part 1, the following steps are in SI units.

Open project NearFld01, and save it as NearFld02. Change the exterior temperatures to –20°C. Edit the compartment geometry to width X = 9.14 m, depth Y = 10.4 m, and height Z = 2.59 m. Relocate the fire to the center of the room which is X = 4.57 m, Y = 5.2 m, and Z = 0 m. Change the dimensions of the horizontal flow connection to the following. Sill: 0.914 m, soffit: 2.13 m, and width: 1.28 m. Press the Save button, and press the Run button.

Data of interest is in the spreadsheet file NearFld02.n. The results of this simulation are also shown in Figure 19.2.

fire. Based on available oxygen, CFAST simulated a smaller lobby fire with a peak of only about 850 Btu/s (900 kW) as shown in Figure 19.2a. The condominium unit had an open window through which oxygen was supplied to the fire. Thus CFAST was able to simulate the prescribed fire for the condominium unit as shown in Figure 19.2a. The layer temperatures and layer heights are shown in Figures 19.2b and c.

Adapting Zone Fire Model Results

To adapt the zone model results for use in CONTAM, average temperatures of the near filed spaces are used in CONTAM. The upper and lower layer temperatures from a zone fire model simulation are converted to an average temperature for use in CONTAM as

$$T_{av} = \frac{T_u(H-z)+T_l z}{H} \tag{19.1}$$

where

T_{av} = weighted average temperature, °F (°C),

T_u = upper layer temperature, °F (°C),

T_l = lower layer temperature, °F (°C),

H = compartment height, ft (m),

z = layer height, ft (m).

For the simulations described in Example 19.1, the average temperatures were calculated by a spreadsheet program, and plotted in Figure 19.3. Example 19.2 develops average temperature data that can be used in CONTAM.

Average temperatures calculated from Equation 19.1 are used as input to CONTAM so that CONTAM can calculate the flows in and out of the near field throughout the simulation. An alternate approach consists of using mass flows in and out of the near field based on a zone fire model simulation as input to CONTAM. This alternate approach is not discussed here.

Modeling with CONTAM

The building is modeled in CONTAM as discussed in Chapter 14, but the following additional tasks must be done: (1) average temperatures must be specified for the near-field spaces discussed previously, (2) two-way flow paths must be specified at some locations, (3) contaminant generation must be defined, and (4) output of contaminant concentrations must be specified.

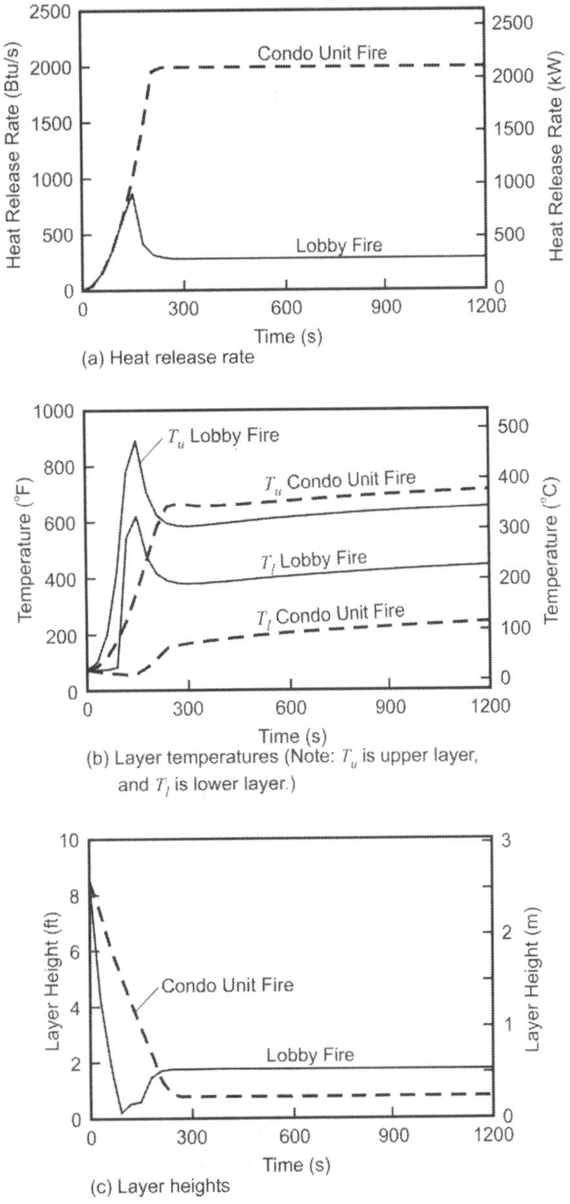

Figure 19.2 *CFAST results of Example 19.1.*

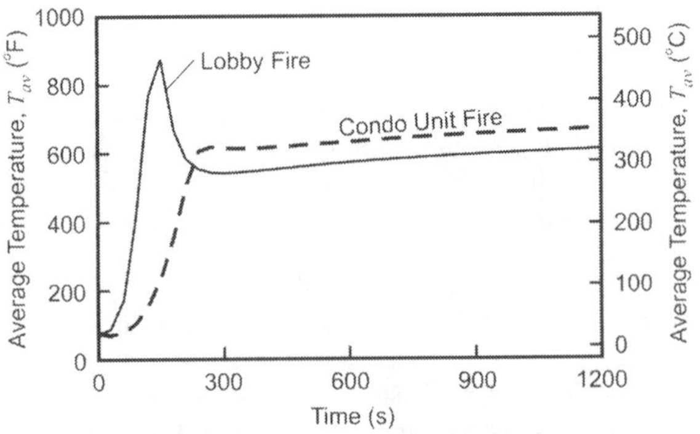

Figure 19.3 *Average temperatures of Example 19.1.*

TWO-WAY FLOW PATHS

For the CONTAM simulations of pressurization smoke control systems discussed in Chapter 14, bidirectional flow was not addressed. For the applications discussed in this chapter, bidirectional flow can be significant. In CONTAM, bidirectional flow is called two-way flow. The driving force of two-way flow is the temperature difference across the flow path. For detailed information about this kind of flow, see Chapter 3.

In CONTAM, a flow path is made bidirectional by choosing the One Opening, Two-way Flow Model when defining the element. For this two-way flow path, the height and width of the opening are specified.

In many applications, the buoyancy forces across flow paths are not clearly dominant, but the use of two-way flow paths is still recommended for the doors and open windows at the boundaries of the near field. Because the wall leakage is usually relatively small, these paths can be modeled as orifice flow paths. Because the flow at floor and roof leakage paths is one dimensional, these paths need to be modeled as orifice flow paths.

The doors in the near field are subjected to elevated temperatures, and the possibility of door warping needs to be considered. The extent of door warping depends on (1) the temperature of the gases near the door, (2) door materials and (3) door fabrication methods. However, there are limited data on this subject (Fire International 1968; Van Geyn 1994).

It is well known that door warping in fire situations can be significant, and door warping of single doors is the subject of Example 19.3. Door warping is also included in the tenability examples discussed later. Because of the large flow areas and complex flow paths associated with elevator doors, elevator door warping is not addressed in the tenability examples discussed later.

CONTAMINANT GENERATION AND FLOW

In CONTAM, the term *Species* is used for substances that can be used as contaminants during a simulation, and the species in a simulation are designated by the user. The flow of species can be handled as either trace or nontrace contaminants. Trace contaminants are those that exist at concentrations that do not cause a "significant" change in the density of air.

Nontrace contaminants are those that can affect the density of the air. With nontrace contaminants, it is possible to simulate the flow of all the relevant species (soot, oxygen, carbon dioxide, carbon monoxide, etc.) involved with a fire, and use the concentrations to evaluate tenability.

The approach described in this chapter is to simulate a single trace contaminant which is the mass density of fuel burned. From mass density of fuel burned, visibility and toxicity can be evaluated. The user defines the generation rate of the mass of fuel burned, and CONTAM calculates the mass density of fuel burned. The maximum generation rate of mass of fuel burned in a fire is

$$G_{max} = \frac{Q_{max}}{\Delta H_{ch}} \tag{19.2}$$

Example 19.2. Average Temperatures

Develop a set of average temperatures for the near field condominium unit fire of Part 2 of Example 19.1. The data in the spreadsheet file NearFld02.n and Equation 19.1 were used to calculate the average temperatures for the near field spaces as shown in Figure 19.3. This spreadsheet has data for every 30 s of both simulations, and this is more data than is needed for the CONTAM simulations.

The smaller set of average temperatures in the table below was developed using engineering judgment, and a different data set could have been selected. Because CONTAM needs the time in the hour, minute, second (hms) format, that is included here.

Condominium Unit Fire

Time, s	h	m	s	T_{av}, °F
0	00:	00:	00	73
90	00:	01:	30	90
240	00:	04:	00	610
1200	00:	20:	00	670

Example 19.3. Door Warping

Calculate door warping for the main door of the condominium in Part 2 of Example 19.1. For a door that warps proportional to the upper layer temperature rise to the fourth power, the door leakage area is $A = A_o + B(T_u - T_o)^4$ where A is the leakage of the warped door in ft² (m²), A_o is door leakage at ambient temperature, B is a constant, T_u is the upper layer temperature in °F (°C), and T_o is the ambient temperature in °F (°C). Use $A_o = 0.17$ ft² (0.0158 m²), and use $T_o = 73$°F (23°C). The door is considered to warp such it has an area of 1.25 ft² (0.116 m²) when the upper layer is 800°F (444°C).

For $A = 1.25$ ft² and $T_u = 800$°F, $B = (A - A_O)/(T_u - T_O)^4 = (1.25 - 0.17)/(800 - 73)^4 = 2.64\times10^{-12}$. A spreadsheet program is used with the upper layer temperatures from Part 2 of Example 19.1, and the area of the warped door is calculated as a function of time for the lobby fire. These values were examined and smaller sets of areas were selected for use in CONTAM.

From the table below, the maximum area is 0.614 ft² for the condominium door. In CONTAM, two-way flow paths need a height and width. For the maximum area, the height is 6.7 ft, and width of 0.092 ft. In CONTAM, the above factors also are needed in the hms format. A factor, F, is used by CONTAM to calculate the door area, and the door area and this function are listed below.

Warped Door of Condominium Unit Fire

Time, s	h	m	s	A, ft²	F
0	00:	00:	00	0.170	0.28
150	00:	02:	30	0.186	0.30
240	00:	04:	00	0.460	0.75
1200	00:	20:	00	0.614	1.00

where

G_{max} = maximum generation rate of fuel burned, lb/s (kg/s),

Q_{max} = maximum heat release rate, Btu/s (kW),

ΔH_{ch} = chemical heat of combustion Btu/lb (kJ/kg).

For values of the chemical heat of combustion, see Chapter 6. As the name implies, the maximum heat release rate is the largest HRR during a CONTAM simulation. In CONTAM, the user enters the value of G_{max}, and a schedule of factors F_i is defined such that the generation rate at time i is the product of the maximum generation rate and F_i

$$G_i = G_{max} F_i \tag{19.3}$$

where

G_i = generation rate of fuel burned at time i, lb/s (kg/s),

F_i = factor at time i, dimensionless.

Example 19.4 develops the maximum generation rate and factors for the condominium fire of Example 19.1. For information about HRR and design fires, see Chapter 5.

TENABILITY CALCULATIONS

The threats to life are reduced visibility, toxic gas exposure, heat exposure, and thermal radiation exposure. In thick smoke, people see poorly and walk slowly or become disorientated, which prolongs exposure to smoke. In many applications the primary threat results from reduced visibility, but the other threats still need to be considered. For smoke control systems using passive barriers or pressurization at barriers, it is not feasible for the smoke control system to protect life in the fire room or spaces open to the fire room.

Visibility in terms of mass optical density is

$$S = \frac{K}{2.303\delta_m m_f} \tag{19.4}$$

where

S = visibility, ft (m),

K = proportionality constant,

δ_m = mass optical density, ft²/lb (m²/g),

m_f = mass concentration of fuel burned lb/ft³ (g/m³).

The proportionality constant K is 8 for illuminated signs and 3 for reflecting signs. For building components that are seen with reflected light, a value of $K = 3$ is used often. The mass concentration of fuel burned is

Example 19.4. Generation Rate

Develop generation data for CONTAM for the condominium unit fire of Example 19.1. The fuel is flexible polyurethane foam with a chemical heat of combustion of ΔH_{ch} = 7570 Btu/lb (17,600 kJ/kg).

The fire has a fast t-squared growth stage up to 2000 Btu/s (2110 kW). The equation for this growth stage is $Q = 1000(t / t_g)^2$ where t_g is 150 s. Solving this equation for time results in $t = t_g (Q/1000)^{1/2}$. The time at which Q is 2000 Btu/s is $t = 150 (2000/1000)^{1/2} = 212$ s.

From Equation 19.2, the maximum generation rate is at 2000 Btu/s (2110 kW) where generation is 0.264 lb/s.

The values of F for this fire were calculated from HRR values calculated from the t-squared equation, but values of F also could have been based on HRR data from spreadsheet NearFld-02.n. These factors are listed in the table at the right.

Generation of Condominium Fire

Time (s)	h	m	s	F
0	00:	00:	00	0.00
30	00:	00:	30	0.02
60	00:	01:	00	0.08
150	00:	02:	30	0.50
212	00:	03:	32	1.00
1200	00:	20:	00	1.00

Example 19.5. Visibility

Part 1: For smoke from a fully developed fire of flexible polyurethane foam, the mass optical density is δ_m = 1600 ft²/lb (0.33 m²/g). Calculate the mass concentration that would result in visibility of 300 ft (91 m) for objects seen with reflected light.

$$m_f = \frac{K}{2.303\delta_m S} = \frac{3}{2.303(1600)(300)} = 2.71 \times 10^{-6}\ \mathrm{lb/ft^3}\ (4.34 \times 10^{-5}\ \mathrm{kg/m^3})$$

Part 2: For the conditions above, what would m_f be for a visibility of 25 ft?

$$m_f = \frac{K}{2.303\delta_m S} = \frac{3}{2.303(1600)(25)} = 3.26 \times 10^{-5}\ \mathrm{lb/ft^3}\ (5.22 \times 10^{-4}\ \mathrm{kg/m^3})$$

calculated by CONTAM. For a particular visibility, Example 19.5 shows how to calculate the corresponding mass concentration of fuel burned. The fire in this example is a fully developed fire of flexible polyurethane foam.

The fractional effective dose (FED) model can be used to evaluate exposure to toxic gases. An FED greater than or equal to one indicates fatality. FED was not developed to predict incapacitation, but sometimes an FED of 0.5 has been used as a conservative level above which incapacitation can happen. For a smoke control system where the visibility criterion is met, the maximum possible FED is

$$FED_{max} = \frac{Kt}{2.303\delta_m S_c LCt_{50}} \quad (19.5)$$

where

FED_{max} = maximum fractional effective dose, dimensionless,

K = proportionality constant,

t = exposure time, min,

δ_m = mass optical density, ft²/lb (m²/g),

S_c = visibility criterion, ft (m),

LCt_{50} = lethal exposure dose from test data, lb·min/ft³ (g·min/m³).

For smoke control systems, Equation 19.5 can simplify evaluation of toxic gas exposure. If the smoke control system meets the visibility criterion and FED_{max} is less than 0.5, toxic gas exposure is not a concern.

Criteria for visibility have been suggested ranging from 13 to 46 ft (4 to 14 m), and it depends on a number of factors. For systems designed to meet most visibility criteria, the other threats are often insignificant. However, the other threats should be evaluated.

Example 19.6 shows how the Equation 19.5 can be used. The fire in this example also is a fully developed fire of flexible polyurethane foam, and this fuel and kind of fire are used in the other examples in this chapter. For more information about tenability and values of δ_m and LCt_{50} for other fuels, see Chapter 6.

Example 19.6. Maximum FED

If the visibility criterion of 25 ft (7.6 m) for reflecting signs is met, calculate the maximum possible FED. The duration of the smoke control system is 20 minutes. The fuel is flexible polyurethane foam with a mass optical density of $\delta_m = 1600$ ft²/lb (0.33 m²/g). For a fully developed fire, the LCt_{50} of flexible polyurethane foam is 0.012 lb·min/ft³ (200 (g·min/m³).

The parameters are: $K = 3$; $t = 20$ min; $\delta_m = 1600$ ft²/lb; $S_c = 25$ ft; $LCt_{50} = 0.012$ lb·min/ft³.

$$FED_{max} = \frac{Kt}{2.303\delta_m S_c LCt_{50}} = \frac{3(20)}{2.303(1600)(25)(0.012)} = 0.0054$$

This FED indicates that this exposure is very small. For a smoke control system meeting the above criterion with a fuel of flexible polyurethane foam, exposure to toxic gases is not an issue.

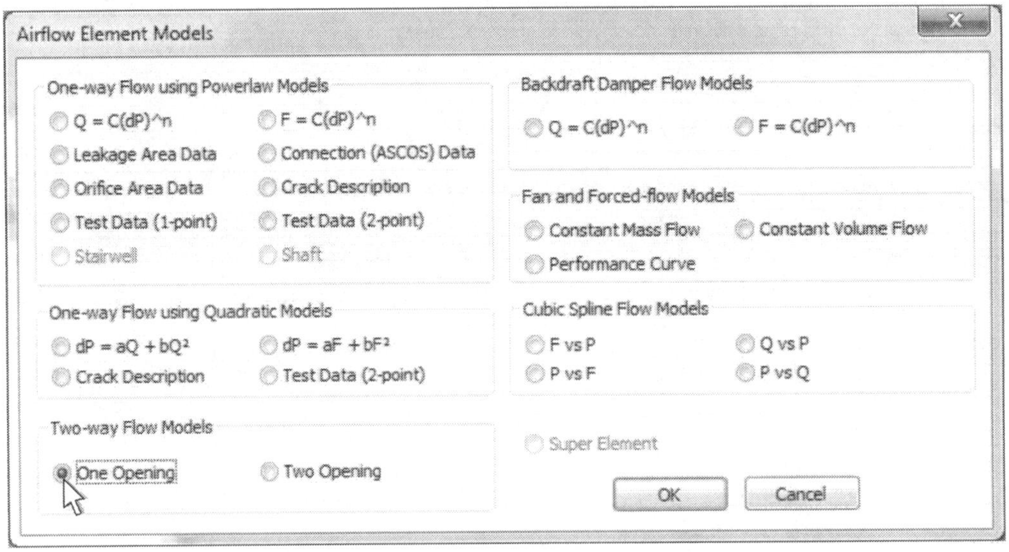

Figure 19.4 *Airflow properties window.*

USE OF CONTAM

This section discusses how to use CONTAM to simulate smoke flow, and how to use the results of the simulations to evaluate tenability. The information presented here on this topic is not exhaustive, and complete information about simulating the flow of containments with CONTAM is provided by Walton and Dols (2005).

CONTAM Input

Before starting smoke transport simulations, the building network should be modeled with CONTAM as described in Chapter 14. The following sections describe additional CONTAM features that are needed for smoke transport and tenability analysis, but the calculations for the tenability analysis are not done by CONTAM. The tenability calculations can be done with a spreadsheet using the CONTAM results.

Two-Way Flow Path

Defining a two-way flow path is somewhat like defining an orifice flow path which is described in Chapter 14. For example, the following steps to define a two-way flow path for a warped door named DOOR-SC2 that is 6.7 ft (2.04 m) high by 0.092 ft (0.028 m) wide with a discharge coefficient[1] of 0.65.

Move the caret to the desired location on the wall (or over an existing flow path), and right click. Click Flow Path. Double click on the red flow path icon that appears. Click New Element. The Airflow Element Models window will pop up as shown in Figure 19.4. Under Two-way Flow Models, select One Opening, and click OK. The two-way model window will pop up. Enter the name

1. The discharge coefficient used in CONTAM is the same as the flow coefficient discussed in Chapter 3.

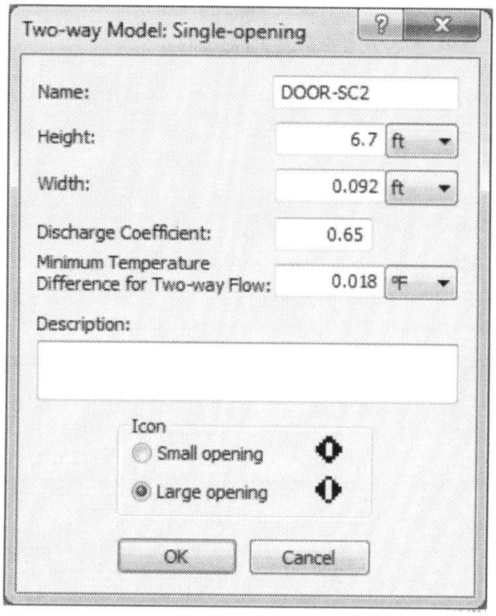

Figure 19.5 *Two-way model window.*

DOOR-SC2, the height of 6.7 ft, the width of 0.092 ft, and the discharge coefficient of 0.65. Select the large opening icon, and the two-way model window should look like Figure 19.5. Click OK.

Dimensionless Schedule

This schedule is used to allow changes during a simulation such as the amount of a source that is generated or the extent to which a door is warped. Schedules consist of week schedules and day schedules. From a physical standpoint, the CONTAM uses discussed here last much less than a day, but the week and day schedules are part of CONTAM because it was developed for indoor air quality analysis. These schedules need to be filled out so that the program will run. The day schedules need data supplied for 24 hours, and the week schedules need a day schedule for each day of the week plus five additional days.

The following steps are to define the schedules for the areas of warped door due to the condominium fire of Example 19.1. With the air properties window open, click Filters and Schedule, click New Schedule, and the week schedule window pops up. Name the schedule Warped-Door. Click New Day Schedule, and the Day window pops up. Name this schedule Warped-Door, and select trapezoidal. Enter the times and factors for the warped door of the lobby fire from Example 19.3. Enter a factor of 1 for time 24:00:00 to complete the day schedule. The Day Schedule should look like Figure 19.6. Click OK. Assign the day schedule,

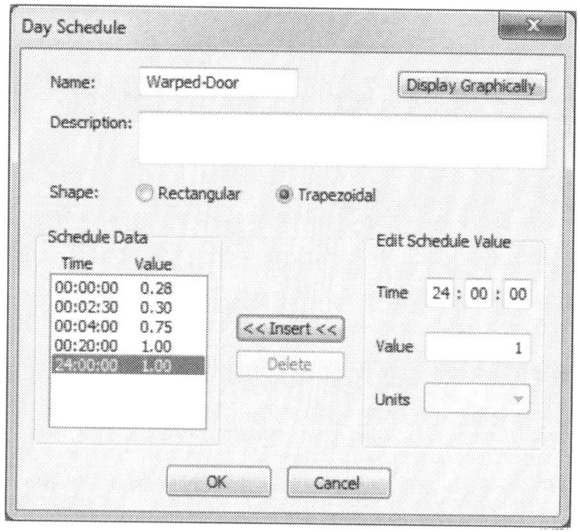

Figure 19.6 *Day schedule for factor of warped door.*

Warped-Door, to each of the days on the week schedule, and click OK.

For the warped door, the same name was used for both the week schedule and the day schedule. Because these schedules are separate things in CONTAM, they can have the same name. A day schedule can be either rectangular or trapezoidal. A rectangular schedule looks like a bar chart, and a trapezoidal one has straight lines connecting the data points. For the applications of this chapter, the day schedules are trapezoidal.

Temperature Schedule

In Chapter 14, the temperature schedule was used in a simple way to make it easy to change the temperatures of spaces like pressurized stairwells, pressurized elevators and parking garages with the season of the year. Temperature schedules can also be used to change the temperature of a space during a simulation. This can be used to change the temperature of near field spaces during the simulation. Temperature schedules are a type of day schedule, and they are assigned to week schedules, the same as dimensionless schedules.

The following is an example of using a temperature schedule for the condominium fire of Example 19.1. The temperatures for this schedule are listed in Example 19.2. Click on the zone icon in the condominium unit 1, select scheduled, click on New Schedule. The Week Schedule window will pop up. Name the week schedule Lobby-Fire, and click on New Day Schedule. The day schedule window will pop up. Name the day schedule Lobby-Fire, and select

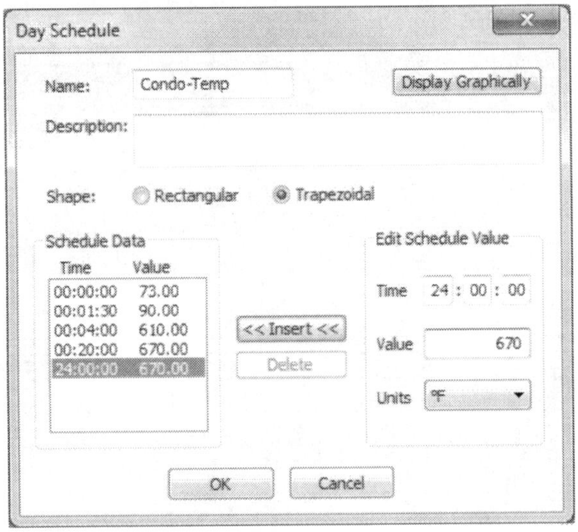

Figure 19.7 *Day schedule for temperature.*

trapezoidal. Enter the temperatures and times from Example 19.2 for the lobby fire. Enter 610°F for time 24:00:00 to complete the day schedule. The day schedule window should look like Figure 19.7. Click OK. Assign the new day to each day on the week schedule. Click OK.

Define a Contaminant

Dealing with contaminants consists of: (1) defining the contaminant and (2) defining the source of the contaminant. The following steps define a contaminant called fuel that consists of the mass of fuel burned per unit volume in units lb/ft^3. Click on Data, followed by Contaminants. The Project Species window pops up (Figure 19.8). Click on New, and the Species Properties window pops up. Enter the name fuel, select a default concentration of lb/ft^3, select Use in Simulation. For a description, enter "Mass of material burned in lb per cubic foot." This window should look like Figure 19.9. Click OK.[2]

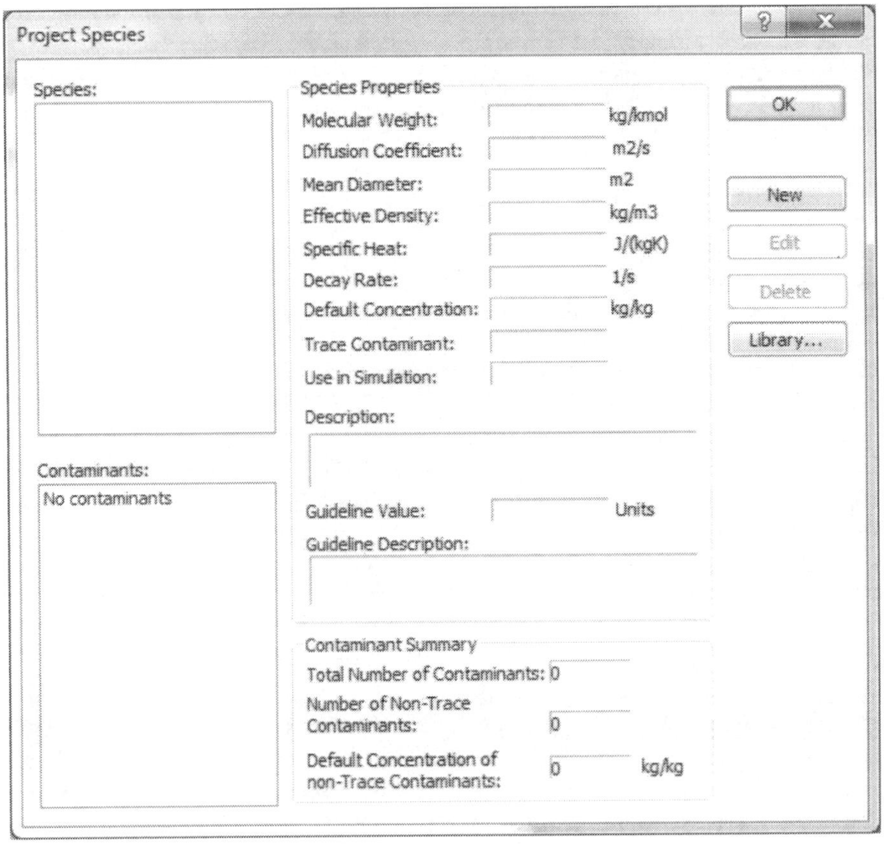

Figure 19.8 *Project species window.*

2. The user needs to click on OK two separate times. The first time the window closes, and another opens where OK is clicked on again.

In the species properties window, a description of the species was entered. Usually adding a description provides no useful information, but the added information could be useful when the species is the mass of fuel burned.

Define a Source

The following steps define a contaminant source with the generation rate of that of the condominium fire of Example 19.4. Move the caret to a location in condominium unit 1, right click, click on Source/Sink. Click on the red source/sink icon, and the Source/Sink Properties window pops up. Click New Element, select constant coefficient. Click OK. The Constant Coefficient Model window pops up. Enter the name Condo-Fire, enter a generation rate of 0.264 lb/s, and click OK. Click Multiplier, Schedule and Location. Click New Schedule. This is a dimensionless schedule like the one described. Complete the day schedule with the times and factors from Example 19.4, and fill out the week schedule. Click OK.

Simulation Parameters

The simulation parameters are used by CONTAM to define a number of specifics about a simulation. The following describes setting these parameters for the simulations described in this chapter. Click Simulation, click Set Simulation Parameters, and the Simulation Parameters window pops up. For airflow, select transient. For contaminants, select transient. For Transient Simulation Start, enter 00:00:00. For Transient Simulation Stop, enter 00:20:00. For each of the 3 Simulation Time Steps, enter 00:00:05. The Simulation Parameters window should look like Figure 19.10. Click OK. Depending on the application, a different simulation stop time and different simulation time steps may be needed.

Examining Results

There are a number of ways to examine contaminant concentrations. The easiest is by using the Plot Contaminants feature. This feature is used when CONTAM is in the results mode. Move the caret inside the room of interest, and right click, followed by clicking on Plot Contaminants. The Chart Control window pops up. Click on Plot Data. The Transient Contaminant Results window pops up with a graph of the contaminant concentrations in the room of interest. The plot looks like Figure 19.11.

With the Plot Contaminants feature, users can only look at concentrations one room at a time. The Export Results feature allows users at one time to produce a text file of all the concentrations for the spaces in a simulation. The Export Results feature is used when CONTAM is in the results mode. Click on Simulation, and click on Export Results. The export data window pops up. Select Report Transient Concentrations of All Zones as shown in Figure 19.12. Click Save As, and enter the name of the file to be exported.

The text files of concentrations can be imported into a spreadsheet program. The text file is read into a spreadsheet where the data is considered delimited by spaces.

Figure 19.13 shows sections of a spreadsheet program using the export results feature. Columns A, B, C, and D are the output data from CONTAM. Column E is the concentration adjusted to prevent dividing by zero. This is done by an If function that assigns the value of 2.71×10^{-6} for concentrations less than 2.71×10^{-6}. From Example 19.5, this concentration results in a visibility of 300 ft (91 m). Column F is the visibility in ft calculated from Equation 19.4.

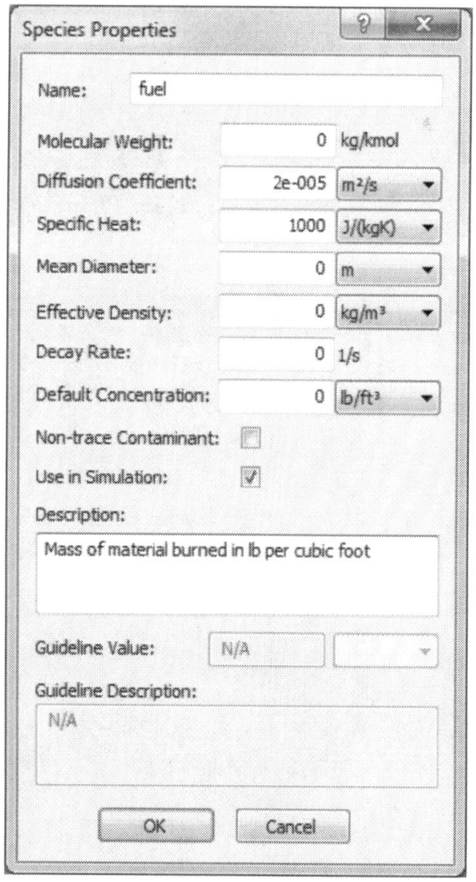

Figure 19.9 *Species properties window.*

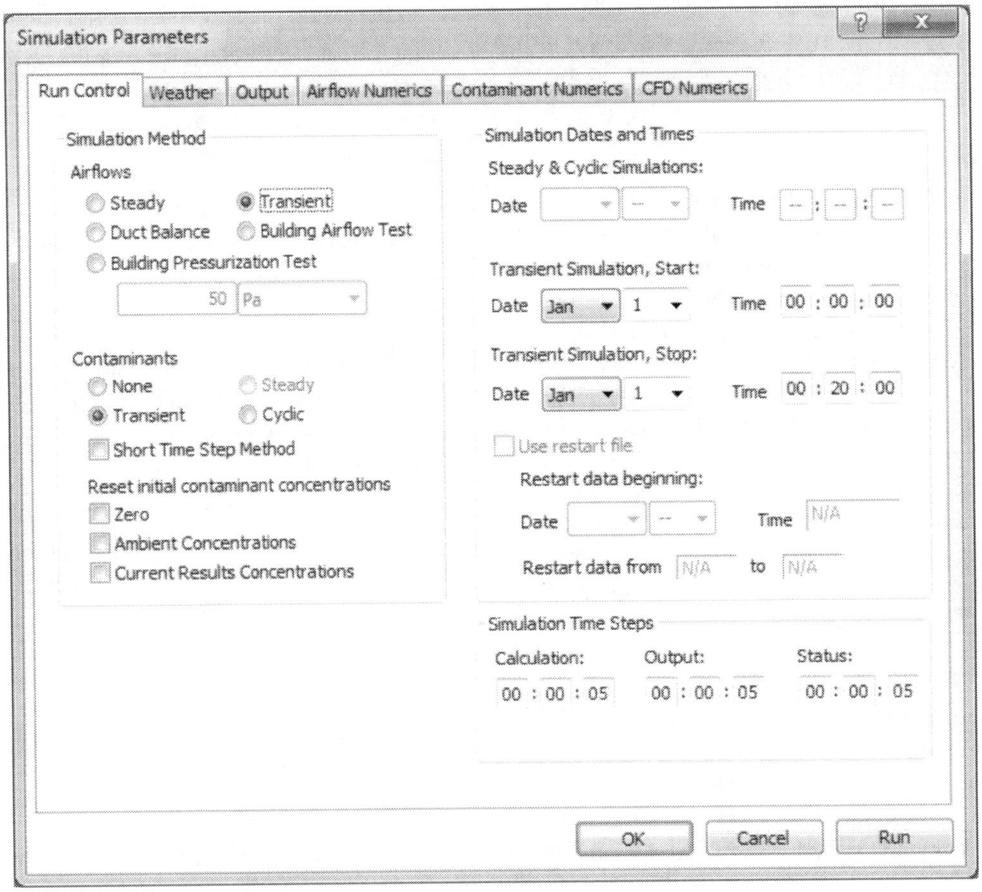

Figure 19.10 *CONTAM simulation parameters window.*

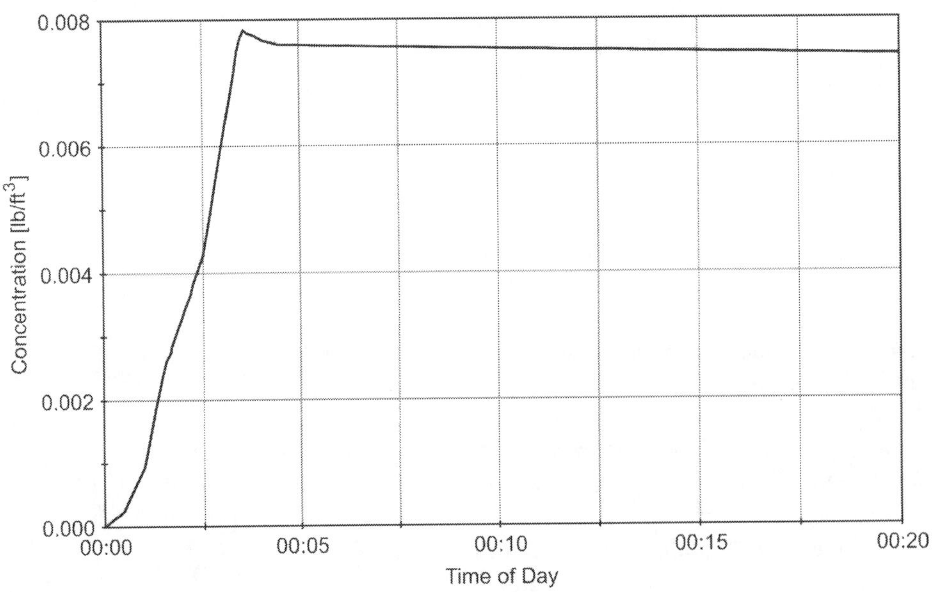

Figure 19.11 *Plot of contaminant concentrations in Unit 1 of Example 19.7.*

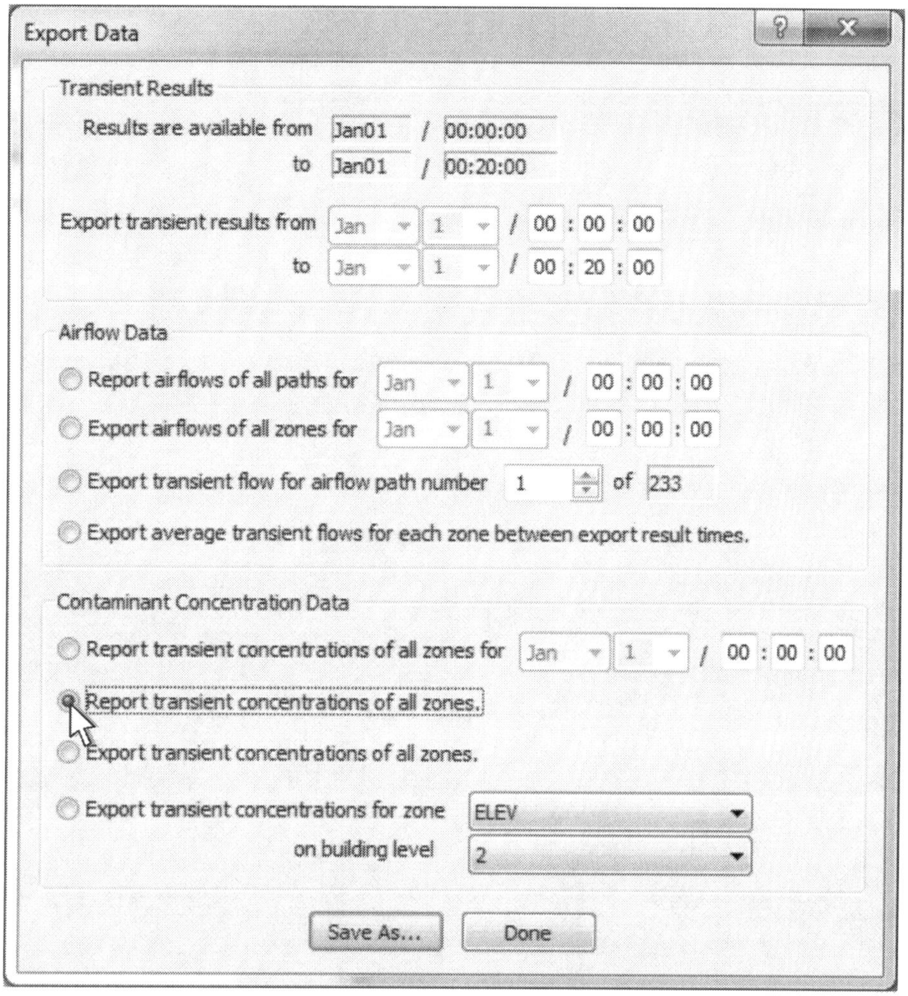

Figure 19.12 *CONTAM export data window.*

TENABILITY EXAMPLES

The examples in this section are simple ones selected to illustrate how CONTAM can be used to simulate smoke transport and how the results of those examples can be used to evaluate tenability. While these examples limit tenability to calculation of visibility, engineering applications should also address exposures to toxic gases, heat and thermal radiation.

Generally, it is beyond the capability of smoke control technology to provide protection in the near field, and elevated temperatures are mostly limited to the near field. This means that for many applications, exposure to heat and thermal radiation is not relevant. As mentioned previously, the FED_{max} can simplify evaluation of toxic gas exposure for smoke control systems. This approach is not appropriate for all applications, and Chapter 6 discusses methods of evaluating toxic gas exposure.

Example 19.7 examines tenability in a condominium building with a fire in one of the units. This building has pressurized stairwells. A visibility criterion of 25 ft (7.6 m) was chosen for this example. From Figure 19.14, it can be seen that this criterion is met for the stairwells and most of the building. As expected, the unit with the fire and several units above the fire do not meet the criterion. Air from the stairwells produces a low level of pressurization in the lobbies relative to units. This air also dilutes what smoke leaks into the lobbies. These mechanisms work together to reduce smoke movement in the building.

There are many other scenarios that can be examined. If the stairwells were not pressurized in Example 19.7, more smoke would have migrated to other spaces, and tenability would have been worse in many locations. If the fire had been in the elevator lobby, the situation would have been much different.

	A	B	C	D	E	F
1	project:	Condo-Fire-01	Sat	Aug		
2						
3	description:					
4						
5	date	time	time	fuel		
6	of	day	[s]	[lb/ft^3]		
7					Adjusted	
8	level:	G	zone:	ELEV	Fuel	Visibility
9					lb/ft^3	ft
10	1-Jan	0:00:00	0	0.00E+00	2.71E-06	300
11	1-Jan	0:00:05	5	0.00E+00	2.71E-06	300
12	1-Jan	0:00:10	10	0.00E+00	2.71E-06	300
13	1-Jan	0:00:15	15	0.00E+00	2.71E-06	300
14	1-Jan	0:00:20	20	0.00E+00	2.71E-06	300
2204	level:	2	zone:	UNIT1		
2205						
2206	1-Jan	0:00:00	0	0.00E+00	2.71E-06	300
2207	1-Jan	0:00:05	5	4.07E-05	4.07E-05	20
2208	1-Jan	0:00:10	10	8.11E-05	8.11E-05	10
2209	1-Jan	0:00:15	15	1.21E-04	1.21E-04	7
2210	1-Jan	0:00:20	20	1.61E-04	1.61E-04	5
2211	1-Jan	0:00:25	25	2.01E-04	2.01E-04	4
2212	1-Jan	0:00:30	30	2.40E-04	2.40E-04	3

Note: The data in columns A to D are from a text file produced by the export results feature of CONTAM. Columns E and F were calculated by the user. The adjusted concentration was calculated using an If function to prevent division by zero when calculating visibility.

Figure 19.13 *Sections of spreadsheet with CONTAM results and calculated visibility.*

Example 19.7. CONTAM and a Condominium Unit Fire

This example is a fire in unit 1 on level 2 of the six-story condominium building of Example 14.3. Unit 1 has an open window that is 4.2 ft (1.28 m) wide by 4 ft (1.22 m) high with a flow coefficient of 0.7. The near field is shown in Figure 19.1c, and a CFAST simulation of the fire is described in Part 1 of Example 19.1. The fuel is flexible polyurethane foam with a mass optical density of δ_m = 1600 ft^2/lb (0.33 m^2/g). The building temperature is 73°F (23°C), and the outdoor temperature is –4°F (–20°C). Atmospheric pressure is 14.3 psi (98.6 kPa).

Part 1: Use CONTAM to calculate contaminant concentrations in this building. Open a CONTAM window, and do the following steps.

- Open project file Condo-01, and save it as Condo-Fire-01.
- Edit the opening flow path between unit 1 and the lobby. This new path is two-way flow path that is 6.7 ft high, 0.092 ft wide, with a discharge coefficient of 0.65. The day schedule needs to be edited with the factors from the condominium fire of Example 19.3.
- Define a window in an exterior wall of unit 1 on level 2. This is a two-way flow path that is 4.2 ft wide by 4 ft high with a discharge coefficient of 0.7.
- Change the temperature of the condominium on level 2 to a schedule with the values from Example 19.2. Name the day schedule Condo-Temp, and select Trapezoidal.
- Define the trace contaminant called "fuel" with the default properties except that the default concentration is in lb/ft^3, select Use in Simulation, and enter the description "Mass of material burned in lb per cubic foot." The species properties window should look like Figure 19.9.
- In unit 1 on level 2, define a source with a generation rate of 0.264 lb/s with a schedule with the condominium fire factors from Example 19.4. On the day schedule, select trapezoidal.
- Set the simulation parameters to: (1) Airflows—transient, (2) Contaminants—Transient, (3) Transient Simulation Start—00:00:00, (4) Transient Simulation Stop—00:20:00. For each of the three Simulation Time Steps use 00:00:05. The simulation parameters window should look like Figure 19.10.
- Run the simulation, and export the results of the concentrations in a text file.

Part 2: Use the text file produced above to calculate the visibility, and examine that visibility. Consider the visibility criterion to be 25 ft (7.6m).

From Example 19.5, the concentration corresponding with the visibility criterion is 3.26×10^{-5} lb/ft^3. Use the Plot Contaminants feature to get a graph of the contaminants for the fire room, and the plot should look like Figure 19.11. It is apparent from this figure that early in the fire the concentration gets much greater than 3.26×10^{-5} lb/ft^3, and this means that the visibility in the fire room quickly drops below the criterion.

Read the text file into a spreadsheet where the data is considered delimited by spaces. Calculate the adjusted concentration of fuel and visibility. The spreadsheet should look like Figure 19.13. The times to reach untenable conditions are shown in Figure 19.14. It can be seen that tenable conditions are maintained in the spaces except for the fire room (unit 1 on level 2) and several of the units above it for several floors.

Level:	Unit 1	Stair 1	Lobby	Elev Shaft	Stair 2	Unit 2
6	OK	OK	OK	OK	OK	OK
5	220 s	OK	OK	OK	OK	OK
4	170 s	OK	OK	OK	OK	OK
3	165 s	OK	OK	OK	OK	OK
2	5 s	OK	OK	OK	OK	OK
G	OK	OK	OK	OK	OK	OK

Notes:
1. The stairwells are pressurized.
2. The fire is an unsprinklered upholstered furniture fire in unit 1 on level 2.
3. The numbers in the spaces above are the times for visibility to be less than 25 ft (7.6 m).
4. OK indicates that visibility is never less than 25 ft (7.6 m).
5. The elevator shaft at the roof level is not shown.

Figure 19.14 *Time to reach untenable conditions for Example 19.7.*

NOMENCLATURE

FED_{max} = maximum fractional effective dose, dimensionless
F_i = factor at time i, dimensionless
G_i = generation rate of fuel burned at time i, lb/s (kg/s)
G_{max} = maximum generation rate of fuel burned, lb/s (kg/s)
H = compartment height, ft (m)
K = proportionality constant
LCt_{50} = lethal exposure dose from test data, lb·min/ft^3 (g·min/m^3)
m_f = mass concentration of fuel burned lb/ft^3 (g/m^3)
Q_{max} = maximum heat release rate, Btu/s (kW)
S = visibility, ft (m)
S_c = visibility criterion, ft (m)
t = exposure time, min
T_{av} = weighted average temperature, °F (°C)
T_l = lower layer temperature, °F (°C)
T_u = upper layer temperature, °F (°C)
z = layer height, ft (m)
ΔH_{ch} = chemical heat of combustion Btu/lb (kJ/kg)
δ_m = mass optical density, ft^2/lb (m^2/g)

REFERENCES

Ferreira, M.J. 1998. Analysis of smoke control system design using a computer-based airflow analysis. Pacific Rim Conference, Society of Fire Protection Engineers, Bethesda, MD.

Ferreira, M.J. 2002. Use of multizone modeling for high-rise smoke control design. *ASHRAE Transactions* 108(2).

Fire International. 1968. The distortion of doors in fires due to asymmetric heating. *Fire International*, 19:36–39.

Floyd, J.E., et al. 2005. A network fire model for the simulation of fire growth and smoke spread in multiple compartments with complex ventilation. *Journal of Fire Protection Engineering* 15(3).

Hadjisophocleous, G., et al. 2011. Algorithm for smoke modeling in large multicompartmented buildings—Implementation of the hybrid model. *ASHRAE Transactions* 117(1).

Hadjisophocleous, G., Z. Fu, and G.D. Lougheed. 2002. Computational and experimental study of smoke flow in the stair shaft of a 10-story tower. *ASHRAE Transactions* 108(1).

Kashef, A., et al. 2011. Algorithm for smoke modeling in large multicompartmented buildings—Development of a hybrid model. *ASHRAE Transactions* 117(1).

Klote, J.H. 1990. Fire experiments of zoned smoke control at the Plaza Hotel in Washington, DC. *ASHRAE Transactions* 96(2).

Klote, J.H. 2004a. Hazards due to smoke migration through elevator shafts—Vol. I: Analysis and discussion. Final Report, NIST GCR 04-864-I, National Institute of Standards and Technology, Gaithersburg, MD.

Klote, J.H. 2004b. Tenability and open doors in pressurized stairwells. *ASHRAE Transactions* 110(1).

Klote, J.H., H.E. Nelson, S. Deal, and B.M. Levin. 1992. Staging areas for persons with mobility limitations. NISTIR 4770, National Institute of Standards and Technology, Gaithersburg, MD.

Van Geyn, M. 1994. National fire door test project—positive pressure furnace fire tests. Technical Report, National Fire Protection Research Foundation, Quincy, MA.

Walton, G.N., and W.S. Dols. 2005, revised 2010. CONTAM 2.4 user guide and program documentation. NISTIR 7251, National Institute of Standards and Technology, Gaithersburg, MD.

Wang, L. 2007. Coupling of multizone and CFD programs for building airflow and contaminant transport simulations. PhD Dissertation, Purdue University, Lafayette.

Wang, L., W.S. Dols, and Q. Chen. 2010. Using CFD capabilities of CONTAM 3.0 for simulating airflow and contaminant transport in and around buildings. *HVAC&R Research* 16(6).

CHAPTER 20

Computational Fluid Dynamics

John H. Klote

In the 1970s, computational fluid dynamic (CFD) modeling was developed at the Imperial College in the United Kingdom (Launder and Spalding 1974). Today, there are many CFD models that can be used for smoke control analysis. Fire Dynamics Simulator (FDS) is a CFD model that was developed at the National Institute for Standards and Technology (NIST) specifically for fire applications (McGrattan et al. 2008a, 2008b). FDS is available from NIST at no cost. Because FDS is extensively used around the world for fire applications, it is the focus of much of this chapter.

This chapter provides general information about CFD. Often, a CFD analysis of smoke control systems is done along with a tenability analysis, and the soot yield of the fire needs to be correctly specified to assure the applicability of the tenability calculations. Most of the equations in this chapter are used for the purpose of explaining concepts, and units are not given for the variables. These equations are valid for SI units (Chapter 1).

TENABILITY ANALYSIS

Smoke is commonly recognized as the major killer in building fires. The threats to life are toxic gas exposure, heat exposure, thermal radiation exposure, and reduced visibility. Reduced visibility is an indirect threat because people exposed to thick smoke become disorientated, which prolongs exposure to smoke. An additional threat associated with reduced visibility is that of falls from balconies and other high places.

When visibility criterion is met for most smoke control systems, the other threats are usually not an issue. When visibility criterion is met, it usually is because the airborne products of combustion are diluted by air. This dilution reduces the smoke temperature and concentrations of toxic gases. It follows that the threats of toxic gas exposure, heat exposure, and thermal radiation exposure are also reduced. However, these exposure threats should be checked for each project to be sure that they are not of concern. For methods to calculate these threats to life and information about tenability criteria, see Chapter 6.

CFD modeling can simulate the smoke flow due to fires, and tenability calculations can be made based on the CFD simulation. FDS has some features that can help with tenability calculations.

CFD CONCEPT

The idea of CFD modeling is to divide a space into a large number of small spaces called cells, and use a computer to solve the governing equations for the flows, pressures, and temperatures throughout the space. The space being modeled is called the *domain*. Most flows involved with smoke control are turbulent, and it is important that simulated flows include turbulence. Turbulence larger than the cell size can be simulated directly by solution of the governing equations, but turbulence on a smaller scale cannot be simulated. Turbulence modeling is used to account for the small scale turbulence, and this is discussed later.

There are many input parameters including physical properties, boundary conditions, and initial conditions. Initial conditions consist of the properties, primarily temperature and pressure, at the beginning of a simulation.

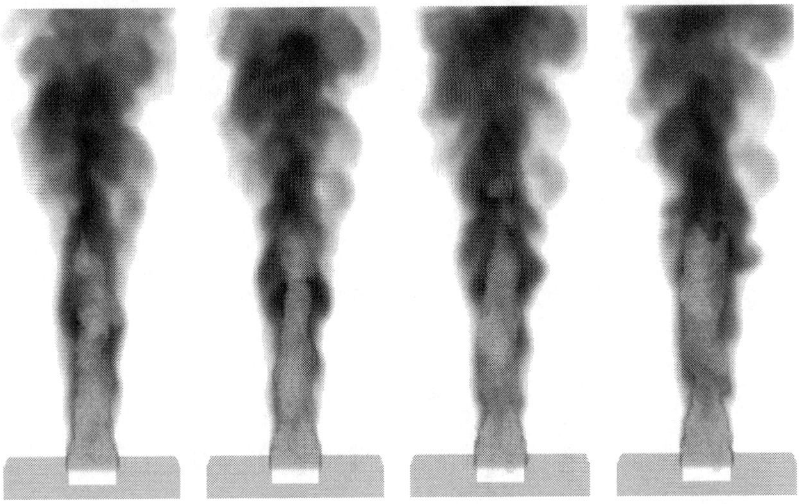

Figure 20.1 *FDS simulation of a plume from a pool fire.*

EXAMPLE APPLICATIONS

CFD modeling has been used for a wide range of applications in many areas including research and design analysis. Analysis of smoke flow in atria is possibly the most extensive fire related application of CFD modeling, but there are many other fire applications of CFD modeling.

An early example of CFD analysis of an atrium[1] smoke control system is the Lloyds Building in London, UK (Waters 1989). An example of FDS modeling of atrium smoke control system is the Newseum in Washington, DC (Klote 2005). A few other fire applications of CFD modeling are: analysis of smoke detector activation (Davis, Forney, and Klote 1991), simulation of flame blow-down in a Navy firefighter training facility (Forney and Davis 1992), and stairwell ventilation systems. Stairwell ventilation systems and atrium smoke control by natural venting are discussed later in this chapter.

Boundary Conditions

The conditions at the boundaries of the domain need to be stipulated, and CFD models have a large number of boundary options. For smoke control applications, the most common boundary conditions are (1) solid surface, (2) velocity boundary, and (3) open boundary. The solid surface boundary is used to define walls, floors, ceilings, and other objects inside the domain. At solid surfaces, the component of velocity tangential to the surface generally is considered to be zero. This is referred to as the no slip condition.

Both velocity and open boundaries can be used where mass flows into or out of the domain. The velocity boundary is particularly useful for defining smoke exhausts that have a constant volumetric flow. Open boundaries also are called pressure boundaries, because the pressures outside the domain are constant. Open boundaries are useful for defining makeup air vents.

Realism

A representation of an atrium or other space can be built inside a computer using CFD modeling to simulate fire-induced flows to a level of detail not possible with any other mathematical model. This allows CFD simulations to account for building features that cannot be accounted for with zone fire modeling of Chapter 18. For example, CFD modeling can simulate smoke flows for arrangements that the balcony spill equations of Chapter 16 were not intended. Also, CFD modeling can account for irregularly shaped ceilings that can only be dealt with roughly by the other methods.

FDS is capable of simulating fire-induced flows with realistic vortices and eddies as can be seen from the simulated fire plume form a pool fire[2] shown in Figure 20.1. Baum, McGrattan, and Rehm (1997)

1. In this handbook, the term atrium is used in a generic sense to mean any large-volume that is at least two stories high, such as an atrium, enclosed shopping mall, sports arena, or an airplane hangar.
2. A pool fire happens when a liquid hydrocarbon is spilled and the resulting "pool" of fuel is ignited. For fire testing, a pool fire can consist of liquid hydrocarbons burning in a metal pan.

demonstrated that FDS is capable of simulating plume flow that agrees well with the plume correlations of McCaffrey (1983).

Model Evaluation

Evaluation of a CFD model consists of verification and validation. Verification consists of checking the correctness of the CFD solution, and validation consists of comparing the results of CFD simulations with experimental measurements. In simple terms, verification is checking the mathematics, and validation is checking the physics.

CFD modeling is so complex that complete verification and validation is impossible. There probably has been more effort in this area for FDS than any other CFD model. Verification efforts included analytical tests, numerical tests, sensitivity analysis and code checking with efforts focused on the basic flow solver, thermal radiation, species, combustion, heat conduction, pyrolysis, and droplets (McGrattan et al. 2008c). Validation efforts included comparison with experiments in the areas of fire plumes, pool fires, compartment fires, tunnel fires, combustion modeling, growing fires, flame spread, air and gas movement in the absence of fire, wind engineering, suppression by sprinklers, suppression by mist systems, and smoke detection (McGrattan et al. 2008d).

FDS and two zone fire models were verified in a study of fire modeling for nuclear power plant applications conducted by the Nuclear Regulatory Commission and the Electric Power Research Institute (NRC 2007). The NRC study involved a number of full scale fire tests relevant to fire safety in nuclear power plants. Based on this study, Figure 20.2 is a comparison of measured and predicted smoke layer temperature rise by different methods of analysis (Salley et al. 2007). The dashed lines indicate the level of uncertainty of the measurements. This figure is an indication of how accurate CFD modeling can be.

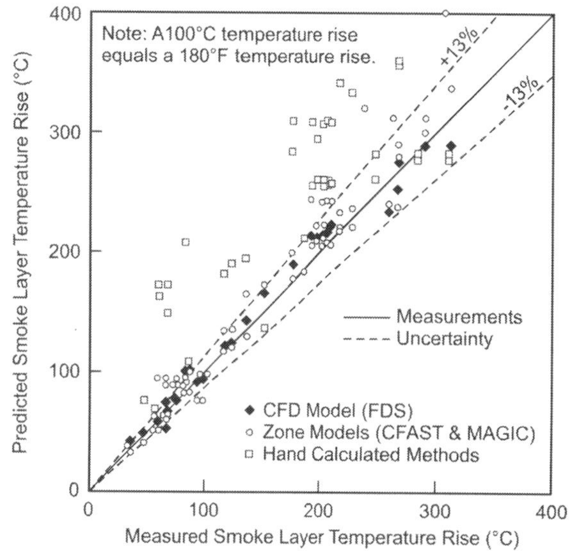

Figure 20.2 *Measured and predicted smoke layer temperature rise by NRC.*

GOVERNING EQUATIONS

The governing equations consist of at least the equation of state and equations for conservation of mass, conservation of momentum, and conservation of energy. For most applications the ideal gas law is the appropriate equation of state.

The derivation of the conservation equations is one of the most intellectually challenging subjects in engineering. This subject is treated in graduate level texts on viscous fluid flow (Papanastasiou, Georgiou, and Alexandrou 2000; Schlichting 2000; White 2005). A detailed development of these equations is presented by Aris (1962). The conservation equations are nonlinear partial differential equations. In addition to expressing these equations in partial differential equation form, they are frequently expressed in vector form and in tensor notation form. Tensor notation was used by Albert Einstein to develop the general theory of relativity. To give an idea of the complexity of topic, the following conservation equations are in vector form.

Conservation of Mass:

$$\frac{D\rho}{Dt} + \rho\nabla \cdot \boldsymbol{u} = 0 \tag{20.1}$$

Conservation of Momentum in the x Direction:

$$\begin{aligned}\rho\frac{Du}{Dt} &= \rho X_x - \frac{\partial p}{\partial x} + \frac{\partial}{\partial x}\left[\mu\left(2\frac{\partial u}{\partial x} - \frac{2}{3}\nabla \cdot \boldsymbol{u}\right)\right] \\ &+ \frac{\partial}{\partial y}\left[\mu\left(\frac{\partial u}{\partial y} + \frac{\partial v}{\partial x}\right)\right] + \frac{\partial}{\partial z}\left[\mu\left(\frac{\partial w}{\partial x} + \frac{\partial u}{\partial}\right)\right]\end{aligned} \tag{20.2}$$

Conservation of Momentum in the y Direction:

$$\begin{aligned}\rho\frac{Dv}{Dt} &= \rho X_y - \frac{\partial p}{\partial y} + \frac{\partial}{\partial y}\left[\mu\left(2\frac{\partial v}{\partial y} - \frac{2}{3}\nabla \cdot \boldsymbol{u}\right)\right] \\ &+ \frac{\partial}{\partial z}\left[\mu\left(\frac{\partial v}{\partial z} + \frac{\partial w}{\partial y}\right)\right] + \frac{\partial}{\partial x}\left[\mu\left(\frac{\partial u}{\partial y} + \frac{\partial v}{\partial x}\right)\right]\end{aligned} \tag{20.3}$$

Conservation of Momentum in the z Direction:

$$\rho\frac{Dw}{Dt} = \rho X_z - \frac{\partial p}{\partial z} + \frac{\partial}{\partial z}\left[\mu\left(2\frac{\partial w}{\partial z} - \frac{2}{3}\nabla \cdot \boldsymbol{u}\right)\right] + \frac{\partial}{\partial x}\left[\mu\left(\frac{\partial w}{\partial x} + \frac{\partial u}{\partial z}\right)\right] + \frac{\partial}{\partial y}\left[\mu\left(\frac{\partial v}{\partial z} + \frac{\partial w}{\partial y}\right)\right] \tag{20.4}$$

Conservation of Energy:

$$\rho\frac{D(C_pT)}{Dt} - \frac{D\rho}{Dt} = \dot{q}''' + \frac{\partial}{\partial x}\left(k\frac{\partial T}{\partial x}\right) + \frac{\partial}{\partial y}\left(k\frac{\partial T}{\partial y}\right) + \frac{\partial}{\partial z}\left(k\frac{\partial T}{\partial z}\right) + \Phi \tag{20.5}$$

where

$$\Phi = -\frac{2}{3}\mu(\nabla \cdot \boldsymbol{u})^2 + 2\mu\left[\left(\frac{\partial u}{\partial x}\right)^2 + \left(\frac{\partial v}{\partial y}\right)^2 + \left(\frac{\partial w}{\partial z}\right)^2\right] + \mu\left[\left(\frac{\partial u}{\partial y} + \frac{\partial v}{\partial x}\right)^2 + \left(\frac{\partial v}{\partial z} + \frac{\partial w}{\partial y}\right)^2 + \left(\frac{\partial w}{\partial x} + \frac{\partial u}{\partial z}\right)^2\right]$$

$\boldsymbol{u}$ = velocity vector,
u = velocity component in x direction,
v = velocity component in y direction,
w = velocity component in z direction,
ρ = density,
p = pressure,
μ = dynamic viscosity,
k = thermal conductivity,
T = absolute temperature,
C_p = constant pressure specific heat,
$\dot{q}'''$ = heat release rate per unit volume,
t = time,
X_j = body force in j direction.

The conservation of momentum equations are also called the Navier-Stokes equations, and sometimes this term is also used in a broader sense to mean all the conservation equations plus the equation of state.

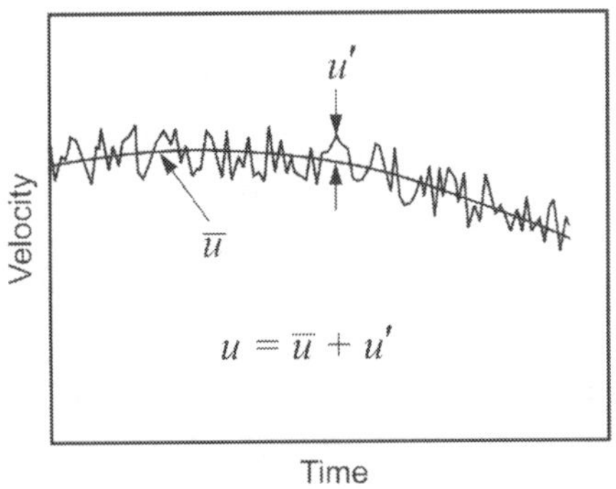

Figure 20.3 *Velocity components in* x *direction.*

TURBULENCE MODELING

The conventional approach for dealing with small scale turbulence consists of Reynolds averaging and turbulence modeling. Reynolds averaging is based on the assumption that the fluctuations associated with turbulence are random. With this averaging, variables are considered to consist of an average component and a random fluctuating component. For example, the velocity in the x direction is written

$$u = \bar{u} + u' \tag{20.6}$$

where

u = velocity in x direction,
$\bar{u}$ = average velocity in x direction,
u' = fluctuating component of velocity in x direction.

These velocity components are shown in Figure 20.3. Considering all the variables being made up of average quantities plus random fluctuating components, a Reynolds averaged form of the governing equations can be developed.

The first turbulence model developed is called the k-ε model where k is the kinetic energy of turbulence and ε is the turbulence dissipation rate. This model uses two empirical partial differential equations (PDEs) to define k and ε.

Other turbulence models have been developed, and turbulence models are classified by the number of PDEs in the model. The k-ε model is a two equation model. The Prandtl model is algebraic with no PDEs, and thus it is classified as a zero equation model. The Reynolds stress models are three to five equation models. Turbulence modeling results in averaged flows without turbulent fluctuations, but the utility of this treatment has been proven for a wide range of applications.

Rather than using the conventional approach to turbulence, FDS uses large eddy simulation (LES). LES does not use the Reynolds averaged equations, and the turbulence modeling in FDS can be thought of as being a zero order model. By using LES, the FDS model is capable of producing realistic simulations like the plume of Figure 20.1.

FIRE MODELING

Because FDS was specifically developed for fire applications, it has many ways to simulate a fire. A burning material can be a solid, liquid or gas. It is possible to prescribe the heat release rate (HRR) of the fire or have it calculated by FDS. Fires can spread out as they grow. Fires can be simulated where the fuel burns away as in a furniture fire or a liquid fuel fire.

For analysis of atrium smoke control systems, the fire is often simulated as a gas burner with a prescribed heat released rate. With this approach, the burning surface, fire reaction, and the HRR are defined by the user. For atrium smoke control purposes, the reaction needs to include the soot yield and the chemical heat of combustion. The chemical composition of the fuel can also be included, but it is not necessary. In order for the visibility calculation of FDS to be accurate, the soot yield and the chemical heat of combustion must be appropriate for the fuel being burned. For fuels that are predominately one material, data from Chapter 5 can be used.

Fuel Mixtures

Some fuels are made of a number of materials, and the following equations can be used for well-mixed fuels. For a mixture of n components, the soot yield and the heat of combustion are

$$y_{p,m} = \sum_{i=1}^{n} f_i y_{p,i} \tag{20.7}$$

$$\Delta H_m = \sum_{i=1}^{n} f_i \Delta H_i \tag{20.8}$$

where

$y_{p,m}$ = soot yield of mixture,
$y_{p,i}$ = soot yield of component i,
ΔH_m = heat of combustion of mixture,
ΔH_i = heat of combustion of component i,
f_i = mass fraction of component i.

MODELING THE SPACE

In FDS, the domain is made up of one or more meshes. A mesh is a parallelepiped, which is a rectangular box. The cells in a mesh are also rectangular. Figure 20.4a shows a domain that consists of one uniform mesh. A uniform mesh is one where all the cells are the same size and shape. Figure 20.4b shows a domain that consists of one nonuniform mesh. Figure 20.4c shows a domain that consists of two uniform meshes with the same size cells. Figure 20.4d also shows a domain that consists of two uniform meshes, but the sizes of cells in the meshes are different.

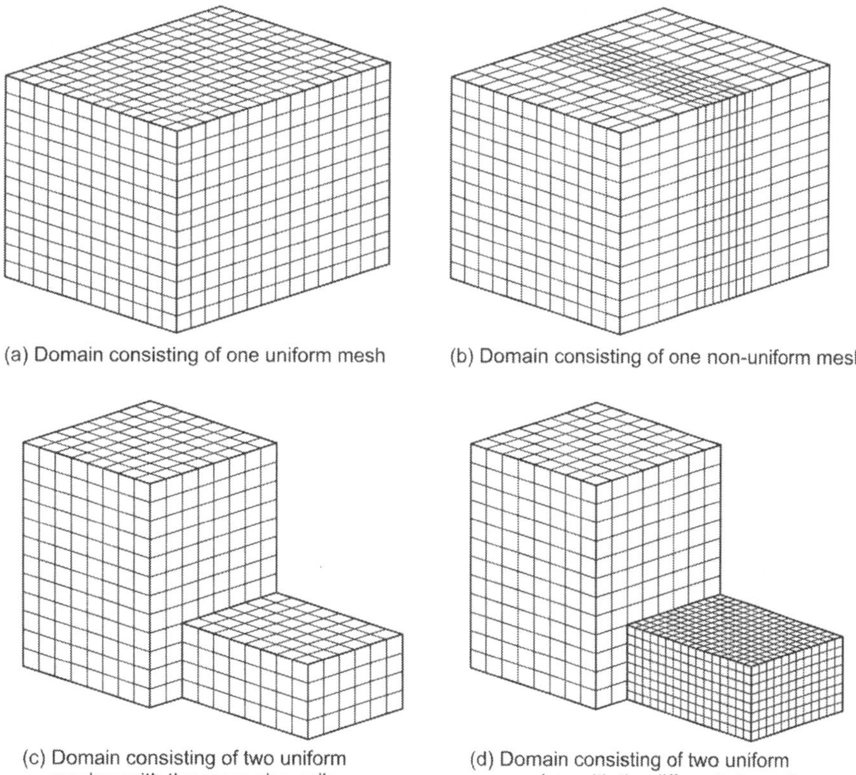

(a) Domain consisting of one uniform mesh

(b) Domain consisting of one non-uniform mesh

(c) Domain consisting of two uniform meshes with the same size cells

(d) Domain consisting of two uniform meshes with the different size cells

Figure 20.4 *Domain consisting of various arrangements of meshes.*

Groups of cells in the domain can be designated as obstructions to represent objects in the domain such as furniture, walls, and floors. The obstructions are assigned thermal properties, and they are included in the heat transfer calculations. Thin objects such as glass walls and railings can be simulated as two-dimensional solids.

Nonrectangular Geometry

Because cells are rectangular, FDS must approximate diagonal and curved surfaces. Some CFD models are capable of using nonrectangular cells, and these models are better for dealing with complex shapes. For example, Figure 20.5a shows how well a grid using nonrectangular cells can model a vaulted ceiling. With FDS, the vaulted ceiling has to be approximated with steps as shown in Figure 20.5b. FDS has a feature (SAWTOOTH=.False.) that can be used to prevent vorticity from being generated at sharp corners. While this feature is not a complete solution to the stepping issue, it prevents the flow field around nonrectangular obstructions from being inhibited by extra drag created at sharp corners.

The stepping issue is also reduced by the large number of cells that can be handled with FDS. The requirement for rectangular cells results from the numeric routine used by FDS to solve the governing equations. This routine is Poisson solver based on fast Fourier transforms (FFTs), which runs much faster than conventional solvers that can accommodate nonrectangular cells. While no studies have been conducted, it is estimated that FDS may be able to deal with as many as ten times the number of cells as CFD models using conventional solvers. With more cells, the size of the steps becomes smaller.

VISUALIZATION

CFD software generates such enormous quantities of data that conventional methods of presenting and analyzing data are inadequate. Postprocessing software is used to produce graphic representations of simulations, and many CFD models have postprocessing software specifically written for them. Smokeview was written for FDS (Forney 2008).

Smokeview produces a 3-D representation of the space being simulated in a window on the user's computer screen. Figure 20.6 is an FDS representation of an atrium where obstructions are used for the balconies and the glass railings, and there is an open doorway on the second floor to a conference room. The user can change the direction from which the scene is being viewed. Smokeview can generate realistic images of smoke flow, and Figure 20.7 shows smoke flowing from the conference room doorway to form a balcony spill plume. Pictures of the smoke flow throughout a simulation can be assembled to make a movie. Smokeview can produce these figures, but third party software is needed to make the movie.

Smokeview also can produce two-dimensional slices showing properties including temperature, velocity, and visibility. Files can also be produced of a property at a point.

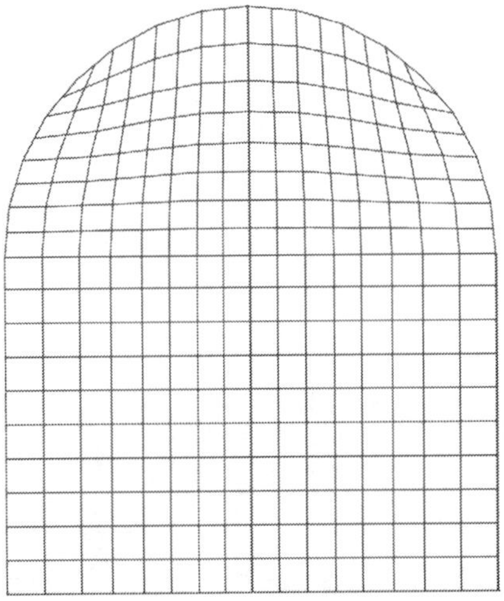

(a) Grid using non-rectangular cells

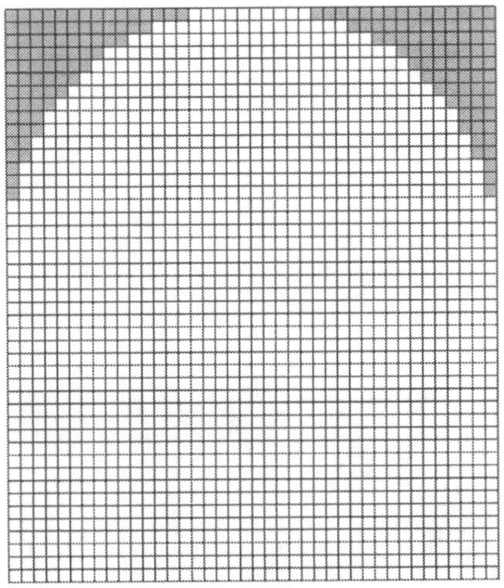

(b) Grid using rectangular cells

Note: FDS can only use rectangular cells.

Figure 20.5 *Grids used for vaulted ceiling.*

MODELING TECHNIQUE

As stated, CFD modeling is capable of accurately and realistically simulating flow, but this depends on modeling technique. Because poor modeling technique can result in simulations that differ significantly from reality, CFD modelers need to develop good modeling techniques. Modeling technique consists of validation and verification plus the choices of input parameters. Good choices for one application may be poor choices for another application.

There are dozens of input parameters including physical properties of solids and gases, boundary conditions, initial conditions, combustion parameters, soot yield, and the size of the cells. The relative importance of each parameter depends on the application, but the size of the cells (or grid size) is probably the most important parameter for any application. For atrium smoke control applications that include a tenability analysis, soot yield is very important. Soot yield is discussed in the previous section.

In general, the smaller the grid, the better the numerical solution of the governing equations. When all other things are the same, computer running time is approximately related to grid size as

$$t_{c2} = t_{c1}\left(\frac{S_1}{S_2}\right)^4 \qquad (20.9)$$

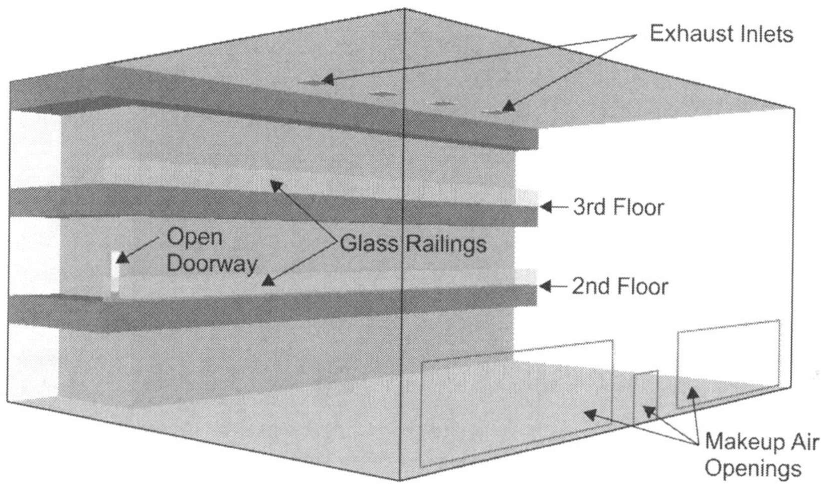

Figure 20.6 *FDS representation of three-story atrium.*

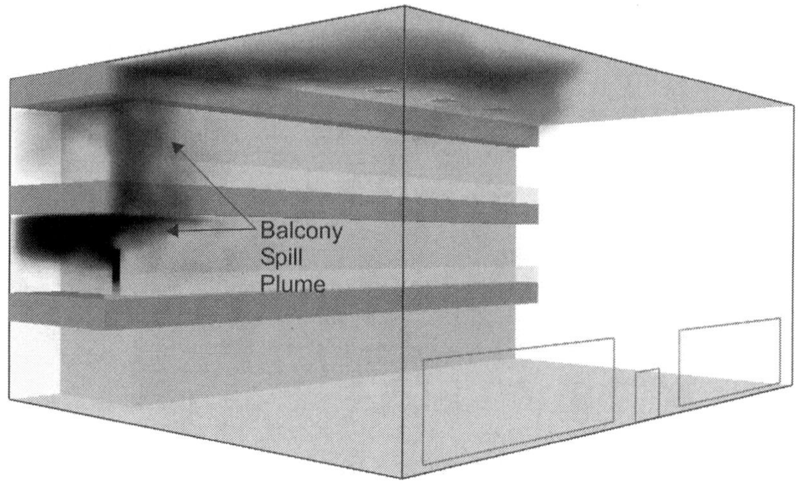

Figure 20.7 *FDS-simulated balcony spill plume two minutes after ignition.*

where

t_{c1} = computer time of simulation 1,

t_{c2} = computer time of simulation 2,

S_1 = grid size of simulation 1,

S_2 = grid size of simulation 2,

Example 20.1 shows that decreasing the grid size from 0.3 to 0.2 m increases the computer running time by a factor of about 5. The temptation to simply use large cell sizes needs to be avoided. To find out the effect of cell size on the solution, a grid sensitivity study can be used.

Figure 20.8 shows the grid sensitivity of an FDS-simulated plume that was part of the NRC verification study. This figure shows FDS predicted plume temperatures for grids of 2, 3, and 4 in. (0.05, 0.075, and 0.10 m). The temperature with 2 in. (0.05 m) grid agrees much better with experimental data, because the air entrained into the plume is simulated more realistically with the 2 in. (0.05 m) grid. For this application, making the grid smaller than 2 in. (0.05 m) grid would not result in a noticeable difference in the results.

Hadjisophocleous and McCartney (2005) present guidelines for the use of CFD simulations for fire and smoke modeling, and Ferreira (2008) discusses atrium design with FDS.

ATRIUM SMOKE CONTROL

CFD modeling is used extensively for atrium smoke control. The algebraic equation method for analysis of atrium smoke control is discussed in Chapter 16. For atrium smoke control systems that use fan-powered exhaust, the equation method provides starting points for CFD modeling.

The equation method does not take into account the transition zone. In systems based on the equation method, occupants may be exposed to smoke in the transition zone. Hopefully, this smoke exposure will not be fatal, but it is beyond the capability of the equation

Example 20.1. Computer Time and Grid Size

An FDS simulation requires 12 h of computer time and has a grid size of 0.3 m. Approximately how much computer time would be needed if the grid size were changed to 0.2 m?

$$t_{c2} = t_{c1}\left(\frac{S_1}{S_2}\right)^4 = 12\left(\frac{0.3}{0.2}\right)^4 = 61 \text{ h}$$

The reduction in grid size increases the computer running time from half a day to about two and a half days. This is only an approximation, and the running time may also depend on other things. In this example, the user should not be surprised if running time is as little as two days or as much as three days.

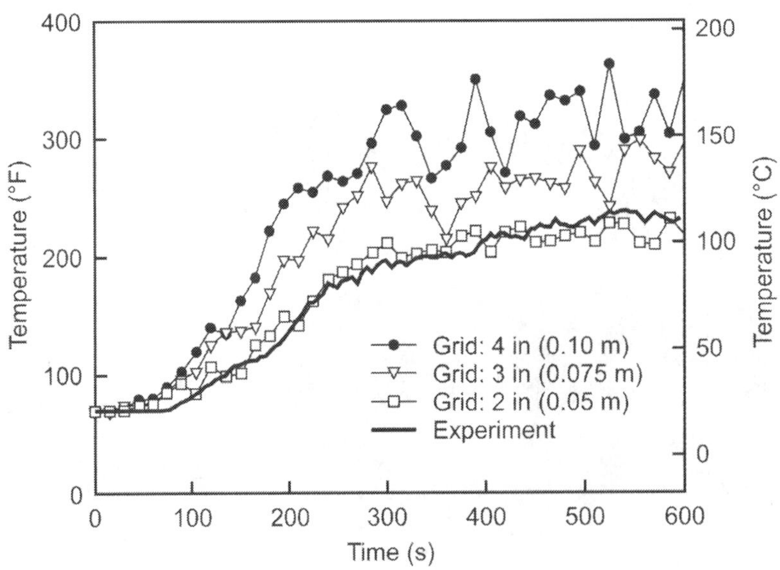

Figure 20.8 *Grid sensitivity of FDS-simulated plume.*

method to evaluate tenability in the transition zone. A major advantage of CFD is that it can be used to evaluate these exposures.

CFD modeling is attractive because it is capable of simulating flows in ways that algebraic equations cannot. Plume contact with walls and the resulting impact on system performance can be realistically simulated with CFD modeling. The impact of makeup air velocity on plume formation can be realistically simulated by CFD modeling, and sometimes CFD simulations can justify a makeup air velocity that exceeds the stipulated limit. CFD modeling can realistically simulate plugholing, and for some applications it is possible to use fewer exhaust inlets than the number required by the equation method.

The balcony spill plume portion of the equation method is only applicable to specific geometries, but almost any possible geometry of balcony spill plume can be simulated by CFD modeling. The stringent smoke layer thickness requirements can sometimes be relaxed based on CFD modeling with a tenability analysis. The equation method is inappropriate for shielded fires, but these fires and the resulting plumes can be simulated with CFD modeling. CFD modeling is capable of analyzing the impact of atrium geometry on system performance beyond anything that can be done with any other form of mathematical modeling.

CFD modeling can be used to simulate the impact of wind on system performance. It can analyze the impact of wind on makeup air velocity to assure that this velocity does not exceed stipulated limits. CFD can model smoke flow out of doors to develop designs that are not likely to have smoke feedback into the makeup air. When there is some smoke feedback, CFD modeling can be used to evaluate the impact on tenability.

Natural Venting

Much of the previous discussion also applies to natural venting. The impact of wind on natural venting systems is much greater than on systems with fan-powered exhaust. It is possible for wind to force smoke downward in a natural venting system. In some situations, wind speeds below the normally used design values may have a significant negative impact. For these reasons, wind effects need to be analyzed with CFD modeling. In hot weather the buoyancy of the smoke may not be sufficient to make natural venting systems work as intended, but this can be analyzed by CFD modeling. The natural venting equation in Chapter 15 can be used to get a starting point for the size of the smoke vent. For more information about CFD modeling of natural venting systems, see Sinclair and Xiangdoing (2012).

STAIRWELL VENTILATION SYSTEMS

CFD modeling can be used with CONTAM to analyze stairwell ventilation systems. These systems are not intended to maintain stairwell pressurization, but they rely on supplying air to and exhausting it from stairwells to protect the stairwells from small amounts of smoke that could leak in through the gaps around a closed door. The idea is that the products of combustion are diluted such that a tenable environment is maintained in the stairwell. This approach has the potential to provide smoke protection for very tall stairwells in very complex buildings.

The smoke leaks through the gaps around the door, and the smoke flows some distance away from the door until it becomes well mixed into the airflow. The concentrations of the products of combustion are greatest at locations in the vicinity of the stair door on the fire floor.

With CONTAM, each level of a stairwell is treated as a separate zone, and the properties in each zone are uniform throughout the zone. Tenability calculations with CONTAM are discussed in Chapter 19, but CONTAM is not capable of simulating detailed smoke flow in the vicinity of this door, and CONTAM cannot evaluate tenability on the stairwell landing of the fire floor where tenability conditions would be the worst. Ferreira and Cutonilli (2008) used CONTAM for analysis of stair ventilation, but their study did not consider the detailed smoke conditions in the vicinity of the stair door on the fire floor.

CFD modeling has the ability to simulate smoke flow in detail on the stairwell landing of the fire floor, and it can be used to evaluate tenability at this location. A CFD model can be used to determine the minimum design flow necessary to maintain tenable conditions for a specific design scenario. CONTAM can be used to simulate airflow in buildings as discussed in Chapter 14. This capability of CONTAM can be used to design a ventilation system that provides the minimum design flow across stair doors under design conditions. For this approach, the CFD simulations and the CONTAM simulations are done separately.

To evaluate the minimum flow needed past stair doors for stringent design conditions, Klote (2011) conducted FDS simulations including tenability analysis for a four-story section of stairwell. The cell size was about 3.7 in. (0.094 m), which was selected based on a sensitivity analysis. The design conditions were (1) a fully developed fire outside the stairwell near the stair door and (2) a warped stair door opening was considered to be 1 in. (25 mm) at the top side away from the hinges. In a fully developed fire, everything in the space that can burn is burning. While there is only limited research on

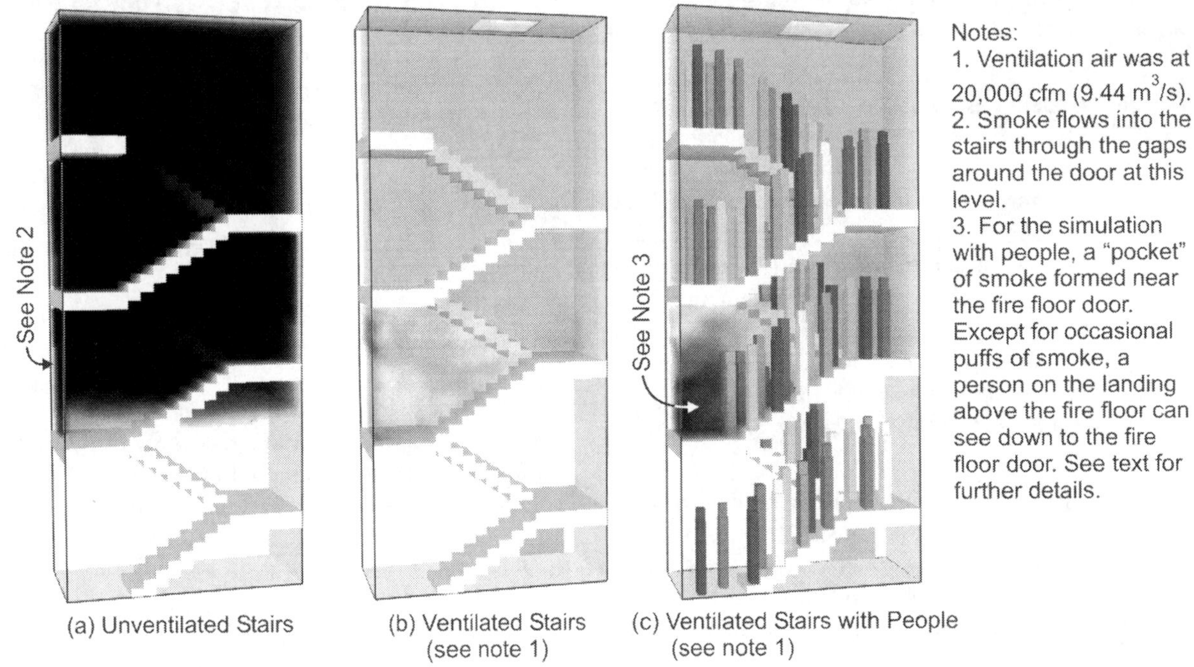

Figure 20.9 *FDS simulations showed that ventilation can provide tenable conditions.*

door warping (Fire International 1968), it is well known that doors often become warped when subjected to hot fire gases. The energy flowing through the door gaps was calculated. Smoke leakage into the stairs was approximated by smoke generation at the gap at the top and side of the stair door.

Figure 20.9a shows simulated smoke flow in the stairwell without forced ventilation. The stairs were filled with dense smoke resulting in visibility of only 2 to 4 ft (0.6 to 1.2 m). Such smoke would make it impossible for a people to see down to their feet. This visibility is definitely unacceptable.

Klote made FDS simulations at ventilation rates of 10,000, 15,000, and 20,000 cfm (4.72, 7.08 and 9.44 m^3/s). At the highest rate, a tenable environment was maintained in the stairwell, and the following discussion is about conditions in the stairs at 20,000 cfm (9.44 m^3/s). Figures 20.9b and 20.9c show smoke at this ventilation rate with and without people. In Figure 20.9c, a "pocket" of smoke forms in front of the door on the fire floor.

With the exception of the space with the "pocket" of smoke, the arrangement of people on the stairs in Figure 3c is that used by Achakji and Tamura (1988) for full people loading of stairs. It is human nature to avoid such a smoky space, and the simulations were made without people in the "pocket" of smoke. Even avoiding the "pocket" of smoke, people would have sufficient room to walk past the fire floor landing. Achakji and Tamura also experimentally determined that they could use cylinders in place of people in the stairs when measuring the flow resistance. Because of the requirements of FDS, it was not feasible to model people by cylinders, and people were modeled in FDS by rectangular solids.

The visibility criterion for this analysis was chosen for the section of stairs near the fire floor because at this location smoke conditions are the worst. The criterion is that a person on the landing above the fire floor can see down to the stair door on the fire floor. Figure 20.10 shows the visibility on a path from the landing above the fire floor to the stairwell door on the fire floor. The path was 5 ft (1.52 m) above the floor. The path length was 14.8 ft (5.4 m), and this length is shown on Figure 20.10 as a dashed line. For visibilities greater than the dashed line, a person can see the fire floor. For all of the simulation without people, visibility is above the dashed line. This means that without people, the visibility criterion is met.

With people in the stairs, it can be seen from Figure 20.10 that most of the time the visibility is above the dashed line. This means that most of the time a person on the landing can see the fire floor door. Visibility is below this line only for short periods of time. What is happening is that occasional puffs of smoke in front of the door prevent a person from seeing the door along the

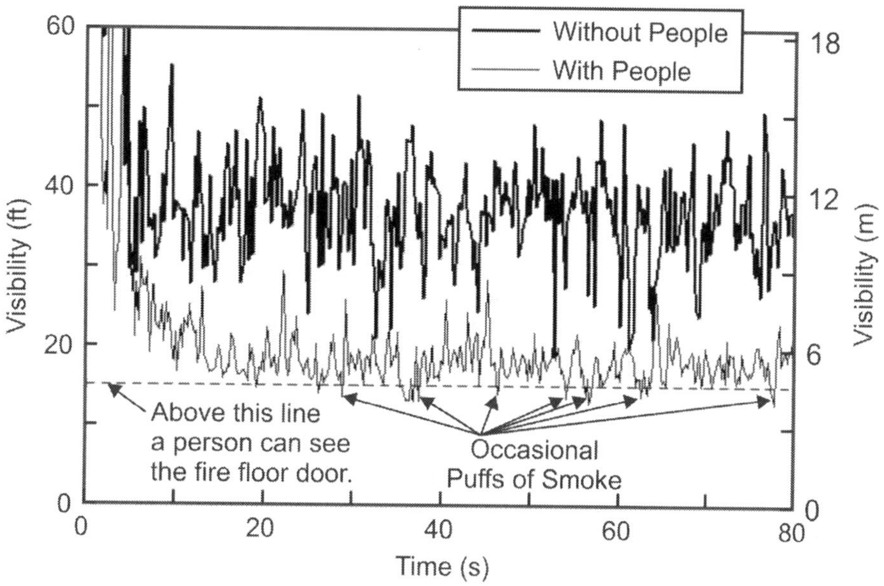

Figure 20.10 *Visibility looking down from the landing above the fire floor.*

path length. But a real person on the landing would be able to see the door on the sides of the puffs of smoke. This means that, with the stairs full of people, the visibility criterion is also met. Also, a person on the fire floor landing would be able to see to the landing above and to the landing below.

For a person in a wheelchair waiting an hour in the stairs on the fire floor, the fractional effective dose (FED) was 0.02. As discussed in Chapter 6, the FED model evaluates toxic gas exposure, and an FED greater than or equal to one indicates fatality. For an FED of 0.02, fatality would not be an issue, but it is possible that this exposure might cause some nausea for some people. The temperature in the stairwell on the fire floor ranged from 74°F to 98°F (23°C to 37°C). This temperature does not result in any threat concerning either heat exposure or thermal radiation exposure. Thus, a tenable environment was maintained in the stairs on the fire floor landing. The results of the CFD simulations showed that the conditions were worst in the stairs on the fire floor landing, and a tenable environment was maintained throughout the stairs.

The FDS simulations discussed demonstrate that stairwell smoke control by ventilation is feasible. These simulations were based on stringent conditions of design fire and door warping, and it showed that 20,000 cfm (9.44 m^3/s) of ventilation air was needed to maintain a tenable environment. A design based on less stringent conditions would be expected to need much less ventilation air.

NOMENCLATURE

c_p = constant-pressure specific heat
f_i = mass fraction of component i
k = thermal conductivity or kinetic energy of turbulence
$\dot{q}'''$ = heat release rate per unit volume
p = pressure
S_1 = grid size of simulation 1
S_2 = grid size of simulation 2
T = absolute temperature
t = time
t_{c1} = computer time of simulation 1
t_{c2} = computer time of simulation 2
u = velocity component in x direction
$\mathbf{u}$ = velocity vector
$\bar{u}$ = average velocity in the x direction
u' = fluctuating component of velocity in the x direction.
u = velocity in the x direction
v = velocity component in y direction
w = velocity component in z direction
X_j = body force in the j direction
$y_{p,i}$ = soot yield of component i
$y_{p,m}$ = soot yield of mixture
μ = dynamic viscosity
ΔH_i = heat of combustion of component i
ΔH_m = heat of combustion of the mixture

ρ = density

ε = turbulence dissipation rate

REFERENCES

Achakji, G.Y., and G.T. Tamura. 1988. Pressure drop characteristics of typical stairshafts in high-rise buildings. *ASHRAE Transactions* 94(1):1223–1236.

Aris, R. 1962. *Vectors, Tensors, and the Basic Equations of Fluid Mechanics*. New York: Dover.

Baum, H.R., K.B. McGrattan, and R.G. Rehm. 1997. Three dimensional simulations of fire plume dynamics. Fire Safety Science Proceedings, 5th International Symposium, March 3–7, Melbourne, Australia.

Davis, W.D., G.P. Forney, and J.H. Klote. 1991. Field modeling of room fires. NISTIR 4673, National Institute of Standards and Technology, Gaithersburg, MD.

Ferreira, M.J., and J. Cutonilli. 2008. Protecting the stair enclosure in tall buildings impacted by stack effect. Proceedings of the CTBUH 8th World Congress, March 3–5, Dubai.

Ferreira, M.J. 2008. Fire dynamics simulator: Ensure your software provides the safest atrium design for real world enforcement. *NFPA Journal* 102(1).

Fire International. 1968. The distortion of doors in fires due to asymmetric heating. *Fire International*, 19:36–39.

Forney, G.P. 2008. User's guide for Smokeview, version 5—A tool for visualizing fire dynamics simulation data. NIST Special Publication 1017-1, National Institute of Standards and Technology, Gaithersburg, MD.

Forney, G.P., and W.D. Davis. 1992. Analyzing strategies for elimination of flame blow-down occurring in the Navy's 19F4 fire fighting trainer. NISTIR 4825, National Institute of Standards and Technology, Gaithersburg, MD.

Hadjisophocleous, G.V., and C.J. McCartney. 2005. Guidelines for the use of CFD simulations for fire and smoke modeling. *ASHRAE Transactions* 111(2).

Klote, J.H. 2005. CFD analysis of atrium smoke control at the Newseum. *ASHRAE Transactions* 111(2):567–574.

Klote, J.H. 2011. Stairwell smoke control by ventilation. *ASHRAE Transactions* 117(1).

Launder, B.E., and D.B. Spalding. 1974. The numerical computation of turbulent flows. *Computer Methods in Applied Mechanics and Engineering* 3:269–289.

McCaffrey, B.J. 1983. Momentum implications for buoyant diffusion flames. *Combustion and Flame* 52(2)149–167.

McGrattan, K.B., et al. 2008a. Fire Dynamics Simulator (version 5) user's guide. NIST Special Publication 1019-5, National Institute of Standards and Technology, Gaithersburg, MD.

McGrattan, K.B., et al. 2008b. Fire Dynamics Simulator (version 5) technical reference guide, Volume 1: mathematical model. NIST Special Publication 1018-5, National Institute of Standards and Technology, Gaithersburg, MD.

McGrattan, K.B., et al. 2008c. Fire Dynamics Simulator (version 5) technical reference guide, Volume 2: verification. NIST Special Publication 1018-5, National Institute of Standards and Technology, Gaithersburg, MD.

McGrattan, K.B., et al. 2008d. Fire Dynamics Simulator (version 5) technical reference guide, Volume 3: validation. NIST Special Publication 1018-5, National Institute of Standards and Technology, Gaithersburg, MD.

NRC. 2007. Verification and validation of selected fire models for nuclear power plant applications. Vols. 1–7, NUREG-1824, U.S. Nuclear Regulatory Commission, Washington DC.

Papanastasiou, T.C., G.C. Georgiou, and A.N. Alexandrou. 2000. *Viscous Fluid Flow*. Worcester Polytechnic Institute, Worcester, Massachusetts: CRC Press.

Salley, M.H., et al. 2007. Verification and validation—how to determine the accuracy of fire models. *Fire Protection Engineering* (34):34–44.

Schlichting, H. 2000. *Boundary Layer Theory*, 8th ed, J. Kestin, Translator. New York: Springer-Verlag.

Sinclair, R. and D. Xiangdoing. 2012. Atrium smoke management natural venting challenges. *ASHRAE Transactions* 118(1).

Waters, R.A. 1989. Stansted terminal building and early atrium studies. *Journal of Fire Protection Engineering* 1(2):63–76.

White, F.M. 2005. *Viscous Fluid Flow*, international ed. New York: McGraw.

CHAPTER 21

Scale Modeling

John H. Klote

In today's world of powerful computers and computational fluid dynamics (CFD), many people forget how extensively scale modeling is used. Because scale modeling is done in the physical world, it has a reality that cannot be equaled by any computer simulation. Scale modeling of smoke movement can be used for (1) research, (2) design analysis, (3) verification of CFD simulations, and (4) fire reconstruction.

DIMENSIONLESS GROUPS

Dimensionless groups are essential for scale modeling, and the following discussion is intended to provide an understanding of source and relative importance of the various groups. For this reason, units are not given for the variables in this section, but the equations in this section are valid for the SI system (Chapter 1). The dimensionless groups that are of interest for fire applications including smoke control can be developed for the one-dimensional form of the governing equations of fluid dynamics (Quintiere 1989).

Conservation of Mass:

$$\frac{\partial \rho}{\partial t} + \frac{\partial (\rho u)}{\partial x} = 0 \qquad (21.1)$$

Conservation of Momentum in Vertical Direction:

$$\rho\left(\frac{\partial u}{\partial t} + u\frac{\partial u}{\partial x}\right) = \frac{\partial p'}{\partial x} + g(\rho_o - \rho) + \frac{4}{3}\mu\frac{\partial^2 u}{\partial x^2} \qquad (21.2)$$

where

$$p' - p - p_o \qquad (21.3)$$

and

$$\frac{dp_o}{dx} = -\rho_o g \qquad (21.4)$$

p_o is the ambient pressure distribution

Conservation of Energy:

$$\rho c_p\left(\frac{\partial T}{\partial t} + u\frac{\partial T}{\partial x}\right) = k\frac{\partial^2 T}{\partial x^2} - 4\kappa\sigma T^4 + \int_0^{4\pi}\kappa I\, d\omega + \dot{Q}''' + \frac{\partial p}{\partial t} \qquad (21.5)$$

Equation of State:

The equation of state for an ideal gas is

$$p = \rho R T \qquad (21.6)$$

Variables in the preceding governing equations are:

c_p = specific heat,
g = acceleration of gravity,
k = thermal conductivity,
T = temperature,
p = pressure,
p_o = ambient pressure,
u = x component of velocity,
x = position,
$\dot{Q}'''$ = rate of chemical energy per unit volume,
ρ_o = ambient density,

I = radiant intensity,
κ = absorption coefficient,
σ = Stefan-Boltzman constant,
μ = dynamic viscosity.

For an ideal gas, the gas constant R is

$$R = c_p - c_v \tag{21.7}$$

where c_v is the constant-volume specific heat.

Dimensionless variables are defined here:

$$\hat{x} = \frac{x}{l}, \tag{21.8}$$

$$\hat{u} = \frac{u}{U}, \tag{21.9}$$

$$\hat{p} = \frac{p}{p_*}, \tag{21.10}$$

$$\hat{p}' = \frac{p'}{p_*}, \tag{21.11}$$

$$\hat{\rho} = \frac{\rho}{\rho_o}, \tag{21.12}$$

$$\hat{T} = \frac{T}{T_0}, \tag{21.13}$$

$$\hat{t} = \frac{t}{\tau}, \tag{21.14}$$

$$\hat{Q} = \frac{Q'''l}{\rho_o V C_p T_o}, \tag{21.15}$$

$$\hat{I} = \frac{I}{\sigma T_o^4}, \tag{21.16}$$

where

l = geometric length scale,
U = characteristic velocity,
τ = characteristic time,
T_o = ambient temperature,
p_o = ambient pressure,
ρ_o = ambient density,
p_* = characteristic pressure defect ($p_* = \rho_o U^2$).

By substituting the dimensionless variables of Equations 21.8 to 21.16 into the governing equations, the following nondimensional form of the governing equations can be developed:

Mass:

$$\Pi_1 \frac{\partial \hat{\rho}}{\partial \hat{t}} + \frac{\partial(\rho \hat{u})}{\partial \hat{x}} = 0, \tag{21.17}$$

Momentum:

$$\hat{\rho}\left(\Pi_1 \frac{\partial \hat{u}}{\partial \hat{t}} + \hat{u}\frac{\partial \hat{u}}{\partial \hat{x}}\right) = -\Pi_2 \frac{\partial \hat{p}'}{\partial \hat{x}} + \frac{1}{\Pi_3}(1-\hat{\rho}) + \frac{4}{3\Pi_4}\frac{\partial^2 \hat{u}}{\partial \hat{x}^2}, \tag{21.18}$$

Energy:

$$\hat{\rho}\left(\Pi_1 \frac{\partial \hat{T}}{\partial \hat{t}} + \hat{u}\frac{\partial \hat{T}}{\partial \hat{x}}\right) = \frac{1}{\Pi_3 \Pi_5}\frac{\partial^2 \hat{T}}{\partial \hat{x}^2} + \Pi_3\Pi_5\Pi_6\Pi_7\left(\int_0^{4\pi} \hat{I}\, d\omega - 4\hat{T}^4\right) + \hat{Q} + \Pi_8 \frac{\partial \hat{p}}{\partial \hat{t}}, \tag{21.19}$$

State:

$$\hat{p} = \left(\frac{\Pi_9 - 1}{\Pi_8 \Pi_9}\right)\hat{\rho}\hat{T}. \tag{21.20}$$

where

$$\Pi_1 = \frac{1}{U\tau},$$

$$\Pi_2 = \frac{p_*}{\rho_o U^2},$$

$$\Pi_3 = \frac{U^2}{gl},$$

$$\Pi_4 = \frac{lU\rho_o}{\mu},$$

$$\Pi_5 = \frac{\mu c_p}{k},$$

$$\Pi_6 = \kappa l,$$

$$\Pi_7 = \frac{\sigma T_o^3 l}{k},$$

$$\Pi_8 = \frac{l p_*}{\rho_o c_p U T_o \tau},$$

$$\Pi_9 = \frac{c_p}{c_v}.$$

Groups Π_1 and Π_2 both have a value of one, and they can be ignored. Group Π_3 is the Froude number, F_r, which can be considered the ratio of inertial forces to buoyancy gravity forces. The Froude number[1] is

$$F_r = \frac{U^2}{gl} \tag{21.21}$$

where

F_r = Froude number,
U = velocity,
g = acceleration of gravity,
l = characteristic length.

Group Π_4 is the Reynolds number, R_e, which can be considered the ratio of the inertial forces to the viscous forces. In addition to the previous expression (Group Π_4) for the Reynolds number, it can be expressed in terms of kinematic viscosity, which is the dynamic viscosity divided by the density ($\nu = \mu/\rho$).

$$R_e = \frac{lU}{\nu} \tag{21.22}$$

where

R_e = Reynolds number,
l = characteristic length,
U = average velocity in flow path,
ν = kinematic viscosity.

Group Π_5 is the Prandtl number, which is nearly constant with respect to temperature. Because smoke is air mixed with a relatively small amount of combustion products, the Prandtl number can be neglected for modeling done in air.

Groups Π_6, Π_7, and Π_8 are heat transfer groups. Group Π_9 is the ratio of specific heats. Because Π_9 is nearly constant for gases, it can be neglected for modeling done in air.

SIMILITUDE

The basic concept of scale modeling is that tests are conducted with a scale model such that the Π groups are preserved. Preserving a Π group means that at a particular location in the model, that Π group has the same value in both the model and the full scale facility. It is not possible to preserve all the groups, but care needs to be taken so that the impact of the groups not preserved is not significant.

Froude modeling, saltwater modeling, and pressure modeling have all been used to simulate smoke movement in fire applications. Froude modeling has probably been used most, and is discussed later. The idea of saltwater modeling is to submerge the scale model in a tank of fresh water and inject salt water to simulate a heat source. The salt water is colored so that it is easily visible. Because the saltwater has a higher density than fresh water, the salt water tends to flow down whereas smoke tends to flow upward. This is accommodated by turning the model upside down in the tank. Saltwater modeling is similar to Froude modeling in that the Froude number is preserved. Because water and salt water are used to simulate the flow of air and smoke, saltwater modeling is sometimes called analog modeling.

Chow and Siu (1993) conducted smoke filling visualization experiments on several atria using saltwater modeling. Yii (1998) conducted a series of saltwater modeling experiments of balcony spill plumes. For general information about saltwater modeling, see Steckler et al. (1986).

In pressure modeling, both the Froude number and the Reynolds number are preserved. To preserve the Reynolds number, the model needs to be in a pressure vessel. However, Froude modeling is done without a pressure vessel, and Froude modeling can be done so that the impact of the Reynolds number is not significant.

FROUDE MODELING

Froude modeling is probably the most common approach to scale modeling of smoke movement. A scale model of the atrium or other facility is built. Tests are conducted in the model in air at normal atmospheric conditions. Scaling relations are used in the design of the tests and to convert measurements from the model to the full-scale facility. These scaling relations are such that temperatures are the same in the models as they would be in the full-scale facility. The Froude number is preserved, and the Reynolds number and the heat transfer groups need to be taken into account.

The scaling relations are

$$x_m = x_f\left(\frac{l_m}{l_f}\right) \tag{21.23}$$

1. An alternate form of the Froude number is $F_r = U/(gl)^{1/2}$. This is simply the square root of the Froude number that is used here, and basic concepts concerning the Froude number and the scaling relations for Froude modeling are the same, regardless of which form of the Froude number is used.

$$T_m = T_f \qquad (21.24)$$

$$\rho_m = \rho_f \qquad (21.25)$$

$$U_m = U_f \sqrt{\frac{l_m}{l_f}} \qquad (21.26)$$

$$\dot{V}_m = \dot{V}_f \left(\frac{l_m}{l_f}\right)^{5/2} \qquad (21.27)$$

$$\dot{m}_m = \dot{m}_f \left(\frac{l_m}{l_f}\right)^{5/2} \qquad (21.28)$$

$$t_m = t_f \sqrt{\frac{l_m}{l_f}} \qquad (21.29)$$

$$Q_m = Q_f \left(\frac{l_m}{l_f}\right)^{5/2} \qquad (21.30)$$

$$Q_{c,m} = Q_{c,f} \left(\frac{l_m}{l_f}\right)^{5/2} \qquad (21.31)$$

$$\Delta p_m = \Delta p_f \left(\frac{l_m}{l_f}\right) \qquad (21.32)$$

where

x = position in model or facility, ft (m),
l = length in the model or facility, ft (m),
T = temperature of gas, °F (°C),

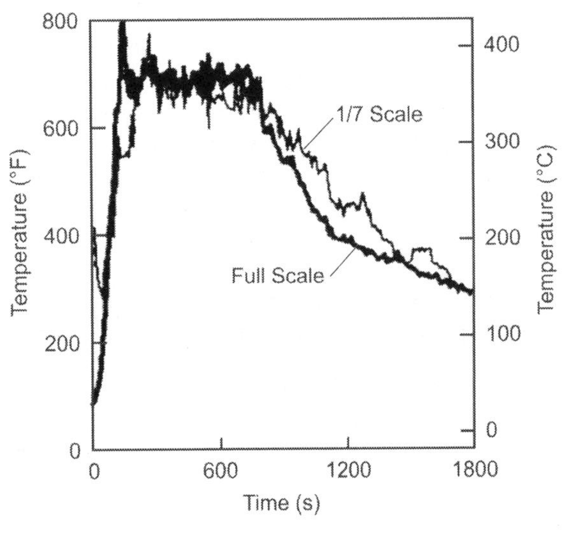

Figure 21.1 *Comparison of corridor temperatures using Froude modeling at NIST.*

ρ = density of gas, lb/ft^3 (kg/m^3),
U = velocity, ft/s (m/s),
V = volumetric flow, ft^3/s (m^3/s),
m = mass flow, lb/s (kg /s),
t = time, s,
Q = heat release rate, Btu/s (kW),
Q_c = convective heat release rate, Btu/s (kW),
Δp = pressure difference, in. H_2O (Pa),

with subscripts of m for the model and f for the full-scale facility.

Research at the National Institute of Standards and Technology (NIST) verified Froude modeling for smoke flow in a corridor (Quintiere et al.1978). Figure 21.1 shows temperature comparison from this NIST study of smoke flow in the full-scale corridor and a one-seventh scale model. As expected from Equation 21.24, the temperatures are nearly the same for the model and the full-scale facility. In Figure 21.1, the time from the scale model was adjusted to the full-scale time. Example 21.1 illustrates the use of the scaling relations.

NIST conducted a series of scale model compartment fires with the purpose of generating a database of comprehensive and accurate measurements that can be utilized both for a better understanding of and improved modeling for compartment fires, especially in the ventilation-limited regime (Johnsson et al. 2007). At the University of Canterbury in New Zealand, one-tenth scale modeling was used to gain a qualitative understanding of the behavior of the balcony spill plume in a six-story atrium (Tan 2009).

Tsujimoto et al. (1990) conducted experiments that verified the preceding scaling equations for smoke movement in atria. Chow and Lo (1995) used Froude modeling to simulate smoke movement and smoke filling in an atrium. Quintiere and Dillon (1997) used scale modeling for fire reconstruction of an atrium fire. The Reynolds number and heat transfer need to be taken into account as discussed next.

Reynolds Number

As discussed in Chapter 3, fluid flow consists primarily of laminar flow, transitional turbulent flow and fully rough turbulent flow. Because of the sizes of building spaces and the magnitudes of air and smoke velocities, the flows of interest in building spaces are fully rough turbulent flow. Provided that the flow in the scale model is fully rough turbulent, the impact of the Reynolds number is not significant. The model needs to be large enough so that the flow will be fully rough turbulent at the locations of interest. The general rule is that the smallest length that can support such turbulent flow is about 1 ft (0.3 m).

Heat Transfer

Approximate heat transfer scaling should be done when heat transfer is significant. For a semi-infinite surface, wall and ceiling materials can be scaled as

$$(k\rho c)_{w,m} = (k\rho c)_{w,f}\left(\frac{l_m}{l_f}\right)^{0.9} \qquad (21.33)$$

where $(k\rho c)_w$ is the thermal inertia of the wall or ceiling material (Btu^2 in h^{-1} ft^{-5} °F^{-2} or kW^2 m^{-4} K^{-2}s), and the subscripts m and f are for the model and the full scale facility respectively. For the thermal inertia of a number of materials, see Chapter 1.

When the smoke temperature is relatively low, heat transfer scaling is not very important. For example, the smoke in contact with an atrium fire is relatively low when the flame height is well below the atrium ceiling. In such an atrium, the smoke temperature drops significantly as it rises above the fire to the ceiling. In such an atrium fire, it is sufficient that the thermal inertias of the model walls and ceilings be roughly within an order of magnitude of the value calculated from the previous equation[2].

When gases are hot, heat transfer scaling is important. If hot smoke flowing under a ceiling or along a wall is important, heat transfer scaling would be important. Froude modeling is appropriate for smoke temperature away from the flame. Froude modeling cannot be expected to model flames realistically, and it cannot model fully developed fires in rooms realistically. However, Froude modeling can realistically model smoke flow away from the flames and away from fully developed room fires.

Construction of Model

Sometimes it is stated that the scale model needs to be built such that every dimension is an exact fraction of the full-scale facility, but not every small detail of the full-scale facility needs to be replicated. Little objects like small light fixtures, light switches, doorknobs, moldings, smoke detectors, and sprinklers would not be expected to impact the gross flow of smoke, and these objects can be neglected. In the absence of well developed criteria about the size of such little objects, it is suggested that objects less than about 9 in. (0.23 m) in the full-scale facility be neglected.

Because of the importance of visualizing smoke flow, the use of glass walls in models is common, regardless of considerations of approximate heat transfer scaling.

Instrumentation

The type and number of instruments used in a scale model depends on the purpose of the project. For projects intended to provide qualitative information about smoke transport, little instrumentation may be needed, but video or photography would be very important. For other projects, extensive instrumentation may be needed. The energy from lights is converted to heat when it is absorbed by solid surfaces such as the walls, ceilings, and floors of the model, and with bright photographic lights this heat can result in air currents in the model. Such air currents can impact smoke flow in the model, and lighting for video or photography should be chosen that does not cause any significant air currents.

The instrumentation needed for such scale model projects is like that used for full-scale fire tests. For information about instrumentation in full-scale tests, see Chapter 22. With any instruments, it is important that the instrument not unduly interfere with the experiment. Because of the small size of scale model experiments, this concern about interference is especially important. For example, a smoke meter that would be appropriate for a full-scale test would interfere with smoke flow in a reduced scale model. In scale modeling, smoke meters can be built into the model such that the only part of the meter in the smoke flow is the light beam.

Example

Example 21.1 illustrates considerations of scale modeling of an atrium with a fire. The scale of the model is chosen based on considerations of the Reynolds number. The materials that the model is made of are chosen based on considerations of thermal inertia and ease of fabrication. The fire size in the model is determined by the appropriate scaling relationship. This example illustrates how to locate a thermocouple in the model and how to scale the time of the thermocouple data to the full-scale facility. In general, there usually is a number of instruments in the model, and the data from these instruments are converted to corresponding values in the full-scale facility by the appropriate scaling relationships.

2. To be in an order of magnitude is to be within a factor of ten. For example, order of magnitude of 2 is about from 0.2 to 20.

Example 21.1. Scale Modeling of Atrium Smoke Movement

It is desired to conduct a series of scale model smoke flow tests in an atrium. The walls and ceiling are primarily concrete, but one wall consists of windows. The atrium is shown in the section in Figure 21.2. The design fire is 3000 Btu/s (3170 kW).

Part 1: What scale should be the scale of the model, and what will be the floor-to-ceiling height of the atrium?

To keep the impact of the Reynolds number insignificant, the smallest dimension of interest should not be less than about 1 ft (0.3 m). For this application, the smallest places that smoke is a concern are the balconies. The smallest relevant dimension of the balconies is the width which is 8 ft (2.44 m). The model atrium will be built with 1 ft (2.44 m) wide balconies. The atrium scale is one-eighth, and the floor-to-ceiling height in the model is 58(1/8) = 7.25 ft (2.21 m).

Part 2: The model should be made of what materials?

Heat transfer is not very important for this application, but approximate heat transfer scaling still should be examined. From Chapter 1, the thermal inertia $(k\rho\, c)_w$ of concrete is 25 Btu^2 ft^{-4} °F^{-1} h^{-1}. Ideally the thermal inertia of the walls, floors, and ceiling of the model would be $(k\rho\, c)_{w,m} = (k\rho\, c)_{w,f}(l_m/l_f)^{0.9} = 25\,(1/8)^{0.9} = 3.8$ Btu^2 ft^{-4} °F^{-1} h^{-1}. Because heat transfer is not very important, the thermal inertia of the surfaces can be roughly within an order of magnitude of this thermal inertia (about 0.4 to 40 Btu^2 ft^{-4} °F^{-1} h^{-1}). Gypsum wallboard has a thermal of 1.5 Btu^2 ft^{-4} °F^{-1} h^{-1} which is acceptable. Gypsum wallboard will be used for this reason and because it lends itself to fabrication.

Glass has a thermal inertia of 14 Btu^2 ft^{-4} °F^{-1} h^{-1}. Similar scaling for the glass windows gives a $(k\rho C)_{w,m}$ of 2.2, and windows should be in the range of about 0.2 to 20 Btu^2 ft^{-4} °F^{-1} h^{-1}. Glass will be used for the windows of the model.

Part 3: What should be the level of detail of the model?

The dimensions of the model need to be one-eighth of those of the full-scale facility, except that small objects less than about 9 in. (0.23 m) in the full-scale facility like light fixtures, light switches, doorknobs, and smoke detectors can be neglected.

Part 4: What is the size of the design fire in the model?

Q_f = 3000 Btu/s. The fire in the model will be $Q_m = Q_f(l_m/l_f)^{5/2} = 3000\,(1/8)^{5/2} = 16.6$ Btu/s (17.5 kW).

Part 5: In the full-scale facility, the temperature 6 ft above the floor of the top balcony is of interest. How high above the corresponding floor does a thermocouple need to be located?

This height is $x_m = x_f(l_m/l_f) = 6(1/8) = 0.75$ ft (0.23 m).

Part 6: At four minutes into a test, the thermocouple of Part 5 has a temperature of 110°F (43°C). How does this correspond to the full-scale facility?

Temperatures are the same for the model and the full scale facility. Time is scaled by the equation $t_f = t_m\,(l_f/l_m)^{0.5} = 4\,(8/1)^{0.5} = 11.3$ min. The test in the model predicts that at 11.3 min into a design fire in the full scale facility, the temperature at the location in Part 5 is 110°F (43°C).

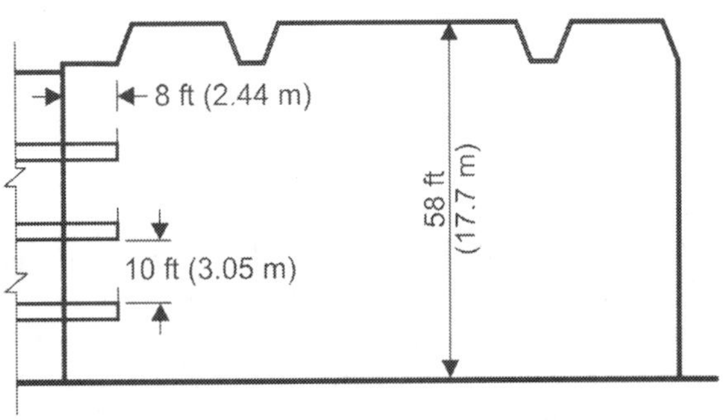

Figure 21.2 *Section of atrium for Example 21.1*

NOMENCLATURE

c_p = constant-pressure specific heat
c_v = constant-volume specific heat
F_r = Froude number
g = acceleration of gravity
I = radiant intensity
k = thermal conductivity
$k\rho\, c$ = thermal inertia
l = characteristic length, or length in the model or facility
m = mass flow
p = pressure
p_o = ambient pressure
Q = heat release rate
Q_c = convective heat release rate

$\dot{Q}'''$ = rate of chemical energy per unit volume
R_e = Reynolds number
T = temperature
t = time
T_o = ambient temperature
U = velocity, characteristic velocity, or average velocity
u = x component of velocity
V = volumetric flow
x = position
ν = kinematic viscosity
k = absorption coefficient
τ = characteristic time
ρ = density
m = dynamic viscosity
s = Stefan-Boltzman constant
ρ_o = ambient density
Δp = pressure difference

REFERENCES

Chow, W.K., and A.C.W. Lo. 1995. Scale modelling studies on atrium smoke movement and the smoke filling process. *Journal of Fire Protection Engineering* 7(2).

Chow, W.K., and W.M. Siu. 1993. Visualization of smoke movement in scale models of atriums. *Journal of Applied Science* 3(2).

Johnsson, E.L., M.F. Bundy, and A. Hamins. 2007. Reduced-scale ventilation-limited enclosure fires—heat and combustion product measurements. International Interflam Conference, 11th Proceedings, Volume 1, September 3–5, London, England.

Quintiere, J.G. 1989. Scaling applications in fire research. *Fire Safety Journal* 15(1):3–29.

Quintiere, J.G., and M.E. Dillon. 1997. Scale model reconstruction of fire in an atrium. 2nd International Symposium on Scale Modeling, June 23-27, University of Kentucky, Lexington, Kentucky.

Quintiere, J.G., B.J. McCaffrey, and T. Kashiwagi. 1978. Scaling study of a corridor subject to a room fire. *Combustion Science and Technology* 18(1).

Steckler, K.D., H.R. Baum, and J.G Quintiere. 1986. Salt water modeling of fire induced flows in multicompartment enclosures. NBSIR 86-3327, National Bureau of Standards, Gaithersburg, MD.

Tan, F. 2009. Physical scale modelling of smoke contamination in upper balconies by a balcony spill plume in an atrium. Fire Engineering Research Report 09/3, University of Canterbury, Christchurch, New Zealand.

Tsujimoto, M., T. Takenouchi, and S. Uehara. 1990. A scaling law of smoke movement in atrium. 11th Joint Panel Meeting of the UJNR Panel on Fire Research and Safety, National Institute of Standards and Technology, Gaithersburg, MD.

Yii, E.H. 1998. Exploratory salt water experiments of balcony spill plume using laser induced fluorescent technique. Fire Engineering Research Report 98/7, University of Canterbury, New Zealand.

CHAPTER 22

Full-Scale Fire Testing

John H. Klote

This chapter discusses full-scale fire testing intended to provide information about smoke control systems or to study related phenomena. This applies to tests that are part of a general research project and to tests that are intended to study a specific smoke control system. ASTM E603 is a guide for room fire tests intended to evaluate the fire-test-response characteristics of materials and assemblies under fire conditions (ASTM 2007). There is no similar guide for fire tests related to smoke control systems, but the information in this chapter should be helpful to those considering such fire tests and those who have to evaluate the results of fire tests. For a general history of fire testing, see Lawson (2009).

RESEARCH AND TESTING

Research is the search for knowledge with an open mind, using both theory and experiment. The two components of theory and experiment are the basis of science and engineering from which humanity has benefited so much. Also, these two components are essential to smoke control research. Smoke control experiments can be bench scale laboratory tests, scale model fire tests, full-scale fire tests or a combination of these. Smoke control tests need to be based on smoke control theory to assure that the tests are well designed and that the test results can be intelligently analyzed.

Full-scale fire tests need to be conducted by people who have an appropriate background. For full-scale fire tests in smoke control, the project leader or members of the project team needs to have a level of expertise in smoke control theory, smoke control design, and full-scale fire testing.

The project team needs to include at least one technician with experience in full-scale fire testing. Typical areas in which technicians need experience are fire hardening, video equipment, thermocouples, pressure transducers, gas analysis, and data acquisition. During project planning, it should be determined what abilities the technicians will need. If necessary, some new abilities will have to be acquired by training or bringing in additional help.

The difference between full-scale fire testing and demonstration fires needs to be addressed. As discussed above, full-scale fire testing is based on smoke control theory and conducted by engineers and technicians who have appropriate experience. It is possible to have a demonstration fire conducted this way, but often demonstration fires are done by people without the appropriate qualifications.

Poorly designed and conducted demonstration fires are a special concern. People can be misled by results of such demonstrations. Worse still is a written report of poorly designed and conducted demonstration fire tests that makes unjustified conclusions and recommendations.

Avoid Poorly Designed and Conducted Fire Tests: Reports of poorly designed and conducted fire tests can make unjustified conclusions and recommendations. Engineers and code officials should evaluate fire tests to determine if the project was properly conducted by experienced professionals. Potential research sponsors should make sure that projects they fund are properly conducted by experienced professionals.

DOCUMENTATION

Documentation should consist of at least a project plan, a safety plan, and a final report. For some projects, there could be additional reports, such as periodic progress reports and a preliminary project report.

Project Plan

The extent of a project plan depends on the complexity of the project. For some extensive and complex projects, the plan might be a formal published report providing information to sponsors and interested parties about the project. For a small and simple project, the plan could be a document of only a few pages intended only for the sponsor and project team. The project plan should address (1) the purpose of the project, (2) the test facility, and (3) instrumentation.

Many services are provided by established laboratories such as building security, electrical power, telephone service, potable water, compressed air, data acquisition systems, gas analysis equipment, smoke meters, pressure transducers, and velocity probes. Organizations with established laboratories often provide dry ice, wire, tubing, and other hardware. The project plan should identify any instruments, equipment, or supplies that are not supplied by the laboratory, so that these can be obtained in a timely manner.

For projects in buildings scheduled for demolition, the project plan should indicate how the necessary services, instruments, equipment, and supplies will be obtained. Items that need to be obtained before the project should be identified so that they can be purchased in a timely manner.

During development of the project plan, any additional abilities needed by the project team should be identified, and steps to provide those abilities should be undertaken.

Security can be especially important when tests are done in a building scheduled for demolition. For a building that is available for only a few days, there may not be enough time to recover from vandalism or theft. Even when a building is available for an extended period, recovering from vandalism or theft can use up significant amounts of project money. In addition to squatters and ordinary thieves, former and current employees with building keys are a concern. The project plan needs to address the security issue. For a building scheduled for demolition, it is recommended that the exterior building locks be replaced before any work related to fire tests begins.

Safety Plan

Because of the nature of fire, the safety plan is especially important. The safety plan can be delegated, but safety is ultimately the responsibility of the project leader. Often a person is appointed as the safety officer. The safety plan includes routine things like (1) the location of the first aid kit, (2) the names of the team members trained in CPR, and (3) what to do in medical and fire emergencies. In the U.S., the universal emergency telephone number is 911, but many organizations have their own fire departments and health care professionals. Arrangements should be made about who to call in medical and fire emergencies. In many organizations, technicians have firefighting training and equipment, and they are capable of dealing with some fire emergencies.

People need to be kept out of the fire and smoke spaces during the tests, and this is especially important when these areas are large. In buildings scheduled for demolition, homeless people can be squatters. Squatting consists of occupying such a space without the right to be there. In buildings where squatters are likely, fire and smoke spaces need to be checked before each test. In one fire test project, police with dogs were used to search for squatters.

Final Report

For some projects, the final report may just be a report of test provided by a laboratory to the client. Such a report of test may only consist of a description of the (1) test facility, (2) instrumentation, (3) schedule of tests, and (4) test results in graphical form. Test results in digital form may also be provided. In this situation, data analysis and conclusions need to be done by the client.

For a significant project that has general interest, a more formal report may be published. In addition to the material in the report of test, a formal final report should describe the purpose of the project, discuss data analysis, and provide conclusions and possibly recommendations.

TEST FACILITY

The test facility consists of the building where the tests are run, fire hardening, instruments, data acquisition system, and supporting materials. The supporting materials include instrument wiring, tubing, and cooling water. Test facilities can be permanent or temporary. There are numerous permanent full-scale fire facilities around the world, and a few are discussed here.

Modern fire tests include a data acquisition system that is controlled by a computer. Instruments are connected to the data acquisition system, and the system has the ability to scan many instruments a number of times each second. These scanned readings are stored in the computer.

The fire research facility at the National Institute of Standards and Technology (NIST) in Maryland operates three hoods of different size capable of determining the heat release rate (HRR) from oxygen consumption measurements (Stroup 2000). Oxygen consumption calorimeters are discussed in Chapter 5.

The Experimental Fire Tower of the National Research Council (NRCC) outside of Ottawa, Canada is a permanent, 10-story building specifically built for smoke control research. Some of the research projects at this tower are mentioned in Chapter 9. Next to this fire tower is a large burn hall which has been used for a variety of fire tests.

In 1989, full-scale fire tests of pressurization smoke control were conducted at the Plaza Hotel building before it was demolished. To measure temperature, pressure difference, gas concentration, smoke obscuration, and wind speed and direction, over 2.5 miles (4.6 km) of wire were installed between the instruments and the data acquisition system. The number of instruments varied with tests, but there were approximately 120 instruments. This test series is mentioned in Chapter 9. Fire tests in buildings scheduled for demolition do not need to be as extensively instrumented as those at the Plaza Hotel.

Fire Test Setup

Figure 22.1 is an example test setup showing the test areas, fuel, instruments, smoke detectors, telltale sprinklers, and video cameras. In this figure, the test areas consist of a burn room and a corridor. For these areas, the floor-to-ceiling height is 8 ft (2.44 m). This setup is inside a large fire laboratory building, and the figure shows an outline of a hood that is part of an oxygen depletion calorimeter. In Figure 22.1, the fuel is a wood crib, which is discussed later.

The number and kinds of instruments for fire tests depend on the purpose of the tests, and Table 22.1 lists instruments that could be used with the example setup. This table includes a channel number for each instrument. The channel numbers indicate the order in which the instruments are connected to the data acquisition system. In this figure, the heights of most of the instruments are in terms of distance from the ceiling, and in terms of distance below the top of the doorway. This convention is often used because smoke flows under ceilings and out the top part of doorways. As with most fire tests, Table 22.1 lists many thermocouples. This indicates the importance of temperature, but also because thermocouples are inexpensive and easy to repair.

A telltale sprinkler is one that is not connected to water, and it has no impact on the fire, but it provides information about when a sprinkler would have actuated. Wires can be connected to parts of the sprinkler such that a circuit is broken on actuation, or a light beam can be broken on actuation. The smoke detectors and the telltale sprinklers can be connected to the data acquisition system. Alternatively, these devices can be connected to a clock board such that a clock stopping indicates actuation of a particular device.

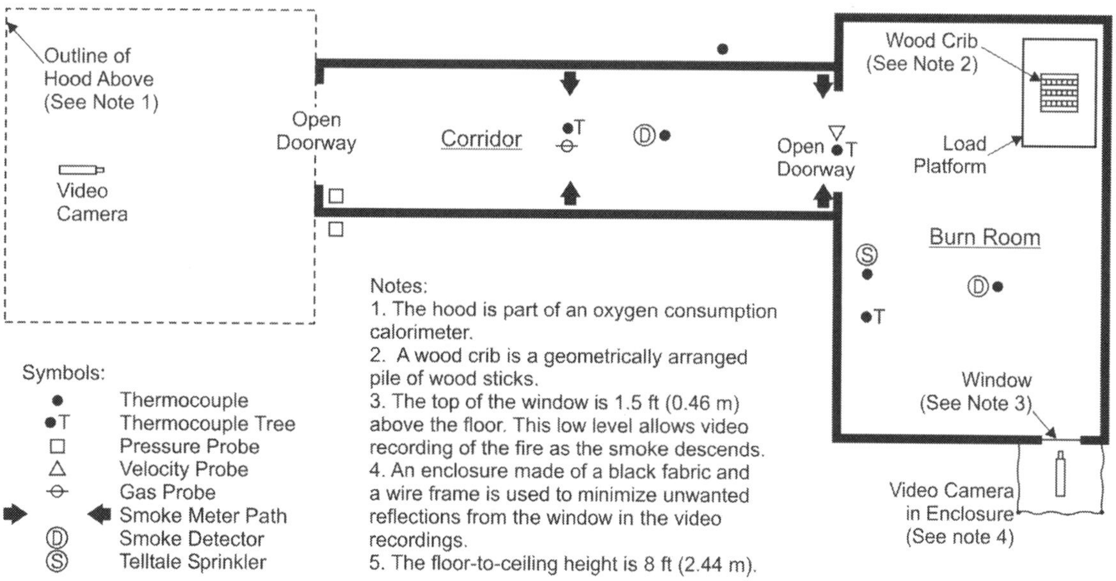

Figure 22.1 *Example fire test setup.*

Table 22.1: List of Instrumentation for Example Plan of Figure 22.1

Channel		Channel	
	Thermocouples:		**Pressure differences:**
00	Adjacent to telltale sprinkler	24	1.51 ft (0.46 m) from ceiling
01	Adjacent to corridor smoke detector	25	4.49 ft (1.37 m) from ceiling
02	Adjacent to burn room smoke detector	26	7.51 ft (2.29 m) from ceiling
03	Inside load platform (not shown)		**Velocity in burn room doorway:**
04	Outside burn room and corridor	27	0.43 ft (0.13 m) below top
05	Outdoors (not shown)	28	0.98 ft (0.30 m) below top
	In burn room doorway	29	2.17 ft (0.66 m) below top
06	0.43 ft (0.13 m) below top	30	3.35 ft (1.02 m) below top
07	0.98 ft (0.30 m) below top	31	4.49 ft (1.37 m) below top
08	2.17 ft (0.66 m) below top	32	6.27 ft (1.91 m) below top
09	3.35 ft (1.02 m) below top		**Gas concentrations:**
10	4.49 ft (1.37 m) below top		*3 ft (0.91 m) below ceiling*
11	6.27 ft (1.91 m) below top	33	Carbon monoxide
	In burn room	34	Carbon dioxide
12	1.51 ft (0.46 m) from ceiling	35	Oxygen
13	2.49 ft (0.76 m) from ceiling		*5 ft (1.52 m) below ceiling*
14	3.51 ft (1.07 m) from ceiling	36	Carbon monoxide
15	4.49 ft (1.37 m) from ceiling	37	Carbon dioxide
16	6.50 ft (1.98 m) from ceiling	38	Oxygen
17	7.51 ft (2.29 m) from ceiling		**Smoke meters:**
	In corridor		*At burn room doorway*
18	1.51 ft (0.46 m) from ceiling	39	5 ft (1.52 m) above floor
19	2.49 ft (0.76 m) from ceiling		*In corridor*
20	3.51 ft (1.07 m) from ceiling	40	5 ft (1.52 m) above floor
21	4.49 ft (1.37 m) from ceiling		Load platform
22	6.50 ft (1.98 m) from ceiling	41	*In burn room*
23	7.51 ft (2.29 m) from ceiling		

Notes: The instruments listed above do not include the smoke detectors, the telltale sprinkler, or those of the oxygen consumption calorimeter. The smoke detectors and the telltale sprinkler are connected to individual clocks that stop on actuation. The instruments of the oxygen consumption calorimeter are connected to a separate data acquisition system.

Fire Hardening

The extent of fire hardening needed depends on the severity of the fire exposure, and fire hardening often needs to be repaired between tests. The following are some approaches that can be used to provide protection for materials during fire tests.

For tests that have a fully developed fire, the walls and ceilings can be protected with calcium silicate board. This board is brittle. Even with experienced fire technicians, breakage is high during installation. Often, two layers of 0.5 in. (12.7 mm) thick calcium silicate boards are used. The first layer of boards is screwed in place, then the second layer is screwed over the first, and then the screw heads are covered with joint compound to protect them. The boards in the second layer are located so that the joints between boards of the second layer do not coincide with those of the first layer. To save costs, sometimes the first board is type X gypsum wallboard.

To maintain realistic thermal conditions at room boundaries, the walls and ceilings can be protected type X gypsum wallboard. For fully developed fires, the gypsum wallboard needs to be replaced after testing.

Because of the buoyancy of fire gases, floors typically need less protection than walls and ceilings, but floors still need protection for fully developed fires. Depending on the extent of the fire exposure, a layer of calcium silicate board or type X gypsum wallboard can be used over floors. Often this board is not fastened to the floor so that it is easy to place instrument wiring under the board.

Alumina silicate blanket insulation can be used to protect many objects such as structural members, pipes, and tubing. This blanket insulation can be secured with steel wires which can withstand many fire exposures.

Video

Videos of fire tests are valuable in many ways. The project engineer can study the video of the test for details that might have been missed during testing. The video can be used in presentations about the project.

Figure 22.1 shows a video camera set up to make a recording of the smoke flow in the corridor through the open doorway. In addition to the fire, the video from this camera will show the instruments in the doorway. Provided that the camera is far enough from the doorway, it does not need to be protected from the heat of the fire. A camera at a location subject to elevated temperatures can be protected by a small enclosure with a glass window.

Figure 22.1 also shows a video camera at a window in the burn room wall. Such a window should be made close to the floor, so that the camera can see the fire under the smoke layer for most of the fire. This camera is in an enclosure made of a black fabric and a wire frame is used to minimize unwanted reflections from the window.

FIRES AND FUELS

Some common fuels for fire tests are (1) materials normally in buildings, (2) wood cribs, (3), liquid hydrocarbons and (4) gaseous hydrocarbons. The kind of occupancy determines the kind of materials in a building, and these materials can be new or used. For example, a fire test of a clothing store can have new garments or garments purchased from a second hand shop or a charity such as the Salvation Army. Old garments have a cost advantage, but new garments can result in more reproducible fires. If old garments of out of date materials are burned, the impact of this on the applicability of the test results should be discussed in the final report. The cost of new garments can be reduced by purchasing second quality garments in quantity. Another example is furniture in a fire test of an office building, and this also can be purchased new or used.

If the early development of fire is not important, gas burners can be used to ignite upholstered furniture and other materials. The high initial heat output from a burner provides reproducible ignition and fire growth. However, early fire development is important for some applications, and other ignition sources can be used.

Waste baskets and cardboard boxes are common ignition sources of larger objects when burning materials that are normally in buildings. These containers are filled with a reproducible quantity of newspaper or other material and ignited by match, pocket lighter or a remote ignition source. An electric match is a remote ignition source made of a wooden match wrapped with nichrome wire. When a voltage is applied to the nichrome wire, it heats up and ignites the match. Sofas and upholstered chairs often are ignited by an electric match located between a cushion and one of the arms.

An advantage of using the kinds of materials normally in buildings is that the interaction between oxygen concentration and the HRR of the fire is like that of real building fires. This can be important for some projects that involve the performance of smoke control systems.

Wood cribs are geometrically arranged piles of wood sticks as shown in Figure 22.1. These fires are discussed in Chapter 5, and they have been used in many research projects where reproducible fires were needed. Wood cribs are often ignited by a small pool fire. An advantage of wood cribs is that the interaction between oxygen concentration and HRR is much like that of materials normally in buildings. To account for the large amounts of plastics in modern buildings, a combination of upholstered furniture and wood cribs can be burned.

The burning characteristics of many solid materials are affected by humidity, and many permanent fire facilities have rooms that maintain solid fuels at constant temperature and humidity before testing. Such humidity control is usually impractical when testing in buildings scheduled for demolition. The moisture in light weight objects such as fabrics changes quickly, and the air humidity at the time of the test is a good indicator. The moisture in larger objects such as furniture and wood cribs changes slowly, and air humidity at the time of the test does not indicate moisture content of these objects. Electronic devices are commercially available to measure the moisture content of wood.

A pool fire consists of liquid hydrocarbons burning in a metal pan, and these fires have the advantage of reproducibility. The HRR depends on the particular hydrocarbon, the shape of the pan and the area of the pan. Round or square pans are most common. Pans often are made of pieces of steel plate welded together. For large pool fires, plate of 0.25 in. (6.4 mm) thickness often is used. For safety reasons, large pool fires should not be ignited by handheld matches or pocket lighters. The person lighting the fire needs to be a safe distance away from the pan.

Gas burners often burn natural gas or propane, and gas fires have the advantage of reproducibility. The flow of gas to the burner is measured by a rotameter or a mass flow meter, and the HRR of the fire is calculated from the flow rate and the heating value of the fuel. For some fire scenarios, there is the possibility that not all the gas will be burned, and pockets of unburned gaseous fuel can pose explosion hazards. To ensure burning of all the fuel, combustion air is sometimes mixed with the fuel before it is burned. Such premixed fires do not have the same interaction between oxygen concentration and HRR of other fires, and this needs to be taken into account when planning the project. Gas burners have a major safety advantage in that they can be turned off in an emergency. A shut off valve to the gaseous fuel should be located away from the burn room and spaces open to it.

INSTRUMENTATION

An instrument is any type of equipment, apparatus, or device that is developed to measure a physical quantity with some degree of accuracy. For an instrument with one input and one output, calibration consists of varying the input over some range of constant values which causes the outputs to vary over some range of constant values. The input-output relationship developed this way comprises the static calibration of the instrument. However, few instruments have only one possible input.

For example, the inputs for a differential pressure transducer are pressure difference and instrument temperature. In theory, a family of input-output relationships for various temperatures could be developed, but this is not done. The instrument temperature is held constant, and the input-output relationship is developed for this temperature. Often, a correction term is developed for other instrument temperatures. For a differential pressure transducer, this correction term should be small. If the transducer were calibrated at 68°F (20°C), it typically could be used at normal room temperatures without concern for the small correction.

The calibration processes usually establishes the accuracy and reproducibility of an instrument by measuring some traceable reference standard. Such standards are defined in the International System of Units and maintained by national standards organizations such as NIST. For many instruments, calibration is either done by such a standards organization or it is traceable to such a standards organization.

In science and engineering, the accuracy of a measurement is the degree of closeness of the measurement to the quantity's "true" value. Because of the nature of measurement, the true value cannot be known, but there is a practical way around this limitation. The true value can be considered the value that would be measured by an exemplar method which is one that is agreed on by experts as being sufficiently accurate for the purposes to which the data ultimately will be put. For the purposes of smoke control fire tests, the calibration processes of national standards organizations such as NIST are exemplar methods.

The repeatability or reproducibility of a measurement is the degree to which repeated measurements under unchanged conditions show the same results. Repeatability is also called precision. In colloquial use, the terms accuracy and precision have basically the same meaning, but in technical use these terms are very different. It is possible to have high repeatability and low accuracy, and it is possible to have low repeatability and high accuracy. The idea of repeatability is illustrated in Figure 22.2a where the data points are close together, but they are some distance from the true value. Accuracy is illustrated in Figure 22.2b where the data points are farther apart, but they are all closer to the true value.

The error of a measurement is the difference between the measurement and the corresponding true value, and knowledge of the characteristics of the measurement process allows bounds to be put on the error of a measurement.

Instruments such as smoke meters, heat flux gages, load cells and load platforms are often located in test spaces that reach elevated temperatures. When in such spaces, it is common to water cool the instruments and insulate them with either calcium silicate board or alumina silicate blanket. Using warm water keeps the instrument temperature near the calibration temperature, and warm water also minimizes the potential for instrument damage due to condensation.

Some instruments are subject to drift which means that the performance of these instruments can move away from the performance at the time of calibration. Instruments that are subject to drift should be checked periodically to determine if their calibration is appropriate. For more information about instrumentation, see Doebelin (2003).

Instrument Wiring

The voltmeters used in almost all data acquisition systems are guarded voltmeters, and proper use of the guarding feature can eliminate extraneous electrical noise that can happen with instrument wiring (HP 1970). There are many ways to connect an instrument to a guarded voltmeter, and most of these involve extraneous noise that sometimes can result in serious errors.

When problems with extraneous noise are suspected with an instrument, extra effort should be expended to connect it in the best way. In general, the best way to connect an instrument to a guarded voltmeter is shown Figure 22.3a.

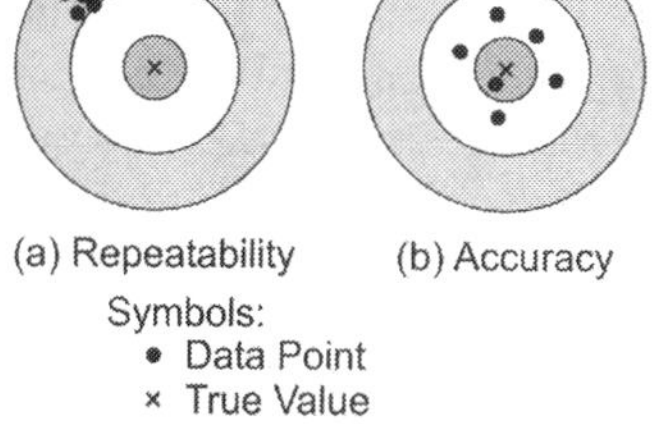

Figure 22.2 *Repeatability and accuracy.*

When an instrument and the guarded voltmeter are far apart, the potential for extraneous voltages in the wire can be high, and these extraneous voltages can be minimized by using a pair of shielded wires. A pair of shielded wires is two insulated conductors enclosed by a common conductive layer. The common conductive layer is the shield which may be made of braided strands of copper (or other metal), a nonbraided spiral winding of copper tape, or a layer of metal foil. The shield can be covered with a jacket. The shield is connected to the low post of the instrument and to the guard post of the guarded voltmeter as in Figure 22.3b.

Prefire Check

Before a fire test, instruments should be checked to determine if they are in working order. This check can be done efficiently by two technicians. One technician causes an input to a specific instrument, and the other technician uses the data acquisition system to determine if there is an appropriate output from that instrument. This is done for all the instruments connected to the data acquisition system. The input to thermocouples is often from a match or a pocket lighter, and the input to differential pressure transducers and velocity probes often consists of the technician blowing on them.

Ideally, instruments found during the prefire check not to be in working order should be repaired or

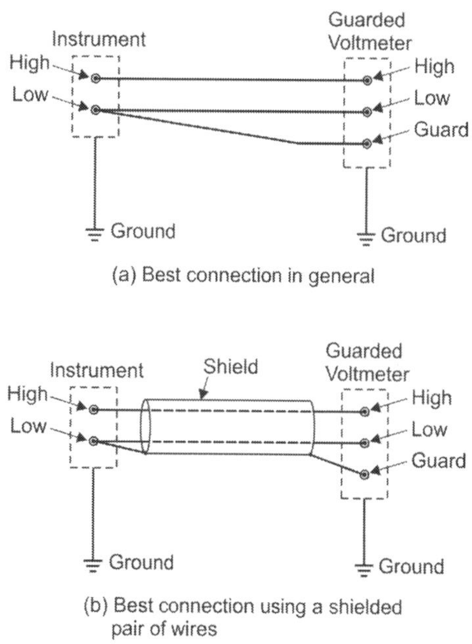

Figure 22.3 *Best connection of instrument to a guarded voltmeter.*

replaced. For instruments that are easy to repair, such as thermocouples, this can be done quickly before the test. If an instrument that is not working is difficult to repair or replace, it may be necessary to consider conducting the test without that instrument. After instruments have been repaired or replaced, they need to be checked to determine if they are in working order. When the data acquisition system is remote from the fire test area, walkie-talkies can be useful for the pre-fire check.

Temperature

There are many kinds of instruments that can measure temperature. Because thermocouples are extensively used to measure temperature in fire tests, the following discussion is limited to thermocouples.

Thermocouples work because of the Seebeck effect, which states that in a circuit made of two dissimilar metals, a current will flow if the two junctions are maintained at different temperatures. The junctions of the circuit in Figure 22.4a are at temperatures T_1 and T_2. This circuit consists of two wires of different metals *A* and *B*, and an ammeter that detects a current, *I*. The magnitude of the current, *I*, depends on the two materials and the temperatures T_1 and T_2.

Voltmeters are used more for data acquisition than are ammeters, and Figure 22.4b is the voltage mode of the thermocouple circuit. If the resistance of the voltmeter is infinite, there is no current flowing in this circuit, and the voltage is the Seebeck electromotive force (emf). The magnitude of the emf depends on the materials and the temperatures T_1 and T_2. The resistance of the voltmeter does not need to be infinite, but it needs to be very large for the voltage mode circuit to work as intended.

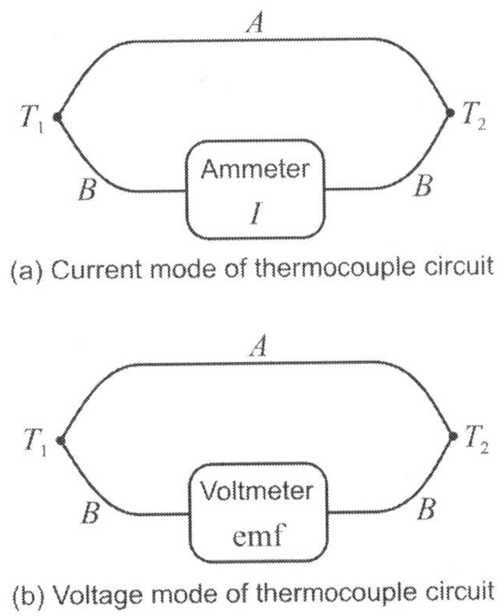

Figure 22.4 *Thermocouple circuits.*

The laws of thermocouples can be stated as follows.

1. The emf of a thermocouple with junctions at T_1 and T_2 is unaffected by the temperature elsewhere in the circuit if the two metals are homogeneous (Figure 22.5a).
2. If a third homogeneous metal, *C*, is inserted into either *A* or *B*, as long as the two new junctions are at like temperatures, the net emf of the circuit is unchanged, regardless of the temperature of *C* away from the junctions (Figure 22.5b).
3. If metal *C* is inserted between *A* and *B* at one of the junctions, the temperature of *C* at any point away from the *AC* and *BC* junctions does not matter. As long as the junctions *AC* and *BC* are both at temperature T_1, the net emf is the same as if *C* had not been inserted (Figure 22.5c).
4. If the emf of metals *A* and *C* is E_{AC}, and that of metals is *B* and *C* is E_{CB}, then the thermal emf of metals *A* and *B* is $E_{AC} + E_{CB}$ (Figure 22.5d).
5. If a thermocouple produces emf E_1 when its junctions are at T_1 and T_2, and E_2 when the junctions are at T_2 and T_3, then the thermal emf will be $E_1 + E_2$ when the junctions are at T_1 and T_3 (Figure 22.5e).

These laws are important for practical applications. The first law states that the lead wires connecting two junctions may be exposed to an unknown or varying temperature environment without affecting the voltage produced. Laws 1 and 2 make it possible to insert a voltmeter into the circuit to measure the voltage. The metal *C* represents the internal circuit of the voltmeter. The voltmeter circuit can be copper or another metal, but the connections to the voltmeter need to be at the same temperature. The voltmeter can be connected in two ways as shown in Figures 22.5b and c. Law 4 is relevant to thermocouple calibration. It is not necessary to calibrate all possible pairs of thermocouple metals. At NIST, the procedure is to calibrate individual metals with platinum, and calculate the other combinations.

In order to measure the temperature at a thermocouple junction, the temperature at the other junction needs to be known. This known temperature is called the reference temperature, and the junction with this temperature is called the reference junction. The normally used reference temperature is 32°F (0°C), and this temperature can be achieved by placing the reference junction in an ice bath. The fifth thermocouple law shows that calibrations based on one reference temperature can be adjusted to other reference temperatures. For fire applications, 32°F (0°C) is used as the reference temperature.

Accurate electronic compensation reference junctions are commercially available, and these should be used for fire tests. These compensation reference junctions adjust the voltage to yield an apparent reference junction of 32°F (0°C). These compensation reference junctions are part of some data acquisition systems.

These electronic compensation reference junctions are recommended because they are more accurate than most ice baths. An ice bath consists of a mixture of water and crushed ice in an insulated vessel, and the construction of an accurate ice bath requires a high level of equipment, skill, and attention to detail. Ice baths constructed by people lacking these skills can have an error of 2°F (1°C) or more.

NIST maintains the ITS-90 database for many types of thermocouples (Croarkin et al. 1993). The most commonly used thermocouples are chromel-constantan

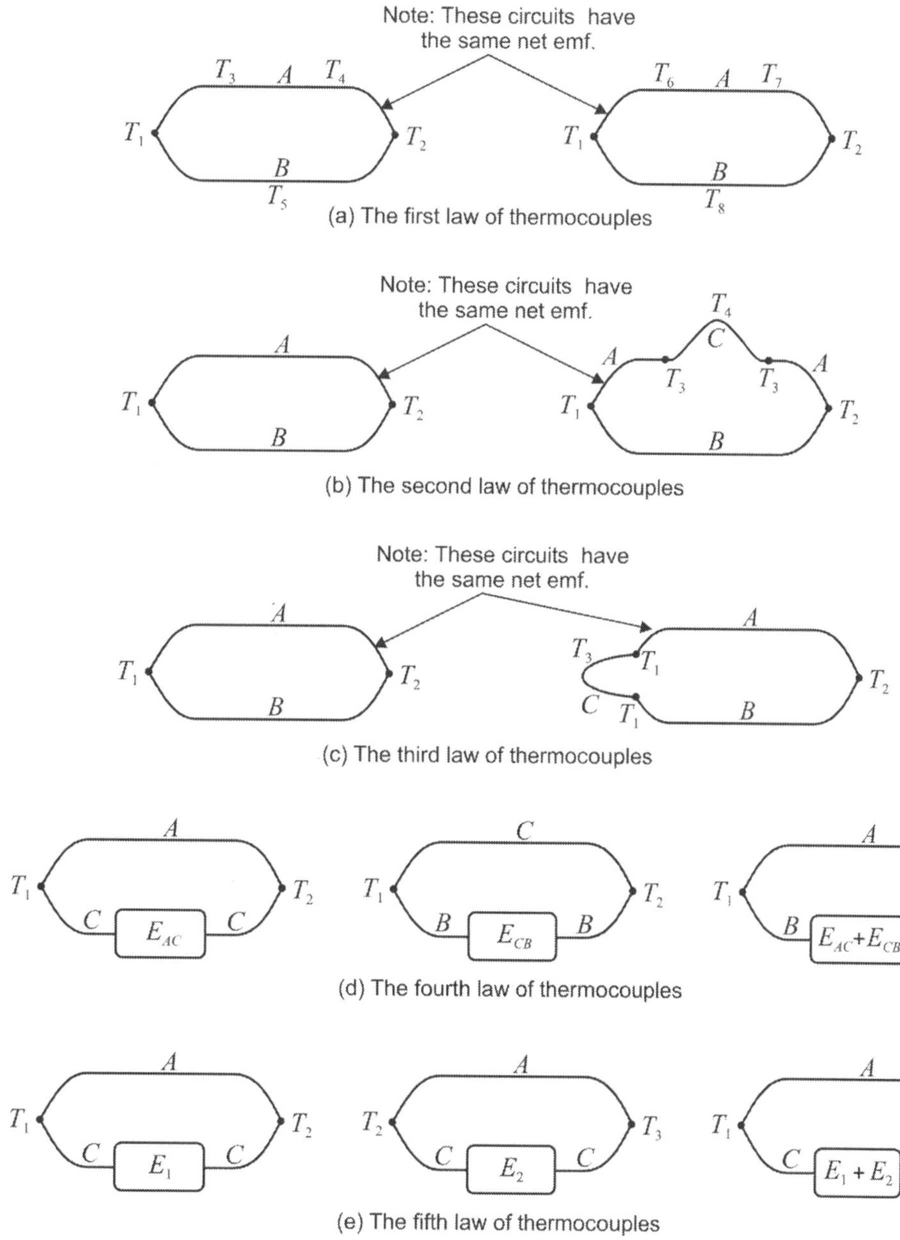

Figure 22.5 *Thermocouple laws illustrated.*

(type E), iron-constantan (type J), chromel-alumel (type K), and copper-constantan (Type T). Platinum 13% rhodium-platinum (type R) thermocouples and platinum 10% rhodium-platinum (type S) thermocouples are used for high temperature and corrosive applications. The temperature-voltage curves of these thermocouples are shown in Figure 22.6.s

Reputable companies who are regularly engaged in the manufacture of thermocouple wire produce wire that meets the composition requirements of NIST ITS-90. Thermocouples made from such wire do not need to be calibrated, and temperatures can be calculated from the equations provided by NIST ITS-90. When thermocouple wire is purchased, the following are suggested: (1) request that all the wire purchased be from the same manufacturing batch and (2) make one thermocouple from the wire and check it against an accepted instrument to assure that the wire is what was expected.

Type K thermocouples are used for full-scale fires by most fire laboratories throughout the world. The use of thermocouples made with platinum is considered a waste of money for practically all fire testing. Type K thermocouples can be used up to about 2500°F (1370°C), but they have an upper limit of about 2280°F (900°C) for continuous operation. Fire tests have relatively short durations, and this upper limit can be exceeded during fire tests. Thermocouples subject to high temperatures need to have high-temperature insulation.

Figure 22.7 shows types of thermocouple junctions. Thermocouple sheaths are made of stainless steel or other alloys. For grounded junctions and ungrounded junctions (Figures 22.7a and 22.7b), the physical protection provided by sheaths is needed only when burning fuels that produce especially corrosive combustion gases. Thermocouple probes with sheaths are commercially available. A disadvantage of sheathed junctions is that they have such large response times that they are not normally used for fire tests.

In most fire tests, thermocouples with fast response are needed to measure changes in temperature accurately. The exposed junction shown in Figure 22.7c is also called a *bare wire bead.* Exposed junctions have faster response times than sheathed junctions, and fire technicians can readily make bare wire beads from thermocouple wire by either gas welding or electric welding. Small thermocouple beads have fast response times and relatively low radiation errors. Small thermocouple beads can be made from small diameter wire. Thermocouples of adequate mechanical strength and fast response time can be made from thermocouple wire of 0.02 in. (0.51 mm) diameter.

Because high temperatures are expected in the burn room and spaces open to the burn room, thermocouple wire in these high-temperature spaces needs to have high-temperature insulation. Outside these high-temperature spaces, thermocouple extension wire with low-temperature insulation can be used. When

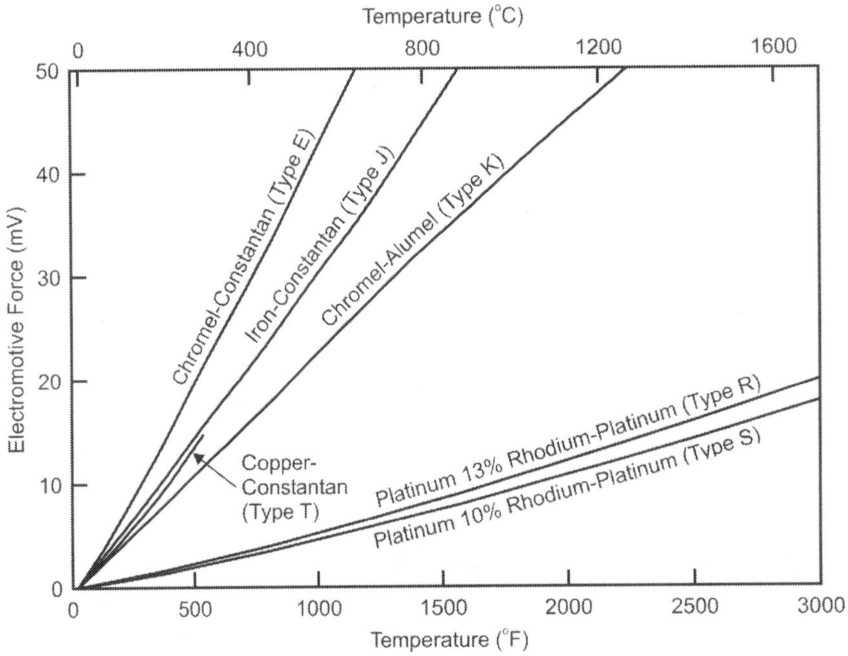

Figure 22.6 *Thermocouple temperature-voltage curves.*

installing thermocouples, extra wire should be used to simplify repairs (Figure 22.8).

Heat Flux

The data from heat flux gages are useful in evaluating the growth and spread of fire, and they are useful in evaluating the threat to life due to thermal radiation. A heat flux sensor is a transducer that generates an electrical signal proportional to the heat rate applied to the surface. The measured heat rate is divided by the surface area of the sensor to determine the heat flux. For fire applications, the heat rate consists of radiative and convective heat. Sensors that measure both radiative and convective heat are called total heat flux gages.

The sensor can be filled with a special air-cooled window to filter out convective heat so that only radiative heat flux is measured, and this kind of gage is called a radiative heat flux gage. There are two types of heat flux gages: the Gardon gage and the Schmidt-Boelter gage. These gages look the same (Figure 22.9), and they both are commercially available. A variety of devices have been made by fire laboratories to calibrate heat flux gages, and all these devices calibrate the gages against a secondary standard which itself has been calibrated by a national standards organization (Pitts 2003).

The sensor of a Gardon gage consists of a foil connected to the sensor body at its external radius, and connected to a thin wire at the center. The foil center and side form the hot and cold junctions of a thermocouple. When the temperature of the sensor increases, the sensor generates a signal. The sensor of the Schmidt-Boelter gage has a plated constantan wire wrapped around an insulating chip. Gardon gages can withstand extremely high flux levels. The range for Schmidt-Boelter technology is more limited, but it can be more sensitive.

In fire tests, both types of heat flux gage are water cooled. The gages, wires, and cooling-water tubes are generally wrapped with alumina silicate blanket insulation.

Pressure Difference

The pressure differences involved with most fire tests range from –0.08 to 0.08 in. H_2O (–20 to 20 Pa), and a sensitivity of about 0.008 in. H_2O (2 Pa) is needed. The only electronic instruments that meet these requirements are variable reluctance differential pressure transducers. These instruments need to be calibrated. Because these instruments are subject to drift, they should be checked periodically, and if needed they should be recalibrated. These transducers can be checked with an inclined liquid manometer.

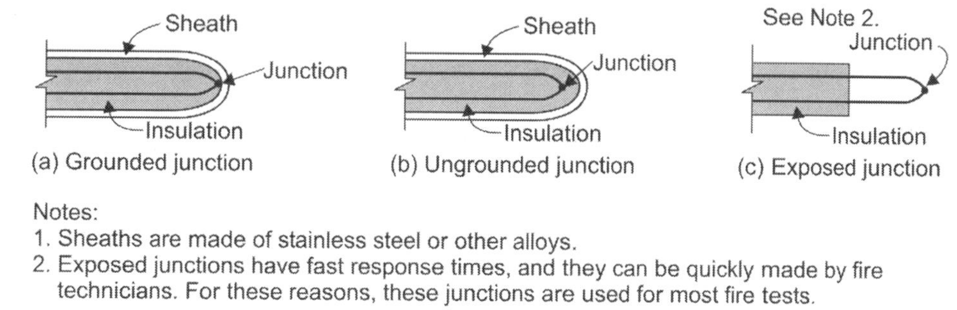

Figure 22.7 *Types of thermocouple junctions.*

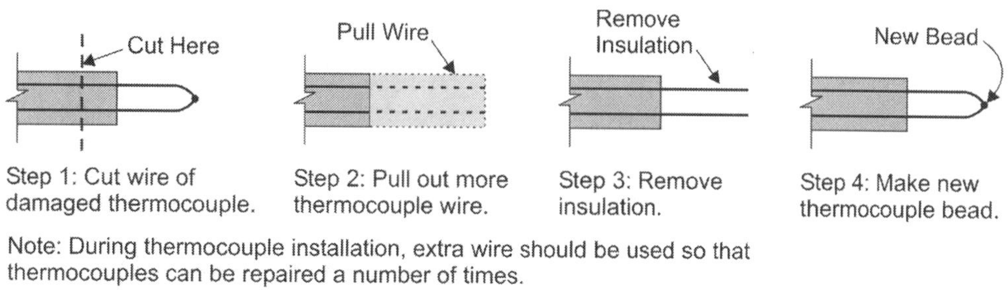

Figure 22.8 *Repair of exposed junction thermocouples.*

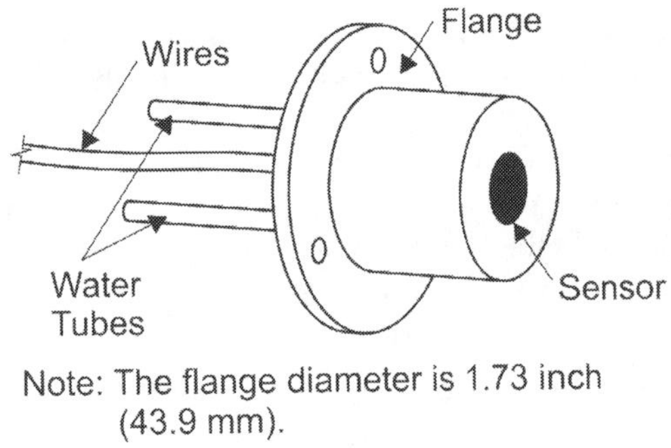

Figure 22.9 *Flux gage.*

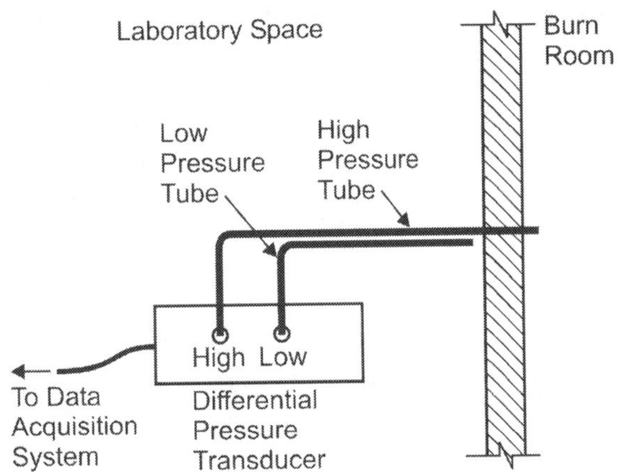

Figure 22.10 *Setup for a differential pressure transducer.*

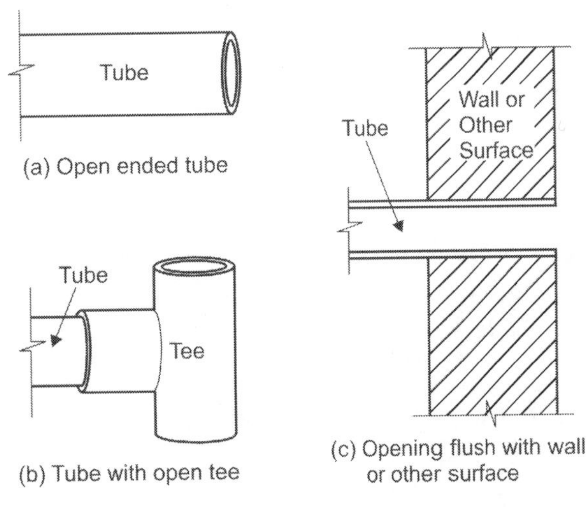

Figure 22.11 *Pressure probes.*

Figure 22.10 shows a differential pressure transducer set up to measure the pressure difference across the wall of the burn room. The low- and high-pressure ports of the transducer are connected to tubes that extend to the location across which the pressure difference measurement is to be made. The low- and high-pressure ports are connected to the low- and a high-pressure tubes. This designation only indicates the order of connection, and both tubes are subjected to very low pressures.

In Figure 22.10, a wire transmits a voltage to the data acquisition system. The convention used here is that a positive voltage indicates a pressure difference from the burn room to the laboratory space, and a negative voltage indicates the opposite. To simplify data analysis, it is suggested that one convention be used for all the differential pressure transducers.

In Figure 22.10, the open ends of the tubes are a kind of pressure probe that is shown in Figure 22.11a, and this open tube should be at a place of negligible air flow to minimize error due to velocity pressure. At locations with some airflow, a tee can be used on the end of the tube to help reduce velocity pressure errors as shown in Figure 22.11b. Pressure probes can also be located flush with walls, floors, and ceilings as shown in Figure 22.11c, and this also helps reduce errors associated with velocity pressure.

The fluid in tubes between probes and differential pressure transducers is almost completely motionless, and the inside diameter of the tube can be very small. In practice, inside diameters usually range from about 0.1 to 0.4 in. (25 to 100 mm). It is suggested that pressure probes subjected to fully developed fires be flush with the wall or other surface (Figure 22.11c), and the tube should be steel pipe or steel tubing. For less intense fires, other metals can be used. For locations outside of the fire test area, plastic tube often is used. Tube needs to be secured to stationary objects to the extent that vibration does not cause significant measurement errors.

Errors introduced into pressure difference measurements by tube temperatures need to be minimized. These errors can be estimated by

$$\Delta p_{Err} = 0.86\frac{g p_{atm} h}{R}\left(\frac{1}{T_l} - \frac{1}{T_h}\right)$$

$$\Delta p_{Err} = \frac{g p_{atm} h}{R}\left(\frac{1}{T_l} - \frac{1}{T_h}\right) \text{ for SI} \tag{22.1}$$

where

Δp_{Err} = pressure difference error due to tube temperature, in. H_2O (Pa),

g = acceleration of gravity, ft/s^2 (m/s^2),

p_{atm} = atmospheric pressure, psi (Pa),

h = height between probes and transducer, ft (m),
R = gas constant of air, 53.34 (287),
T_l = absolute temperature in low-pressure tube, °R (K),
T_h = absolute temperature in high-pressure tube, °R (K).

At sea level, Equation 22.1 becomes

$$\Delta p_{Err} = 7.36h\left(\frac{1}{T_l} - \frac{1}{T_h}\right)$$
$$\Delta p_{Err} = 3460h\left(\frac{1}{T_l} - \frac{1}{T_h}\right) \text{ for SI} \qquad (22.2)$$

The temperatures T_l and T_h are averaged values over the lengths of the tubes. Equations 22.1 and 22.2 are for tubes that are in a straight line from the pressure probes to the transducer. These equations provide information about eliminating Δp_{Err}. It follows that

- If both tubes are horizontal, h is zero, and there is no error due to tube temperature. This is so regardless of the temperature of the tubes.
- If T_l is the same or nearly the same as T_h, then there is no error due to tube temperature. This is so regardless of how large h is.

Tubes that extend considerable vertical distances in building can pose difficulties that may not be immediately apparent. This can be seen in an example of tubes installed through several stories of mechanical shaft in a building scheduled for demolition. Unknown to the test engineer, one of the tubes is located adjacent to an insulated hot-water pipe. That tube has a much higher temperature than the other, resulting in significant errors in the pressure difference measurements. For tubes installed on the building exterior, uneven solar radiation can have similar unwanted consequences.

Example 22.1 shows that solar radiation on one of the two tubes 45 ft (13.7 m) long can result in an error of 0.04 in. H_2O (10 Pa). Considering the maximum pressure difference in a fully developed room fire is about 0.08 in. H_2O (20 Pa), the error estimated in this example is extremely large. It is likely that an observant project engineer would notice this error and take steps to correct it.

Example 22.2 shows that the tube contacting hot- and cold-water pipe over distance of 45 ft (13.7 m) can result in an error of 0.024 in. H_2O (6.0 Pa). Errors of this size are a particular concern because they are small enough to go

Example 22.1. Δp Error due to Solar Radiation

Fire tests are conducted in a building scheduled for demolition. The burn rooms are four stories above the location of the instruments and the data acquisition system. The tubes are installed on the building exterior with a vertical distance of 45 ft (13.7 m). The air temperature is 70°F (21°C), and the low pressure tube is in the shade and at that temperature. The dark-colored high pressure tube is in direct sun light, and that tube reaches an average temperature of 105°F (41°C).

The temperatures are: $T_l = 70 + 460 = 530$°R; $T_h = 105 + 460 = 565$°R.

$$\Delta p_{Err} = 7.63h\left(\frac{1}{T_l} - \frac{1}{T_h}\right) = 7.63(45)\left(\frac{1}{530} - \frac{1}{565}\right) = 0.04 \text{ in. } H_2O \text{ (10 Pa)}$$

For fire testing, this error is extremely large, and it makes the data from this transducer useless. This demonstrates that attention to detail is needed to keep the temperature in these tubes the same. Correcting such an error after the tests is almost impossible in most situations because only guesses can be made about the temperatures in the tubes.

Example 22.2. Δp Error due to Water Pipes

Fire tests are conducted in another building scheduled for demolition, and a set of pressure probes is 45 ft (13.7 m) above the differential pressure transducer. The tubes are installed in a plumbing shaft, and the tubes are attached to uninsulated water supply pipes. The low pressure tube contacts the cold water pipe, and it is at 60°F (16°C). The high pressure tube contacts the hot-water pipe, and it is at 80°F (27°C).

The temperatures are: $T_l = 60 + 460 = 520$°R; $T_h = 80 + 460 = 540$°R.

$$\Delta p_{Err} = 7.63h\left(\frac{1}{T_l} - \frac{1}{T_h}\right) = 7.63(45)\left(\frac{1}{520} - \frac{1}{540}\right) = 0.024 \text{ in. } H_2O \text{ (6.0 Pa)}$$

This error is significant. Errors of this size are a particular concern because they are small enough to go unnoticed, but large enough to have an adverse impact on the conclusions of the project.

Example 22.3. Δp Error with Transducers on the Fire Floor

The differential pressure transducers are located on the fire floor, but they are remote from the fire spaces. One low-pressure tube is 70°F (21°C), and the corresponding high pressure tube is 77°F (25°C). For this transducer, $h = 6$ ft (1.8 m).

The temperatures are: $T_l = 70 + 460 = 530$°R; $T_h = 77 + 460 = 537$°R.

$$\Delta p_{Err} = 7.63h\left(\frac{1}{T_l} - \frac{1}{T_h}\right) = 7.63(6)\left(\frac{1}{530} - \frac{1}{537}\right) = 0.001 \text{ in. } H_2O \text{ (0.25 Pa)}$$

This error is not significant.

unnoticed, but large enough to have an adverse impact on the conclusions of the project. Example 22.3 shows that with the transducer on the fire floor, the error is negligible, provided that some care is taken to keep the temperatures relatively close to each other.

CAUTION

Avoid the Absolute Pressure Blunder: Some engineers have used absolute pressure transducers in place of differential pressure transducers, but this is a mistake. To get a pressure difference, the reading from one absolute pressure transducer is subtracted from that of another. Absolute pressure is about 400 in. H_2O (100,000 Pa), and the pressure differences of interest are at or below 0.08 in. H_2O (20 Pa). This amounts to subtracting two large numbers to get a very small difference which is poor practice because it magnifies error. Also, absolute pressure transducers magnify fluctuations in atmospheric pressure which are typically in the range of 0.02 to 0.04 in. H_2O (5 to 10 Pa). With modern data acquisition systems, the readings are made some time apart at different atmospheric pressures.

Solenoid Valves

To minimize the number of expensive differential pressure transducers needed, arrangements of solenoid valves have been used so that the pressure differences at many locations can be measured with one transducer. The idea is that, after the data acquisition system has taken a reading for one location, the two solenoid valves for that location close, and two solenoid valves for the next location open. After waiting for pressure transients in the tubes to drop off, the next reading is taken. This is done over and over to make measurements at all locations. A computer can be used to control the solenoid valves. This approach requires effort and dedication to detail.

Velocity

In fire applications, pitot-static tubes have significant problems associated with water droplet formation and solid particulate deposition. Smoke environments can cause difficulties with laser Doppler velocimetry, particle image velocimetry, and hot-wire anemometry. Stereoscopic particle image velocimetry has been used to measure buoyancy induced flow through a doorway (Bryant 2005). The bidirectional probe shown in Figure 22.12a was developed by McCaffrey and Heskestad (1976) with a large inlet to overcome the problems associated with smoky environments.

Bidirectional probes can be located where knowledge of flow direction is not known, and bidirectional probes respond appropriately when the flow reverses. McCaffrey and Heskestad tested probes with diameters D ranging from 0.5 to 1.0 in. (12.7 to 25.7 mm). They found that velocity is relatively insensitive to direction for an angle of about 100 degrees centered on the axis of the probe (Figure 22.12b). The velocity varies over this angle by only about 10%.

Bidirectional probes based on McCaffrey and Heskestad research have been used extensively for fire research and testing. Kent and Schneider (1987) have used a glass coating to protect probes from particularly corrosive environments. For additional information about bidirectional probes, see Sette (2005).

Thermocouples need to be located near bidirectional probes, because the temperature near a probe is needed to calculate the velocity. Bidirectional probes are connected to differential pressure transducers, and velocities are calculated from the pressure differences. The same kind of variable reluctance differential pressure transducers discussed should be used with bidirectional probes. The previous discussion about pressure difference measurement also applies here.

Gas Concentration

Fire tests that include considerations of tenability often include measurements of gas concentrations. The

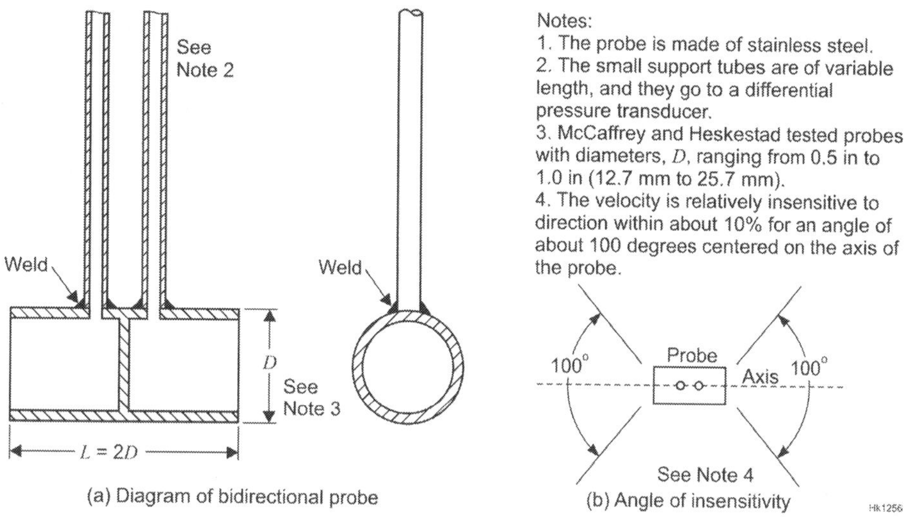

Figure 22.12 *Bidirectional velocity probe.*

two approaches to gas analysis are the batch method and the continuous method. For reasons to be discussed, the batch method is not recommended for serious fire testing. The batch method consists of people collecting samples of gas in small bottles or syringes, and the gases are analyzed later. People need to record where and when the samples were taken. With the batch method in a half hour test, the number of samples taken at each location can range from 3 to 10. When taking a small number of samples, the probability of missing important information is high, as is illustrated by Figure 22.13. Ideally, particulates and water vapor should be removed before analysis, but this can be difficult with the batch method.

The continuous method is an automated version of the batch method. A small gas pump continuously pulls gas from a probe. This gas is continuously treated to remove particulates and water vapor before it enters a gas analyzer. Data is collected from the analyzer by the data acquisition system at regular time intervals. The sampling rate with the continuous method ranges from 1 to 20 s depending on (1) objectives and requirements of the test, (2) number of instruments, (3) duration of the test, and (4) capabilities of the data acquisition system.

For tenability calculations, concentrations of at least CO, CO_2, and O_2 are needed. Sometimes CO and CO_2 concentrations are measured, and the concentration of O_2 is calculated by oxygen depletion. Gas analyzers are commercially available to measure concentrations of a number of gases in a flow stream. Infrared cells measure low concentrations of CO, CO_2, and other gases by determining the absorption of an emitted infrared light source. Infrared cells do not analyze O_2 concentration, but lambda sensors can measure oxygen.

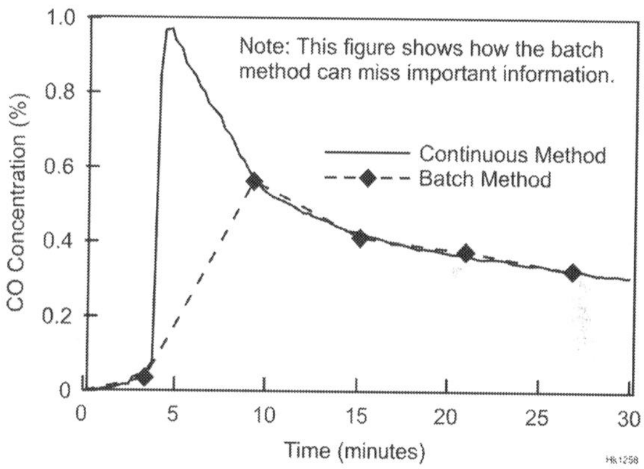

Figure 22.13 *Comparison of batch and continuous methods for CO.*

Gas analyzers are calibrated with gases of certified concentration, and cylinders of certified gases can be purchased commercially. Such a cylinder comes from the manufacturer with label attached that states the composition of the gas. To calibrate an analyzer for a specific gas, at least a certified *zero* gas and a certified *span* gas are used. For example, consider calibration of a CO analyzer for a test where the maximum CO to be measured is estimated at 0.5% by volume. The zero gas would have no CO or it would have only a small trace of CO. A span gas would be chosen with a concentration somewhat above the highest concentration anticipated in the test.

Because many gas analyzers are subject to drift, it is suggested that gas analyzers be calibrated before each fire test. Use and maintenance of gas analyzers requires

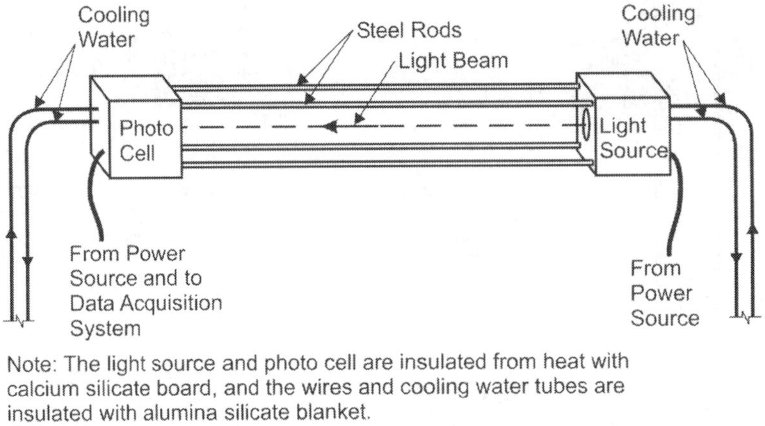

Figure 22.14 *Smoke meter used to measure smoke obscuration.*

special skills, and some manufactures provide classroom instruction for their equipment.

As with differential pressure transducer, an arrangement of solenoid valves can be used to allow one gas analyzer to take data from a number of probes during a test. The use of solenoid valves in this way requires considerable effort and dedication.

Smoke Obscuration

Smoke meters are used to measure smoke obscuration. Figure 22.14 is an illustration of a smoke meter. A beam of light is projected from a light source. After passing through a path filled with smoke, the beam falls on the photo cell. The data acquisition system reads the signal from this cell, and the smoke obscuration can be calculated. Smoke meters are commercially available, but they also can be made by fire technicians.

Lasers have been used as the light source, but a laser emits light of one wavelength. Because the human eye functions over a spectrum of wavelengths, incandescent bulbs are used in most smoke meters. A photovoltaic diode is used in the photo cell to detect the light and convert it to a voltage. Because the performance of the light source and the photo cell are temperature dependent, both need to be water cooled. Smoke meters are calibrated by placing a neutral density light filter in the light beam.

Because smoke meters are often subjected to high temperatures, they are protected. Typical protection consists of insulating the light source and photo cell with calcium silicate board, and insulating the wires and cooling water tubes with alumina silicate blanket. For smoke obscuration terminology and equations, see Chapter 6.

Load Cells and Load Platforms

Load cells and load platforms are used to measure the mass loss of burning materials. These devices are commercially available in a range of sizes and capacities, and the sensors in them are strain gages. Figure 22.15a shows a load cell located above the burn room ceiling weighing a mattress and bedding. The load cell needs to be protected from high temperatures or water cooled.

Figure 22.15b shows a load platform weighing a chair. Load platforms are commercially available in a range of sizes and capacities, and the sensors in them also are strain gages. For large fires such as in a fully involved fire, such protection is particularly challenging. Fire laboratories commonly calibrate load cells and load platforms with calibration weights made of steel or cast iron as shown in Figure 22.15c. These weights are available commercially in a wide range of sizes.

NONFIRE MEASUREMENTS

The instruments and measurements discussed so far have been for use during fire tests, but measurements are often needed before or after fire tests. For many such measurements the instruments and the data acquisition system can be used. For measurements that cannot be done with instruments associated with the data acquisition system, handheld instruments are often used. For example, temperatures can be measured by thermometers and handheld electronic sensors. Often such measurements are taken one or two minutes before the start of the test.

This section discusses ways to measure pressure difference, air velocity, and volumetric flow. The methods discussed here can also be used for research involved with field testing and acceptance testing, but acceptance testing of smoke control systems is usually done by air-balancing contractors using specialized instruments beyond the scope of this chapter.

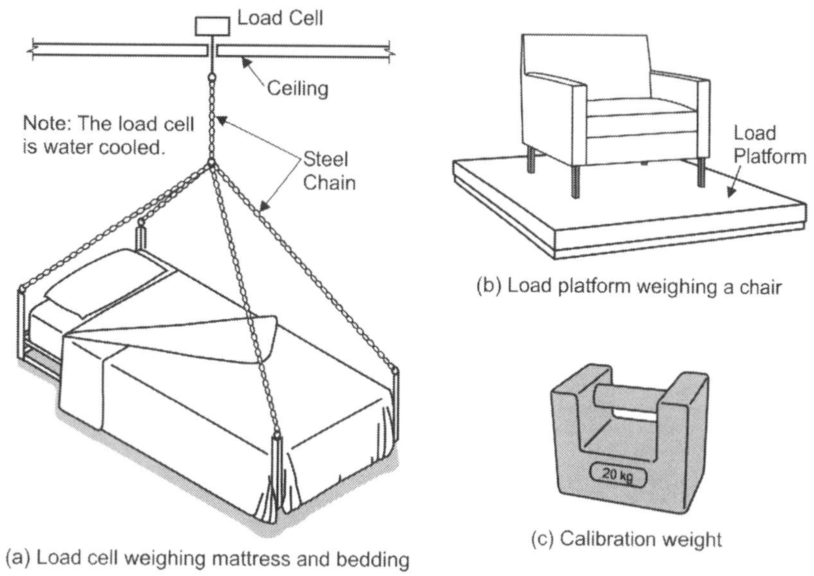

Figure 22.15 *Load cell and load platform.*

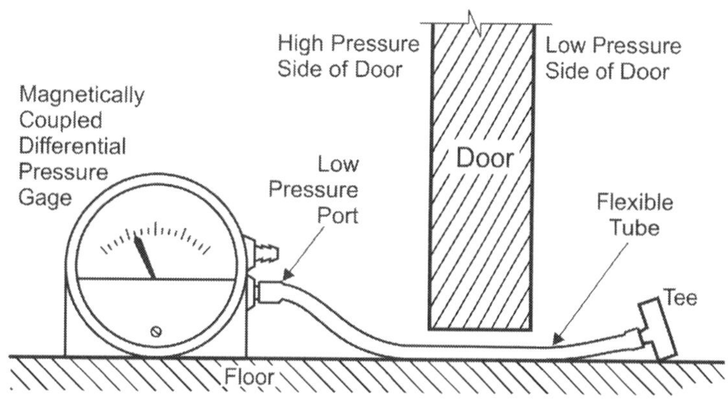

Figure 22.16 *Setup for measuring pressure difference across a door.*

Pressure Difference

Figure 22.16 shows a setup that has been extensively used for measuring the pressure difference across a closed door. The magnetically coupled differential pressure gage in this setup would typically have a sensitivity of about 0.01 in. H_2O (2.5 Pa) and a range from 0 to 0.25 in. H_2O (0 to 62 Pa). This kind of gage should be calibrated periodically. In this setup, the gage is located on the high pressure side of the door. Flexible tube connected to the low pressure port of the gage goes through a gap underneath and is terminated with a tee on the low pressure side of the door. The tee minimizes errors of pressure difference due to air velocity, but the tee can be omitted provided that the end of the tube is in relatively still air. An alternate setup could be used with the gage on the low pressure side of the door. Rubber or flexible plastic tube can be used, but caution needs to be exercised so as not to pinch the tube when the gap under the door is small. Small diameter metal tubing can sometimes be used in such cases, particularly through the gaps of some gasketed doors.

Bourdon tube gages are the most common type of pressure gages, but the friction of the mechanical linkages of these instruments limits sensitivity. No Bourdon tube gage is known with sufficient sensitivity for smoke control application.

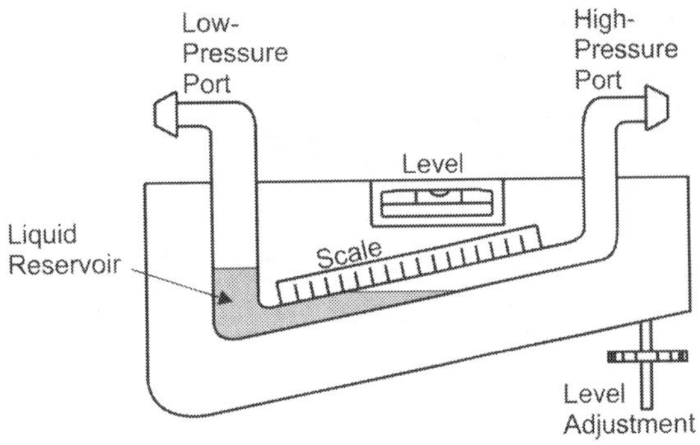

Figure 22.17 *Inclined liquid manometer.*

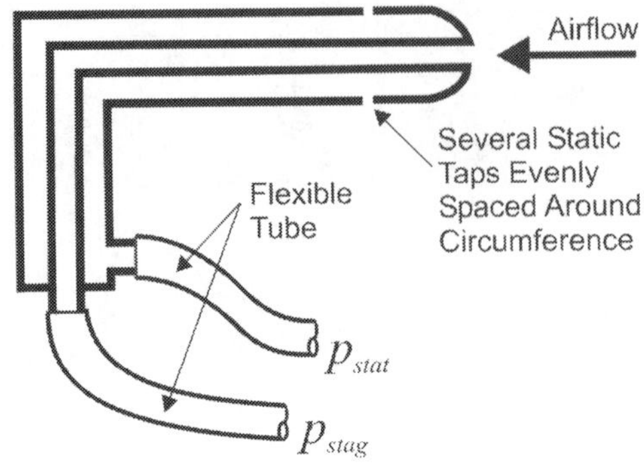

Figure 22.18 *Pitot-static tube.*

Pressure difference can also be measured with an inclined liquid manometer (Figure 22.17). Before use, this instrument needs to be adjusted so that it is level. The scales of inclined manometers should be made so that pressure difference can be read directly. Because the measurement principle of these devices is so fundamental, it is believed that commercially available inclined manometers are of sufficient accuracy for smoke control testing without independent calibration. Pressure difference can also be measured with an electronic differential pressure transducer of the type used for fire testing.

Velocity

Instruments for velocity measurement in smoke control applications include thermal anemometers and pitot-static tubes. Velocity measurements need to be made with as little interference with the flow as possible, and people making such measurements needs to keep away from the flow. These instruments can be used with a rod that can be held from outside the flow area.

Thermal anemometers are also called hot-wire anemometers and hot-film anemometers. These anemometers are available in two types: constant-current and constant-temperature, and both types have a velocity probe with a fine wire called a filum. For the constant-current type, the filum is subjected to a constant electrical current and the temperature of the filum depends upon the convective cooling of air flowing past it. Thus with the constant-current type, velocity is determined from the temperature of the filum. The constant-temperature type uses the same principle in a different way. The electrical current through a filum is adjusted so that its temperature remains constant. For this instrument, velocity is determined from the current. Handheld, battery-powered, thermal anemometers are commercially available with ranges of about 10 to 5000 fpm (0.05 to 25 m/s) and accuracies of about 5%.

Figure 22.18 is a schematic of a pitot-static tube, and it needs to be used with a differential pressure gage or transducer previously discussed. For a specific pressure difference, the velocity can be calculated as

$$U = 1097\sqrt{\frac{\Delta p}{C_{pt}\rho}}$$
$$U = 4.427\sqrt{\frac{\Delta p}{C_{pt}\rho}} \quad \text{for SI} \tag{22.3}$$

where

U = velocity, fpm (m/s),
Δp = pressure difference from manometer, in. H_2O (mm H_2O),
ρ = density of air, lb/ft^3 (kg/m^3),
C_{pt} = correction factor, dimensionless.

Manufacturers of Pitot-static tubes frequently supply information about the correction factor as a function of flow velocity or of Reynolds number. When used with an electronic differential pressure transducer, a pitot-static tube is capable of a range of 200 to 3000 fpm (1 to 15 m/s). When used with an inclined liquid manometer, the range can be 400 to 2000 fpm (2 to 10 m/s).

Volumetric Flow

The velocity in ducts, open doorways and other building spaces is characterized by turbulence often including large stationary vortices. Care needs to be taken to assure accurate flow measurements. Two approaches to flow measurement are (1) using a flow hood and (2) making a velocity traverse.

Figure 22.19 shows a flow hood being used to measure the flow of a ceiling supply. These devices are commercially available, and they can be used to mea-

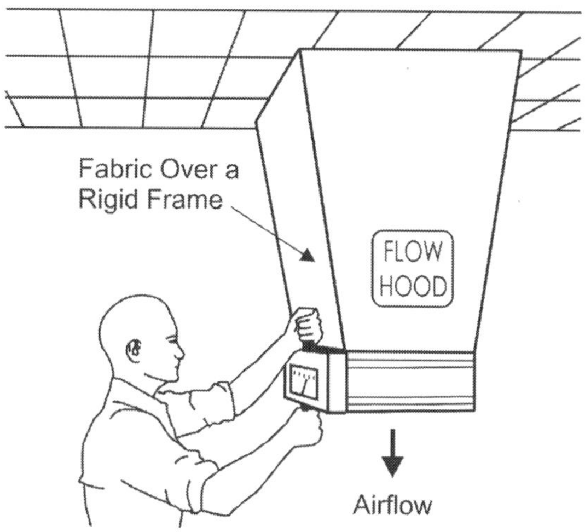

Figure 22.19 *Flow hood being used to measure volumetric flow of ceiling mounted supply.*

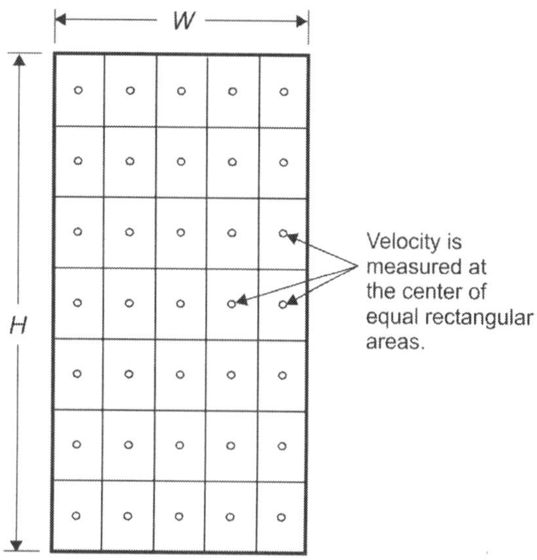

Figure 22.20 *Flow measurement traverse for corridors and open doorways.*

sure exhaust, as well as supply. The orientation of the hood can be changed to measure the flow at wall-mounted inlets and outlets. The accuracy of flow hoods is believed to be in the range of 10% to 15%, provided that the pressure loss through the hood is small compared to that through the ducts.

A velocity traverse consists of making velocity measurements in a systematic way so that the volumetric flow can be calculated. Figure 22.20 shows a velocity traverse of a corridor or open doorway, and the approach can also be used for ductwork. Because of the likely variations of velocity in doorways and corridors, it is recommended that these openings be divided into 30 to 64 equal spaces. Velocity readings should be taken in the center of equal areas over the cross section, and this can be done with thermal anemometers or pitot-static tubes. Stationary vortices need to be identified, and care needs to be taken to assign negative values to flows that are in the opposite direction as the main flow. Because stationary vortices are common in stair doors and corridors during smoke control operation, care needs to be taken at these locations. Punk sticks can be helpful in locating stationary vortices. The volumetric flow for a rectangular duct or other opening is calculated from the formula

$$V = HWU_{av} \tag{22.4}$$

where

V = volumetric flow, cfm (m^3/s),

H = height of opening, ft (m),

W = width of opening, ft (m),

U_{av} = average velocity, fpm (m/s).

DATA REDUCTION AND ANALYSIS

The process of data reduction and analysis is a process of inspecting, cleaning, transforming, and modeling data with the goals of highlighting useful information, developing empirical equations, and developing conclusions. Sometimes noise in data is excessive, and data is smoothed. Data smoothing is not universally accepted, but it is often done. Smoothing is addressed later, but when data has been smoothed it should be so stated in the final report of the project.

The first step in this process is to convert the raw data from the data acquisition system to data in terms of physical units. Raw data was collected in volts from instruments, and this is converted to appropriate physical units such as °F (°C), in. H_2O (Pa), and ft/min (m/s). For each instrument, there is a conversion equation. Some laboratories have developed their own general purpose data reduction programs to do this conversion. Alternatively, this conversion can be done with a spreadsheet program.

Once the data is in physical units, it is graphed and examined. Even with modern instrumentation and data acquisition systems, data errors can happen. Figure 22.21 shows a temperature curve with an erroneous datum of –3.216E+5°F (–1.786E+5°C). This temperature has no physical meaning, and it is an obvious error. It might seem that it would be best to simply omit the erroneous datum, but this can cause problems because much of the data reduction and analysis software requires data points at uniform intervals. To overcome this, the erroneous datum can be

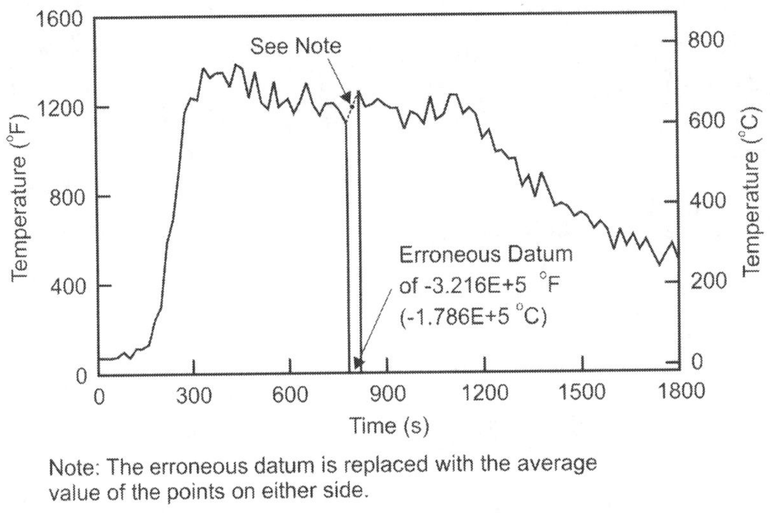

Figure 22.21 *Adjustment for erroneous datum.*

replaced with the average value of the points on either side as was done in Figure 22.21.

The extent to which data is analyzed depends on the purpose of the tests. For tests conducted to examine the performance of a specific smoke control system or a portion of a smoke control system, analysis can be as simple as presenting data in graph form and explaining in detail how the system did or did not meet some level of performance criteria. Such criteria are often project specific. For example, a criterion might be that the visibility on the floor above the fire floor is not to be less than 35 ft (10.7 m) for 20 minutes after ignition. Analysis could consist of presenting the data from a number of smoke meters on that floor, and discussing if the criterion is met. For information about analysis of fire test data, see Peacock and Babrauskas (1991).

For smoke control projects that have a more pronounced research component, data analysis is usually much more complex. The purpose of a series of tests can be categorized as (1) validation of a theoretical equation, (2) validation for a computer model, (3) study a phenomenon, and (4) development of an empirical equation for a phenomenon. Each of these presents unique challenges, and data analysis for such projects requires a high level of expertise in smoke control theory and practice.

Often, least squares curve fitting is used to develop empirical equations. The least squares method can be used to develop straight lines, polynomials, and exponential equations. Sometimes a phenomenon is divided into regions, and different empirical equations are developed for each region. Error analysis can be important for such efforts. For information about numerical methods that can be used for data analysis, see Hoffman (2001) and Press et al. (2007). For information about error analysis see Bevington and Robinson (2003) and Mandel (1985).

Data Smoothing

Often data is smoothed with the intent of removing a large amount of noise and keeping the essential character of the data. The noise can be from fluid turbulence, variations in atmospheric pressure, fluctuations in wind, and electrical noise. Electrical noise should be minimized by use of proper instrument wiring as discussed.

Smoothing of experimental data is often done using the following floating average method. Smoothing tends to reduce the peaks and valleys in data, and care needs to be taken not to destroy any important information. The following discussion is based on data that has been collected at uniform time intervals ($\Delta t_1 = \Delta t_2 = \Delta t_3$, etc.). A three-point floating average is written as

$$y_{Si} = \frac{1}{3}(y_{i-1} + y_i + y_{i+1}) \text{ for } i = 1, N \tag{22.5}$$

where

y = original data for $i = 1$ to N,

y_{Si} = smoothed data for $i = 1$ to N,

N = number of data points.

For any odd number of points averaged, the general equation for the floating average is

$$y_{Si} = \frac{1}{k}\sum_{j=1}^{k} y_{i-m+j} \text{ for } i = 1, N \tag{22.6}$$

Example 22.4. Data Smoothing

Temperature data from a half hour fire test has been collected at 20 s intervals, and the first five temperatures are 22.33°C, 22.34°C, 22.30°C, 24.15°C, and 22.32°C. A five-point average is to be calculated, and the equation is $y_{Si} = (y_{i-2} + y_{i-1} + y_i + y_{i+1} + y_{i+2})/5$ which is calculated for data points $i = 1$ to N. N is the number of data points, which is 91.

Part 1: Calculate the smoothed temperature for $i = 1$. For Point 1, $y_{S1} = (y_{-1} + y_0 + y_1 + y_2 + y_3)/5$, but there are no data for points y_{-1} and y_0. For these points, the value of y_1 is used.

$$y_{S1} = (22.33 + 22.33 + 22.33 + 22.34 + 22.30)/5 = 22.33°\text{C}.$$

It should be noted that in data smoothing, each smoothed datum point is calculated from original data, and smoothed data is never to be calculated from any smoothed data at other times.

Part 2: Calculate the smoothed temperature for $i = 3$.

$$y_{S3} = (y_1 + y_2 + y_3 + y_4 + y_5)/5 = (22.33 + 22.34 + 33.30 + 24.15 + 22.32) = 22.69°\text{C}.$$

Part 3: Calculate smoothed data for all the 91 points. The original data and the smoothed data are listed in the table below, and they are shown in Figure 22.22. The smoothed data was calculated using a spread sheet program. As in Part 1, the original data for Point 1 was used for points 0 and –1. A similar situation happens at the end of the data when original data are needed for Points 92 and 93, and the value at Point 91 was used for those at 92 and 93.

i	Original Data	Smoothed Data	*i*	Original Data	Smoothed Data	*i*	Original Data	Smoothed Data	*i*	Original Data	Smoothed Data
1	22.33	22.33	24	740.76	680.59	47	689.35	607.90	70	476.59	431.57
2	22.34	22.69	25	777.25	709.89	48	640.34	623.58	71	427.26	427.95
3	22.30	22.69	26	602.03	705.40	49	517.68	614.20	72	321.38	419.34
4	24.15	22.77	27	678.96	659.66	50	681.01	614.58	73	479.71	397.21
5	22.32	27.74	28	728.01	623.69	51	542.63	608.12	74	391.76	399.07
6	22.73	28.67	29	512.04	603.74	52	691.25	643.47	75	365.92	408.05
7	47.20	34.88	30	597.39	573.01	53	608.03	617.85	76	436.60	369.32
8	26.95	53.26	31	502.31	522.07	54	694.44	651.63	77	366.27	374.00
9	55.22	78.31	32	525.31	538.92	55	552.89	647.64	78	286.05	363.94
10	114.22	130.41	33	473.29	525.50	56	711.56	655.01	79	415.15	362.70
11	147.94	197.60	34	596.28	558.64	57	671.29	624.17	80	315.64	364.52
12	307.72	283.61	35	530.33	551.92	58	644.87	650.30	81	430.40	365.96
13	362.90	387.29	36	667.97	582.23	59	540.24	610.19	82	375.38	361.20
14	485.27	497.36	37	491.73	601.91	60	683.56	592.73	83	293.24	352.80
15	632.60	552.29	38	624.86	610.29	61	510.98	555.33	84	391.33	343.58
16	698.32	628.33	39	694.68	615.34	62	584.00	563.59	85	273.66	335.35
17	582.34	655.83	40	572.20	653.83	63	457.89	509.02	86	384.29	325.14
18	743.12	675.21	41	693.21	639.44	64	581.53	523.40	87	334.23	291.39
19	622.76	686.09	42	684.21	642.97	65	410.71	483.31	88	242.18	301.15
20	729.52	676.12	43	552.91	649.04	66	582.87	491.30	89	222.60	275.05
21	752.73	677.59	44	712.32	628.31	67	383.56	461.96	90	322.46	258.96
22	532.46	701.19	45	602.57	629.34	68	497.84	475.13	91	253.78	261.28
23	750.46	710.73	46	589.54	646.82	69	434.80	444.01			

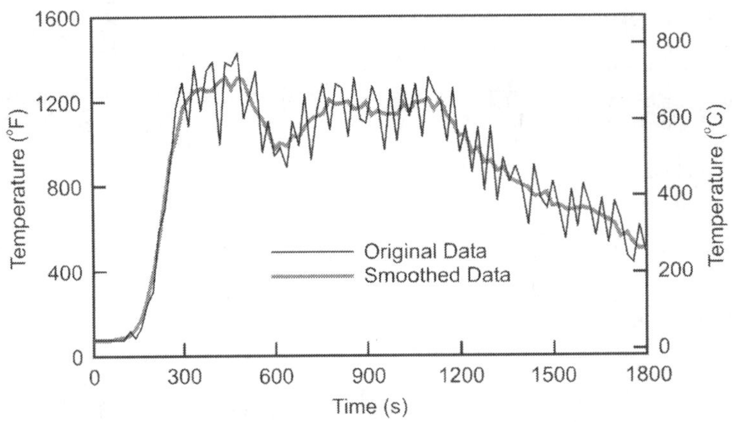

Figure 22.22 *Example of data smoothing by five point floating average.*

where

k = odd number of data points averaged,

m = $k/2 + 0.5$.

From Equation 22.5, the first point of smoothed data is $y_{S1} = (y_0 + y_1 + y_2)/3$, but there is no value for y_0 because y_1 is the first data point. In this situation, the value of y_1 is used for y_0 when calculating y_{S1}. A similar situation can occur at the end of the data, and a similar approach can be used. Alternatively, at the end of the fire test, a few additional scans can be made to provide original data points at the test end for data smoothing.

Example 22.4 is data smoothing using a five-point floating average, and this example deals with the data points at the beginning and end of the test. Figure 22.22 shows the original data and the smoothed data from Example 22.4. From this figure, it can be seen how smoothing can flatten out the peaks and valleys of the original data.

The number of points averaged, k, is usually 3, 5, or 7. As the number of points averaged becomes larger, the smoothed curve becomes smoother, but the potential for destroying important information increases. There is no consensus regarding the value of k that should be used, and there is no consensus regarding when smoothing is needed. Data smoothing is a matter of professional judgment.

NOMENCLATURE

C_{pt} = correction factor, dimensionless

g = acceleration of gravity, ft/s^2 (m/s^2)

H = height of opening, ft (m)

h = height between the probes and the transducer, ft (m)

k = odd number of data points averaged

m = $k/2 + 0.5$

N = number of data points

p_{atm} = atmospheric pressure, psi (Pa)

R = gas constant of air, 53.34 (287)

T_h = absolute temperature in the high-pressure tube, R (K)

T_l = absolute temperature in the low-pressure tube, R (K)

U = velocity, fpm (m/s)

U_{av} = average velocity, fpm (m/s)

V = volumetric flow, cfm (m^3/s)

W = width of opening, ft (m)

y = original data for $i = 1$ to N

y_{Si} = smoothed data for $i = 1$ to N

ρ = density of air, lb/ft^3 (kg/m^3)

Δp = pressure difference from manometer, in. H_2O (mm H_2O)

Δp_{Err} = pressure difference error due to tube temperature, in. H_2O (Pa)

REFERENCES

ASTM. 2007. Standard guide for room fire experiments, E 603. American Society for Testing and Materials, Philadelphia.

Bevington, P.R., and D.K. Robinson. 2003. *Data Reduction and Error Analysis for the Physical Sciences*, 3rd ed. New York: McGraw Hill.

Bryant, R.A. 2005. Particle image velocimetry measurements of buoyancy induced flow through a doorway. NISTIR 7252, National Institute Of Standards and Technology, Gaithersburg, MD.

Croarkin, M.C., et al. 1993. Temperature-electromotive force reference functions and tables for the letter-designated thermocouple types based on the ITS-90. Monograph 175, National Institute of Standards and Technology, Gaithersburg, MD.

Doebelin, E. 2003. *Measurement Systems Application and Design*, 5th ed. New York: McGraw Hill.

Hoffman, J.D. 2001. *Numerical Methods for Engineers and Scientists*, 2nd ed. New York: McGraw Hill.

HP. 1970. Floating measurements and guarding. Application Note 123, Hewlett Packard.

Kent, L.A., and M.E. Schneider. 1987. The design and application of bi-directional velocity probes for measurements in large pool fires. *ISA Transactions* 26(4):25–32.

Lawson, R.J. 2009. A history of fire testing. NIST Technical Note 1628, National Institute of Standards and Technology, Gaithersburg, MD.

Mandel, J. 1984. *Statistical Analysis of Experimental Data*. Mineola, NY: Dover.

McCaffrey, B.J., and G. Heskestad. 1976. Robust bidirectional low-velocity probe for flame and fire application. *Combustion and Flame* 26(1):125–127.

Peacock, R.D., and V. Babrauskas. 1991. Analysis of large-scale fire test data. *Fire Safety Journal* 17(5):387–414.

Pitts, W.M., et al. 2003. Round robin study of total heat flux gauge calibration at fire laboratories. NIST Special Publication 1031, National Institute of Standards and Technology, Gaithersburg, MD.

Press, W.H., et al. 2007. *Numerical Recipes*, 3rd ed. New York: Cabbridge University Press.

Sette, B.J G. 2005. Critical considerations on the use of a bi-directional probe in heat release measurements. *Fire and Materials* 29(5):335–349.

Stroup, D.W., et al. 2000. Large fire research facility (building 205) exhaust hood heat release rate measurement system. NISTIR 6509, National Institute of Standards and Technology, Gaithersburg, MD.

CHAPTER 23

Commissioning and Special Inspections

Michael J. Ferreira and John H. Klote

Commissioning is the means to demonstrate to an owner that the smoke control system installed in a project meets the smoke control system design for the project. Commissioning is the process for verifying and documenting that the performance of facilities, systems, and assemblies meets defined objectives and criteria. Commissioning refers to the process of examining, comparing, testing, and documenting the installation and performance of a smoke control system to ensure that it functions according to an approved design.

Special inspections are a means that an authority having jurisdiction (AHJ) uses to determine that a smoke control system meets the code requirements. The International Building Code (IBC) has requirements for a special inspection and describes the qualifications required for a special inspector (ICC 2012).

COMMISSIONING PROCESSES

The commissioning process begins at the start of the project and continues throughout the project. ASHRAE Guideline 5 provides methods for verifying and documenting that the performance of smoke control systems conforms with respect to the intent of the design (ASHRAE 2012). NFPA 3 is a recommended practice on commissioning that can be adapted for smoke control (NFPA 2012a). For smoke control systems, an AHJ such as a building official or fire marshal typically enforces a combination of building codes, fire codes, and local standards. The intent of the smoke control system commissioning testing is to determine that the system meets the owner's project requirements (OPR), including code requirements and inspections by the AHJ throughout the delivery of the project.

To achieve successful commissioning of a system, a number of different people will typically be involved in the process. In addition to the building owner and AHJ, the system designer, general contractor, subcontractors, fire protection engineering consultants, and test and balance technicians can be involved. At the end of the testing, documentation is provided that the system is working properly according to the design.

Commissioning activities can occur at multiple stages during the construction process. Duct inspections, duct leakage testing, and barrier inspections are activities that typically occur early in the construction process when the ducts and barriers are readily visible. Component testing, including airflow measurement, can occur at a midpoint in construction where power is provided to individual devices, but central monitoring and control has not yet been provided. Sequence of operations and final performance testing typically occurs when construction is nearly complete, often just before the building is intended to obtain its permits and open to the public.

The remainder of this chapter concerns matters related to special inspections according to the IBC. They relate to portions of commissioning. For details of commissioning, one is referred to the aforementioned ASHRAE guideline.

Roles and Responsibilities

The people conducting the testing can vary depending on the complexity of the system design. For some simple systems, the installing contractor can test and certify the system at the completion of construction and then the AHJ can test the system. For complicated systems or when mandated by the local requirements, independent testing can be performed by a third party to certify the proper operation of the system. Certain activities require the participation of the mechanical, electrical, and controls contractors, as well as the general contractor involved in the project.

Independent inspection can require the services of test and balance contractors or licensed engineers, depending on how the AHJ enforces local requirements. While NFPA 92 (NFPA 2012b) does not specify the requirements for the people performing the testing, the IBC specifically requires that this role be performed by a special inspector.

The IBC specifies that a special inspector charged with the inspection of smoke control systems must have expertise in fire-protection engineering, mechanical engineering, and certification as air balancers. It is important to note that the required qualifications do not need to be fulfilled by a single individual but rather by a team that collectively possess all of the required skills (i.e., a fire protection engineer, mechanical engineer, and a certified air-balancing contractor). It is often the case that engineers work in conjunction with air balancers to perform the testing.

Because the intent of the special inspection per the IBC is to confirm that the contractor follows proper installation and construction methods as detailed in approved construction documents (Klote and Evans 2007), it can be desirable to have the design engineer directly involved in commissioning. However, some jurisdictions require that the special inspector not be previously involved with the design or installation, to avoid potential conflicts of interest.

Recommended Documentation

Before testing is started, a written plan should be submitted to project stakeholders (architects, engineers, and building owners) and the AHJ in order to obtain concurrence on the extent and details of testing. The plan should include (1) an outline of all the testing, (2) a description of all the types of testing, (3) a list of the prerequisite states of construction required for the tests when appropriate, and (4) sample data sheets.

In addition, a detailed testing sequence plan is recommended that can be helpful to the project team to prepare for testing, including pretesting of the equipment before the test personnel arrive on site. This can minimize delays during testing that have the potential to delay the opening of the building.

At the completion of testing, a comprehensive test report is typically prepared that documents the inspections, tests, and results of the tests. The IBC requires such a report. The methodology used for testing during commissioning must be clearly documented explaining what constitutes as pass or fail condition. The required format of the commissioning test report varies among jurisdictions, but many reports share common features (Klote and Evans 2007):

- Summary of the results obtained
- A compilation of all inspection reports and any noncompliance issues
- The collection of testing and inspection logs
- Data sheets for all of the inspected components
- Signatures of the special inspection team members

SPECIAL INSPECTION PHASES

The two phases of special inspection are inspection and testing. The goal of inspection is to determine that the specified system components have been installed, and that the installation of these components is according to the manufacturer's instructions. Testing is intended to establish that the system design achieves the accepted performance criteria. In practice, the process of evaluating a smoke control system is often done in many stages of inspection and testing. It is important to note that inspection can be performed at different phases of construction rather than directly prior to building occupancy (Klote and Evans 2007).

Installation and Component Verification

Installation and component verification involves inspecting all components of the smoke control system and making sure they are present and are installed in accordance with design specifications. The purpose of equipment installation verification and component identification is to determine that the installed equipment is as specified in the design documents. This stage can be conducted as soon as the physical installation of the equipment is complete and consists of the following activities:

- All components and subsystems that are part of the smoke control system should be identified by manufacturer, model number, and building-specific mark number. The installation of each smoke control system component should be checked against the product data sheet to verify proper installation.
- Determine that all smoke control fans and related components are labeled in accordance with local building codes and agree with labeling on design documentation and the firefighter's smoke control station (FSCS).
- Passive smoke barriers should be inspected and compared to original design documentation. Penetrations should be properly sealed.
- Outdoor air inlets and outlets should be located and inspected to determine that the risk of smoke and fire being reintroduced into the building is minimal.

- Standby power systems should be visually inspected to determine their compliance with the applicable building codes.

An FSCS is a system for use by the fire service that provides graphical monitoring and manual overriding capability over smoke control systems and equipment at designated locations within a building. In some standards and specifications, the FSCS is also called the *firefighter's smoke control panel and the firefighter's control panel.* Typically an FSCS is designed and built specifically for a particular building, and Figure 23.1 is an example of an FSCS. For more information about FSCS, see Chapter 8.

Inspection and Equipment Functional Testing

The goal of equipment functional testing is to determine that the smoke control system is operational, properly supervised, and gives an accurate status indication on the FSCS. As with component inspections, this stage of testing can be conducted as soon as the smoke control components of interest are installed, powered by a permanent source, and connected to the proper controls. Testing in this phase is conducted using normal power.

During this stage, the smoke control equipment is usually manually activated from the FSCS, and verification of each fan and damper status is by visual inspection. Upon activation of a component, the functionality of the FSCS can be verified by observing the correct indication of the status of all smoke control equipment. In many cases, smoke control inspectors may wish to make a checklist of all the components to be tested to facilitate the process.

Regardless of the performance objectives of a smoke control system (atrium[1] exhaust, zoned smoke control by pressurization), the system will use a number of typical basic components that can be a part of a building's mechanical, electrical, or architectural systems. Each of the critical components contributing to the operation of a smoke control system must be identified and tested accordingly in order to determine the long term operability of the system.

Smoke Barriers

As noted in Chapter 9, a smoke barrier is a continuous wall, floor, or ceiling assembly that is designed and constructed to restrict the movement of smoke in conjunction with a smoke control system. Smoke barriers may or may not also have a fire resistance rating. Smoke-tight construction is used to limit the potential for air movement across the smoke barrier. Openings such as doors, HVAC ductwork penetrations, and other utility penetrations are protected by supplemental means to inhibit the passage of smoke.

Smoke barrier doors are provided with supplemental gaskets and drop seals to make them tighter than non-smoke rated fire doors. HVAC ductwork penetrations have a smoke damper installed where the ductwork passes through the smoke barrier, and these penetrations will have a combination fire/smoke damper if the barrier is also a fire barrier. Other utility penetrations are typically fire-stopped where they pass through the smoke barrier to prohibit the passage of smoke.

Barrier components that require testing and inspection include (1) door latches, gaskets and drop seals, and (2) penetration seals (e.g., ductwork, conduit, and cable trays). Common problems with smoke barriers identified during testing include

- excessive leakage causes design pressures not to be met;
- door closers or other operators not designed to overcome system design pressures, resulting in the door "hanging open"; and
- doors or windows not properly gasketed as a smoke-protected opening penetration.

Fans

Mechanical fans are critical components for most smoke control systems. Roof-mounted supply and exhaust fans should be oriented to minimize the potential to reintroduce smoke into the building, because the smoke exhaust outlet is too close to a pressurization fan inlet. Proper fan orientation should be confirmed by inspection.

Fan components that require inspection and testing include the following:

- Fan blades
- Belts (for belt-driven fans)
- Power sources (normal and emergency)
- Variable-frequency drives
- Maintenance disconnects
- Verification devices (pressure or current transducers, flow switches)

Fans used for smoke control should be tested for airflow, current, and voltage by the test and balance contractor. The special inspector should review these reports. This review should check that the contractor has measured the fan airflow in smoke control mode, and that the following are in agreement with the design

1. In this handbook, the term atrium is used in a generic sense to mean any large-volume that is at least two stories high such as an enclosed shopping mall, sports arena, or an airplane hangar.

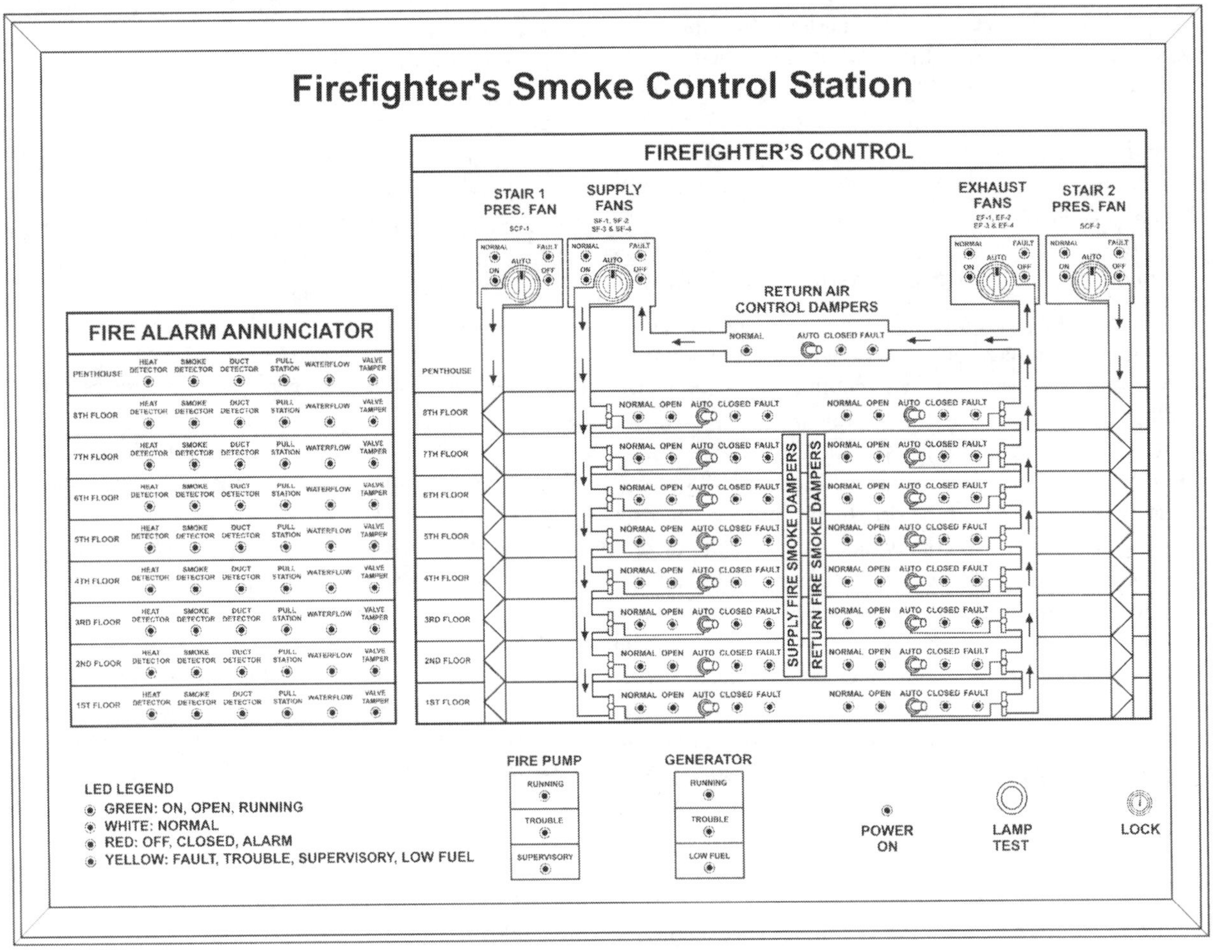

(a) FSCS for a building with pressurized stairwells and zoned smoke control (See Note 2)

Notes:
1. The Firefighter's Smoke Control Station (FSCS) is also called the *Firefighter's Smoke Control Panel* and the *Firefighter's Control Panel.*
2. The purpose of this figure is to give readers an overall idea of what an FSCS can look like, but much of the wording on this figure is too small to be legible.

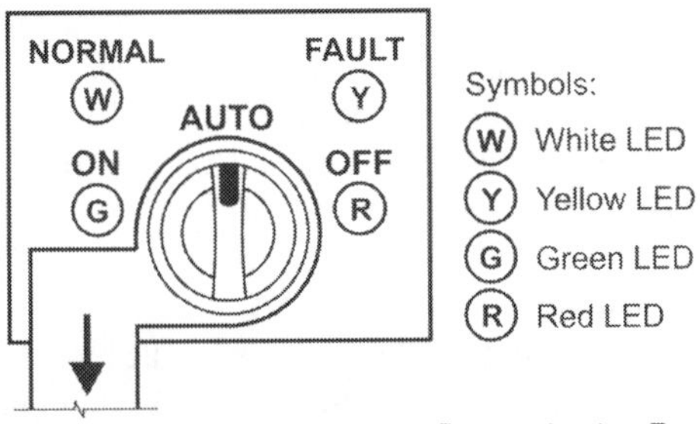

(b) Status indication for Stair 1 Pressurization Fan

Figure 23.1 *Example firefighter's smoke control station (FSCS).*

documents and local code: (1) impeller rotation, (2) motor speed, (3) number of belts, and (4) belt tension.

Testing should be performed to verify the proper supervision of fans by disconnecting power to the fans while they are running and confirming that the proper "Fault" condition is displayed on the FSCS. It should be noted that disconnecting power while the fan is not running is not expected to result in a fault indication, as discussed regarding end-to-end verification in Chapter 8. It should be verified that the fan can be controlled by the FSCS and that the correct indication of the fan's status is given. Finally, it should be verified that fans are activated within time required by the local building regulations.

Common problems with fans identified during commissioning testing include the following:

- Fans wired backward (results in reversed flow direction for axial fans and reduced flow for centrifugal fans)
- Maintenance disconnect not properly monitored to indicate fault if fan turned off locally by the disconnect (power monitored downstream of all disconnects)
- Flow indication status not provided by a flow sensor, pressure sensor, or current transducer
- Exhaust inlets located in too close a proximity to stair pressurization or other supply inlets designed to remain on during smoke exhaust; Excessive leakage causes design pressures not to be met

Dampers

As discussed in Chapter 7, dampers are used for one or more of the following purposes: (1) balance flow, (2) control flow, (3) resist the passage of fire, and (4) resist the passage of smoke. Dampers intended to resist the passage of fire are called fire dampers, and dampers intended to resist the passage of smoke are called smoke dampers. Dampers that intended to resist the passage of both fire and smoke are combination dampers.

Combination dampers are often used where the damper is located at a penetration through a barrier that is both a rated fire barrier and a smoke barrier. Combination dampers have an integrated thermal element that is capable of causing closure of the damper when exposed to a specified temperature threshold. Smoke dampers can also have an integrated smoke detector to cause closure of the damper when smoke is present without requiring the damper to be remotely-controlled to close using some other initiating device. Smoke dampers or combination can also be used at shaft wall enclosures or in air transfer grilles between smoke control zones to close off an air transfer opening.

Damper components that require inspection and testing include the following:

- Damper blades and gaskets
- Damper motor
- Power sources (normal and emergency)
- Integrated smoke detectors
- Temperature switches or fusible links
- Verification devices (end switches)

Test 1: Dampers and doors that are important to smoke control system operation must be inspected to determine that they operate properly. The response of each damper, door, etc. should be visually checked to confirm that it has operated properly and that the appropriate status indication is provided on the FSCS. The time to operation of the devices should conform to the requirement of the local building code.

Test 2: Verify that dampers and doors are properly supervised by lifting a lead or by intentionally misaligning a damper/door at a few random locations. The FSCS should provide the proper fault indication for each test.

Common problems with dampers identified during commissioning testing include the following:

- End switches not provided to indicate open/closed status
- Damper blades warped or broken during installation, resulting in improper seal
- Perimeter of damper not properly sealed where dampers penetrate walls
- Thermal element with temperature lower than expected smoke temperature for dampers designed to be open to facilitate exhaust

Operable Doors and Windows

There are two ways that operable doors, windows, and shutters can be utilized in a smoke control system design. Door, windows, and shutters can be designed to close upon an alarm to maintain the smoke-tightness of a designated smoke barrier. Doors and windows can also be designed to open to provide a source of outdoor makeup air for an atrium smoke control system.

Operable doors designed to close in the event of an alarm are typically held open by magnetic hold open devices that close the door when current is cut off to the magnet. Operable shutters can also be used to close off vertical openings or other large openings to create a smoke barrier separation. These devices are relatively easy to test and commission.

Doors, windows, and exterior louvers can be opened for makeup air upon alarm. Doors are typically operated with devices similar to handicapped access door-opening devices. Windows can be either center-pivot or side-hinged. Various louver types are possible, from the simplest counterweighted louvers to electrically operated louvers.

For all operable devices used for makeup air, the position of the opening must be capable of being monitored to ensure that it opens properly in the event of an alarm. Monitoring also allows these devices to be incorporated into the automatic weekly self-test. All operable doors/windows should be tested during commissioning.

Common problems with doors and windows identified during commissioning testing include the following:

- Excessive door-closer force contributes to excessive door-opening force under pressure
- Door and window operators not designed to overcome system design pressures
- Door and window position not monitored
- Testing of doors and windows not incorporated into weekly self-test or other periodic testing

Verification of Self-Test Feature

To evaluate the performance of the weekly self-test feature of the UUKL listed smoke control equipment (discussed in Chapters 8 and 24), manipulate the system time to force the automatic actuation of the testing sequence. Verify the proper operation of the self-test by first running the test with all components in proper working order. The self-test should then report completion of the test with no faults identified. The test should then be repeated with multiple components purposely put in a fault condition, verifying that the test correctly identifies the faults. Test the reporting function of the system by viewing the generation of a report and, if required, the transmittal of a trouble signal to the central station.

Firefighter's Smoke Control Station (FSCS)

To test an FSCS, it should first be verified that all devices that are part of the smoke control system are properly identified on the panel, and provided with the proper status indication. All devices should then be manually manipulated into the possible states to verify that all lights show the proper status. This testing is typically performed with one test participant at the panel controlling the switches and verifying indication and another test participant at the device being manipulated to visually verify status.

Common problems with FSCS panels identified during testing include the following:

- Status lights on FSCS not designed per typical convention (nonintuitive color schemes, fault lights not shown).
- Status lights lit based on switch position, not positive confirmation of status from device (the light is essentially wired to the switch).
- Status of fans or dampers does not occur within the allotted time frame.
- Status improperly indicated. This can occur when a device such as a damper end switch is wired improperly.
- Connection between panels not supervised.

Sequence of Operations Testing

The sequence of operations is the documented sequence of component actions that are programmed to happen in the response to a given change of state event. The purpose of the sequence of operations testing is to verify that the automatic functions of the smoke control system function as designed. The smoke control system must properly align in response to a representative fire event triggered by an automatic initiating device.

Smoke control systems are usually comprised of many components that require a sequence of operation for proper performance. A testing matrix (also called an activation schedule in some standards) is often created to facilitate the process, and a testing matrix needs to include all of the components to be tested. Figure 23.2 shows an example sequence of operations matrix. The report of sequence of operations testing should include a test number, the expected outcome of the test per sequence of operations, and whether the test was a pass or fail.

Sequence of operations testing is intended to demonstrate that the smoke control system responds properly to the various types of alarm inputs received. Before the sequence of operations testing can be conducted, all initiating devices and fire alarm components necessary to test the automatic operation of the smoke control system must be installed and operational. The sequence of operations testing is performed under normal and standby power supplies. Testing under normal power should be conducted first, and testing with standby power should follow when the system has been shown to function properly under normal power. Often, sequence of operations tests require the presence of mechanical, electrical, and controls contractors to aid in the testing process.

To test each sequence of operations, the input specified on the sequence of operations matrix is activated or simulated, and then operation of each fan, damper, door, and other devices contained in the sequence of operations matrix is visually verified to confirm that the system has responded correctly. Proper indication of device status should also be confirmed at the FSCS. Testing should be performed for each activation signal shown on the sequence of operations matrix. Altering the status of randomly selected smoke control equipment should result in a "fault" condition on the FSCS. It should also be verified that the activation of an alarm in other areas of the building does not change the status of the smoke control system, unless specifically intended to do so per the design.

During sequence of operations testing, it should also be verified that the FSCS properly overrides the smoke control system after automatic activation. Repeat a sequence of operations test to ensure proper system performance when normal power is shut down while in alarm, transitioning the smoke control system over to standby power. Repeat a sequence of operations test to verify proper system performance when an alarm is initiated while under standby power.

System Performance Testing

Commonly, testing and balancing is required before formal acceptance testing to achieve the expected performance of all the components. Testing and balancing refers to the process where the as-built performance of smoke control systems is tested in the field and compared to the required design conditions. Adjustments to the installed system, such as refining the supply airflow rates, are made to ensure that the smoke control system is functioning as intended in the approved design documentation.

System performance testing is the phase where the code-specified performance parameters appropriate to the smoke control design are measured. For example, building codes require that a minimum pressure difference exist between a pressurized stairwell and other zones in the building, and that door-opening force must not exceed a specified amount. In this case, performance testing would focus on measuring the pressure difference across stairwell doors and door-opening forces. Some common parameters measured during smoke control system performance testing are (1) exhaust/supply airflow quantities, (2) airflow velocities at atrium or other large open space perimeters, (3) door-opening forces, and (4) pressure differences between zones.

Zoned Smoke Control

For zoned smoke control systems, one zone should be put into the smoke control mode, and the pressure differences at the boundaries of that zone should be measured. After smoke control operation in that zone has been deactivated, another zone should be tested in the same manner. This should be repeated until all smoke zones have been tested. Systems with automatic activation should be activated by putting an appropriate initiating device into alarm. All of the potential sequences of operations need to be tested.

For some zoned systems, certain devices (e.g., manual pull station) can only activate a portion of the system, such as stair pressurization. An additional device

Fire Floor / *Device*	Actuate Off-Site Reporting	Release Automatic Door Hold-Open Throughout Building	Initiate Elevator Recall	Initiate Evacuation Alarm				Actuate Stair Pressurization Fans Throughout Building	Floor Supply Fans	Floor Supply Dampers	Floor Return Fans	Floor Exhaust Dampers			
				1st Floor	2nd Floor	3rd Floor	4th Floor					1st Floor	2nd Floor	3rd Floor	4th Floor
1st Floor															
Sprinkler Water Flow	X	X		X	X			X	Off	Closed	On	Open	Closed	Closed	Closed
Smoke Detector	X	X		X	X			X	Off	Closed	On	Open	Closed	Closed	Closed
Elevator Lobby Smoke Detector	X	X	X	X	X			X	Off	Closed	On	Open	Closed	Closed	Closed
2nd Floor															
Sprinkler Water Flow	X	X		X	X	X		X	Off	Closed	On	Closed	Open	Closed	Closed
Smoke Detector	X	X		X	X	X		X	Off	Closed	On	Closed	Open	Closed	Closed
Elevator Lobby Smoke Detector	X	X	X	X	X	X		X	Off	Closed	On	Closed	Open	Closed	Closed
3rd Floor															
Sprinkler Water Flow	X	X		X	X	X	X	X	Off	Closed	On	Closed	Closed	Open	Closed
Smoke Detector	X	X		X	X	X	X	X	Off	Closed	On	Closed	Closed	Open	Closed
Elevator Lobby Smoke Detector	X	X	X	X	X	X	X	X	Off	Closed	On	Closed	Closed	Open	Closed
4th Floor															
Sprinkler Water Flow	X	X			X	X	X	X	Off	Closed	On	Closed	Closed	Closed	Open
Smoke Detector	X	X			X	X	X	X	Off	Closed	On	Closed	Closed	Closed	Open
Elevator Lobby Smoke Detector	X	X	X		X	X	X	X	Off	Closed	On	Closed	Closed	Closed	Open

Figure 23.2 *Sequence of operations matrix.*

(e.g., corridor smoke detector) may be needed to activate the full smoke control sequence for a zone. When this is the case, pressure differences must be verified under each condition. The impact of auxiliary systems (e.g., laboratory hood exhausts constantly ON) on pressures should also be assessed.

During testing, it is not acceptable to measure pressure differences produced by subsystems operating alone. For example, it is not acceptable to activate one stair's pressurization systems, measure the resulting pressure differences, then deactivate this system and test another stair system, followed by a smoke zone exhaust subsystem. While the stair pressures may be within the desired range, unacceptable door-opening forces can be produced when all these subsystems operate at the same time. The smoke control subsystems interact with each other, and the pressure differences need to be measured with all the systems operating as they would during a fire. Pressure differences are additive, and all pressurization systems within a given building will work with or against the other systems with respect to pressure differences.

Pressurized Stairwells

With all stairwell doors closed, pressure differences across each stairwell door should be measured. Then one door should be opened and pressure difference measurements made at each closed stairwell door. This should be repeated until the number of doors opened equals the number of doors required by the code authority to be opened. If the design is based on all doors being closed, this testing may not need to be performed.

As discussed in Chapter 10, a compensated stairwell pressurization system is one that adjusts pressurization to account for opening and closing doors. The intent of a compensated system is to prevent loss of pressurization when one or more stair doors are open. VAV compensated stairwell pressurization systems use VAV fans controlled by pressure sensors installed to measure the pressure differences between the stair and adjacent spaces (usually the corridor) on multiple floors.

When testing a VAV compensated stair system, it is important to test all potential failure modes. In general, it is more acceptable to have a lower pressure difference in the stair than to overpressurize the stair, causing excessive door-opening forces that impede occupant access to the stairs.

Sometimes compensated stairwell pressurization systems are not designed to account for all combinations of open stairwell doors. For example, holding a single door open for a prolonged period of time can cause the stairwell pressurization fan to ramp up due to the loss of pressure in the stairwell. Particularly for those cases where the stair fan is conservatively oversized, the increased airflow into the stair can cause doors to slam shut, which can potentially cause injury to building occupants. Once the doors are closed, the overpressure will cause excessive door-opening forces until the pressure is relieved from the stair.

Elevator Smoke Control

The tests for elevator smoke control systems depend on the type of elevator smoke control system installed. In general, the design pressure differences should be measured at the appropriate locations for the particular design. If the intent of the system is to pressurize enclosed elevator lobbies, pressure differences across closed lobby doors to the building should be measured. If the intent of the system is to pressurize the elevator shaft to prevent smoke flow through it, the pressure differences across the elevator doors should be measured.

With elevator shaft pressurization, the shaft should be pressurized after elevator recall. When a smoke detector in an elevator lobby goes into alarm, the elevator goes into recall mode in which cars are moved to the exit landing and removed from service. In the event of a fire on the exit floor, the cars are recalled to an alternate floor. Where elevator hoistway venting is required by local code, one of the following must be done: (1) the elevator pressurization system can account for pressurization air lost out the vent, (2) the vent can be closed upon approval of the AHJ, or (3) the vent can be eliminated upon approval of the AHJ.

It is important to verify the proper operation of the elevator doors under the maximum design pressure for the system, including the impact of stack effect. Manufacturer data on the operation of elevator doors under pressure is not readily available, thus it is important to ensure that doors will operate properly under pressure to allow occupants to exit the elevator cars.

Atrium Smoke Control

As discussed in Chapter 15, there are many design approaches for atrium smoke control, and the most commonly used approach in North America is steady mechanical smoke exhaust. The makeup air velocity for smoke exhaust systems and natural venting systems must not exceed a specified limit. The exhaust flows and makeup air velocities need to be measured. Upper layer air temperature of the space can be measured to check that design considerations about smoke stratification in the atrium are appropriate.

For each design approach, measuring system performance is impractical because this would require that a design fire be built in finished the atrium. Therefore, testing and balancing is typically simply to verify that the air flows provided by the fans are in accordance with the design.

MEASURING PERFORMANCE

There are many times during commissioning and testing where inspectors need to know the flow direction, velocity, and volumetric flow in order to determine if the smoke control system is functioning correctly. For some tasks, expensive devices are not always necessary. Simple flow indicators such as a piece of paper placed in front of a vent or hung from the air grill will readily indicate the direction of airflow. Other methods can employ smoke from punk sticks or other smoke-producing materials to visually trace the airflow direction.

Performance testing is often done by a certified test and balance contractor, and these contractors use specialized instruments for the measurement of velocity and volumetric flow. Chapter 22 discusses some commonly used instruments that can be used to measure pressure difference, velocity, and volumetric flow.

Door-Opening Forces

Door-opening forces are typically measured using push-type or pull-type spring-loaded gages. Push-type gages have been specifically designed for measurement of door forces, and push-type gages are generally preferred for commissioning testing. For more information about door-opening forces and gages, see Chapter 9.

The following discusses measuring the opening force for a door that has a push-bar release as opposed to a door knob. When measuring door-opening force it is important to consider whether the moment-driven door-opening force is being measured or the latch release force. Depending on the door-opening mechanism, the two types of measurements can differ. Some types of push-bar release latches require a higher force to release the latch, which can be more easily provided by using weight to push down on the bar. Once the latch is released, the force needed to open the door can then be measured. It is common for the latch to be released and then gently held down, while the door-opening force gage is placed just above the latch and pushed with the other hand to measure the force required push the door open to 90°. Figure 23.3 shows this action using a spring-loaded push-type gage. The manner for measuring the door-opening force should be reviewed with the AHJ before testing.

Automatic Sensors

There are a number of different types of automatic sensors used to obtain feedback regarding the operating status of smoke control system devices during their operation. These devices are also used during commissioning to verify proper operation of equipment, usually by monitoring status at the FSCS.

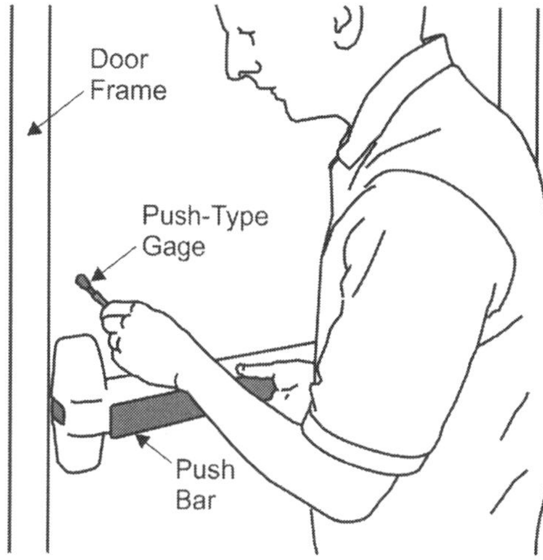

Figure 23.3 *Use of push-type door force gage on door with push-bar.*

Automatic sensors are the instruments providing primary system feedback during the weekly self-test and any other periodic testing performed for the system. Automatic sensors are discussed in Chapters 8 and 24.

CHEMICAL SMOKE

Smoke bombs (also called smoke candles) produce a chemical smoke that is very different from the smoke that would be produced by a design fire. The major differences are (1) visibility through chemical smoke is very different from visibility through real smoke and (2) chemical smoke lacks the buoyancy of smoke from a design fire.

A smoke control test is very different from a smoke control demonstration. In a smoke control test, the performance of the system or systems is measured as just described. In a smoke control demonstration, the operation of the system or systems is displayed. This can be stated in simple terms. In tests, people find out if systems can perform as intended. In demonstrations, people see the systems operating. Because chemical smoke is so different from real smoke, it is not possible to use chemical smoke to determine if a smoke control system will operate under design conditions.

Sometimes, an AHJ will require smoke bomb tests of smoke control systems, but it should be realized that these are only demonstrations, not tests. It is not possible to use a smoke bomb test to learn how the system will perform under design conditions, but smoke bomb tests can give occupants a false sense of security.

The following discussion of smoke bomb tests for zoned smoke control systems and atrium smoke control systems also applies to testing with other forms of chemical smoke, including the smoke from theatrical smoke generators.

Zoned Smoke Control

For zoned smoke control systems, a major problem with most smoke bomb tests of these systems is that they are intended to evaluate some improvement of smoke conditions in the zone where the fire is located. This is based on the mistaken belief that smoke control is capable of producing a significant improvement in tenable conditions within the zone where the fire is located. This belief is far from reality. In fact, zoned smoke control is basically a containment system, and it is not the intent of zoned smoke control to maintain tenable conditions in the zone with the fire.

Smoke bomb tests of zoned smoke control systems are inappropriate, except as described in the section Other Uses of Smoke Bombs. Smoke bomb tests for zoned smoke control are described here in general terms so that the reader can recognize this type of test and understand the problems with it. The smoke control system is put in operation. In the zone which is being exhausted, a number of smoke bombs are ignited. The smoke bombs produce all their smoke in a few minutes and the zone rapidly fills with smoke. Because the smoke control system is exhausting air and chemical smoke from this zone, the concentration of chemical smoke decreases with time. However, with no air supplied to the zone, there is minimal airflow or dilution. The smoke control system is mistakenly declared a success if at some specific time after ignition, a particular object (such as an exit sign) is visible by a human observer at specific distance (such as 20 ft).

The problems with this kind of smoke bomb test are numerous, and the unrealistic nature of these tests was illustrated by the smoke control experiments at the Plaza Hotel. The criterion for successful operation is not objective. Further, the potential danger of exposing the observer or other people to toxic chemical smoke must dealt with. The obscuration of smoke from a building fire is much different from that of chemical smoke. Most flaming fires produce a hot, dense, black smoke while most smoke bombs produce a cool, white smoke. Because a zoned smoke control system is intended to maintain pressure differences at the boundaries of the smoke zone, the system should be tested by measuring pressure differences.

A very serious problem with this type of smoke bomb test is that it can give building occupants and fire service officials a false sense of the security. The test can lead people to wrongly think that zoned smoke control is capable of achieving a significant improvement in tenable conditions within the fire space, or it may cause them to believe that a properly functioning smoke control system will not provide the intended life-safety function.

For zoned smoke control, DO NOT USE SMOKE BOMB TESTS. These tests are unrealistic, subjective, and tend to give occupants a false sense of security. Zoned smoke control systems should be tested by measuring the pressure differences that the system produces.

Atrium Demonstration Testing

Demonstration testing consists of observing the smoke control system operating with visible smoke from a smoke generator or smoke bomb. The term demonstration testing is incorrect because such tests cannot measure or evaluate system performance for design fire for reasons discussed in the following text. Demonstration testing provides little, if any, assurance that a smoke control system will operate as intended, but it is addressed here because it is sometimes required by an AHJ.

The problems with using smoke bombs to test atrium smoke control are due to the lack of buoyancy of the chemical smoke. The simplest approach to these tests is to put a number of smoke bombs in a container and ignite them. The resulting smoke has very little buoyancy, and it often stagnates before it can reach the ceiling. Smoke can stagnate even when the atrium smoke exhaust is operating. Using a cold-smoke apparatus like the one shown in Figure 23.4a may improve buoyancy slightly, but the chemical smoke still has a tendency to stagnate.

Smoke can be heated to increase buoyancy. Because of concerns about property damage, the smoke at the ceiling must not reach temperatures that can happen with design fires. For this reason, hot-smoke bomb tests cannot test an atrium smoke control system under design fire conditions.

When smoke bomb demonstrations are required by the AHJ, the use of a hot-smoke apparatus is recommended so that the chemical smoke will reach the ceiling of the atrium. Designers should make it clear to the involved parties that such demonstrations do not test the atrium smoke control system under design fire conditions.

For such demonstrations, an apparatus using a gaseous or liquid fuel can be used with a fire sized so that the smoke will reach the ceiling of the atrium. Gaseous

CAUTION: Smoke bomb tests and other chemical smoke tests are not recommended, because they do not produce realistic smoke.

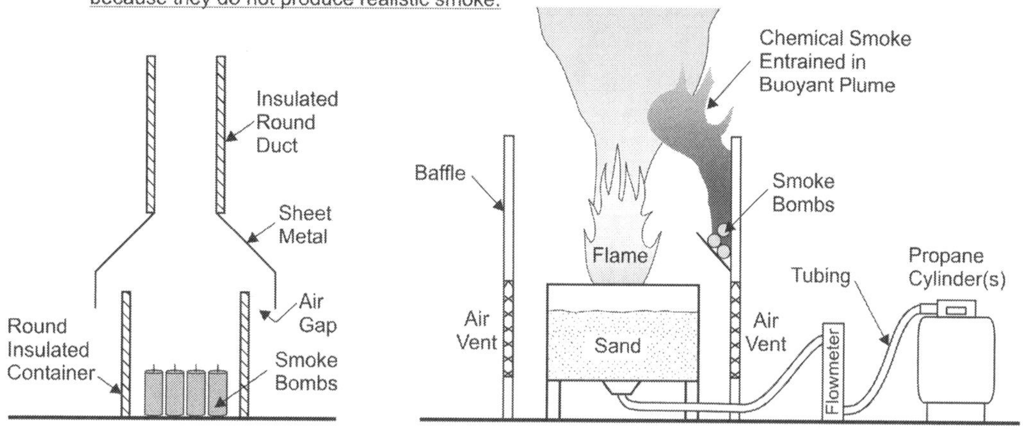

(a) Cold smoke apparatus (see note 1) (b) Hot smoke apparatus (see note 2)

Notes:
1. The cold smoke apparatus produces smoke with little buoyancy that may stagnate and not reach the ceiling.
2. The hot smoke apparatus produces smoke with more buoyancy that may reach the ceiling. The flowmeter measures the flow of propane. Using this flow and the heating value of the propane, the heat release rate of the fire can be sized to maintain a desired centerline plume temperature at the atrium ceiling.
3. When smoke bomb tests are required by an AHJ, the hot smoke apparatus is suggested.

Figure 23.4 *Cold- and hot-smoke demonstration apparatuses.*

fuel is preferred for safety reasons, because the apparatus can be designed so that the fuel can be cut off in an emergency. Dillon (1994) conducted demonstrations of the smoke control system in an enclosed stadium. He used propane to heat chemical smoke from smoke bombs. To prevent property damage, Dillon sized his fires so that the centerline plume temperature at the ceiling would not exceed 115°F (46°C).

Figure 23.4b shows such a hot-smoke apparatus that uses a clean-burning propane burner as the hot, buoyant smoke source, along with smoke bombs as the visible smoke component for the test. Smoke candles are ignited in a tray that is part of the apparatus. The cold, visible smoke from the smoke bombs will be entrained in the upward rising buoyant plume and smoke will rise to the upper levels of the atrium. Figure 23.5 shows a hot smoke demonstration in an atrium.

For atrium smoke control, SMOKE BOMB TESTS ARE NOT RECOMMENDED. The smoke from smoke bombs lacks the buoyancy of smoke from a design fire even when done with a hot-smoke apparatus. When smoke bomb tests are required by the AHJ, the use of a hot-smoke apparatus is recommended such that the smoke will reach the ceiling of the atrium.

(Photograph courtesy of Hughes Associates, Inc.)

Figure 23.5 *Hot-smoke demonstration in an atrium.*

Other Uses of Smoke Bombs

Chemical smoke from smoke bombs can be used to test for smoke feedback into supply air. A general procedure for this testing is described here. A number of smoke bombs are placed in a metal container, and all bombs are simultaneously ignited. The container is located near an exhaust inlet in the smoke zone being tested so that all of the chemical smoke produced by the bombs is drawn directly into the exhaust air stream. If chemical smoke is detected in the supply air, its path should be determined, the path should be blocked, and then the smoke feedback test should be conducted again.

Smoke bombs or other tracers can be useful in locating the leakage paths that sometimes defeat a smoke control system. For example, if the construction of a stairwell is unusually leaky, pressurization of that stairwell may not be possible with fans sized for construction of average tightness. Chemical smoke generated within the stairwell will flow through the leakage paths and indicate their location so that they can be caulked or sealed. Smoke bombs may also be used to locate unintended leakage paths in a zoned smoke control system. In this case, smoke bombs would be ignited to simulate a zone with a fire, and the zoned smoke control system would be activated. If any smoke is found in other parts of the building, then an unintended leakage path exists that needs to be sealed.

REFERENCES

ASHRAE. 2012. ASHRAE Guideline 5-2012, *Commissioning Smoke Management Systems*. Atlanta: ASHRAE.

Dillon, M.E. 1994. Case study of smoke control system testing for a large enclosed stadium. *ASHRAE Transactions* 100(2).

ICC. 2012. *International Building Code*® (*IBC*®). Country Club Hills, IL: International Code Council.

Klote, J. and D. Evans. 2007. *A Guide to Smoke Control in the 2006 IBC*. International Code Council, Country Club Hills, IL.

NFPA. 2012a. NFPA 3, *Recommended Practice on Commissioning and Integrated Testing of Fire Protection and Life Safety Systems*. Quincy, MA: National Fire Protection Association.

NFPA. 2012b. NFPA 92A, *Standard for Smoke Control Systems*. Quincy, MA: National Fire Protection Association.

CHAPTER 24

Periodic Testing

Michael J. Ferreira and Paul G. Turnbull

After a smoke control system has been commissioned, testing must still be performed periodically so that the system is in the proper operating condition in the event of a fire. Periodic testing needs to be performed over the life of a building to determine that the installed smoke control systems are capable of operating as designed. Periodic testing includes (1) manual testing involving ongoing inspection and maintenance and (2) automatic testing to determine that integral equipment is functional and operational. Automatic testing is often performed at a higher frequency than manual testing. Continued inspection and testing helps so that adjustments and repairs can be made to account for unforeseen changes to the building or failure of components.

Until recently, smoke control system reliability has been somewhat compromised because periodic testing was limited to manual testing. Inspections performed years after commissioning showed that some smoke control systems were inoperable, turned off, or made ineffective due to modifications to equipment or the building.

It is expected that the reliability of smoke control systems will be significantly improved by the use of automatic weekly self-testing of system components, afforded by Underwriters' Laboratories listed equipment carrying the UUKL product designation. Weekly self-testing will be discussed at length in this chapter.

FACTORS IMPACTING TESTING

It is important that a record of the design basis for the smoke control system be maintained with a record of the results of previous periodic tests that have been performed, to provide the person(s) doing the testing with a basis for conducting the current tests. A number of factors can prevent a smoke control system from working as designed, including architectural changes to the building and equipment and sensor malfunction.

Architectural Changes

Architectural changes that impact periodic testing are most often considered minor changes, such as those involved in a tenant improvement (TI) project, because major building renovations usually trigger a reevaluation of the building and its installed systems per the prevailing building code at the time of construction.

Seemingly minor architectural changes can significantly alter the function of a smoke control system. Experience has shown that the following can frequently occur:

- Changes to a smoke barrier's leakage rate due to unintended penetrations (e.g., new unsealed utility penetrations above a suspended ceiling), or modifications/adjustments to door hardware (e.g., removal of door sweeps, change in door closer)
- Addition of a major leakage path between two adjacent smoke control zones (e.g., addition of a vertical circulation stair connecting two floors in a high-rise building)
- Changes to a floor layout during renovations to relocate or remove a smoke barrier (e.g., conversion of a multitenant office floor served by a pressurized exit access corridor to a single floor tenant with an open office configuration)
- Addition/removal of stair vestibules or elevator lobbies

Architectural changes should be evaluated at the time of performing manual periodic testing, first by inspection and then my measuring system performance (e.g., required airflows/pressures/door-opening forces). Changes to smoke barriers that impact the leakage across the barrier will most likely have only local impact that would require sealing penetrations or increasing system airflows to get the desired pressure difference and/or door-opening force. More major changes, such as adding a convenience stair between two smoke control zones in a high-rise building or removal of a pressurized exit access corridor due to a change to an open office configuration, could require a rethinking of the smoke control design basis for those zones.

Equipment Maintenance

Routine maintenance can inadvertently impact the ability of equipment to properly operate in the smoke control mode, and the resulting impairment cannot be identified without testing.

One of the most common causes of equipment impairment are local disconnect switches that are left in the OFF or HAND position after maintenance is complete, and not returned to the ON or AUTO position. This condition can remedied by (1) either monitoring the proper operation of the equipment (e.g., fan airflow), (2) monitoring for presence of power downstream of the disconnect, or (3) monitoring the position of the disconnect switch. Using any of these methods allows for the annunciation of a trouble or system fault that will not clear until the maintenance disconnect is returned to the proper position to allow for automatic operation of the system component.

Unless a full functional test is performed, monitoring the position of a disconnect or presence of power downstream of the disconnect alone will not verify proper operation of the equipment. For example, the operation of fire/smoke dampers can be impacted by drywall screws inadvertently introduced into the path of the damper blade travel during barrier modifications, which would impact the ability of the damper to open/close properly. A fan belt can break or be left off during maintenance, which is another condition that would not be identified until the fan is called on to operate.

Manual inspection or testing alone is not sufficient to identify equipment maintenance issues due to the long intervals between required testing. It could take as long as six months to a year to identify problems with a smoke control system component if left solely to manual periodic testing. Therefore, impairments to individual pieces of equipment are more readily identified by way of automatic testing, whether this be by way of continuous monitoring where possible or by frequent automatic self-tests.

Sensors and Instrumentation

For complex smoke control systems, components can be controlled by automatic sensors or other instrumentation that continuously monitors conditions within the building. These devices are typically used to either detect a fire and activate the system or modify the system configuration subsequent to activation due to the resulting building conditions.

Devices used to activate a smoke control system consist of sprinkler water flow switches, smoke detectors, and heat detectors that are continuously monitored by the fire alarm system. The means for determining reliability of detection devices has been incorporated in the various codes and standards pertaining to fire alarm system design. A similar degree of monitoring is not usually provided for devices that modify the smoke control system configuration.

The most common type of devices used to configure smoke control systems during their operation are pressure transducers. Current transducers, voltage monitors, end switches and position switches are other devices that can be used to monitor equipment status during operation. These devices can indicate the need to (1) modulate fan speed, (2) modulate damper position, or (3) perform complicated control functions based on pressure differences between smoke control zones. For example, a common method of designing pressurized stairwells in high-rise buildings is to design a system that modulates the speed of pressurization fans as a function of the measured pressure difference between the stair and the floor served. This is typically done at multiple floors in the building.

For all types of sensors used, the calibration of the sensor must be maintained over time in accordance with the manufacturer's recommendations. For certain types of sensors (in particular, pressure transducers), sensor drift can occur over time. Unless the sensor is periodically calibrated, the system can modulate based on incorrect data resulting in the pressures differences or door-opening forces outside the bounds of either the design basis, or in the worst case, outside of safe limits.

Unfortunately, checking the calibration of sensors is a step that is often overlooked during the periodic testing process.

Environmental Factors

A smoke control system is designed for a range of environmental conditions that typically consist of outdoor temperatures, wind speeds, and wind directions. Acceptance testing is done over a relatively short period of time just before the building is completed, and it is impossible to test over the range of environmental conditions that are encountered during the life of the building. However, environmental conditions can be considered when performing periodic testing.

The indoor-to-outdoor temperature difference results in stack effect that can have a significant impact on pressure differences and door-opening forces, particularly in stairwells. A building that is commissioned during mild weather conditions can experience unacceptable pressure conditions during extreme winter or summer periods, and the resulting impact on interior pressure differences or door-opening forces cannot be identified until a seasonal periodic test is performed. It is at this point that the system can be modified to address the previously unidentified condition. For buildings in extreme hot or cold climates, it is often advisable to schedule a manual periodic test during the extreme conditions, particularly if acceptance testing occurred during a more favorable time of year.

Wind typically has the greatest impact on exterior openings to a building, especially openings used to provide makeup air for an atrium[1] smoke control system. Experience has shown that wind can cause some types of makeup air doors/windows to hang up or not open due to high wind pressures. The design cannot fully anticipate high winds that can occur at certain times of year or due to a funneling effect created by adjacent buildings at the site that were not considered in the design.

The impact of extreme temperature or wind conditions can sometimes be identified during an automatic self-test. For example, if an exterior makeup air door was reported to not operate properly during an automatic test but upon inspection appears fully functional, investigation can show that the door did not open fully (and thus register open by its end switch) due to extreme wind conditions at the time of testing.

RECOMMENDED TESTING

To determine the proper operation of a smoke control system over the life of the building, a program of periodic testing needs to be adopted that includes both manual and automatic testing of the system. Periodic testing should be performed at frequencies that are both practical and sufficient to confirm that the system as a whole and its components remain operational.

Automatic testing needs to be performed at a relatively high frequency, to verify that the system components are operational and capable of performing in a smoke control mode. Manual testing can be performed less frequently, with the purpose of verifying that changes to the building's architecture or systems do not have an adverse impact on the performance of the system. For both types of testing, it is important that the appropriate stakeholders have a role in performing and monitoring the testing and responding rapidly to impairments or deficiencies that are identified by the testing. It is important that testing results be properly documented to ensure long-term continuity of the testing program for the smoke control system.

Manual Testing

In spite of the cost and some possible interruptions, manual testing is important for the long-term functionality of the system, and life-safety protection afforded by the system. Manual testing should be performed by persons who are familiar with the intended operation of the system, either through training or by way of review of design intent reports, sequences of operations, and system drawings. Otherwise, testing can be incomplete and could miss system deficiencies. **Testing data (e.g., test procedure, measured pressure differences and door-opening forces, environmental conditions at the time of testing) should be recorded for all manual tests in order to provide a baseline for comparison for future manual tests.**

The type of testing that should be performed can vary by system type, but should always include witnessed operation of all system components, usually performed by manipulating devices (e.g., fans, dampers, operable doors/windows) from the firefighter's smoke control station (FSCS). Printed test reports from the automatic weekly self-testing should be reviewed to identify inoperable or problematic system components. Other type of testing recommended by system type is discussed in the sections that follow.

Frequency of Manual Testing

Guidance on the frequency for which manual testing should be performed varies. NFPA 92 (NFPA 2012) recommends that dedicated systems be tested at least semiannually while nondedicated systems are only recommended to be tested annually. NFPA 92 requires that periodic testing operates the smoke control system for each control sequence in the current design while verifying that the correct outputs are observed for each given input. Dedicated systems are smoke control systems and components that are installed for the sole purpose of providing smoke control, and upon activation these systems operate specifically to perform the smoke control

1. In this handbook, the term atrium is used in a generic sense to mean any large-volume that is at least two stories high, such as an enclosed shopping mall, sports arena, or an airplane hangar.

function. Nondedicated systems are smoke control systems and components that are used both to provide normal HVAC functions as well as smoke control functions. Because dedicated systems should only be activated during a fire event or when tested, they require more frequent testing. Less frequent testing requirements exist for nondedicated systems based on the assumption that problems are more likely to be detected by building occupants who notice a failure in the HVAC system that is used on a daily basis.

Zoned Smoke Control

As discussed elsewhere in this book, zoned smoke control typically involves the creation of pressure differences across defined smoke barriers between smoke control zones. For complex zoned smoke control systems, manual testing can be time consuming when performed properly, due to the need for testing the often complex interactions of devices for alarms in different smoke control zones.

The first step in testing a zoned smoke control system is to review the system documentation to determine the location of smoke control zone boundaries in the building. The sequence of operations (in table or narrative form) should then be reviewed to verify the intended operation of devices for an alarm in a given zone.

Once an understanding is developed of the intended operation of the system, the smoke barriers at all of the smoke control zone boundaries need to be inspected. The first thing to verify is whether all of the boundaries still exist in their intended locations. If a boundary has changed, it would be necessary to flag this and determine if the smoke control system function has been compromised. An example of this would be a convenience stair installed for a multifloor tenant in an office building. If the zoned smoke control system's exhaust fan is only sized to maintain the required pressure difference for a single floor, opening the zone to a second floor effectively doubles the size of the zones and can cause pressure differences to be lower than required. Changes would then have to be made to the system to account for the change in zone boundaries.

When inspecting the smoke barriers at the boundaries of smoke control zones, doors should be checked to determine if they close and latch properly. Utility penetrations in these barriers should be inspected where possible to determine if the penetrations remain properly sealed. For doors that release by magnetic hold-open devices, the doors should be reinspected during the sequence of operations testing to make sure that the doors close properly and do not hang open either due to a hardware issue or to the pressures differences produced by the smoke control system. Door-closing failure is a common problem identified when testing zoned smoke control systems.

Once system devices are manually verified to be operable by the FSCS, individual sequences for each smoke control zone should be executed to verify that all devices align properly. This can be performed by activating a system by a smoke detector or another initiating device. For a complex smoke control system where zoned system activation is provided on the FSCS, the sequence should be verified both by an initiating device and manually by the FSCS. When it has been verified that a zone is operating according to its proper sequence, verification of pressures and door-opening forces can occur by locally measuring the pressure and door-opening forces at multiple locations at the zone boundary, including all doors leading to the exterior or to adjacent smoke control zones.

Pressurized Stairwells

For many buildings, the only smoke control systems installed in the building are pressurized stairwells. For simple constant-speed pressurization systems, manual testing can be performed by initiating an alarm and (1) verifying fan and damper operation, (2) measuring pressure differences, and (3) measuring door-opening forces. This type of testing can easily be performed by a building engineer without any special qualifications.

In some buildings, the stair pressurization systems are more complex modulating systems that can require knowledge of the programmed sequence to test. Care must be given not only to verify that the pressure differences are in the required range when doors are closed, but that fans do not overpressurize or underpressurize the stair when a door is held open for a prolonged period of time as can occur during occupant evacuation. Overpressurization can cause doors to inadvertently slam with the potential for occupant injury, or to create excessive door-opening forces for a period of time once the door is closed. Underpressurization can allow smoke to migrate into the stairwell, making the stairwell unusable for egress.

Where pressurized stairwells exist in combination with other pressurized stairwells and/or elevator pressurization systems, or as part of a zoned smoke control system, pressure differences and door-opening forces need to be verified for all possible sequences of operations. For example, it is common to activate stair pressurization on all alarms, including manual pull stations. Elevator pressurization can wait for activation of an elevator lobby smoke detector to initiate this system. Zoned smoke control requires additional location-specific detection to activate. Because of this hierarchy of sequences, a number of different pressure states can exist. The stair pressurization system alone exerts one set of pressures on the stair doors. Other systems can exert additional complementary or opposing forces on doors that impact whether the pressure exceeds the required minimums or allowable maximums on the door.

For modulating systems where pressure sensors are used to control fan speeds to deliver the desired pressure, sensor calibration must be verified during manual testing of pressurized stairwells. An on-site verification of the sensor can be performed by comparing the pressure difference measured by the sensor during a test to pressure readings taken manually with a handheld measuring gage near to the location of the installed sensor. To avoid inadvertent creation of adverse pressure conditions within a stairwell due to the faulty sensor, sensors that are not measuring pressure correctly should be replaced. Sensors that are out of calibration can be sent to an accredited laboratory for recalibration.

Elevator Smoke Control

Elevator pressurization systems are tested similarly to stair pressurization systems with regard to the testing of individual pieces of equipment or the hierarchy of sequences. Operation of devices is verified along with the pressure difference at each elevator door. Unlike stair doors, there is no definitive maximum pressure or door-opening force, as elevator manufacturer data for maximum operating pressures are generally not available. Therefore, once the elevator shafts are pressurized it is important to individually verify that elevator doors are capable of opening and closing properly when subject to the design overpressure.

Atrium Smoke Control

For atrium smoke control systems, the first step in performing manual testing is to review the system documentation to determine the location of smoke control zone boundaries in the building. The sequence of operations (in table or narrative form) should then be reviewed to verify the intended operation of devices for the atrium zone(s). As with a zoned smoke control system, the boundary smoke barriers surrounding the atrium smoke control zone should be inspected for integrity.

The performance of an atrium smoke control system is primarily dependent on the exhaust and supply airflows from and to the zone. These are verified during the initial commissioning process (Chapter 15) by a certified test and balance (TAB) contractor. It is not be necessary that every periodic test be done by a TAB contractor. It is often sufficient to review the printed results from the automatic self-tests, which should verify that all fans are operating within their intended range.

System devices should be manually verified to be operable by the FSCS, and then the sequence for the atrium smoke control zone should be executed to verify that all devices operate properly. This can be performed by activating the system by a smoke detector or another initiating device. Proper operation of all exterior doors/windows used to provide makeup air from the exterior should also be visually verified.

Automatic Testing

For smoke control systems to provide their intended life-safety function it is important to verify that the system will be capable of operating during a fire event. To provide this implied level of reliability, some sort of constant monitoring of the system would ideally be provided, analogous to that provided for a fire alarm system. The codes and standards governing smoke control use words like *supervision* and *verification* to describe this process. Requirements for monitoring "positive confirmation of actuation, testing, or manual override" and "presence of power downstream of all disconnects" are applied to smoke control systems, sometimes with minimal guidance on how this is to be accomplished.

Fire alarm systems are capable of providing continuous electrical supervision and verification of devices. If a power/monitoring wire leading to a smoke detector is cut, or if a smoke detector becomes dust-logged and thus becomes impaired for its intended function, a trouble signal will be indicated at the Fire Alarm Control Panel (FACP). Unlike fire alarm systems, dedicated smoke control system components (e.g., fans and dampers) are dormant until the need arises that requires their function. Faulty wiring or inoperative equipment adversely affecting the intended smoke control objective can go undetected until the system is directed to operate. Electrical monitoring methods do not work for mechanical, pneumatic, hydraulic, and nonaddressable (4–20 ma and 0–10 volt) equipment typically employed to activate smoke control dampers, fans, etc. In addition, the fans and dampers themselves are listed only according to safety concerns rather than the according to reliability standards applied to fire alarm equipment.

To address this concern, the Underwriters Laboratories (UL) Standard 864 (UL 2003) includes a section for control equipment for smoke control applications. The equipment listed for smoke control applications carries the four letter UUKL product classification. Equipment that is listed under category UUKL is evaluated to the same hardware reliability requirements as fire alarm equipment, and is evaluated against operational requirements similar to those of NFPA92 regarding signal prioritization when multiple activation signals are received, and the ability to provide an automatic self-test function.

The self-test function allows performance testing of devices periodically to provide the end-to-end verification of component functionality. UL 864 specifically requires that for dedicated equipment, the automatic self-test function shall be capable of being performed at least on a weekly basis. Other test intervals can also be available. The self-test shall automatically activate each associated function, either on a system, subsystem, zone, or individual equipment basis. For example, a stairwell pressurization system can be self-tested on a system basis to confirm required pressurization is reached, or each fan and damper individually activated and operation confirmed by "end-to-end verification." UL 864 specifies that both audible and visual trouble signals shall be annunciated at the FSCS identifying any function that failed to perform as intended. No specific annunciation is mandated when each function operated as intended. The supervision and verification requirements for equipment controlled by UUKL listed equipment is shown in Figure 24.1.

The codes and standards vary on how the requirement for UUKL weekly self-testing is invoked. NFPA 92 (2012) requires smoke control systems to have a means for verifying correct operation when activated, including positive confirmation of activation, testing, manual override, and the presence of operating power downstream of all circuit disconnects. The annex material describes multiple methods of end-to-end verification that are capable of ascertaining the desired result. NFPA 92 specifically references the UUKL listing requirement and specifies a weekly self-test interval for dedicated smoke control equipment, and for nondedicated equipment, states that the means and frequency of verification varies according to the complexity and importance of the system. The IBC (ICC 2012), while not specifically using the term UUKL, specifically states that control systems must comply with UL 864 and be listed as smoke control equipment. The IBC states verification is to include positive confirmation of actuation, testing, manual override, the presence of power downstream of all disconnects and, through a preprogrammed weekly test sequence, report abnormal conditions audibly, visually, and by printed report.

Nondedicated smoke control systems and subsystems are not required to have the capability of an automatic self-test function, because the equipment is activated on a routine basis and any inoperability would be noted by a lack of environmental control or comfort level. For dedicated equipment, a test is performed weekly on a predetermined day at a predetermined time, usually selected to cause minimal interruption to the normal operation of the building. A printed report should be made as a record of the self-test, and such a report is required in the event of abnormal conditions. The printed report notes any devices that failed the self-test and a system fault is registered so that the device(s) not operating properly can be investigated and repaired. As shown in Figure 24.2, the printed report ideally provides detailed information as to the device that failed so that the device in question can be readily investigated. Grouping of devices only causes additional work, as multiple devices would have to be inspected to identify which device is the problem device. To facilitate long-term maintenance, it is important that the weekly self-test reports be retained and kept in a log book or file so that the performance of problem devices can be tracked over time.

In order for the weekly self-test to be performed, sensors need to be in place to enable monitoring of the performance of the component being tested. The type of sensor used depends upon the type of component being

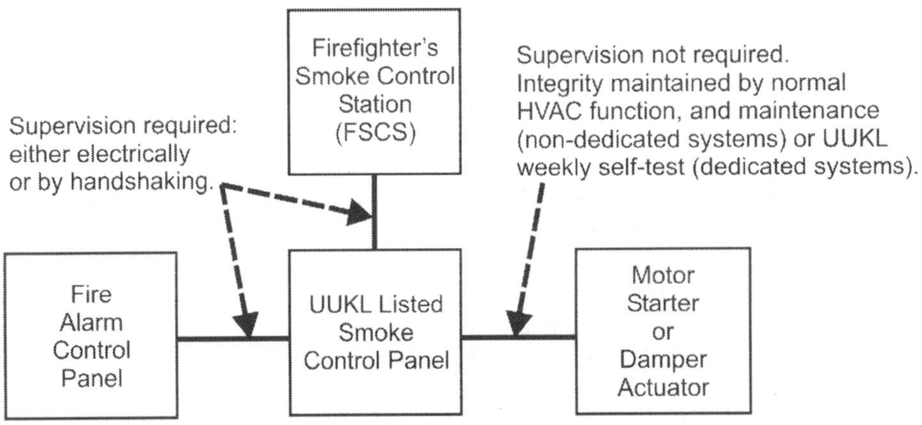

Figure 24.1 *Monitoring of smoke control equipment.*

monitored, as discussed in the sections that follow. For fans, dampers, and automatic doors/windows, these sensors serve as the means for providing supervision of the device, and meet the requirement for verification and presence of power, consistent with the intent of the IBC, NFPA 92A, and NFPA 92B.

Fans

The primary purpose of sensors provided for fan monitoring is to verify that the fan airflow is consistent with the design. Multiple fault conditions can prohibit fans from performing as designed. Power failure, broken fan belts, and locked rotors will prevent a fan from operating. Clogged filters, blocked duct, and adverse wind conditions can cause the airflow of the fan to be outside of its intended range. In addition, the fan can be inadvertently turned off either by a local maintenance disconnect switch or HOA switch at a remote motor control center.

Voltage monitors, current transducers, differential pressure transducers, and flow monitors are sensors that can be used to monitor fan status as shown in Figure 24.3. These sensors can be used in various combinations. At a minimum, it is typical to monitor voltage upstream of the local motor controller to verify that power is present up to the controller at all times. Downstream of the motor controller, power is not present unless the fan is called upon to operate, either during a fire event or during the preprogrammed weekly self-test. Monitoring fan function downstream of the motor controller should be accomplished using sensing devices that directly indicate that airflow is present, such as a flow monitor, pressure transducer, or sail switch in the airstream from the fan.

Figure 24.3 shows a generic "flow monitor" at this location. Other types of devices, such as current transducers, have been proposed for fan monitoring, but only a few of these devices are sensitive enough to detect the difference between a fan running normally and one that is running with a broken belt or blocked filter, as described in the following paragraph. While any of these monitoring methods are also capable of determining loss of power due to the local disconnect switch being in the OFF position at the time of the self-test, it is desirable to also provide electrical supervision of the switch so that a fault condition is immediately indicated when the switch is turned off.

Some codes and standards have limited performance verification strictly to flow and pressure verification downstream of the fan. Where this limitation does not exist, newer state-of-the-art current transducers equipped with both high and low limit settings have become a popular choice to provide fan monitoring. Low-current conditions would occur either when the fan is left off or when rotors freely spin due to fan belt breakage. A high-current condition would occur due to

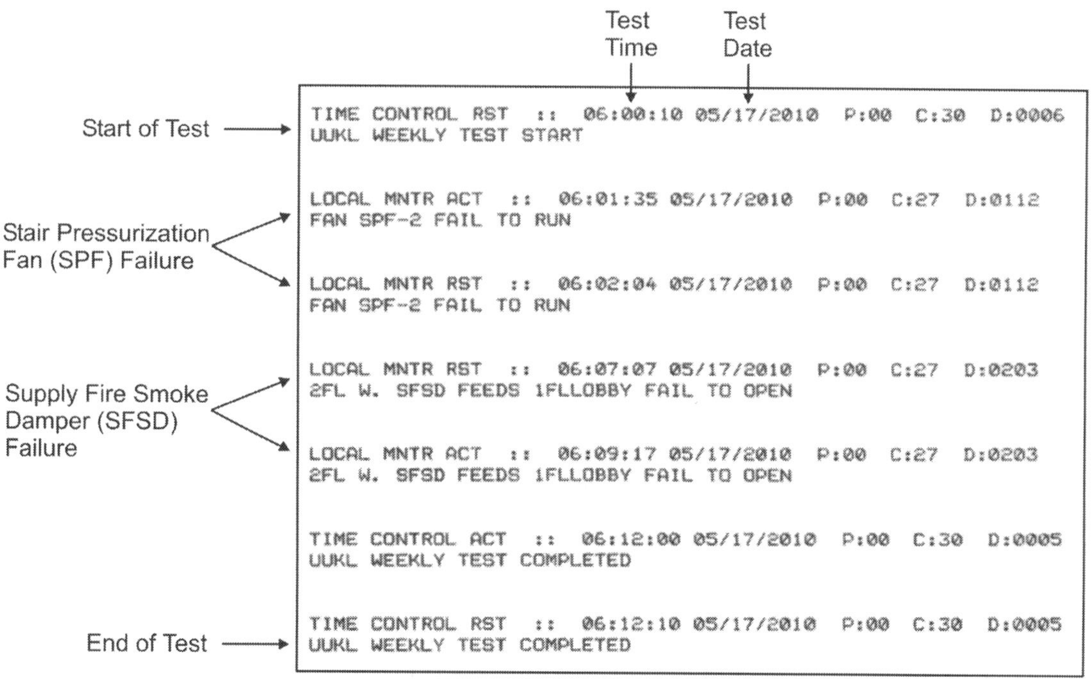

Figure 24.2 *An example self-test report.*

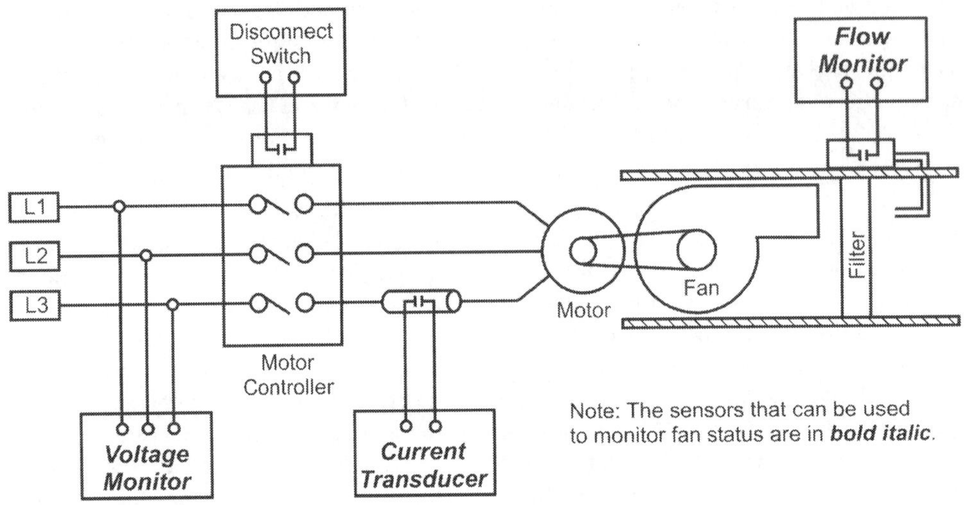

Figure 24.3 *Sensors that can be used to monitor fan status.*

a locked rotor. Even among these state-of-the-art devices, only a select few are sensitive enough to detect reduced airflow due to a blocked filter in the airstream.

Current transducers are not as susceptible to breakage as a sail switch, nor are they as susceptible to calibration issues as pressure transducers. When using current transducers, manufacturer listings should be consulted to determine whether their use is specifically prohibited for life-safety purposes, as is the case for some models. While it could be argued that the current transducers are being used for a monitoring rather than life-safety/control function, this could prove problematic with respect to liability should a system fail to operate in a fire event.

Dampers

The most common sensors used to monitor damper function and status are end switches, limit switches, or proximity sensors. These devices can be placed either on the damper blades, or on the frame of the damper. Optical switches and mechanical switches are available to provide this function.

When direct indication of the damper blade position is required, the most reliable method for monitoring damper function is to place the sensor on the damper blade or frame, to make sure that the damper has properly open or closed. Sensors placed on the actuator arm are often held in place by a set screw that can loosen over time with the operation of the damper and provide a false indication of damper status. This is a common problem that should be avoided when monitoring a damper integral to a smoke control system.

Commercially available dampers, including fire/smoke dampers (FSDs) commonly used in smoke control applications, typically ship with integrated sensors to monitor damper positions. In some instances, however, end switches are added to a damper after installation to monitor damper function. The return damper in typical packaged air-handling units (AHUs) is often not monitored for position, for example. Often, these units are used in a 100% outdoor air mode to provide smoke control functions. An end switch is often added to the return damper to ensure that it is closed and not allowing recirculation of smoke by the supply side of the package unit. During periodic testing, end switches installed as a system retrofit should be inspected to ensure that they are not impeding the ability of the damper to properly open and close.

Automatic Doors/Windows

As with dampers, automatic doors/windows are monitored using end switches, limit switches or proximity sensors. End switches are not commonly used to monitor doors/window as it is often difficult to place a sensor in the door/window frame to verify open or closed status. Doors and window used to provide makeup air can be located in a curtain wall where it is not desirable to run power and control monitoring out to an end switch either due to practical or aesthetic concerns. As such, automatic doors and windows are most often monitored by limit or proximity sensors built into the actuator. For example, doors in exterior walls are often equipped with a door operator/closer that has a proximity sensor monitoring the position of the closer, and not the actual position of the door. Should the connection between the actuator arm and the door break or become disconnected, the proximity sensor can indicate that a door has properly opened when it in fact remains closed. During periodic testing, it is desirable to operate doors and windows manually and to visually verify that the door/window sensor is indicating the correct status of that opening.

ROLES AND RESPONSIBILITIES

When a building is first built and is being readied to open, commissioning testing is performed to establish the functionality of the installed life safety systems. Third party inspection is often required by code and is performed by an independent special inspector or commissioning agent. Final testing is then witnessed by the local authorities having jurisdiction who are often a combination of building department and fire department personnel.

After the building is occupied, long-term functionality and reliability of a smoke control system is provided only when periodic testing is routinely performed to confirm that the system continues to operate as designed. Unlike commissioning, the responsibility for this testing often relies solely with the building owners and operators, as local jurisdictions simply do not have the resources to witness periodic testing for all smoke control systems within their jurisdiction.

It is critical that building owners and operators recognize the importance of periodic testing and plan to perform this testing on a regular basis, even though there is a cost due to manpower involvement and a potential for business interruption during the tests. To facilitate a long-term testing program, it is important to maintain copies of the basis of design documents as well as logs and test reports for individual tests for the purposes of establishing a baseline against which to compare future results.

Manual Testing

Manual testing should be performed either annually or semiannually depending upon whether the system primarily uses nondedicated or dedicated equipment. Manual testing is often performed by a building engineer or other building maintenance staff. However, for more complicated systems, manual testing can be contracted out to a third party (i.e., a fire protection engineer, mechanical engineer, or certified air-balancing contractor). Should manual testing be performed by a third party, the basis of design documents should be made available to enable the third party to develop an understanding of the proper function of the system. A log of the tests performed and results (including all pressure, flow, and door-opening force measurements made during testing) should be maintained at the building. Reports should be kept of deficiencies found during the manual tests and the action taken to correct them.

To facilitate manual testing, it is beneficial if a test plan/procedure be developed at the time of design, and that this procedure should be included in the basis of design documents. This is so all of the required testing necessary to verify the proper operation of the system is being performed during each manual periodic test, regardless of who is performing the testing.

Automatic Testing

Automatic testing of dedicated smoke control equipment should have been set up at the time of commissioning, according to the requirements for the UUKL weekly self-test. Subsequent to commissioning, the results of each self-test should be reviewed weekly to identify inoperable devices and establish a plan for returning the system to operation. This is typically the responsibility of a building engineer assigned to the building. The responsible engineer should become familiar with the location of all devices integral to the function of the smoke control system and should maintain a log of self-test reports in order to be able to track problem devices over time. Reports should be kept of deficiencies found during the automatic tests and the action taken to correct them.

REFERENCES

NFPA. 2012. NFPA 92, *Standard for Smoke Control*. Quincy, MA: National Fire Protection Association.

ICC. 2012. *International Building Code*® (*IBC*®). Country Club Hills, IL: International Code Council.

UL. 2003 (rev. 2008). Standard 864, *Control Units and Accessories for Fire Alarm Systems*, 9th ed. Northbrook, IL: Underwriters Laboratories.

Appendix A

Derivations of Equations

John H. Klote

This appendix has the derivations of many of the equations used in smoke control engineering. Because the equations are in base units and coherent derived units of the SI units, no units are given in this appendix. For information about the SI system, see Chapter 1.

1.0 FUNDAMENTAL EQUATIONS OF ENGINEERING

The following are fundamental equations of engineering from which the other equations in this appendix are derived.

1.1 Hydrostatic Pressure

The hydrostatic pressure equation is

$$p = p_o - \rho g z \qquad \text{(A1.1.1)}$$

where

p = pressure at elevation z,

p_o = pressure at reference elevation z_o,

g = acceleration of gravity,

z = elevation.

Equation A1.1.1 is exact for the pressure in a fluid at rest, but it is a good approximation when the velocity is relatively low. It is applicable for fires in rooms and flows in building spaces including shafts.

1.2 Orifice Equation

The orifice equation flow written is

$$m = CA\sqrt{2\rho\Delta p} \qquad \text{(A1.2.1)}$$

where

m = mass flow through path,

C = flow coefficient, dimensionless,

A = flow area (or leakage area),

Δp = pressure difference across path,

ρ = density gas in path.

This is extensively used for flow in buildings. When this equation is used for an orifice flow meter, C is called the discharge coefficient.

The volumetric flow is related to the mass flow as

$$m = \rho V s \qquad \text{(A1.2.2)}$$

where V is the volumetric flow. By substituting Equation A1.2.2 into Equation A1.2.1 and rearranging, the orifice equation can be written in terms of volumetric as

$$V = CA\sqrt{\frac{2\Delta p}{\rho}}. \qquad \text{(A1.2.3)}$$

1.3 Ideal Gas Law

The density of air and smoke are expressed by the ideal gas law which is

$$\rho = \frac{p}{RT} \qquad \text{(A1.3.1)}$$

where

ρ = density,

p = pressure,

R = gas constant,

T = absolute temperature.

2.0 BIDIRECTIONAL FLOW

Bidirectional flow is also called two-way flow, and it happens between two spaces at different temperatures as shown in Figure A2.1.

2.1 Pressure Differences

This section derives the equations for pressure difference related to bidirectional flow. The hydrostatic pressure equation, Equation A1.1.1, can be written for two spaces as

$$p_1 = p_o - \rho_1 g z \qquad \text{(A2.1.1)}$$

and

$$p_2 = p_o - \rho_2 g z \qquad \text{(A2.1.2)}$$

where the subscripts 1 and 2 refer to spaces 1 and 2. At the neutral plane, the elevation z is chosen to be zero, and the pressure at the neutral plane is p_o. The pressure difference from space 1 to space 2 is $\Delta p_{12} = p_1 - p_2$. Substituting Equations A2.0.1 and A2.0.2 into the equation for Δp_{12} results in

$$\Delta p_{12} = g(\rho_2 - \rho_1)z. \qquad \text{(A2.1.3)}$$

The densities in spaces 1 and 2 can be written in terms of the ideal gas law (Equation A1.3.1), and they can be substituted into the previous equation to give

$$\Delta p_{12} = \frac{g p_{atm}}{R}\left(\frac{1}{T_2} - \frac{1}{T_1}\right)z \qquad \text{(A2.1.4)}$$

where p_{atm} is the absolute atmospheric pressure. This is the pressure difference shown in Figure A2.1c.

2.2 Continuous Opening

This section derives the equation for the height of the neutral plane H_n for bidirectional flow through a continuous rectangular opening between two spaces at different temperatures (Figure A2.1a). The height of the opening is H, and the width is w. The neutral plane is at a height of H_n. The temperature in space 1 is greater than that in space 2 which results in flow into space 1 at the bottom and out of space 1 at the top.

The orifice equation can be written in differential form as

$$dm_{12} = Cw\sqrt{2\rho_1 \Delta p_{12}}dz. \qquad \text{(A2.2.1)}$$

In this equation, the density of space 1 is used because that is the density of the air in the path. Substituting Equation A2.1.3 into Equation A2.2.1 yields

$$dm_{12} = Cw\sqrt{2\rho_1 g(\rho_2 - \rho_1)}dz. \qquad \text{(A2.2.2)}$$

Integrate this equation from z = 0 to $(H - H_n)$.

$$m_{12} = \frac{2}{3}Cw(H - H_n)^{3/2}\sqrt{2\rho_1 g(\rho_2 - \rho_1)} \qquad \text{(A2.2.3)}$$

By analogy, the flow in the opposite direction is

$$m_{21} = \frac{2}{3}CwH_n^{3/2}\sqrt{2\rho_2 g(\rho_2 - \rho_1)}. \qquad \text{(A2.2.4)}$$

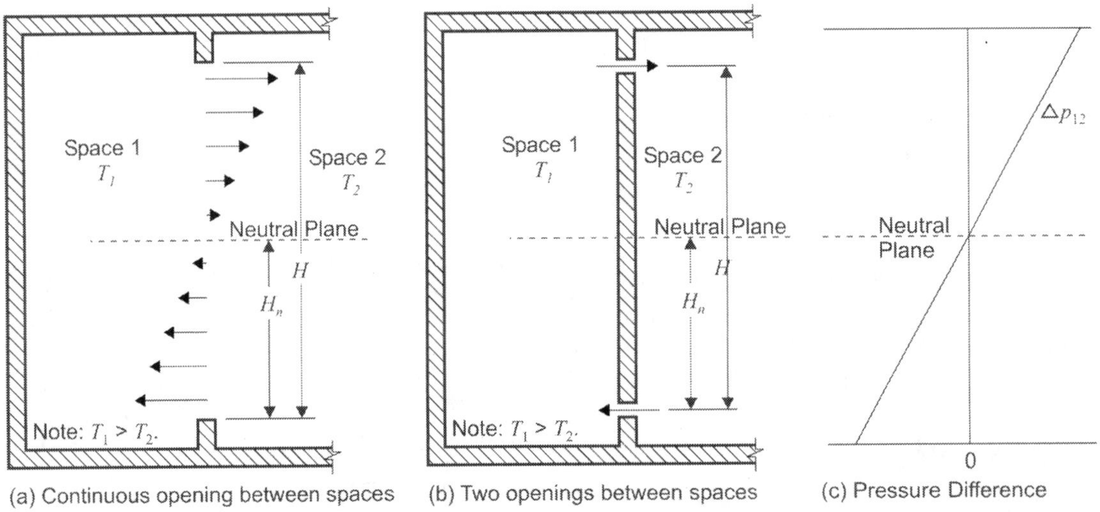

Figure A2.1 *Bidirectional flow between two spaces.*

If there are no other flows into or out of the two spaces, $m_{12} = m_{21}$. Setting Equations A2.2.3 and A2.2.4 equal to each other results in

$$\begin{aligned}&\frac{2}{3}Cw(H-H_n)^{3/2}\sqrt{2\rho_1 g(\rho_2-\rho_1)}\\&=\frac{2}{3}CwH_n^{3/2}\sqrt{2\rho_2 g(\rho_2-\rho_1)}\end{aligned}. \qquad \text{(A2.2.5)}$$

Canceling like terms and rearranging results in

$$H_n = \frac{H}{1+(\rho_2/\rho_1)^{1/3}}. \qquad \text{(A2.2.6)}$$

The densities in spaces 1 and 2 can be written in terms of the ideal gas law (Equation A1.3.1), and they can be substituted into the previous equation to give

$$H_n = \frac{H}{1+(T_1/T_2)^{1/3}}. \qquad \text{(A2.2.7)}$$

2.3 Two Openings

This section derives the equation for the height of the neutral plane H_n for bidirectional flow through two openings between two spaces at different temperatures (Figure A2.1a). The height between the openings is H, which is much larger than the sizes of the openings. Thus the variation of pressure with height for the openings can be neglected. The temperature in space 1 is greater than that in space 2 which results in flow into space 1 at the bottom and out of space1 at the top. The areas above and below the neutral plane are A_a and A_b respectively.

From Equation A2.1.3, the pressure difference at the bottom opening is

$$\Delta p_{12} = g(\rho_2-\rho_1)H_n. \qquad \text{(A2.3.1)}$$

From the orifice equation and the ideal gas law, the flow from space 1 to space 2 below the neutral plane is

$$m_{12} = CA_b\sqrt{2\rho_2 g(\rho_2-\rho_1)H_n}. \qquad \text{(A2.3.2)}$$

In the same manner, the flow above the neutral plane is

$$m_{21} = CA_a\sqrt{2\rho_1 g(\rho_2-\rho_1)(H-H_n)} \qquad \text{(A2.3.3)}$$

If there are no other flows into or out of the two spaces, $m_{12} = m_{21}$. Setting Equations A2.3.2 and A2.3.3 equal to each other results in

$$\begin{aligned}&CA_b\sqrt{2\rho_2 g(\rho_2-\rho_1)H_n}\\&=CA_a\sqrt{2\rho_1 g(\rho_2-\rho_1)(H-H_n)}\end{aligned}. \qquad \text{(A2.3.4)}$$

Canceling like terms and rearranging results in

$$H_n = \frac{H}{1+(\rho_2/\rho_1)(A_b/A_a)^2}. \qquad \text{(A2.3.5)}$$

The densities in spaces 1 and 2 can be written in terms of the ideal gas law (Equation A1.3.1), and they can be substituted into the previous equation to give

$$H_n = \frac{H}{1+(T_1/T_2)(A_b/A_a)^2} \qquad \text{(A2.3.6)}$$

3.0 EFFECTIVE FLOW AREAS

The following sections develop equations for the effective flow areas for a system of flow paths in parallel and one in series as shown in Figure A3.1.

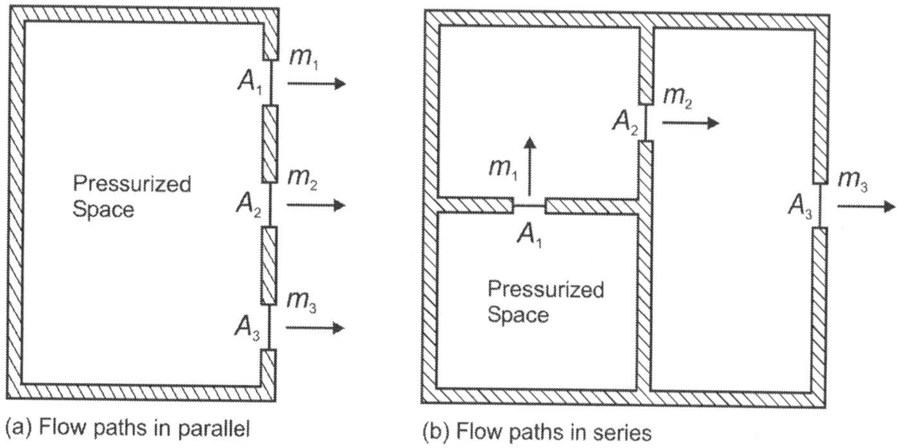

Figure A3.1 *Systems of flow areas.*

3.1 Parallel Paths

This section derives the equation for the effective area of three flow paths in parallel as shown in Figure A3.1a. The paths and rooms between them are considered to be at the same temperature, and the flow coefficients are considered the same for all the paths. For paths in parallel, the pressure difference across each path is the same as that across the other paths. For three paths in parallel, the total flow m_T from the room is the sum of the flows through each path.

$$m_T = m_1 + m_2 + m_3 \qquad \text{(A3.1.1)}$$

where m is the mass flow, and the subscripts 1, 2, and 3 refer to flow paths 1, 2, and 3 respectively. The orifice equation can be used to describe the flow through the system in terms of the effective area as

$$m_T = CA_e\sqrt{2\rho\Delta p} \qquad \text{(A3.1.2)}$$

where

C = flow coefficient,

A_e = effective area of system,

ρ = density of air in flow paths,

Δp = pressure difference across paths.

In the same way, the flow through path 1 is

$$m_1 = CA_1\sqrt{2\rho\Delta p}. \qquad \text{(A3.1.3)}$$

The flow through the paths 2 and 3 can be written the same way. Substituting these flows in Equation A3.1.1 results in

$$\begin{aligned} CA_e\sqrt{2\rho\Delta p} &= CA_1\sqrt{2\rho\Delta p} \\ &+ CA_2\sqrt{2\rho\Delta p} + CA_3\sqrt{2\rho\Delta p} \end{aligned}. \qquad \text{(A3.1.4)}$$

Canceling like terms in Equation A3.1.4 yields

$$A_e = A_1 + A_2 + A_3. \qquad \text{(A3.1.5)}$$

This can be extended to any number of paths.

3.2 Series Paths

This section derives the equation for the effective area of three flow paths in series as shown in Figure A3.1b. For these paths, the total pressure difference, Δp_T, from the pressurized room to the outside is the sum of the pressure differences Δp_1, Δp_2, and Δp_3 across each of the respective flow areas, A_1, A_2, and A_3:

$$\Delta p_T = \Delta p_1 + \Delta p_2 + \Delta p_3 \qquad \text{(A3.2.1)}$$

The flow, m, is the same for each flow path. The orifice equation written for the flow through the entire system of paths results is $m = CA_e\sqrt{2\rho\Delta p_T}$. Solving this equation for pressure difference results in

$$\Delta p_T \frac{1}{2\rho}\left(\frac{m}{CA_e}\right)^2. \qquad \text{(A3.2.2)}$$

Again, flow coefficients and the temperatures are considered that the same for all the flow paths. Similar equations can be written for each of the three paths, and substituting these into Equation A3.2.1 yields

$$\begin{aligned} \frac{1}{2\rho}\left(\frac{m}{CA_e}\right)^2 &= \frac{1}{2\rho}\left(\frac{m}{CA_1}\right)^2 \\ &+ \frac{1}{2\rho}\left(\frac{m}{CA_2}\right)^2 + \frac{1}{2\rho}\left(\frac{m}{CA_3}\right)^2 \end{aligned} \qquad \text{(A3.2.3)}$$

Canceling like terms and rearranging results in

$$A_e = \left(\frac{1}{A_1^2} + \frac{1}{A_2^2} + \frac{1}{A_3^2}\right)^{-1/2}. \qquad \text{(A3.2.4)}$$

This can be extended to any number of paths.

Different Temperatures and Coefficients

If the temperatures and flow coefficients are different for the flow paths, the following equations can be derived in the same manner as the previous equations.

$$A_e = \frac{T_e^{1/2}}{C_e}\sum_{i=1}^{n} C_i A_i T_i^{1/2}$$

for parallel paths (A3.2.5)

and

$$A_e = \frac{T_e^{1/2}}{C_e}\left(\sum_{i=1}^{n} T_i (C_i A_i)^{-2}\right)^{-1/2}$$

for series paths. (A3.2.6)

4. PRESSURIZED STAIRWELLS

This section applies to pressurized stairwells in an idealized building that (1) has no vertical leakage through the floors and shafts, and (2) has leakage that is the same from floor to floor.

4.1 Pressures and Flows for an Idealized Building

This section derives equations for the pressure differences of a pressurized stairwell in an idealized building. When the pressure in a stairwell is hydrostatic, Equation A1.1.1 can be written for the stairwell as

$$p_S = p_{Sb} - \rho_S g z \qquad \text{(A4.1.1)}$$

where

p_S = absolute air pressure in stairwell at elevation z,

p_{Sb} = absolute air pressure in stairwell at stairwell bottom,

g = acceleration of gravity,

ρ_S = air density within stairwell,

z = elevation above stairwell bottom.

Equation A4.1.1 is appropriate when the pressure losses due to friction in the stair shaft are negligible. In a pressurized stairwell, this usually happens when all the doors are closed. When some doors in a pressurized stairwell are open, pressure losses due to friction can be significant. For a stairwell ventilation system where air is both supplied to and exhausted from the stair shaft, pressure losses due to friction can be significant. When pressure losses due to friction are significant, equation A4.1.1 is not appropriate.

The pressure outdoors is hydrostatic, and it is

$$p_O = p_{Ob} - g\rho_O z \qquad \text{(A4.1.2)}$$

where

p_O = absolute air pressure at elevation y, (Pa),

p_{Ob} = absolute air pressure at stairwell bottom, (Pa),

ρ_O = air density outdoors, (kg/m^3).

The pressure difference from the stairwell to the outdoors is $\Delta p_{SO} = p_S - p_O$. Using Equations A4.1.1 and A4.1.2, the pressure difference from the stairwell to the outdoors can be written as

$$\Delta p_{SO} = (p_{SBb} - g\rho_S z) - (p_{Ob} - g\rho_O z) \qquad \text{(A4.1.3)}$$

where Δp_{SOb} is the pressure difference from the stairwell to the outdoors at the bottom of the stairs. Rearranging Equation A4.1.3 yields

$$\Delta p_{SO} = \Delta p_{SOb} + gy(\rho_O - \rho_S). \qquad \text{(A4.1.4)}$$

Substituting the density from the ideal gas law into this equation results in

$$\Delta p_{SO} = \Delta p_{SOb} + gz\left(\frac{p_{atm}}{RT_O} - \frac{p_{atm}}{RT_S}\right) \qquad \text{(A4.1.5)}$$

where

p_{atm} = atmospheric pressure,

T_O = absolute temperature outdoors,

T_S = absolute temperature in stairwell.

Rearranging Equation A4.1.5 yields

$$\Delta p_{SO} = \Delta p_{SOb} + Bz \qquad \text{(A4.1.6)}$$

where

$$B = \frac{g p_{atm}}{R}\left(\frac{1}{T_O} - \frac{1}{T_S}\right) \qquad \text{(A4.1.7)}$$

The pressure difference from the stairs to the outdoors can be written as

$$\Delta p_{SO} = \Delta p_{SB} + \Delta p_{BO} \qquad \text{(A4.1.8)}$$

where

Δp_{SB} = pressure difference from stairs to building,

Δp_{BO} = pressure difference from building to outdoors.

Using the orifice equation, the mass flow from the from the stairs to the building is

$$m_{SB} = C_{SB} A_{SB} \sqrt{2\rho_S \Delta p_{SB}}. \qquad \text{(A4.1.9)}$$

The mass flow from the building to the outdoors is

$$m_{BO} = C_{BO} A_{BO} \sqrt{2\rho_B \Delta p_{BO}} \qquad \text{(A4.1.10)}$$

The mass flow from the stairs to the building equals that from the building to the outdoors ($m_{SB} = m_{BO}$). This can also be written as

$$C_{SB} A_{SB} \sqrt{2\rho_S \Delta p_{SB}} = C_{BO} A_{BO} \sqrt{2\rho_B \Delta p_{BO}} \qquad \text{(A4.1.11)}$$

Consider C_{SB} equal to C_{BO}, then cancel the like terms in Equation A4.1.11 to get

$$A_{SB} \sqrt{\rho_S \Delta p_{SB}} = A_{BO} \sqrt{\rho_B \Delta p_{BO}}. \qquad \text{(A4.1.12)}$$

Squaring both sides of Equation 13 and rearranging yields

$$\Delta p_{BO} = \Delta p_{SB} \frac{A_{SB}^2 \rho_S}{A_{BO}^2 \rho_B}. \qquad \text{(A4.1.13)}$$

Substituting the ideal gas law into the above equation produces

$$\Delta p_{BO} = \Delta p_{SB} \frac{A_{SB}^2 T_B}{A_{BO}^2 T_S} \qquad \text{(A4.1.14)}$$

Substituting the previous equation into Equation A4.1.8 and rearranging yields

$$\Delta p_{SO} = \Delta p_{SB}\left(1 + \frac{A_{SB}^2 T_B}{A_{BO}^2 T_S}\right) \qquad \text{(A4.1.15)}$$

Let $F_R = 1 + \frac{A_{SB}^2 T_B}{A_{BO}^2 T_S}$, and Equation A4.1.15 becomes

$$\Delta p_{SO} = F_R \Delta p_{SB} \quad \text{(A4.1.16)}$$

Equation A4.1.16 can be written for the bottom and top of the stairs as

$$\Delta p_{SOb} = F_R \Delta p_{SBb} \quad \text{(A4.1.17)}$$

and

$$\Delta p_{SOt} = F_R \Delta p_{SBt} \quad \text{(A4.1.18)}$$

4.2 Average Pressure Difference for an Idealized Building

This section derives the pressure differences for a pressurized stairwell in an idealized building. The average pressure difference is defined as the pressure difference that will result in the same total flow as a nonuniform pressure profile. For the mass flow to the stairwell to the outside, this can be written as

$$m_{SO} = C A_{SO} \sqrt{2\rho \Delta p_{SOav}} \quad \text{(A4.2.1)}$$

Using a differential form of the orifice equation, the mass flow from the stairwell to the outside is

$$dm_{SO} = C W_{SO} \sqrt{2\rho \Delta p_{SO}}\, dz \quad \text{(A4.2.2)}$$

where W_{SO} is the effective leakage width from the stairwell to the outside. This effective leakage width can be written as $W_{SO} = A_{SO}/H$ where H is the stairwell height. Substituting this and Equation A4.1.6 into Equation A4.2.1 results in

$$dm_{SO} = \frac{C A_{SO}}{H} \sqrt{2\rho(\Delta p_{SOb} + Bz)}\, dz \quad \text{(A4.2.3)}$$

Integrating this equation from $z = 0$ to $z = H$ gives the flow from the stairwell to the outside.

$$m_{SO} = \frac{2}{3} C A_{SO} \sqrt{2\rho} \left(\frac{\Delta p_{SOt}^{3/2} - \Delta p_{SOb}^{3/2}}{\Delta p_{SOt} - \Delta p_{SOb}} \right) \quad \text{(A4.2.4)}$$

Equating Equations A4.2.1 and A4.2.4 results in

$$C A_{SO} \sqrt{2\rho \Delta p_{SOav}} = \frac{2}{3} C A_{SO} \sqrt{2\rho} \left(\frac{\Delta p_{SOt}^{3/2} - \Delta p_{SOb}^{3/2}}{\Delta p_{SOt} - \Delta p_{SOb}} \right). \quad \text{(A4.2.5)}$$

Canceling like terms and rearranging yields

$$\Delta p_{SOav} = \frac{4}{9} \left(\frac{\Delta p_{SOt}^{3/2} - \Delta p_{SOb}^{3/2}}{\Delta p_{SOt} - \Delta p_{SOb}} \right)^2 \quad \text{(A4.2.6)}$$

The equation for Δp_{SBav} can be derived in a similar manner.

4.3 Height Limit for an Idealized Building

This section derives the height limit for a pressurized stairwell in an idealized building. The pressure difference from the stairwell to the building at the top of the stairs can be expressed as

$$\Delta p_{SBt} = \Delta p_{SBb} + \frac{BH}{F_R}. \quad \text{(A4.3.1)}$$

Solve this equation for H:

$$H = \frac{F_R}{B}(\Delta p_{SBt} - \Delta p_{SBb}) \quad \text{(A4.3.2)}$$

Substitute Equation A4.1.7 into Equation A4.3.2:

$$H = \frac{R F_R (\Delta p_{SBt} - \Delta p_{SBb})}{g p_{atm} \left(\frac{1}{T_O} - \frac{1}{T_S} \right)} \quad \text{(A4.3.3)}$$

In winter, when Δp_{SBb} is the minimum design value Δp_{min} and Δp_{SBt} is the maximum design value Δp_{max}, the stairwell is as tall as it can be for the idealized building. Under these conditions, the height is called the height limit H_m. For winter conditions, the height limit is

$$H_m = \frac{R F_R (\Delta p_{\max} - \Delta p_{\min})}{g p_{atm} \left(\frac{1}{T_O} - \frac{1}{T_S} \right)} \quad \text{(A4.3.4)}$$

To generalize this equation for both winter and summer, an absolute value term can be used as

$$H_m = \frac{R F_R (\Delta p_{\max} - \Delta p_{\min})}{g p_{atm} \left| \frac{1}{T_O} - \frac{1}{T_S} \right|} \quad \text{(A4.3.5)}$$

5. DOOR-OPENING FORCES

This section derives the equation for the door opening force for a hinged door in a smoke control system that relies on pressurization as shown in Figure A5.1.

The sum of the moments about the hinge is

$$M_r + A\Delta p\left(\frac{W}{2}\right) - F(W - d) = 0 \quad \text{(A5.1)}$$

where

F = total door-opening force,

M_r = moment of door closer and other friction,
W = door width,
A = door area,
Δp = pressure difference across door,
d = distance from doorknob to knob side of door.

The moment to overcome the door closer and friction consists of all moments about the hinge due to the door closer or friction forces such as friction in the hinges or rubbing of the door against the door frame. The component force F_{dc} at the knob to overcome the door closer and other friction is

$$F_{dc} = \frac{M_r}{W-d}. \quad \text{(A5.2)}$$

Solve Equation A5.2 for M_r, substitute it into Equation A5.1, and rearrange to get

$$F = F_{dc} + \frac{WA\Delta p}{2(W-d)}. \quad \text{(A5.3)}$$

6. AVERAGE PLUME TEMPERATURE

This section derives the equation for the average temperature of a smoke plume. Ideally, the plume is a steady flow process with the control volume shown in Figure A6.1.

Neglecting the small amount of mass added to the plume flow due to combustion, the first law of thermodynamics for the plume is

$$Q_g + Q_t = m(h_e - h_i + \Delta KE + \Delta PE) + W \quad \text{(A6.1)}$$

where

Q_g = heat generated within control volume,
Q_t = heat transferred from surroundings into control volume,
m = mass flow rate,
h_i = enthalpy of flow entering control volume,
h_e = enthalpy of flow leaving control volume,
ΔKE = change in kinetic energy,
ΔPE = change in potential energy,
W = work done by system on surroundings.

The heat generated is the heat release rate of the fire ($Q_g = Q$ where Q is the total heat release rate of the fire). For practical purposes, there is no heat transfer from the plume by thermal conduction. Heat is transferred from the plume to the surrounding almost completely by radiation. This means that the heat transferred from surroundings into the control volume can be considered equal to the negative of the radiative heat transfer ($Q_t = -Q_r$ where Q_r is the radiative heat transfer). The changes in kinetic and potential energy are negligible compared to the heat release rate of the fire ($\Delta KE = 0$, $\Delta PE = 0$). The steady plume does no work on its surroundings ($W = 0$).

Substituting these into Equation A6.1yields

$$Q - Q_r = m(h_e - h_i). \quad \text{(A6.2)}$$

The convective heat release rate is $Q_c = Q - Q_r$. Specific heat is considered constant, which means the enthalpies entering and leaving the control volume are $h_i = C_pT_o$ and $h_e = C_pT_p$ where C_p is the specific heat, T_o is ambient temperature, and T_p is the plume temperature at elevation z. Substituting these equations into Equation A6.2 and rearranging yields

$$T_p = T_o + \frac{Q_c}{mC_p}. \quad \text{(A6.3)}$$

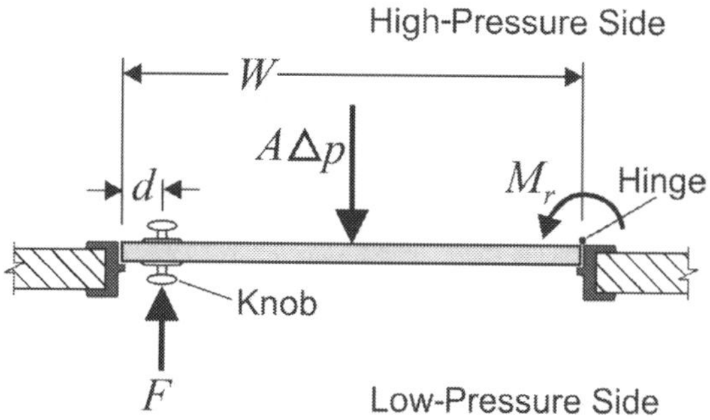

Figure A5.1 *Forces on side-hinged swinging door.*

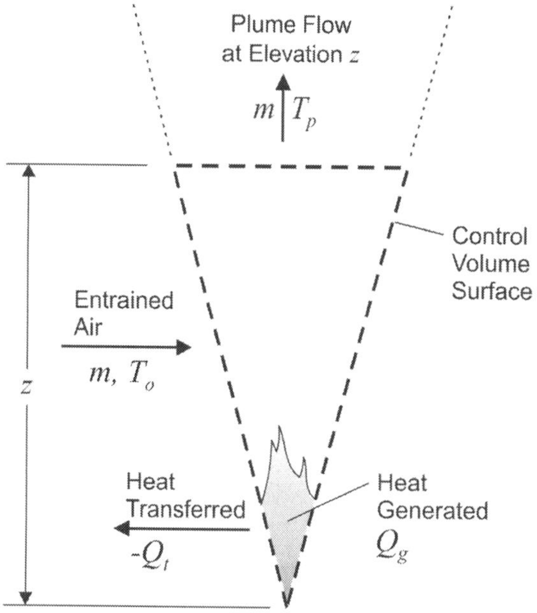

Figure A6.1 *Control volume for idealized plume.*

7. SCALE MODELING RELATIONS

This section derives the scaling relations for Froude modeling. The idea of Froude modeling is to keep the Froude number the same between a scale model and a full-scale facility in such a way that the temperature in the model equals that in the full-scale facility.

The Froude number is

$$F_r = \frac{U^2}{gl} \tag{A7.1}$$

where

U = velocity,
l = length,
g = acceleration of gravity.

The Froude number in the model equals that in the full-scale facility. This can be written as

$$\frac{U_m^2}{gl_m} = \frac{U_f^2}{gl_f} \tag{A7.2}$$

where

U_m = velocity in model,
U_f = velocity in full-scale facility,
l_m = length in model,
l_f = length in full-scale facility,
g = acceleration of gravity.

The model is built to scale, which means that positions in the model are scaled as

$$x_m = x_f\left(\frac{l_m}{l_f}\right) \tag{A7.3}$$

where x_m and x_f are positions in the model and full-scale facility, respectively. These scaling relations are such that temperatures are the same in the models as they would be in the full-scale facility.

$$T_m = T_f \tag{A7.4}$$

Equation A7.2 is rearranged to give the scaling relation for velocity

$$U_m = U_f\left(\left(\frac{l_m}{l_f}\right)\right)^{1/2}. \tag{A7.5}$$

By squaring both sides of Equation A7.3, a scaling relation for area is

$$A_m = A_f\left(\frac{l_m}{l_f}\right)^2 \tag{A7.6}$$

where

A_m = area in model,
A_f = area in full-scale facility.

Volumetric flow is velocity multiplied by area ($V = UA$). Multiply the left side of Equation A7.5 by the left side of Equation A7.6, and doing a similar thing to the right sides, this results in

$$V_m = V_f\left(\frac{l_m}{l_f}\right)^{5/2} \tag{A7.7}$$

where

V^m = volumetric flow in model,
V_f = volumetric flow in full-scale facility.

The density is defined by the ideal gas law ($\rho = p/RT$ where ρ is density, p is pressure, R is the gas constant and T is absolute temperature). Combining this law with Equation A7.4 results in

$$\rho_m = \rho_f \tag{A7.8}$$

where

ρ_m = density of gas in model,
ρ_f = density of gas in full-scale facility.

Mass flow rate is volumetric flow multiplied by density; thus, combining Equations A7.7 and A7.8 results in

$$m_m = m_f\left(\frac{l_m}{l_f}\right)^{5/2} \tag{A7.9}$$

where

m_m = mass flow in model,
m_f = mass flow in full-scale facility.

Velocity is length per unit time. Substituting $U_m = l_m / t_m$ and $U_f = l_f / t_f$ into Equation A7.5 and rearranging results in

$$t_m = t_f\left(\frac{l_m}{l_f}\right)^{1/2} \tag{A7.10}$$

where

t_m = time in model,
t_f = time in full-scale facility.

The heat convective portion of the heat release rate can be considered as an enthalpy flow ($Q_c = mc_p\,\Delta T$ where c_p is the specific heat, ΔT is temperature difference). Because the model is in air, the specific heat in the model equals that in the full-scale facility. Because of Equation A7.4, the temperature difference in the model equals that in the full-scale facility. Multiplying both sides to Equation A7.9 by $c_p\,\Delta T$ results in

$$Q_{c,m} = Q_{c,f}\left(\frac{l_m}{l_f}\right)^{5/2} \tag{A7.11}$$

where

$Q_{c,m}$ = convective heat release rate in model,

$Q_{c,f}$ = convective heat release rate in full-scale facility.

The pressure difference due to velocity is $\Delta p_m = \rho\, U_m^{\,2}/2$ and $\Delta p_f = \rho\, U_f^{\,2}/2$, and substituting these into Equation A7.5 yields

$$\Delta p_m = \Delta p_f \left(\frac{l_m}{l_f}\right) \quad \text{(A7.12)}$$

where

Δp_m = pressure difference in model,

Δp_f = pressure difference in full-scale facility.

8. PLUGHOLING: SEPARATION DISTANCE

This section derives the minimum distance between exhaust inlets so that they can be considered separate inlets for the purpose of preventing plugholing. Figure A8.1 shows an exhaust inlet in a ceiling. This analysis is based on the concept of inviscid flow which is the flow of an ideal fluid that has no viscosity or insignificant viscosity. Sources and sinks are common topics in inviscid flow theory. The flow of a real fluid towards the exhaust inlets behaves like a sink. As a digression, the flow from an air supply does not behave like a source. This analysis does not account for the differences in density between a smoke layer and the air below, but the results are considered useful for this application.

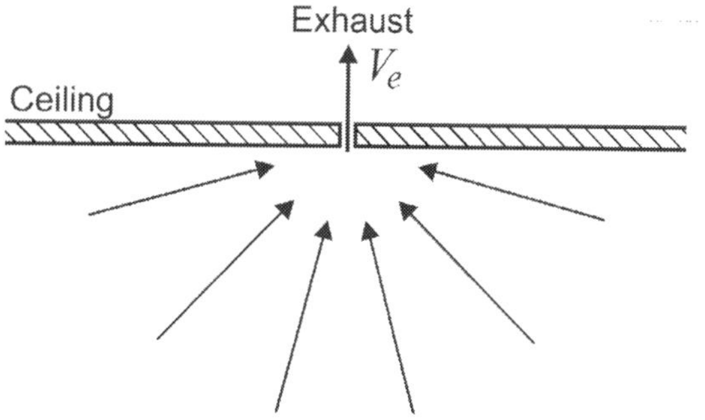

Figure A8.1 *Flow toward an exhaust inlet.*

For a sink in a solid surface like a ceiling, the velocity is

$$u = \frac{V_e}{2\pi R} \quad \text{(A8.1)}$$

where

u = velocity,

V_e = volumetric exhaust rate,

R = radius from exhaust inlet.

It is arbitrarily considered that where the velocity is 0.2 m/s (40 fpm), the inlets are far enough apart to be considered separate. At this velocity, the radius equals the minimum separation of S_{min}. Substituting $R = S_{min}$ and $u = 0.2$ into Equation A8.1) and rearranging results in

$$S_{min} = 0.9 V_e^{1/2} \quad \text{(A8.2)}$$

where

S_{min} = minimum edge-to-edge separation between inlets,

V_e = volumetric flow rate of one exhaust inlet.

INDEX